Earth's Evolving Systems

The History of Planet Earth

Ronald Martin, Ph.D.

University of Delaware

Newark, Delaware

JONES & BARTLETT
LEARNING

World Headquarters
Jones & Bartlett Learning
5 Wall Street
Burlington, MA 01803
978-443-5000
info@jblearning.com
www.jblearning.com

Jones & Bartlett Learning books and products are available through most bookstores and online booksellers. To contact Jones & Bartlett Learning directly, call 800-832-0034, fax 978-443-8000, or visit our website, www.jblearning.com.

Substantial discounts on bulk quantities of Jones & Bartlett Learning publications are available to corporations, professional associations, and other qualified organizations. For details and specific discount information, contact the special sales department at Jones & Bartlett Learning via the above contact information or send an email to specialsales@jblearning.com.

45716-2

Production Credits

VP, Executive Publisher: David D. Cella
Executive Editor: Matthew Kane
Associate Editor: Audrey Schwinn
Senior Production Editor: Nancy Hitchcock
Marketing Manager: Lindsay White
Production Services Manager: Colleen Lamy
Manufacturing and Inventory Control Supervisor: Amy Bacus

Composition: Cenveo® Publisher Services
Cover Design: Scott Moden
Rights & Media Specialist: Jamey O'Quinn
Media Development Editor: Shannon Sheehan
Cover Image: Shutterstock, Inc./Francesco R. Iacomino
Printing and Binding: LSC Communications
Cover Printing: LSC Communications

Library of Congress Cataloging-in-Publication Data
Names: Martin, Ronald E.
Title: Earth's evolving systems : the history of planet Earth / Ronald
    Martin, PhD, University of Delaware, Newark, Delaware.
Description: Second edition. | Burlington, Massachusetts : Jones & Bartlett
    Learning, [2018] | Includes index.
Identifiers: LCCN 2016037413 | ISBN 9781284108293
Subjects: LCSH: Earth (Planet)—History. | Geodynamics—History.
Classification: LCC QE501 .M2864 2018 | DDC 551.7—dc23
LC record available at https://lccn.loc.gov/2016037413

6048

Printed in the United States of America
20 19 18 17 16   10 9 8 7 6 5 4 3 2 1

This book is again dedicated to the late Dr. Allan Thompson (Department of Geological Sciences, University of Delaware), who did not shrink from learning something new and then teaching it. And to all those instructors who, like Al did, teach about the importance of the science of geology by transporting students to the other-worlds of Earth recorded in the rocks of geologic time.

BRIEF CONTENTS

CONTENTS

PREFACE TO THE SECOND EDITION: FOR THE INSTRUCTOR AND STUDENT

As the title indicates, *Earth's Evolving Systems* attempts to bridge the gap between traditional historical geology texts and the study of Earth's systems. The response to the first edition of *Earth's Evolving Systems* has been quite gratifying, especially given the recent emphasis by a National Science Foundation–sponsored webinar by the American Geophysical Union and American Geological Institute in October 2015 entitled "Geoscience Workforce and the Future of Undergraduate Geoscience Education." The respondents to this webinar emphasized at the outset the complex, dynamic linkages among Earth's systems, the role of "deep time" (and thus the role of the scale of time in understanding process), the origin and evolution of life, climate change, and energy resources. All of these topics were emphasized in the first edition of *Earth's Evolving Systems* and continue to be emphasized in the second.

Nevertheless, there is always room for improvement, and I have attempted to respond positively to reviewers' comments on the first edition. This has of course involved some compromises, given each instructor's approach to his or her particular course and research and teaching interests. Chapters have been updated with information on significant advances that have been reported in the literature over the past several years. Themes stated at the beginning of each chapter are now restated or rephrased, in some cases as "big concept" questions, which are highlighted at relevant points in the text margins of the chapters. As before, each chapter is followed by a summary that provides a detailed overview of the chapter.

The following key points about the second edition are applicable to all chapters:

- As in the first edition, a major theme of the text is the method of multiple working hypotheses and debates, among them the origin of the theory of plate tectonics, the origins of the atmosphere and life, the tectonics of the western United States, human evolution, and the recognition of Milankovitch cycles.
- Discussion and contributions and photos of some major women scientists to the earth sciences, such as Marie Tharp and Lynn Margulis, have been included in the relevant chapters.
- An extensive list of references is provided at the end of each chapter, along with a list of key terms and review questions. In addition, a second set of questions, called "Food for Thought," is provided to stimulate students to think beyond the chapter material.

Part I: Earth Systems: Their Nature and Their Study

Major changes were made to Chapters 1–6 to improve the flow of the material in Part I:

- **Chapter 1**: A brief discussion of Vladimir Vernadsky, the founder of Earth systems science, has been added. The discussion on the nature of historical sciences such as geology has been improved by eliminating Chapter 18 from the first edition and incorporating certain elements of that chapter into Chapter 1.
- **Chapter 2**: As before, much of the discussion of Earth's history revolves around the framework of the tectonic cycle. Plate tectonics has therefore been moved from Chapter 6 to Chapter 2.
- **Chapter 3**: The discussion of the interactions among Earth's systems has been simplified, and the introduction and discussion of specific stable isotopes have been pushed back to the chapters where they are explicitly tied to the geologic record. A new section has been added to this chapter, "How Does the Tectonic Cycle Affect Other Earth Systems?" which describes the effects of the tectonic cycle on sea level, ocean circulation, the hydrologic cycle, and major lithologies.
- **Chapters 5 and 6**: Chapter 5, which presents evolution, remains largely unchanged, but it now precedes Chapter 6, which deals with geologic time and stratigraphy. Discussion of iterative evolution has been moved from Chapter 14 to the section on marine organisms during the Paleogene.

Part II: The Precambrian: Origin and Early Evolution of Earth's Systems

- **Chapter 7**: Chapter content has been updated to reflect the most recent research.
- **Chapter 8**: A few reviewers questioned the relevance of a chapter on the origins of life in an Earth science text. However, I believe that life's origins are among the most fascinating chapters in Earth's history and that this is when the initial, fundamental interactions among all of Earth's systems began to occur. Life has been a geologic force throughout much of Earth's history, as emphasized throughout the text. The study of the interactions between life and Earth therefore serves as a bridge

between the biologic and inorganic worlds. Furthermore, like evolutionary theory, origin of life studies present viable alternatives to Creationism. A new paragraph at the beginning of the chapter now reiterates the rationale for retaining Chapter 8.

- **Chapter 9**: Chapter content has been updated to reflect the most recent research.
- **Chapter 10**: The discussion of the origins of various important fossil phyla has been augmented.

Part III: The Phanerozoic: Toward the Modern World

- **Chapters 11–15**: Chapters on the Phanerozoic continue to use the tectonic cycle as a basic framework for understanding the history of the Earth. Many figures in these chapters have been replaced and sections on various taxa augmented with multiple photos and new artwork.
- **Chapter 15**: The section on human evolution in Chapter 15 has been completely revised and reviewed by two professional paleoanthropologists.

Part IV: Humans and the Environment

- **Chapter 16**: As before, Chapter 16, which is on rapid climate change, sets the stage for the Gordian knot of natural versus anthropogenic climate change and its sociopolitical implications for future climate and energy resources, which are discussed in Chapter 17.
- **Chapter 17**: As explained in Chapter 1, the initial study of Earth systems was a response to anthropogenic effects. Humans are now a major, if not the most important, geologic force on the planet. The emphasis on the environment and "sustainability" at many academic institutions, including my own, does not diminish the importance of historical sciences, such as geology, in addressing these problems. In fact, the inclusion of chapters on anthropogenic impacts and their potential resolution is a prime opportunity to make historical geology not just an exercise in the "past" but to make it "contemporary" and "relevant" and to potentially awaken students' latent interest in the history of Earth and its lifeforms. Consequently, I have occasionally tied certain portions of Chapters 16 and 17 to examples from the geologic record.

Ron Martin
Newark, Delaware
August 10, 2016

THE STUDENT EXPERIENCE

The second edition of *Earth's Evolving Systems: The History of Planet Earth* was designed with numerous features to create an engaging learning environment for students and to enhance their experience with the text:

■ **Major Concepts and Questions Addressed in This Chapter**—Every chapter opens with a list of questions that will be addressed throughout the chapter. Students should review this list prior to digging into the chapter to help guide their focus. The new text design also incorporates icons identifying where in the chapter each concept is addressed to help guide study and review.

CHAPTER

1

Introduction: Investigating Earth's Systems

MAJOR CONCEPTS AND QUESTIONS ADDRESSED IN THIS CHAPTER

- Ⓐ Why study the history of Earth?
- Ⓑ How did the science of Earth systems arise?
- Ⓒ What is a system, and how does it work?
- Ⓓ What are Earth's systems, and what are their basic characteristics?
- Ⓔ Why is geologic time important to understanding how Earth's systems interact?
- Ⓕ How do different processes act on different durations of time?
- Ⓖ How do we study Earth's systems and the history of their interactions?

CHAPTER OUTLINE

1.1 Why Study the History of Earth?
1.2 What Are the Major Earth Systems, and What Are Their Characteristics?
BOX 1.1   The Origins of the Science of Earth Systems
1.3 Geologic Time and Process

1.4 Directionality and the Evolution of Earth Systems
1.5 Geology as an Historical Science
1.6 Method and Study of Earth's Evolving Systems

The Anasazi cliff dwellings at Mesa Verde National Park, Colorado. The Anasazi (the "Ancient Ones") civilization vanished suddenly, possibly as a result of prolonged drought. The Anasazi civilization once encompassed an area the size of New England in the Four Corners region where Colorado, Utah, New Mexico, and Arizona meet today. Based on archeological evidence, the Anasazi civilization flourished during what is called the Little Climate Optimum from about 900–1300, and traded with other civilizations as far south as Mexico and Central America. The Anasazi adopted an agricultural lifestyle and built extensive cities in the sides of cliffs. However, the Anasazi began to disperse from about 1280–1300, leaving behind their dwellings, and their civilization disappeared. Similarly, increasing evidence indicates that modern global change—due to the combustion of fossil fuels—will alter precipitation patterns, leading to more intense heat waves and prolonged drought in different regions all over the world, including North America.

Courtesy of National Park Service.

3

1.1 Why Study the History of Earth?

Ⓐ *Earth's Evolving Systems* is about the history of the Earth, the natural processes that have shaped it, and the history of these processes and their interactions through vast intervals of time. *Geology* is the science that studies the history of the Earth and its life preserved as fossils. Why should we be concerned about Earth's history? Because understanding how the Earth changes and has changed tells us about how natural processes affect humans and how humans affect natural processes. The history of the Earth confronts us with events and possibilities that we cannot imagine. Many natural processes act so slowly we would be unaware of them except for the geologic record of their activities preserved by rocks and fossils. Most people are unaware that Earth's environments are constantly changing. We assume that landscapes—mountains, valleys, rivers and streams, and coasts—do not change because the changes are typically so slow and subtle they take place over time spans equivalent to many, many human generations; from many millions of years down to millennia and centuries. Also, some processes are so infrequent or sudden, we would not know they occur except, again, to look at the geologic record.

Ⓑ Scientists have only recently begun to appreciate just how strongly changes in Earth's environments have affected—and still affect—humankind, from our evolutionary beginnings through the origins of ancient settlements and civilizations—and perhaps their collapse—right up to

XV

- **Featured Boxes**—Many chapters contain boxes providing greater depth on special topics.

BOX 13.3 Late Cretaceous Extinctions and the Scientific Method

Most mass extinctions appear to be somehow related to the tectonic cycle. However, the Late Cretaceous extinctions involved—and may well have resulted primarily from—an impact, as indicated by the occurrence of shocked mineral assemblages (**Box Figure 13.3A**). Whereas the impact hypothesis certainly arouses our imaginations, how the hypothesis came to be widely accepted by the scientific community is also a prime example of how scientific investigation works (see Chapter 1). Moreover, the corroboration of the hypothesis paved the way for the acceptance of extraterrestrial impacts as important—even extraordinary—agents of geologic, climatic, and biospheric change. it also radically altered—once and for all—earth scientists unquestioned acceptance of Lyell's dogma of slow, gradual change to a broader doctrine that recognized that Earth systems processes vary through time and in rate (see Chapter 1).

Initially, a dark sedimentary layer containing a high concentration of the element iridium was found near Gubbio, Italy, almost by accident (see Chapter 1). The iridium layer also occurred at the time of the mass extinction at the end of the Cretaceous Period about 65 million years ago, during which dinosaurs and many other organisms became extinct. Iridium is not normally found in rocks of Earth's crust and could have come from only two sources: volcanoes fed by the mantle, which is enriched in iridium, or from an extraterrestrial body. The hypothesis was that the iridium layer was generated by a meteor enriched in iridium. The impact presumably threw a gigantic dust cloud into Earth's stratosphere that suddenly cooled the planet, causing extinction; the blockage of sunlight also shut down marine photosynthesis causing a **Strangelove Ocean** (named after the character of the same name in a famous movie) in which there was a sudden, strong shift in carbon isotope ratios to much lower values (see Chapter 9; **Box Figure 13.3B**).

A prediction made from the hypothesis was that if an impact were responsible for the Late Cretaceous extinctions, an iridium layer should be found all over the world in rocks of exactly the same age. Scientists tested the hypothesis by exploring for the iridium layer all over the world, on land and in deep-sea cores, where the rocks were of the right age. The hypothesis was corroborated: the Late Cretaceous iridium layer is now known not only from Gubbio, Italy, but also from Stevns Klint (Steven's Cliff) near Copenhagen, Denmark; El Kef, Tunisia, in north Africa; and El Mimbral, Mexico (to name only a few of the more famous and intensively studied localities), as well as in many deep-sea cores (see Box Figure 13.3B).

BOX FIGURE 13.3A An artist's visualization of the impact of an asteroid with Earth.
© Gl0k/Shutterstock, Inc.

- **Concept and Reasoning Checks**—As students progress through the chapter they will encounter these questions, which will encourage them to pause and assess their grasp of the material.

CONCEPT AND REASONING CHECKS

1. Diagram the hydrologic cycle.
2. How are the hydrologic cycle and atmospheric circulation related?
3. What drives surface ocean circulation?
4. What causes the deep oceans to circulate?
5. How do the oceans influence Earth's albedo?

CONCEPT AND REASONING CHECKS

1. What is the evidence for the solar nebula hypothesis as opposed to the original Kant-Laplace hypothesis?
2. How do the inner planets, including Earth, differ from the outer planets?
3. Why might carbonaceous chondrites have been an important source of water for early Earth?

CONCEPT AND REASONING CHECKS

1. Volcanism has been implicated in several mass extinctions. Which ones?
2. Diagram the test of a meteor impact as the causal agent of the Late Cretaceous mass extinction in terms of the scientific method diagrammed in Chapter 1 (see Box 13.3).

- **Summaries**—Each chapter concludes with a bulleted list of the key concepts addressed in the chapter.

SUMMARY

- The theory of plate tectonics really began with early ideas about orogenesis, or mountain building. Hypotheses and theories of mountain building changed radically over the past two centuries, and their development is a prime example of how scientists work and think.
- The discovery of radioactivity led to more modern theories of mountain building. Of these, it is Alfred Wegener's hypothesis of continental drift—based on a variety of evidence—that paved the way for the modern theory of plate tectonics. Initially, Wegener's hypothesis was roundly criticized because he could not identify a mechanism to make continents drift. Consequently, continents (or at least many geologists' minds) remained "fixed" until the work of Harry Hess in the 1950s, which proposed the process of seafloor spreading as a mechanism to move continents.
- In the 1960s, the detection of magnetic seafloor stripes corroborated seafloor spreading and provided the mechanism of continental drift that had eluded Wegener. Seafloor spreading also corroborated Hess's views about the formation of guyots, heat flow beneath mid-ocean ridges, and the destruction of seafloor in trenches. Rearranging the continents into different positions also began to make sense of apparent polar wandering curves.
- Consequently, what had been known as continental drift was wedded to seafloor spreading to produce the theory of plate tectonics.
- Today, plate tectonics is recognized as an integral component of Earth's systems. We know that Earth's lithosphere (the crust and uppermost mantle) consists of about 15 large and small plates that are moved by the production of new seafloor at mid-ocean ridges. Forming portions of the plates are continents. The plates move over the asthenosphere of the mantle. Beneath the mantle are an outer fluid and a solid inner iron and nickel-rich core that generate Earth's magnetic field.
- Although convection cells are widely viewed as moving the plates, several hypotheses have been proposed to explain how the seafloor actually moves: (1) slab-pull, in which a descending slab pulls the rest of slab behind it downward; (2) ridge-push, in newly formed ocean crust as spreading centers pushes the slab ahead of it; (3) gravity slide, in which a slab slowly "slides" down the side of a spreading center, pushing the slab ahead of it; and (4) suction from the descending portion of a plate.
- Based on plate tectonics, different features of the planet can be arranged into a sequence of stages called the tectonic cycle: East African Rift Valley, Red Sea, Atlantic Ocean, Pacific Ocean, and suture (Himalayas). Not all rift valleys become seaways, however; many have become failed rift valleys or aulacogens, down which some of the world's major rivers such as the Amazon flow. The tectonic cycle has occurred a number of times during Earth's history, each cycle spanning several hundred million years.
- Based on the tectonic cycle, continental margins and plate boundaries can change through time. There are two basic types of continental margins: active and passive. Passive continental margins, like those along the Atlantic Ocean, accumulate sediment along their margins. Active margins, like those along the Pacific Ocean's ring of fire, are sites of subduction, volcanism, and earthquakes.
- Plate boundaries are classified into three basic categories: convergent (associated with sea floor trenches), divergent (associated with rifting), and transform, which are associated with offsets of mid-ocean ridges.
- Convergent boundaries are themselves of three types: volcanic island arc (for example, Japan), continental volcanic island arc (for example, the Cascades), and collisional island arc (for example, the Cascades), and collisional island arc (for example, the Himalayas).
- The three types of convergent plate boundaries parallel the different types of orogenesis and the formation of major geologic structures such as faults and folds: island arcs only, plate collisions without continents, and continent–continent collisions.
- As orogenesis occurs, smaller pieces of crust with distinctive geologic features (rock type, fossils, paleomagnetic directions) called microcontinents or exotic terranes can be sandwiched between the larger continents.
- No one has ever observed the tectonic cycle because of the immense amounts of geologic time involved in its completion, but it can be pieced together based on observations of modern tectonic settings.

- **Key Terms List**—A list of the key terms from each chapter is provided to help students review new vocabulary.

- **Review Questions**—These end-of-chapter questions are great for homework assignments or self-guided study.

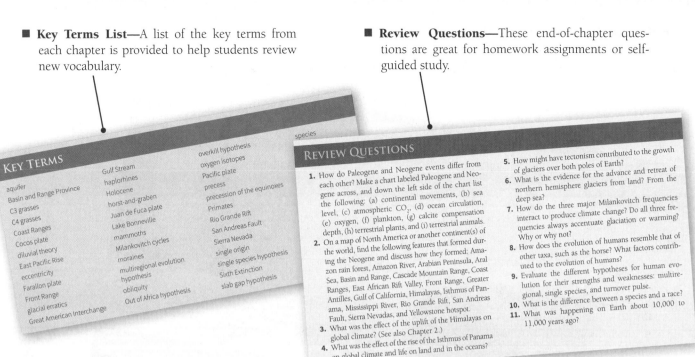

KEY TERMS

aquifer	Gulf Stream	overkill hypothesis
Basin and Range Province	haplorhines	oxygen isotopes
C3 grasses	Holocene	Pacific plate
C4 grasses	horst-and-graben	precess
Coast Ranges	Juan de Fuca plate	precession of the equinoxes
Cocos plate	Lake Bonneville	Primates
diluvial theory	mammoths	Rio Grande Rift
East Pacific Rise	Milankovitch cycles	San Andreas Fault
eccentricity	moraines	Sierra Nevada
Farallon plate	multiregional evolution hypothesis	single origin
Front Range	obliquity	single species hypothesis
glacial erratics	Out of Africa hypothesis	Sixth Extinction
Great American Interchange		slab gap hypothesis

REVIEW QUESTIONS

1. How do Paleogene and Neogene events differ from each other? Make a chart labeled Paleogene and Neogene across, and down the left side of the chart list the following: (a) continental movements, (b) sea level, (c) atmospheric CO_2, (d) ocean circulation, (e) oxygen, (f) plankton, (g) calcite compensation depth, (h) terrestrial plants, and (i) terrestrial animals.

2. On a map of North America or another continent(s) of the world, find the following features that formed during the Neogene and discuss how they formed: Amazon rain forest, Amazon River, Arabian Peninsula, Aral Sea, Basin and Range, Cascade Mountain Range, Coast Ranges, East African Rift Valley, Front Range, Greater Antilles, Gulf of California, Himalayas, Isthmus of Panama, Mississippi River, Rio Grande Rift, San Andreas Fault, Sierra Nevadas, and Yellowstone hotspot.

3. What was the effect of the uplift of the Himalayas on global climate? (See also Chapter 2.)

4. What was the effect of the rise of the Isthmus of Panama on global climate and life on land and in the oceans?

5. How might have tectonism contributed to the growth of glaciers over both poles of Earth?

6. What is the evidence for the advance and retreat of northern hemisphere glaciers from land? From the deep sea?

7. How do the three major Milankovitch frequencies interact to produce climate change? Do all three frequencies always accentuate glaciation or warming? Why or why not?

8. How does the evolution of humans resemble that of other taxa, such as the horse? What factors contributed to the evolution of humans?

9. Evaluate the different hypotheses for human evolution for their strengths and weaknesses: multiregional, single species, and turnover pulse.

10. What is the difference between a species and a race?

11. What was happening on Earth about 10,000 to 11,000 years ago?

- **Food for Thought**—More in-depth than the Review Questions, the Food for Thought activities are great for individual or group assignments in or out of the classroom. They will challenge students to think critically about the material presented in the chapter.

- **Sources and Further Reading**—The list of references for the chapter is a great place for students to begin additional research into special topics.

FOOD FOR THOUGHT:
Further Activities In and Outside of Class

1. Construct a table of the hypotheses described in the text for the origin of the Basin and Range. List the hypotheses down the left side and place a heading at the top titled "Evidence." Include in that column both the geologic evidence and the forces inferred from the evidence. Then, to the right place a column titled "Success of the Hypothesis" with two columns underneath for each of the main regions of the Basin and Range emphasized in the text: Northern (N) and Southern (S). For each hypothesis, indicate whether it satisfactorily explains the evidence within the region (+), does not explain it one way or the other (0), or contradicts it (−). Discuss your results in terms of the method of multiple working hypotheses (see Chapter 1).

2. Which normally causes sea level to change faster: the advance and retreat of glaciers or the movements of continents?

3. Describe the tectonics and sedimentation of the western United States in terms of the method of multiple working hypotheses (see Chapter 1).

4. How did preadaptation in early primates later affect the evolution of humans?

5. Why is the fossil record of humans so hotly debated?

6. Of the different hypotheses for human evolution, which one do you favor and why?

7. What is the significance of the finding of *Plesiadapis* in both North America and Europe?

8. Quite frequently, new species of humans are based on a single fossil fragment such as a jaw fragment. How can an entire new species be inferred from a single fragmentary fossil? (Hint: See Chapter 4 for Cuvier's correlation of parts.)

SOURCES AND FURTHER READING

Alroy, J. 2001. A multispecies overkill simulation of the end-Pleistocene megafaunal mass extinction. *Science*, 292, 1893–1896.

Asfaw, B. et al. 2002. Remains of *Homo erectus* from Bouri, Middle Awash, Ethiopia. *Nature*, 416, 317–320.

Baldridge, W. S. 2004. *Geology of the American Southwest: A journey through two billion years of plate-tectonic history.* Cambridge, UK: Cambridge University Press.

Bower, B. 1999. DNA's evolutionary dilemma. *Science News*, 155(6, February 6):88–90.

Bower, B. 2001. Fossil skull diversifies family tree. *Science News*, 159(12, March 24), 180.

Bower, B. 2002. Evolution's surprise: Fossil find uproots our early ancestors. *Science News*, 162(2):19.

Brunet, M. et al. 2002. A new hominid from the Upper Miocene of Chad, central Africa. *Nature*, 418, 145–151.

Bloch, J. E., Chritz, K. L., Jablonski, N. G., Leakey, M. G., Sukanthi, F. K. 2013. Diet of Theropithecus from 4 to 1 Ma in Kenya. *Proceedings of the National Academy of Sciences*, www.pnas.org/cgi/doi/10.1073/pnas 1214209110

Bloch, J. G. B., et al. Oldest known euarchontan tarsals and affinities of Paleocene *Purgatorius* to Primates. *Proceedings of the National Academy of Sciences*, 104:11487–11492.

Bond, G. Mass extinctions pinned on ice age harbingers.

deMenocal, P. B. 2011. Climate and human evolution. *Science*, 331, 540–542

Gibbons, A. 2011. Skeletons present an exquisite paleo-puzzle. *Science*, 333, 1370–1372.

Guo, Z. T. et al. 2002. Onset of Asian desertification by 22 Myr ago inferred from loess deposits in China. *Nature*, 416, 159–163.

Imbrie, J., and Imbrie, K. P. 1979. *Ice ages: Solving the mystery.* Short Hills, NJ: Enslow Publishers.

Leakey, R. and Lewin, R. 1995. *The Sixth Extinction: Patterns of life and the future of humankind.* New York: Doubleday Dell.

Leakey, M. G. et al. 2001. New hominid genus from eastern Africa shows diverse middle Pliocene lineages. *Nature*, 410, 433–440.

Leakey, M. G., Spoor, F, Dean, M. C., Feibel, C. S., Antón, S. C., Kiarie, C., and Leakey, L. N. 2012. New fossils from Koobi Fora in northern Kenya confirm taxonomic diversity in early *Homo*. *Nature*, 488, 201–204.

Leonard, W. R. 2002. Food for thought: Dietary change was a driving force in human evolution. *Scientific American*, 287(6):106–115.

Monastersky, R. 1996. Out of arid Africa. *Science News*, 150(5, August 3), 74–75.

Monastersky, R. 1999. The killing fields: What robbed the Americas of their most charismatic mammals? *Science News*, 156(23):360–361.

Murphy, J. B., Oppliger, G. L., Brimhall, G. H., and Hynes, A. 1999. Mantle plumes and mountains. *American Scientist*, 87(2):146–153.

Ni, Xijun, et al. "The oldest known primate skeleton and early haplorhine evolution." *Nature*, 498.7452 (2013):60–64.

Pastor, J., and Moen, R. A. 2004. Ecology of ice-age extinctions. *Nature*, 431, 639–640.

Roberts, R. G. et al. 2001. New ages for the last Australian megafauna: Continent-wide extinction about 46,000 years ago. *Science*, 292, 1888–1892.

Ruddiman, W. F., and McIntyre, A. 1976. Northeast Atlantic paleoclimate changes over the past 600,000 years. In R. M. Cline and J. D. Hays (eds.), *Investigation of Late Quaternary paleoceanography and paleoclimatology* (pp. 111–146). Boulder, CO: Geological Society of America.

Sonder, L. J., and Jones, C. H. 1999. Western United States: How the west was widened. *Annual Review of Earth and Planetary Sciences*, 27, 417–462.

Stringer, C. 2003. Out of Ethiopia. *Nature*, 423, 692–695.

Stuart, A. J., Kosintsev, P. A., Higham, T. F. G., and Lister, A. M. 2004. Pleistocene to Holocene extinction dynamics in giant deer and woolly mammoth. *Nature*, 431, 684–689.

Tattersall, I. 1997. Out of Africa again ... and again? *Scientific American*, 279(4):60–67.

Tattersall, I. 2000. Once we were not alone. *Scientific American*, 282(1):56–62.

Tavaré, S., Marshall, C. R., Will, O., Soligo, C., and Martin, R. D. 2002. Using the fossil record to estimate the age of the last common ancestor of extant primates. *Nature*, 416, 726–729.

Templeton, A. R. 2002. Out of Africa again and again. *Nature*, 416, 45–51.

Vekua, A. et al. 2002. A new skull of early *Homo* from Dmanisi, Georgia. *Science*, 297, 85–89.

Wilson, E. O. 1992. *The diversity of life.* Cambridge, MA: Belknap Press.

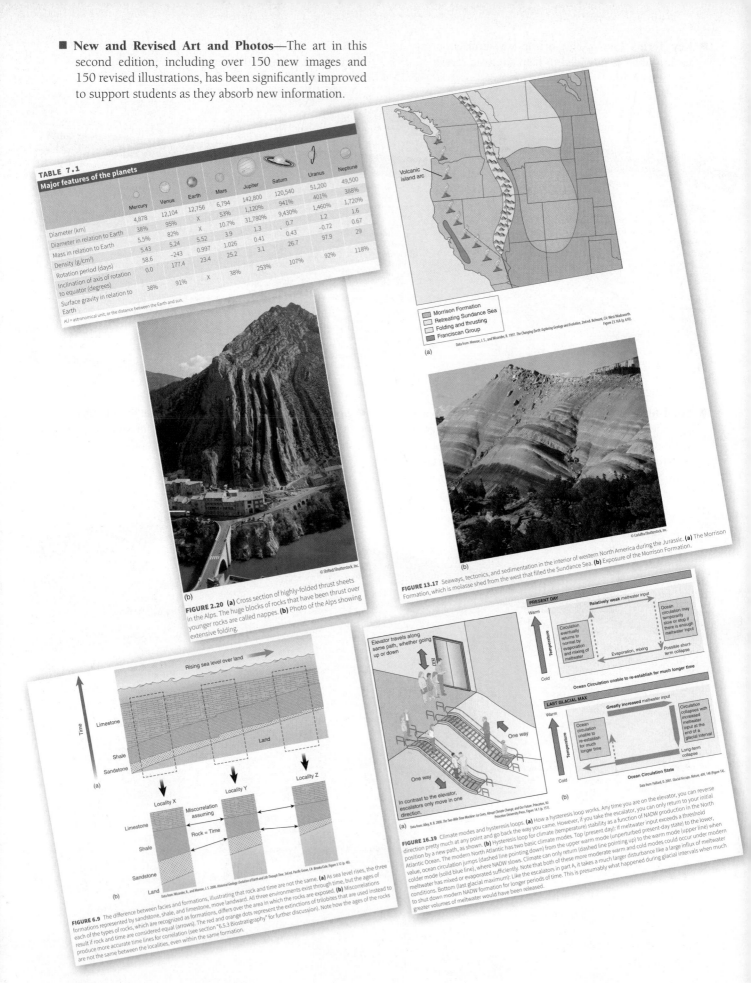

■ **New and Revised Art and Photos**—The art in this second edition, including over 150 new images and 150 revised illustrations, has been significantly improved to support students as they absorb new information.

TABLE 7.1
Major features of the planets

	Mercury	Venus	Earth	Mars	Jupiter	Saturn	Uranus	Neptune
Diameter (km)	4,878	12,104	12,756	6,794	142,800	120,540	51,200	49,500
Diameter in relation to Earth	38%	95%	X	53%	1,120%	941%	401%	388%
Mass in relation to Earth	5.5%	82%	X	10.7%	31,780%	9,430%	1,460%	1,720%
Density (g/cm³)	5.43	5.24	5.52	3.9	1.3	0.7	1.2	1.6
Rotation period (days)	58.6	−243	0.997	1.026	0.41	0.43	−0.72	0.67
Inclination of axis of rotation to equator (degrees)	0.0	177.4	23.4	25.2	3.1	26.7	97.9	29
Surface gravity in relation to Earth	38%	91%	X	38%	253%	107%	92%	118%

AU = astronomical unit, or the distance between the Earth and sun.

FIGURE 2.20 (a) Cross section of highly-folded thrust sheets in the Alps. The huge blocks of rocks that have been thrust over younger rocks are called nappes. (b) Photo of the Alps showing extensive folding.

© Shifted/Shutterstock, Inc.

FIGURE 13.17 Seaways, tectonics, and sedimentation in the interior of western North America during the Jurassic. (a) The Morrison Formation, which is molasse shed from the west that filled the Sundance Sea. (b) Exposure of the Morrison Formation.

Volcanic island arc

Morrison Formation
Retreating Sundance Sea
Folding and thrusting
Franciscan Group

Data from: Monroe, J. S., and Wicander, R. 1997. *The Changing Earth: Exploring Geology and Evolution*, 2nd ed. Belmont, CA: West/Wadsworth, Figure 23.16A (p. 610).

© CataFba/Shutterstock, Inc.

FIGURE 6.9 The difference between facies and formations, illustrating that rock and time are not the same. (a) As sea level rises, the three formations represented by sandstone, shale, and limestone, move landward. All three environments exist through time, but the ages of each of the types of rocks, which are recognized as formations, differs over the area in which the rocks are exposed. (b) Miscorrelations result if rock and time are considered equal (arrows). The red and orange dots represent the extinctions of trilobites that are used instead to produce more accurate time lines for correlation (see section "6.5.3 Biostratigraphy" for further discussion). Note how the ages of the rocks are not the same between the localities, even within the same formation.

Data from: Wicander, R., and Monroe, J. S. 2000. *Historical Geology: Evolution of Earth and Life Through Time*, 3rd ed. Pacific Grove, CA: Brooks/Cole, Figure 3.12 (p. 46).

FIGURE 16.19 Climate modes and hysteresis loops. (a) How a hysteresis loop works. Any time you are on the elevator, you can reverse direction pretty much at any point and go back the way you came. However, if you take the escalator, you can only return to your initial position by a new path, as shown. (b) Hysteresis loop for climate modes. Top (present day): The modern North Atlantic has two basic climate modes. Top (present day): If meltwater input exceeds a threshold value, ocean circulation jumps (dashed line pointing down) from the upper warm mode (unperturbed present-day state) to the lower, colder mode (solid blue line), where NADW slows. Climate can only return (dashed line pointing up) to the warm mode (upper line) when meltwater has mixed or evaporated sufficiently. Note that both of these more moderate warm and cold modes could occur under modern conditions. Bottom (last glacial maximum): Like the escalators in part A, it takes a much larger disturbance like a large influx of meltwater to shut down modern NADW formation for longer periods of time. This is presumably what happened during glacial intervals when much greater volumes of meltwater would have been released.

Data from: Alley, R. B. 2000. *The Two-Mile Time Machine: Ice Cores, Abrupt Climate Change, and Our Future*. Princeton, NJ: Princeton University Press, Figure 14.1 (p. 111).

Data from: Pollard, D. 2001. Glacial hiccups. *Nature*, 409, 148 (Figure 1A).

TEACHING TOOLS

A variety of Teaching Tools are available for qualified instructors to assist with preparing for and teaching their courses. These resources are accessible via digital download and multiple other formats:

■ **Lecture Outlines in PowerPoint format**—The *Lecture Outlines in PowerPoint format* provide lecture notes and images for each chapter of *Earth's Evolving Systems: The History of Planet Earth, Second Edition*. Instructors with Microsoft PowerPoint can customize the outlines, art, and order of presentation and add their own material.

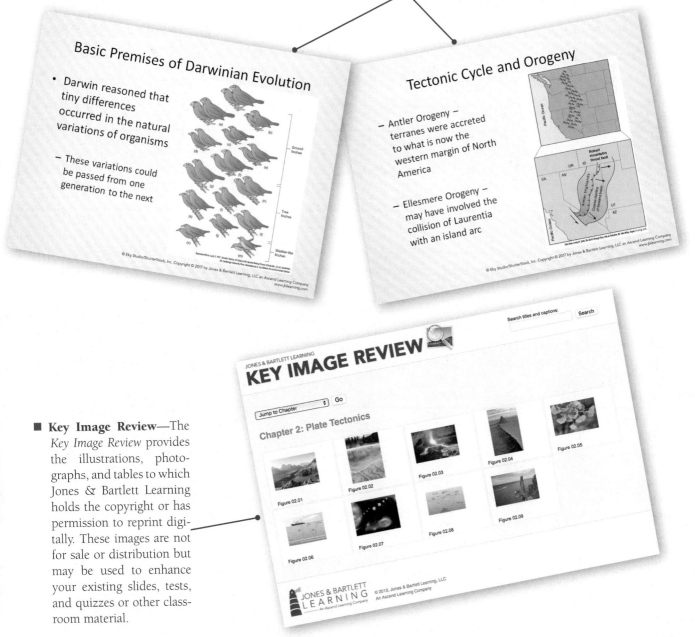

■ **Key Image Review**—The *Key Image Review* provides the illustrations, photographs, and tables to which Jones & Bartlett Learning holds the copyright or has permission to reprint digitally. These images are not for sale or distribution but may be used to enhance your existing slides, tests, and quizzes or other classroom material.

- **Test Bank Material**—The author has provided 500+ multiple-choice questions, including true-false, matching, and identifications. Each chapter has approximately 30 to 40 questions. The author of this text has used some—but certainly not all—of these questions in his introductory course. Many questions ask for basic factual information, others are intended to make students "think about it." In some cases, essentially the same questions are worded differently. Alternative wordings and answers are suggested for some questions. Some questions refer to specific figures in the text. Instructors are welcome to modify the questions as they see fit. Short and long essay questions can be developed from the Review Questions and Food for Thought exercises at the end of each chapter and the Concept and Reasoning Checks embedded throughout. These could be used in smaller classes as writing assignments. Students could be assigned the questions ahead of time or given a list to choose from. These questions are available as an instructor download.

- **Instructor's Manual**—An Instructor's Manual containing an instructor's overview, instructional aids, answers to Review and Food for Thought questions, and suggestions for homework or in-class projects and assignments is available for each chapter.

Earth's Evolving Systems had its beginnings in my book *One Long Experiment: Scale and Process in Earth History* (1998, Columbia University Press), the reviews of which were encouraging.

Many individuals contributed to the publication of this work. I would like to thank Stan Wakefield for putting me in touch with Jones & Bartlett Learning regarding the manuscript. I would also like to thank Editor Audrey Schwinn who guided the text through its preproduction phase; Rights and Media Specialist Jamey O'Quinn; and Media Development Editor Shannon Sheehan. I would also like to thank Senior Production Editor, Nancy Hitchcock, whose careful eyes for detail have much improved the book and kept it on track for publication.

A number of recent undergraduate geology majors at the University of Delaware have contributed to this book with their enthusiasm during the courses I have taught, especially Emily Cahoon, Mary Cassella, Steve Cinderella, Lauren Cook, Laura Dodd, Kevin Gielarowski, Josh Humberston, Deon Knights, Kelsey Lanan, Sherri Legg, Amanda Lusas, Briana Lyons, Suzie McCormick, Livia Montone, Steve Mulvry, Sharon Nebbia, Marc Roy, Nick Spalt, Justin Walker, Jessie Wenke, Dave Wessell, and Erika Young. So, too, have many students in my introductory course. I hope their enthusiasm validates my approach with the readers. Jean Self-Trail read portions of Chapter 14. I also thank my running buddies for many years of physical and mental exertion: Al, Dick, and Sandy.

My sincere thanks to Drs. Karen Rosenberg and Thomas Rocek of the Department of Anthropology at the University of Delaware for their review of the section on human evolution in Chapter 15; any errors are, however, mine.

Jones & Bartlett Learning would also like to thank and acknowledge Dr. Amanda Julson of Blinn College for her work on revising the Lecture Outlines in PowerPoint format and the Web Links, and for creating the Instructor's Manual for this edition. In addition, we sincerely appreciate the assistance of Professor Ann Harris of Eastern Kentucky University and Dr. A. M. Hunt of University of Cincinnati in creating the online assessment questions that accompany this edition.

I express my gratitude to the reviewers of the first edition, whose feedback helped to shape the text in many ways:

Rick Batt, Buffalo State College
Alan Benimoff, College of Staten Island–CUNY
Walter S. Borowski, Eastern Kentucky University
Robert Cicerone, Bridgewater State College
Joshua C. Galster, Montclair State University
William Garcia, University of North Carolina at Charlotte
Tamie J. Jovanelly, Berry College
Matthew G. Powell, Juniata College
Steven H. Schimmrich, SUNY Ulster County
   Community College
Greg W. Scott, Lamar State College–Orange

Comments from the following reviewers helped to shape this second edition:

Alan I. Benimoff, College of Staten Island
Harry Dowsett, U.S. Geological Survey
Antony N. Giles, Nicholls State University
Danny Glenn, Wharton Junior College
Warren D. Huff, University of Cincinnati
Takehito Ikejiri, University of Alabama
Arthur C. Lee, Roane State Community College
Margaret Karen Menge, Delgado Community College
Jill Mignery, Miami University
Donald Neal, East Carolina University
Cynthia L. Parish, Lamar University
Carrie E. Schweitzer, Kent State University at Stark
David Richard Schwimmer, Columbus State University

Finally, I thank my wife, Carol, for her encouragement throughout the writing and production of this book. Watching our daughter, Dana, grow up has perhaps contributed more to my teaching and to this book than either she or I will ever know or understand.

Ronald Martin
Newark, Delaware

Ron Martin is Professor of Geological Sciences at the University of Delaware. He grew up in southwestern Ohio, where world famous assemblages of Late Ordovician fossils drew his attention to paleontology. He received a B.S. degree in Geology and Paleontology from Bowling Green State University (Ohio), M.S. in Geology from the University of Florida, and Ph.D. in Zoology from the University of California at Berkeley. He worked as an operations micropaleontologist and biostratigrapher for Unocal in Houston (Texas) from 1981–1985 before coming to the University of Delaware. He has taught introductory courses in physical geology and Earth history (upon which *Earth's Evolving Systems* is based), Paleontology, Paleoecology, Sedimentology, and Stratigraphy, and Advanced (Sequence) Stratigraphy, among others and he has been nominated several times for the university-wide Best Teacher Award. His research interests include the taphonomy (preservation) and biostratigraphy of microfossil assemblages and, most recently, the role of phytoplankton evolution in the diversification of the marine biosphere. He is the author or co-author of more than 60 papers; in addition to *Earth's Evolving Systems*, he has also authored *One Long Experiment: Scale and Process in Earth History* (Columbia University Press) and *Taphonomy: A Process Approach* (Cambridge University Press) and edited *Environmental Micropaleontology: The Application of Microfossils to Environmental Geology* (Kluwer/Plenum Press). He received the Best Paper Award in 1996 from the journal *Palaios* for "Secular Increase in Nutrient Levels Through the Phanerozoic: Implications for Productivity, Biomass, and Diversity of the Marine Biosphere"; his work was also featured as the cover article in the June (2013) issue of *Scientific American*: "Tiny Engines of Evolution," which was translated into French, German, Spanish, and Japanese sister publications. He is past president of the North American Micropaleontology Section of the Society of Sedimentary Geology, former Editor of the *Journal of Foraminiferal Research,* and Associate Editor of *Palaios*. He was Visiting Professor at the Université de Lille (France) in 2014.

PART I

EARTH SYSTEMS: THEIR NATURE AND THEIR STUDY

Part I of *Earth's Evolving Systems* examines the principles and concepts critical to the study of the processes of each of the basic Earth systems: the solid Earth, the hydrosphere, the atmosphere, and the biosphere. In examining these systems, Part I emphasizes the following:

1. Why study Earth history?
2. Basic components and behavior of each system and how they evolve
3. How changes in the distributions of the continents and oceans affect the other systems
4. How the interactions of Earth's systems regulate climate
5. The importance of geologic time to the study of physical and biological processes
6. How we study Earth's systems

Introduction: Investigating Earth's Systems

MAJOR CONCEPTS AND QUESTIONS ADDRESSED IN THIS CHAPTER

A Why study the history of Earth?

B How did the science of Earth systems arise?

C What is a system, and how does it work?

D What are Earth's systems, and what are their basic characteristics?

E Why is geologic time important to understanding how Earth's systems interact?

F How do different processes act on different durations of time?

G How do we study Earth's systems and the history of their interactions?

CHAPTER OUTLINE

The Anasazi cliff dwellings at Mesa Verde National Park, Colorado. The Anasazi (the "Ancient Ones") civilization vanished suddenly, possibly as a result of prolonged drought. The Anasazi civilization once encompassed an area the size of New England in the Four Corners region where Colorado, Utah, New Mexico, and Arizona meet today. Based on archeological evidence, the Anasazi civilization flourished during what is called the Little Climate Optimum from about 900–1300, and traded with other civilizations as far south as Mexico and Central America. The Anasazi adopted an agricultural lifestyle and built extensive cities in the sides of cliffs. However, the Anasazi began to disperse from about 1280–1300, leaving behind their dwellings, and their civilization disappeared. Similarly, increasing evidence indicates that modern global change—due to the combustion of fossil fuels—will alter precipitation patterns, leading to more intense heat waves and prolonged drought in different regions all over the world, including North America.

Courtesy of National Park Service.

1.1 Why Study the History of Earth?

A *Earth's Evolving Systems* is about the history of the Earth, the natural processes that have shaped it, and the history of these processes and their interactions through vast intervals of time. *Geology* is the science that studies the history of the Earth and its life preserved as fossils.

Why should we be concerned about Earth's history? Because understanding how the Earth changes and has changed tells us about how natural processes affect humans and how humans affect natural processes. The history of the Earth confronts us with events and possibilities that we cannot imagine. Many natural processes act so slowly we would be unaware of them except for the geologic record of their activities preserved by rocks and fossils. Most people are unaware that Earth's environments are constantly changing. We assume that landscapes—mountains, valleys, rivers and streams, and coasts—do not change because the changes are typically so slow and subtle they take place over time spans equivalent to many, many human generations; from many millions of years down to millennia and centuries. Also, some processes are so infrequent or sudden, we would not know they occur except, again, to look at the geologic record.

B Scientists have only recently begun to appreciate just how strongly changes in Earth's environments have affected—and still affect—humankind, from our evolutionary beginnings through the origins of ancient settlements and civilizations—and perhaps their collapse—right up to the present (refer to this chapter's frontispiece). Humans have now begun to affect Earth's environments at rates much faster than the rates of natural processes. The rapid growth of human populations (**Figure 1.1**) has led to the spread of agriculture and deforestation, heavy industry and power plants fired by fossil fuels, and the dependence on petroleum (oil and gas) to power automobiles for transportation (**Figure 1.2**).

The burning of fossil fuels releases **greenhouse gases**, especially carbon dioxide, into the atmosphere. Greenhouse gas traps solar radiation as heat in Earth's atmosphere, causing the atmosphere and surface to warm (**Figure 1.3**). Without carbon dioxide in the atmosphere, Earth's average surface temperature would be about −18°C (0.5°F) instead of its current (and more comfortable!) temperature of +15°C (59°F). But humans have begun to burn fossil fuels at an unprecedented rate, and no one really knows what the outcome will be of the rapid accumulation of carbon dioxide in the atmosphere. In fact, carbon dioxide levels in the atmosphere have increased about 30% since the beginning of the Industrial Revolution (**Figure 1.4**). We know this based on carbon concentrations in gas bubbles found in core samples taken through the glacial ice of Greenland and Antarctica. The bubbles are a record of the composition of ancient atmospheres. As the use of fossil fuels has increased, so too has Earth's average surface temperature, so that the greenhouse effect is no longer considered by most scientists to be purely natural. As far as scientists can tell, these changes will continue through the 21st century and beyond, potentially affecting future human generations, environments, and ecosystems.

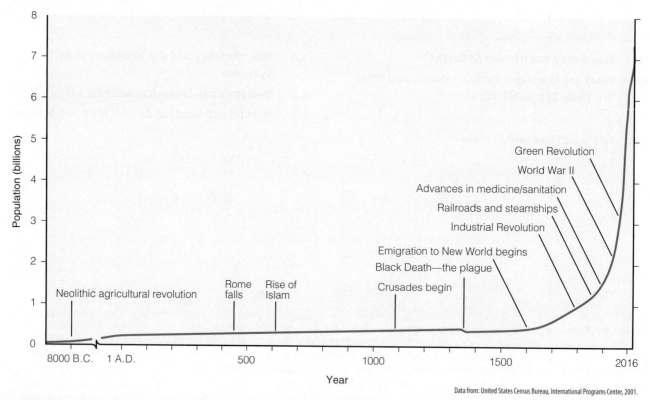

Data from: United States Census Bureau, International Programs Center, 2001.

FIGURE 1.1 Human population growth. Global human population growth since 8,000 B.C. Note the steep rise of population growth beginning in the 1800s in response to the Industrial Revolution and advances in medicine and sanitation.

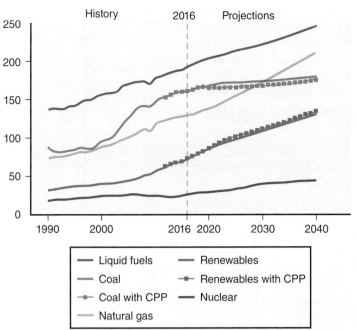

FIGURE 1.2 Historical and projected trends of world energy consumption. Given the projection, society might become increasingly dependent on fossil fuels unless other energy sources are found.

History   2016   Projections

Legend:
- Liquid fuels
- Coal
- Coal with CPP
- Natural gas
- Renewables
- Renewables with CPP
- Nuclear

Data from: the United States Department of Energy, Energy Information Administration, *International Energy Annual*, 2002, 2003 (May–July 2005), 2005, and *System for the Analysis of Global Energy Markets*, 2005 and 2006.

(a)

© Studio 1a Photography/Shutterstock, Inc.

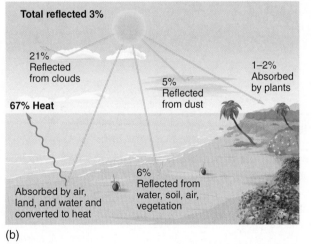

Total reflected 3%

21% Reflected from clouds

5% Reflected from dust

1–2% Absorbed by plants

67% Heat

6% Reflected from water, soil, air, vegetation

Absorbed by air, land, and water and converted to heat

(b)

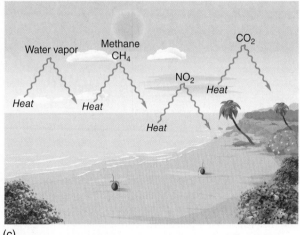

Water vapor   Methane CH₄ CO_2

Heat   Heat   NO₂   Heat

Heat

(c)

FIGURE 1.3 (a) How a greenhouse works. Solar energy penetrates through the glass and is reflected by the floor of the greenhouse as infrared radiation. The infrared radiation is trapped by the glass ceiling and warms the interior of the greenhouse. **(b)** The atmospheric greenhouse effect works in the same way; atmospheric carbon dioxide acts like the glass ceiling of the greenhouse by trapping solar energy that has been reflected by **(c)** the Earth's surface as infrared radiation; this warms the atmosphere. Other gases like methane (CH_4) can combine with oxygen to produce carbon dioxide (CO_2) and greatly exacerbate global warming. Water vapor also contributes to warming. But, scientists have focused on carbon dioxide because of the rapid increase of carbon dioxide concentrations in Earth's atmosphere during the past half century.

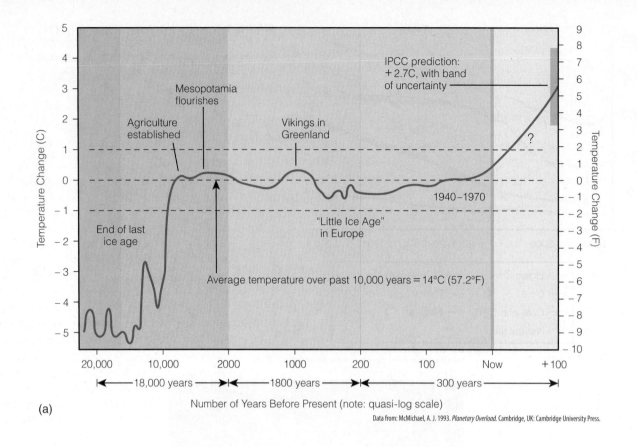

(a)

Number of Years Before Present (note: quasi-log scale)

Data from: McMichael, A. J. 1993. *Planetary Overload*. Cambridge, UK: Cambridge University Press.

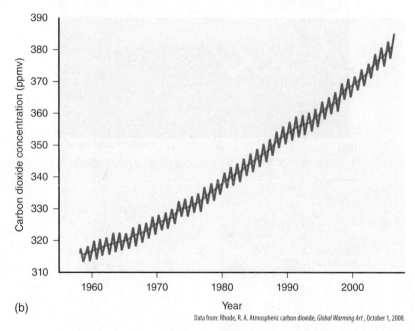

(b)

Year

Data from: Rhode, R. A. Atmospheric carbon dioxide, *Global Warming Art*, October 1, 2008.

FIGURE 1.4 **(a)** Earth's average surface temperature increased during the last half century as compared to the previous 20,000 years. Note that some civilizations or settlements flourished during times of very mildly increased temperatures that were much lower than those of the past half century and projected into the future. There were also far smaller human populations during earlier times. (Compare Figure 1.1.) **(b)** Since carbon dioxide concentrations in the Earth's atmosphere began to be measured in 1958, atmospheric carbon dioxide has risen steadily. These measurements were taken at a station at Mauna Loa, Hawaii. Another station takes measurements in Antarctica. The oscillations reflect seasonal changes in photosynthesis.

C The fact that Earth's environment has likely affected past civilizations and that **anthropogenic**—or human-generated—activities such as fossil fuel combustion are thought by almost all scientists to be affecting Earth has led to the study of Earth as a series of systems. A **system** can be viewed as a series of parts or components that interact together to produce a larger, more complex whole.

Geology, then, is not just about describing and naming rocks and fossils. The entire Earth can be viewed as a system, and it is the record of the interactions of its component systems that we will study in this book. Geology is the science that examines the evolution of the natural processes on Earth, the evolution of life, and the evolution of these interactions and how they have caused Earth to evolve toward its present state. It is the geologic record of rocks and fossils that preserves the history of these interactions. This is what geology studies. Humans are now the primary geologic force on our planet because of the rapidity of anthropogenic change. It is the science of Geology that provides the clues as to how the Earth and its life have behaved—often unpredictably from a modern, anthropogenic view.

What are these systems and how do they interact? Over what scales of time do these systems and their processes interact? What methods do we use to study the history of these systems, and how do we determine the durations of time over which these systems interact? It is these questions to which we devote the rest of this chapter.

1.2 What Are the Major Earth Systems, and What Are Their Characteristics?

D Earth's surface environments are regulated by four major systems and their component subsystems (**Figure 1.5**). The **solid Earth system** consists of the nonliving, solid Earth, from its center to its surface, including the continents and the seafloor. The **atmosphere** comprises the gaseous envelope surrounding Earth, whereas the **hydrosphere** consists of the oceans, rivers and streams, lakes, and ice contained in mountain glaciers and polar ice caps. Glaciers and related environments are sometimes grouped into a separate system called the **cryosphere**. The **biosphere** consists of all living organisms and their dead remains.

In this chapter, we consider the traits systems share in common. First, each major Earth system consists of a series of parts or components that comprise a larger integrated and complex whole. Each of these components in turn consists of smaller parts with their own systems. Some compartments may serve as **reservoirs**, in which certain types of matter (e.g., carbon from photosynthesis) can be stored—or sequestered—for some length of time ranging from perhaps days or weeks to tens of millions of years or more. Second, most natural systems, both living and nonliving, are **open systems** (**Figure 1.6**). This means that the reservoirs of the

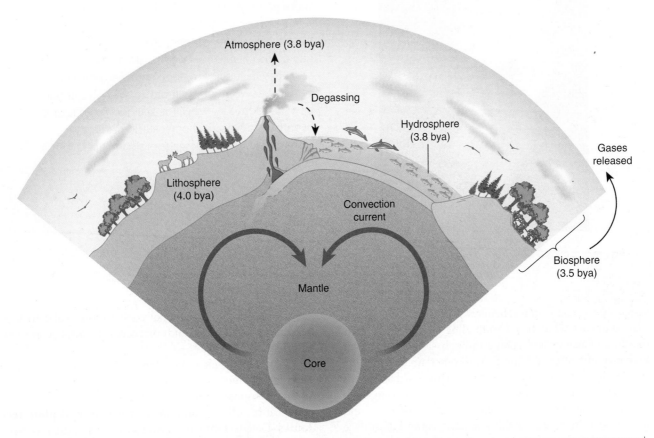

FIGURE 1.5 The four major systems of the Earth and the basic processes within each. Note that each system has its own components and that each system is cyclic. The approximate ages of the systems are shown as billions of years ago (bya).

BOX 1.1 The Origins of the Science of Earth Systems*

Like so many other scientific disciplines, the science of Earth systems originated through the work of more than one person. But if its origins can be traced to one person in particular, it would likely be the Russian scientist Vladimir Ivanovich Vernadsky (1863–1945), whose work and integrity are a source of national pride in both Russia and the Ukraine (**BOX Figure 1.1A**). Vernadsky is so famous that he has had streets and monuments erected in his honor and his image reproduced on postage stamps. In perhaps his greatest work, *The Biosphere*, Vernadsky stated that life is not only *a* geologic force, it is *the* geologic force, and that the role of life on the Earth has increased through time. Indeed, ***humans have become the prime geologic agent on Earth.***

But Vernadsky also had very broad, interdisciplinary interests, stimulated by the work of the French chemist Louis Pasteur, one of the founders of medical microbiology, and the French physiologist Claude Bernard on the concept of **homeostasis** (the tendency of a system to maintain its internal stability in response to external disturbances). Vernadsky also conducted experimental studies on minerals with the French chemist Henri Louis Le Châtelier in Paris.

Vernadsky's *The Biosphere*, first published in Russian in 1926, was translated into French in 1930, but, unfortunately, was not translated into English until 1998; thus, many western scientists remained unaware of it until after the Second World War, when some of Vernadsky's major themes were summarized by the ecologist G. Evelyn Hutchinson (1903–1991). Vernadsky's work then began to give rise to various fields such as geomicrobiology and biogeochemistry: the study of geologically and biologically important chemical cycles like those discussed in this text.

Similar themes began to be advocated in the West during the 1970s by the British scientist James Lovelock in his Gaia hypothesis (named after the Greek goddess of mother Earth), which advocated a "physiology" of the Earth involving not only life but also temperature, atmospheric composition, and ocean chemistry. Nevertheless, many western scientists, including Lovelock, remained unaware of Vernadsky's work until after the Gaia hypothesis had been developed. As we will see, although Lovelock's hypothesis advocated homeostasis, the Earth's systems have also undergone profound directional change, or evolution.

*Summarized from Margulis, L. Foreward, and Grinevald, J. 1998. Introduction: The Invisibility of the Vernadskian Revolution, in Vernadsky, V. I. *The Biosphere.* Copernicus/Springer-Verlag. New York, NY. (Translated by D. B. Langmuir; M. A. McMenamin, Editor), pp. 14–32.

© RIA Novosti/Science Source.

BOX FIGURE 1.1A Vladimir Ivanovich Vernadsky.

---

systems exchange matter (chemical substances) and energy (like sunlight) with their surrounding environment. It is the flow of matter and energy through systems and their exchange of matter and energy with other systems and the surrounding environment—termed **fluxes**—that keeps open systems functioning. For example, **convection cells** (like those found in a pot of boiling water) transfer heat and molten rock from deep within the Earth toward its surface during the process of seafloor spreading. The heat is radiated from Earth's surface into the surrounding environment

(space) as the molten material cools to form solid rock to produce the ocean floor and continents of Earth's outer shell, or **crust**.

Convection cells and seafloor spreading are also responsible for the movement of the continents and large pieces of the lithosphere (the crust and uppermost mantle of the solid Earth), called plates, in what is known as **plate tectonics**. Plate tectonics (or tectonism) refers to the processes that cause the movement of these plates. These processes have produced mountain chains, ocean basins, and other

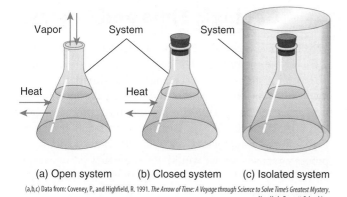

(a) Open system   (b) Closed system   (c) Isolated system

(a,b,c) Data from: Coveney, P., and Highfield, R. 1991. *The Arrow of Time: A Voyage through Science to Solve Time's Greatest Mystery.* New York: Fawcett Columbine.

FIGURE 1.6 **(a)** An open system exchanges both matter and energy with its surroundings. **(b)** A closed system exchanges energy (by temperature changes) with its surroundings. **(c)** An isolated system does not exchange matter and energy with its surroundings. Natural systems are typically open systems.

features on Earth's surface, all the while interacting with the other Earth systems, which have profoundly affected Earth's climate through geologic time.

Convection cells also occur within Earth's atmosphere. Atmospheric convection cells result from the differential heating of Earth's surface and distribute heat and moisture over Earth's surface, thereby affecting surface temperatures and the precipitation patterns of the hydrosphere. Water is critical to life as we know it on Earth; most organisms consist of more than 60% water (and some more than 90%). Water also provides habitats for organisms. Like carbon dioxide, water vapor also acts as a greenhouse gas, affecting Earth's temperature and habitability.

The energy of sunlight penetrating the atmosphere is also used by plants during photosynthesis to produce simple sugars from carbon dioxide and water. These plants are then eaten by herbivores and their stored energy is in turn consumed by predators higher in food pyramids. The biosphere has had a profound impact on the evolution of the Earth. In fact, *life may be viewed as a geologic force.* Without the evolution of photosynthesis on Earth and the storage of carbon dioxide in plant matter, the carbon dioxide levels of Earth's atmosphere would more nearly resemble those of Mars or Venus and there would be little or no oxygen present for respiration (**Figure 1.7**). Plants also profoundly affect the physical and chemical breakdown—or **weathering**—of the rocks of Earth's crust. Weathering processes are critical to the long-term, or *geologic,* cycle of carbon that occurs over tens of millions of years, as we will see in coming chapters.

E These examples illustrate another important point about Earth's systems, namely that they interact to regulate Earth's climate and maintain it in a relatively stable state (homeostasis). These interactions occur through **feedback** (**Figure 1.8**). Positive feedback promotes an effect, whereas negative feedback counters an effect. Most of us are all too familiar with one type of positive feedback: audio feedback. In a sound system, the microphone converts sound (vibrations in the air) into electrical impulses that are

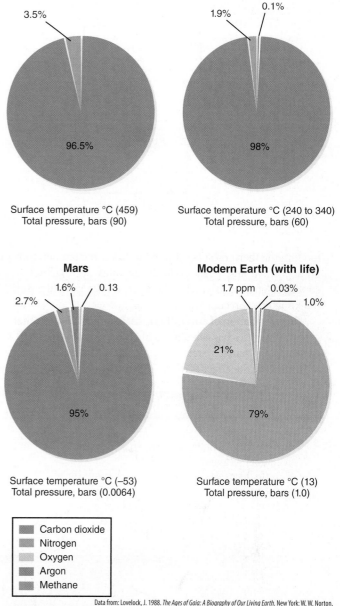

Data from: Lovelock, J. 1988. *The Ages of Gaia: A Biography of Our Living Earth.* New York: W. W. Norton.

FIGURE 1.7 The carbon dioxide concentration of the Earth's atmosphere as compared to that of Mars and Venus if life had not evolved on Earth.

sent to the amplifier, which enhances the signal and sends it to a speaker. The speaker then converts the amplified electrical signals back into vibrations (sound), which are picked up by the microphone and sent to the amplifier, producing an ear-piercing sound. Negative feedback in this system would have just the opposite effect: the reduction of sound.

Typically, however, positive and negative feedback act together to maintain homeostasis within a system; otherwise, the system would spiral out of control (positive feedback) or go to extinction (negative feedback). Another example of homeostasis produced by feedback is the temperature control of a house. If the house becomes too cold, a thermostat is triggered that turns on the furnace to warm the house to the desired level set on the thermostat. If the house becomes

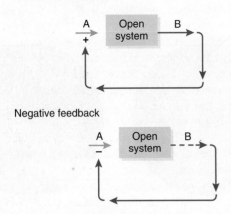

Positive feedback

Negative feedback

FIGURE 1.8 Positive and negative feedback are involved in temperature regulation.

too hot, the thermostat is triggered to lower the temperature. Both positive and negative feedback must act together to counterbalance each other; otherwise, a system would shift too far in one direction.

The regulation of atmospheric carbon dioxide and other greenhouse gases has acted in a similar manner (**Figure 1.9**). Nevertheless, Earth systems may act in a nonlinear manner, meaning that the effect is much greater than the cause.

CONCEPT AND REASONING CHECKS

1. What are the characteristics of open systems?
2. Are natural systems open systems?
3. What are the basic components of each system just described?

1.3 Geologic Time and Process

Our discussion of feedback has raised the related issues of time and process. The processes relevant to the study of Earth's systems vary according to the duration of time involved. *Many processes occur so slowly they are imperceptible to us on human time scales.* Nevertheless, these processes have had a profound impact on Earth's environments over long spans of many millions of years. On the other hand, some processes are so rare, we would not know they occur except, again, to look at the geologic record. In either case, though, *the only way to detect the existence of these processes, their potential impact on humanity, and humanity's potential impact on them is to examine the geologic record of these changes as they are preserved by the rocks and fossils.*

To come to grips with this important realization, one must wrestle with the enormity of **geologic time**. As the first Earth scientists began to study the planet's history in earnest, they quickly recognized that rocks occur in layers that lay on top of one another. In the 1600s this basic observation by Nicolaus Steno led to the **Principle of Superposition**, which states that in a sequence of rocks, younger rocks lie on top of older rocks (**Figure 1.10**). Thus, the sequence of rocks provides what are called **relative ages**, meaning that something is older or younger than something else. Still, most scientists refrained from speculating on how old the rocks were and generally assumed the Earth was only a few thousand years old, based on biblical interpretation. During the 18th and early 19th centuries, however, scientists began to realize that Earth was far older than previously imagined. One of the primary pieces of evidence for this conclusion was the relationship of stratified rocks like those at Siccar Point, along the northeast coast of England. Here, horizontally layered

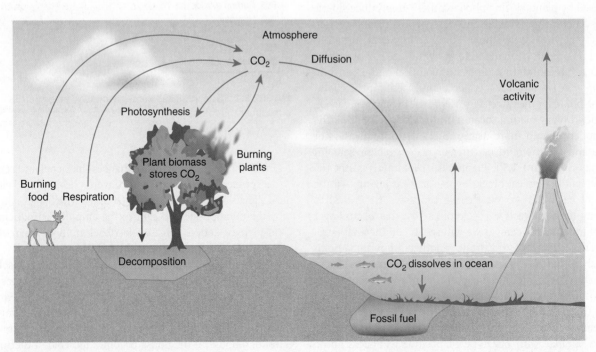

FIGURE 1.9 The carbon cycle. Photosynthesis, respiration, and organic decay in the carbon cycle transfer matter and energy between reservoirs within ecosystems.

Courtesy of Ronald Martin, University of Delaware.

FIGURE 1.10 The Principle of Superposition states that younger rocks are laid down on top of older rocks. Thus, the oldest rocks occur at the bottom of the section and the youngest at the top. These stratified (layered) rocks are exposed in the Grand Canyon.

(or **stratified** [from "stratum" for layer]) rocks had originally been laid down and then tilted and eroded, after which more stratified rocks were deposited on top (**Figure 1.11**). Because processes that could produce such phenomena are not observed, the inference was that the processes must occur that produced the rocks themselves and their angular relationship must occur unimaginably slowly as compared to human time scales. The recognition of the enormity of time is generally attributed to the so-called father of geology, James Hutton (1726–1797), a Scottish farmer who was a member of a small circle of Enlightenment naturalists and scientists of the 1700s who periodically met in the pubs of Edinburgh, Scotland, to discuss natural history and science (or "natural philosophy," as it was called in those days) (**Figure 1.12**).

About this time, fossils were also coming to be accepted as evidence for prehistoric life. The presence of fossils as evidence of ancient life eventually led to the recognition in 1815 of the **Principle of Faunal Succession**, which states that different groups of fossils follow each other in a characteristic upward sequence, that is, through time. Faunal succession was originally established in England, but with its recognition scientists began to look more closely at the occurrence of fossil assemblages elsewhere. This led to the development of the geologic time scale, the subdivisions of which are largely based on the occurrence of distinctive fossil assemblages through time (**Figure 1.13**). For example, the Cretaceous Period, when Earth was quite warm and the seas quite high, is named for the widespread chalks ("creta") that consist of the microscopic remains of ancient plankton that floated in the seas during this period and which are integral to the geologic cycle of carbon that occurs on durations of many millions of years (**Figure 1.14**). The geologic time scale is, then, a record of the history of life itself. What the overall succession of life forms tells us is that life on Earth evolved during a time span of 4.5 to 5 billion years from the relatively simple to the, seemingly, more "advanced." Because the biosphere has interacted with and influenced the other Earth systems through this span, it is paramount that the student knows the geologic scale of time and how it evolved (Figure 1.13; see also **Chapter 6**).

The initial time scale represented relative ages. Much later, during the 20th century, **absolute ages** (dates in years) began to be determined from the rates of **radioactive decay** of certain minerals. Because radioactive elements decay at known rates, they can be used to calculate the ages of the rocks in which they occur. The modern time scale, which incorporates both changes in fossil assemblages through time and absolute ages, is routinely used in studies of Earth systems, and we will refer to it frequently throughout this book.

The initial recognition that Earth was quite old raised several profound scientific and philosophical questions: Are the processes that we observe and measure today on human time scales necessarily representative of the geologic past? In

(a)

Courtesy of Clifford E. Ford.

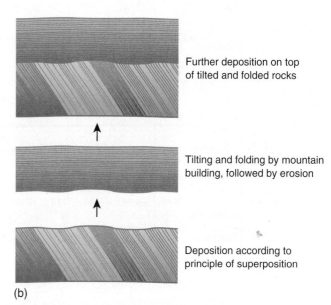

Further deposition on top of tilted and folded rocks

Tilting and folding by mountain building, followed by erosion

Deposition according to principle of superposition

(b)

FIGURE 1.11 (a) James Hutton recognized the enormity of geologic time based on his observations of soil formation and the relationships of stratified rocks like those at Siccar Point, Scotland. **(b)** Interpretation of the events affecting the relations of the strata.

(a)

(b)

FIGURE 1.12 Two founders of geology. **(a)** James Hutton (1726–1797). **(b)** Sir Charles Lyell (1797–1875), who is credited with formulating the Principle of Uniformitarianism.

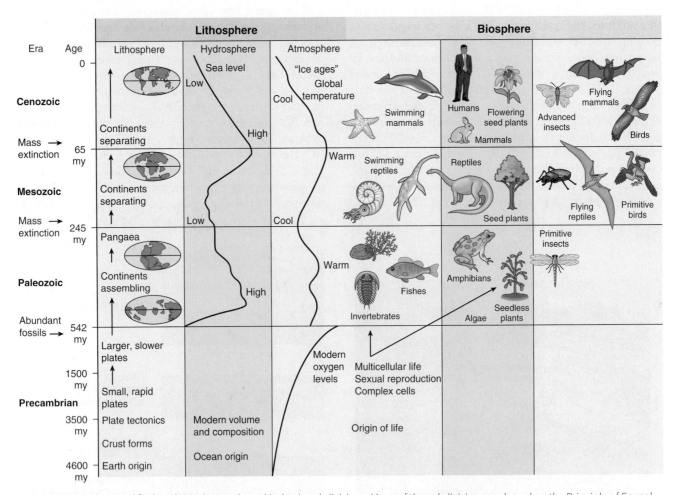

FIGURE 1.13 The simplified geologic time scale and its basic subdivisions. Many of the subdivisions are based on the Principle of Faunal Succession, which recognizes that the Earth's biota has changed through time. Some of the major changes in the biotas are shown. The absolute ages come from radioactive elements that decay at known rates. The eras and periods of the geologic time scale were largely recognized in the 19th century. "Paleozoic" refers to "ancient life," "Mesozoic" as "middle life," when more modern forms began to greatly diversify, and "Cenozoic" means "recent life," many forms of which have persisted to the present.

other words, does nature act in a uniform manner? If so, it must take a long time for many geologic processes to have an effect. Or, can the *rates* of the processes vary, and can the *processes* themselves vary through time?

The guiding principle that leads Earth scientists through the maze of Earth processes and systems is the **Principle of Uniformitarianism**. This principle is frequently stated as "the present is the key to the past." In other words, the processes we observe today are the same as those that have always operated. The origins of uniformitarianism are typically traced to James Hutton's *Theory of the Earth, with Proofs and Illustrations* (1795), which discussed his observations and those of others. Some earlier scientists had begun to think along similar lines, even going so far as to believe Earth was quite old, but Hutton was the first to have synthesized his ideas into a theory based on observation. Hutton had been trained as a physician and, as a student, had defended a thesis on the circulation of human blood. Hutton's studies in medicine undoubtedly influenced his view of Earth's processes. After he inherited a small farm, Hutton became

interested in how soils form and are replenished. Among his conclusions was that natural processes are self-renewing, or cyclic, like the circulation of the blood (and the processes of Earth's systems introduced previously in this chapter). Hutton further concluded that these processes are typically quite slow and so the Earth's history must span an unimaginable amount of what we now call "geologic time."

He also became interested in how rocks form. Hutton was well aware of the power of heat because of other studies of how to make steam engines more efficient for the ongoing Industrial Revolution. He ultimately concluded that mountain building results from the activity of heat and that Earth therefore had to have an internal source of heat. In Hutton's theory, volcanoes acted as valves to release the heat and pressure generated within the Earth (we now know that the source of Earth's internal heat is radioactive decay, which drives the processes of plate tectonics).

Hutton's work laid the foundation for more serious geologic inquiry in the 19th and 20th centuries. Among these works, those of Sir Charles Lyell (1797–1875) have perhaps

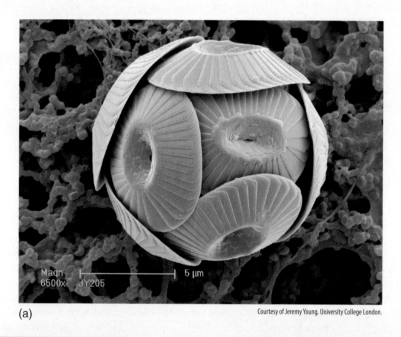

(a)

Courtesy of Jeremy Young, University College London.

(b)

Courtesy of Michele Miller.

FIGURE 1.14 **(a)** The skeleton of a modern unicellular alga called a coccolithophore. **(b)** The White Cliffs of Dover consist of chalk produced by vast numbers of these kinds of algae during the geologic period called the Cretaceous ("creta" means chalk) period, when the seas flooded the continents (refer to Figure 1.13).

(continues)

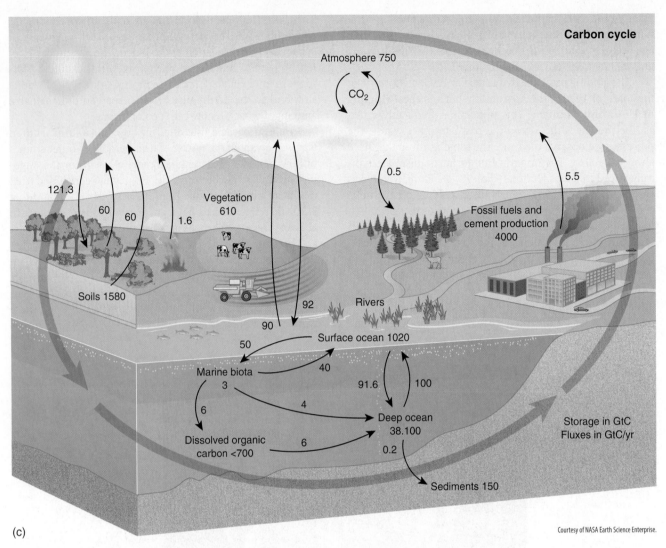

Carbon cycle

Atmosphere 750

CO_2

121.3

Vegetation 610

60 60 1.6

Soils 1580

0.5

Fossil fuels and cement production 4000

5.5

92

Rivers

90

50

Surface ocean 1020

Marine biota
3

40

91.6   100

6

4

Deep ocean
38.100

Dissolved organic carbon <700

6

0.2

Storage in GtC
Fluxes in GtC/yr

Sediments 150

(c)

Courtesy of NASA Earth Science Enterprise.

FIGURE 1.14 (Continued) **(c)** Coccolithophorids and other calcareous plankton are involved in the geologic cycle of carbon. Modern volcanoes pump carbon dioxide into the atmosphere; feedback by increasing levels of carbon dioxide in the atmosphere resulting from ancient eruptions must have similarly spewed enormous volumes of the same gas into the atmosphere. The carbon dioxide then remained in the atmosphere for long periods of time, spanning tens of millions of years. Based on the greenhouse effect, the carbon dioxide warmed the Earth, speeding up—through positive feedback—the chemical reactions involved in the weathering of rocks on land. As these reactions were speeded up, they in turn began to act as negative feedback: much of the carbon dioxide pumped into the atmosphere by volcanism was used up by weathering, eventually lowering atmospheric concentrations of carbon dioxide and lowering Earth's surface temperature. These weathering reactions produce a type of sedimentary rock called limestone. The carbon dioxide in the atmosphere is transferred to and sequestered (or stored) in limestone by coccolithophorids that secrete microscopic platelets of calcium carbonate ($CaCO_3$) to produce a form of limestone. In the geologic past these particular organisms produced vast amounts of limestone on the continents, but if the limestones are deposited in the deep sea, they can eventually be carried into deep-sea trenches, where they are subducted and heated. The carbon dioxide stored in the limestone (as carbonate) is released and vented back to the atmosphere through volcanoes. These processes therefore act as positive feedback on atmospheric carbon dioxide levels.

had the greatest influence on geology (Figure 1.12). Lyell developed Hutton's views further by gathering enormous amounts of evidence from his travels in Europe. Lyell's synthesis of his observations and conclusions resulted in the publication of his three-volume treatise, titled *Principles of Geology*, in England in 1830; 11 more editions of the *Principles* followed into the 1870s.

Lyell is usually credited with developing the Principle of Uniformitarianism (although he did not coin the term), which he used to counter the rival doctrine of what he derisively called **catastrophism**. Lyell did not deny that

catastrophic changes such as earthquakes or volcanic eruptions occurred on a *local basis*, but he condemned the concept of *global* catastrophes that he attributed to the so-called catastrophists. According to Lyell, catastrophism viewed the Earth's geologic record as having resulted from a series of sudden global catastrophes. Lyell also argued that catastrophism was an attempt to make Earth's age fit with a chronology (time scale) derived from the Bible. This portrayal of catastrophism by Lyell was grossly inaccurate, but Lyell was a lawyer and he used his skills at argumentation and rhetoric to build a convincing argument for his own case.

According to Lyell, the Principle of Uniformitarianism meant that the natural processes of today have *always remained the same and are therefore typically very slow and gradual*. Indeed, given that Earth is now known to be approximately 4.5 to 5 billion years old (based on absolute ages), there has been plenty of time for very slow processes to produce major changes on Earth, just as Lyell (and Hutton) said. Lyell also concluded, however, that not only are the rates of geologic processes typically slow and gradual, but *the rates of modern processes are exactly the same as those that have acted in the past*. Lyell's concept of slow, gradual change dominated geologic thought until well into the 20th century, and his influence on the earth sciences up to the present cannot be overestimated.

The uniformitarian approach is sometimes referred to as **actualism**. According to actualism, processes observed in modern environments, using modern **analogs**, or examples, can be extrapolated to the interpretation of ancient environments. Thus, if a past phenomenon can be explained as a result of processes *actually* observed today, we need not invent or search for other processes to explain the phenomenon. In this view the Principle of Uniformitarianism is the broader assumption that the behavior of nature has been uniform through time. Geologists and paleontologists (those who study the history of life) are especially proud to claim the Principle of Uniformitarianism as their own because of its implications for Earth history. Nonetheless, *the Principle of Uniformitarianism underlies all scientific inquiry*. No form of scientific research could be conducted if nature did not behave—and if we did not *assume* that nature behaves—in a relatively uniform manner. For example, chemical elements and the compounds they form must have always behaved the same way chemically; otherwise, a chemist could not conduct meaningful research, even under the controlled conditions of a laboratory experiment. Normally, actualism and the Principle of Uniformitarianism seem to work quite well, as attested to by, say, obtaining similar results in repeated laboratory experiments.

G What is typically left unsaid about uniformitarianism and actualism is *that the past is often the key to the present*. As we have already noted, the geologic record demonstrates that there are processes acting on Earth that either act so slowly or so infrequently as to be imperceptible on human time scales. Moreover, scientific reasoning from rocks and fossils often indicates that, given enough time, unusual processes or events not observed today have also occurred. In fact, we now know that present-day conditions were often not representative of the past and that a number of unusual—even catastrophic—changes on Earth have occurred. For example, we know from the geologic record that movement of the continents on geologic scales of time has interacted with the other Earth systems to drastically alter climate. The geologic record also tells us that at times meteors have collided with Earth to produce certain **mass extinctions** (**Figure 1.15**). Thus, Earth scientists have broadened their view of the Principle of Uniformitarianism to mean that (1) processes have largely remained the same but the *rates* of the processes can vary, (2) processes vary according to the

Courtesy of Billy Glass, University of Delaware.

FIGURE 1.15 The geologic record of a mass extinction. The hammer spans a dark layer enriched in the chemical element iridium preserved in sedimentary rocks in Colorado. The iridium layer formed about 65 million years ago at the end of the Cretaceous Period as a result of the collision of an extraterrestrial body with the Earth. The collision sent enormous volumes of dust and iridium into the atmosphere and then settled to produce a global layer marking the extinction. This particular extinction is thought to have resulted in the final demise of the dinosaurs.

duration of time, and (3) sometimes highly unusual conditions and phenomena occur, including catastrophic ones like meteor impacts and mass extinctions.

1.4 Directionality and the Evolution of Earth Systems

An important aspect of Lyell's *Principles of Geology* was that he did not discuss what he called Earth's "cosmogony," or origins; to do so, according to Lyell, exceeded the bounds of uniformitarianism. This is because Lyell envisioned Earth as being in **equilibrium**, meaning that a system exhibits no *net* change. Thus, according to Lyell, Earth had undergone no net change through time. To be sure, Lyell envisioned Earth as changing, but he envisioned no overall, or net, change to Earth. In Lyell's view, for example, mountains forming in one place were counterbalanced by erosion somewhere else.

Although Earth's systems do behave in a cyclic manner, as Lyell and Hutton envisioned, there is substantial evidence that Earth has also evolved; that is, that the seemingly equilibrium states of Earth (no net change) have actually shifted through time in particular directions. This is because of the flux—the inflow and outflow—of matter and energy through the open systems and subsystems of the Earth (Figure 1.6). Open systems are often said to be in "equilibrium" with their surroundings because they appear to be undergoing no net change. This is because we do not observe these systems over sufficiently long durations of time, *geologic* time. Such open systems are more accurately said to be in **steady state**, but even this term is a bit of a misnomer. Systems in steady state actually exist far from the *true* equilibrium of an isolated system. *True* equilibrium is tantamount to the death of a living organism; the exchange of matter and energy by the living organism with its environment has ceased. Open systems can undergo directional change because the flux of matter and energy maintains them far from this true equilibrium. (You might want to consult von Bertalanffy's book. For more details about it, go to the section "Sources and Further Reading.")

Indeed, as indicated by the title of this book, *Earth's systems have evolved and continue to evolve.* The word **evolution** means to develop or unfold. In the case of Earth as a whole, with all its interacting systems, the evolution of Earth has resulted in **directionality**—or **secular change**—through time. Thus, even though Earth's systems act in a cyclic manner, they also tend to change, inexorably, in certain directions through geologic time. *This secular change in Earth's systems is another fundamental reason for studying Earth's history.* Slow, directional change means that a system can appear to exhibit no net change, like the environments we observe on human time scales, when in fact environments might be changing in a directional manner on geologic scales of time. Furthermore, what we observe or measure today is by no means always representative of Earth's distant past because conditions on Earth have changed. Therefore, we cannot fully understand Earth's systems simply by looking at how they behave at present, because Earth's systems have evolved. Indeed, Earth's history has really consisted of a succession of vastly different worlds leading up to ours.

Directional changes in Earth's systems have resulted from the flow of matter and energy between Earth's systems through geologic time. Among the numerous examples of directionality in the history of Earth's systems that we will consider are (1) the formation of the solid Earth by cooling from a molten ball during the time the solar system formed, (2) the formation and evolution of the continents by the continued transfer of molten material and heat from within Earth's interior to the lithosphere, (3) the early evolution of the atmosphere during the degassing of the primitive Earth by volcanoes, and (4) the origin and evolution of life, which led first to photosynthesis and the oxygenation of the atmosphere and then ultimately to the origin of more complex organisms. The biosphere eventually began to be integrated with Earth's other systems and acted as a geologic force to produce new cycles, such as the geologic cycle of carbon.

Despite the flow of matter and energy through natural systems, however, natural systems are often assumed by earth scientists to exist in a state of equilibrium, oscillating about an "average" long-term condition. In the short term, the system might change, but in the long term its average behavior presumably does not. This view of nature is widely accepted today, by which it is often referred to as the state of **dynamic equilibrium**. If the system is disturbed away from its so-called equilibrium state, it is assumed that it will return to its original state, given enough time. In other words, there is presumably no net change to the system because not enough time has passed to have allowed net change to occur.

1. What distinguishes the Principle of Uniformitarianism as Lyell recognized it from the Principle of Uniformitarianism as it is now recognized?
2. What sorts of evidence indicate directionality in Earth's history?
3. What has caused directional change in Earth's history?

1.5 Geology as an Historical Science

Geology and the other Earth sciences are often termed **historical sciences**. What is meant by the term "historical science"? So far in this chapter, we have dealt with Earth systems as if they were components of a larger entity resembling a machine. The idea that nature and its components can be treated as machines is not new. It dates at least to the time of Galileo (1564–1642) and René Descartes (1596–1650; a French mathematician and philosopher for whom the Cartesian coordinate system of geometry is named). Even Hutton and other scientists of his time viewed the Earth as machine-like, in part because of Sir Isaac Newton's earlier formulation of the law of gravitation based on the behavior of the planets.

With the doctrine that nature is a machine came another view: that nature can be understood by taking it apart, like a machine, into smaller parts that are more easily analyzed. This view has had a tremendous impact on scientific thinking and is referred to as **reductionism**. For example, we broke the larger whole Earth system into four basic component systems: atmosphere, biosphere, hydrosphere, and solid Earth. Admittedly, these are very large parts, each of which is itself a system! But, by breaking a system (in this case, the whole Earth) down into smaller components, we can presumably examine the structure and behavior of the smaller components and identify particular cause and effect relationships. Each component can in turn consist of smaller compartments with their own systems and fluxes. However, as we have already seen, Earth's systems are not isolated but open to each other, and we cannot fully understand them without understanding their interactions.

Much of science is also based on the assumption that each effect (observation, outcome) has a *particular* cause and that we can predict an effect given a particular cause; this view of science is called **determinism**. Many of the phenomena and processes studied by scientists, including those who study Earth systems, occur on sufficiently short time scales that the processes and their effects can be observed or measured. Determinism is associated with the simplest notion of cause and effect, namely that an effect has a single cause, like that observed in a laboratory setting, which, unlike natural systems, typically represent isolated or closed systems in which all factors thought to affect the laboratory system are held constant and the factor of interest is allowed to vary. The notion that the simplest explanation of a phenomenon is the one that is most likely to be correct is called **Ockham's Razor**, or the **Principle of Parsimony**, after William of Ockham (ca. 1297–1347). This principle has become an integral part of the scientific method because it emphasizes that one should seek simplicity in scientific explanation.

Nevertheless, Earth's phenomena—its processes and their outcomes—are not strictly deterministic because in natural settings the same effect might have multiple or overlapping causes. On seasonal time scales, for example, atmospheric carbon dioxide is regulated by seasonal changes in sunlight and photosynthesis (Figure 1.4b), but when taken on the scale of many millions of years, continental weathering also affects atmospheric carbon dioxide. Another prime example of the notion of **multiple causation** is the mass extinction of Earth's biota. Mass extinctions occurred a number of times during Earth's history, but not all have resulted from meteor impacts. The exact cause of each episode has varied, and in some cases there might have been more than one cause that overlapped in time. The processes that occurred during the ancient past have also undoubtedly varied with historical circumstances, or contingency. **Contingency** means that historical circumstances influence the outcome of one or more processes or events. In other words, processes and events in Earth's history are conditional—or contingent—on what happened before. Historical processes set the stage for, and can even constrain, the future behavior of Earth's systems. For example, although the chemical reactions themselves involved in weathering have remained the same (based on the Principle of Uniformitarianism), their rates have been affected by the kinds of rocks exposed at Earth's surface (which in turn depends on processes like plate tectonics and mountain building), the presence or absence of land plants (the roots of which accelerate weathering), and the concentration of atmospheric carbon dioxide, which, as already discussed, also affects rates of weathering.

1.6 Method and Study of Earth's Evolving Systems

Many students have been introduced to "the" scientific "method," but the scientific method in earth sciences differs somewhat from that which is normally taught in the classroom. Earth scientists, especially geologists and paleontologists, tend to think in a way that differs from that of other scientists, in what to some degree resembles that of historians. This is because *they are historians*. The scientific method as it is used in historical sciences like geology is a recurring theme throughout this book.

The scientific method is *iterative,* meaning it might need to be repeated. This also means the method is *self-correcting.* In other words, the scientific method feeds back on itself in a kind of feedback loop that is supposed to keep the thinking reasonable and on the right track (**Figure 1.16**). Through this process of finding out what *doesn't* work, we presumably come to a better understanding of what *does* work. In

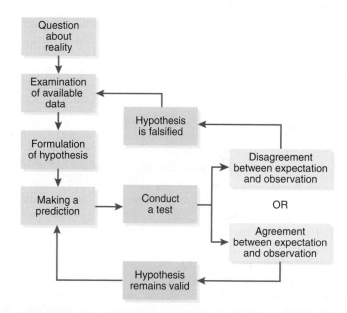

FIGURE 1.16 The iterative process of the scientific method.

other words, scientists attempt to *falsify* everything they can, and whatever is left presumably explains the phenomenon of interest, *at least for the time being.* The development of the modern scientific method as outlined so far is largely attributed by scientists to the 20th century philosopher, Karl Popper. Popper first published his highly influential work *The Logic of Scientific Discovery* in 1934. Popper stated that *the best hypotheses are the ones that can be easily falsified* because then one knows for sure that the hypothesis is false and can move on to other hypotheses. However, although the hypotheses that are not falsified gain support, they too might someday be falsified as new data appears or new tests develop, for example by new technologies.

This approach to science is best exemplified by experimental sciences such as physics and chemistry. Experiments usually involve a laboratory system (or a field setting) in which all conditions are held constant except one to determine the role of that particular condition, known as a *variable*, in the system; this makes prediction much easier because the system is essentially being treated as a closed system, "cut out" from its surroundings and thus its interconnections with other systems. Each experiment must be accompanied by what is called a **control**, which is identical to the other experimental setups, except that none of the conditions is varied. The control allows one to distinguish any other potential change that might have occurred while a particular variable was being manipulated during the experiment. Earth scientists can also use experiments to formulate and test hypotheses about what happens in nature. These experiments might be done in the laboratory (for example, on the melting and crystallization of rocks) or in the field (such as experiments on rates of weathering under different conditions). The results are then applied to understanding the Earth under more complex conditions. So far, this is the scientific method into which students have been indoctrinated.

More often than not, however, geologists use data that already exists in nature and that are the end products of the natural processes of open systems with all their interactions, feedbacks, and contingencies. Geologists typically test hypotheses by first looking for rocks, sediments, or fossils of the appropriate age—on land, in cores taken in the oceans, or both—and make observations or measurements on the variables of interest. Then, geologists use what some have termed *retrodiction* or *postdiction* (to distinguish it from *prediction*), meaning *geologists must frequently reason backward from geologic data such as rocks and fossils to infer the processes and the rates of those processes that produced what is being observed or measured.* This is why the Principle of Uniformitarianism is so important, because without it we would not be able to justify such a procedure.

On the other hand, *it is not uncommon for the geologic record to indicate unusual, perhaps even unique, conditions not observed by humans.* For example, when iridium was found to be associated with the end-Cretaceous extinctions, the hypothesis appeared that all mass extinctions had been caused by meteor impacts, and there was an immediate search for iridium layers in end-Cretaceous sections associated with other mass extinctions. When was the last time you observed a meteor impact?

Obviously, there is a greater amount of uncertainty involved in testing most hypotheses in earth sciences because earth scientists are less assured of the constancy of conditions in the natural environment through time than in a laboratory. Earth scientists, especially geologists and **paleontologists**, are thus forced to reconstruct past phenomena through postdiction. It is primarily for these reasons that historical sciences are often viewed as being less "scientific" and as being derivative sciences based ultimately on physics and chemistry. Hutton and, especially, Lyell felt this way: Hutton, perhaps because he was associated with the Enlightenment and Lyell because of his extreme form of uniformitarianism. But the argument that experimental sciences are somehow superior to historical ones is flawed for several reasons. First, any experimental setting makes certain assumptions, consciously or subconsciously, and the underlying assumptions can be so numerous that they cannot all possibly be taken into account. Assumptions underlying an experiment might be false and, if so, could alter the outcome of an experiment, so the hypothesis is rejected when it is actually true, or vice versa. Second, even if the results are unambiguous and the hypothesis is rejected, scientists frequently do not discard the hypothesis completely. They either modify it or test it under different conditions. So-called anomalous results or phenomena (such as the iridium layer discovered near Gubbio, Italy) can even lead to new discoveries undetectable on human time scales. When iridium was not detected at all other extinction boundaries, scientists did not abandon the impact hypothesis; instead, they focused on the extinctions associated with iridium or searched for new sections that might be more complete and contain iridium layers.

Ideally, scientists are supposed to formulate multiple hypotheses to explain natural phenomena. If scientists favor a "pet" hypothesis over another, they are, according to conventional wisdom, more likely to overlook important phenomena or data bearing on the question of interest. This **method of multiple working hypotheses** was proposed in 1897 by the American geologist T. C. Chamberlin. The method of multiple working hypotheses is especially apropos to earth sciences, in which multiple factors acting on different scales of time can interact to produce a final effect (see the previous discussion on multiple causation).

However, rather than each scientist formulating multiple hypotheses, what typically happens is that scientists working on a particular research problem tend to form "social" groups that favor a particular hypothesis. Although this is perhaps undesirable and can seem odd, it is understandably human. Each scientist has had different training and different experiences than others, so it is not surprising that each one brings different viewpoints to a problem and different ways of thinking about it and attempting to solve it. Those with similar viewpoints on a particular question tend to proceed by consensus, more or less mutual agreement. Very importantly, however, different parties or individuals might strongly disagree with the consensus on a particular question; "group think" can therefore stifle independent thought and scientific research. In fact we are far from knowing everything there is to know about Earth and its history. If this were not true, we would not have the multiplicity of hypotheses about certain events and processes on Earth!

Furthermore, as previously stated, the Principle of Uniformitarianism itself underlies all of science. The process of scientific induction itself cannot be proven to be the best scientific method. In the process of induction, premises established by observation, measurement, or calculation are used to infer the unknown of the past, present, and future. This problem is sometimes referred to as **Hume's problem**, after the Scottish philosopher who formulated it, and has long plagued philosophers. Briefly stated, Hume posed the question: how do we know induction is the best method? To establish unequivocally that induction is the best method, we would have to use deduction. But a deductively valid argument does not make conclusions that go beyond the premises because the conclusion states things we already know. Furthermore, the premises of deductive arguments are typically based on certain assumptions. These assumptions include the principle of the uniformity of nature! What, then, about the future, which is unknown? How could we use a deductively valid argument to infer that induction will work in the future in a laboratory setting, much less for an ancient historical one? We are, Hume argued, reduced to using induction to validate induction. We really have no other choice than to *assume the uniformity of nature* embodied in the Principle of Uniformitarianism. According to the argument, induction has worked in the past, so it will work in the future because nature's behavior is presumably uniform. Although many philosophers have attempted to subvert Hume's problem by logical means (Popper's method of falsification was just one attempt), no one has ever satisfactorily demonstrated another approach that thoroughly replaces induction. But nature is not always uniform. Iridium layers and mass extinctions are prime examples but, as we will see throughout this text, by no means the only ones!

As we will also see, although induction *seems* to work *most* of the time, the "experts" have frequently been wrong. In other chapters, for example, we will examine how the hypothesis of "continental drift" was formulated about 1915. This hypothesis was constructed using an inductive argument, the premises of which were the "jigsaw" fit of the continents and matching rocks and fossils on continents now located on opposite sides of the Atlantic Ocean. From these premises, it was concluded by Alfred Wegener that the continents must have originally been joined together into a larger supercontinent named **Pangaea** ("all land"; **Figure 1.17**). It was this hypothesis that would ultimately lead to the theory of plate tectonics. However, in 1915 there was no known force (which we now know to be related to convection cells and seafloor spreading) that moved the continents. As a result, almost all other scientists rejected the hypothesis. Not until the second half of the 20th century did scientists begin to obtain data that indicated the existence of convection cells deep within the Earth and their role in driving plate tectonics and the movement of the continents.

Then, the hypothesis began to be regarded as part of a larger theory now called plate tectonics. The value of a theory, like the theory of plate tectonics, is that it takes observations and data and synthesizes them into a larger, more coherent explanatory picture. Hutton's *Theory of the Earth* (see the section "Sources and Further Reading") and Lyell's *Principles of Geology* are examples of geologic theory. The prime value of a theory is that it suggests further hypotheses and tests to substantiate or refute the theory; thus, a theory might be modified, as was Lyell's original view of uniformitarianism, or even overturned. Another way in which

theory develops is when scientists formulate conceptual frameworks that produce hypotheses that can be tested. This kind of theory might originate as "thought experiments" in a scientist's mind when they imagine, "What if . . .?" In either case, the formulation of a theory represents a scientifically valid approach. If a scientific relationship is called a theory, there must be a reason. Thus, you must be wary when you hear phrases like "it's only a theory," or "it's just theoretical." What are the data for the theory? Have hypotheses stemming from the theory been tested and corroborated?

More much certain relationships between phenomena, which are considered fundamental truths, are called **laws**. Laws are embedded within a larger body of theory. Laws note strikethrough are frequently "empirical" or derived from observation and experiment. Such laws attempt to explain repeated observations without really understanding why the phenomena occur. Sir Isaac Newton's laws of motion, which can be used to predict eclipses and planetary movements both in the future and the past, are examples of such laws (but in Newton's time, like Wegener centuries later, some scientists could not understand how a force called gravity could act across the great distances separating planets). Partly because of the power of these laws, Newton is often considered the quintessential scientist of all time. For a relationship to be a law, it must be repeatable and hold under a variety of conditions. Newton's laws of motion hold, for example, in other solar systems; otherwise, they would not be laws. But laws are *abstractions*. They help explain what has happened or will happen, irrespective of other factors such as contingency or historical circumstances, but they do not tell us *exactly what happened*.

Earth scientists deal with history, which is about changing conditions through time. Historical conditions are never

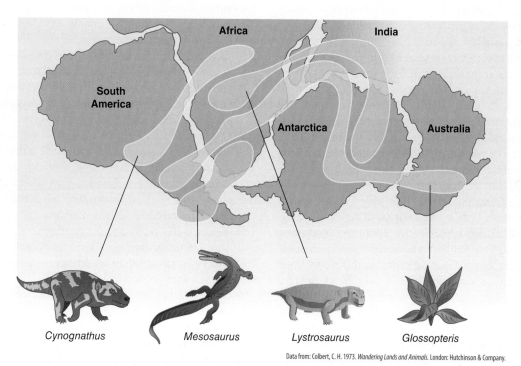

Data from: Colbert, C. H. 1973. *Wandering Lands and Animals*. London: Hutchinson & Company.

FIGURE 1.17 The hypothesis of continental drift was based on the "jigsaw" fit of the continents and matching rocks and fossils on continents now located on opposite sides of the Atlantic Ocean.

exactly the same because of contingency. So, instead of laws, earth scientists tend to rely on theory, usually backed up by observation and measurement, and, to a lesser extent, experiment. Instead of laws, general truths in geology are usually referred to as **principles**. A principle is like a good "rule of thumb" because it might have some exceptions, whereas laws are supposed to be invariable. The Principles of Superposition and Uniformitarianism are examples. Superposition typically holds unless, for example, rocks have been overturned by mountain building, resulting in their order from top to bottom being reversed.

Despite this outline of basic method, most scientists don't go around thinking continuously about using the scientific method; their thinking is typically much, much looser than that. Instead, scientists tend to look primarily for two things: phenomena that require explanation (i.e., a "good" research question or problem) and ways to test hypotheses or explain phenomena. As you proceed through the text, you should be aware of how hypotheses or theories have been arrived at and tested in earth sciences.

CONCEPT AND REASONING CHECK

1. How does the science of geology resemble or differ from other sciences with regard to processes and their rates, experimental versus natural conditions, causation, contingency, and laws?

SUMMARY

- Earth's climate has influenced human civilization for thousands of years and will continue to do so in the future. During the 20th century, the expanding human population, agriculture, and industrialization all began to significantly affect Earth's environments. The view among scientists that human activities are dramatically influencing Earth's climate and biota led to the science of Earth systems.

- A system is a series of parts or compartments that comprise a larger integrated whole. The major Earth systems are the solid Earth, atmosphere, hydrosphere, and biosphere. Each system is an open system characterized by the cyclic flow of matter and energy.

- Earth's systems share a number of features: The systems maintain themselves far from equilibrium by exchanging matter and energy with their environment, *which includes other systems and their products*. The systems interact with one another and might do so nonlinearly. Because of the flow of matter and energy through the systems and the interactions of the systems with one another, the complexity of each system—or the number of its components—increases. Each system is typically hierarchical, with different processes acting and interacting on different time scales. A sixth trait of Earth's evolving systems is that the direction and magnitude of change is affected by preexisting conditions, or contingency. The geologic record strongly indicates that much of what has occurred on this planet often depended on contingency and was the result of a

unique (or nearly so) conjunction of circumstances, making the systems essentially irreversible and giving them a history.

- Earth's systems interact with one another through positive and negative feedback to maintain a relatively constant environment. Positive feedback promotes an effect, whereas negative feedback counters an effect.

- Earth scientists use the same basic approach as other scientists in seeking cause and effect relationships: the Principle of Uniformitarianism.

- Many Earth processes have largely remained the same through time, acting imperceptibly, whereas other processes have acted suddenly and catastrophically, such as extraterrestrial impacts. In fact, there has been plenty of time for both slow processes and unusual ones to have caused enormous change on Earth.

- Earth has not existed strictly in an equilibrium state of no net change through time. It has evolved, and its history exhibits secular change, or directionality.

- Given the variety of processes that occur over long spans of time, geologists and other earth scientists typically cannot reduce the scientific study of Earth's systems to simple cause and effect relationships like those observed in controlled laboratory settings. We really cannot understand all processes of Earth or humankind's rapid impact on Earth's environments without understanding Earth's geologic record and processes, which occur on different scales of time.

absolute age

actualism

anthropogenic

biosphere

catastrophism

contingency

convection cells

cryosphere

determinism

directionality (secular change)

dynamic equilibrium

feedback

flux

geologic time

greenhouse gases

historical science

homeostasis

hydrosphere

method of multiple working hypotheses

multiple causation

Ockham's Razor

open system

Pangaea

plate tectonics

Principle of Faunal Succession

Principle of Parsimony

Principle of Superposition

Principle of Uniformitarianism

radioactive decay

reductionism

relative age

reservoirs

steady state

stratified

system

REVIEW QUESTIONS

1. What are the four basic Earth systems?
2. What are the basic characteristics of a natural system?
3. Diagram the flow of matter and energy that occurs in the atmosphere, lithosphere, biosphere, and hydrosphere.
4. What is the difference between true equilibrium and steady state? Is steady state a kind of equilibrium? Explain.
5. How do each of Earth's systems interact to control atmospheric carbon dioxide on geologic scales of time?
6. How does the biosphere regulate atmospheric carbon dioxide on different scales of time?
7. Describe positive and negative feedback. Give examples using Earth's systems.
8. Give examples of how life has acted as a geologic force.
9. What is meant by the term "geologic time"?
10. Why is the Principle of Uniformitarianism important to the study of Earth's systems?
11. How does the modern view of uniformitarianism differ from that of Lyell's?
12. What is the importance of the following to the history of Earth's systems: (a) contingency and (b) directionality.
13. What are some key reasons for studying the history of Earth?
14. How are geologists both scientists and historians?

FOOD FOR THOUGHT:
Further Activities In and Outside of Class

1. Do you think the Earth is in equilibrium? Why or why not? On what time scales might it appear to be in equilibrium?
2. Why do you suppose the measurements of Figure 1.4b were taken in Hawaii?
3. As described in the text, the impact of anthropogenic activities on Earth greatly accelerated during the 20th century. Based on absolute ages, Earth is about 5 billion years old. To get some idea of how fast anthropogenic impacts have occurred, calculate the percentage of Earth's existence represented by the 20th century.
4. Assume that anthropogenic carbon dioxide emissions actually cause global warming. Are there any plausible negative feedback mechanisms that would decrease atmospheric carbon dioxide in coming decades that would keep atmospheric carbon dioxide at relatively constant levels?
5. Distinguish between the different meanings that have been given to the term "uniformitarianism."
6. What is a control, and why is it important to a laboratory experiment? How could "natural" conditions recorded in rocks and sediments serve as controls?
7. James Hutton likened the Earth to a machine, whereas some later workers have likened it to a giant organism. Is one of these descriptions better than the other? Why?
8. Distinguish between laws as they are used in experimental sciences and the use of principles like those used in geology.

SOURCES AND FURTHER READING

Ager, D. 1993. *The new catastrophism: The importance of the rare event in geological history*. Cambridge, UK: Cambridge University Press.

Chaisson, E. J. 2001. *Cosmic evolution: The rise of complexity in nature*. Cambridge, MA: Harvard University Press.

Chamberlin, T. C. 1897. The method of multiple working hypotheses. *Journal of Geology*, 5, 837–848.

Collier, M., and Webb, R. H. 2002. *Floods, droughts, and climate change*. Tucson, AZ: University of Arizona Press.

Coveney, P., and Highfield, R. 1991. *The arrow of time: A voyage through science to solve time's greatest mystery*. New York: Fawcett Columbine.

Dott, R. H. 1998. What is unique about geological reasoning? *GSA Today* (October), 1.

Frodeman, R. (1995) Geological reasoning: geology as an interpretive and historical science. *Geol Soc Am Bull,* 107:960–968.

Gorst, M. 2001. *Measuring eternity: The search for the beginning of time*. New York: Broadway Books.

Hutton, J. 1795. *Theory of the Earth*.

Lovelock, J. 1988. *The ages of Gaia: A biography of our living Earth*. New York: W. W. Norton.

Mackenzie, F. T., and Lerman, A. 2006. *Carbon in the geobiosphere: Earth's outer shell*. Dordrect, The Netherlands: Springer.

Martin, R. E. 1998. *One long experiment: Scale and process in Earth history*. New York: Columbia University Press.

Peters, E. K. 1996. *No stone unturned: Reasoning about rocks and fossils*. New York: W. H. Freeman.

Pielou, E. C. 2001. *The energy of nature*. Chicago: University of Chicago Press.

Schumm, S. A. 1991. *To interpret the Earth: Ten ways to be wrong*. Cambridge, UK: Cambridge University Press.

Simpson, G. G. 1963. Historical science. In C. C. Albritton (Ed.), *The fabric of geology* (pp. 24–48). Reading, MA: Addison-Wesley.

von Bertalanffy, L. 1968. *General system theory*. New York George Braziller.

Westbroek, P. 1991. *Life as a geological force*. New York: W. W. Norton.

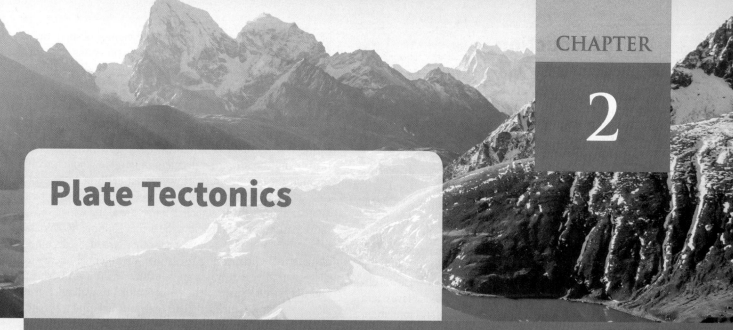

Plate Tectonics

MAJOR CONCEPTS AND QUESTIONS ADDRESSED IN THIS CHAPTER

Ⓐ What is the theory of plate tectonics?

Ⓑ What is the internal structure of Earth, and how do we know it?

Ⓒ How did plate tectonics develop from a hypothesis to a theory?

Ⓓ How is radioactivity involved in plate tectonics?

Ⓔ How does seafloor spreading occur?

Ⓕ What are the different types of plate margins?

Ⓖ What are the different "styles" of mountain building?

Ⓗ What is the tectonic cycle, and how does it help us understand plate tectonics?

CHAPTER OUTLINE

The Himalayan Mountains separate India from China and began to form in the Cenozoic Era. The Himalayas are themselves part of a much larger chain of mountains extending across Asia through Turkey and southern Europe, including the Alps. This massive chain of mountains has formed as a result of the collision of the Indian, Arabian, and African plates with Asia and southern Europe.

© NIRUN NUNMEESRI/Shutterstock, Inc.

2.1 Introduction

A Some of the most important processes of Earth's systems involve those of **plate tectonics**. The outermost rigid lithosphere of Earth is subdivided into a series of gigantic fragments called **plates** (**Figure 2.1**); **tectonics**, or tectonism, refers to the large-scale processes that cause the movement of the plates and continents.

The flux of heat from the Earth's interior drives the movement of the plates and the continents over the Earth, producing ocean basins and closing them when continents collide to produce mountains chains. Understanding the theory of plate tectonics is critical to the study and understanding of the other Earth systems on geologic scales of time. Plate tectonics drives the formation of various types of rocks in what is called the **rock cycle**, and affects Earth's climate by effecting broad changes in carbon dioxide levels and sea level, and affecting atmospheric and ocean circulation. We will study these effects in subsequent chapters.

The theory of plate tectonics is an outstanding example of how a hypothesis develops into a theory. As we examine the processes of plate tectonics, we also examine how the modern theory of plate tectonics itself evolved.

2.2 Structure of Earth

B A basic knowledge of the structure of deep Earth serves as the foundation for understanding how the processes of plate tectonics work. However, we cannot observe deep Earth directly because of the tremendous pressures and temperatures. But, we can infer Earth's structure indirectly from the behavior of seismic waves emitted by earthquakes.

During the 20th century, a series of earthquake monitoring stations was established throughout the world. Each station records shock waves sent through Earth when an earthquake occurs. **Earthquakes** can be generated by volcanic activity, the upward movement of **magma** (molten rock) through the Earth's crust. However, an earthquake is typically generated when one block of rock suddenly moves past another along a **fault**, which is a fracture along which rocks move past one another. Before the actual faulting, two blocks of rock push or pull against each other, gradually accumulating stress; the growing stress imparts potential energy to the system. Eventually, the stress exceeds a critical threshold and the system "snaps." The process of faulting is much like slowly bending a small wooden stick or pencil until it breaks. When faulting finally occurs, the potential energy that has accumulated in the rocks as a result of the stress is suddenly released as a burst of kinetic energy (**Figure 2.2**). The site of energy release is called the **focus**, whereas the surface location above the focus is called the **epicenter** (**Figure 2.3**). The foci and epicenters of earthquakes tend to occur in certain parts of the world, particularly along plate boundaries where plates move past one another.

Typically, the displacement along a fault is no more than a few feet to a few tens-of-feet. Nevertheless, the displacement

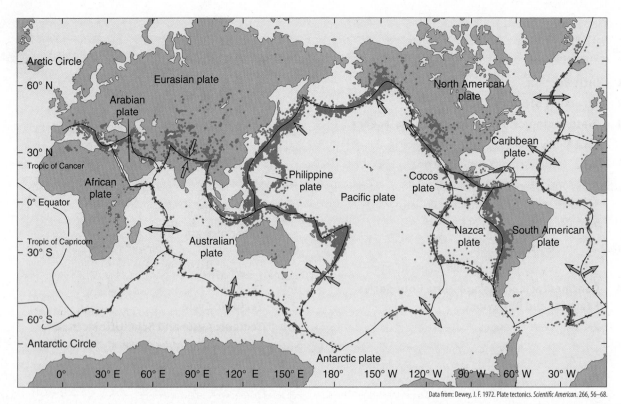

Data from: Dewey, J. F. 1972. Plate tectonics. *Scientific American.* 266, 56–68.

FIGURE 2.1 The lithospheric plates of modern Earth. Arrows indicate relative plate motions and blue dots indicate locations of volcanoes and earthquakes. Note that the distribution of earthquakes and volcanoes over the Earth reflects the boundaries between plates. The ring of volcanoes around the Pacific Ocean basin is commonly referred to as the "Ring of Fire." The continents are moved about by seafloor spreading because the continents themselves are parts of their respective plates.

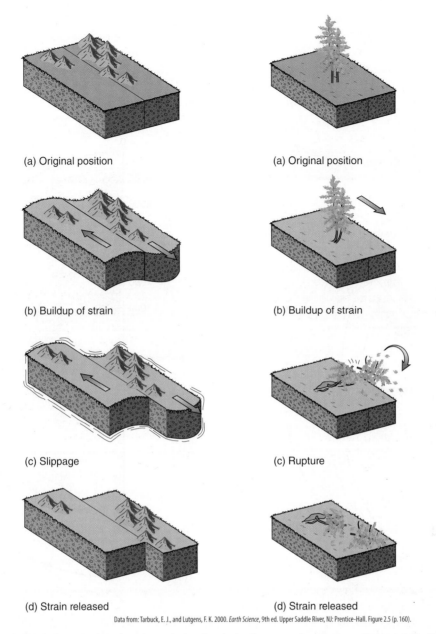

(a) Original position

(a) Original position

(b) Buildup of strain

(b) Buildup of strain

(c) Slippage

(c) Rupture

(d) Strain released

(d) Strain released

Data from: Tarbuck, E. J., and Lutgens, F. K. 2000. *Earth Science*, 9th ed. Upper Saddle River, NJ: Prentice-Hall. Figure 2.5 (p. 160).

FIGURE 2.2 Faulting and the production of earthquakes. Blocks of rock slowly accumulate stress until the rock finally ruptures like a tree bending in response to a strong storm until the tree trunk shears off.

can release a tremendous amount of energy that is sent out in all directions through the Earth as shock waves. It is these shock waves that produce an earthquake. Two earthquakes vie for the largest ever recorded: one centered at Prince William Sound in Alaska in 1964 that registered 9.2 on the earthquake scale, and one that occurred in 1960 off the coast of Chile that was initially measured at 8.6 but was later revised upward to 9.5. Earthquakes of this magnitude release energy equivalent to *more than 1 billion tons of TNT!*

The behavior of the shock, or seismic, waves generated by earthquakes has been used to infer the internal structure of Earth (Figure 2.3). These studies have inferred a distinct increase in the velocity of seismic waves at the **Mohorovičić discontinuity** (or **Moho**), named for its discoverer, Croatian seismologist Andrija Mohorovičić. The Moho separates the crust of Earth from the upper mantle. The Moho varies in depth from about 6 km beneath the ocean floor to about 30 to 40 km beneath the continents but can increase in depth to 70 to 90 km beneath mountains on some continents. Seismic waves also reveal the boundary between the **lithosphere** ("lithos," rigid), which consists of the crust and uppermost mantle. The lithosphere ranges up to about 250 km in thickness beneath continents but thins to only a few kilometers in the oceans. The plates of plate tectonics consist of the lithosphere. The boundaries between the modern plates are recognizable based on the distribution of volcanoes and earthquakes like that of the ring of volcanoes called the "Ring of Fire" around the perimeter of the Pacific Ocean that is associated with plate boundaries (Figure 2.1).

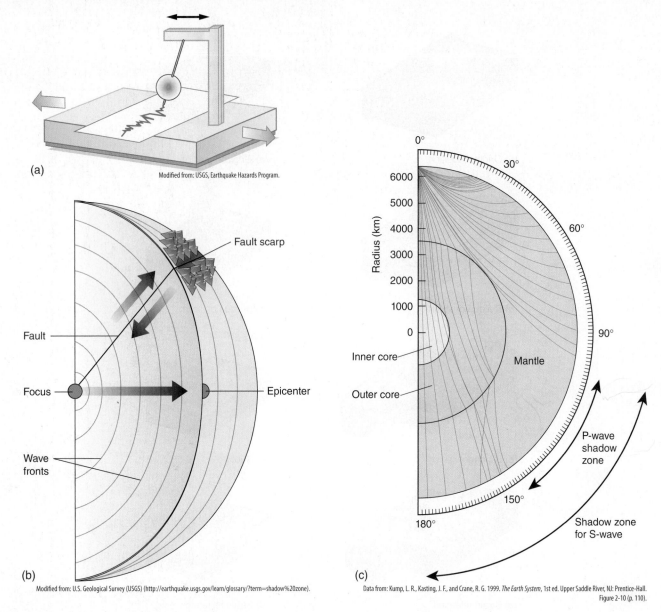

(a)

Modified from: USGS, Earthquake Hazards Program.

(b)

Fault scarp

Fault

Focus

Epicenter

Wave
fronts

Modified from: U.S. Geological Survey (USGS) (http://earthquake.usgs.gov/learn/glossary/?term=shadow%20zone).

(c)

0°

30°

60°

90°

6000
5000
4000
3000
2000
1000
0

Radius (km)

Inner core

Outer core

Mantle

P-wave
shadow
zone

150°

180°

Shadow zone
for S-wave

Data from: Kump, L. R., Kasting, J. F., and Crane, R. G. 1999. *The Earth System*, 1st ed. Upper Saddle River, NJ: Prentice-Hall.
Figure 2-10 (p. 110).

FIGURE 2.3 **(a)** How a seismograph records seismic waves. Seismic waves cause the base of the seismograph to move, in turn causing the pen to move back and forth across the paper. **(b)** The focus and epicenter of an earthquake. **(c)** The basic internal structure of the Earth inferred from seismic waves. P-waves emitted by the earthquake travel through both solids and liquids, but S-waves travel only through solids. Based on seismic waves, the outer core is thought to be fluid and rich in iron and nickel, whereas the inner core is thought to be solid. The P-wave "shadow zone" is the zone in which no P-waves arrive directly. P-waves received within this zone arrive later than would be expected because they have first been reflected by either the Earth's surface or some boundary within the Earth.

The lithospheric plates move over the portion of the mantle beneath them called the **asthenosphere** ("asthenos," soft; **Figure 2.4**). The movement of the plates is driven by the loss of heat from radioactive decay within Earth. Because of the increasing temperature and pressure within Earth, the upper 150 km or so of the asthenosphere weakens, becoming plastic. This means the asthenosphere is neither truly rigid like the lithosphere nor molten like magma but that it moves and deforms very slowly (although it deforms faster than the lithosphere). As a result, the upper portion of the asthenosphere is characterized by a low-velocity zone, in which seismic waves tend to slow down. It is the plastic nature of this low-velocity zone that allows the lithospheric

plates to move over the asthenosphere. It is also the asthenosphere into which the lithospheric plates descend in seafloor trenches.

The changes in seismic velocity within the lithosphere and asthenosphere correspond to significant changes in the rocks. The boundary between ultramafic (those highly enriched in iron and magnesium) and mafic (those somewhat less enriched in these elements) and mafic igneous rocks is interpreted as the base of the oceanic crust (see Chapter 3). Seafloor crust consists of a particular kind of mafic igneous rock called **basalt**, whereas continents are considered to be more "granitic" in composition (see Chapter 3). Coarse-grained (exhibiting obvious crystals) mafic igneous

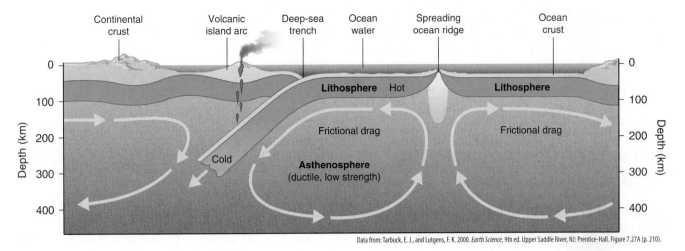

Continental crust   Volcanic island arc   Deep-sea trench   Ocean water   Spreading ocean ridge   Ocean crust

Lithosphere   Hot   **Lithosphere**

Frictional drag   Frictional drag

Cold

Asthenosphere
(ductile, low strength)

Data from: Tarbuck, E. J., and Lutgens, F. K. 2000. *Earth Science*, 9th ed. Upper Saddle River, NJ: Prentice-Hall. Figure 7.27A (p. 210).

FIGURE 2.4 Modern interpretation of seafloor spreading and subduction.

rocks called **gabbros**, which are chemically equivalent to basalt, are probably also important in the lower continental crust. By contrast, the mantle portion of the lithosphere and the asthenosphere is composed of other ultramafic rocks called *peridotites*. When erupted at the Earth's surface, peridotites are called *komatiites*. Ultramafic volcanics such as komatiites solidify at relatively high temperatures greater than about 1,600°C. Komatiites were a prominent component of the early Earth's crust billions of years ago, suggesting that Earth's surface was still quite hot and that the lithosphere had not yet fully formed. The asthenosphere extends downward to about 660 km, which is close to the maximum depth of earthquake foci (**Figure 2.5**). At the base of the asthenosphere is the layer of the mantle called the *mesosphere*, which is thought to be more rigid than the asthenosphere and to behave passively. The mesosphere is thought to be composed of yet another ultramafic rock that is a high-density form of shallower rocks called *perovskite*. The increase in pressure with depth in the mantle causes a phase change, in which the atomic arrangement is altered without altering its chemical composition.

Beneath the asthenosphere, lying at the center of the Earth, is the **core**. The core is thought to be rich in iron and nickel (Figure 2.5). Because Earth's crust is iron-poor, iron and perhaps other dense metals sank to Earth's center billions of years ago when early Earth was still molten. **P-waves** emitted by earthquake travel through both solids and liquids but **S-waves** only travel through solids. Based on the behavior of S-waves, the outer core is thought to be liquid, whereas the inner core is inferred to be solid because of the internal pressure of Earth. The core is also thought to be responsible for the generation of Earth's magnetic field because of its iron-rich composition. As we will see in Chapter 7, the development of Earth's magnetic field was very significant because it protected Earth from cosmic radiation and allowed an atmosphere initially to develop and then evolve through processes such as volcanic outgassing and photosynthesis.

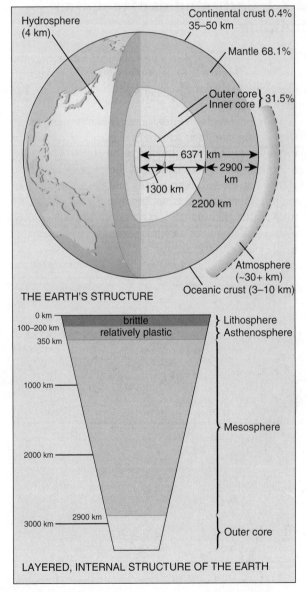

FIGURE 2.5 The internal structure of the Earth. Percentages indicate the relative composition of each component of the entire Earth.

CONCEPT AND REASONING CHECKS

1. What do P- and S-waves tell us about Earth's internal structure?
2. How can two different minerals have the same chemical composition?

2.3 Plate Tectonics: From Hypothesis to Theory

Today, the theory of plate tectonics is almost universally accepted by earth scientists. However, this has not always been so. Despite the accumulating knowledge of Earth's structure, the theory of plate tectonics took well more than half century to gain wide acceptance by the late 1960s.

2.3.1 Early studies of mountain building

One of the most obvious features at Earth's surface is mountain chains (**Figure 2.6**). Mountain building is known as **orogenesis**, and a particular episode of orogenesis is referred to as an **orogeny**.

Long before plate tectonics, early earth scientists studied mountains and their structures in places such as Scotland and the Alps. They recognized that orogenesis could reveal the inner workings of Earth and how rocks are formed, uplifted, and eroded. Both Hutton and Lyell (see Chapter 1) concluded that there was an ongoing release of heat from within Earth. Hutton also believed individual volcanoes served as "safety valves" to release excess heat from Earth. He maintained that deformation was part of a slow process of gradual uplift that maintained Earth's surface in equilibrium with the counter process of erosion. Erosion was viewed as relieving the pressure of rocks overlying the continents and allowing uplift, whereas sedimentation caused depressions in the crust where sediments were buried, deformed, and melted, producing volcanoes.

Debate about orogenesis began somewhat later in the United States. Based on the proximity of the Appalachian Mountains to the east coast of North America, some workers hypothesized that lateral forces from the ocean side of the Appalachians had caused their uplift. According to these workers this unidentified force caused a long, linear trough called a **geosyncline** to first deepen, allowing great thicknesses of sediment to accumulate. The sediments were later deformed and metamorphosed by the same lateral force, while volcanoes spewed forth. All these processes—the deposition of thick sequences of sediment, volcanism, metamorphism, and deformation—had been documented on the eastern side of the Appalachians by previous field observations of rock outcrops (for further discussion of the debate about the geosynclinal theory, you might want to see Greene, cited in the section "Sources and Further Reading").

2.3.2 Hypothesis of continental drift

Despite the seeming success of the geosynclinal theory at explaining the formation of mountains, the key question remained unanswered: what force caused the deformation and displacement of rocks during orogenesis? Scientists are normally quite skeptical until a particular mechanism is identified that accounts for a particular effect, in this case, mountains.

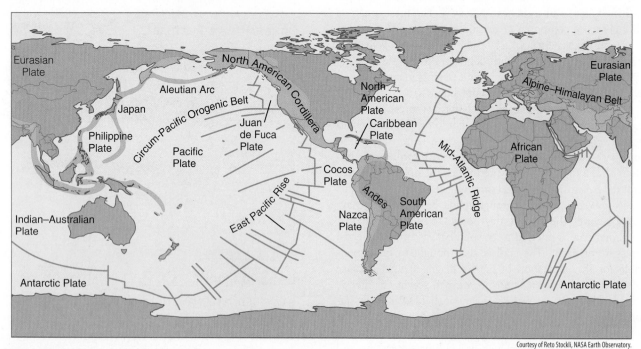

Courtesy of Reto Stockli, NASA Earth Observatory.

FIGURE 2.6 Major mountain ranges and island chains of the world in relation to major plate boundaries.

As a result, a multiplicity of hypotheses about the exact mechanism of orogenesis appeared during the later 19th and early 20th centuries. By about this time the hypothesis of **continental drift**, which stated the continents move over Earth's surface, had appeared. The famous South African geologist, Alexander L. Du Toit, and European geologists had earlier noted the similarity of features of Earth's crust such as mountain belts, lithologies, and fossils on either side of the Atlantic Ocean.

However, the German scientist, Alfred Wegener (1880–1930), was the first to attempt a plausible explanation of continental drift in his famous book, *The Origin of Continents and Oceans,* published in 1915. Wegener hypothesized that the present-day continents had once been joined into a huge supercontinent he called **Pangaea** ("all land"). Pangaea then separated, or rifted, into separate land masses as the continents moved toward their present positions.

Besides the "jigsaw" fit of the continents on either side of the Atlantic (which explained the observations of Du Toit and others), Wegener used a variety of other evidence to build an argument for continental drift (**Figure 2.7**). Among the evidence he used was the occurrence of fossil plants called seed ferns (belonging to what is called the *Glossopteris* flora) and ancient glacial deposits called tillites to bolster his hypothesis. Today, fossil seed ferns and tillites are found on widely separated continents, but in Wegener's view the deposits would have been adjacent to one another when the continents had been joined together. Wegener reasoned that it would have been impossible for the seeds of these particular plants to have been transported by air or ocean currents from one continent to another because the seeds were dense and unlikely to float for long; the seeds would probably not have survived prolonged exposure to seawater, either. Similarly, tillites of the same age on different continents could only have formed if the glaciers (which appeared separated based on the distribution of tillites) had once been united. He also noted the occurrence in South Africa of fossil reptiles named *Mesosaurus* that are found in rocks of similar age in South America. *Mesosaurus* lived in fresh water and would not have been able to swim across the Atlantic.

Wegener suggested what he called a "working hypothesis" about how the continents moved apart. He reasoned the continents were composed of less dense (more buoyant) rock that remained solid at depth, forming deep "roots" beneath the continents extending down as much as 100 km into Earth. These roots were thought to allow the "granitic" continents to "float" on top of the more plastic rocks of the mantle, like ships on an ocean. But Wegener came up against the same problem that had plagued other geologists: he could not suggest a force capable of moving the continents. Consequently, many scientists dismissed his idea as implausible, and his hypothesis was ridiculed.

D One of the few scientists who seemed to favor Wegener's hypothesis was the British geophysicist Sir Arthur Holmes of Edinburgh University (Scotland). Holmes had the imagination to suggest that continents moved as the result of convection cells generated by heat from within Earth's mantle

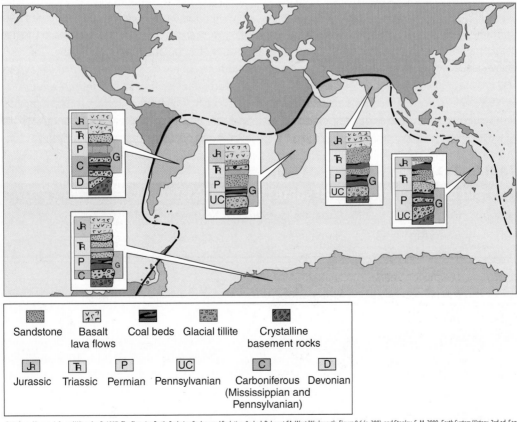

Sandstone   Basalt lava flows   Coal beds   Glacial tillite   Crystalline basement rocks

JR Jurassic   TR Triassic   P Permian   UC Pennsylvanian   C Carboniferous (Mississippian and Pennsylvanian)   D Devonian

Data from: Monroe, J. S., and Wicander, R. 1997. *The Changing Earth: Exploring Geology and Evolution*, 2nd ed. Belmont CA: West/Wadsworth. Figure 9.6 (p. 200), and Stanley, S. M. 2009. *Earth System History*, 3rd ed. San Francisco, CA: W. H. Freeman. Figure 8–7 (p. 181).

FIGURE 2.7 Wegener used a variety of evidence to infer that continents on either side of the Atlantic had once been joined, including fossils, geologic structures, and the matching sequences of rocks shown here.

(Figure 2.4). One of the most important scientific discoveries to occur during the late 19th century was the discovery of radioactivity, which releases heat. Later, in 1908, it was proposed by the Irish physicist John Joly that the radioactive decay of uranium provided an important heat source that drives mountain building. Thus, radioactivity was the source of heat that Hutton and Lyell had noticed much earlier. Holmes was well aware that radioactive decay generates heat because he had been involved in developing the use of radioactive isotopes to date Earth. Holmes therefore suggested that heat produced by radioactive decay also generated convection cells within Earth that moved the continents.

BOX 2.1 Radioactive Decay and Production of Heat: Processes That Drive Plate Tectonics

Radioactive decay is the spontaneous breakdown of the atomic nuclei of certain chemical elements. Radioactivity generates the heat that drives seafloor spreading and plate tectonics and the generation of igneous and metamorphic rocks.

The phenomenon of radioactivity in uranium ores was reported by the French scientists Henri Becquerel and Marie and Pierre Curie, for which the three were awarded the Nobel Prize in Physics in 1903. Marie Curie later isolated the radioactive elements radium and polonium from uranium ore, for which she received the Nobel Prize in Chemistry in 1911. She died of leukemia in 1934 as a result of prolonged exposure to radiation.

The theory of radioactive decay was subsequently developed by the physicist, Ernest Rutherford, based on atomic theory. Elements with the same atomic number (number of protons) but differing numbers of neutrons are called isotopes. Normally, the protons and neutrons of an atomic nucleus are held together by the so-called strong force and are stable. However, very large atoms or ones with too many or too few neutrons are unstable and break down. For example, the element uranium (U), which has an atomic number of 92, has two main isotopes: U-238 (designated as ^{238}U) and U-235 (^{235}U). ^{238}U decays to ^{206}Pb, whereas ^{235}U decays to ^{207}Pb (**Box Table 2.1A**).

The radioactive isotopes of an element emit one or more types of subatomic particles from their unstable nuclei. In alpha (α) decay, an alpha particle consisting of two protons and two neutrons is released. In beta (β) decay, a neutron breaks down into a proton and an electron, and the proton is retained, whereas the electron is ejected. Hence, alpha decay results in a decrease of four in atomic mass and two in atomic number, whereas beta decay results in an increase of one in atomic number but no change in atomic mass (the mass of an electron is quite small compared with that of a proton or a neutron). The alpha and beta decays were so named because they were the first and second events in the decay sequence to be understood, but other decays happen in other sequences. The third type of decay, electron capture, transforms a proton into a neutron.

The half-lives of various isotopes in the table indicate how long it takes for any amount of that isotope to decay to its daughter product.

BOX TABLE 2.1A

Half-lives and daughter products of some radioactive isotopes			
Parent Isotope	Half-Life*	Daughter Isotope	Source Materials
Carbon-14	5730 years	Nitrogen-14	Organic Matter
Uranium-238	4.47 billion years	Lead-206	Zircon, uraninite, pitchblende
Uranium-235	704 million years	Lead-207	
Uranium-234	248 thousand years	Thorium-230	Lavas, coralline limestones
Thorium-232	14 billion years	Lead-208	
Rubidium-87	48.8 billion years	Strontium-87	Muscovite mica, potassium feldspar, biotite, glauconite, whole metamorphic or igneous rock
Potassium-40	1251 million years (1.251 billion years)	Argon-40 (and calcium-40)*	Muscovite, biotite, hornblende, whole volcanic rock, glauconite, potassium feldspar[†]

*Half-life data from: Steiger, R. H., and Jäger, E. 1977. Subcommission on geochronology: convention on the use of decay constants in geo- and cosmochronology, *Earth and Planetary Science Letters* 36:359–362.
[†]Although potassium-40 decays to argon-40 and calcium-40, only argon is used in the dating method because most minerals contain considerable calcium-40, even before decay has begun.

2.3.3 Hypothesis of seafloor spreading

Despite Holmes's support, Wegener's hypothesis lay dormant until the 1950s. Earlier Robert Dietz had noticed the chain of submarine mountains called the Emperor Seamounts extended off to the northwest from Hawaii to the Midway chain of islands (**Figure 2.8**) and thought that they had somehow been moved there. A former naval officer named Harry Hess began to analyze bathymetric charts that

he had made while serving in the Pacific theater during World War II. Hess noticed that some of these **seamounts** on his profiles had flat tops. He concluded the flat tops had formed by wave erosion when the tops of the seamounts were much closer to the ocean surface (**Figure 2.9**) and named these particular seamounts **guyots** (after the 19th century geographer, Arnold Guyot). However, the tops of seamounts were now so deep (thousands of meters below the ocean surface) that falling sea level could not be responsible for the erosion. Some other process must have been involved. Hess also

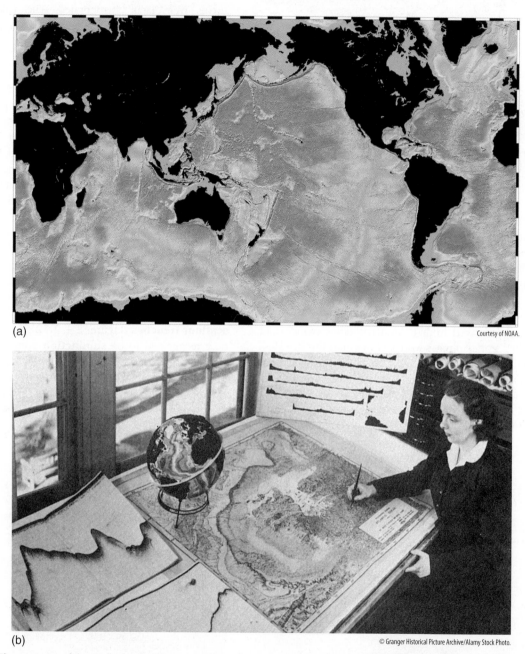

(a)

Courtesy of NOAA.

(b)

© Granger Historical Picture Archive/Alamy Stock Photo.

FIGURE 2.8 (a) The anatomy of the ocean basins. Note extensive submarine mountain ranges such as the mid-Atlantic Ridge and East Pacific Rise, seafloor trenches, faults called transforms that offset the ridges (described in the text), and numerous seamounts. Compare with Figure 2.1. Note the distinct "kink" in the island chain leading northwest from Hawaii toward Midway. Robert Dietz suspected this island chain was formed by some sort of movement, which we now know as seafloor spreading. The kink in the chain appears to have formed by a change in the direction of plate movement. **(b)** Marie Tharp (1920–2006), who constructed the first map of the ocean basins. Very importantly, construction of the initial map helped lead to the acceptance of continental drift, seafloor spreading, and ultimately the theory of plate tectonics.

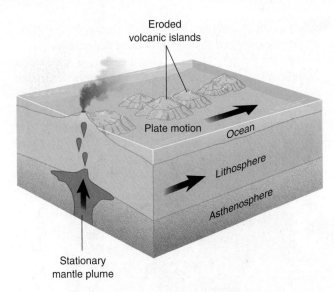

Eroded
volcanic islands

Plate motion

Ocean

Lithosphere

Asthenosphere

Stationary
mantle plume

FIGURE 2.9 The formation of guyots according to Hess.

reasoned that *if the ocean basins had remained fixed in size through geologic time*, there should be far more seamounts and guyots in the ocean basins than the approximately 10,000 he counted. Based on the approximate rate of sediment deposition on the seafloor (generally a few centimeters per thousand years), he further calculated the thickness of sediment on the seafloor should be about 20 kilometers, which is much greater than its current thickness of roughly 1 km.

Thus, all Hess's evidence pointed to the Pacific Ocean basin being much younger than the age of Earth, which by this time was known to be billions of years old based on absolute ages derived from radiometric dates (which had also resulted from Holmes's work). How could this be? Hess had noted that heat flow from within the Earth was especially high beneath the long mid-ocean ridges. Hess therefore hypothesized that hot mantle material was moving to the surface of the ocean floor as convection cells beneath

the mid-ocean ridges, similar to Holmes's earlier hypothesis. Hess suggested as the mantle material moved to the surface, it solidified to form ocean crust that spread away in opposite directions from a ridge; after this the crust cooled, became more dense, and sank as it moved away from the ridge. In the process volcanoes fed by mantle material near the ridges were also transported away from their source, cooled, and sank. As they sank the volcanoes were eroded to form guyots at the ocean surface before descending into the deep sea away from the ridges.

Hess's hypotheses about the origins of seamounts and guyots were later corroborated. The volcanic Hawaiian Islands occur almost in the middle of a huge piece of oceanic crust known as the Pacific plate and so are located far from any trench where subduction and melting of ocean crust might occur. The islands actually occur near what is now referred to as a **hotspot** (**Figure 2.10**), which consists of magma piped to the surface by a **mantle plume** from deep within Earth (**Box 2.2**). The magma that forms the plume is thought to originate within the mantle or in some cases at the core–mantle boundary. Here, heat released from the core is thought to cause lower mantle rocks to melt, making them less dense and able to rise through the surrounding mantle (Figure 2.10). Today, hotspots are known from all over the world (**Figure 2.11**) and serve as "windows" on Earth's interior.

Like Wegener, Hess further hypothesized that the granitic (felsic) continents, which were relatively buoyant, moved over the top of the more mafic mantle. He also reasoned that if ocean crust forms by seafloor spreading at **mid-ocean ridges**, ocean crust must be descending somewhere else, most likely at the features already known as deep-sea or **seafloor trenches** (Figures 2.4, **2.12**). Seafloor trenches are now recognized to be **subduction zones**, into which the leading edge of a plate descends.

Convection cells and seafloor spreading not only provided Wegener's mechanism to move continents but also the mysterious lateral force that deformed and pushed up

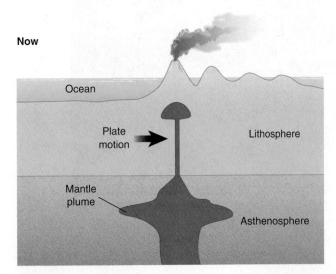

Now

Ocean

Plate
motion

Lithosphere

Mantle
plume

Asthenosphere

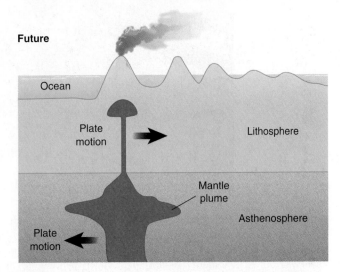

Future

Ocean

Plate
motion

Lithosphere

Mantle
plume

Asthenosphere

Plate
motion

FIGURE 2.10 The formation of a mantle plume and hotspot. Magma is thought to originate within the mantle or in some cases at the core–mantle boundary, where heat released from the mantle or core causes rocks to melt, making them less dense and able to rise through the mantle and crust.

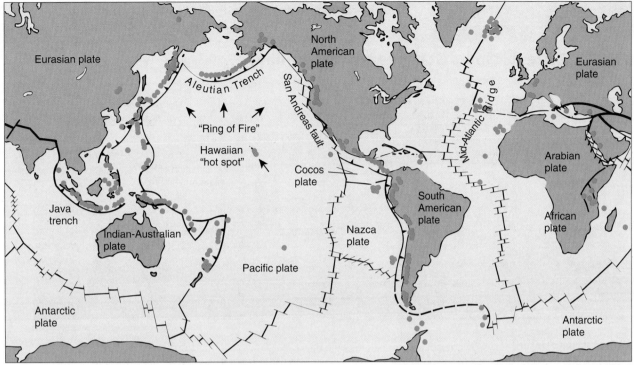

Data from: USGS (http://vulcan.wr.usgs.gov/Glossary/PlateTectonics/Maps/map_plate_tectonics_world.html.)

FIGURE 2.11 Location of hotspots throughout the world. Some of the hotspots, such as Hawaii, appear to occur over deep mantle plumes, whereas others might be associated with shallower plumes.

BOX 2.2 Are Hotspots Stationary?

Hotspots typically have been assumed to be stationary. For example, it has long been assumed that the Hawaiian hotspot is fixed because of the ages of the islands. The age of the Hawaiian Islands increases away from the island of Hawaii (700,000 years) to 27 million years at Midway Island, and then to 65 million years in the Emperor Seamount Chain near the Aleutian Islands, off Alaska (**Box Figure 2.2A**). The increasing ages were thought to result from the movement of the Pacific plate over the presumably stationary hotspot. Each island is thought to be carried away from the hotspot by the spreading ocean crust, as if on a conveyor belt, so that volcanism eventually shuts down. Erosion then beveled the extinct volcanoes to produce guyots as the volcanoes sank beneath the ocean surface. A stationary hotspot is also thought to underlie Yellowstone National Park in northwest Wyoming. The progressively younger appearance of volcanism eastward across the Snake River Plain of southern Idaho is thought to result from the westward movement of the North American plate over the Yellowstone hotspot (**Box Figure 2.2B**).

Despite the explanatory power of the hotspot concept, the stationary positions of at least some hotspots have come into question. The changes in the angles of the Hawaiian and Emperor seamount chains, for example, have generally been thought to reflect changes in the directions of movement of the lithospheric plates over the Hawaiian hotspot (**Box Figure 2.2C**); however, recent reconstructions of plate movements fail to predict the bend in the Hawaii–Emperor seamount chain. When it is assumed that the Hawaiian hotspot is fixed, its observed path does not coincide with the predicted path. Recent evidence also suggests the position of the mantle plume beneath the Hawaiian Islands might have been distorted over millions of years in response to mantle flow and seafloor spreading. The predicted and observed paths can be made to agree more closely by using a combination of plate motions elsewhere (which can deform and affect the motion of the Pacific plate) and distortion of the Hawaiian hotspot by mantle flow.

(continues)

BOX 2.2 Are Hotspots Stationary? (Continued)

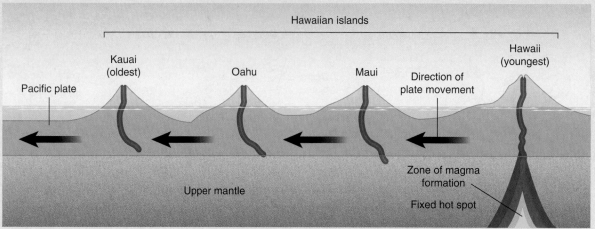

Data from: U.S. Geological Survey (http://pubs.usgs.gov/gip/dynamic/hotspots.html).

BOX FIGURE 2.2A The Hawaiian hotspot. The formation of the Hawaiian islands occurs as the ocean crust moves over a hotspot that was once thought to be stationary. The stationary position of the hotspot was suggested by the increasing ages of the islands away from the apparent site of the hotspot.

© Photos.com.

BOX FIGURE 2.2B The progressive appearance of volcanism across southern Idaho during the last 17 million years to northwest Wyoming is thought to have resulted from the movement of the North American plate over a stationary hotspot presumed to lie beneath Yellowstone National Park.

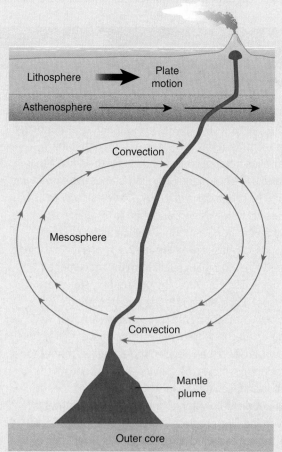

Data from: Tarduno, J. A. 2008. Hotspots unplugged. *Scientific American, 298*, 88–93.

BOX FIGURE 2.2C The Hawaiian hotspot might, however, not be stationary.

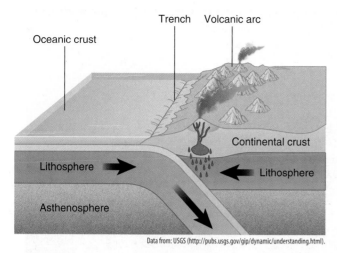

FIGURE 2.12 Structure of a seafloor trench.

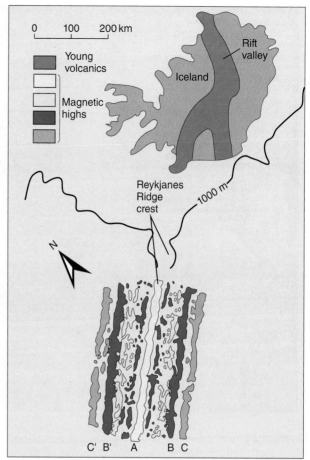

Data from: Heirtzler, J. R., LePichon, X., and Baron, J. G. 1966. Magnetic anomalies over the Reykjanes Ridge. *Deep Sea Research, 13*, 427–433.

FIGURE 2.13 Magnetic anomaly stripes. Magnetic anomaly stripes run parallel and are symmetrically arranged on both sides of the mid-ocean ridge axis. For example, anomalies B and C on the eastern ridge flank have counterparts B' and C' on the western flank that are the same distance from the crestal anomaly A.

sediments and other rocks to produce mountain chains from geosynclines. This mechanism explained not only the occurrence of mountain ranges along other continental margins, such as the Rockies and the Andes, but also the location of mountain ranges within the interiors of present-day continents such as the Urals of Russia, which formed when separate continents collided hundreds of millions of years ago.

2.3.4 Corroboration of seafloor spreading

Still, continental drift and seafloor spreading remained only interesting—but intriguing—hypotheses because there was no proof seafloor spreading actually occurred. Only a pattern had been demonstrated but not a mechanism for how continents might move. This came in the 1960s in the form of a series of ingenious experiments that changed many scientists' view of plate tectonics from an interesting hypothesis to an overarching theory.

As igneous rocks cool from their molten state, iron-rich mineral particles present in the magma align themselves with Earth's magnetic field. Earth's magnetic field has also flip-flopped back and forth between normal and reversed states through geologic time (see Chapter 6). Before the acceptance of plate tectonics, however, many scientists were skeptical that magnetic reversals actually occurred.

Fred Vine and Drummond Matthews (Cambridge University, England) combined the hypothesis of magnetic reversals with Hess's hypothesis of seafloor spreading. In a series of papers, they tested Hess's hypothesis of seafloor spreading by towing an instrument called a magnetometer back and forth over mid-ocean ridges (see Giere in "Sources and Further Reading," which provides a fuller description of the actual experiments). They found the rocks on either side of the ridges occurred in paired polarity reversals, or "magnetic stripes," with one member of each pair occurring on either side of the ridge. Moreover, both members of each pair exhibited the same polarity, but the polarity of different

pairs alternated back and forth between normal and reversed (**Figure 2.13**). This suggested the polarity stripes in the ocean crust had first formed near the ridge and then moved away. Thus, Vine and Matthews simultaneously corroborated the hypothesis of magnetic reversals *and* the hypothesis of seafloor spreading. Similar symmetric striped patterns have since been documented on the sides of spreading centers all through the ocean basins (**Figure 2.14**). Based on radiometric dates and the positions of the stripes, rates of seafloor spreading are calculated to be on the order of a few centimeters per year, although rates vary widely (see Figure 2.1).

Another type of paleomagnetic evidence also suggested the continents could move. The angle at which the magnetic particles in a rock point toward the magnetic north pole depends on the latitude at which the particles were when they aligned themselves with the magnetic field (the Earth's magnetic and geographic poles do not coincide; see Chapter 6). The latitude, and therefore the angle, of the magnetic particles varies from one continent to the next (**Figure 2.15**). When the positions of the poles were drawn for different times for a

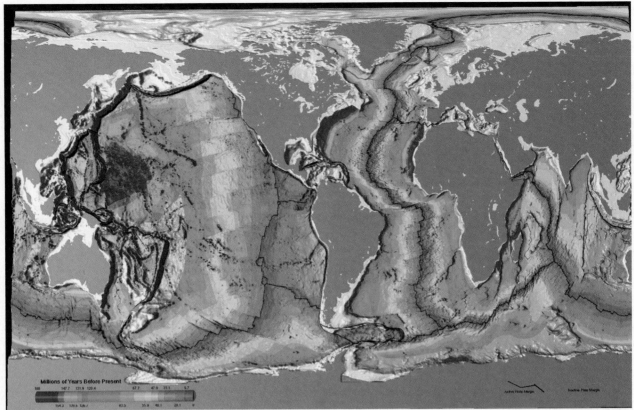

Courtesy of NGDC/NOAA.

FIGURE 2.14 Map of magnetic seafloor stripes in the ocean basins. Stripes of approximately the same age are shown in the same color. The youngest stripes are red and lie closest to seafloor spreading centers.

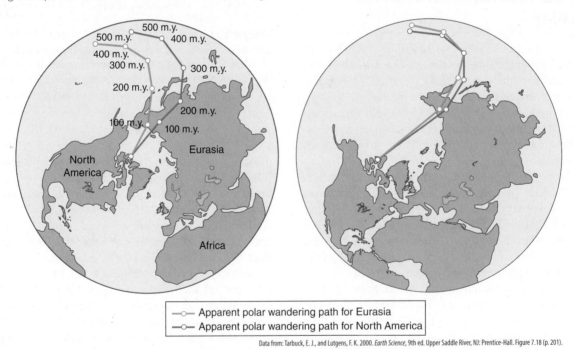

—○— Apparent polar wandering path for Eurasia
—○— Apparent polar wandering path for North America

Data from: Tarbuck, E. J., and Lutgens, F. K. 2000. *Earth Science*, 9th ed. Upper Saddle River, NJ: Prentice-Hall. Figure 7.18 (p. 201).

FIGURE 2.15 Polar wandering curves indicate the former positions of continents. Apparent polar wandering curves are based on the angle between magnetic particles and the Earth's surface. Magnetized particles in the rocks align themselves with the magnetic force field. The angle of the particles with respect to the Earth's surface indicates the approximate latitude, or how far north or south of the equator the continent originally occurred. However, the angle of the magnetic particles says nothing about the longitude, or east-west position, of the continent. The same angle would be found on separate continents as long the continents lay anywhere on a circle at the same latitude. Thus, two continents could yield the same paleomagnetic data (angle) but have lain very far apart. This is why other types of data—rock types, geologic structures, fossils—must also be integrated with the paleomagnetic data to reconstruct the jigsaw puzzle of ancient continental positions, much as Wegener did originally.

single continent, they showed the magnetic poles wandered a great deal. The "pro-drift" scientists believed this showed the continents must move about independent of one another. However, the modern magnetic poles are known to move, so the "fixists" (who rejected plate tectonics) seized on these results and argued that the curves reflected the movement of the magnetic poles beneath Earth's surface and not the movement of the continents. As a result the curves were often referred to as **apparent polar wandering curves**. Mostly, though, the curves were simply regarded as being unreliable.

With the acceptance of plate tectonics, it was realized that polar wandering curves actually agreed with drifting continents when the continents were moved to their earlier positions. When this was done, the alignment of the magnetic particles agreed with the alignment that should have occurred when the continents were in their earlier configurations. Thus, scientists achieved more consistent and detailed reconstructions of continental movements. In fact, moving the continents back to their original positions shows that a number of different kinds of data from continents now separated on either side of the Atlantic Ocean actually "matched," as Wegener had originally suggested.

The modern theory of plate tectonics recognizes several important concepts. First, the continents, along with seafloor crust, comprise larger units called plates. Earth's lithosphere is divided into approximately eight large and seven smaller rigid plates of various sizes whose margins move past one another. Second, the continents do not simply float on the mantle but are *parts of the plates and move with them.* Third, we know that mid-ocean ridges correspond to *seafloor spreading centers,* that seafloor spreading causes the plates to move as units over mantle rocks, and that these movements can result in orogenesis. We also now know why the ocean basins are so young. The oldest known seafloor crust dates from the Jurassic Period, and older seafloor crust, along with its overlying sediment, has largely been subducted.

For these and other reasons, plate tectonics has sometimes been hailed as a scientific "revolution" (although the development of the theory really took about half a century!). Plate tectonics showed that Earth was far more dynamic than had previously been thought, and it helped to change our view of how Earth works. All the evidence for continental drift, when drawn together and eventually confirmed by the evidence for seafloor spreading, quickly replaced the older geosynclinal theory of orogenesis. In fact, the term "geosyncline" has been abandoned.

CONCEPT AND REASONING CHECKS

1. What do guyots tell us about the behavior of Earth?
2. How were Hess's ideas about the behavior of Earth corroborated?
3. The age of seafloor crust increases away from mid-ocean ridges. Why?

2.4 Continental Margins and Plate Boundaries: Features and Behavior

The initial theory of plate tectonics was quite "elegant." The elegance of a theory is something that is prized by scientists because it implies we thoroughly understand something of very fundamental importance about nature (although what we think we know can still turn out to be wrong!). However, the updated theory of plate tectonics is far more complicated than Wegener or Hess envisioned it. As it turns out the margins, or boundaries, of the continents and plates vary. So, too, do the processes with which they are associated.

2.4.1 Types of margins

The theory of plate tectonics recognizes two major types of margins. The margins of continents not presently associated with subduction or other types of large-scale tectonic activity such as orogenesis are called **passive margins** (**Figure 2.16**). Passive margins are associated with the initial rifting of continents and the subsequent opening of ocean basins along spreading centers. As the warm ocean crust moves away from the spreading centers, it cools and sinks below sea level as sediment accumulates on top. As sedimentation continues, the lithosphere can begin to subside, allowing great thicknesses of sediment—up to many thousands of meters—to accumulate. Volcanism and earthquake activity are also relatively infrequent. An example of a passive continental margin is the Atlantic seaboard of the United States (Figure 2.16).

According to plate tectonics, as one ocean basin is opening, another must close. Today, as the Atlantic Ocean continues to open (following the rifting of Pangaea), the Pacific Ocean is closing. The margins of the continent along which plate boundaries move are called **active margins** (Figure 2.4). Active margins are often—but not always—associated with subduction and frequent volcanism and earthquakes, like that around the rim of the modern Pacific Ocean's "Ring of Fire" (Figure 2.1), along with faulting and other forms of rock deformation.

2.4.2 Tectonic features of Earth's surface

Before describing the different types of plate boundaries and the processes that form them, we must first examine a number of tectonic and orogenic features with which they are associated. These features can be used to infer past episodes of mountain building even when the mountains are no longer present.

The distribution of igneous and metamorphic rocks often provides clues to ancient plate boundaries. For example, igneous rocks such as granites can be found in or near the cores of some mountain ranges. Different types of volcanic rocks are also found associated with different plate margins. Other types of rocks that form under different temperatures

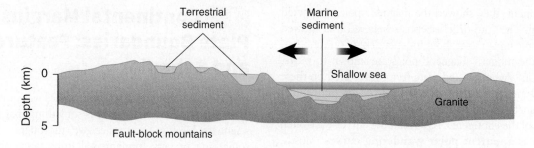

Terrestrial sediment

Marine sediment

Shallow sea

Granite

Fault-block mountains

(a) **Initial rifting (Triassic period: 200 million years B.P.)**

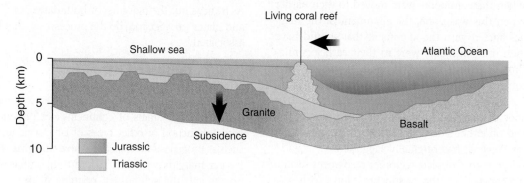

Living coral reef

Shallow sea

Atlantic Ocean

Granite

Subsidence

Basalt

Jurassic
Triassic

(b) **Jurassic margin (150 million years B.P.)**

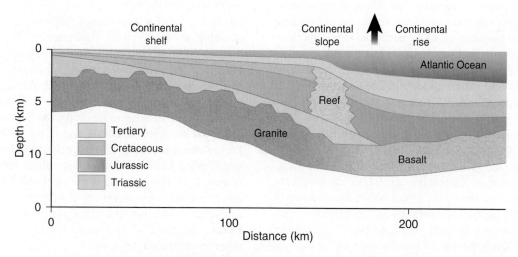

Continental shelf

Continental slope

Continental rise

Atlantic Ocean

Reef

Tertiary
Cretaceous
Jurassic
Triassic

Granite

Basalt

Distance (km)

(c) **Present-day margin southeast of Cape Cod**

(a, b, c) Data from: Watkins, J. R. 1979. *Geological and Geophysical Investigations of Continental Margins.* Tulsa, OK: American Association of Petroleum.

FIGURE 2.16 Passive spreading or rifting plate margins. **(a)** Passive margins are associated with the opening of ocean basins along spreading centers and the initial spreading or rifting (horizontal arrows) of continents. **(b, c)** After the ocean basin has formed, passive margins form along the margins of continents, such as the east coast of North America today, where deposition (red upward-pointing arrow) and subsidence (downward-pointing arrows) have built a thick sedimentary sequence upward and outward into the ocean basin. No subduction zone is present.

and pressures (called metamorphic rocks) along an orogenic belt are also indicative of past orogenesis.

Large-scale features found within sedimentary rocks are also indicative of deformation due to plate tectonics. These features include faults and folds. Several types of faults are associated with tectonism. **Normal faults** occur when a block of rock—called the "hanging wall"—moves

downward *relative* to the block on the other side of the fault, called the "footwall" (so named because you can imagine standing on it as opposed to the hanging wall which "hangs" over you; **Figure 2.17**). Note the use of the word "relative" in describing the movement of rocks on either side of the fault. This is because we cannot tell whether the hanging wall moved down or the footwall moved up. In either case,

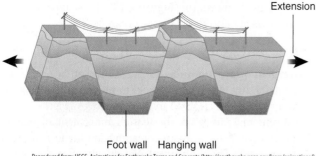

Extension

Foot wall   Hanging wall

Reproduced from: USGS, Animations for Earthquake Terms and Concepts (http://earthquake.usgs.gov/learn/animations/).
Accessed July 19, 2010.

FIGURE 2.17 Horst-and-graben structures formed by extension and normal faulting. Arrows indicate extension.

the net effect is the same: the hanging wall moved down *relative* to the footwall. Normal faults tend to be associated with tectonic regimes undergoing *extension*. In **extensional** ("pull-apart") **tectonic regimes** pieces of crust are pulling apart (Figure 2.17). In such a regime it is more likely that blocks of crust will move downward relative to adjacent blocks. A prime example of an extensional regime is the Basin and Range of the western United States, especially Nevada. Here, an entire series of north–south trending fault-bounded basins—called grabens—are separated by uplifts called horsts (Figure 2.16). The grabens have moved downward along normal faults *relative* to the horsts because of the overall stretching of the crust. The repeated occurrence of these **horst-and-graben** structures has been used to infer that the crust in this part of the world is pulling apart.

Reverse faults are the opposite of normal faults. In reverse faults the hanging wall moves up relative to the footwall. Reverse faults are associated with tectonic regimes undergoing compression. **Compressional tectonic regimes** commonly occur when pieces of crust are pushed together or over one another, such as during subduction or orogenesis. Reverse faults generally occur at relatively high angles to the horizon. A **thrust fault** is a particular type of reverse fault in which the angle of the fault is quite low or nearly horizontal. Thrust faults are commonly associated with orogenesis and can be involved in the movement of large slabs of rock over many tens to hundreds of kilometers.

Not all faults involve the vertical movement of rocks along the fault. **Strike-slip faults** involve horizontal movement of rocks relative to each other along the fault (**Figure 2.18**). The San Andreas Fault is a strike-slip fault marking the boundary between the Pacific and North American plates. The San Andreas is termed a right-lateral strike-slip fault; this is because as one imagines standing on the Pacific Plate looking to the east, the North American Plate moves to the right relative to the Pacific Plate (Figure 2.18). Left-lateral strike-slip

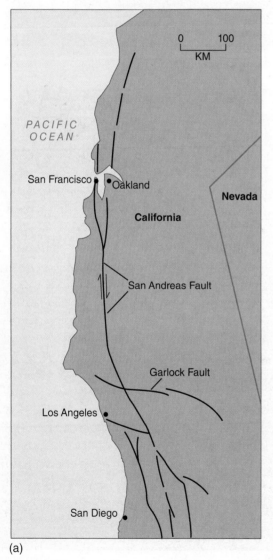

(a)

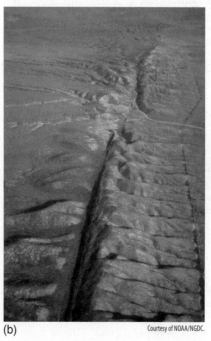

(b)

Courtesy of NOAA/NGDC.

FIGURE 2.18 (a) The San Andreas Fault is a right-lateral strike-slip fault, in which the rocks move past one another horizontally. The sense of movement along left-lateral strike-slip faults is just the opposite to that of the San Andreas fault. Given its current rate of motion, movement along the San Andreas will eventually send California and the Baja Peninsula of Mexico into the Pacific Ocean. This gave rise to fears many years ago that California was "falling into the sea." **(b)** Aerial photo of a portion of the San Andreas Fault. Note that, as shown in part b, there can be some vertical movement along strike-slip faults.

2.4   Continental Margins and Plate Boundaries: Features and Behavior   **39**

faults exhibit relative movement in the opposite direction, that is, the rocks on the other side of the fault from where one is standing move to the left.

Another phenomenon is also commonly associated with tectonism and the deformation of rocks, especially by large-scale thrust faulting during orogenesis: **folding**. Folds are of two types: anticlines and synclines. In **anticlines** the strata dip away from the main axis of the fold, with the oldest strata found in the core of the anticline, and the youngest strata on its crest and sides. In **synclines** just the opposite occurs: the strata dip toward the center of the fold, and the strata found along the axis of the syncline are younger than strata farther away from the axis (**Figure 2.19**). Anticlines and synclines frequently occur adjacent to one another because both form during folding.

Sometimes folds become oversteepened and come to lie on their sides. Thrust faults are commonly associated with this type of folding. Under these conditions the compressional forces finally become so great that instead of continued folding, the rocks fracture and a hanging wall moves over a footwall at low angles. Spectacular folds of this type are seen in mountain ranges all over the world; in the Alps, they are referred to as **nappes** (**Figure 2.20**).

2.4.3 Types of plate boundaries

Ⓖ Ⓗ The **plate boundaries** associated with active continental margins can be categorized as divergent,

convergent, or transform (**Figure 2.21**). These types of plate boundaries are also found on the ocean floor.

Divergent plate boundaries

Divergent plate boundaries are associated with rifting. Initially, rifting of the crust occurs to produce rift valleys on land or along the crests of mid-ocean ridges (Figure 2.21). Divergent margins mark the *upward* movement of mantle material involved in seafloor spreading and the formation of seafloor crust. Thus, divergent plate boundaries are extensional, or pull-apart, in nature and tend to be associated with

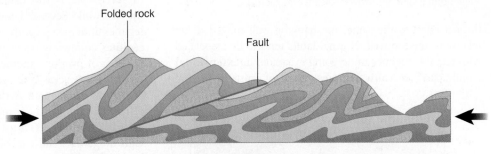

(a) **Sedimentary rocks squeezed by compression**

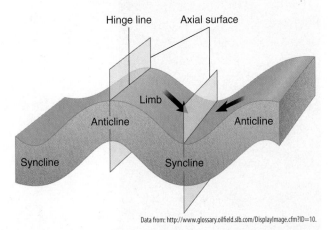

Data from: http://www.glossary.oilfield.slb.com/DisplayImage.cfm?ID=10.

FIGURE 2.19 The features of anticlines and synclines. The central portion of an anticline or syncline is known as its axis. In anticlines, the strata dip away from the axis of the anticline, with the oldest strata being found in the core of the anticline and the youngest on its crest and limbs. In synclines, just the opposite occurs: the strata dip toward the center of the syncline and the strata found along the axis of the syncline are younger than strata farther away. The axis of an anticline or syncline might remain parallel to the Earth's surface, or it might plunge beneath the surface at an angle.

(b) © Shifted/Shutterstock, Inc.

FIGURE 2.20 (a) Cross section of highly-folded thrust sheets in the Alps. The huge blocks of rocks that have been thrust over younger rocks are called nappes. **(b)** Photo of the Alps showing extensive folding.

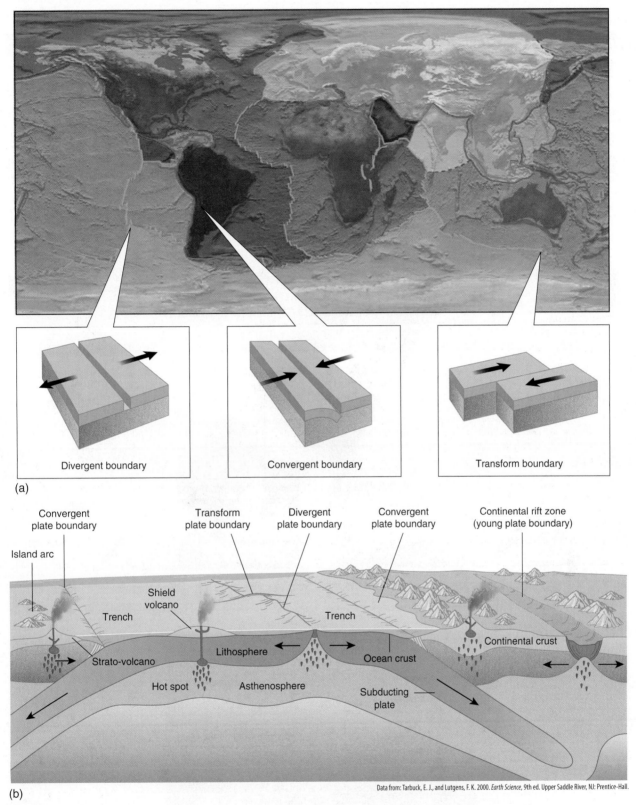

(a)

(b)

Data from: Tarbuck, E. J., and Lutgens, F. K. 2000. *Earth Science*, 9th ed. Upper Saddle River, NJ: Prentice-Hall.

FIGURE 2.21 Types of plate boundaries. Divergent plate boundaries are caused by extensional forces, whereas convergent plate boundaries (Figure 2.22) are caused by compressional forces. Transforms offset mid-ocean ridges and magnetic stripes.

normal faults. Divergent plate boundaries, especially spreading centers, are commonly associated with earthquakes and volcanism because of the proximity of mantle rocks (**Table 2.1**).

Convergent plate boundaries

By contrast, **convergent plate boundaries** mark the subduction of seafloor crust into subduction zones (**Figure 2.22**). Convergent plate boundaries are compressional in nature and

TABLE 2.1

Types of active plate boundaries and their features

Geologic Feature	Relative Plate Motion	Geologic Phenomena	Topography	Examples
Mid-ocean ridge	Divergent	Seafloor spreading, submarine volcanism	Ocean ridge, volcanoes	Mid-Atlantic Ridge, Iceland
Rift valley	Divergent	Continental rifting, volcanism	Rift valley	East African Rift Valley
Trench	Convergent Ocean-ocean collision Ocean-continent collision Continent-continent collision	Subduction, volcanic island arc formation Subduction, volcanic arc formation, orogeny Orogeny	Deep-sea trench and volcanic arc Deep-sea trench and volcanic arc Mountain belt	Western Aleutian Islands (off Alaska) Cascades (U.S. Pacific Northwest), Andes Mountains Himalaya, Japan
Transform	Strike slip	Crustal deformation, earthquakes	Offsets in mid-ocean ridges	Mid-ocean ridges in general (e.g., Mid-Atlantic Ridge, East Pacific Rise)

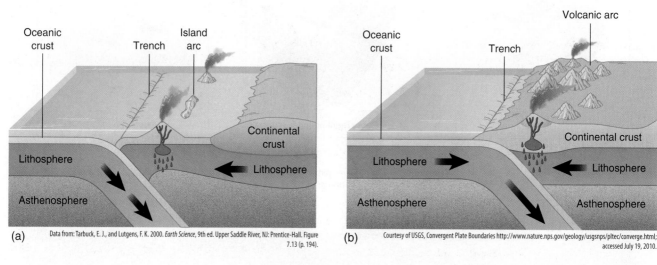

(a) Data from: Tarbuck, E. J., and Lutgens, F. K. 2000. *Earth Science*, 9th ed. Upper Saddle River, NJ: Prentice-Hall. Figure 7.13 (p. 194).

(b) Courtesy of USGS, Convergent Plate Boundaries http://www.nature.nps.gov/geology/usgsnps/pltec/converge.html; accessed July 19, 2010.

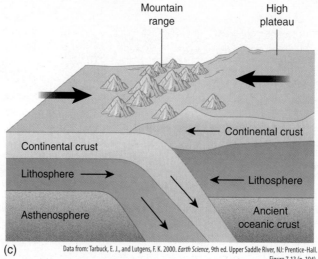

(c) Data from: Tarbuck, E. J., and Lutgens, F. K. 2000. *Earth Science*, 9th ed. Upper Saddle River, NJ: Prentice-Hall. Figure 7.13 (p. 194).

FIGURE 2.22 Convergent plate boundaries. **(a)** Oceanic–oceanic plate boundaries produce island arcs like that of Japan. **(b)** Oceanic–continent plate boundaries produce continental volcanic arcs, like the Cascades of the Pacific Northwest. **(c)** Continent–continent plate boundaries produce mountain chains like the Himalayas.

tend to be associated with earthquakes and thrust faulting. Examples include the Pacific Ocean's "Ring of Fire." During subduction, the components of the plate and surrounding rock with relatively low melting points become molten. This tends to produce magma that might eventually reach the surface to feed volcanoes. The sediment and water carried with the subducting plate lower the melting temperatures of the subducted materials and the surrounding rocks.

There are three types of convergent boundaries (Figure 2.22). The first type, **oceanic–oceanic plate boundaries**, forms by subduction of the leading edge of an oceanic crustal plate beneath the margin of another plate comprised of ocean crust. Subduction in this manner produces volcanic island arcs. Examples of oceanic–oceanic plate boundaries include the Aleutian Islands off Alaska, the islands that comprise the country of Japan, and the Philippines and Indonesia (Figure 2.1). The basin between the island arc and the trench is called a **fore-arc basin**. The width of the fore-arc basin depends on the angle of plate subduction. If the angle of subduction is relatively steep, the island arc occurs closer to the trench. By contrast, if the angle of subduction is shallow, the subducting plate must descend further beneath the overriding plate before sufficient melting begins to generate magma that feeds volcanoes (Figure 2.22); this results in the location of the volcanic chain farther from the trench.

If the rate of subduction is faster than the movement of the overriding plate, the lithosphere behind the volcanic island arc can be subjected to tensional forces that pull it apart. A **back-arc basin** is thought to be produced in this way (Figure 2.22). The back-arc basin might continue to spread apart and grow if magma continues to break through the thinned crust to produce new ocean crust. The Sea of Japan, which separates the volcanic arc of Japan from mainland China, represents a modern back-arc basin. Depending on the angle at which the plate is subducted into the mantle along oceanic–oceanic plate boundaries, partial melting produces magmas up to several hundred kilometers from the trench where subduction is occurring.

The second type of convergent plate boundary is called **oceanic–continent** (Figure 2.22). In this type of plate boundary a plate whose leading edge consists of ocean crust is subducted beneath another plate whose leading edge consists of continental crust. This produces continental volcanic arcs. Prime examples of oceanic–continent boundaries are the Andes in South America and the Cascades of the Pacific Northwest. The Cascades extend from southern Canada into northern California and include Mount St. Helens in Washington state (which last erupted in 1980; **Figure 2.23**), Mount Mazama in Oregon (which erupted about 5,000 BC, the remnant of which is Crater Lake), and Mount Lassen in northern California (which last erupted about 1915).

The third type of convergent boundary is **continent–continent** (Figure 2.22). Ocean crust of the leading edge of one plate bearing a continent is subducted beneath the margin of a second plate (also bearing a continent) as a volcanic island arc forms on the second plate. As the continent on the first plate draws near the subduction zone, subduction slows because the continent is too buoyant to be subducted. Then,

© Photodisc.

FIGURE 2.23 The eruption of Mt. St. Helens in 1980 in the Cascade Range of the Pacific Northwest.

the margin of the subducting plate detaches and descends. The descending slab continues into the asthenosphere, where it might melt (in some cases the slabs might persist in the mantle after subduction).

Subduction of the second plate then begins in the reverse direction. As subduction in the reverse direction proceeds, the second continent approaches the first, whereas a continental volcanic arc may form on the first continent. Eventually, the continents collide to produce mountain ranges such as the Himalayan Mountains (refer to this chapter's frontispiece). Volcanism tends to be less common at continent–continent boundaries but may be associated with the production of andesitic and rhyolitic magmas during subduction.

Transform boundaries

The third type of plate boundary, called a **transform plate boundary** (also called transform faults or just transforms), marks the location where plates or segments of plates move past one another *horizontally* (Figure 2.21). Transforms link offsets of different segments of mid-ocean ridges. Rocks on either side of the fault offsetting adjacent ridge segments move horizontally in *opposite* directions *relative* to each other, just as they do in strike-slip faults. However, *beyond* the offset segments, the rocks on either side of the fault move in the *same* direction (Figure 2.21). An alternative explanation for transforms is that relative movement on each side of the fault is *always* in the same direction (Figure 2.21). However, although this can account for the offsets in mid-ocean ridges, it cannot explain the offsets in magnetic stripes. In fact, this is how transforms were first recognized by J. Tuzo Wilson, a Canadian geophysicist who helped transform continental drift into the modern theory of plate tectonics. Thus, transforms account for *both* the offsets of mid-ocean ridges and magnetic stripes. Transforms also indicate that spreading rates along different ridge segments are the same; otherwise, the ridge segments would not be offset.

Transforms are typically associated with ocean basins. Occasionally, though, transforms extend onto land. The San Andreas Fault, for example, is thought to have been derived from a transform that now separates the North American and Pacific plates (**Figure 2.24**). Today, a fault links the modern spreading center known as the East Pacific Rise in the Gulf of California to the Juan de Fuca spreading center located off the coast of the Pacific Northwest. Offset segments of the spreading center in the Gulf of California extend all the way up the Gulf and then seem to disappear beneath western North America (Figure 2.24).

It therefore appears that an ancient plate—called the Farallon plate—was being subducted beneath the North

BOX 2.3 New Hypotheses About Plate Tectonics

Despite our understanding of the internal structure and function of Earth, like all science, good questions spawn answers that breed more questions. Even the now widely accepted theory of plate tectonics is subject to modification. We have already seen one example of this: the stationary or nonstationary nature of hotspots (Box 2.2). And we still do not fully understand the forces that drive the movements of the plates. The principal hypothesis presented in this book, involving the passive movement of the plates on top of the convective flow of the mantle, stems from the work of Holmes. However, the movement of the plates is constrained because of their relatively "tight fit" against one another. This limits our ability to infer the directions of convection beneath them.

A recently developed technique called **seismic tomography**, which is an outgrowth of medical technology, reveals that the internal workings of the mantle are far more complex than just the movement of large convection cells. We now know, for example, that although heat flow is highest along mid-ocean ridges (seafloor spreading centers) and decreases away from them, the heat flow varies dramatically from one ridge or portion of a ridge to the next. This suggests the amount of heat moved by convection cells is not uniform. Indeed, as we have already seen, transforms indicate that spreading rates along different ocean ridge segments are not uniform.

As a result of these studies, several other hypotheses have been suggested for the actual driving force of plate movements. Some scientists believe as the leading edge of a plate cools, it becomes sufficiently dense to descend back into the mantle. If

the descending portion of the plate remains attached to the rest of the plate, it may pull the rest of the plate behind it (*slab pull hypothesis*). Other scientists believe plates move by pushing from either magma upwelling at spreading centers (*ridge push hypothesis*) or by the plates sliding down the side of the mid-ocean ridge by gravity (*gravity slide* hypothesis). Still others believe if the descending slab becomes detached from the rest of the plate, it might exert a kind of *suction* on the rest of the plate (like fluid moving through a straw but at very slow rates). And still others believe that as mantle plumes move upward, they eventually begin to spread apart like convection cells, causing seafloor spreading.

Other data offer tantalizing clues as to the mechanisms of plate motion, but they are not conclusive. For example, the velocity of plate movement is inversely related to the continental mass on the plate. Plates bearing continents move far more slowly than oceanic plates, suggesting a deep continental *root* exerts a drag on plate movement. Also, plates bounded by long subduction zones tend to move faster than those with shorter subduction zones. However, no relation has been found between the length of divergent boundaries and plate velocities or with the length or angle of subducted slabs.

Other fundamental questions are associated with plate movement: what causes a trench to start in the first place, and why does a trench occur where it does? Is the initiation of a trench basically a function of the cooling of the leading edge of the plate, so the plate begins to descend *wherever* it first becomes sufficiently dense to sink?

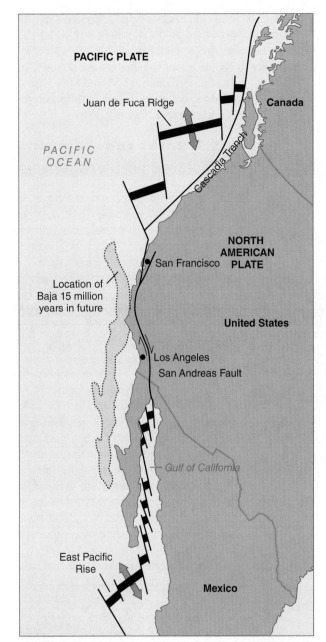

FIGURE 2.24 A hypothesis for the formation of the San Andreas Fault. The San Andreas Fault is a former transform that now links the Gulf of California spreading center to the Juan de Fuca spreading center in the Pacific Ocean. The San Andreas fault system is thought to have formed as a result of the North American plate overriding the ancient Farallon plate. As this happened, transforms offsetting segments of the East Pacific Rise are thought to have "linked up" to produce the San Andreas fault system.

American plate along the margin of western North America. Most of the Farallon plate has in fact been subducted, leaving behind two small plates: the Juan de Fuca plate along the Pacific Northwest and the Cocos plate along southern Mexico and Central America (Figures 2.1 and Figures 2.24).

As the North American plate moved westward, it eventually collided with the Pacific-Farallon Ridge. This ridge represents the spreading center that originally bounded

the Farallon and Pacific plates; the Farallon Plate moved east from the ridge, whereas the Pacific Plate moved to the west-northwest (Figure 2.24). As the North American plate collided with the Pacific-Farallon Ridge, it encountered one or more transforms. As this happened the North American and Pacific plates began to slide past one another horizontally, giving rise to the San Andreas Fault system. Eventually, the Baja Peninsula also began to rift away from the Mexican mainland to form the Gulf of California.

CONCEPT AND REASONING CHECKS

1. What is the difference between a transform fault and a strike-slip fault?
2. Do convergent or divergent plate boundaries occur along a passive margin?
3. Given your knowledge of the evolution of the San Andreas Fault system, do you suppose plate boundaries change through time? For example, can a passive margin become active, or vice versa?

2.5 Orogenesis

2.5.1 Types of orogenesis

Orogenesis occurs in different ways along convergent plate boundaries. The styles of orogenesis parallel convergent plate boundaries: oceanic–oceanic, oceanic–continent, and continent–continent.

The first two categories of convergent plate boundaries reflect what is called orogenesis without continental collision. In the first case the mountains that form are really volcanic island arcs like those of Japan in which one piece of ocean crust is subducted beneath another. The second style of orogenesis is exemplified by the modern Andes in which mountains, including volcanoes, form on land. In this type of orogenesis the leading edge of a plate composed of oceanic crust is subducted beneath another plate, which bears a continent. As subduction of the oceanic crust occurs beneath the margin of the continent, enormous volumes of magma can be generated, producing a continental volcanic arc. Because temperature and pressure tend to decline away from the main axis of intrusive activity, belts of rock reflecting differing intensities of metamorphism called metamorphic grades can form more or less parallel to the main axis of orogenesis (see Chapter 3). However, these patterns can be obscured by folding and thrust faulting associated with orogeny. The third category of orogenesis reflects the collision of continents, originally found on separate plates, to produce mountain chains. Orogenesis along continent–continent convergent plate boundaries occurs when the continents finally collide. This results in enormous volumes of sediment being shed to produce enormous delta complexes or submarine fans.

In many cases, volcanic island arcs, small **microcontinents**, or both can be accreted to the margins of plates at convergent boundaries. If the newly accreted rocks possess distinctive traits such as unusual lithologies, fossils, magnetic particle orientations, or some combination, they are recognizable as distinct **terranes** (**Figure 2.25**). These terranes are sometimes referred to as **exotic terranes** because their particular lithologic, fossil, or paleomagnetic traits differ dramatically from those of the continent with which they collided. Such a scenario would occur if, say, islands of New Zealand were to collide with a larger continent (Figure 2.1).

The entire state of Alaska and large parts of Washington, Oregon, and California consist of such terranes (Figure 2.25).

2.5.2 Rocks and sediments associated with orogenesis

Various types of rocks and sediments are associated with orogenesis. In the trench itself a jumbled mass of rock and sediment referred to as mélange (French for "mixture") forms in deep water. **Mélange** consists of volcanic debris, muds, graywackes, turbidites, and pieces of ocean crust (**Figure 2.26**). During subduction the mélange is folded and thrust-faulted to produce an **accretionary wedge** or prism consisting of thick "slices" of rock stacked on top of one another. Metamorphic rocks called blueschists can also occur in the accretionary wedge.

The mélange might also include **ophiolites**. Ophiolites are slabs of ocean crust and sometimes portions of the upper mantle, along with sedimentary rocks deposited in deep water nearer the top of the ophiolite (**Figure 2.27**). Ophiolites are very important geologically because they represent the remains of ancient ocean crust and possibly upper mantle not otherwise exposed at Earth's surface.

The idealized ophiolite consists of a distinct bottom-to-top sequence of layered ultramafic rocks and layered gabbros (Figure 2.27). The basal ultramafics formed by the settling of crystals to the bottom of the magma, where they solidified to form rock, whereas the contact between the layered ultramafics and layered gabbros probably represents the base of the oceanic crust. Sheet-like intrusive ("injected") rocks called dikes composed of the mafic rock diabase follow upward in the sequence. The sheet-like dikes formed as new dikes were intruded one after the other, as the earlier dikes split to allow the intrusion of later ones. Above this appear basaltic "pillow lavas" (submarine eruptions) that were fed by the dikes from the deeper magma chamber. As the lavas were extruded into seawater, they were quenched to form their characteristic billowy appearance.

As the sediments and rocks adjacent to orogenic belts are uplifted, they are highly folded to produce anticlines and synclines and thrust-faulted great distances as a result

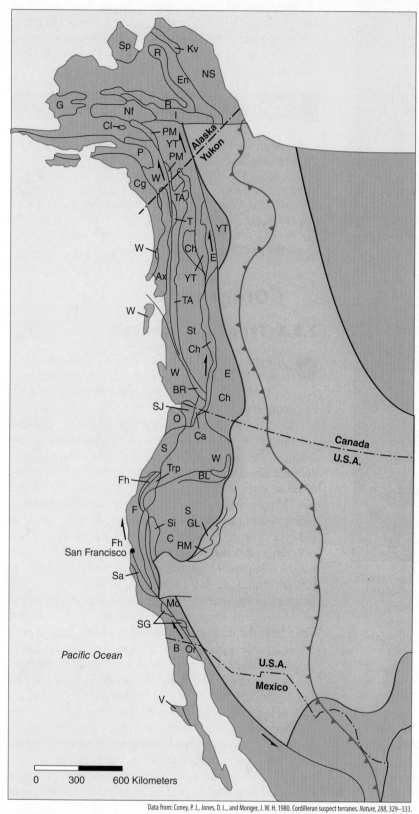

Data from: Coney, P. J., Jones, D. L., and Monger, J. W. H. 1980. Cordilleran suspect terranes. *Nature, 288*, 329–333.

FIGURE 2.25 Exotic terranes of the western United States. Letters indicate various terranes and small arrows indicate sense of strike-slip movement. Blue line indicates eastward extent of deformation by tectonism.

(a)

(b)

FIGURE 2.26 **(a)** Mélange consists of jumbled muds, graywackes ("dirty sandstones" often containing rock fragments), turbidites (sediments deposited by submarine landslides), and pieces of ocean crust that are folded and faulted into huge slices of sediment stacked on top of one another as they are "bulldozed" and "scraped up" by a moving plate. **(b)** Repeated layers of deep-sea microfossil oozes can also be found among melanges. These photographs are of the Franciscan mélange along the west coast of California north of San Francisco.

of the compressive forces associated with subduction and orogenesis. At the same time subduction of the plate beneath the continental margin and the weight of the growing mountain range cause the crust along the landward side of the volcanic arc to gradually sink and form a deep **foreland basin**. The sediments that are uplifted are shed into the deep waters of the trench and foreland basin as shales and turbidites and are referred to collectively as **flysch** (a term coined by geologists working in the Alps during the 19th century).

Eventually, so much sediment is shed from the uplifting mountains into the foreland basin that the seaway is driven out. This sediment consists of nonmarine sediments shed directly from the mountains onto river floodplains, swamps, and coastal and lagoon environments that can extend landward for hundreds of kilometers. Such sediments are often referred to as **molasse** (a term also coined by geologists working in the Alps). The sediments are often so thick that they form enormous wedges of sediment consisting of poorly

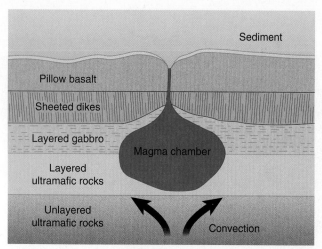

Data from: Levin, H. 2006. *The Earth Through Time*, 8th ed. Hoboken, NJ: John Wiley. Figure 7-47 (p. 191).

FIGURE 2.27 The idealized sequence of lithologies in an ophiolite. Note the relation of the sequence to the structure and formation of ocean crust from the upper mantle.

sorted sediments such as conglomerates (gravels) nearest the uplift, grading into sands, shales (muds), and limestones farther away.

CONCEPT AND REASONING CHECKS

1. Look at a map of the Pacific Ocean basin, and locate possible islands or island chains that might eventually be accreted as microcontinents or exotic terranes.
2. How long would it take for this to occur given current rates of seafloor spreading?

2.6 Isostasy

After orogenesis has begun, it can continue for many tens of millions of years. This is not due just to the continuation of compressive forces that push the mountains up. Ironically, it is the erosion of the mountains that causes the mountains to continue to rise.

The coupling of uplift and erosion is the result of the phenomenon called **isostasy**. According to isostasy, vertical adjustments of Earth's lithosphere cause portions of Earth's crust to occur at levels determined by their thickness and density. In other words, isostasy is much like flotation (**Figure 2.28**). Imagine pushing a piece of wood beneath the surface in a pan of water and then letting go; the wood bobs upward. Similarly, a great thickness of sediment or glacial ice will push the underlying crust downward, but when the overlying weight is removed, the crust will move upward. In fact, isostasy was first recognized in the 1800s in Scandinavia, where ancient coastlines were exposed. As the overlying ice sheets of the last glacial age retreated, the release of the weight allowed the underlying crust to move upward, or rebound, slowly.

The same principle applies to the erosion of mountain ranges. As overlying sediment is eroded, the deep crustal root beneath the mountain slowly rebounds upward because it is less dense (Figure 2.28). Geologists now recognize erosion plays a much more prominent role in orogenesis than even Lyell imagined. This is because the erosion of mountain chains provides *positive feedback* on mountain building. As the weight of the overlying rocks is removed, mantle heat lies closer to Earth's surface, intensifying metamorphism there. *Thus, differential erosion of mountains feeds back on isostatic adjustments of the lithosphere by affecting the site and intensity of metamorphism beneath the mountain range.*

If erosion occurs for sufficiently long periods of time, once spectacular mountain chains several kilometers or more high are eroded down to their roots. Nevertheless, many ancient mountain chains, including those that formed billions of years ago in the Precambrian, have left their traces in the form of igneous intrusions, metamorphic rocks, and large-scale folds and faults exposed on the modern continents.

CONCEPT AND REASONING CHECKS

1. Explain isostasy in your own words.
2. How does erosion cause mountain building?

2.7 Tectonic Cycle

 With the acceptance of plate tectonics, geologists began to search Earth's surface for modern examples of how plates and continents behave through time. Although many of the sites they examined had long been regarded as geologically interesting, their true significance for plate tectonics was not understood until *after* acceptance of the theory of plate tectonics. In other words, the theory of plate tectonics provided the framework for ordering the isolated observations into a coherent framework.

In re-examining these sites, geologists began to piece together the **tectonic cycle** (**Figure 2.29**). The tectonic cycle is frequently referred to as the **Wilson cycle** in honor of J. Tuzo Wilson. Based on the different types of modern plate boundaries, Wilson proposed a cycle of continental rifting, the opening of ocean basins, the disappearance of the oceans by subduction, and finally the collision of continents. Each of these cycles takes several hundred million years to complete. Thus, the processes involved are so slow that geologists were not even aware these cycles occurred until the advent of plate tectonics theory.

A modern example of the initial stage of the tectonic cycle is exemplified by the East African Rift Valley, which is an elongate valley (graben) associated with rivers, lakes, and the volcanic eruptions of Mount Kilimanjaro. The rift valleys presumably form over hotspots produced by mantle plumes, which pipe magma from deep within Earth and push up the lithosphere. As the lithosphere is elevated, Y-shaped sets of

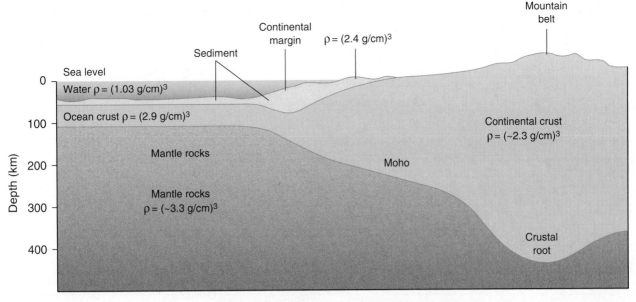

FIGURE 2.28 Isostatic uplift of a mountain range occurs as erosion removes overlying rocks of a mountain. The mountain has a relatively low density (ρ) granitic "root" that is buoyant and "bobs" up and down on the mantle, much like an ice cube pushed down in a pan of water. If sufficient sediment is eroded, it may cause subsidence at the site of deposition.

valleys produce **triple junctions** to relieve the stress on the crust (**Figure 2.30**). Because the crust is relatively cool, it is brittle and easily broken compared with the warmer and more plastic mantle beneath it. The valleys of the triple junctions form the initial rift valleys. If rifting continues, two of the arms of the triple junction continue to spread and eventually widen to produce an elongate, deep seaway such as the Red Sea between Africa and Saudi Arabia, whereas the third arm stops spreading. The arms that continue to spread eventually expand into an ocean basin, such as the Atlantic Ocean. The final stage of the tectonic cycle is the collision of continents, one on each plate. Continental collision results in orogenesis and the **suturing** of one continent to another. Today, this process is exemplified by the collision of India with Asia to form the Himalayas.

Not all tectonic cycles run to completion. The failed rift valleys of the triple junctions are referred to as

aulacogens. Although they are mostly buried by much younger sediments, aulacogens are frequently associated with depressions at Earth's surface that form river valleys. Examples are the valleys of the Amazon (which flows through a foreland basin), the Niger, and the Hudson rivers. These particular aulacogens were associated with the initial rifting of Pangaea (Figure 2.30). Another example is the Mississippi River, which flows down an aulacogen that formed during rifting of another supercontinent during the late Precambrian.

Although rifting has long ceased along aulacogens, occasionally movement along the faults that bound the ancient basins occurs, producing earthquakes. These sorts of earthquakes are unusual because they occur within plates, not along their margins. In 1811 and 1812 earthquakes struck near New Madrid, Missouri, located near the aulacogen down which the Mississippi flows. The earthquakes

Stage	Motion	Physiography	Example
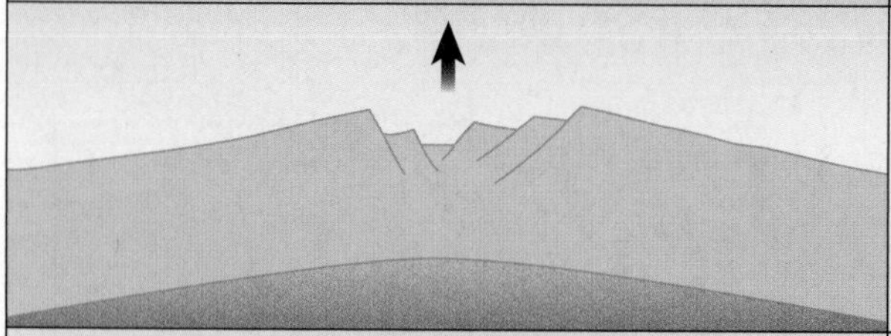	Uplift	Complex system of linear rift valleys on continent	East African Rift Valley
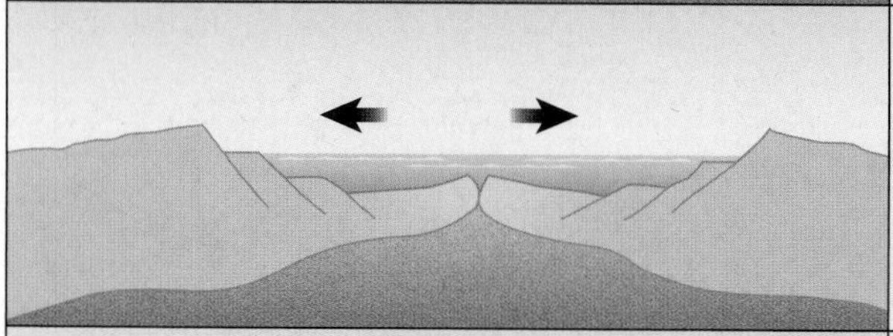	Divergence (spreading)	Narrow seas with matching coasts	Red Sea
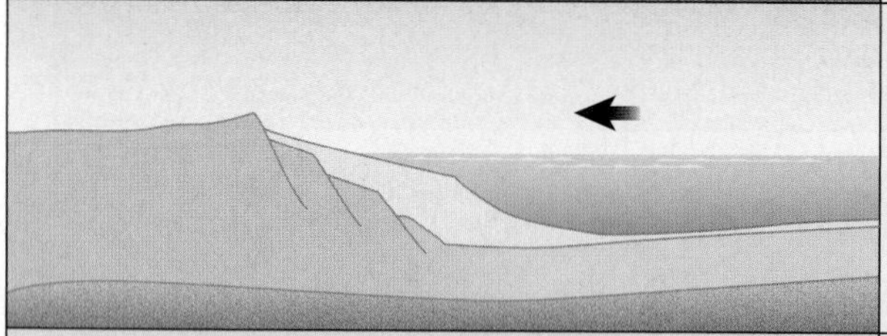	Divergence (spreading)	Ocean basin with continental margins	Atlantic, Indian, and Arctic oceans
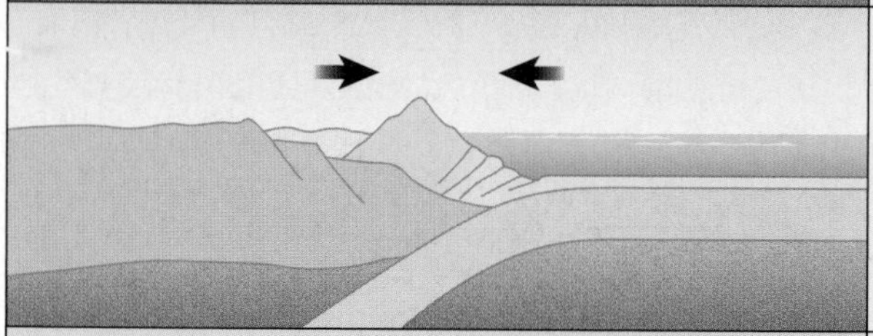	Convergence (subduction)	Island arcs and trenches around basin edge	Pacific Ocean
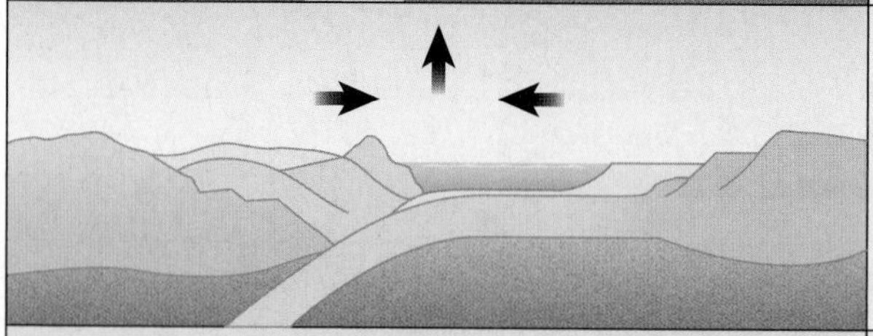	Convergence (collision and uplift)	Narrow, irregular seas with young mountains	Mediterranean Sea
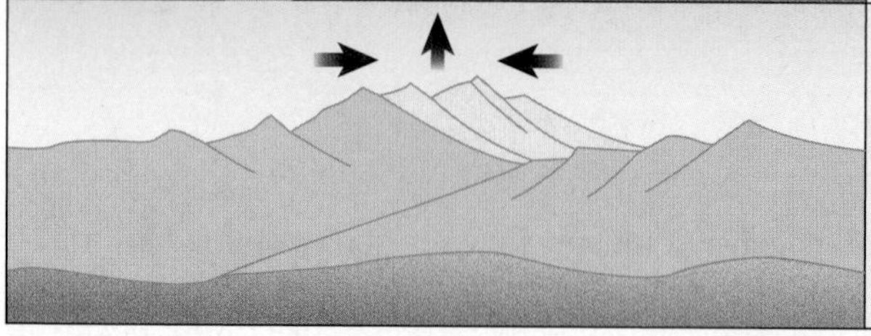	Convergence (and uplift)	Young to mature mountain belts	Himalayas

Data from: Wilson, J. T. American Philosophical Society Proceedings 112 (1968): 309–320; Jacobs, J. A., et al. *Physics and Geology*. McGraw-Hill, 1974.

FIGURE 2.29 Stages of the tectonic cycle with modern examples. Arrows indicate general directions of tectonic forces of seafloor spreading and compression and uplift during orogeny. Arrows indicate general directions of tectonic forces of seafloor spreading and compression and uplift during orogeny.

were so strong they were felt all over the eastern two-thirds of the United States, from the East Coast to the Rockies. The strong earthquakes resulted from the transmission of the shock waves through rigid continental crust rather than the warmer crust associated with subduction zones, where earthquakes normally occur. Fortunately, much of North America was still sparsely inhabited at this time, minimizing the damage. The earthquakes killed an estimated 10 to 20 people in New Madrid, but nearly destroyed the town. Much smaller earthquakes still occur quite frequently in the vicinity of New Madrid but attract little attention. Another earthquake struck Charleston, South Carolina, in 1886, killing 60 persons and causing $23 million in damage. Although the east coast of the United States is now a

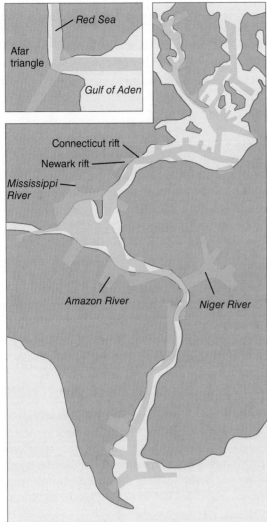

Data from: Burke, K., and Wilson, J. T., Hot spots on the earth's surface: *Scientific American, 235*(2), 46–60.

FIGURE 2.30 Map of aulacogens, or failed rift valleys, associated with a reconstruction of Pangaea. Note that a number of major rivers flow down aulacogens because these failed rift valleys are still expressed at the Earth's surface as low-lying areas.

passive margin, this earthquake resulted from movement along a fault bounding a Triassic rift basin that dates from the breakup of Pangaea.

 CONCEPT AND REASONING CHECKS

1. Diagram and describe the tectonic cycle.
2. Piece together a tectonic cycle using modern examples other than those just described.

2.8 Tectonic Cycle and Scientific Method

Plate tectonics act in a cyclic manner: the tectonic cycle (Figure 2.29). However, the tectonic cycle has never actually been observed to occur as a continuous cycle because of the immense amounts of geologic time involved to complete the cycle; seafloor spreading rates are on the order of a few centimeters per year. Instead, the tectonic cycle has been pieced together from modern examples.

Although this might seem odd, geologists frequently use the approach of piecing sequences together from stages that have been inferred separately. For example, the rock cycle has been inferred from the particular features of different rocks, not from direct, continuous observation of all stages of the cycle. Hess inferred seafloor spreading from the sequence of active volcanoes to seamounts to guyots in the Pacific. Similarly, Charles Darwin made similar inferences about the growth of ring-like coral reefs called atolls on the sides of seamounts (see Chapter 4). Scientists in other fields also use this approach. Biologists infer the stages of the life cycles (birth, growth, reproduction, and death) of different species (such as the reproductive cycles of plants) from observations of different stages. Similarly, astronomers have used observations of different modern stars to infer the "life cycles" of stars from birth to death.

The recognition of the tectonic cycle has other important implications for the study of Earth systems. The tectonic cycle has presumably acted through time. Thus, plate tectonics and the tectonic cycle affect climate by affecting the configuration of continents, the opening and closing of ocean basins, sea level, and the distribution of oceanic and atmospheric currents. As mountains are uplifted, they also affect rainfall patterns and erosion. All these factors in turn affect the distribution of organisms, environments, and rocks in the marine realm and on land, as we will examine later.

CONCEPT AND REASONING CHECKS

1. Was the tectonic cycled pieced together from observations of modern environments alone or were modern observations put together in a sequence based on a theory of the cyclic rifting and collision of the continents?
2. Based on your answer to question 1, which comes first in the so-called "scientific method" (Chapter 1): observation or theory? Or can both occur simultaneously?

SUMMARY

- The theory of plate tectonics really began with early ideas about orogenesis, or mountain building. Hypotheses and theories of mountain building changed radically over the past two centuries, and their development is a prime example of how scientists work and think.

- The discovery of radioactivity led to more modern theories of mountain building. Of these, it is Alfred Wegener's hypothesis of continental drift—based on a variety of evidence—that paved the way for the modern theory of plate tectonics. Initially, Wegener's hypothesis was roundly criticized because he could not identify a mechanism to make continents drift. Consequently, continents (or at least many geologists' minds) remained "fixed" until the work of Harry Hess in the 1950s, which proposed the process of seafloor spreading as a mechanism to move continents.

- In the 1960s, the detection of magnetic seafloor stripes corroborated seafloor spreading and provided the mechanism of continental drift that had eluded Wegener. Seafloor spreading also corroborated Hess's views about the formation of guyots, heat flow beneath mid-ocean ridges, and the destruction of seafloor in trenches. Rearranging the continents into different positions also began to make sense of apparent polar wandering curves.

- Consequently, what had been known as continental drift was wedded to seafloor spreading to produce the theory of plate tectonics.

- Today, plate tectonics is recognized as an integral component of Earth's systems. We know that Earth's lithosphere (the crust and uppermost mantle) consists of about 15 large and small plates that are moved by the production of new seafloor at mid-ocean ridges. Forming portions of the plates are continents. The plates move over the asthenosphere of the mantle. Beneath the mantle are an outer fluid and a solid inner iron and nickel-rich core that generate Earth's magnetic field.

- Although convection cells are widely viewed as moving the plates, several hypotheses have been proposed to explain how the seafloor actually moves: (1) slab-pull, in which a descending slab pulls the rest of slab behind it downward; (2) ridge-push, in newly formed ocean crust as spreading centers pushes the slab ahead of it; (3) gravity slide, in which a slab slowly "slides" down the side of a spreading center, pushing the slab ahead of it; and (4) suction from the descending portion of a plate.

- Based on plate tectonics, different features of the planet can be arranged into a sequence of stages called the tectonic cycle: East African Rift Valley, Red Sea, Atlantic Ocean, Pacific Ocean, and suture (Himalayas). Not all rift valleys become seaways, however; many have become failed rift valleys or aulacogens, down which some of the world's major rivers such as the Amazon flow. The tectonic cycle has occurred a number of times during Earth's history, each cycle spanning several hundred million years.

- Based on the tectonic cycle, continental margins and plate boundaries can change through time. There are two basic types of continental margins: active and passive. Passive continental margins, like those along the Atlantic Ocean, accumulate sediment along their margins. Active margins, like those along the Pacific Ocean's ring of fire, are sites of subduction, volcanism, and earthquakes.

- Plate boundaries are classified into three basic categories: convergent (associated with sea floor trenches), divergent (associated with rifting), and transform, which are associated with offsets of mid-ocean ridges.

- Convergent boundaries are themselves of three types: volcanic island arc (for example, Japan), continental island arc (for example, the Cascades), and collisional (Himalayas).

- The three types of convergent plate boundaries parallel the different types of orogenesis and the formation of major geologic structures such as faults and folds: island arcs only, plate collisions without continents, and continent–continent collisions.

- As orogenesis occurs, smaller pieces of crust with distinctive geologic features (rock type, fossils, paleomagnetic directions) called microcontinents or exotic terranes can be sandwiched between the larger continents.

- No one has ever observed the tectonic cycle because of the immense amounts of geologic time involved in its completion, but it can be pieced together based on observations of modern tectonic settings.

accretionary wedge

active margins

anticlines

apparent polar wandering
  curves

asthenosphere

aulacogens

back-arc basin

basalt

compressional tectonic
  regimes

continental drift

continent–continent

convergent plate boundaries

core

divergent plate boundaries

earthquakes

epicenter

exotic terranes

extensional tectonic regimes

fault

flysch

focus

folding

fore-arc basin

foreland basin

gabbros

geosyncline

guyots

horst-and-graben

hotspot

isostasy

lithosphere

magma

mantle plume

mélange

microcontinents

mid-ocean ridges

Mohorovičić discontinuity
  (Moho)

molasse

nappes

normal faults

oceanic–continent

oceanic–oceanic plate
  boundaries

ophiolites

orogenesis

orogeny

P-waves

Pangaea

passive margins

plate boundaries

plates

plate tectonics

reverse faults

rock cycle

S-waves

seafloor trenches

seamounts

seismic tomography

strike-slip faults

subduction zones

suturing

synclines

tectonic cycle

tectonics

terranes

thrust fault

transform plate boundary

triple junctions

Wilson cycle

REVIEW QUESTIONS

1. Draw and label a cross-section of the Earth from core to crust.
2. Identify active and passive plate margins at various places on Earth.
3. Draw cross-sections of each of the following plate boundaries and indicate the relative plate motions: (a) convergent (all three types), (b) divergent, (c) transform.
4. What was the evidence used by Wegener to infer continental drift?
5. What was the evidence used by Hess to conclude that seafloor trenches exist?
6. How did the recognition of seafloor "stripes" corroborate plate tectonics?
7. What sorts of rocks are found in each of the following settings: (a) convergent (all three types), (b) divergent, (c) transform?
8. What is the difference between the geographic and magnetic poles of Earth?
9. How could one test the accuracy of reconstructed continental configurations based on polar wandering curves?
10. Distinguish between the different possible mechanisms of plate movement.
11. What sorts of features are used to distinguish an exotic terrane?
12. What are the stages of the tectonic cycle? Give modern examples.
13. How is isostatic rebound involved in orogeny?

1. Why is the theory of plate tectonics superior to the geosynclinal theory?
2. Describe all the different ways geologists infer the structure and composition of Earth's interior.
3. Identify possible sites in the world other than those discussed in the text that might represent stages of the tectonic cycle.
4. How do we know that the tectonic cycle occurs?
5. Is the east coast of North America today an active or a passive margin? Explain. What about the west coast?
6. Does the type of plate boundary always remain the same along the active margin of a continent?
7. Can more than one subduction zone occur along the margin of a continent? What are some examples?
8. Why does Yellowstone Park have lots of geysers, hot springs, and huge lava flows?
9. Why do you suppose earthquake activity extends across Asia and the Mediterranean?
10. The location of the Ural Mountains in the interior of Russia was long considered an anomaly of geosynclinal theory because geosynclinal theory could not explain the presence of these mountains in the interior of a continent. Why are the Urals there?
11. Assume that a seafloor spreading center is located 1000 km from a trench and that seafloor is formed at the spreading center at the rate of 5 cm per year. How long will it take for new crust created at the spreading center to reach the trench?
12. What island nation(s) would western California perhaps resemble in, say, 30 million years, if current tectonic processes continue at their present rate?
13. Orogenic episodes in North America and Europe were given different names prior to the acceptance of the theory of plate tectonics. Explain.
14. Why is plate tectonics considered a theory and not a hypothesis or law?
15. (a) Review the evidence used by Harry Hess to infer the ocean basins are geologically young. (b) Now arrange Hess's observations regarding heat flow at, and features of, mid-ocean ridges as a series of predictions from your conclusion(s) to part a. Are you now using inductive or deductive logic (see Chapter 1)? (c) What are the basic differences between inductive and deductive logic?
16. How does the "elegance" of a theory resemble Ockham's Razor (see Chapter 1)?

SOURCES AND FURTHER READING

Albritton, C. C. 1980. *The abyss of time: Changing conceptions of the earth's antiquity after the sixteenth century.* San Francisco: W. H. Freeman.

Conrad, C. P., and Lithgow-Bertelloni, C. 2002. How mantle slabs drive plate tectonics. *Science, 298,* 207–209.

DePaolo, D. J., and Manga, M. 2003. Deep origin of hotspots—the mantle plume model. *Science, 300,* 920–921.

Foulger, G. R., and Natland, J. H. 2003. Is "hotspot" volcanism a consequence of plate tectonics? *Science, 300,* 921–922.

Giere, R. N. 1999. *Science without laws.* Chicago: University of Chicago Press.

Glen, W. 1982. *The road to Jaramillo: Critical years of the revolution in earth science.* Stanford, CA: Stanford University Press.

Greene, M. T. 1982. *Geology in the nineteenth century: Changing views of a changing world.* Ithaca, NY: Cornell University Press.

Gurnis, M. 2001. Sculpting the Earth from inside out. *Scientific American, 284*(3), 40–47.

Hess, H. H. 1962. History of ocean basins. In Engel, A. E. J. (Ed.), *Petrologic studies—A volume in honor of A. F. Buddington* (pp. 599–620). Boulder, CO: Geological Society of America.

Hodges, K. 2006. Climate and the evolution of mountains. *Scientific American, 295*(2), 73–79.

Hughes, S. et al. 1999. Mafic volcanism and environmental geology of the eastern Snake River Plain, Idaho. In Hughes, S. S., and Thackray, G. D. (Eds.), *Guidebook to the geology of Eastern Idaho* (pp. 143–168). Pocatello, ID: Idaho State University Press.

King, P. B. 1977. *The evolution of North America*. Princeton, NJ: Princeton University Press.

Murphy, J. B., and Nance, D. 2004. How do supercontinents assemble? *American Scientist, 92*(4), 324–333.

Pinter, N., and Brandon, M. T. 1997. How erosion builds mountains. *Scientific American, 276*(4), 74–79.

Shirey, S. B., and Richardson, S. H. 2011. Start of the Wilson cycle at 3 Ga shown by diamonds from subcontinental mantle. *Science, 333*, 434–436.

Steinberger, B., Sutherland, R., and O'Connell, R. J. 2004. Predictions of Emperor-Hawaii seamount locations from a revised model of global plate motion and mantle flow. *Nature, 430*, 167–173.

Takeuchi, H., Uyeda, S., and Kanamori, H. 1970. *Debate about the earth: Approach to geophysics through analysis of continental drift*. San Francisco: Freeman, Cooper, & Co.

van Andel, Tj. H. 1992. Seafloor spreading and plate tectonics. In Brown, G. C. et al. (Eds.), *Understanding the earth: A new synthesis* (pp. 167–186). Cambridge, UK: Cambridge University Press.

Wegener, A. 1929. *The origin of continents and oceans* (1966 reprint of 4th revised edition). New York: Dover Publications.

Wilson, J. T. 1963. Continental drift. *Scientific American, 208*, 86–100.

Wilson, J. T. 1965. A new class of faults and their bearing on continental drift. *Nature, 207*, 343–347.

Earth Systems: Processes and Interactions

MAJOR CONCEPTS AND QUESTIONS ADDRESSED IN THIS CHAPTER

A What is the rock cycle, and how is it related to plate tectonics and the tectonic cycle?

B What are the major types of rocks, and how do we infer how and where they formed?

C What are the major patterns of atmospheric circulation?

D How does atmospheric circulation affect the distribution of heat and moisture over Earth's surface?

E How fast do the oceans circulate?

F How do the atmosphere and continental configuration affect ocean circulation?

G What accounts for the distribution and diversity of plants and animals over Earth's surface?

H How is the availability of energy related to the structure of biologic communities?

I How are nutrients and other elements recycled, and why is nutrient recycling important?

J How has the tectonic cycle affected the atmosphere and hydrosphere?

CHAPTER OUTLINE

Yellowstone Gorge, Yellowstone National Park, Wyoming. The bright colors of the rocks result from lava flows and ash falls.

© Filip Fuxa/Shutterstock, Inc.

3.1 The Solid Earth System: Components and Processes

A Chapter 3 examines each of Earth's major systems in greater detail. As we proceed, recall the basic features of natural systems introduced in Chapter 1: (1) each major Earth system consists of a series of parts or compartments that comprise a larger integrated and complex whole, (2) each system is an open system that exchanges matter and energy with the environment, and (3) each system behaves in a cyclic manner because of the flow of matter and energy through the system.

The flows of matter and energy within and between systems are known as fluxes. This chapter emphasizes two major features of the fluxes of matter and energy:

1. Fluxes of matter and energy within systems and between systems are cyclic.
2. Systems interact with one another through the fluxes of matter and energy.

Recall, for example, that plate tectonics is driven by the flow of heat, which is itself produced by radioactive decay (Chapter 2). We will concentrate on how plate tectonics interacts with each of the other major Earth systems beginning with the rock cycle, and then move on to the behavior of the modern atmosphere, the hydrosphere, and biosphere. This will allow us to then examine how the tectonic cycle has broadly influenced the other systems through geologic time.

3.2 Rock Cycle

B The major cycle operating within the solid Earth system, especially the lithosphere, is the rock cycle (**Figure 3.1**). The rock cycle involves the formation and destruction of the three major rock types, or *lithologies*: **igneous**, **sedimentary**, and **metamorphic**. Igneous rocks are those that have cooled and solidified from magma (from the Latin, for "characterized by fire").

In an idealized example of the rock cycle, igneous rocks erode to produce sedimentary rocks that are later metamorphosed and then melted to produce igneous rocks (Figure 3.1). Usually, though, the rock cycle does not function this simply. Depending on conditions, all preexisting rocks, whether they are igneous, sedimentary, or metamorphic, can be subjected to any one or more of the processes of the rock cycle out of this sequence. Igneous rocks can, for example, be metamorphosed without first having been eroded; metamorphic rocks can erode to produce sedimentary rocks; sedimentary rocks can be metamorphosed; or sedimentary rocks can become caught up in a "sedimentary loop" in which they are recycled through the processes of erosion, transport, deposition, and lithification all over again to produce new sedimentary rocks.

3.2.1 Igneous rocks

Intrusive igneous rocks: occurrence, texture, and composition

Intrusive igneous rocks form beneath Earth's surface. Bodies of solidified magma beneath Earth's surface are referred to as **plutons** and vary substantially in size and shape. Gigantic plutons are called *batholiths*, whereas smaller dome-shaped

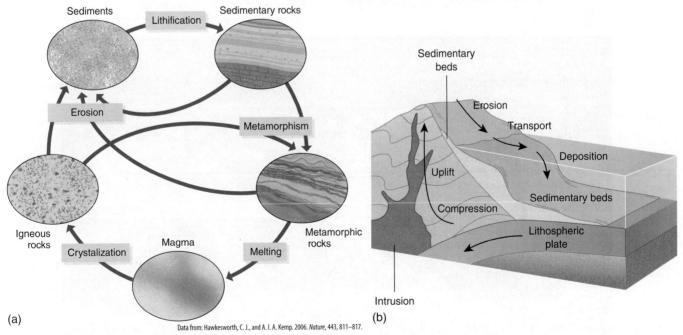

(a)

Data from: Hawkesworth, C. J., and A. I. A. Kemp. 2006. *Nature*, 443, 811–817.

(b)

FIGURE 3.1 The rock cycle in relation to plate tectonics. The rock cycle involves the formation of molten magma and its intrusion into surrounding rocks or extrusion onto the Earth's surface as volcanoes; uplift, weathering and erosion, and redeposition to form sedimentary rocks; and metamorphism of preexisting rocks. Note the similarity of plate tectonics and the rock cycle.

ones are called *laccoliths* (**Figure 3.2**). The dome shape of laccoliths occurs because the magma is still relatively thick and tends to collect in one spot, resulting in an igneous body with a relatively flat base and domed upper surface. Intrusive rocks can also occur as relatively thin bodies cutting across surrounding rocks (*dikes*) or injected parallel to strata (*sills*). One of the most spectacular sills is the Palisades Sill, located along the Hudson River, north of New York City; the Palisades are composed primarily of igneous rocks emplaced as the supercontinent Pangaea rifted apart (**Figure 3.3**).

Although no one has ever seen an intrusive igneous rock form, their emplacement beneath Earth's surface can be inferred from their coarse-grained, or **phaneritic**, texture ("phaneritic" means visible, referring to the visible mineral faces in the rock). The term *texture* refers to the size, shape, and arrangement of the grains in a rock. Phaneritic textures occur when minerals in the rock exhibit relatively large, blocky mineral faces that make them easily visible. This happens when the flux of heat from the magma to the surrounding environment occurs very slowly, giving the crystal faces sufficient time to grow into visible surfaces (**Figure 3.4**).

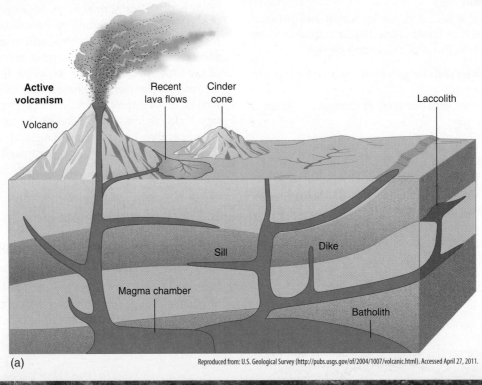

(a)

(b)

FIGURE 3.2 **(a)** Intrusive igneous rock bodies: batholiths, laccoliths, dikes, and sills. **(b)** Igneous dikes cut across one another in this outcrop. We can use cross-cutting relationships like this to date rocks (see Chapter 5).

Courtesy of Fred Wehner, www.tug44.org.

FIGURE 3.3 The Palisades Sill along the Hudson River, north of New York City. These rocks are enriched in the mineral olivine and were intruded during the early rifting of the supercontinent Pangaea (see the section "Tectonic Cycle" in Chapter 2).

Intrusive igneous rocks also vary according to their chemical (mineralogic) composition. Rocks that are highly enriched in magnesium and iron, such as those within the mantle, are referred to as **ultramafic**. Rocks of the mantle are thought to consist mainly of peridotite. At shallower depths in the mantle, where lower pressures and temperatures are found,

ultramafic rocks of the mantle give rise to mafic intrusive igneous rocks called **gabbros** (Figure 3.4). Gabbros are very dark in color because they contain relatively large amounts of iron and magnesium. Gabbros are probably common at the base of continental crust, and the boundary between ultramafic and gabbroic rocks is typically interpreted as the base of the oceanic crust. Diabase is a more medium-grained mafic igneous rock that is often found in dikes and sills.

Most continental crust is composed of granite, or rocks of "granitic" composition (Figure 3.4). The term "granitic" means the continental crust has an overall chemical composition similar, but not necessarily identical, to granite. Granites consist predominantly of the minerals potassium feldspar and silica-rich minerals such as quartz, and paper-like micas; hence, granitic rocks are sometimes referred to as being **felsic**.

Between gabbro and granite are igneous rocks of intermediate chemical composition. Like granite, these rocks also tend to be associated with continental crust. One of the most important of these rocks is **diorite**. Unlike granite, diorite does not have visible quartz crystals (Figure 3.4). Its white and dark grains tend to impart a salt-and-pepper appearance to it. Although diorite is sometimes involved in mountain building or other tectonic activity, granite is more commonly intruded as large felsic bodies. So, too, are **granodiorites**, which are grayish rocks with a composition intermediate between diorite and granite (Figure 3.4). Granodiorites, in part, form the enormous batholiths of the Sierra Nevada of

(a) © Tyler Boyes/Shutterstock, Inc.
(b) © Tom Grundy/Shutterstock, Inc.
(c) Courtesy of NASA/JPL.
(d) Courtesy of NASA/JPL.

FIGURE 3.4 Types of phaneritic or coarse-grained igneous rocks. These kinds of igneous rocks are emplaced deep within the Earth's crust, allowing them to cool slowly and develop their coarse-grained crystalline texture. **(a)** Gabbro. **(b)** Granite. **(c)** Diorite. **(d)** Granodiorite.

Courtesy of National Park Service.

FIGURE 3.5 Yosemite Valley in the Sierra Nevada of California. The Sierra Nevada consists of enormous batholiths of granodiorite that were later exposed by uplift and erosion. The U-shaped Yosemite Valley seen here was carved out much later by glaciers.

California (**Figure 3.5**). These granodiorites were originally emplaced within the Earth and brought to the surface by later uplift and erosion that formed the Sierra Nevada.

Extrusive igneous rocks: occurrence, texture, and composition

The magmas that form intrusive igneous rocks can also be extruded onto Earth's surface. Thus, granite, diorite, and gabbro each have an extrusive counterpart that is their chemical equivalent: rhyolite, andesite, and basalt, respectively (**Figure 3.6**). **Rhyolites**, **andesite**, and **basalt** can all occur on land, although basalts are more likely to erupt onto the seafloor. For this reason ocean crust is commonly referred to as being basaltic in composition. The texture of all three types of extrusive rocks is said to be fine-grained, or **aphanitic** ("without visible appearance"); aphanitic textures result from relatively rapid cooling and solidification on Earth's surface, leaving insufficient time for visible crystal faces to grow (Figure 3.6).

Extrusive igneous rocks form in association with volcanic activity and can form distinctive features at Earth's surface indicative of their mode of origin. On land and in the sea, extrusive igneous rocks are represented by **lava** and lava flows (**Figure 3.7**). In the sea, lava flows produce distinctive, bulbous *pillow lavas* as the hot magma is rapidly quenched by the much cooler seawater (Figure 3.7). Volcanoes can also pump large amounts of *volcanic ash* and gaseous *aerosols* high into the atmosphere, blocking sunlight and cooling the Earth, but the climate change is normally temporary, lasting only a few years (**Figure 3.8**). Eventually the ash rains down, sometimes blanketing large areas. Ash deposits are instantaneous in terms of geologic time and if sufficiently widespread are used to correlate, or "match," deposits in widely separated areas (see Chapter 6). In the case of *pumice*, the rock is so filled with vesicles, it floats (Figure 3.6)!

(a) © Tyler Boyes/Shutterstock, Inc.

(b) © Tyler Boyes/Shutterstock, Inc.

(c) © Tyler Boyes/Shutterstock, Inc.

(d) Courtesy of Willie Scott/USGS.

FIGURE 3.6 Types of aphanitic or fine-grained igneous rocks. These kinds of igneous rocks are extruded at or near the Earth's surface, allowing them to cool rapidly so that visible crystals do not have time to form. **(a)** Basalt. **(b)** Andesite. **(c)** Rhyolite. **(d)** Pumice.

(a)
© AbleStock.

(b)
Courtesy of NOAA/OAR/National Undersea Research Program (NURP).

FIGURE 3.7 Extrusive igneous rocks. **(a)** A lava flow on land. **(b)** Modern pillow lavas form during submarine volcanic eruptions.

If the volcanic vents or pipes that brought the magma to the surface fill with solidified magma, the resulting volcanic necks and dikes can form distinctive features at Earth's surface after erosion of the surrounding rock (**Figure 3.9**). Some magmas might also reach the surface through cracks or fissures in Earth's crust as relatively gentle *fissure eruptions*. Volcanic eruptions sometimes produce terrains that can be quite colorful, such as those of Yellowstone National Park (refer to this chapter's frontispiece). As we will see in coming chapters, enormous volcanic (including fissure) eruptions are thought to have been important agents of catastrophic climate change and mass extinction in Earth's ancient past. These fissure eruptions are thought to have injected enormous volumes of carbon dioxide into the atmosphere, rapidly altering Earth's surface temperature.

The styles of volcanic eruption—relatively gentle or explosive—differ because of the viscosity of the magma. Viscosity refers to the ability of a fluid to flow; molasses, for example, is much more viscous than water. The most important factor in determining viscosity is silica content; the greater the silica content, the more viscous the magma and the slower the flow. As a result, outpourings of basalt are common at Earth's surface, whereas the eruption of granitic

magmas is less so. Granitic magmas have a greater concentration of silica than basaltic ones; thus as they near the surface and cool, they become less mobile. Consequently, granitic magmas are more likely to form large intrusions. If a volcano does spew magma of granitic composition, the magma will tend to block the release of gases (like carbon dioxide and water vapor) until they come near the surface. When it reaches the surface, the gas rapidly expands (because of the lowered pressure), producing explosive eruptions.

How do different kinds of magmas form?

The differing mineralogy and texture of igneous rocks leads to a fundamental question: how does granite, or at least crust with a granitic chemical composition, form? For that matter,

Courtesy of Ronald Martin, University of Delaware.

FIGURE 3.8 Volcanic ash layer at hammer approximately 400 million years old exposed in northern Pennsylvania.

© Tom Bean/Alamy Stock Photo.

FIGURE 3.9 Shiprock, New Mexico, a volcanic vent exposed by erosion. The ridge is a feeder dike leading to the vent.

given that many magmas are thought to originate in the mantle, which is ultramafic to mafic in composition, how do magmas with andesitic or felsic compositions form?

Several processes are involved and are referred to collectively as **magmatic differentiation**. One of the most important is called **fractional crystallization** (**Figure 3.10**). For example, as basaltic magma ascends toward Earth's surface, it moves into zones of lower pressure and temperature. As the magma rises, the minerals in the magma with the highest melting points begin to crystallize and settle to the bottom, because as the magma temperature falls, those minerals with the highest melting points reach their crystallization temperature first. Thus, minerals with the lowest melting points are those that crystallize last from a magma. The minerals crystallize out in a definite sequence that has been demonstrated in laboratory experiments (called Bowen's Reaction Series after N. L. Bowen, who determined the sequence). As iron-, magnesium-, and calcium-rich minerals crystallize from magma and settle toward the bottom of the magma body, the magma left behind becomes progressively enriched in sodium-plagioclase and then potassium feldspar, micas, and leftover silica. Thus, the magma left behind becomes more felsic in composition and has a higher amount of silica and water than the original basaltic source.

Many magmas are thought to form through the process of **partial melting**. Partial melting occurs because of the different melting points of minerals and changing temperatures within the lithosphere. Ultramafic rocks consist predominantly of minerals with high melting points that are relatively low in silica, whereas more felsic rocks consist of minerals that melt at lower temperatures and have higher silica contents (Figure 3.10). Ultramafic rocks dominate in

the mantle. Mantle rocks such as peridotites are normally under enormous temperatures and pressures from the overlying rocks, but the high pressure normally keeps them from melting. However, as mantle rocks move beneath mid-ocean ridges by upward convection, the pressure is released and melting begins. Minerals with lower melting points, like those in basalt, melt first, separate, and rise from the remaining magma, producing basalt.

Andesitic and more felsic magmas also likely form by other processes. Magma may **assimilate** (incorporate) more felsic rock—if present—through which it is rising, altering the chemical composition of the magma. **Magma mixing** might also occur when one body of magma overtakes another as it rises toward the surface and mixes with it.

More felsic magmas that generate granites are probably too enriched in silica to have been derived directly from mafic magmas. Rather, granitic magmas are probably generated by the differentiation of andesitic magmas or the partial melting of silica-rich continental rocks adjacent to the magma. The heat to melt the rocks likely comes from magma that has risen through the rocks after being generated by plate subduction.

The composition of extrusive igneous rocks—rhyolite, andesite, and basalt—also varies according to where volcanism occurs. We can use this feature to infer past tectonic activity and the types of plate margins. Ocean crust, for example, is composed of basalt, so volcanic eruptions associated with mid-ocean ridges are typically basaltic in composition. However, most of the 600 or so modern volcanoes occur where one lithospheric plate descends beneath another. If subduction of ocean crust occurs beneath the margin of another piece of ocean crust, magmas of either basaltic or andesitic (more silica-rich) composition are often produced. These magmas

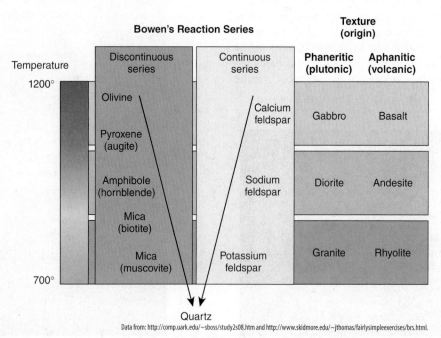

Data from: http://comp.uark.edu/~sboss/study2s08.htm and http://www.skidmore.edu/~jthomas/fairlysimpleexercises/brs.html.

FIGURE 3.10 Bowen's Reaction Series shows the fractional crystallization of minerals and the resulting igneous rocks. The pattern of crystallization is Y-shaped. The branch called the continuous series represents the continuing enrichment of magma in sodium (Na) as calcium (Ca)-rich rocks crystallize out.

feed volcanic island arcs like the Aleutian Islands off Alaska. Because the volcanoes are fed by less viscous basalt, the volcanoes tend to occur at lower elevations and have a shield-like appearance because the lava flows more easily.

On the other hand, if subduction occurs beneath a continental margin, the remelting of crustal rocks and water-laden sediments tends to produce magmas of andesitic composition. These types of volcanoes include the Cascade Range of the Pacific Northwest and the Andes of South America ("andesite" derives from the fact that it is common in the Andes). If the magmas associated with plate subduction become sufficiently granitic in composition, rhyolites result. However, rhyolites are not as common as andesites because granitic magmas are less likely to be erupted; the magma that feeds volcanic chains on land is more silica-rich, so that the magmas are more viscous and flow less easily. Thus, volcanoes on land tend to build to higher elevations, giving them a steep, cone-like appearance.

By contrast to volcanism along plate margins, intra-plate volcanism occurs within a plate. Examples include the Hawaiian islands. These islands also appear to be associated with a hotspot, where magma has reached the surface as if it had flowed through a pipe (see Chapter 2 for further discussion of the existence and behavior of hotspots). The types of extrusive rocks produced by intraplate volcanism vary from basalts to rhyolites. The magmas erupting in Hawaii are basaltic in composition, indicating they originate from the mantle or the base of the lithosphere. The low silica content of the basaltic magmas lowers their viscosity, making them flow more easily, accounting for the gentler slopes and shield-like appearances of these volcanoes.

CONCEPT AND REASONING CHECKS

1. Draw a table indicating the major types of intrusive igneous rocks and their extrusive counterparts.
2. What processes cause magmas to change composition?

3.2.2 Sedimentary rocks

A large proportion of Earth's surface is covered by **sedimentary rocks**, which form at Earth's surface. Sedimentary rocks are typically layered, or stratified. Rocks that are uplifted and exposed to the atmosphere slowly undergo physical and chemical weathering. Physical weathering transforms larger rocks into smaller grains, whereas chemical weathering attacks chemical bonds within minerals, breaking them down further. Physical and chemical weathering produce materials that are more easily removed and transported by the processes of erosion, which occurs by the action of landslides, streams and rivers, wind, and glaciers. The eroded grains are eventually deposited as loose sediment, which can eventually undergo **lithification** (consolidation or hardening) during the processes of burial, compaction, and cementation to form a sedimentary rock (Figure 3.1).

Not all sedimentary rocks form by erosion and deposition, however. Some form by chemical precipitation or

from the accumulation of shells of dead organisms. Thus, sedimentary rocks frequently contain fossils, which are the remains (shell, bone) or traces (tracks, trails) of preexisting organisms. Sedimentary rocks and their preserved fossils represent an enormous archive of past interactions of Earth's systems because they are indicative of surface conditions such as sea level and related climatic feedbacks. For these reasons, we will postpone discussion of sedimentary rocks to Chapter 4, where we can more fully discuss their features and contained fossils.

3.2.3 Metamorphic rocks

All types of rocks can be subjected to changing temperatures and pressures that physically and chemically transform the rocks from one type to another. These processes are referred to as metamorphism ("meta," change; "morphos," form). Unlike igneous rocks, metamorphism might involve, at most, only partial remelting and rearrangement of grains in the rock. Metamorphism is frequently associated with the intense pressures and temperatures generated during mountain building, but metamorphism also occurs in or adjacent to subduction zones (where intense pressures and temperatures are also generated), in the solid rocks adjacent to magma, during the burial of rocks, or by the injection of hot fluids into rocks.

Metamorphic rocks: texture and mineralogy

During metamorphism, new textures and minerals form that are indicative of the intensity of the temperatures, pressures, and chemical conditions that occurred. Metamorphism occurs in a variety of settings: the localized alteration of rock adjacent to hot magma; the production of hot fluids by magma that are injected into the surrounding rock; or much broader, or regional, metamorphism associated with the generation of high temperatures and pressures during mountain building

The intensity, or grade, of metamorphism is reflected by the texture and mineralogy of the resulting rock. Let's consider texture first. A basic distinction between different types of metamorphic rocks is whether they are **foliated** or **nonfoliated**. Metamorphic rocks that exhibit a preferred orientation of their mineral grains are said to be foliated ("folium," leaf); foliation occurs in response to the increased pressure and other stresses of mountain building. Metamorphic rocks not exhibiting an orientation of mineral grains are said to be nonfoliated.

The lowest grade of metamorphism in foliated metamorphic rocks occurs in **slates**, which form from fine-grained, clay-rich sedimentary rocks called *shales* at relatively low temperatures and pressures (**Figure 3.11**). Slightly increased temperatures and pressures cause the tiny clay minerals in shales to react chemically. This in turn causes the minerals to reorganize and produce parallel arrangements of small mica flakes that cause the slate to break along preferred planes, or *slaty cleavage*. A **phyllite** forms from shale under conditions of higher temperature and pressure; phyllite is distinguished from slate by its shinier appearance, which is caused

(a)

© Tyler Boyes/Shutterstock, Inc.

(b)

© Tyler Boyes/Shutterstock, Inc.

(c)

© Tyler Boyes/Shutterstock, Inc.

(d)

Courtesy of William S. Schenck, Delaware Geological Survey.

FIGURE 3.11 Foliated metamorphic rocks and the rocks from which they form. Different types of foliated sedimentary rocks form under progressively more extreme conditions. **(a)** Slate forms under the least extreme conditions, when the platy clay minerals of a shale are compressed together and realigned. **(b)** In schists, which may develop from shales and sandstones, mica crystals grow in size to give the rock a scaly appearance. The small, red-colored dots are garnets, which recrystallize from the original minerals during metamorphism. **(c)** In gneisses, which can form from a variety of sedimentary and igneous rocks, including granites, higher-grade metamorphism causes the minerals to segregate into distinct bands. **(d)** Migmatites have a wavy, banded appearance and form by partial melting.

by slightly larger grain sizes and the presence of the micas muscovite and chlorite.

Under somewhat higher temperatures and pressures, minerals such as micas grow to produce larger crystals visible to the naked eye, giving the rock a scaly appearance. This kind of foliation is called *schistosity* and the rocks are called **schists**. There are a variety of schists, but most are mica schists consisting of muscovite and biotite (Figure 3.11). The mineral garnet, which only forms by metamorphism, can also form from the minerals present in the original rock.

Even more intense metamorphism causes the minerals in the original rock to recrystallize into distinct bands to produce **gneisses** (Figure 3.11). Gneisses can form from different rocks, including shales and alternating shales, sandstones, and even granite. Gneisses typically consist of light bands of quartz, feldspars, and muscovite alternating with darker bands of amphiboles and biotite. Garnet is often found in gneisses, as well (Figure 3.11).

Even more intense alteration produces **migmatites**. Migmatites (meaning "mixed igneous and metamorphic") form when partial—but not complete—melting occurs, producing a wavy, layered appearance (Figure 3.11). The wavy appearance results from the fact that light-colored silicate minerals like quartz and feldspar melt before more mafic minerals with higher melting points such as hornblende.

Thus, the light-colored minerals with the lower melting points tend to segregate from the darker-colored minerals that have not melted, imparting a distinct "wavy" banding to migmatites.

Another type of foliated metamorphic rocks form under somewhat different conditions than those above. **Blueschist** forms in the thick sedimentary wedges associated with subduction zones, where relatively high pressures but low temperatures occur. In subduction zones, the ocean crust initially begins to cool as it descends, so that relatively little heat rises to the base of the rocks and sediments found in the subduction zone. By contrast, the pressures due to compression and burial increase dramatically.

Nonfoliated metamorphic rocks typically form from parent rocks that consisted of essentially one mineral. **Marble** is a nonfoliated metamorphic rock that forms during the recrystallization and interlocking of the individual calcite grains of limestones. Similarly, **quartzites** form by the recrystallization of individual quartz grains in sandstones. Minor impurities in the original rocks can impart distinctive coloration to marbles and quartzites.

The intensity of metamorphism is also indicated by the predominant minerals that compose the rocks. Chlorite is a greenish mica that is associated with greenschist, whereas sillimanite is associated with the highest metamorphic grade (not coincidentally, sillimanite is used to make temperature-resistant porcelains like those used in spark plugs). Various types of garnets characterize the different metamorphic grades between greenschist and sillimanite-bearing metamorphic rocks.

Types of metamorphism

Most metamorphic rocks are generated by **regional metamorphism** (**Figure 3.12**). As the term implies, regional metamorphism occurs over broad—or regional—scales, such as that associated with increased temperature and pressure and large-scale deformation during orogeny, or mountain building. We can use the intensity, or grade, of metamorphic rocks associated with regional metamorphism to infer the intensity of metamorphism, even after the mountains have long vanished. In the simplest, ideal case of regional metamorphism the metamorphic grades are expressed as successive zones of decreasing metamorphism away from the centers of mountain belts where the deformation is greatest and magma is most likely to have been emplaced. Thus, migmatites tend to occur closest to the center of mountain belts, with gneisses, schists, phyllites, and slates tending to occur progressively farther away (Figure 3.12). The mineralogic composition of the rocks parallels the gradation in foliation, with schist and gneiss occurring in the highest grade metamorphism and slate and phyllite occurring in the lower grades. Blueschist is associated with the high pressures that occur in the sedimentary wedges of subduction zones. However, in most cases, the patterns of metamorphism are more complex because of movement and deformation of the rocks during mountain building and erosion thereafter (see Chapter 2).

Another type of metamorphism important to the history of Earth's systems is **hydrothermal metamorphism**, or **hydrothermal weathering**, that occurs at seafloor spreading centers. Here, seawater percolates through hot ocean crust, altering the concentrations of certain dissolved ions in the seawater (**Figure 3.13**). Changes in the concentrations of these dissolved ions are recorded in the calcareous shells of fossil organisms and are used to infer past changes in rates of seafloor spreading and continental weathering.

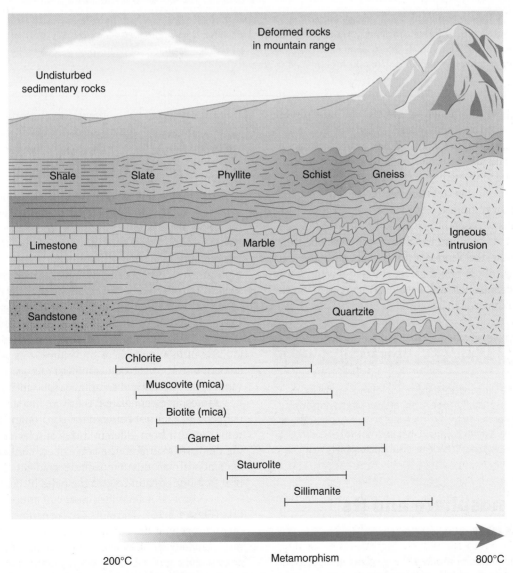

FIGURE 3.12 Changes in lithology and mineralogy that result during the metamorphism. Certain minerals are indicative of metamorphism and certain associations of these minerals are indicative of the intensity of metamorphism.

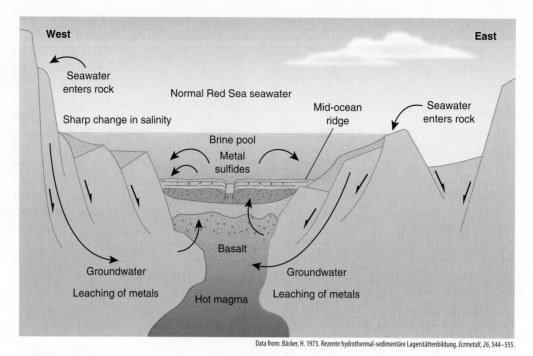

West

East

Seawater
enters rock

Normal Red Sea seawater

Sharp change in salinity

Mid-ocean
ridge

Seawater
enters rock

Brine pool
Metal
sulfides

Basalt

Groundwater

Groundwater

Leaching of metals

Hot magma

Leaching of metals

Data from: Bäcker, H. 1973. Rezente hydrothermal-sedimentäre Lagerstättenbildung. *Erzmetall, 26,* 544–555.

FIGURE 3.13 Hydrothermal weathering occurs when seawater percolates through hot ocean crust at seafloor spreading centers, such as this one in the Red Sea. This mechanism alters the ionic composition of seawater. Magnesium (Mg^{2+}) ions from ocean crust dissolve into seawater as calcium (Ca^{2+}) ions move from seawater into the crust. The ratio of Mg^{2+}/Ca^{2+} ions appears to be related to the mineralogy of the shells (calcite or aragonite) secreted by organisms living in the oceans.

Finally, there is **impact** or **shock metamorphism**, which occurs when an extraterrestrial body such as an asteroid hits the Earth (**Figure 3.14**). When an impact occurs, there is a tremendous and instantaneous increase in pressure and temperature to thousands of degrees. Projectiles consisting of melted rock called **tektites** can be ejected thousands of kilometers away from the site of impact (Figure 3.14). Shocked mineral assemblages also form from impact. Shocked minerals such as quartz have well-developed thin layers, or laminae, that represent rearrangements of the crystalline structure of the mineral (Figure 3.14). The only way known to generate such well-developed laminae is by the impact of an extraterrestrial object.

CONCEPT AND REASONING CHECKS

1. Diagram the sequence of metamorphic grades (phyllite, schist, gneiss, etc.) that one might encounter moving away from a batholith.
2. What are the different types of metamorphism?
3. Diagram the basic rock cycle on a subducting plate margin.

3.3 Atmosphere and Its Circulation

The atmosphere comprises the gaseous envelope surrounding Earth. On the scale of the Earth, the atmosphere is only a thin veil that separates life on the planet from destruction, but without the atmosphere life as we know it on Earth would not exist. The atmosphere helps warm the Earth through the greenhouse effect (otherwise, life would freeze) and also protects Earth's surface from harmful cosmic radiation.

The lowest layer of the atmosphere, called the troposphere, is about 10 to 15 kilometers in thickness. Although the troposphere is quite thin, it is where about 80% of the gases and almost all water vapor are concentrated. As a result, it is the troposphere where most of our "weather" occurs. The top of the troposphere, for example, is marked by the anvil shapes of storm clouds. The composition of the modern atmosphere is dominated by several gases. Nitrogen (78%) and oxygen (21%) are by far the two most prominent components (**Figure 3.15**). Although carbon dioxide comprises only 0.038% of the atmosphere by volume, this relatively small amount of carbon dioxide is sufficient to warm Earth. Water vapor is also present and contributes substantially to warming.

Atmospheric circulation is driven, fundamentally, by the temperature contrast between the equator and the poles and results in turn from different fluxes of solar radiation reaching the Earth's surface. The atmosphere constantly attempts to "smooth out" this temperature gradient by transferring heat from the equator toward the poles by convection.

Let's look at a simplified model of atmospheric circulation (**Figure 3.16**). The sun's rays penetrate a thinner layer of atmosphere at the equator than at the poles because the rays penetrate the atmosphere at more-or-less right angles nearest the equator but more tangentially nearer the poles. Consequently, the sun's rays entering the atmosphere over the equator are less likely to collide with air molecules and

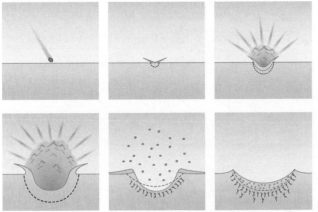

(a) Reproduced from: Christian Koeberl, "El'gygytgyn: a very special meteorite impact crater," Website of the FWF (Austrian Science Fund), project: P21821-N19 (http://lithosphere.univie.ac.at/impactresearch/elgygytgyn-crater/). Accessed April 19, 2011.

0.5 mm

(b) Courtesy of Billy P. Glass, University of Delaware.

(c) Courtesy of Billy P. Glass, University of Delaware.

FIGURE 3.14 Impact metamorphism and its evidence. **(a)** Impact or shock metamorphism occurs when an extraterrestrial body such as an asteroid hits the Earth. When an extraterrestrial object hits the Earth, there is a tremendous and instantaneous increase in pressure and temperature to thousands of degrees. **(b)** Projectiles consisting of melted rock called tektites can be ejected thousands of kilometers. **(c)** Shocked quartz. The parallel laminations represent the rearrangement of the mineral's crystalline structure in response to sudden intense pressure and temperature generated by an impact. Well-developed shocked mineral assemblages are much better indicators of impact than iridium alone (see Chapter 1).

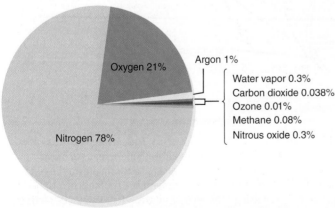

Oxygen 21%

Argon 1%

Nitrogen 78%

Water vapor 0.3%
Carbon dioxide 0.038%
Ozone 0.01%
Methane 0.08%
Nitrous oxide 0.3%

Data from: Mackenzie, F. T., and Mackenzie, J. A. 1995. *Our Changing Planet: An Introduction to Earth System Science and Global Environmental Change.* Englewood Cliffs, NJ: Prentice-Hall.

FIGURE 3.15 The composition of the modern Earth's atmosphere.

be scattered back to outer space than the rays that penetrate the atmosphere nearer the poles. The more intense heating at the equator causes the air to warm and rise there; because it is warm, the air can also hold more moisture. As the warm air rises it loses heat energy and cools, releasing its moisture as rain. It is for this reason rain forests occur in the tropics near the equator. As the air masses rise into the atmosphere, they begin to cool, so their density increases. Unlike the simple diagram in Figure 3.16a, however, the air masses stop rising and the air begins to spread parallel to Earth's surface toward either pole (Figure 3.16b and **Figure 3.17**). This air cools further as it flows parallel to Earth's surface. These air masses eventually cool sufficiently to further increase their density, causing them to descend back toward Earth before reaching the poles. Because these descending (high pressure) air masses have already lost their moisture, the land

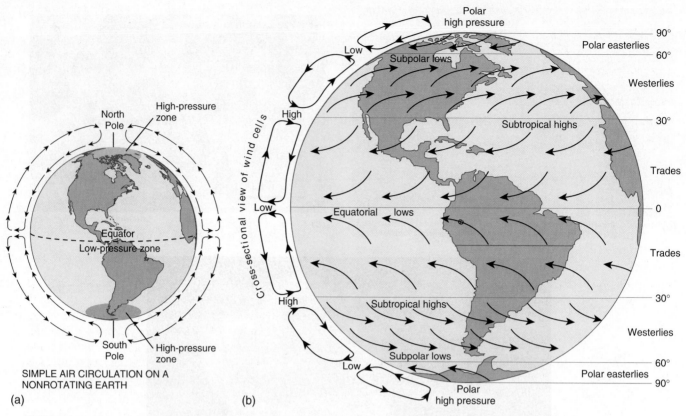

FIGURE 3.16 The structure and basic circulation of the atmosphere. **(a)** Cross-section of the Earth's atmosphere showing the basic components of highly simplified convection within the atmosphere as initially described in the text. Low-pressure zones are regions of ascending moist air and rainfall, whereas high-pressure zones consist of descending dry air masses. **(b)** Differences in heating between the equator and poles actually cause the formation of multiple convection cells in the atmosphere that determine the broad patterns of rainfall and dryness, as further discussed in the text.

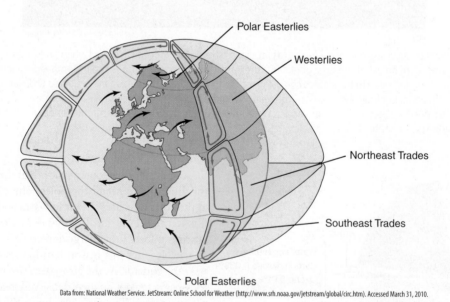

FIGURE 3.17 Major wind systems of the Earth resulting from atmospheric convection and the Coriolis effect. The curvature of the airflow imparted by the Coriolis effect establishes the major wind patterns. As air descends at 30°E north and south latitude, it tends to move from east to west over the Earth's surface to produce the trade winds. Trade winds in the northern hemisphere flow from the northeast to southwest, whereas their counterparts in the southern hemisphere flow from southeast to northwest; northern and southern hemisphere trade winds meet at the Intertropical Convergence Zone above the equator. Warm air rising at about 60°E north and south latitude tends to flow from west to east (again because of the Coriolis effect) to produce the westerlies ("coming from the west") that move major weather systems over the United States and Europe. The polar easterlies ("coming from the east") lie toward the poles and move from east to west like the trade winds, meeting the warmer westerlies along a polar front.

beneath them often consists of deserts, such as the Sahara Desert in northern Africa.

After its descent, the air flows parallel to the Earth's surface. Some air flows back toward the equator to be incorporated into the rising air there, whereas the rest flows toward higher latitudes. As it does so, the dry air picks up moisture and warms once again, eventually rising into another convection cell. This pattern is repeated yet again in a third set of convection cells closest to the poles.

The flow of air within the convection cells is not simply straight up and down; it is actually curved. The curvature results from the **Coriolis effect**, which in turn results from the rotation of the Earth. The curvature of the airflow imparted by the Coriolis effect establishes the major wind patterns in Earth's atmosphere (Figures 3.16 and 3.17). The Coriolis effect exists because the atmosphere and the Earth move together as the Earth rotates eastward around its axis. As the Earth rotates, a point near the north pole moves around a circle that is much smaller in diameter than a point at the equator. In other words, during the same interval of time, the point near the north pole travels a shorter distance than the point at the equator. Thus, the point near the north pole moves more slowly than the point at the equator. Now imagine that an air molecule at the north pole is displaced toward the equator. Because of the physical phenomenon of inertia (which states that an object continues to move with the same speed and direction unless a force acts upon it), the air molecule maintains the original speed and direction that it had when it began to move south toward the equator. Consequently, the molecule moving from the north pole toward the equator will lag behind the one at the equator because the point at the equator is moving faster than the point coming from the north pole; thus, the point at the equator will have moved eastward away from where the point from the north pole will arrive. Conversely, a point moving from the equator toward the north pole will move ahead of the point at the pole. A similar phenomenon occurs in the southern hemisphere, but the movements are the reverse, or mirror image, of those in the northern hemisphere.

Because of the Coriolis effect, as air descends within the atmospheric convection cells at 30° north and south latitudes, the air masses tend to move from east to west over Earth's surface to produce the **trade winds**. Trade winds in the northern hemisphere flow from the northeast to southwest, whereas their counterparts in the southern hemisphere flow from southeast to northwest. Trade winds in the northern and southern hemispheres converge near the equator. Similar phenomena account for the well-known **westerlies** (coming from the west) that move major weather systems over the United States and Europe and the **polar easterlies** (which come from the northeast in the northern hemisphere and the southeast in the southern hemisphere) nearer the poles. As we will see below, the basic pattern of atmospheric circulation determines not only the broad pattern of precipitation over the planet but also the broad distribution of Earth's biotas while driving the major surface currents of the ocean. These patterns are in turn influenced by the distribution of the continents

(discussed later in this Chapter) and the presence or absence of mountain ranges (**Box 3.1**).

3.4 The Hydrosphere

The **hydrosphere** is critical to maintaining Earth's climate and life. The presence of water in sediments lowers the melting point of rocks in subduction zones and is therefore critical to maintaining the processes of plate tectonics. Water also provides habitat for countless numbers of organisms and is necessary for life as we know it; most organisms consist of more than 60% water, and some more than 90%. Water is also critical to life because water vapor is—like carbon dioxide—a greenhouse gas, affecting Earth's temperature and habitability.

3.4.1 Hydrologic cycle

Like other Earth systems, the hydrosphere is cyclic. The **hydrologic cycle** involves the flux of water through several **reservoirs** (**Figure 3.18**). **Precipitation** reaches the ground as rain or snow if the air just above the land is sufficiently cold, such as at high latitudes near the poles or at high elevations (mountains). Some precipitation may undergo **evaporation** or flow over Earth's surface as **runoff** in streams and rivers; as is described in Chapter 2, the occurrence of some of the major river systems of the world is determined by the presence of deep valleys within the Earth's crust that represent failed rift systems. Much of this water will of course reach the oceans, whereas some infiltrates the ground and flows through subterranean rocks and sediments as groundwater. The remaining water in the cycle is used by plants, which are technically part of the biosphere (**Figure 3.19**). Most land plants lose tremendous amounts of water out the bottom of the leaves through the process of **transpiration**. The water is lost through countless numbers of microscopic openings called stomata that allow the exchange of carbon dioxide and oxygen between the leaves and the environment during photosynthesis. The process of transpiration is tremendously important to the hydrologic cycle, because without transpiration many tropical regions would suffer from drought.

Some workers recognize a separate Earth system referred to as the **cryosphere**, that includes glaciers, both those of mountains (called alpine glaciers) and polar ice caps. The development of polar glaciers depends on the supply of moisture that is part of the overall hydrologic cycle.

Box 3.1 Asian Monsoon: Influence of Large Land Masses on Atmospheric Circulation and the Hydrologic Cycle

Large land masses can profoundly influence the circulation of the atmosphere through the phenomenon of monsoons. The word "monsoon" is usually associated with torrential rainfall but actually refers to reversing airflow (**Box Figure 3.1A**).

Approximately 50 to 60 million years ago, the continent of India began to collide with the continent of Asia, resulting in the Himalayan Orogeny and leading to the Asian monsoon over Tibet and India. What we see today is that, as the summer sun warms the Tibetan Plateau to the north of India, warm air begins to rise, drawing warm moist air from the Indian Ocean over India. When this warm, moist air encounters the south side of the Himalayas, which form a narrow rampart along the southern margin of the Tibetan Plateau, the air is forced upward by the orographic effect. As the air rises it cools, and precipitation results on the southern side of the Himalayas, leaving Tibet high and dry. During the winter the reverse conditions hold: cold air, which is dense, descends over Tibet and southward over India. This airflow is also considered part of the monsoon.

But what exactly were the effects of the Himalayan Orogeny on climate and the Asian monsoon through

geologic time? How, in other words, did the Asian monsoon evolve toward its present state? How can we tell what happened? One way to determine the effects of the Himalayan Orogeny on Asian climate, especially given the complex interactions of Earth's systems, is to use computer models of climate change. A model is a kind of sophisticated hypothesis (see Chapter 1) that tries to take into account all the components or processes of a system that are important to the system while excluding those that are not considered important. Using climate models, we can run experiments on complex systems under controlled conditions, like running experiments in a laboratory. This allows earth scientists to identify the mechanisms that cause variation in global climate because many factors can be held constant, whereas others are varied to examine their effect on the behavior of the model.

Based on climate models, as India collided with Asia the fluctuations in the climate of the Tibetan Plateau became more extreme because land was being uplifted higher into the atmosphere. Because of the increasing elevation of the plateau, the atmosphere over the plateau became progressively thinner. As a result, the sun warmed the air near

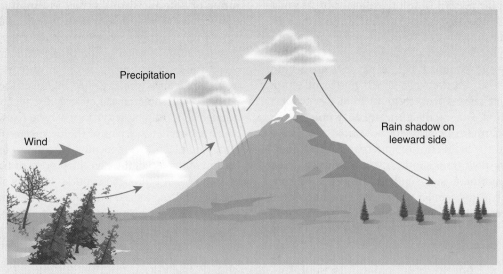

BOX FIGURE 3.1A A monsoon like that of the Indian subcontinent.

Box 3.1 Asian Monsoon: Influence of Large Land Masses on Atmospheric Circulation and the Hydrologic Cycle (Continued)

Earth's surface more rapidly during the summers, decreasing the air's density and allowing it to rise more easily. By contrast, during winters air masses cooled more readily, increasing their density and causing them to descend over the plateau. Because the air currents that feed into and out of the air masses over the plateau also flow over the Indian subcontinent, the slow rise of the plateau began to intensify the monsoonal airflow over India and Tibet (**Box Figure 3.1B**).

During this time, rainfall shifted from the north side of the Himalayas to the south side. Accordingly, over tens of millions of years, what was a warm, wet lowland in Tibet became a cool, desert-like plateau. The shift in rainfall to the south side of the Himalayas might in turn have fed back positively on erosion, uplift, and metamorphism in the region.

The results of a model—like those of a hypothesis—should be tested or at least constrained by the real data of the geologic record (see Chapter 1). If the results of the model agree with geologic evidence, we can be reasonably sure the model simulates what actually happened. If not, the model must be reexamined and modified or possibly even discarded. This approach is the same iterative one used in the scientific method.

One way is to construct arguments based on the record of rocks and fossils, reasoning backward to make predictions, and then look for evidence in the geologic record to corroborate or refute the predictions. In fact, the model results generally agree with the fossil record. Fossils collected from rock exposures in Tibet indicate that vegetation in Tibet before 30 million years ago was that of a subtropical-to-tropical forest like that of the southeastern United States today. However, by 5 to 10 million years ago these forests had given way to deciduous forest like that of temperate latitudes and eventually to grasses and scrubby vegetation. The changes in vegetation through time indicate that rainfall was decreasing.

The thinner atmosphere over uplifted areas intensifies airflow driven by temperature changes during the winter and summer. In the winter, cool air

descends over the uplift and flows away, whereas in the summer, air warms and rises over the uplift, drawing air toward and over the uplift. Changes in the precipitation patterns in turn cause changes in vegetation.

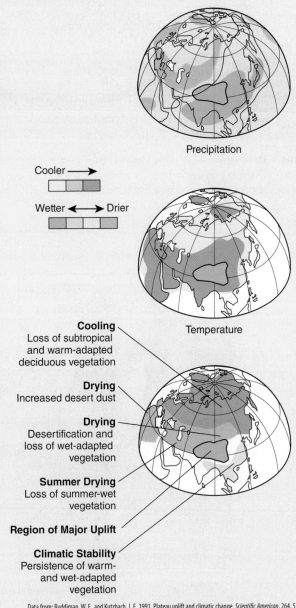

Cooler →

Wetter ←→ Drier

Precipitation

Temperature

Cooling
Loss of subtropical and warm-adapted deciduous vegetation

Drying
Increased desert dust

Drying
Desertification and loss of wet-adapted vegetation

Summer Drying
Loss of summer-wet vegetation

Region of Major Uplift

Climatic Stability
Persistence of warm- and wet-adapted vegetation

Data from: Ruddiman, W. F., and Kutzbach, J. E. 1991. Plateau uplift and climatic change. *Scientific American*, 264, 5.

BOX FIGURE 3.1B Effect of the rise of the Himalayan Mountains on the Asian monsoon.

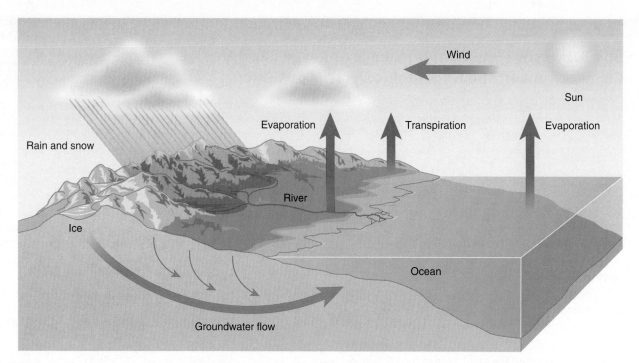

FIGURE 3.18 The hydrologic cycle. Note the involvement of the biosphere through the process of transpiration (water loss) through the microscopic stomata on the undersides of leaves. Note also the overlap of the hydrosphere and atmosphere: the atmosphere contains water vapor and also affects surface ocean circulation through the production of wind currents, as discussed in the text.

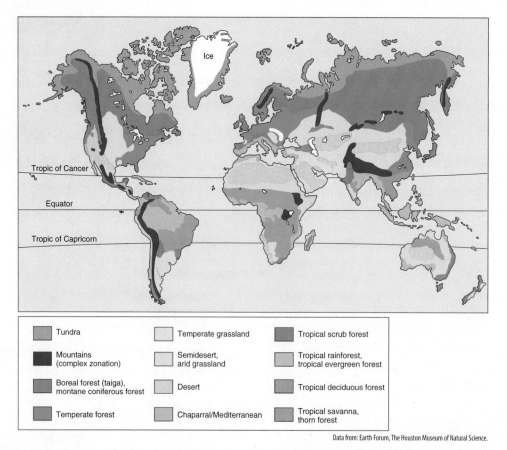

Tundra	Temperate grassland	Tropical scrub forest
Mountains (complex zonation)	Semidesert, arid grassland	Tropical rainforest, tropical evergreen forest
Boreal forest (taiga), montane coniferous forest	Desert	Tropical deciduous forest
Temperate forest	Chaparral/Mediterranean	Tropical savanna, thorn forest

FIGURE 3.19 Convection cells within the atmosphere determine the major belts of precipitation, vegetation, and deserts over the Earth. Compare to Figure 3.17.

Glaciers in turn affect the return of moisture to the atmosphere because they store water. Today, about 90% of the world's freshwater is stored in Antarctic ice. Glaciers—at least polar ice caps—have not always been present during Earth's history, but when they have been present, they have exerted a profound impact on Earth's climate (such as during the Pleistocene "Ice Ages"; see Chapter 15).

3.4.2 Ocean circulation

E Unquestionably, the most dominant components of the hydrosphere are the oceans. The modern ocean basins have an average depth of 3.8 km and represent more than 70% of Earth's surface. The oceans store 96.5% of Earth's water, account for 86% of all evaporation, and receive 78% of all precipitation. Because of their area and volume, the oceans exert a tremendous influence on climate. Only about 2% of the heat generated by the greenhouse effect at Earth's surface is used to drive atmospheric circulation and those portions of the hydrologic cycle that occur in the atmosphere. Much of the rest of the heat is stored in the oceans. In fact, because of water's high specific heat capacity (the heat stored per unit volume), the oceans represent the largest reservoir and regulator of heat on Earth. In other words, after ocean waters have warmed up, they release the heat only slowly. This property allows ocean currents to retain heat for long periods of time, permitting them to redistribute heat over long distances—even to colder high latitudes—and affecting climate.

The major wind systems of Earth discussed earlier determine the patterns of major oceanic surface currents by pushing and dragging the water ahead of them (**Figure 3.20**). The oceanic surface currents normally flow down to about 100 to 200 meters. Some of the most pronounced surface currents belong to large, circular **gyres**, which result in part from the Coriolis effect. Surface water masses that flow from high to low latitudes in the gyres move along the western margins of continents. These currents flow more slowly than the water already at the equator, so these currents are "deflected" westward by the Coriolis effect relative to the currents already at the equator (compare Figures 3.17 and 3.20a). When it reaches the equator, the water continues to move westward, driven by the trade winds, and travels across the ocean. Upon reaching the eastern margins of continents, the water is deflected away from the equator back toward high latitudes. Because they are now flowing faster than the water already at higher latitudes, these currents are deflected eastward by the Coriolis effect. As they reach higher latitudes, the surface currents are moved eastward by the prevailing westerlies.

F **G** The behavior of the modern Earth's surface water masses must be distinguished from that of its deep water masses. The deep water masses circulate in what has been called an oceanic conveyor (**Figure 3.21**). To understand the oceanic conveyor, try to imagine releasing a colored water molecule at a particular spot in the ocean that is easily distinguished from all other water molecules. Assume that you could repeat this experiment a number of times by re-releasing the colored molecule each time it returned to the site of release, allowing the molecule to circulate through the oceans again.

On average, it would take about 1,000 years for the colored molecule to return to the site of release. During this time the molecule, typically, would have circulated through the entire world ocean before returning back to its site of release. This rate of ocean circulation is relatively rapid and might even occur on the scale of centuries in separate ocean basins.

We know about the patterns and rates of modern ocean circulation based on the concentration of anthropogenic radioactive isotopes such as ^{14}C (carbon-14) and ^3H (tritium). Although ^{14}C and ^3H are both produced naturally in Earth's atmosphere by cosmic ray bombardment, their production peaked in the early 1960s because of nuclear weapons testing during the Cold War. Because they are radioactive, the amount of these isotopes in a water mass decreases as soon as the supply of radioactive parent in the atmosphere is cut off by the descent of surface water masses into the deep ocean. When the supply of radioactive parent in the atmosphere is cut off to the descending water mass, the amount of radioactive parent in the water mass begins to decline, giving a measure of time since the water mass became isolated from the atmospheric source.

The circulation of the modern oceanic conveyor has often been referred to as **thermohaline circulation**, meaning the circulation was thought to be driven by differences in the temperature and salinity of the different water masses. The deep water masses of the ocean tend to form in cold, high latitudes, especially near glaciers. Because cold water is denser than warm water, surface waters chilled by the glacial ice tend to sink. The deepest ocean water is quite cold, on the order of about 4°C. Also, the water taken up from the oceans into glacial ice is freshwater. This leaves the salt behind in the remaining ocean water near the ice, which also increases the water's density and causes it to sink.

There are two basic deep water masses produced in the oceans: **Antarctic Bottom Water (AABW)** and **North Atlantic Deep Water (NADW)** (Figure 3.21). As its name suggests, AABW forms off Antarctica but bathes the floors of ocean basins all over the world. Eventually, though, it appears to flow to the surface in the North Pacific and Indian Oceans and then returns along the ocean surface to Antarctica, where it cools and sinks again. NADW is produced mainly off the coasts of Greenland and Iceland in the North Atlantic (Figure 3.21). As the surface waters sink there, they are replaced by the northward flow of the Gulf Stream. As it flows northward, the Gulf Stream waters cool, releasing their heat to the environment (this is why the climates of countries such as England are relatively mild, despite the fact that they are located at relatively high, cool latitudes). Some of the Gulf Stream's water also evaporates, increasing the salinity. As it cools and becomes saltier, the Gulf Stream water eventually becomes dense enough to sink, thereby perpetuating the production of NADW and the oceanic conveyor. Today, the rate of NADW flow is about equal to that of 100 Sverdrups (named in honor of a famous oceanographer). One Sverdrup is equal to a flow rate of 10^6 cubic meters per second and 100 Sverdrups is about equal to the flow rate of the modern Amazon River! By contrast to AABW, NADW only flows within the Atlantic Basin, eventually returning to

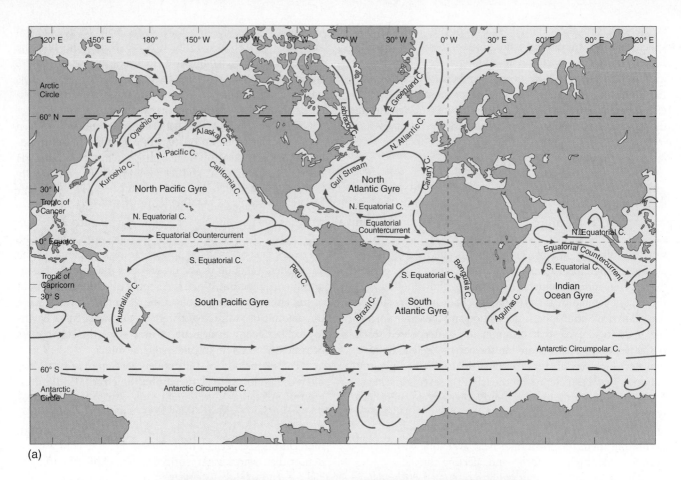

(a)

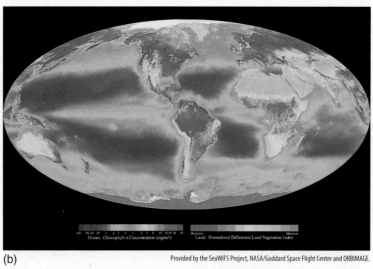

Ocean: Chlorophyll *a* Concentration (mg/m³) Land: Normalized Difference Land Vegetation Index

(b)

Provided by the SeaWiFS Project, NASA/Goddard Space Flight Center and ORBIMAGE.

FIGURE 3.20 Major ocean surface currents and oceanic production. **(a)** The major ocean surface currents are driven by atmospheric circulation. Compare Figure 3.17. **(b)** Primary production on the Earth as indicated by chlorophyll concentration. In the marine realm, the most productive regions are coastal regions shown in red, whereas the least productive regions are shown in deep blue and correspond to the oceanic gyres of Figure 3.20a. On land, productivity declines with lighter shades of green and yellow. The tropical rainforests of Brazil are highly productive, whereas the Arctic and Sahara Desert are among the least productive.

the ocean surface off Antarctica by the process of **upwelling**. It takes only about 200 years for NADW to circulate in the Atlantic Ocean basin based on the rates of radioactive decay of ^{14}C and other isotopes produced by nuclear bomb testing in the 1950s and 1960s. When it reaches the surface at Antarctica, NADW mixes with water around Antarctica, some

of which will again descend as AABW, which takes much longer to circulate through the oceans.

Although density differences no doubt play a role in deep-ocean circulation, the concept of thermohaline circulation has been modified in recent years by the concept of **meridional ocean circulation**, so named because of

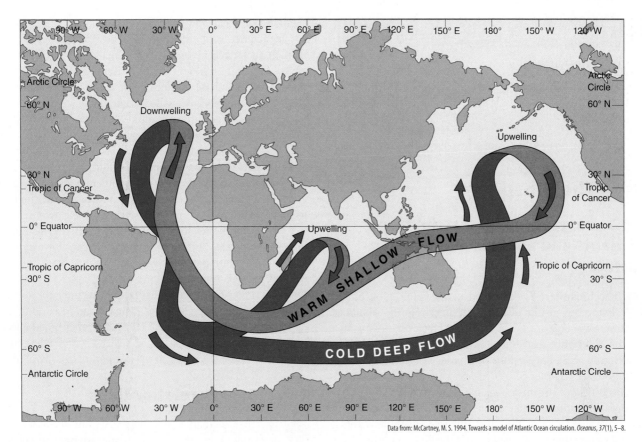

Data from: McCartney, M. S. 1994. Towards a model of Atlantic Ocean circulation. *Oceanus, 37*(1), 5–8.

FIGURE 3.21 The oceanic conveyor involves the thermohaline circulation of deep water masses due to differences in temperature and salinity. Two of the main water masses produced are Antarctic Bottom Water (AABW) and North Atlantic Deep Water (NADW).

the movement of water masses north and south through the ocean basins parallel to the meridians of longitude. Although density differences due to temperature and salinity are still thought to be involved in the movement of these deep water masses, some workers now think that the density differences are not necessarily themselves the prime driver of deep-ocean circulation. Instead, winds are thought to produce the density differences in upper water masses of the top several hundred meters by evaporation and heat loss. These upper-water masses then become sufficiently dense to sink, driving the movement of deeper water masses.

The oceans also affect Earth's climate in a third way: by influencing Earth's **albedo**, or surface reflectivity, which in turn affects global temperatures. The oceans are dark compared with land, and darker surfaces tend to radiate the sun's energy back into the atmosphere as heat, which is trapped by carbon dioxide and water vapor, warming the Earth via the greenhouse effect (see Chapter 1). Thus, when oceans have been widespread—at many times even flooding most of the continents—Earth has tended to warm in part because of its decreased albedo. On the other hand, when the oceans have receded from land, more bare land increased Earth's albedo and cooled the planet. Other types of surfaces on land (ice, forests) also affect Earth's albedo (**Figure 3.22**). We examine the processes that cause sea level to rise and fall in Chapter 6.

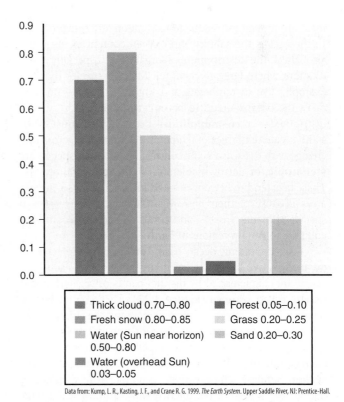

Data from: Kump, L. R., Kasting, J. F., and Crane R. G. 1999. *The Earth System*. Upper Saddle River, NJ: Prentice-Hall.

FIGURE 3.22 Albedo (reflectivity) of various types of natural surfaces.

Legend:
Thick cloud 0.70–0.80
Fresh snow 0.80–0.85
Water (Sun near horizon) 0.50–0.80
Water (overhead Sun) 0.03–0.05
Forest 0.05–0.10
Grass 0.20–0.25
Sand 0.20–0.30

3.5 The Biosphere

3.5.1 Biogeography: distribution of plants and animals over Earth's surface

The **biosphere** is composed of the total living biota (living organisms) of Earth, ranging from the smallest bacteria to the largest trees and whales. However, organisms are not uniformly distributed over the planet. This basic observation was a cornerstone of Charles Darwin's theory of evolution (see Chapter 5).

Each type of organism—or *species*—has a certain range of physical environmental factors within which it can live and reproduce its own kind. It is this tolerance to environmental factors that largely determines the biogeographic distribution of different species. Temperature is considered to be the most important factor determining the distribution of different species because it affects the rates of biochemical pathways such as photosynthesis and respiration; such processes tend to be characteristic of each species. On land, rainfall is also important because animals and plants must make up for water loss due to respiration and transpiration. In the oceans the salinity and oxygen content of the water also affect the distribution of organisms, especially close to shore where large rivers enter. Modern species that, for example, live in temperate and higher latitudes—on land or in the oceans—tend to be **eurytopic** and more widely distributed—or **cosmopolitan**—because they must withstand greater extremes of temperature and other factors that change with the seasons. By contrast, many other species are **stenotopic**, or narrowly tolerant of environmental change. These taxa tend to be concentrated in, or endemic to, certain areas because of their narrower tolerances. Many tropical species of plants, insects, and corals are stenotopic and are endemic to certain regions of Earth.

Each continent, ocean basin, or sea tends to have its own distinctive biota. On land, larger regions of each continent are characterized by distinct bands of vegetation and deserts called **biomes** (**Figure 3.23**). These broad bands, which tend to circle the globe on land, tend to correspond to differences in the amount of solar radiation and precipitation reaching Earth's surface, as described earlier in the chapter. Each continent in turn represents a particular biogeographic **region**. Similarly, there is a steep temperature gradient in the surface waters of the open oceans. Surface water temperatures at the equator can reach 30°C, whereas those near

the poles can be as low as 0°C, so that different species tend to characterize each water mass. The oceanic realm is in turn characterized by one or more smaller scale **provinces**, especially where different ocean currents encroach on land masses. Thus, in the marine realm species living closer to land occur in provinces whose boundaries correspond to changes in temperature and salinity between adjacent water masses impinging on the coasts.

These distinctive biotas have also changed through time as continents have changed position during tectonic cycles. In the process, populations of many different kinds of plants and animals have become separated from one another for tens to hundreds of millions of years, evolving their own distinctive faunas and floras, for example, the distinctive marsupial fauna of Australia. Rifting of a continent can, for example, lead to the evolution of what was once a single biota into two or more regions separated by new ocean basins. It was this type of evidence that was initially used to infer the former existence of the supercontinent Pangaea and to support the hypothesis of continental drift. These and other broad effects of the tectonic cycle on Earth systems and sedimentary environments will be discussed in the last part of this chapter.

3.5.2 Energy relationships

The flux of matter and energy within ecosystems is determined by the characteristic **niches** of species within the ecosystem. The term "niche" refers to the particular role or function of a species within an ecosystem and is defined primarily by the **trophic (food, energy) relationships** of the species, or "who eats whom." The niches of the different species of a community can be diagrammed as "links" in a **food chain** or food web. It is these energetic relationships that organize the species inhabiting an area into communities and ecosystems.

The energetic relationships of food chains and webs can also be arranged into **food (or energy) pyramids** (**Figure 3.24**). The different levels of the food pyramid show how food (energy, matter) is passed from one group of organisms to another in a community. Plants (**autotrophs or producers**) occur at the base. During photosynthesis energy from light is used by plants (or producers) to combine carbon dioxide from the atmosphere with water to produce one molecule of sugar and oxygen, which is released as a byproduct into the atmosphere. Energy from sunlight is stored in the chemical bonds of the sugar, whereas carbon dioxide is taken out of the atmosphere and sequestered in the reservoirs of living and dead organic matter. **Herbivores** occur at the level immediately above producers. **Secondary consumers (carnivores)** occur at the next higher level above herbivores and **top carnivores** at the apex of the pyramid. Although the conversion of light energy to plant biomass varies, it is roughly about 10% efficient. Moreover, as the energy stored in plant biomass is passed from one level to the next of the pyramid, energy is lost, so that only so many levels can be supported. This means that only 10% of the original light energy received by the plant is converted to plant biomass. Now assume that each level of a food pyramid is only 10%

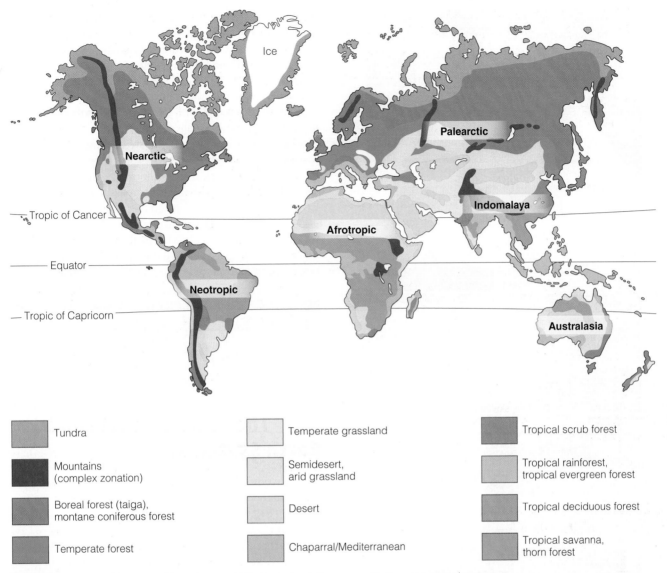

Ice

Palearctic

Nearctic

Indomalaya

Tropic of Cancer

Afrotropic

Equator

Neotropic

Tropic of Capricorn

Australasia

Tundra

Temperate grassland

Tropical scrub forest

Mountains
(complex zonation)

Semidesert,
arid grassland

Tropical rainforest,
tropical evergreen forest

Boreal forest (taiga),
montane coniferous forest

Desert

Tropical deciduous forest

Temperate forest

Chaparral/Mediterranean

Tropical savanna,
thorn forest

FIGURE 3.23 Major biogeographic regions and biomes on land. Compare with Figures 3.17 and 3.19.

efficient in converting energy from the next lower level. Thus, only 10% of the energy stored in plant biomass (or about 1% of the energy originally received from the sun) at the base of the pyramid is converted into animal (herbivore) biomass at the next higher level; the rest of the energy is lost to the locomotion, body heat, and so on of the herbivores. At the next higher level only 10% of herbivore biomass is converted to carnivore biomass, or 0.1% of the original energy captured by plants (one-tenth of 1%). Ten percent of this primary carnivore biomass is converted to secondary carnivore biomass or 0.01% of the original energy in plant biomass (one-tenth of 0.1%). At this point the apex of the food pyramid has typically been reached, presumably because there is insufficient energy to support higher levels of the pyramid.

The level at which populations of organisms are no longer sustainable is called the **carrying capacity** of the environment. Carrying capacity therefore refers to the maximum population sizes (or biomass) that can be sustained by the environment. Carrying capacity is largely a function of energy

(food availability) but also reflects other factors such as available habitat, mates, and nutrients (as you'll see in a moment).

3.5.3 Biogeochemical cycles

The biosphere is intimately involved in the cycles of various chemical elements on different scales of time. Such cycles are called **biogeochemical cycles** because they involve the interplay of several of Earth's major systems. In Chapter 1, we were primarily concerned with the biogeochemical cycle of carbon.

During the cycle of organic carbon, as in other systems, matter and energy are transferred between reservoirs. The initial production of organic carbon by photosynthesis is referred to as gross primary production. Much of the gross primary production is used as an energy source at the base of food pyramids, as just described; that which is left is called net primary production. Net primary production can be incorporated into the reservoirs of living biomass or

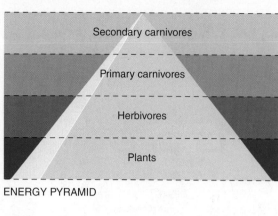

ENERGY PYRAMID

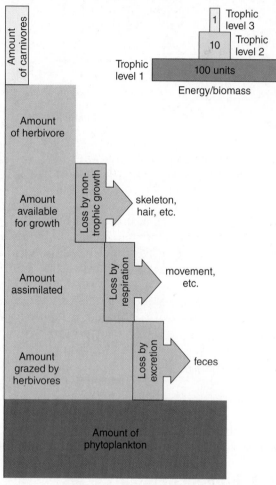

Data from: Russell-Hunter, W. D. 1979. *Aquatic Productivity*. New York: MacMillan Publishers.

FIGURE 3.24 Most energy that is transferred upward through the levels of a food pyramid is lost as heat, thereby limiting the number of levels in a food pyramid or links in a food chain.

dead biomass, where it is stored until it is eventually broken down. The breakdown of dead organic matter and the release of energy stored in the chemical bonds occur by the process of respiration, which is essentially the reverse of photosynthesis. Bacteria, along with fungi and burrowing organisms like worms, act as **decomposers**, which initially break down—or **remineralize**—dead organic matter and **recycle** the nutrients trapped in the organic matter back to the ecosystem. **Nutrient** is any substance required by

organisms for normal growth and maintenance (**Figure 3.25**). Without this recycling of nutrients from dead organic matter, ecosystems would quickly shut down.

This illustrates a very important point about ecosystems, communities, and the biosphere in general: they are all typically nutrient limited. Many nutrients like phosphorus (which is a critical component of DNA and other molecules of biologic systems) are originally derived from the weathering of continental rocks. However, the weathering of continental rocks is very slow, so that nutrient recycling is necessary to sustain ecosystems. Near the sediment surface, organic matter might be broken down by decomposers. Burrowers pump oxygen into the sediment, which oxidizes the organic matter to carbon dioxide, causing it to decay and release nutrients. As we will see in subsequent chapters, the availability of various nutrients has changed through time, affecting the flux of carbon between different reservoirs in the carbon cycle.

CONCEPT AND REASONING CHECKS

1. What factors determine the distribution of organisms over Earth's surface?

3.6 The Tectonic Cycle and Earth Systems

Earlier, we used the rise of the Himalayas to examine how continental movements and orogeny associated with the tectonic cycle can dramatically affect climate, the hydrologic cycle, and biotas over enormous regions—specifically, in case of the Himalayas, the origins of the Asian monsoon (Box 3.1).

The tectonic cycle has exerted similar effects on *global* scales by shifting the positions of the continents beneath atmospheric convection cells. The positions of the atmospheric cells have presumably remained relatively constant through geologic time, so we will move the continents beneath them and see what happens. For example, the movement of a portion of a continent beneath a low pressure cell will increase rainfall because as the air within the cell rises, it cools and loses its moisture, resulting in forests in temperate latitudes and tropical rainforests nearer the equator (**Table 3.1**). Conversely, moving a continent beneath a high pressure cell will likely result in a drier continental interior because the air within the high pressure cell picks up moisture as it descends toward the Earth's surface, potentially resulting in deserts or grasslands. Such conditions occurred in what is now the southwestern United States during the Jurassic Period, when thick, cross-bedded desert sands were deposited. On the other hand, the movement of a continent near or over a pole will result in the production of glaciers, increasing the latitudinal temperature gradient from poles to the equator while impacting the distribution of terrestrial vegetation, as occurs today.

The rifting and assembly of a supercontinent during a tectonic cycle also exert a tremendous impact on sea level,

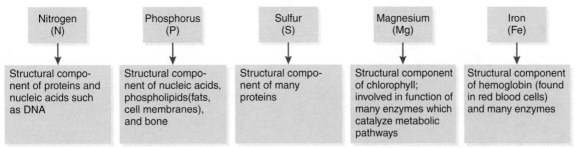

Nitrogen (N)	Phosphorus (P)	Sulfur (S)	Magnesium (Mg)	Iron (Fe)
Structural component of proteins and nucleic acids such as DNA	Structural component of nucleic acids, phospholipids(fats, cell membranes), and bone	Structural component of many proteins	Structural component of chlorophyll; involved in function of many enzymes which catalyze metabolic pathways	Structural component of hemoglobin (found in red blood cells) and many enzymes

Data from: Ricklefs, R. E., and Miller, G. L. 2000. *Ecology*, 4th ed. New York: W. H. Freeman.

FIGURE 3.25 The roles of nutrients essential to plants and animals.

patterns of surface ocean currents, and rates of deep-ocean circulation. Each tectonic cycle lasted approximately 250–300 million years and consisted of an initial "greenhouse" phase followed by an "icehouse" phase.

Let's begin with the rifting of a supercontinent like Pangaea during the Triassic Period, about 250 million years ago (see Chapter 13). A supercontinent insulates the Earth, slowing radiogenic heat loss from the core and mantle. Eventually heat buildup between continental crust becomes sufficient to cause doming, rifting, and breakup of the continent, followed by rapid seafloor spreading. Seafloor spreading centers (or mid-ocean ridges) amount to vast submarine mountain ranges, dwarfing even the Himalayas. During the Mesozoic, for example, the equatorial Tethyan Seaway (after Tethys, a daughter of the gods Uranus and Gaea) flowed around the Earth largely unimpeded. The Tethyan Seaway was a warm seaway because its currents, driven by the trade winds, flowed around the Earth largely unimpeded, all the while exposed to the heat of the sun in the tropics and subtropics (**Figure 3.26a**; **Table 3.2**). As mid-ocean ridges grew in volume, they also began to displace the oceans out of their basins onto the continents, much like when you sit in a bathtub full of water (also known as Archimedes' principle). At the same time, the warm crust forming at the spreading centers cooled as it spread away from the ridge and began to sink relative to the height of the oceans; this would have also caused the seas to flood the continents. In fact, during the Mesozoic Era, and especially the Cretaceous Period, sea levels were roughly several hundred meters higher than today (**Figure 3.27**). Increased rates of seafloor spreading would have also been associated with increased rates of subduction; thus, volcanism would have increased, as well, pumping more carbon dioxide into the atmosphere and promoting "greenhouse" conditions. These conditions would have been accentuated by decreased albedo related to the spread of relatively shallow epicontinental (or epeiric) seas. Warm conditions would have also intensified the hydrologic cycle on land; the increased heat likely increased the energy and rates of circulation of atmospheric convections cells.

These conditions continued until about the middle of the Cenozoic Era (late in the Paleogene period), when the Earth began to enter a prolonged "icehouse" phase (see Chapters 14 and 15). Why did the conditions begin to reverse themselves? Because the continents were now moving back toward one another and even colliding in some cases. The Himalayas, the Alps, and mountain ranges in western North and South America all began to form. This widespread uplift drove the seas off the continents as volcanism became less pronounced. By this time, too, Antarctica had moved the South Pole and massive ice sheets began to form, drawing sea level down and increasing Earth's albedo yet further.

A similar cycle of "greenhouse-to-icehouse" conditions occurred during the Paleozoic Era (see Chapters 11 and 12).

TABLE 3.1

Effects of the tectonic cycle on the hydrologic cycle and the distribution of terrestrial floras		
Conditions	Supercontinent Present (Continents Clumped Together)	Continents Rifted (Continents Far Apart)
Tectonic style	Collision, orogeny, closing of ocean basins	Rifting
Relative latitudinal temperature gradient	High	Low
Relative area of continental interiors	High	Low
Relative climatic extremes	High	Low
Terrestrial vegetation: High (subpolar to polar) latitudes	Glaciers, taiga, and tundra	Deciduous forests
Terrestrial vegetation: Mid (temperate) latitudes	Deciduous forests and steppe	Tropical wet forests
Terrestrial vegetation: Low (subtropical to tropical) latitudes	Savannahs and deserts more likely present	Rainforests more likely, along with savannahs and deserts

From van Andel, Tj. H., *New Views on an Old Planet*. © Cambridge, UK: Cambridge University Press 1985, 1994. Reprinted with the permission of Cambridge University Press.

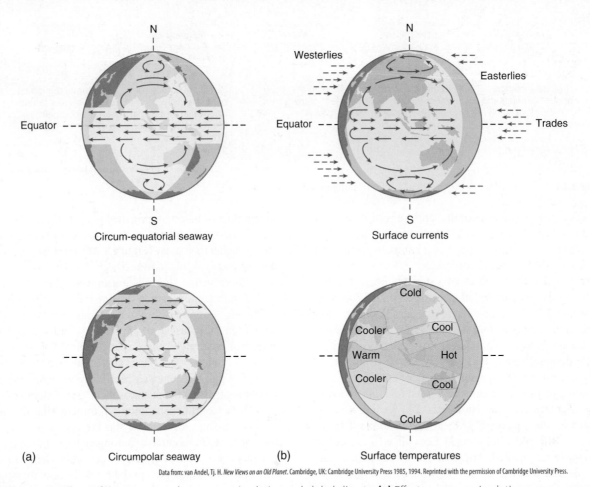

N

Westerlies

Easterlies

Equator

Trades

Equator

S

S

Circum-equatorial seaway

Surface currents

Cold

Cooler

Cool

Warm

Hot

Cooler

Cool

Cold

(a)   Circumpolar seaway   (b)   Surface temperatures

Data from: van Andel, Tj. H. *New Views on an Old Planet*. Cambridge, UK: Cambridge University Press 1985, 1994. Reprinted with the permission of Cambridge University Press.

FIGURE 3.26 The effects of the tectonic cycle on ocean circulation and global climate. **(a)** Effects on ocean circulation as a supercontinent rifts apart. A more even distribution of warm temperatures results if equatorial waters flow around the Earth more than once, warming the currents before they are eventually deflected to the north and south. As we will see, this was the basic situation during the existence of the Tethyan Seaway during the Mesozoic Era. If a continent is isolated over a pole by a surface current, then ice caps may start to grow, as eventually happened on Antarctica during the Neogene period. **(b)** Effects on ocean circulation and global temperature distributions as a supercontinent is assembled. Note how the temperatures of the surface waters mimic the flow of surface currents. Water in the equatorial current is moved westward by the trade winds, warming along the way because of exposure to the sun. Much of this water is deflected to the north and south by Pangaea and cools as it forms two large gyres. The rest of the water returns as an equatorial countercurrent and warms even further. Consequently, water in the western portion of the ocean is warm and that in the eastern portion is even warmer.

A different supercontinent called **Pannotia** began to rift during the Cambrian, the continents began to disperse, and the proto-Atlantic Ocean called the "Iapetus Ocean" (after Iapetus, a son of Oceanus and Gaea, and a brother of Tethys) formed between what was then North America (Laurentia) and Europe (Baltica). This particular greenhouse phase lasted until about the middle of the Paleozoic Era or so (Devonian Period). Then, continental collisions and associated orogenies began to assemble Pangaea during the last half of the Paleozoic Era (Mississippian-to-Permian periods). Supercontinent assembly is associated with the production of enormous mountain ranges and extremely large continental interiors that are prone to warming; during this time, Pangaea also lay astride the equator, where it was subject to warm, moisture-laden winds brought by the trade winds. Together, these conditions likely resulted in even more intense monsoons than those experienced over the Indian subcontinent today (see Chapter 12). Such conditions are indicated during the Late Paleozoic era, when widespread

coal-forming swamps occurred in what is now the eastern half of North America and portions of Europe as the supercontinent Pangaea was being assembled (see Chapter 12).

So, what happened to ocean circulation when Pangaea began to block the warm ocean circulation of the earlier Paleozoic? The warm tropical surface currents would have been diverted to higher latitudes to the north and south, making it more likely that they would cool (**Figure 3.26b**). And like the Cenozoic Era, an enormous glacier formed on the Gondwana continents of the Southern Hemisphere, as the sea level gradually fell. This in turn likely imparted greater "continentality" to Pangaea, meaning greater seasonality and climatic extremes on the Earth.

The broad patterns of sedimentary environments responded to the broad cycles of sea-level change associated with the tectonic cycle, as we will see in later chapters. Both greenhouse phases were characterized by warm, and extensive, shallow seas several hundred meters or more higher than today's. These are prime conditions for the formation of

TABLE 3.2

Effects of the tectonic cycle on climate, sea level, and ocean circulation

Conditions	Icehouse State (Late Paleozoic; Oligocene-Recent)	Greenhouse State (Early-to-Middle Paleozoic; Cretaceous-Eocene)
Tectonic style	Collision, orogeny, closing of ocean basins	Rifting
Relative rates of seafloor spreading	Slow	Fast
Relative rates of ocean ridge formation, subduction, volcanism, and release of CO_2 to atmosphere	Low	High
Relative latitudinal temperature gradient	High	Low
Relative sea level	Low, due to decreased production of mid-ocean ridges; narrow continental shelves	High, sometimes hundreds of meters, due to increased mid-ocean ridge production; with broad epeiric seas covering the continents producing enormous continental shelves
Degree of stratification of oceans into distinct water masses	Ocean is highly stratified because of differences in temperature and salinity of distinct water masses	Much less stably stratified than icehouse state
Surface ocean temperatures	Great range due to large latitudinal temperature gradient: Surface water temperatures range from less than 2°C (at poles) to greater than 25°C at equator	Smaller range due to small latitudinal temperature gradient: Surface water temperatures not much greater than icehouse phase at equator but much greater (12–15°C) at high latitudes
Deep ocean temperatures	Cold, ranging from about 1°C to 2°C, due to glaciation and production of deep-water masses nearer poles	Quite warm compared to present, ranging from 15°C near equator to about 10°C at poles, due to absence of glaciers
Rates of ocean circulation and oxygenation of deep-water masses	Relatively fast because of density differences between water masses; faster circulation moves oxygenated surface waters deeper into basins	Low-density surface water leads to slow deep water circulation and low oxygen content due to warm temperatures, often with widespread anoxia and black shales (warm water holds less oxygen than cold water)
Organic matter remineralization and nutrient cycling	Vigorous flow of strongly oxygenated bottom water, so relatively little organic matter preservation	Little remineralization of dead organic matter due to low oxygen conditions, so nutrient recycling may be much reduced
Primary productivity (marine photosynthesis)	Numerous, diverse marine environments and provinces, sometimes with high productivity due to upwelling	Low marine productivity, but good petroleum source beds because of organic matter preservation (due to low oxygen conditions)
Relative rates of erosion and mechanical versus chemical weathering	Fast, mechanical weathering tends to dominate	Slow, chemical weathering tends to dominate

From van Andel, *New Views on an Old Planet.* © Cambridge, UK: Cambridge University Press 1985, 1994. Reprinted with the permission of Cambridge University Press.

carbonate (limestone) platforms, in which reefs—formed by various taxa through time—thrived. Much of the world's petroleum also comes from rocks of the Cambrian-to-Devonian and Mesozoic eras; these intervals are characterized by widespread, carbon-rich black shales, which could have served as source beds for petroleum. These conditions have led some workers to suggest that ocean circulation was quite sluggish and conditions highly reducing ("anoxic" or without oxygen) during greenhouse phases because such conditions would have allowed the preservation of large amounts of dead organic matter in sediments to produce petroleum (Table 3.2). On the other hand, icehouse conditions would have been characterized by more rapid rates of ocean circulation and better oxygenation because of the presence of ice sheets at one or both poles.

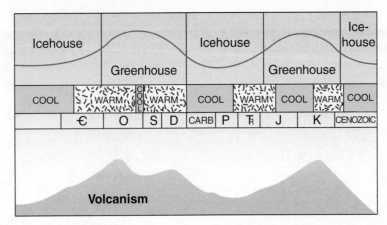

FIGURE 3.27 Cycles of continental configuration, sea level, and climate during the Phanerozoic Eon. During "greenhouses" (Cambrian through Devonian periods; Jurassic through Paleogene periods), continents dispersed and sea level rose as a result of increased sea floor spreading rates and mid-ocean ridge (MOR) volume. Increased volcanism (based on the volume of igneous rocks emplaced) resulted in increased atmospheric CO_2 levels that presumably caused the Earth's average surface temperature to rise during these intervals. Just the opposite conditions apparently prevailed during "icehouses" (Mississippian through Permian periods; Paleogene period to the recent): atmospheric CO_2 levels and sea level declined, presumably as a result of decreased spreading rates and MOR volume. *This figure and the associated discussion in the text serve as a framework for later discussion of Earth history in Chapters 11–15.*

SUMMARY

- The solid Earth system involves the rock and tectonic cycles. The release of heat generated by radioactive decay from within the Earth drives the processes of plate tectonics: seafloor spreading, the movement of continents, mountain building, and the rock and tectonic cycles.

- As heat is brought toward Earth's surface by convection cells, igneous and metamorphic rocks are formed and mountain chains are uplifted. The mountains undergo weathering, erosion, deposition, and lithification to produce sedimentary rocks. Igneous and sedimentary rocks can change through the processes of metamorphism, as can metamorphic rocks.

- Circulation within the atmosphere results from differential heating of Earth's surface and the transfer of heat and moisture across latitudes via large convection cells. Coupled with the Coriolis effect, this results in the major wind systems of the Earth: trades, westerlies, and polar easterlies.

- The hydrosphere involves the hydrologic cycle. Precipitation produces rain, snow, and ice. Runoff from the land flows into rivers and streams, infiltrates the ground, evaporates to the atmosphere, and is lost by transpiration from plants during photosynthesis.

- The biosphere affects the carbon cycle and Earth's climate through photosynthesis and respiration, both of which help to regulate levels of the greenhouse gas carbon dioxide in the atmosphere.

- The biogeographic distribution of organisms is determined in part by their tolerance to environmental change. Widely distributed species tend to be eurytopic, or broadly tolerant of environmental change. By contrast, stenotopic species are less tolerant of environmental change and tend to be endemic to certain areas.

- Biomes are caused by differential heating of Earth's surface by the sun, which affects surface temperature and rainfall. Each continent is characterized by a particular biogeographic region. In the sea, subdivisions called provinces reflect changes in water masses, especially temperature related to ocean currents caused by atmospheric circulation and upwelling.

- The distribution of modern organisms is also determined by historical factors such as the movement of plates, which affect atmospheric and oceanic currents via rifting and orogeny.

- A biologic community coupled with its physical environment is called an ecosystem. The flow of energy from the sun to photosynthetic plants to consumers (herbivores and carnivores) establishes the trophic relationships and niches of food chains. Trophic relationships in turn link food chains into food webs that comprise biologic communities.

- Eventually, the availability of energy appears to run out because of the laws of thermodynamics, and further links in food chains and higher levels in food pyramids cannot be supported.

- Biogeochemical cycles transfer elements such as carbon, oxygen, and nutrients from one reservoir of Earth's systems to another. The reservoirs may be in the same system or in different systems. Biogeochemical cycles are critical to the functioning of Earth's systems because they help to recycle nutrients and other elements critical to life.
- The tectonic cycle has affected the broad patterns of sea level, ocean circulation, and climate of Earth on time scales of several hundred million years.

KEY TERMS

albedo
andesite
Antarctic Bottom Water (AABW)
aphanitic
assimilate
autotrophs or producers
basalt
biogeochemical cycles
biomes
biosphere
blueschist
carrying capacity
Coriolis effect
cosmopolitan
cryosphere
decomposers

diorite
eurytopic
evaporation
felsic
foliated
food chain
food (or energy) pyramids
fractional crystallization
gabbros
gneisses
granodiorites
gyres
herbivores
hydrologic cycle
hydrosphere
hydrothermal metamorphism

hydrothermal weathering
igneous
impact
lava
lithification
magma mixing
magmatic differentiation
marble
meridional ocean circulation
metamorphic
migmatites
niches
nonfoliated
North Atlantic Deep Water (NADW)

nutrient
Pannotia
partial melting
phaneritic
phyllite
plutons
polar easterlies
precipitation
provinces
quartzites
recycle
region
regional metamorphism
remineralize
reservoirs
rhyolites
runoff

schists
secondary consumers (carnivores)
sedimentary rocks
shock metamorphism
slates
stenotopic
tektites
thermohaline circulation
top carnivores
trade winds
transpiration
trophic (food, energy) relationships
ultramafic
upwelling
westerlies

REVIEW QUESTIONS

1. Using labeled diagrams and in your own words, describe the reservoirs, processes, and fluxes of the each of the four basic Earth systems, including their cycles.
2. What are the ways by which different types of magmas form?
3. Diagram the sequence of crystallization of the different igneous rock types and their compositions and textures.
4. What does the texture and mineralogy of each of the following igneous rocks tell about how and where each rock formed: gabbro, basalt, granite, rhyolite, granodiorite, and andesite?
5. What does the texture and mineralogy of each of the following metamorphic rocks tell about how and where each rock formed: slate, phyllite, schist, gneiss, marble, and quartzite?
6. Using labeled diagrams and in your own words, describe how each of the following sets of systems interact via any processes and fluxes:
   Solid Earth-Atmosphere
   Atmosphere-Hydrosphere
   Solid Earth-Biosphere
   Atmosphere-Hydrosphere-Biosphere
   Biosphere-Hydrosphere
7. Diagram and discuss a monsoon.
8. Why is the heat capacity of the oceans important?
9. What factors determine the broad distribution of plants and animals?

10. What traits does an ecosystem share with other kinds of systems?

11. What are nutrients, and why are they important? Identify the different reservoirs and discuss the regeneration of nutrients in the oceans.

12. Construct a chart indicating how the tectonic cycle of greenhouse-icehouse conditions affects each of the following: global temperature, sea level, the hydrologic cycle, ocean currents, albedo, and the distribution of plants and animals? Indicate on the chart when these conditions occurred during geologic time.

FOOD FOR THOUGHT:
Further Activities In and Outside of Class

1. Assume that you are an early geologist (like James Hutton; see Chapter 1) trying to determine how rocks form. Use inductive logic (see Chapter 1) to infer how each of the major types of rocks—igneous, sedimentary, metamorphic—forms. What sorts of evidence would you seek? Where would you go to look? How would you test your inferences?

2. In a few words, infer the conditions under which you believe each of the following rocks formed based on the description (in a few cases, there can be more than one interpretation):
   a. Phaneritic, abundant potassium feldspar, mica, and quartz
   b. Dark, with phaneritic texture
   c. Coarse-grained, banded rock containing quartz, feldspar, micas, and hornblende
   d. Shiny, generally fine-grained but with some coarser amphibole grains
   e. Dark, with aphanitic texture

3. Is the rock cycle a theory (see Chapter 1)? Explain.

4. Why does continental crust tend to be granitic in composition?

5. Granites and granodiorites are often involved in mountain building but diorites less so. Why? (Hint: Think about the processes by which magmas of different compositions form.)

6. The slopes of Hawaiian volcanoes are relatively gentle compared with those of the Cascades. Why?

7. The composition of the extrusives associated with the Snake River Plain west of Yellowstone Park range from basaltic to rhyolitic, but those nearer Yellowstone range from andesitic to rhyolitic. How might the composition of the extrusives located closer to Yellowstone have changed through time?

8. Imagine as you drive along a highway for many tens of miles that the rocks exposed in the outcrops change from granodiorites to gneisses and then phyllites. In what sort of geologic setting did these rocks originally form?

9. If you were looking for evidence of an ancient asteroid impact, for what sorts of evidence would you look?

10. Would the biosphere completely shut down if photosynthesis stopped?

11. The amount of time it takes for a particular substance to be replaced in a particular reservoir is called its residence time. The total inventory of phosphorus dissolved in all of the oceans is 89,910 terragrams (1 terragram [or Tg] = 1,012 grams). The amount of phosphorus delivered to the oceans by runoff (the flux) from land is about 1.9 Tg per year. What is the residence time of phosphorus in the oceans if it is replaced only by runoff from land?

12. The residence time of phosphorus in the surface layer of the ocean is only 2.6 days, which is quite rapid. Explain.

13. The phosphorus inventory of all the deep ocean water masses is 87,100 Tg and the amount of phosphorus coming into this reservoir from all sources is about 60 Tg per year. What is the residence time of phosphorus in this reservoir? Why do you suppose scientists are interested in ocean circulation and its effect on atmospheric carbon dioxide concentrations?

SOURCES AND FURTHER READING

Bowen, N. L. 1928. *The evolution of the igneous rocks.* New York: Dover Publications (reprinted 1956).

Ricklefs, R. E., and Miller, G. L. 2000. *Ecology*, 4th ed. New York: W. H. Freeman and Company.

van Andel, Tj. H. 1994. *New views on an old planet: A history of global change.* Cambridge, UK: Cambridge University Press.

Wunsch, C. 2002. What is the thermohaline circulation? *Science*, 298, 1179–1180.

Sedimentary Rocks, Sedimentary Environments, and Fossils

MAJOR CONCEPTS AND QUESTIONS ADDRESSED IN THIS CHAPTER

A What are sedimentary rocks, and why are they important?

B How can sedimentary rocks tell us about how they formed?

C What are the major types of sedimentary rocks, and how do they form?

D What are the different types of marine and terrestrial environments and their sediments?

E What features of modern environments can be used to recognize their ancient counterparts?

F What factors affect the deposition of sediment in the deep sea?

G What are sedimentary structures, and what do they tell us about how sedimentary rocks were deposited?

H What are fossils, what do they tell us, and how do they form?

CHAPTER OUTLINE

4.1 Introduction to Sedimentary Rocks

A Sedimentary rocks are an important "subsystem" of the rock cycle, which is itself part of the much larger solid Earth system (see Chapter 3). Earth's surface represents the interface between the solid Earth, hydrosphere, atmosphere, and biosphere, and this interface is where sedimentary rocks form. Sedimentary rocks are therefore indicative of surface conditions and related climatic feedbacks. In this chapter, we examine the formation of sedimentary rocks and how we use them (based on the Principle of Uniformitarianism and modern analogs; see Chapter 1) to infer past surface conditions.

There are several common types of sedimentary rocks. Among the most abundant are **terrigenous** ("land-derived") sedimentary rocks. This means the rocks have been derived *from* various types of preexisting rocks by weathering, erosion, transport, and deposition of sediment, such as sand or mud (**Figure 4.1**). Sediment then undergoes the processes of **lithification** (or solidification) to produce sedimentary rocks. A second broad class of sedimentary rocks is biogenic in origin, meaning that the rocks consist predominantly of the shells of dead organisms, or **fossils**. The third class of sedimentary rocks is chemical in origin and is formed by chemical precipitation from water. As we will see, besides sediment grain size and composition and fossils, sedimentary

rocks can preserve certain types of structures that provide clues to their origin and thus past surface conditions.

4.2 Processes of Weathering

B Rocks exposed at Earth's surface break down by the processes of physical and chemical weathering. The processes of physical and chemical weathering act together. **Physical weathering** involves the formation and enlargement of cracks and other surfaces in rocks that increase the surface area available for chemical weathering. Cracks might initially form by mountain building and other forms of tectonism that deform rocks, often producing large sets of fractures. After the rocks are exposed at Earth's surface by erosion, the cracks can be enlarged (if climate or elevation permit) by the process of **freeze-thaw**. Water has the unique property of expanding when it freezes. Thus, if cracks are filled with water that freezes, the cracks are pushed farther and farther apart like a wedge, ever so slowly, exposing the rock surfaces to further weathering. Plant roots that grow into cracks also push fragments of rock farther apart.

As rock surfaces are exposed to air and water, they undergo **chemical weathering** (**Figure 4.2**). Chemical weathering results from weak acids produced by the decay of dead organic matter (for example, **humic acids** from dead plant matter) and from the dissolution of atmospheric

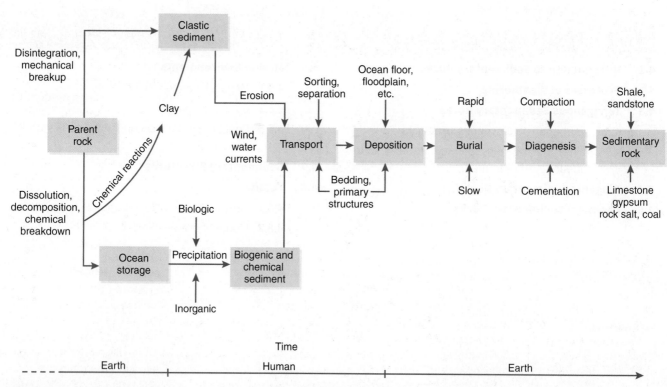

Sedimentary rock cycle

FIGURE 4.1 The processes involved in the formation of sediment and sedimentary rocks.

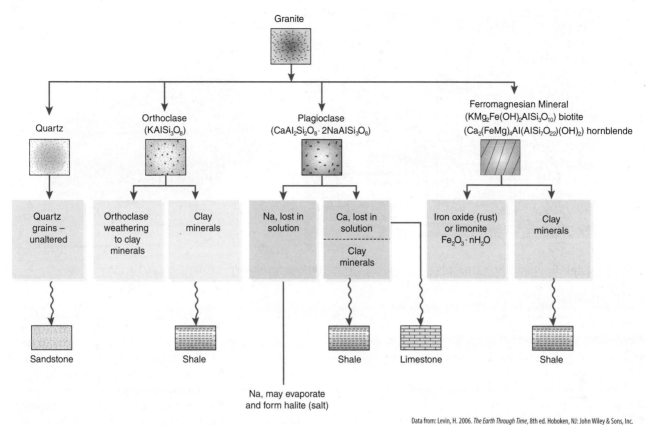

Granite

Quartz

Orthoclase
$(KAISi_3O_8)$

Plagioclase
$(CaAl_2Si_2O_8 \cdot 2NaAlSi_3O_8)$

Ferromagnesian Mineral
$(KMg_2Fe(OH)_2AlSi_3O_{10})$ biotite
$(Ca_2(FeMg)_4Al(AlSi_7O_{22})(OH)_2)$ hornblende

Quartz
grains –
unaltered

Orthoclase
weathering
to clay
minerals

Clay
minerals

Na, lost in
solution

Ca, lost in
solution

Clay
minerals

Iron oxide (rust)
or limonite
$Fe_2O_3 \cdot nH_2O$

Clay
minerals

Sandstone

Shale

Shale

Limestone

Shale

Na, may evaporate
and form halite (salt)

Data from: Levin, H. 2006. *The Earth Through Time*, 8th ed. Hoboken, NJ: John Wiley & Sons, Inc.

FIGURE 4.2 The basic processes of chemical weathering of a granite and some of the basic sedimentary rocks that may result. Feldspars and micas weather to clay, leaving the quartz grains behind, while calcium in solution may contribute to limestone formation.

carbon dioxide (CO_2) into rainwater to produce **carbonic acid** (which is also found in carbonated beverages; the fizzing produced when the container is opened is the carbon dioxide coming out of solution). Ions such as calcium, sodium, and potassium are dissolved from the minerals forming the rock, causing the rocks to break down. The ions can in turn be carried in dissolved form by rivers and groundwater to the oceans. In the case of granite, feldspar (orthoclase and plagioclase) and micas weather to form muds; more highly resistant minerals such as quartz resist chemical weathering but are freed from the surrounding rock to eventually produce sands and gravels.

If the rock weathers in place and there is sufficient input of organic matter (such as leaves) from land plants, soil forms. **Soil** is the accumulation of weathered debris composed of disintegrated and decomposed rock, decayed organic matter or **humus** (primarily from plants), water, and air. Humus is important because it enhances the soil's ability to retain water. The dead organic matter, which derives primarily from plants, also produces organic acids that promote weathering. Thus, the chemical composition and texture of a particular soil depend on a variety of factors, including the parent rock material, the type of organic matter that

forms the humus (for example, forest or grassland), and the amount of time during which the soil has formed.

Several feedback processes can also affect soil composition. Nutrients are released during decay, promoting further plant growth and the production of organic matter that can then decay, promoting positive feedback on weathering. Microbes, including bacteria and fungi, accelerate the breakdown of organic matter, whereas burrowing animals such as earthworms, which also feed on the organic matter, churn and mix the soil. Charles Darwin wrote an entire book on the subject of soil formation; he made a crude calculation that up to about 50,000 worms per acre could transport roughly 18 tons of sediment per acre to the surface in a year!

The factors that affect soil formation also vary with climate—especially temperature and rainfall—which changes across latitude. **Laterites** form in warm, wet tropical climates where chemical weathering is intense (**Figure 4.3**). In the tropics, chemical weathering is so intense that nutrients are leached from the soil so that even silica dissolves, leaving behind a brick-like soil enriched in clays, iron, and aluminum oxides. In some cases the soils are so enriched in aluminum they are mined as the ore, **bauxite** (Figure 4.3).

(a)

© Tahiti/Fotolia.com.

(b)

© Denis Selivanov/Shutterstock, Inc.

FIGURE 4.3 (a) Intense weathering in the tropics produces laterites. **(b)** Bauxite, an aluminum ore produced by intense tropical weathering.

CONCEPT AND REASONING CHECKS

1. Diagram the processes of the sedimentary portion of the rock cycle.
2. What sorts of soils form under conditions of severe weathering in the tropics?
3. Do you suppose the soils in question 2 would be good for farming?

4.3 Terrigenous Sedimentary Rocks

4.3.1 Formation of terrigenous sediments

There are three broad types of terrigenous sediments based on the size of the predominant **sedimentary grains**: gravel, **sand**, and **mud** (**Table 4.1**). These three terms indicate only the size of the sediment particles, not their chemical composition. For example, sand can consist of the quartz grains of stereotypical white, sandy beaches; grains of basalt like those of the black beaches of volcanic islands in the southwest Pacific and Hawaii; or coral and algal fragments of tropical reef environments.

Loose rock and mineral grains of these various sizes that result from weathering can be eroded and transported by landslides, streams, and rivers and eventually deposited, buried, and lithified to form sedimentary rocks of different types. Lithification involves the consolidation of loose sediment into rock by burial, compaction, and cementation. **Compaction** occurs due to the weight of overlying sediment, which progressively destroys the pore spaces between sediment grains while bringing the sediment grains closer together. Compaction is very important in the lithification of muds. As muds are buried by the deposition of overlying

sediment, water is squeezed out and the individual particles, which bear positive and negative charges at their surfaces, come closer together and bond. Mud particles bond together to produce rocks called **mudstones**; mudstones of silt-sized sediment are called siltstones, whereas those composed of clay-sized particles are called claystones. Mudstones exhibiting a layered appearance are called **shales**. By contrast, the larger grains of sands and gravels are mostly incompressible. Instead, the grains of sands and gravels typically undergo **cementation** by the precipitation of chemical substances (typically calcium carbonate or silica) from the waters filling the pore spaces between the sediment grains or from groundwater as it percolates through the sediment.

4.3.2 What do terrigenous sedimentary rocks tell us about how they formed?

Sedimentary rocks have histories. We can decipher those histories using a few simple indicators: the size, shape, and mineral composition of the grains of the rock.

During transport and deposition, sediment grains of different sizes can be deposited together or they might be separated by size according to the amount of environmental energy of waves and currents available to transport them. The separation of sediment grains according to size is called **sorting**. Depending on its velocity, water can transport all sizes of sedimentary grains. Coarser grains such as pebbles and cobbles tend to be deposited closest to the source area because they are the least easily transported, whereas granules and sands tend to be carried farther from the source area. Mud (silt and clay) tends to be transported the farthest because it is easily suspended in water for long periods of time. Muds tend to be deposited in quiet waters, such as those of relatively isolated lagoons or far from shore in lakes or oceans, where there is time for the particles to slowly settle from suspension.

TABLE 4.1

Wentworth grain-scale for classic sediment showing size ranges (in mm) along with corresponding particle, sediment, and rock names

Particle Size (mm)	Particle Name	Sediment Name		Rock Name
256	Boulder	Gravel		Breccia or Conglomerate
128				
54				
32	Cobble			
16				
8				
4	Pebble			
2	Coarse	Sand		Sandstone
1.0				
0.5	Medium			
0.25	Fine			
0.0625 (1/16 mm)	Silt	Mud	Silt	Siltstone
0.0039 (1/256 mm)	Clay		Clay	Shale or Claystone

Data from Ritter & Petersen: *Interpreting Earth History: A Manual in Historical Geology*, Figure 3.1.

The degree of sorting exhibited by terrigenous sedimentary rocks provides important clues to the rocks' origins. By reasoning backward from a terrigenous sedimentary rock's texture, we can infer how far the sediments were transported from their source (and possibly the direction in which they were transported), their original source rocks, and the amount of environmental energy to which the sediments were exposed by currents and waves. Poorly sorted sediments consist of a wide range of grain sizes because they have undergone little transport. This is typical of sediment-choked **braided streams and rivers**, which today are most common in relatively barren mountainous regions and indicate rapid erosion and deposition close to their source area. Typically, gravels tend to be too coarse to be carried very far and, along with sands and muds, are deposited nearer their source to form poorly sorted rocks called **conglomerates** (Table 4.1). Conglomerates can form from landslides adjacent to uplifted regions, but they can also be shed for many hundreds of kilometers depending on the amount of uplift. **Breccias** are like conglomerates, but the grains are angular because there has been insufficient transport to round the sediment grains by grain collisions during erosion and transport (Table 4.1). By contrast, well-sorted sediments consist of similar grain sizes because they have been either transported farther from their source—allowing greater time for sorting of the sediment grains—or deposited under more uniform energy conditions. These conditions are more typical of meandering or winding rivers found at much lower elevations.

If there is sufficient current energy in streams and rivers, sand and mud-sized grains might be further separated as they are transported to the oceans, where they are eventually deposited. Here, the sediments can be reworked and further sorted by waves and currents in shallow water. As waves pass through a column of water, individual water particles move up and down in a circular motion (**Figure 4.4**).

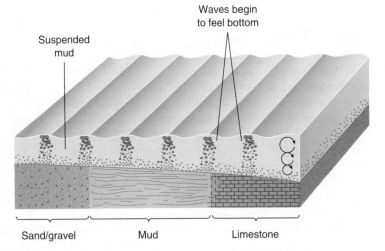

FIGURE 4.4 Sorting of sediment grains by waves. As sediment is transported, the grains tend to separate according to size. After sediment is deposited in a lake or the sea, it is typically reworked and sorted by wave movement and currents (see text for explanation). Coarse-grained sediments like gravels and sands tend to be deposited closer to shore under high-energy conditions, whereas muds are deposited farther from shore in quieter water where waves and currents do not feel bottom. Even farther from shore, after most of the mud has settled from suspension, sediments tend to consist of the accumulation and concentration of shells; these sediments will ultimately form limestones.

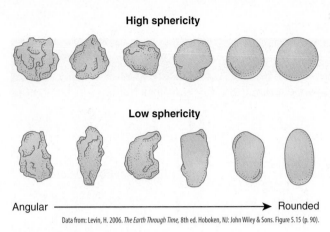

High sphericity

Low sphericity

Angular ——————————————▶ Rounded

Data from: Levin, H. 2006. *The Earth Through Time,* 8th ed. Hoboken, NJ: John Wiley & Sons. Figure 5.15 (p. 90).

FIGURE 4.5 Rounding of an individual quartz grain by abrasion. Rounding refers to the rounding off of the angular corners of the grain, whereas sphericity refers to the overall shape of the grain, ranging from spherical to ellipsoidal.

As the waves approach shore, the circular motions of the water particles eventually begin to "feel bottom," and the circles nearest the bottom begin to flatten into ellipses. As the waves feel bottom, they cause the surface sediment on the bottom to move back and forth slightly. This back-and-forth movement causes the fine-grained muds to be thrown into suspension and carried farther from shore to quieter water where they can settle out (Figure 4.4). The sands left on the shallow bottom are often well sorted, meaning the grains are all about the same size.

The grains of sands and gravels can also undergo rounding as they are transported (**Figure 4.5**). Well-rounded sediment grains indicate prolonged exposure to environmental energy such as that which occurs during transport by rivers and streams and wave or current-swept environments. During this time, abrasion of the grains against one another rounds off the angular corners of individual sand and gravel grains.

The extent of transport, rounding, and sorting of sediment grains in a terrigenous sedimentary rock reflects the rock's **maturity**. Maturity is of two types: **compositional** and **textural**. A rock that is compositionally mature is one in which relatively unstable minerals such as feldspars and micas have been weathered and transported away (as clays), leaving only chemically inert grains such as quartz. Texturally mature rocks are those whose grains are well rounded and well sorted because as soon as this stage is reached, it is unlikely that the grains can be rounded or sorted further. **Sandstones** consisting of pure quartz grains that are well rounded and well sorted (all grains of nearly the same size) are said to be compositionally and texturally mature. It takes a relatively long time to produce a compositionally and texturally mature quartz sandstone. This is because the quartz grains are normally angular when they are first weathered from their source rocks and become more rounded as the grains collide with one another during transport. Thus, the sand grains are likely to have been transported long distances, thereby sorting the grains, as well. By contrast, conglomerates consist of a mixture of gravel, sand, and mud and

are often both compositionally and texturally immature. The compositional and textural immaturity of conglomerates indicates that the individual grains have not been intensely weathered or transported very far.

Two types of terrigenous sandstones of particular importance, **graywackes** and **arkoses**, are considered compositionally and texturally immature (**Figure 4.6**). Graywackes are "dirty sandstones" consisting of poorly sorted, highly angular sediment grains, rock fragments, and mud. Despite their name, graywackes need not be gray and can be brownish or dark-colored. Graywackes sometimes contain significant amounts of volcanic rock fragments when they are associated with tectonically unstable areas, such as deep-sea trenches, which are associated with volcanic activity and submarine landslides called turbidity currents (see later in this chapter). Arkoses are enriched in angular fragments of feldspar and often form close to mountain ranges (**Figure 4.7**). Because they contain abundant feldspar, arkoses often have a pink or reddish appearance, so they can resemble granites. But the sediment grains are not fused and the angular nature of the feldspar grains indicates that the grains have not been transported relatively far; otherwise, the feldspar's cleavage (by which the mineral tends to break apart along planes of weakness) would have allowed it to break apart into smaller grains that would have been weathered or transported away, perhaps as clays.

CONCEPT AND REASONING CHECKS

1. What processes produce textural and compositional maturity of a terrigenous sedimentary rock?
2. Starting with a typical granite, produce a mature quartz sandstone.

4.4 Biogenic Sedimentary Rocks

As their name indicates, biogenic sedimentary rocks are *biologic* in origin; that is, they are generated by organisms. Many creatures use dissolved ions in water to secrete shells that are used for skeletal support and protection. Upon death, the shells are contributed to the sediment and in some cases are so abundant they are the most common components of the rock.

Many biogenic sedimentary rocks are **limestones**, which are composed of calcium carbonate, or $CaCO_3$. Organisms such as algae or clams use dissolved calcium (Ca^{2+}), bicarbonate (HCO_3^-), and carbonate (CO_3^2) in seawater to produce shells of calcium carbonate ($CaCO_3$). The $CaCO_3$ might have been originally precipitated as the mineral calcite or aragonite (in which case, it has usually **recrystallized** to calcite). Upon death, the shells are incorporated into the surrounding sediment and buried (**Figure 4.8**). The resulting sediment particles, or **bioclasts**, range in size from muds to sands and gravel-sized particles, some of which can

FIGURE 4.6 Types of sandstones and their maturity. **(a)** Graywacke. Graywackes often contain significant amounts of volcanic rock fragments and are frequently associated with tectonically unstable areas such as deep-sea trenches. **(b–e)** The figure shows how a granite **(b)**, which is enriched in potassium feldspars, mica, and quartz grains initially weathers to a pure quartz sandstone **(e)**. Initially, micas weather easily so that only feldspar and quartz grains remain to produce a conglomeratic arkose **(c)**, and then an arkose sandstone **(d)**, and finally a pure quartz sandstone **(e)** after the feldspars have completely weathered to mud and been washed away.

Courtesy of Ronald Martin, University of Delaware.

FIGURE 4.7 The Fountain Arkose forms spectacular cliffs on the eastern side of the Rocky Mountains near Boulder, Colorado. These sediments were originally laid down more or less horizontally and later deformed, uplifted, and exposed by erosion at the Earth's surface.

disintegrate into much smaller particles upon death of the organism.

Several features of shallow-water limestones make them useful indicators of past conditions. First, shallow-water limestones tend to form where there is little terrigenous sediment input; otherwise, the terrigenous sediment would dilute the sediment grains composed of calcium carbonate, making it a terrigenous sedimentary rock. Second, unlike terrigenous sediments, shallow-water limestones do not normally consist of sediment grains transported great distances. Instead, these limestones tend to form more or less in place from the hard parts of preexisting organisms, making them

reliable indicators of the conditions under which the sediments accumulated. Some of the most spectacular examples of biogenic limestone formation are the coral reefs found in warm, shallow, tropical waters. In fact, many biogenic limestones form in shallow, warm tropical waters, although occasionally some form in cooler waters of temperate latitudes. Thus, based on the Principle of Uniformitarianism, ancient reefs are usually interpreted to have formed in similar environments.

Biogenic limestones can also form deeper in the oceans, however, where there is also generally little input of terrigenous sediment except for fine-grained muds settling out

Courtesy of Ronald Martin, University of Delaware.

FIGURE 4.8 A limestone composed of fossils, which are the remains of preexisting organisms. These particular fossils are impressions of bivalves (clams) from Miocene rocks outcropping along Chesapeake Bay (see Chapter 10).

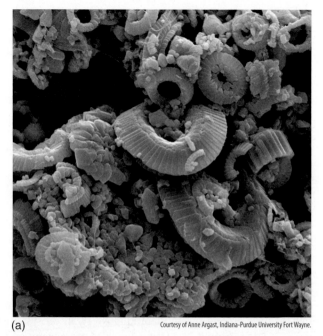
(a)   Courtesy of Anne Argast, Indiana-Purdue University Fort Wayne.

(b)   © Photos.com.

FIGURE 4.9 Calcareous and siliceous oozes and the microfossils which form them. **(a)** Calcareous ooze is composed of unicellular phytoplankton called coccolithophorids (shown here) and unicellular zooplankton called foraminifera. **(b)** Siliceous ooze can be composed of unicellular phytoplankton called diatoms or siliceous ooze called radiolaria (shown here). Chert can form from siliceous oozes during burial and compression.

of the water column (more on this in the next section). These types of limestones are called oozes and form from the calcareous remains of floating microscopic organisms, or **plankton**. The limestones of the White Cliffs of Dover consist of **calcareous oozes**, for example (see Chapter 1; **Figure 4.9**). Another type of biogenic ooze is **siliceous ooze**, which forms from the siliceous remains of other types of plankton (Figure 4.9). Upon burial, siliceous oozes can be transformed into **chert**.

4.5 Chemical Sedimentary Rocks

Chemical sedimentary rocks are generated by precipitation from water. One broad group of chemical sedimentary rocks is the **evaporites**. Evaporites are represented by the minerals **gypsum**, **anhydrite**, and **halite** (salt). Based on field observations and laboratory experiments, these minerals precipitate in a definite order as seawater evaporates in restricted basins and lagoons. Gypsum forms first, when the original volume of water has been reduced to about 20% of the original volume. The precipitation of gypsum is followed by halite when the water volume has dropped to about 10% of the original. Salts enriched in magnesium and potassium ("bitter salts") form when the water volume reaches about 5% of the original.

In some cases $CaCO_3$ precipitates inorganically out of seawater as round or ellipsoidal sand-size sedimentary grains called **ooids** (**Figure 4.10**). Microscopic algae on the surfaces of the ooids can also help promote precipitation. Each ooid consists of a central nucleus (often consisting of

a detrital or biogenic sand grain) around which concentric rings composed of microscopic needles of $CaCO_3$ have accumulated. These oölitic sands are typically found in very shallow, wave and current-swept environments, which seem to promote the formation of ooids. When lithified, ooids form the particular type of limestone called **oölite**. Thus, ancient oölites are indicative of warm, shallow, wave-swept conditions. Such conditions are widespread today in the Bahamas.

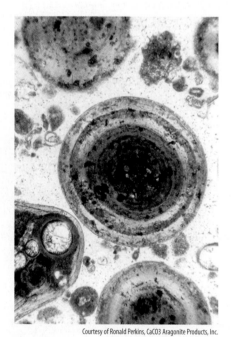
Courtesy of Ronald Perkins, CaCO3 Aragonite Products, Inc.

FIGURE 4.10 Ooids in cross-section under the microscope.

Limestones that form in relatively shallow waters can be altered to **dolostones**. Dolostones are composed of the mineral dolomite and for this reason dolostone is usually referred to as dolomite. Several hypotheses for the formation of dolomites have been proposed that involve the alteration of preexisting limestones exposed to seawater of elevated salinity like those associated with evaporites. The highly saline waters—or **brines**—form when the evaporation of seawater exceeds precipitation, so that ions such as those of magnesium become concentrated in the water. As the seawater percolates through the limestone ($CaCO_3$), it is altered to produce dolomite, or $CaMg(CO_3)_2$.

Another type of chemical sedimentary rock forms as the result of the dissolution of limestones at Earth's surface. This results in **karst topography** (or just **karst**) that is associated with caves and collapse structures called sinkholes. As ground-water percolates through soils it picks up carbon dioxide (CO_2) from decaying organic matter in soil to produce carbonic acid. The carbonic acid in turn dissolves the $CaCO_3$ in limestones. Upon reaching the subsurface CO_2 leaves the groundwater, diffusing into the air of the cave. As a result the dissolved $CaCO_3$ becomes more concentrated in the solution and precipitates as **travertine**, forming the distinctive features known as **stalactites** (which hang from the cave ceiling) and **stalagmites** (which grow upward from the cave floor). Cave deposits of travertine are known collectively as **speleothems**. Speleothems are important because they provide detailed records of climatic fluctuations on the continents.

CONCEPT AND REASONING CHECKS

1. How do biogenic sedimentary rocks and evaporites differ from terrigenous sedimentary rocks?
2. What might biogenic sedimentary rocks and evaporites tell one about the past climate in which they formed?

Ⓔ Ocean environments are referred to collectively as **marine environments**. The sediments transported to the coasts by rivers and streams are typically redistributed by waves and currents before being deposited and buried in marine environments. In some cases sediments can be transported far from shore into deep oceanic environments. Marine sediments are also important sources of fossils that tell us about Earth's past life and surface conditions.

4.6.1 Marginal marine environments

Marginal marine environments are those where land and sea meet (**Figure 4.11**). These environments are associated with coasts, bays, barrier islands, and **estuaries** that are often heavily influenced by the fresh waters and sediment brought by rivers and streams. Together, these environments account for only a relatively small area of Earth's surface, but they are among the most productive of all ecosystems per unit area (**Table 4.2**). Marginal marine systems not only serve as breeding grounds for fisheries, they also trap many pollutants before they reach the ocean, and they slow the erosion of the land.

Marginal marine environments include the **intertidal** (or **littoral**) **zone**, which is the land alternately inundated and exposed by the tides. The tides are caused by the gravitational attraction of the moon, which pulls the surface waters of the oceans away from the Earth, causing the surface ocean to rise and inundate the land periodically and then recede as the Earth continues to rotate. Depending on location the tides might expose large areas, such as mudflats, mangrove swamps, and the lower portions of salt marshes.

Marginal marine environments are stressful to most organisms because of the daily and seasonal fluctuations in temperature, salinity, and sediment input. Normal seawater is approximately 3.5% dissolved salts, mostly sodium chloride. Salinity is typically expressed as parts per thousand; for

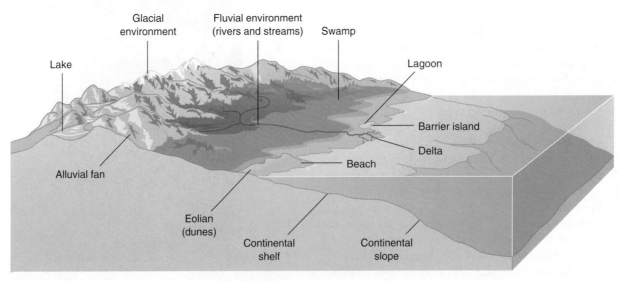

FIGURE 4.11 Marginal marine environments.

TABLE 4.2

The productivity of Earth's marine and terrestrial ecosystems

Ecosystem Type	Swamp, Marsh	Lake, Stream	Cultivated Land	Rock, Ice, Sand	Desert Scrub	Tundra, Alpine Meadow	Temperate Grassland	Woodland, Shrubland	Savannah	Boreal Forest	Temperate Deciduous Forest	Evergreen Forest	Tropical Rainforest (Evergreen)	Open Ocean	Upwelling Areas	Continental Shelf	Algal Beds and Reefs	Estuaries	Total Global Weighted Mean
Area/10^6 km^2	2	3	14	24	18	8	9	8	15	12	7	5	17	332	0.4	27	0.6	1.4	511
Mean plant biomass per unit area/kgC m^{-2}	6.8	0.01	0.5	0.01	0.3	0.3	0.7	2.7	1.8	9.0	14	16	20	0.0014	0.01	0.005	0.90		5.5549
Total plant biomass/10^{12} kgC	14	0	7	0	5	5	6	22	27	110	98	80	340	0.46	0.004	0.14	0.54	0.63	831.76
Mean NPP per unit area/kgC m^{-2} y^{-1}	1.1	0.225	0.290	0.002	0.032	0.065	0.225	0.270	0.315	0.360	0.540	0.585	0.900	0.057	0.225	0.162	0.900	0.810	0.393
Total NPP/10^{12} kgC y^{-1}	2.2	0.6	4.1	0.04	0.6	0.5	2.0	2.2	4.7	4.3	3.8	2.9	15.3	19	0.1	4.3	0.5	1.1	73.3

NPP = Net Primary Productivity.
Data from: Cockell, C. (Ed.). 2008. *An Introduction to the Earth-Life System*. Cambridge University Press, Cambridge, UK, Table 2.1 (p. 69) and Table 2.2 (p. 80).

example, 3.5% is equal to 35 parts per thousand (denoted as 35‰ or 35 ppt). In bays and lagoons and the upper reaches of estuaries, the salinity might be diluted far below normal marine salinity to only a few parts per thousand or even down to that of fresh water (0‰). Oxygen levels can also become quite low, especially in warm weather when the waters are only able to hold less dissolved oxygen. The sediments of many marginal marine environments consist of terrigenous gravels, sands, and muds. The waters are often quite turbid as a result of the input of suspended mud by streams and rivers and its resuspension by waves and currents.

One of the most prominent features of coastal settings are river deltas, so named because their appearance sometimes resembles the shape of the Greek letter "Δ." Deltas form when rivers deliver so much sediment to the coast that the coast builds out by the process of **progradation**. As a river and its sediment flow into the sea, the main channel of the river branches into **distributaries** that tend to be enriched in sand (**Figure 4.12**). Between the distributary channels is found the **floodplain**. **Oxbow lakes**, which are the meanders of former river channels that have been cut off by the river channel, are common on the floodplain. The sediments of the floodplain and oxbow lakes are frequently rich in organic carbon because of the presence of swamps and marshes which contribute dead organic matter to the sediment (Figure 4.12).

Depending on the relative importance of sediment input and waves and currents, deltas are broadly divisible into three different types: river dominated, wave dominated, and tide dominated (Figure 4.12). In river-dominated settings so much sediment is delivered to the delta that deposition dominates, and the delta builds out as irregular lobes. If

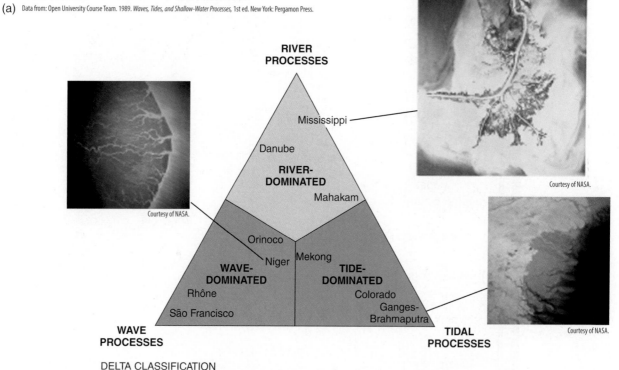

(a) Data from: Open University Course Team. 1989. *Waves, Tides, and Shallow-Water Processes*, 1st ed. New York: Pergamon Press.

DELTA CLASSIFICATION
(b) Data from: Open University Course Team. 1989. *Waves, Tides, and Shallow-Water Processes*, 1st ed. New York: Pergamon Press.

FIGURE 4.12 Deltaic environments and delta types. **(a)** As the river and its sediment flow into the sea, the main channel of the river branches into distributaries that tend to be enriched in sand. In between the distributary channels are located swamps and marshes of the river's floodplain and former river meanders, which have been cut off to produce oxbow lakes. **(b)** Types of deltas. In river-dominated settings, such as the Mississipi River shown here, so much sediment arrives that the deltas build out as irregular lobes, such as the "Bird's Foot Delta" of the Mississippi River. In wave-dominated settings like that of the Niger (and also Nile River of Egypt), there is less sediment input. Consequently, there is a greater chance for reworking of sediments by waves and currents to produce sand bodies that run more or less parallel to the coast. In tide-dominated deltas such as the Ganges-Brahmaputra of India, sediments are distributed more perpendicularly to the coast by tidal currents.

sufficient sediment is deposited on the delta, the shelf where the delta is prograding becomes so filled with sediment that the delta shifts position and can even flip-flop back and forth through time. This process is called **delta lobe-switching**. Lobe-switching is exemplified by the Mississippi River, which has shifted its position a number of times over the past 5,000 to 6,000 years (**Figure 4.13**). Today, the Mississippi delivers an average of 436,000 tons of sediment *per day* to the Gulf of Mexico! However, the present Mississippi's "Bird's Foot Delta" (so named because of its shape) results from human diversion efforts; without these, the Mississippi would flow into the Atchafalaya Bay to the west.

Wave-dominated deltas are exemplified by the Nile River of Egypt. However, unlike river-dominated deltas, wave-dominated deltas have the characteristic delta shape because there is less sediment input. Consequently, the sands and muds brought to the sea are more likely to be reworked, winnowed, and redistributed along the shore by waves and currents. The sediments tend to be sorted and redistributed relatively parallel to the coast by waves and currents. In **tide-dominated deltas**, tidal currents, which can sometimes be quite strong, tend to redistribute sediments in bodies that are oriented more perpendicular to the coast.

4.6.2 Coral reefs

Among the most spectacular of all ecosystems are **coral reefs** (refer to this chapter's frontispiece). Along with tropical rain forests, coral reefs are among the most diverse biologic communities of the world (Table 4.2). Coral reefs thrive in shallow, clear, warm tropical waters of normal marine salinity where there is little if any terrigenous influx. In fact, the reefs themselves are among the prime producers of carbonate sediments in such environments.

The word "**reef**" is actually a nautical term for a barrier to ships. Coral reefs are biogenic barriers made by corals, most of which live in massive colonies. Taken together, the countless numbers of corals secrete enormous amounts of $CaCO_3$, the primary component of limestone, which comprises the reef framework (**Figure 4.14**). The voids in the framework are filled by biogenic sediment produced by other calcium carbonate-secreting organisms such as algae, and the frame and loose sediments are cemented together by calcareous cement.

Depending on water clarity, corals can grow down to only about 50 to 75 meters of water depth within the photic zone because light is required by the corals for **symbiotic algae**, which live in the tissues of the coral. The algae

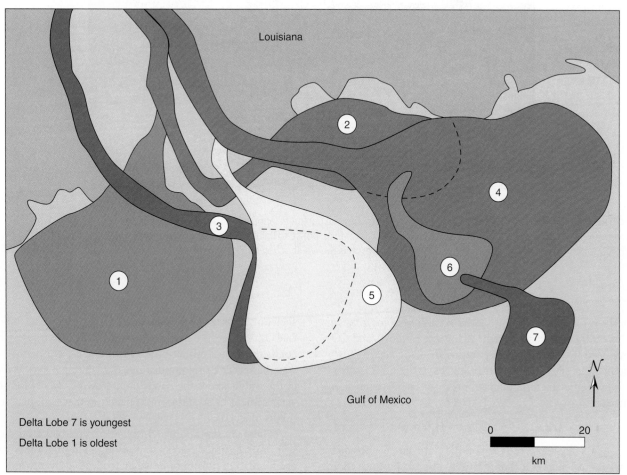

Delta Lobe 7 is youngest

Delta Lobe 1 is oldest

0   20

km

Data from: Walker, H. J., Coleman, J. M., Roberts, H. H., and Tye, R. S. 1987. Wetland loss in Louisiana. *Geografiska Annaler. Series A, 69*(1), 189–200.

FIGURE 4.13 Lobe-switching of the Mississippi Delta during the last few thousand years. If the Mississippi River had not been diverted, it would flow into the Atchafalaya Bay to the west.

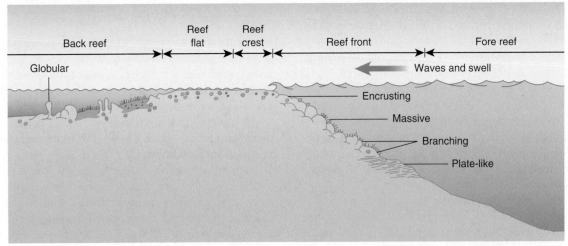

Data from: James, N. P. 1983. *Coral Reefs*. IN: Scholle, P. A., Bebout, D. G., and Moore, C. H. (eds.), Carbonate Depositional Environments. *American Association of Petroleum Geologists Memoir, 33*. Tulsa, OK. Figure 14 (p. 354).

(a)

(b)

FIGURE 4.14 Coral reefs. **(a)** Cross-section of a coral reef showing back-reef lagoon (with small patch reefs), reef crest, and fore reef environments with different types of coral colonies which form in response to light intensity and wave energy. **(b)** Modern coral reef with actively growing coral. The water is almost crystal clear, which is the result of low nutrient levels and small populations of plankton.

produce food and oxygen for the coral by photosynthesis, whereas the algae use nutrient-rich waste products and CO_2 produced by the corals. The CO_2 used by the algae (shown on the right of the equation that follows) is initially dissolved in seawater in the form of bicarbonate, or HCO_{3^-} (shown on the left):

$$Ca^{2+} + 2HCO_{3^-} \rightarrow CaCO_3 \downarrow + H_2O + CO_2 \uparrow \qquad (3.1)$$

As the algae use the CO_2 on the right side of the equation, the calcium (Ca^{2+}) and HCO_{3^-} react to produce more CO_2 to replace that used by the algae. In the process $CaCO_3$ is precipitated (as indicated by the arrow pointing downward) in the tissues of the coral to produce its limestone skeleton. This is why corals are able to secrete so much $CaCO_3$. Symbiosis between the corals and algae is also why corals thrive in nutrient-poor environments. Nutrient-poor waters prevent dense populations of planktonic algae living in the waters above the corals from growing; otherwise, the waters might be so turbid from plankton that light penetration is blocked.

There are several different types of reefs based on their size, shape, and distance from land. **Fringing reefs** can occur adjacent to land or be separated from it by a relatively shallow **back reef** lagoon that is one to a few kilometers in width (Figure 4.14). Small **patch reefs** composed of corals and other organisms can occur in the back-reef lagoon.

© tororo reaction/Shutterstock, Inc.

(a)

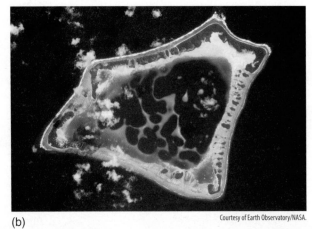

Courtesy of Earth Observatory/NASA.

(b)

FIGURE 4.15 **(a)** Australia's Great Barrier Reef. A fringing reef resembles a barrier reef except that it is located closer to shore and consequently has a much smaller lagoon. **(b)** Atoll in the Pacific Ocean viewed from space. The brown and green areas represent vegetated islands which have formed around the central bluish lagoon formed as the volcano sank beneath the surface of the ocean.

The organisms living on the patch reefs tend to be somewhat more tolerant of the more turbid and sometimes warmer waters of the back reef. The most actively growing part of a fringing reef is the reef crest, where corals usually grow into the surf to obtain oxygen and get rid of waste products (Figure 4.14). Barrier reefs are similar to fringing reefs but are located much farther from land (**Figure 4.15**). The best example is the Great Barrier Reef of Australia, although another smaller one is located off the coast of Belize in the Caribbean.

Atolls are ring-like reefs with a deep central lagoon (Figure 4.15). Based on his observations while serving as the naturalist on the voyage of the *H.M.S. Beagle* in the 1830s (see Chapter 4), Charles Darwin first proposed a hypothesis for the formation of atolls (**Figure 4.16**). He observed that some atolls surrounded volcanoes in different stages of submergence (as indicated by lagoons of different sizes), whereas other atolls had only deep lagoons with no volcanoes at all. Darwin inferred that the reefs initially developed on the shores of volcanoes and that the volcanoes eventually sank beneath the ocean as the reefs continued to grow upward. Although Darwin's hypothesis was criticized and alternative hypotheses were offered by other investigators, Darwin's hypothesis was eventually corroborated by drilling in the 20th century.

As we will see in later chapters, corals are only one of a number of diverse groups that formed reefs or reef-like structures during Earth's history. In all cases, though, ancient reefs seemed to thrive in warm, relatively shallow, and clear waters.

4.6.3 Continental shelves, continental slopes, and the abyss

Several types of environments exist beyond the coast (**Figure 4.17**). The **neritic zone** consists of the water above the **continental shelf**. Continental shelves typically grade outward very gradually from the coasts to an average water depth of about 200 meters at the continental **shelf break**. In fact, the slope of most continental shelves is so gradual that if sea level were to fall and someone were to be transported onto a shelf out of sight of dry land, they would need a compass to orient themselves and return home. The width of continental shelves also varies greatly. The widest continental shelves tend to occur along passive (tectonically inactive) continental margins like those along the east coast of North America. In other cases, such as along the coast of California, there is almost no continental shelf but instead a series of deep-water basins (**Figure 4.18**).

Beyond the continental shelf break lies the **continental slope** (Figure 4.17), the waters above which are typically referred to as the **bathyal zone**. Although the continental slope appears quite steep in diagrams, its inclination is only a few degrees greater than that of the shelf (the slope is exaggerated in textbooks to make the continental slope fit the page). Beyond the lower reaches of the continental slope are large areas of the ocean bottom called **abyssal plains**, which mostly lie below about 4,000 meters water depth (Figure 4.17). Depending on the location, the abyssal plain might grade into the **hadal environment** of seafloor trenches, which are characterized by water depths greater than about 5,000 meters.

All bottom environments of the marine realm are referred to collectively as **benthic** environments and the creatures that live on the bottom are referred to collectively as the **benthos**. The corresponding waters above bottom in these environments are referred to as the **pelagic** environment (Figure 4.17). The uppermost portion of the pelagic zone that is most strongly penetrated by sunlight is called the **photic zone**, which normally extends down to about 200 meters. Here are found many species of plankton, which consist of floating plants and animals ranging in size from

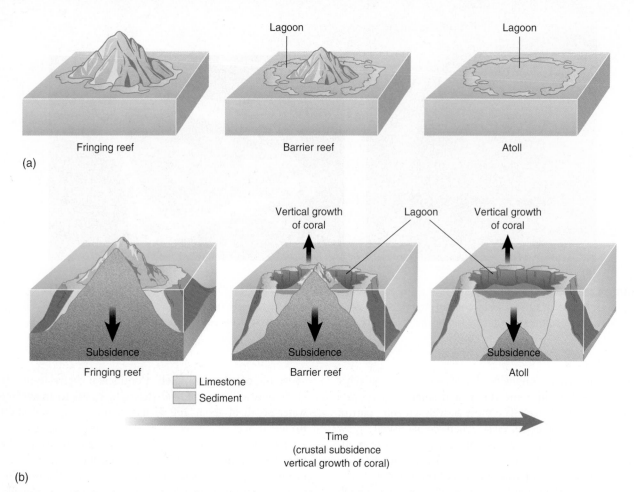

(a)

(b)

FIGURE 4.16 The development of an atoll. Note how fringing and barrier type reefs can form and disappear as the atoll develops.

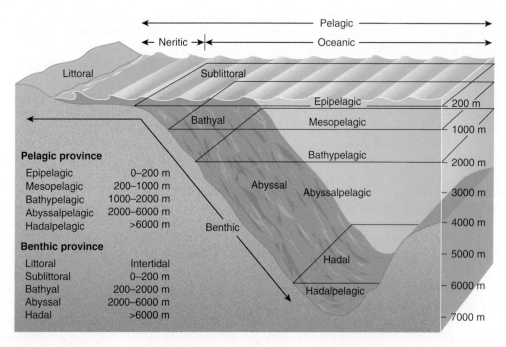

FIGURE 4.17 Environments along a passive continental margin, such as the east coast of North America.

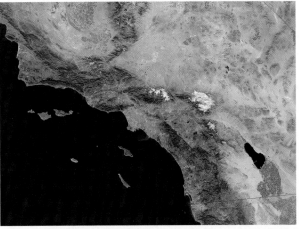

Courtesy of Jacques Descloitres, MODIS Rapid Response Team, NASA/GSFC. Accessed on October 25, 2011, at http://visibleearth.nasa.gov/.

(a)

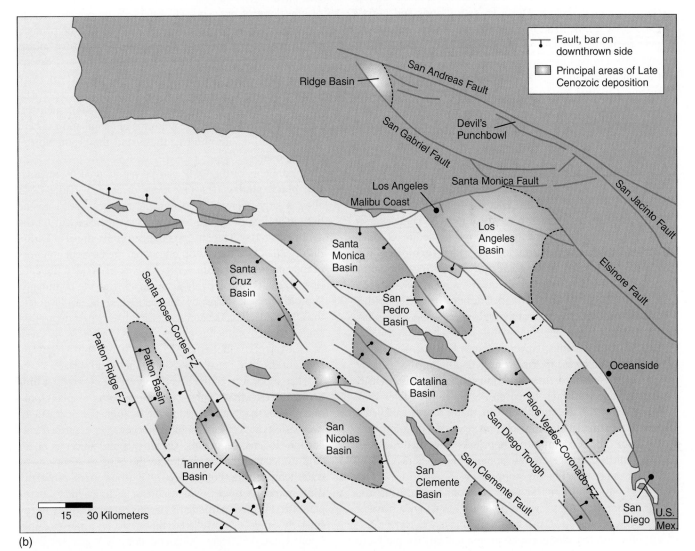

(b)

FIGURE 4.18 Aerial view of the continental margin off southern California. Note the lack of a typical continental shelf and slope. Deep-water basins formed by faulting dominate instead while uplifted portions of crust adjacent to the basins occasionally penetrate the water surface to form islands. The basins are often blocked off from extensive ocean circulation. Such basins are said to be "silled" and contain low-oxygen water because of the poor water circulation. Other examples of silled basins include Norwegian fjords and the ancient Permian basin of what is now west Texas (see Chapter 12).

bacteria to unicellular **phytoplankton** and **zooplankton** and **nekton** (swimming species). Much of the pelagic zone overlying the continental shelf lies within the photic zone. However, most plankton cannot withstand the wide ranges of water temperature and salinity of the inner-to-middle shelf. It is only nearer the shelf edge and beyond that more normal marine conditions, with relatively constant temperatures and salinity, are encountered; here, plankton typically flourish and become quite diverse.

Enormous thicknesses of sediment, sometimes several thousands of meters or more, can be deposited on the shelves and slopes. These sediments were mostly eroded from the continents and deposited at their margins as the continents grew in size and often consist of terrigenous sands and muds. These sediments are sorted and rounded as ocean waves begin to feel bottom on the shallow parts of the continental shelves. However, some sands and gravels exposed at the surface of the continental shelf can be left over from previous falls of sea level caused by the advance and retreat of past glaciers. Extensive coring of sediments on the continental shelves has documented that the complex of marginal marine environments migrates landward and seaward in response to sea-level change. These coarser **relict sediments** therefore represent shallow-shelf sediments that formed when marginal marine environments were located much farther out on the shelf.

Sediments on the continental slope tend to be more fine-grained than those on the shelf, often consisting of muds deposited from suspension. Like the shelf, however, slope sediments can also consist of sands or gravels. These coarser sediments were delivered to the slope by sediment-laden **turbidity currents** to form deposits called **turbidites**. Turbidity currents can be triggered by volcanism, earthquakes, or where river deltas deposit so much sediment that the sediment pile becomes unstable and slides downslope. Turbidity currents can reach remarkably high speeds. On November 18, 1929, for example, transatlantic cables lying on the seafloor off Grand Banks, Newfoundland, broke in succession from the shallowest to the greatest depths on the seafloor. It was later realized the cables must have been broken by a submarine landslide. Based on the times the cables were broken and the distances between them, it was calculated that this particular turbidity current reached 100 kilometers per hour (about 65 miles per hour)! As it reached the deep seafloor, the current slowed to about 24 kilometers per hour (15 miles per hour) before stopping.

Turbidity currents were very prominent during the Pleistocene "Ice Ages", when polar ice caps took up water from the oceans, lowering sea level. This is when river deltas now located along coasts prograded farther out onto continental shelves and deposited large quantities of sands and muds near the shelf edge. Because these deposits were water saturated, they became unstable and frequently flowed downslope. Consequently, continental slopes often consist of huge quantities of turbidite-derived sediment in fan-like structures called submarine fans. Many **submarine canyons (fans)** that cut through the shelf edge and onto to the continental slope are also thought to have been formed by turbidity currents or submarine slumps. Like river deltas, the fans typically consist of overlapping lobes that have "switched" through time.

Normally, a portion of the pelagic zone that impinges on the upper continental slope is occupied by an **oxygen minimum zone (OMZ)** (**Figure 4.19**). The top of the OMZ in modern oceans is normally several hundred meters below the ocean's surface and can lie at depths of 1,000 meters or more. In modern oceans, the OMZ results from the decay of dead organic matter produced by photosynthesis in the photic zone. As the organic matter settles through the water column most of it decays, using up much of the available oxygen in the water column and making the sediments and water above them **anoxic**.

The degree of oxygen loss in the OMZ depends on several factors. The rate of ocean circulation affects the rate of upwelling of nutrient-rich waters of the oceans back to the surface and the photic zone and the rate of primary productivity of the plankton. Productivity in turn affects the amount of dead organic matter that settles and decays through the water column after photosynthesis. An OMZ also typically forms along the continental slope in coastal settings (Figure 4.19). If large deltas are present, such settings can be subject to large amounts of runoff and nutrient input that stimulate photosynthesis and the production of organic matter.

Sediments of the OMZ can be represented by organic-rich **black shales**. The low levels of oxygen in the OMZ allow a larger percentage of organic matter to settle to the bottom and be preserved. Ancient black shales occur on the continents today. This suggests that at certain times in Earth's history, the OMZ moved up onto what are now the interiors of the continents (Figure 4.19). If sea level flooded the continents, the OMZ probably also spread over the continents, depositing black shales; by contrast, when the sea level fell, the OMZ moved off the continents and black shale deposition shifted to the continental slopes. This is quite possible given that the top of the modern OMZ tends to maintain a relatively constant distance between itself and the ocean surface.

Large portions of the ocean bottom are covered by **biogenic oozes**, which consist of the remains of planktonic microfossils that have settled to the bottom. The oozes are of two basic types, calcareous and siliceous, as described earlier for biogenic sediments. Calcareous and siliceous oozes tend to occur in bands parallel to the equator according to water temperature (**Figure 4.20**). Siliceous oozes, especially **diatomaceous oozes**, which consist of unicellular phytoplankton, are often particularly prominent where deep, nutrient-rich waters are upwelled to the surface. A prime example is off Antarctica. Here, Antarctic Bottom Water (AABW), which is part of the oceanic conveyor (see Chapter 3), upwells nutrient-rich waters, stimulating photosynthesis by diatoms.

At ocean depths greater than about 4,000 meters, calcareous oozes are often absent. At these depths sufficient CO_2 dissolves into the water because of the pressure of the overlying water This chemical behavior is very important because most organisms can tolerate only a limited range

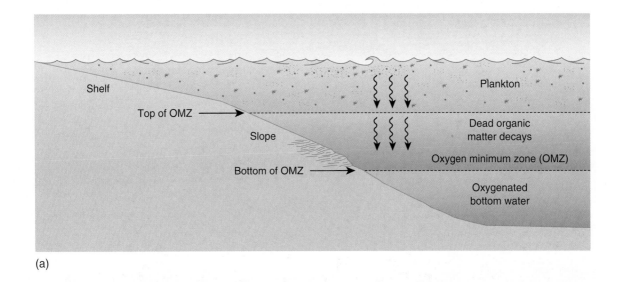

(a)

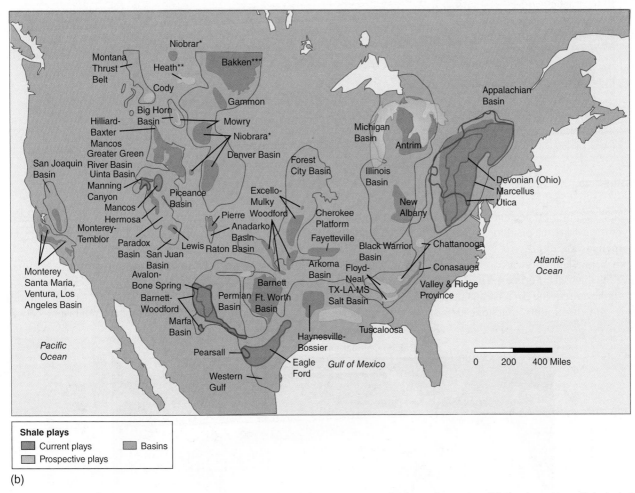

(b)

FIGURE 4.19 **(a)** The modern Oxygen Minimum Zone (OMZ) and the processes involved in its formation. **(b)** The deposition of black shales on the continents occurred during times of high sea level, allowing the Oxygen Minimum Zone to move onto the continents.

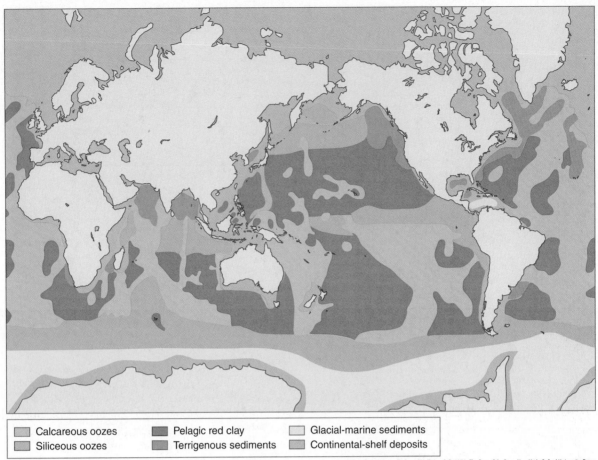

Calcareous oozes

Siliceous oozes

Pelagic red clay

Terrigenous sediments

Glacial-marine sediments

Continental-shelf deposits

Data from: Heezen, B. C., and Hollister, C. D. 1971. *The Face of the Deep*. New York: Oxford University Press.

FIGURE 4.20 The distribution of deep-sea sediments in ocean basins.

of acidity or, in more technical language, **pH**. The acid dissolves the calcareous plankton as it settles below the **calcite compensation depth** (CCD), neutralizing the acid. The depth below which calcareous oozes have completely dissolved is the CCD (**Figure 4.21**). Thus, the CCD represents an equilibrium in the deep oceans between limestone deposition and dissolution; limestone deposition occurs above

the CCD but dissolution occurs below it. Limestone deposition can have a similar effect on the behavior of the CCD. When enough $CaCO_3$ is deposited in the deep oceans (by calcareous ooze), the CCD deepens; however, when insufficient $CaCO_3$ is deposited in the deep ocean, the CCD is "starved," so to speak, of $CaCO_3$ and the CCD shallows to dissolve calcareous ooze to maintain equilibrium between

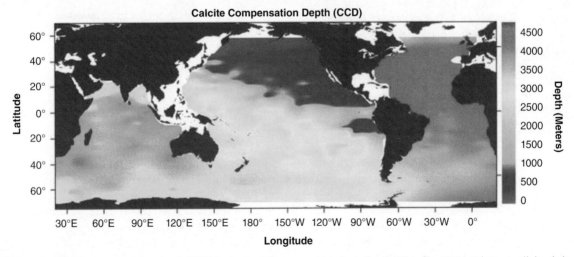

FIGURE 4.21 The calcite compensation depth (CCD) in the world's oceans. Note how the depth to the CCD tends to parallel red clay deposition in Figure 4.20. Red clays in turn tend to correspond to oceanic gyres (Chapter 3), which are regions of very low primary productivity and calcareous ooze deposition. This causes the CCD to shallow, as discussed in the text.

limestone deposition and dissolution. Below the CCD most of what remains is insoluble mud called **red clay** that has settled from suspension (Figure 4.20); the reddish color of red clay is due to the oxidation of iron. Because they do not dissolve in the presence of CO_2, siliceous oozes can be found below the CCD.

4.7 Sedimentary Structures

G We can use various features—called sedimentary structures—as indicators of the processes and environmental conditions involved in the formation of sedimentary rocks. **Mud cracks** form in mud-rich sediments as they dry out, producing a distinctive hexagonal pattern that resembles a tiled floor (**Figure 4.22**). The presence of mud cracks is indicative of shallow water environments, like those along a lake or lagoon or temporary pools of water that form after a desert rainstorm.

We can use other features to infer the directions of the predominant currents—or paleocurrents—responsible for transporting the sediment to the site of deposition. **Ripple marks** form in response to currents of wind or water (Figure 4.22). Asymmetric ripple marks form in sediments exposed to currents coming predominantly from one direction, whereas symmetric ripple marks are generally called *wave ripples* or *oscillation ripples* and are typical of shoreline settings. Wind can also produce large sand dunes, which are really large-scale ripple marks.

Both ripple marks and sand dunes can typically exhibit **cross-bedding** in response to fluid flow (water or wind; Figure 4.22). Water or wind pick up sediment grains from the windward side of the ripple or dune and deposit the grains on the leeward side (Figure 4.22). As this happens, sets of inclined sedimentary layers form on the leeward side. When the fluid changes direction, the sets of cross-beds come to lie at angles to each other. In the geologic record, deserts are characterized by large sets of cross-bedded sediments, which reflect the erosion, transport, and deposition of sand grains by wind. By contrast, the sediments of cross-beds deposited by water tend to be smaller, coarser, and less well sorted than those of sand dunes because wind can only move finer particles (Figure 4.22).

Two other types of structures are indicative of the rapid deposition of sediment-laden currents. **Graded bedding** occurs when sediment grains in a bed become progressively finer upward (Figure 4.22). Graded bedding can occur when flood or storm waters recede and the current energy decreases. Graded bedding is especially common in turbidites. You can recognize each turbidite by its graded bedding. As the turbidity current slows and stops, sediment grains "fine upward." The fining upward sequence of sediments reflects the same processes as those involved in sorting; gravels and sands are less easily transported and so settle out of suspension before muds, which are more easily transported and therefore settle much more slowly. Thus, a complete turbidite consists of sands and gravels at its base, grading upward into mud at its top. However, because more than one turbidite often occurs in a sedimentary sequence, the upper portions of turbidites can be scoured away by later turbidity currents.

4.8 Fossils

4.8.1 Early processes of fossilization

H Fossils are the remains or traces of preexisting organisms. Taphonomy ("laws of preservation") is the science that studies the formation and preservation of fossils—or fossilization. Fossils are normally represented by hard parts such as shells, bones, or teeth. Fossils also include trace fossils, or **ichnofossils**, such as tracks (including those of dinosaurs), trails, burrows, and markings left by plant roots (**Figure 4.23**).

Normally, three basic prerequisites for fossilization are necessary: (1) the dead organism must have hard parts such as shells or bones, (2) the hard parts must be buried rapidly, and (3) the hard parts must be buried in fine-grained sediment. Because coarse-grained sediments are deposited in wave or current-swept environments, these settings are typically less favorable for fossil preservation because of abrasion and breakage. Much of the fossil record has formed in marine environments because sediments are more likely to be deposited and buried in marine environments compared with terrestrial environments, which are more likely to be eroded.

Because biogenic hard parts typically consist of calcium carbonate ($CaCO_3$), we consider what happens to calcareous hard parts, such as clam shells, during the early processes of fossilization. However, the same basic processes occur with skeletons and bones. Initially, the shells lie close to the sediment–water interface. If the shells consist of multiple parts, upon their death the shells and skeletons begin to disaggregate, or **disarticulate**. A clam, for example, consists of two halves, or valves, held together by soft parts such as muscles and special ligaments. Disarticulation occurs because of the decay of the soft parts and because of scavengers seeking food. If sufficiently strong, waves and currents in the marine realm can also cause disarticulation, abrasion, and transport of hard parts to sites where the organisms did not live; on land, streams, and rivers carry out the same process with skeletons.

In the marine realm burrowing, or **bioturbation**, by worms, crabs, and other organisms also helps break down

(a) Courtesy of National Park Service.

(b) © Linda Armstrong/Shutterstock, Inc.

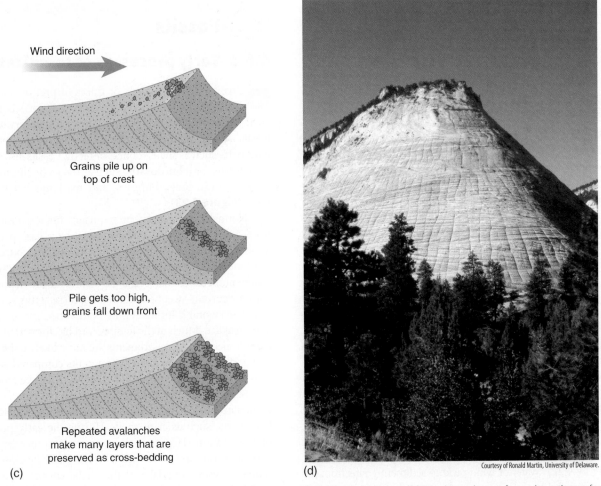

Wind direction

Grains pile up on
top of crest

Pile gets too high,
grains fall down front

Repeated avalanches
make many layers that are
preserved as cross-bedding

(c)

(d) Courtesy of Ronald Martin, University of Delaware.

FIGURE 4.22 Sedimentary structures. **(a)** Mud cracks, which form in muds during dessication. **(b)** Ripple marks can form along the surface of sediment. **(c)** Formation of cross-bedding. Cross-beds can form by the action of either wind or water, which pick up sediment grains from the windward or upstream side of the structure and deposit them on the leeward side. As this happens, sets of inclined layers of sediment form. When the fluid changes direction, the sets of cross-beds come to lie at angles to each other. **(d)** Cross-bedding in the Navajo Sandstone of southern Utah. These beds were formed in an ancient desert about 150–200 million years ago.

(e)

Courtesy of Ronald Martin, University of Delaware.

FIGURE 4.22 **(e)** A turbidite forms by submarine slumping. The beds typically exhibit "fining upward," in which coarse sands and gravels (yellow arrows) of the current give way to finer muds (red arrows) of the current, which settle out of suspension on top of the coarser sediment. Graded beds can stack on top of one another—like those shown here—as a result of repeated turbidity currents but very often the upper portions of the graded beds are eroded by succeeding turbidity flows. Note the uneven, wavy-looking bases of the sand beds. Sedimentary structures such as those resulting from mud flow are sometimes evident in turbidites and also indicate the turbulent nature of deposition.

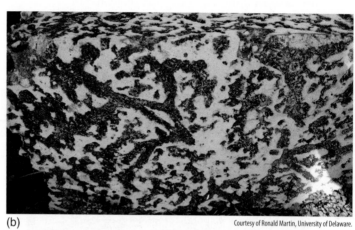

(b)

Courtesy of Ronald Martin, University of Delaware.

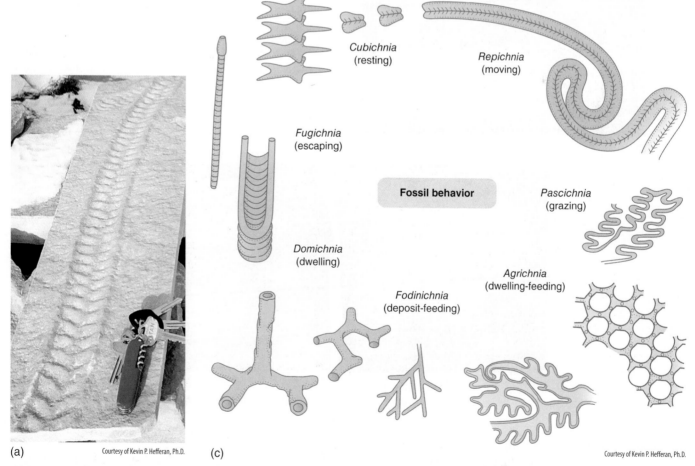

(a)

Courtesy of Kevin P. Hefferan, Ph.D.

(c)

Courtesy of Kevin P. Hefferan, Ph.D.

FIGURE 4.23 Ichnofossils reflect the behavior of organisms and can be quite complex. **(a)** Trilobite tracks. **(b)** Shrimp burrows (dark) in the whitish Key Largo Limestone (Pleistocene) of the Florida Keys. **(c)** Further examples of ichnofossil behavior.

hard parts. Burrows can extend downward from the sediment surface up to about a meter but are usually most abundant near the sediment–water interface. As burrowing occurs, oxygenated seawater is pumped downward into sediment as the burrowers search for food. Some oxygen attacks dead organic matter in the sediment, causing it to decay and release carbon dioxide. The carbon dioxide dissolves into water to produce carbonic acid that dissolves the shells. If the shells manage to be buried beneath this layer, they are more likely to be preserved. Bioturbation also tends to destroy any fine layering in the sediments that formed as the sediments were deposited, often destroying the fine-scale sedimentary record.

This is why anoxic basins are so important. The basins are anoxic because they are cut off from the normal flow of ocean currents, which would otherwise supply oxygen. Such basins are found, for example, in Norwegian fjords, off the coasts of California and Venezuela, and in the Gulf of Mexico. Because burrowing organisms require oxygen, these creatures are absent from such basins, which can then preserve fine-scale layers called **varves** that can be used in paleoclimate studies.

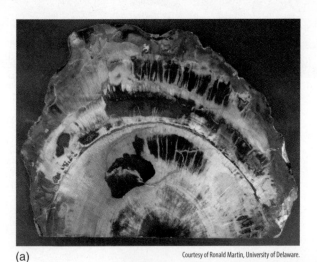

(a)

Courtesy of Ronald Martin, University of Delaware.

4.8.2 Modes of preservation

Even if the hard parts survive the host of physical and chemical reactions that occur in the surface layers of sediment, the hard parts are typically subject to other processes after burial. Many of these processes grade into one another. For this reason, they are sometimes referred to collectively as **petrifaction** ("turning to stone"; **Figure 4.24**). In **permineralization** mineral material—often some form of silica (SiO_2) dissolved in groundwater—fills the voids and pores of hard parts, usually bone and plant material; permineralization sometimes preserves the microscopic structure of fossils in striking detail (Figure 4.24). In other cases, the original hard part material is dissolved and simultaneously replaced by another mineral, often pyrite (iron sulfide, or FeS_2 commonly referred to as "Fool's Gold" in its crystal form). Both dissolved iron and sulfides are typically abundant in the pore waters of terrigenous sediments and precipitate when sufficiently concentrated to produce pyrite. This particular process is called **replacement**, and this particular type of replacement is called **pyritization**.

Often, the original hard part material undergoes the process of **recrystallization**, during which the atomic structure of the mineral alters to a new form. Many organisms that secrete calcareous shells ($CaCO_3$) secrete the shell in the form of the mineral aragonite. Aragonite is unstable, however, and eventually recrystallizes to the more stable mineral calcite. The basic features of the hard parts are typically preserved by recrystallization, whereas the fine anatomic features are lost.

More drastic changes to the hard parts can also occur. If the shells are completely entombed and then dissolve away, they leave a cavity known as an **external mold**. Although the hard parts are completely absent, the external mold

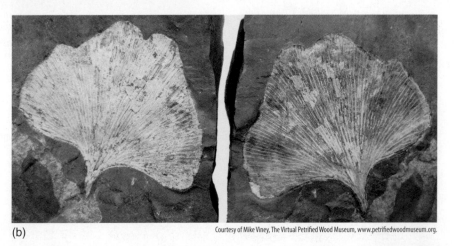

(b)

Courtesy of Mike Viney, The Virtual Petrified Wood Museum, www.petrifiedwoodmuseum.org.

FIGURE 4.24 Modes of preservation. **(a)** Permineralization of plant matter, in which silica (SiO_2) has filled the voids and pores of plant material, can result in exquisite preservation. This particular fossil is a polished section of a tree stump of the Middle Triassic genus *Araucarixylon* (related to the modern "Monkey Puzzle Tree") found in the Petrified Forest National Park, northern Arizona. Note the preservation of the tree rings. **(b)** Plant fossils are often preserved as impressions and corresponding compressions, which consist of a residue of carbon that exhibits the general outline and some surface features of the original organism. Impressions and compressions are often preserved together in sedimentary nodules. The nodules can then be split open, exposing both impression and compression. This particular fossil is an ancient *Ginkgo* leaf.

preserves the exterior features of the hard parts. If the cavity later fills in and the surrounding sediment is then eroded away, a **cast** results. In the case of plant roots, the cast forms a lining around the burrow. Usually, the material forming a cast is different from the original fossil and that of the surrounding sediment. A cast therefore resembles the original shape of the fossil and preserves the external features, but there is no preservation of internal features. The external surfaces of casts can sometimes be quite detailed anatomically, especially if they consist of SiO_2. An **internal mold** results when the original shell fills with sediment and is lithified, often by calcite or pyrite, and the original shell then dissolves away. Internal molds also do not preserve internal features.

Despite their general lack of hard parts, soft-bodied organisms or the nonmineralized portions of various organisms, such as plants, jellyfish, and insects, are sometimes preserved. Plant fossils and insects are often preserved as **impressions**, which often consist of a residue of carbon that exhibits the general outline and some surface features of the original organism (Figure 4.24). Impressions of plant fossils often have a corresponding **compression** that, along with the impression, is preserved together in sedimentary nodules. The nodules can then be split open, exposing both impression and compression.

Although hard parts are normally required for fossilization, soft parts or numerous articulated hard parts (fish skeletons, for example) are sometimes preserved by very rapid burial by turbidity currents or storms. Rapid burial removes the remains from burrowing and associated destructive chemical reactions of the shallow surface layers. For example, carbon films can be derived from soft-bodied animals that have no hard parts and would otherwise completely decay; the rapid burial of the soft parts inhibits scavenging and complete decay of the carcasses. In other cases, the depositional setting facilitates certain chemical reactions that preserve soft materials or the soft parts are encased in virtually inert substances such as **amber**, which is fossilized tree resin (**Figure 4.25**). In still other cases, spectacular fossils, especially of vertebrates, have been preserved in tar, refrigeration in ice (such as mammoths), or **mummification**, in which dry environments inhibit organic decay.

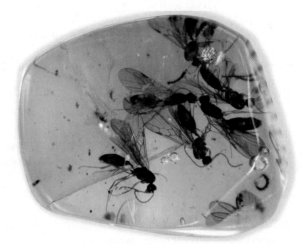
© Ismael Montero Verdu/Shutterstock, Inc.

FIGURE 4.25 Insects preserved in amber (fossilized tree sap).

These sorts of fossil deposits, which are not normally preserved, are called **Lagerstätten** (singular, Lagerstätte). The term, loosely translated from the German, means "mother lode," as in mining. Although once viewed as curiosities, Lagerstätten have been reexamined intensely in recent years because they provide a perspective on Earth's ancient biodiversity and evolution that is normally not preserved. We will encounter a number of Lagerstätten as we study the history of Earth's systems.

CONCEPT AND REASONING CHECKS

1. What do each of the following sedimentary structures indicate about the original depositional environment? (a) mud cracks, (b) cross-bedding, (c) graded beds.
2. How does bioturbation affect (a) stratification and (b) the cycles of carbon and nutrients?
3. What is the difference between an external mold, an internal mold, and a cast?
4. What basic feature of Lagerstätten distinguishes them from all other types of fossil assemblages?

SUMMARY

- Modern sedimentary environments serve as analogs for the interpretation of ancient sedimentary ones based on the Principle of Uniformitarianism.
- Rocks exposed at Earth's surface interact with the atmosphere, hydrosphere, and biosphere and are weathered, eroded, and redeposited to produce sedimentary rocks. Sedimentary rocks are important because the sedimentary processes of weathering, erosion, and deposition are associated with important climatic feedbacks. Sedimentary rocks are also

important because they contain fossils, which record the appearance and evolution of life forms on Earth that have also been integrally involved in the evolution of Earth.
- Terrigenous (detrital) sediments and rocks consist of three main grain sizes: gravel, sand, and mud. These sediments correspond to three main types of sedimentary rocks: conglomerates, sandstones, and shales.
- The degree of compositional and textural maturity indicates the extent of weathering and transport of the

sediments forming a rock. The greater the maturity, the greater the extent of weathering, transport, and size sorting and the greater the amount of time likely required to produce the rock.

■ Other types of sedimentary rocks are biogenic. These types of rocks tend to accumulate where they are not diluted by terrigenous sediments. Most limestones consist of the calcareous ($CaCO_3$) hard parts of organisms. Deep-sea calcareous oozes (also limestones) consist of the remains of calcareous plankton. Siliceous (SiO_2) oozes consist of the remains of siliceous plankton and, upon burial, are transformed into chert.

■ Another category of sedimentary rocks is chemical rocks such as evaporites, which form in restricted basins flooded by the sea. Evaporite minerals include salt and gypsum, which turn to anhydrite upon burial. Limestones can be altered to dolomite by the brines that produce evaporites.

■ Continental shelves extend from land to approximately 200 meters water depth. Continental shelves are formed by huge thicknesses of sediment eroded from the continents and deposited at their margins, which made the continents grow outward through time.

■ Beyond the edge of the continental shelf is the continental slope, or bathyal zone. Beyond the lower reaches of the continental slope are large areas of the ocean bottom called abyssal plains. Depending on the location, the abyssal plain can grade into the hadal environments of seafloor trenches.

■ The environments of the inner continental shelf are referred to as marginal marine environments and are subdivided into the intertidal zone, salt marshes, mangrove swamps, estuaries, lagoons, bays, deltas, and coral reefs. Marginal marine environments tend to be stressful for most organisms because of highly variable temperature and salinity, and so the biotas of most marginal marine environments are normally not very diverse. With the exception of coral reefs, sediments are typically of terrigenous origin and consist of gravels, sands, and muds.

■ River deltas are of three types: river dominated, wave dominated, and tide dominated. River-dominated deltas can be associated with turbidity currents and submarine fans on the continental slope.

■ Among the most spectacular marginal marine environments are coral reefs that, along with tropical rain forests, are some the most diverse ecosystems of the world. Well-developed coral reefs are found in shallow, warm, nutrient-poor waters of relatively constant and normal marine salinity. Waters must also be relatively clear because of the corals' symbiotic relationship with photosynthetic algae, which allows reefs to secrete massive amounts of biogenic limestone.

■ Further from shore, nearer the edge of the continental shelf, temperatures and salinities remain relatively stable so that the diversity of the floating plankton and swimming nekton increases.

■ Distinct transition zones or layers occur in the water column with respect to oxygen concentration (oxygen minimum zone).

■ The sediments of outer shelf environments and deeper waters become more and more dominated by various types of calcareous and biogenic oozes, but in the deep sea below the calcite compensation depth, only red clay and siliceous oozes might remain.

■ Sedimentary rocks can display sedimentary structures indicative of their depositional environment: mud cracks (indicative of temporary bodies of water), ripple marks and cross-bedding (which can indicate the direction of water or wind currents), and graded bedding and sole marks (indicative of rapid deposition, such as by submarine turbidity currents).

■ Most sediments and fossils are preserved in marine environments. Preservation of fossils normally requires hard parts and rapid burial in fine-grained sediments. Otherwise, the hard parts might dissolve in the surface layers of sediment because of dissolution caused by bioturbation and organic decay that produce acids. If the hard parts survive and are buried, they can later be preserved as various types of molds or casts or they might be recrystallized, replaced, or petrified. In exceptional cases called Lagerstätten, soft parts are preserved.

KEY TERMS

abyssal plains	bathyal zone	black shales	cast
amber	bauxite	braided streams and rivers	cementation
anhydrite	benthic	breccias	chemical weathering
anoxic	benthos	brines	chert
arkoses	bioclasts	calcareous oozes	compaction
atolls	biogenic oozes	calcite compensation depth	compositional
back reef	bioturbation	carbonic acid	compression

conglomerates	halite	oöids	sand
continental shelf	humic acids	oölite	sandstones
continental slope	humus	oxbow lakes	sedimentary grains
coral reefs	ichnofossils	oxygen minimum zone	shales
cross-bedding	impressions	patch reefs	shelf break
delta lobe-switching	internal mold	pelagic	siliceous ooze
diatomaceous oozes	intertidal zone	permineralization	soil
disarticulate	karst	petrifaction	sorting
distributaries	karst topography	pH	speleothems
dolostones	Lagerstätten	photic zone	stalactites
estuaries	laterites	physical weathering	stalagmites
evaporites	limestones	phytoplankton	submarine canyons (fans)
external mold	lithification	plankton	symbiotic algae
floodplain	marginal marine environments	progradation	terrigenous
fossils	marine environments	pyritization	textural
freeze-thaw	maturity	recrystallization	tide-dominated deltas
fringing reefs	mud	recrystallized	travertine
graded bedding	mud cracks	red clay	turbidites
gravel	mudstones	reef	turbidity currents
graywackes	mummification	relict sediments	varves
gypsum	nekton	replacement	wave-dominated deltas
hadal environment	neritic zone	ripple marks	zooplankton

REVIEW QUESTIONS

1. How do physical and chemical weathering accelerate overall rates of weathering?
2. Draw and label a cross-section of a continental margin indicating all benthic and pelagic environments.
3. Why are marginal marine environments important?
4. Describe the symbiotic relationship between corals and algae.
5. How is anoxia produced?
6. Describe the action of the CCD as it affects acidity.
7. Sketch the distribution of different sediment types in the ocean basins, and explain their occurrence in terms of deep-ocean circulation, nutrient availability, and dissolution.
8. Why are Lagerstätten important?
9. Reason backward from the following kinds of sediments or rocks to the environments in which they originated. There might be more than one possibility for each, depending on the description. How does the presence of fossils help in making your paleoenvironmental interpretations?
   a. Shale
   b. Shale with planktonic foraminifera
   c. Shale with plant roots
   d. Shale with shallow-water marine clams
   e. Ripple-marked coarse sandstone
   f. Cross-bedded, fine-grained sandstone
   g. Coarse-grained sands with shallow-water clams, grading fining upward into muds with planktonic foraminifera
   h. Diatom ooze
   i. Mud with abundant peat, but no large tree trunks, branches, etc.
   j. Chalk (calcareous ooze)
   k. Calcareous, sandy muds containing abundant algal fragments, burrows, and root casts
   l. Black shale

1. Normally, fine-grained sediments such as muds should be found on the outer continental shelves, but sometimes coarse-grained sediments are found. Explain.

2. What is dominant paleocurrent direction(s) in symmetric versus asymmetric ripple marks?

3. Ripple marks and mud cracks indicate the tops of sedimentary layers. However, these two types of sedimentary structures are sometimes found to be upside down, especially in areas affected by mountain building. Explain.

4. What should happen to the CCD in response to the input of anthropogenic CO_2 into the atmosphere? How long do you believe it would take for the CCD to take up the CO_2? Do you believe the CCD would be important in regulating CO_2 levels on time scales of a few years? Decades? Centuries? Millennia? Why? (Hint: Recall the rates of modern ocean circulation discussed in Chapter 3.)

SOURCES AND FURTHER READING

Bromley, R. G. 1996. *Trace fossils: Biology, taphonomy, and applications.* London: Chapman and Hall.

Garrels, R. M., and Mackenzie, F. T. 1971. *Evolution of sedimentary rocks.* New York: W. W. Norton.

Martin, R. E. 1999. *Taphonomy: A process approach.* Cambridge, UK: Cambridge University Press.

Scholle, P. A., Bebout, D. G., and Moore, C. H. 1983. *Carbonate depositional environments.* Tulsa, OK: American Association of Petroleum Geologists.

Walker, R. G. (Ed.). 1986. *Facies models.* Toronto, Ontario, Canada: Geological Association of Canada.

Evolution and Extinction

MAJOR CONCEPTS AND QUESTIONS ADDRESSED IN THIS CHAPTER

A How did early theories of evolution differ from Darwin's?

B What sorts of evidence led Darwin to his theory of evolution?

C What are the basic premises of Darwinian evolution?

D How is genetic information stored and transmitted from parents to offspring?

E What produces genetic variation?

F What is the evidence for natural selection?

G How do new species originate?

H What is the fossil evidence for evolution?

I What are the patterns of speciation in the fossil record and how are they related to the biological theory of speciation?

J How do wholly new structures and higher taxa arise?

K How does extinction affect evolution?

L What are the major patterns of evolution during geologic time?

CHAPTER OUTLINE

Galápagos tortoises. The occurrence of different tortoises on different islands was among some of the evidence used by Darwin to infer his theory of evolution. Inset: Location of the Galápagos Islands.

© Frans Lanting Studio/Alamy Stock Photo.

5.1 Introduction

Charles Darwin (1809–1882) is rightfully credited with founding the modern theory of evolution because the basic foundation of the theory laid down by him has remained largely intact to the present. Darwin's theory supported, and was supported by, the notion of deep time implicated by Hutton and Lyell. It is important to remember, however, that *Darwin's theory also imparted secular change, or directionality, to the history of the Earth and its life, rather than the cyclic view of Lyell's:* strict (dynamic) equilibrium, with no net change. But, with Darwin's theory came the growing realization that not only has life evolved but so too have Earth's physical systems. Natural systems can exhibit directionality or secular change because they are open systems; that is, they can *evolve*. It is the flow of matter and energy through the biosphere that has moved the biosphere away from equilibrium (i.e., "death") and promoted its evolution. Darwin's theory is therefore not only fundamental to the study of the history and evolution of life but also strengthened the inkling that physical systems—and thus Earth—can evolve, as well.

Like so many theories, however, Darwin's was preceded by others. To fully comprehend the significance of the modern theory of evolution, we must examine these earlier theories. In doing so, we get another glimpse as to how scientists think and how theories are developed and modified, leading to new questions. As we will see, several of these questions stem from a better understanding of the fossil record: How do new species arise? How do new anatomic structures like wings or eyes and wholly new groups of organisms—like reptiles or mammals—arise? How do mass extinctions of the biosphere—well documented in the fossil record—create new evolutionary opportunities?

5.2 Early Theories of Evolution

Two of Darwin's predecessors, both French, stand out for their views on evolution: Jean Baptiste Pierre Antoine de Monet, Chevalier de Lamarck, or just Lamarck (1744–1829), and Baron Georges Cuvier (1762–1839) (**Figure 5.1**). Lamarck originally believed that certain groups of organisms—now called species—were fixed and did not evolve, but began to change his views as his studies progressed (**Box 5.1**). Lamarck based his theory of evolution on spontaneous generation which stated that living matter arises from nonliving matter. Lamarck based his acceptance of spontaneous generation on his observation that the simplest organisms had no obvious organs, so Lamarck reasoned they must have evolved directly from nonliving substances.

Lamarck also believed in the concept of a Chain of Being, which stated that all organisms could be arranged in a continuous hierarchy stretching from the simplest organisms all the way to humans at the top. The origins of the Chain of Being concept can be traced all the way back to Aristotle. According to this view, because only the simplest animals could evolve from nonliving matter, more complex

(a) Courtesy of National Library of Medicine.

(b) Courtesy of National Library of Medicine.

FIGURE 5.1 (a) Jean Baptiste Pierre Antoine de Monet, Chevalier de Lamarck (1744–1829). **(b)** Baron Georges Cuvier (1762–1839).

BOX 5.1 Evolution of Biologic Classification and the Species Concept

A **species** is a group of organisms that interbreed with one another *and* produce fertile offspring. The point of reproduction is to propagate the species. Species are usually recognized by physical features thought to reflect the identity of the species, such as anatomic traits, distinctive coloration, or behavior. Many species have "common" names used by laypeople. However, these terms can be confusing, especially if they are used to refer to different species. To avoid confusion among scientists, each species is given a scientific name consisting of two names: the genus and the species. For example, the scientific name of human beings is *Homo sapiens. Homo* is the genus and *sapiens* is the species. The scientific name for humans means "prescient man," referring to the mental ability of humans to foresee into the future and anticipate the consequences of their actions and those of nature. By international convention the scientific name is always underlined or set off from the surrounding text in some other manner, and the genus name is always capitalized, whereas the species name is not (species is used for both the singular and plural; there is no such word as "specie"). Each genus is represented by one or more species. Thus, when referring to a particular species, we cite both names—genus and species—not just the species name by itself, because there could well be another completely different genus represented by a species with the same species name.

The first naturalist to systemically catalog and describe animals and plants was the Swedish botanist, Carolus Linnaeus (1707–1778), who held that species were fixed. He eventually described about 4,200 species of animals and 7,700 species of plants. There are now about 1.4 million described species, which might represent only about one-tenth of all modern species. Many species have common names, but the same common name might be used in different areas for different species. To avoid confusion Linnaeus gave each species a Latinized description. However, use of the description was quite clumsy, and he decided to give each species a "nickname." This led to the establishment of **binomial nomenclature** in which each species is recognized by two names, as described earlier. Linnaeus' system is still used today. In fact, his monumental work, *Systema Naturae,* first

published in 1758, is taken as the starting point of binomial nomenclature, and any species described before this time are considered invalid. However, species originally described after 1758 can still be redescribed or placed in new groupings based on new data.

As evolutionary theory began to take hold, Linnaeus' system of biologic classification began to take on a new meaning: evolutionary relationships. A classification recognizes similarities or dissimilarities between objects, in this case species. The purpose of a biologic classification is to show evolutionary relationships, or **phylogeny**, between different groups of organisms, or taxa (taxon, singular). Biologic classification and phylogenetic relationships are hierarchies. These hierarchies consist of different levels, or categories—kingdom, phylum, class, order, family, genus, and species—each of which can be further subdivided or lumped together. The actual groups, or taxa (such as particular species), are placed into higher taxa corresponding to the different categories; thus, each taxon shares characteristics with the other members of the same taxon. The science of biologic classification is called **taxonomy**. A biologic classification is therefore like a series of boxes nested within progressively larger boxes. In biologic classification, these boxes are called categories with the categories becoming more and more inclusive. Taxa and categories can also be subdivided (for example, subclasses) or lumped together (for example, superfamilies). For example, *Homo sapiens* is classified as follows:

CATEGORY: TAXON

Kingdom: Animalia

Phylum: Chordata (animals with some form of backbone)

Class: Mammalia (mammals, which are warm-blooded, possess hair, and produce milk for their young)

Order: Primates (monkeys, apes, humans)

Superfamily: Hominoidea

Family: Hominidae

Genus: *Homo*

Species: *sapiens*

(continues)

Besides the Kingdom Animalia, several other kingdoms are recognized: Plantae, Fungi (mushrooms, etc., including microscopic fungi), Protista (single-celled plants and animals), and two kingdoms of bacteria, the Archaea and the Eubacteria.

The modern view of biologic classification and the concept of species have themselves evolved. Biologists recognize that within a species there is typically a broad range of variation in physical features. Sometimes, this range can be quite broad, such as that found among different breeds of dogs. Despite their tremendous differences in size and coloration, different breeds of dogs are thought to be descended from an ancestral wolf lineage. This lineage alone contained the DNA from which all different purebred lines of dogs have been bred. Obviously, this range of variation in physical traits can sometimes make it difficult to decide if we are studying separate species or variations within the same species. This was the basic problem confronted by early evolutionary biologists who began to describe different types of organisms. Lamarck saw the possibility for almost endless variation and the continued production of new forms from nonliving matter; nature was just too "fluid" in his view to have such distinct biologic boundaries that separated species. On the other hand, Cuvier had a very narrow notion of a species, in which a particular specimen was supposed to represent all members of that species; this kind of thinking originated with the Greek philosopher Plato, who viewed all living things as having an "ideal type."

Darwin also wrestled with this problem and eventually recognized that each species can vary within broad limits. In other words, most specimens resemble the "average" appearance of the species, but many specimens deviated from the species' average, sometimes quite strongly. Nevertheless, the existence of species was actually used as an argument *against* Darwinian and earlier theories of evolution. The reasoning was that because species are not observed to change into new species, species do not evolve. Even though Darwin used the results of animal breeding experiments to support the process of natural selection, new species were never produced. Some of Darwin's critics used this evidence against his theory, as well.

This problem is still encountered today in the form of "lumpers" and "splitters." Splitters tend to recognize more species than do lumpers because splitters tend to "split out" species based on traits that lumpers otherwise include within the range of variation of a particular species. This is nowhere more pronounced than in the recognition of new species of ancestral humans (see Chapter 15).

species must be descended from simpler ones in a progressive manner. Thus, in Lamarck's view, new, simple creatures evolved through separate acts of spontaneous generation and then moved up the Chain of Being along a kind of evolutionary "escalator." The dead residues fell back to the base of the escalator to be used again, in a cyclic manner, in the evolutionary process. A particular group's complexity was therefore a measure of its age: the more primitive a group, the more recently it had evolved. Because the Chain of Being emphasized that evolution was progressive, it also emphasized that evolution progressed toward more perfect forms, namely humans.

Finally, Lamarck believed that organisms were capable of evolving the organs they needed and could change their characteristics through an internal "striving" in response to new environmental conditions; organisms evolved new structures according to their "need" through use and disuse. This **theory of the inheritance of acquired characteristics** asserted, for example, that the short-necked ancestors of modern giraffes had lengthened their necks as members of each generation strove to reach the foliage higher in trees; the slightly longer neck of each generation was passed to the next during reproduction. In other words, evolution was **teleological**, meaning it was goal-oriented or purposeful, a view held by many other naturalists before and after Lamarck. But the theory of the inheritance of acquired characteristics begged the question as to how the characteristics are passed on to offspring. Consequently, Lamarck's theory was never widely accepted, and some scientists dismissed him as a crank.

One such scientist was Georges Cuvier. Cuvier developed a classification scheme based on his studies of comparative anatomy for which he was renowned (Box 5.1). By observing a single bone or tooth Cuvier could predict the rest of the animal's skeleton through his knowledge of anatomic relationships (what Cuvier called the "correlation of parts").

Based on anatomic relationships, Cuvier concluded that species and other taxa were so complex that they could only be fixed and unchangeable (Box 5.1). Instead of an escalator-like progression from simple to more complex creatures, different groups of organisms were viewed as being separate branches in a tree-like arrangement. One branch was no more advanced than another; branches were merely different because each was adapted to a different mode of life.

Through his studies of fossil forms, Cuvier also gradually realized a series of extinctions had occurred. Because the changes between fossil assemblages were abrupt, he concluded these extinctions were caused by movements of the land and sea. Cuvier was therefore lumped into the catastrophist camp by Lyell (see Chapter 1). However, contrary to what Lyell said, Cuvier did not explain the repopulation of Earth after an extinction as the work of a Creator, and he refused to equate the last extinction with the biblical deluge. Cuvier did not accept the evolution of species (or transmutation, as it was then called) after extinction but instead suggested that populations repopulated Earth from refuges after extinction.

Despite Cuvier's long-standing influence, however, other workers saw tantalizing patterns in the classification—or groupings—of different plants and animals groups, and the concept of transmutation started to make headway once again in the 1830s (Box 5.1). This was basically the state of evolutionary biology when Charles Darwin came onto the scene.

CONCEPT AND REASONING CHECKS

1. Was Darwin the first person to recognize a theory of evolution?
2. Why were species originally thought to indicate a lack of evolution?

5.3 Charles Darwin and the Beginnings of the Modern Theory of Evolution

To fully appreciate the perspective and power of the evolutionary viewpoint, we need to briefly examine how Darwin arrived at his theory. Charles Darwin (**Figure 5.2**) was the naturalist on the voyage of the *HMS Beagle* from 1831 to 1836. During the expedition Darwin visited the Canary Islands in the Atlantic Ocean, made multiple short expeditions into the interior of South America from both the east and west coasts, and explored the Galápagos Islands about 600 miles west of South America and coral reefs of the South Pacific. Being the keen naturalist that he was, Darwin filled his notebooks with many observations during the trip and sent crate after crate of specimens back to England. At least two fundamental themes run through

(a)
Courtesy of National Library of Medicine.

(b)
Courtesy of National Library of Medicine.

FIGURE 5.2 Charles Darwin **(a)** in 1849 at the age of 40, a decade before publication of *On the Origin of Species*. **(b)** Darwin in 1881, the year before his death.

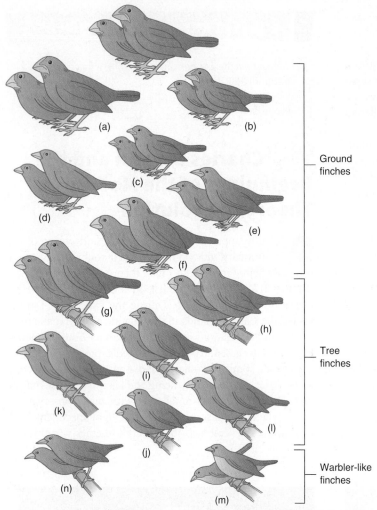

Ground
finches

Tree
finches

Warbler-like
finches

FIGURE 5.3 Darwin's finches.

Like the finches and tortoises, plants and other organisms also varied from one island to the next. Darwin absorbed this information, as well.

Upon his return to England, Darwin devoted the rest of his life to synthesizing the observations recorded in his notebooks from the voyage into a series of books (**Box 5.2**). From his observations, Darwin began to develop general hypotheses about nature. As it turned out, one of the most influential works Darwin ever read was Charles Lyell's *Principles of Geology* (see Chapter 1). He received the first volume before leaving on the expedition of the *Beagle* and received the latter two volumes during the trip, and Darwin and Lyell became close colleagues after Darwin's return.

Although Darwin had long intended to gather his observations into a theory and a book, he procrastinated, despite the warnings of Lyell and other colleagues that someone else might "scoop" him. In science, as in the rest of society, priority means a great deal in terms of recognition and prestige. In fact, Alfred Russel Wallace (**Figure 5.4**), another British naturalist who had also developed a keen interest in beetles as a young man, contacted Darwin in 1858 about his own theory of evolution. Wallace had spent many years in the tropical rain forests of the Amazon and southeast Asia and had developed a theory remarkably similar to that of Darwin's. Wallace

his observations and *still serve as foundations for the modern theory of evolution*: the seemingly infinite variety and variation of plants and animals in nature (Box 5.1) and their biogeographic distribution (see Chapter 3). Why, Darwin asked himself, would a Creator produce so many different kinds of organisms, and why would the Creator distribute them in different places over the planet?

One of the most often-cited examples of Darwin's observations is "Darwin's finches" (**Figure 5.3**). Different islands of the Galápagos, sometimes within sight of one another, are inhabited by different species of finches. The beaks of the finch populations were so different from one another that Darwin did not realize how closely related the species were. After study by a noted ornithologist in England upon Darwin's return, it became clear that the beaks reflected the adaptation and evolution of finch populations on different islands into different species in response to different food sources. For example, finches with robust beaks ate large seeds, whereas others with longer beaks pried insects from underneath bark. Why should this be? Similarly, while visiting the Galápagos, Darwin was informed by one of the inhabitants that one could tell which island he or she was on simply by looking at the tortoises (refer to this chapter's frontispiece).

FIGURE 5.4 Alfred Russel Wallace (1823–1913). Wallace's middle name resulted from a misspelling on his birth certificate.

BOX 5.2 Charles Darwin, the Person

Charles Darwin was the son of Robert Darwin, a prominent physician, who encouraged Charles to follow in his footsteps by attending medical school at Edinburgh. Charles attended the school but was bored by the lectures and horrified by anatomy classes, and he eventually fled. Unfortunately for Robert Darwin, Charles seemed to have had no other interests than beetle-collecting and hunting. Consequently, his father decided that Charles should become a minister, which in those days would have served as a respectable occupation to allow Charles to continue with his naturalistic pursuits.

So, Charles next attended Cambridge University, which was then, along with Oxford University, a bastion of the Anglican Church. There, Darwin met some of the scientific luminaries of the day. These included the famous geologist Adam Sedgwick, who would eventually introduce him to the field of geology in Wales, and John Henslow, who taught mineralogy and botany and who would recommend Darwin for the position of naturalist on the voyage of the *HMS Beagle*. Charles's father at first objected to the trip but eventually relented when his uncle and future father-in-law, Josiah Wedgwood, intervened.

After his return to England, Darwin eventually married his first cousin, Emma Wedgwood, and they settled briefly in London before moving to Downe, located west of London, in 1842. Darwin probably wanted to spend more time thinking and writing and less time on professional activities and the hubbub of London, all of which seemed a distraction to him. Emma and Charles settled into a comfortable existence at Downe, during which Charles spent much of his time writing in his study, conducting experiments or dissecting specimens, and taking long walks in his garden. Darwin developed a deep respect for the organisms he studied and eventually abandoned hunting altogether.

Emma and Charles dearly loved one another and had 10 children, 2 of whom died of disease, and another, who apparently suffered from intellectual disability, died at a young age. However, it was the loss of his beloved daughter, Anne Elizabeth ("Annie"), at the age of 10 that might have finally shaken Darwin's belief in God once and for all; Darwin did not attend Annie's funeral and could never bring himself to visit her grave.

Although he had been in good health as a young man and during most of the *Beagle*'s voyage, Darwin's health suffered through the years, and he frequently visited spas for so-called water cures. It is thought that Darwin might have contracted a parasitic disease while in South America. Also, given his religious background, the implications of his theory of evolution must have also caused Darwin a great deal of mental anguish (compare Figure 5.2, a and b). Emma was quite religious, and she and Charles learned not to speak directly of Charles's work, instead communicating about it by notes when necessary.

Charles Darwin died suddenly at Downe House. He had wished to be buried next to his children, but after a state funeral he was entombed in Westminster Abbey, only a few feet away from Sir Isaac Newton, who was at that time regarded as the preeminent scientist of all time.

---

reported that he first realized his theory while suffering from one of numerous feverish bouts with malaria during an expedition in Indonesia; the idea occurred to him that individuals less resistant to disease, predation, and environmental change would be culled from natural populations.

When Wallace communicated with Darwin about his theory of evolution, Darwin began to panic. At the suggestion of Lyell and other colleagues, Darwin and Wallace copublished a brief abstract, and then Darwin set to work on his seminal work, *On the Origin of Species*, the first edition of which was published in 1859. Twelve hundred fifty copies of the first edition were printed and sold out overnight. Darwin was keenly aware of the implications of his theory for human evolution, but he refrained from tackling this problem in the *Origin* and only published his arguments later in *The Descent of Man*. Nevertheless, many laypeople and scientists immediately understood the implications of Darwinian evolution for humans, and it is therefore not surprising that the *Origin* received many negative reviews when it was first published.

5.4 Basic Premises of Darwinian Evolution

Like Lyell had done in his *Principles of Geology* (see Chapter 1), Darwin marshaled a mountain of evidence to support his theory in *Origin*. Based on this evidence he developed an inductive argument (see Chapter 1) for his theory based on a number of premises. As noted previously, two basic observations led to Darwin's theory of evolution:

the great diversity of plants and animals and their biogeographic distribution. The fundamental questions, then, were how did diversity arise and how did the organisms come to live in different areas? Darwin had to develop a satisfactory mechanism for evolution to occur. It was not enough just to say that evolution occurred; Darwin had to say *how* evolution occurred. Scientists are extreme skeptics until a particular mechanism (cause) can be identified that accounts for a particular effect. Unfortunately for Darwin, he never found the exact mechanism (for reasons we discuss shortly), but Darwin reasoned that tiny differences occurred in the natural variation of organisms and that these variations could be passed from one generation to the next. Another of Darwin's basic premises was that plant and animal populations do not grow unchecked. Darwin therefore reasoned there must be a "struggle for existence" that eliminates "unfit" individuals, those with less suitable variations, whereas fitter individuals survive. Darwin called the struggle for existence **natural selection**; he coined the term based on artificial breeding or selection, which is still used today to produce more productive lines of plants and animals for food. If there is natural selection, with unfit individuals weeded out, then, Darwin reasoned, there must be **differential reproduction**: Survivors live long enough to reproduce and pass their more favorable traits to their offspring so there is **descent with modification**. In this way natural selection for more "fit" individuals occurs over long intervals of time. Thus, depending on how you want to view it, natural selection acts as negative feedback (see Chapter 1) on unfit individuals or as positive feedback on more fit ones (**Box 5.3**).

BOX 5.3 Misuse of Darwin's Theory

Darwin based his thinking in part on Thomas Malthus' *Essay on the Principle of Population,* first published in 1797 (Wallace had also read Malthus' book some time before his bouts with malaria). Malthus' views, along with those of others of the time, have echoed down to the present and have often been misused. Thomas Malthus believed "the passion between the sexes" was too great to overcome; thus, human populations would increase "geometrically" (or exponentially), meaning that after a slow increase, human population would suddenly skyrocket upward (see Chapter 1 for human population growth curve). In contrast, according to Malthus, agricultural food production would increase at a much slower arithmetic rate, behaving like a straight line. Thus, according to Malthus, starvation would always be part of the human condition. Malthus also concluded that poverty was natural and could never be eradicated, and there should be no attempts at state support of the poor. Malthus, not Darwin, first coined the phrase "struggle for existence" in his description of primitive tribes and believed that in his own society competition was best for all.

However, Malthus was perhaps not as ruthless as social reformers of the time portrayed him. Malthus believed the wealthy had not become rich because they possessed superior abilities, and he viewed wealth as a responsibility that required its use to help society in terms of employment and progress. He also advocated educating the poor in an attempt to eliminate poverty.

Today, Malthus is viewed by some as an alarmist. As the human population has increased, technologic breakthroughs have generally increased the ability of food supplies to sustain human populations over much of Earth. The massive starvation predicted by Malthus has not occurred in *developed* countries, but famine has indeed occurred frequently in underdeveloped countries with large populations. Today, many believe biotechnology holds the greatest promise for sustaining human populations, even as expanding populations continue to affect the environment (see Chapter 17).

By all accounts Darwin's sociopolitical views were "Whiggish," meaning he was inclined toward what we now consider a moderate-to-liberal social and political viewpoint. Unfortunately, Darwin's adaptation of Malthus' views continued to be used to espouse what is called **social Darwinism**, in which society is supposed to improve by the action of natural selection on human efforts: some persons naturally succeed, whereas others fail. One of the most notable persons identified with social Darwinism was Herbert Spencer (1820–1903), who emphasized laissez-faire economic views. Spencer certainly believed in societal progress, but he also believed it should occur in a Lamarckian-style "upward striving" of individuals, not through natural selection. Still others claim that the strongly pro-Darwin stance of the famous German zoologist, Ernst Haeckel, planted the seeds of the National Socialist (Nazi) movement in Germany and Austria. These and other movements, not just in Germany but also in Great Britain and America, promoted the movement of racial stereotyping and selective breeding of humans called **eugenics**.

Darwin also argued that natural selection of the small differences between organisms could account for trends seen in the fossil record. Thus, he became keenly aware of the tremendous amount of time required for natural selection to produce evolutionary change. This is not unlike Lyell's view of slow, gradual change through time, except that Darwin's theory resulted in *directional* change, something Lyell initially rejected (see Chapter 1).

CONCEPT AND REASONING CHECKS

1. What were some of the basic observations Darwin made during the voyage of the *HMS Beagle*?
2. What are the basic tenets of the theory Darwin developed after his return to England?

5.5 Inheritance and Variation

To make his theory of evolution credible, Darwin had to provide evolution with a mechanism of inheritance. He already knew, based on the experiments of animal breeders, that agricultural stocks had been improved to feed the expanding human population, so there was obviously some sort of mechanism for parental traits to be passed to the offspring. This same mechanism, he reasoned, would result in the gradual modification of traits over many generations in natural populations. Up to and after Darwin's time, Lamarck's theory of acquired characteristics was one of only two theories of inheritance that had been suggested. The other widely accepted theory of inheritance in Darwin's time was that of **blending inheritance**, which stated that the traits of individuals were simple blends of those of its parents, like mixing a bucket of red and white paint to produce pink. Based on Lamackian theory and blending inheritance, Darwin developed his own theory of inheritance: pangenesis. Darwin suggested that each organ of an individual's body developed special particles called "gemmules," which were transported by the bloodstream to the gonads (testes or ovaries). Because the gemmules originated in the body's cells, they could take on characteristics in the manner Lamarck had described. Each offspring would be a blend of gemmules from both parents, although in some cases the offspring might receive more gemmules from one parent than the other.

The discovery of the basic mechanism of inheritance is rightfully attributed to an Augustinian monk named Gregor Mendel (**Figure 5.5**). About the mid-1860s Mendel conducted experiments involving the crossing of pure-breeding lines of peas in the gardens of a monastery in Brünn (Brno), in what is now the Czech Republic. Based on the results of many different kinds of experimental crosses, Mendel concluded that genetic traits occurred in pairs (one each from the male and female) called alleles. Alleles behaved like particles, and Mendel's theory came to be called the **particulate theory of inheritance**. Alleles were either *dominant*

© National Library of Medicine.

FIGURE 5.5 Gregor Mendel (1822–1884).

or recessive. Mendel reasoned that a recessive trait was only expressed when both alleles were recessive (now called homozygous recessive); otherwise, the dominant trait was expressed if both alleles were dominant (homozygous dominant) or if the dominant allele was present in just one "dose" (heterozygous). Thus, even if organisms appear outwardly identical, they might be different genetically. However, even if organisms are genetically identical, they might still appear slightly different. Genetically identical twins, for example, typically have slightly different appearances because of slight differences in environment (different diets, etc.). We must therefore differentiate between "genotype" and "phenotype." The term genotype refers to the actual genetic makeup of the organism (homozygous dominant, heterozygous, or homozygous recessive); phenotype refers to a particular genotype plus the effects of the environment on the genotype. Mendel's particulate theory of inheritance was highly significant, because it contradicted the blending theory of inheritance. Unfortunately, Mendel published the results of his work in a rather obscure journal, so the significance of his results lay dormant until around the turn of the century. Later, workers discovered that Mendel had sent a copy of his paper to Darwin, but Darwin apparently never read it.

Still, two basic questions remained: what were the particles, and how did they behave when observed under the microscope? If the particles could be identified, perhaps their exact behavior could be understood. During the last half of the 19th century, **chromosomes** ("colored body") were found to duplicate and separate from one another during cell division and were therefore thought to be involved in heredity. Then, Mendel's results were rediscovered about 1900, and their full significance for heredity and the production of variation was realized. Each species has a characteristic number of chromosomes. The number varies substantially between species and does not indicate the species' level of evolutionary complexity. When gametes (sperm and egg) are produced by the process of gametogenesis, the chromosomes are shuffled much like a deck of cards. During gametogenesis one half of *each* duplicated pair of chromosomes has an equal chance of going into one of the two cells resulting from each cell division. This is much like flipping a coin, during which you have a 50:50 chance of obtaining heads or tails. Thus, the potential number of chromosome combinations (and therefore potential genotypes of gametes) is equal to 2 raised to the power of the number of pairs of chromosomes. In humans the chromosome number is 46, so there are 23 pairs of chromosomes. So, the total number of different genotypes of gametes in humans (excluding mutation, discussed later in Section 5.6) is 2^{23}, or about 8.4 million! No wonder there was so much variation within the same species: gametogenesis and sexual reproduction resulted in **genetic recombination** that produced vast numbers of genotypes on which natural selection could act.

5.6 Genetic Code and Mutation

(E) Still, this was not the complete picture of heredity and variation. Breeding experiments established that changes in the appearance of certain portions of chromosomes corresponded to certain alterations in body parts and coloration. Thus, variation also originated from changes, or mutations, to certain portions of the genetic material called genes; at the time it was thought that a particular gene coded for a particular trait. Mutations resulted in even greater amounts of genetic recombination than gametogenesis alone. In fact, mutations are now recognized as providing the "raw material" upon which natural selection acts. But what was the exact genetic material in the chromosomes, its structure, and chemical composition? Answers to these questions would show *how* the genetic information was encoded and passed on to offspring.

The fact that deoxyribonucleic acid, or **DNA**, serves as the hereditary code was not established until the 1940s. The actual structure of DNA was determined shortly thereafter in the 1950s, with the bulk of the credit normally given to James Watson (a geneticist) and Francis Crick (a biochemist). The fundamental building blocks of DNA are called nucleotides (adenine, guanine, thymine, and cytosine); nucleotides are attached to a double-stranded "spiral staircase" composed of sugars and phosphates (**Figure 5.6**).

The sequence of nucleotides along the sugar-phosphate backbone comprises the genetic code. The code is "decoded" to produce other molecules and structures according to the Central Dogma of cell biology. The **Central Dogma** of cell biology states the genetic code stored in DNA is read by the process of **transcription** to produce messenger RNA (mRNA), and the message turned into other molecules and

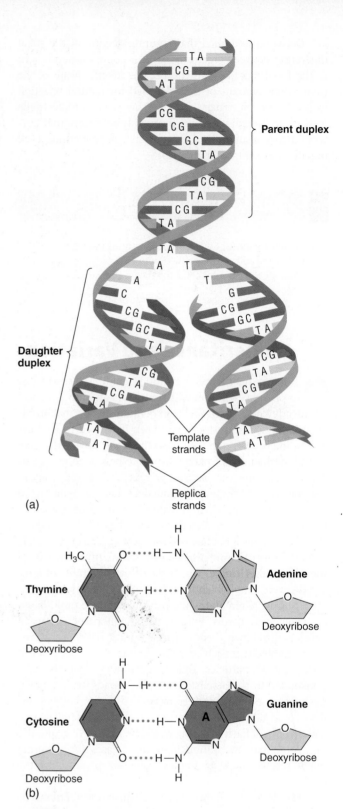

(a)

(b)

FIGURE 5.6 The basic structure of DNA. **(a)** The double helix unwinds in order to duplicate itself or to be transcribed to produce messenger RNA. The nucleotides along the unwound strands of DNA serve as templates which are read to produce the complementary strands. **(b)** Complementary nucleotides along the helices bond through weak hydrogen bonds (red dots). The hydrogen bonds are easily broken and re-made during DNA duplication and transcription.

structures. The mRNA is single stranded, and after it has been synthesized from DNA it is released from the DNA and travels out of the cell nucleus to the cell's cytoplasm. The mRNA is read or **translated** into proteins in the cytoplasm. In diagrammatic form,

$$DNA \rightarrow mRNA \rightarrow protein$$

There are three main types of molecules in all creatures: carbohydrates (sugars), lipids (fats), and **proteins**. Although carbohydrates and lipids are both involved in metabolism and growth, neither is particularly different from one species to another. It is the proteins that give each species its characteristics. Proteins occur as structural proteins that give a cell shape and aid in cellular movement and as organic catalysts called **enzymes**. Each enzyme molecule consists of one or more proteins, which in turn consist of strands of amino acids (**Figure 5.7**). After each strand is initially produced, it "balls up" into a characteristic shape based on the distribution of positive and negative charges and the strength of chemical bonds (Figure 5.7). One or more of the protein balls then form the actual enzyme molecule.

Enzymes are highly specific for certain biochemical reactions. In these reactions an enzyme molecule binds with a substrate molecule for much, much less than a split second; the enzyme catalyzes the chemical change in the substrate (perhaps a chemical bond is broken or made, or a particular chemical grouping changed slightly), and then the altered substrate molecule is released for use in another reaction. The enzyme molecule and its substrate molecule react at the enzyme's **active site**. The active site is highly specific for the substrate molecule because of the distribution of positive and negative charges at the active site and because of the active site's shape. Thus, the substrate behaves like a key inserted into the lock represented by the enzyme (Figure 5.7). If any changes to the active site's structure or the distribution of positive or negative charges occur because of a mutation, it is very likely that the enzyme will be dysfunctional. An important step in a metabolic pathway might not be catalyzed properly, and the organism would likely die.

Thus, mutations are typically lethal. Nevertheless, some mutations are beneficial, or at least "neutral," and so might persist in natural populations rather than being weeded out by natural selection. Mutations, coupled with the process of genetic recombination, are acted on by natural selection to produce evolutionary change that results in new species.

CONCEPT AND REASONING CHECKS

1. What processes produce genetic variation?
2. What is the importance of enzymes, and how do mutations potentially affect enzyme function?

5.7 Evidence for Natural Selection

Despite the voluminous evidence that Darwin marshaled in support of his theory of evolution and despite the theory's wide explanatory power, Darwin regarded natural selection as a hypothesis. Darwin based the hypothesis of natural selection on artificial breeding experiments and the fact that natural populations do not exhibit wild fluctuations in abundance. Darwin further supported his hypothesis through his studies of **sexual selection**. Sexual selection acts on mating success, either through competition of the members of one sex of a species for mates, through choices by members of the opposite sex, or some combination. Sexual selection results in what are seemingly bizarre mating rituals and exaggerated phenotypes such as beautiful coloration in birds or large antlers in elk and other mammals.

Since Darwin's time, a number of cases of natural selection have been well documented that further substantiate his hypothesis. The most famous example is **industrial melanism** in England ("melanic" refers to dark colored). During the industrial revolution in England, pollution controls were unheard of and the countryside, especially around the heavily industrialized city of Manchester, was often blackened with the soot from the burning of coal. Consequently, the background on which the peppered moth, *Biston betularia,* lived began to darken (**Figure 5.8**). A similar phenomenon occurred among other species of insects. The peppered look of the moth had originally served to camouflage it from predators, especially birds. As tree trunks and other substrates were darkened by soot, black moths, which were always present in very small numbers in natural populations, became the dominant form because birds served as a selective agent and were more likely to prey on lighter-colored moths. With the decline of coal use in later years, the original peppered coloration spread once again through most of the population.

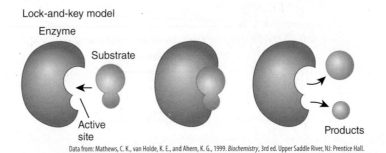

Lock-and-key model

Data from: Mathews, C. K., van Holde, K. E., and Ahern, K. G., 1999. *Biochemistry*, 3rd ed. Upper Saddle River, NJ: Prentice Hall.

FIGURE 5.7 The enzyme lock-and-key mechanism.

© Michael Willmer Forbes Tweedie/Science Source.

FIGURE 5.8 Industrial melanism in the peppered moth, *Biston betularia*, on a **(a)** dark (melanic) and **(b)** pale, "peppered" form.

It is now known that the black allele is a dominant allele and the white is recessive. Thus, the natural populations before the onset of pollution were represented by animals that were homozygous recessive for the white allele. This sort of natural selection is referred to as **directional selection** because the genotypes in the population are dominated by one genotype. Although dominant mutations normally spread quickly, even after many generations, they still do not comprise 100% of genotypes of the population (**Figure 5.9**). On the other hand, even though recessive mutations normally take many generations to spread, they can still spread quite rapidly because of directional selection.

Other examples of natural selection are related to the role of medicine and agriculture in society. For example, natural selection also acts in the case of sickle cell anemia. Sickle cell anemia is found mainly in African-Americans and causes the collapse of red blood cells so they have a crescent (sickle)-shaped appearance rather than a disc-like appearance (**Figure 5.10**). Sickled red blood cells carry less oxygen

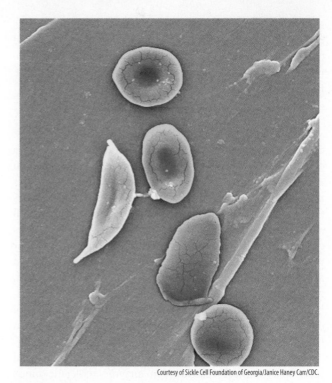

Courtesy of Sickle Cell Foundation of Georgia/Janice Haney Carr/CDC.

FIGURE 5.10 Normal and sickled red blood cells.

to the body's tissues and organs, resulting in fatigue or death. Sickle cell anemia results from a single slight mutation in the DNA code. This mutation affects the ability of the protein hemoglobin in red blood cells to bind oxygen.

Sickle cell anemia was originally widespread in tropical Africa. Malaria is also widespread there because of heavy rainfall and standing water, which mosquitoes (the vectors or carriers of the disease) use for breeding (**Figure 5.11**). Malaria is actually caused by a protist that infects red blood cells and causes them to burst (hence, the fever associated with malaria).

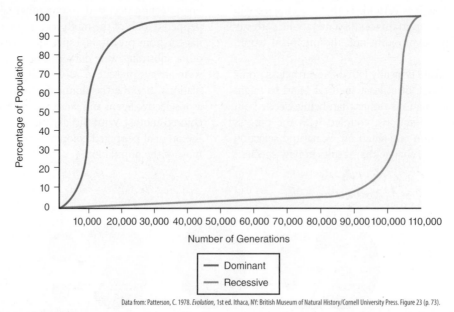

Data from: Patterson, C. 1978. *Evolution*, 1st ed. Ithaca, NY: British Museum of Natural History/Cornell University Press. Figure 23 (p. 73).

FIGURE 5.9 Spread of dominant and recessive alleles in a hypothetical population of organisms. Notice that the dominant mutation spreads quickly, but even after many generations has still not spread completely through the population. On the other hand, the recessive mutation takes many more generations to spread, but eventually spreads quite rapidly.

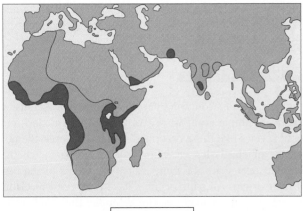

Data from: Strickberger, M. W. 1985. *Genetics*, 3rd ed. New York: MacMillan.

FIGURE 5.11 The occurrence of sickle cell anemia in tropical Africa. The highest occurrence is in tropical west Africa, from where most slaves originated.

However, the protist cannot infect sickled cells. Thus, humans either homozygous dominant for the allele for normal hemoglobin or homozygous recessive for the allele for sickle cell anemia are more likely to die of anemia or malaria, respectively, than individuals who are heterozygous (Figure 5.11). Individuals with one allele of each are more likely to get enough oxygen while avoiding infection of blood cells by malaria. This example is one of **balancing selection**, in which a relatively stable proportion of genotypes is maintained in a population by a selective force (in this case, malaria).

Balancing selection is exhibited in other ways. In **hybrid vigor**, crosses of inbred strains sometimes produce offspring that exhibit greater crop yields than the parent strains. Each parent strain is homozygous dominant or homozygous recessive for a particular allele that produces only one type of protein, whereas the offspring are heterozygous for the allele. The offspring therefore produce two types of proteins and can be at a selective advantage. In humans, inbred strains are frequently homozygous recessive for genes that produce defects that are lethal; this is why laws prohibit the marriage of first cousins, who are too closely related genetically (see Box 5.2).

Another example of natural selection is antibiotic resistance. Bacteria grown in culture and treated with antibiotics such as penicillin quickly become resistant to the antibiotics. Bacteria reproduce quite quickly, often in a matter of hours, and if mutants resistant to antibiotics appear they spread through the culture populations quite rapidly. In recent years antibiotic use has become very widespread and so has antibiotic resistance. Thus, new antibiotics might need to be developed in the future to counteract resistant bacteria.

Natural selection is also put to use in genetic engineering. **Genetic engineering** is the science that manipulates the DNA of viruses, bacteria, and other organisms, including humans, and is used to increase the yield and disease resistance of crops. It is also being used to combat disease and birth defects in humans.

5.8 Speciation

So, we now know how genetic traits are passed from parents to offspring. But how do new species arise? This is really the question that Darwin and others were trying to answer.

Mutations are spread through populations by interbreeding. Thus, speciation could occur by geographic isolation of populations such that they are prevented from interbreeding. This particular type of speciation is known as **allopatric speciation**. Biologic populations are not uniformly distributed over their geographic ranges and typically consist of a main population with smaller isolated populations on their periphery. According to allopatric speciation new species originate by geographic isolation of local populations, called **demes** or **peripheral isolates**, by rivers and streams, mountain chains, or changes in local climate from one side of a valley to another, such as sunlight and moisture. Although the populations are **reproductively isolated**, they begin to diverge genetically from one another. If the populations remain isolated for a long enough time for sufficient genetic divergence to occur, the populations become reproductively isolated from one another and are therefore new species.

Genetic transformation of the demes to new species can occur in several ways. Obviously, the addition of mutations adds more variety to the genotypes of demes, and mutations spread much more quickly through small populations than large ones. Speciation can be accelerated in small populations by two other processes. In the **founder effect**, a new deme is not genetically representative of the original parent population from which it came. Imagine a box full of solid black and solid white balls mixed together, from which 10 balls are randomly chosen. The likelihood that the proportion of black to white balls chosen (the deme) is exactly the same as the proportion in the much larger collection in the box (parent population) is not very high. The founder effect accounts for the establishment of the finch populations in the Galápagos (Figure 5.3). Genetically different populations of finches settled on different islands, so that the founding populations on each island were already different at the outset. Another mechanism involved in speciation involves genetic drift. In **genetic drift**, some genetic traits are simply lost by chance from demes, whereas others are passed on. In this way demes can also change genetically through time.

Given enough time, then, geographic isolation leads to reproductive isolation and allopatric speciation. Reproductive isolation can develop because of different seasonal times of reproduction, behavioral differences associated with mating rituals, and so on. However, if the barriers to isolation are removed before reproductive isolation has been completed, the populations can still interbreed. Populations that have partly diverged from one another but not developed into full-fledged species are called **subspecies**. In humans these differences are less pronounced and are recognized as **races**. Racial differences between humans are therefore a matter of natural biologic evolution. For example, dark skin is the result of the production of the dark pigment, melanin, that protects the skin from excessive sunlight in the tropics.

5.9 Evolution and the Fossil Record

Much of the evidence so far marshaled in support of evolution in this chapter has come from biology. However, as Darwin and his contemporaries recognized, much of the basic evidence for evolution comes from the fossil record.

5.9.1 Comparative anatomy

Some of the evidence for evolution in the fossil record was already well known before Darwin. For example, studies of comparative anatomy recognized two basic kinds of structures. **Homologous structures** have a common evolutionary ancestry but are dissimilar in function, such as the limbs of mammals (**Figure 5.12**). These structures are similar because they share a common ancestry, but they have evolved or diverged for different functions through **divergent evolution**. Conversely, **analogous structures** are similar in function but dissimilar in structure. Examples include the wings of birds, bats (mammals), and butterflies and the streamlined bodies of fish, ichthyosaurs (extinct marine reptiles), and dolphins (mammals) (**Figure 5.13**). These groups are not closely related by evolution. Analogous structures arise through **convergent evolution**, in which natural selection acts on very different taxa to evolve or converge on, for example, wings for flight or streamlined bodies for moving through water quickly. Although analogous structures are not used to determine evolutionary relationships, they serve as yet another example of the action of natural selection, namely that depending on their habitat and niche, very different groups of organisms can evolve similar structures because they are subject to the same selective pressures.

Because biologic classification is supposed to represent true phylogenetic, or evolutionary, relationships, all the members of a particular category must be descended from a common ancestor (see Box 5.1). Such taxa are called **clades**, and for this reason branching or divergent evolution is sometimes referred to as cladogenesis. Clades are considered to be "monophyletic," meaning that each represents a distinct taxon that reflects its true evolutionary relationships. Taxa that do not share a common ancestry but are mistakenly misclassified together are referred to as grades because convergent evolution has resulted in very different taxa attaining the same "grade" or superficial appearance. Such grades are said to be "polyphyletic" because they include taxa not closely related evolutionarily.

5.9.2 Cladistics

For many years, evolutionary relationships were established based on simple comparisons of anatomy of modern and fossil species like those described earlier. However, the new science of **cladistics** classifies taxa according to whether different taxa share the same traits irregardless of the time of their origin. All sorts of traits have been used in cladistics, ranging from anatomic (for example, the number and arrangement of bones in a skeleton) to the sequences of nucleotides in DNA and RNA and amino acids in proteins.

But the basic procedure in cladistics is always the same. The basic assumption of cladistics is that two taxa share the same trait because they have a common ancestry. These relationships are represented in a **cladogram** like that shown for the major groups of vertebrates in **Figure 5.14**. All the vertebrate groups shown possess jaws. Such a trait, which is shared by all the taxa, must have appeared first and is therefore considered **primitive**. **Derived traits**—lungs, claws, scales, fur, and mammary glands—appear later in succession in each of the remaining vertebrate taxa: amphibians, reptiles, and mammals. Birds, which are characterized by feathers, are probably an offshoot of small bipedal dinosaurs (dinosaurs that walked or ran on their hind legs) that appear to have used feathers for insulation (see Chapter 13). Mammals were probably also derived from another group of reptiles that share certain anatomic features with mammals.

Although this particular cladogram was constructed for higher taxonomic categories (some major taxa of vertebrates), the same basic procedure is used to examine the relationships at much lower levels, such as the species belonging to a particular genus or the genera belonging to a particular family. As before, more primitive traits are thought to be shared by different groups, whereas derived traits are shared by fewer groups. A hypothetical, evolutionary tree is shown in **Figure 5.15**. Figure 5.15a implies time on the vertical scale, whereas the relative distances between taxa A, B, and C in the diagram roughly correspond to their anatomic differences, which presumably reflect their evolutionary relationships.

These relationships are shown differently in a cladogram. Figure 5.15b represents the cladogram of the common and derived characters of taxa A, B, and C. According to the cladogram A and C are once again more closely related to each other than they are to B.

Traditional evolutionary "trees" and cladograms do not always yield the same classifications, however. For example, **Figure 5.16** shows a particular diagram of the relationships between humans and related groups. The traditional classification places all humans in the same family, Hominidae, and other forms such as gorillas and chimpanzees in the family Pongidae. In this view, the families Hominidae and Pongidae are considered to be monophyletic. However, this view is undoubtedly highly anthropocentric, or "human centered."

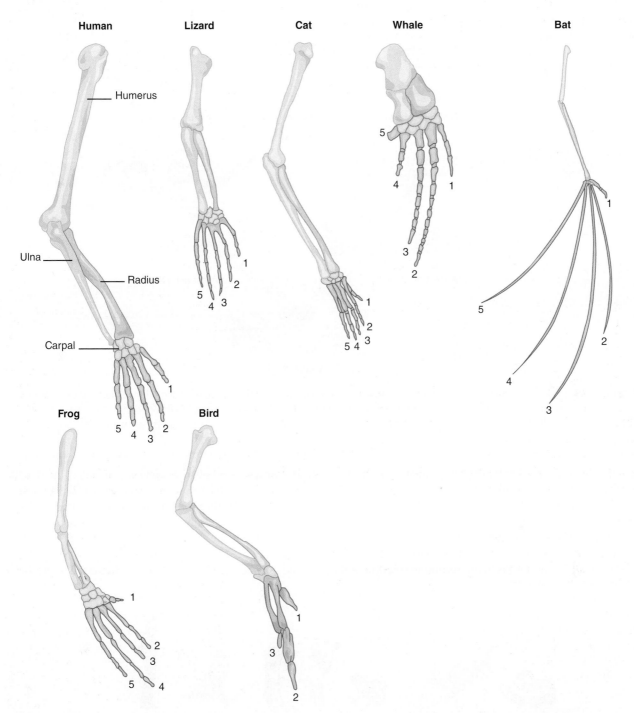

Human · Lizard · Cat · Whale · Bat · Frog · Bird

FIGURE 5.12 Homologous structures of the limbs of different mammals. Numbers and colors refer to bones with a common evolutionary ancestry. Numbers and colors indicate the homologous structures.

Many workers view the family Pongidae as being a "grab bag" or "wastebasket" of all taxa that are not considered "human"; such a taxon is said to be "paraphyletic" because it does not include all of its descendants. In fact, the diagram in Figure 5.16 suggests that humans basically lie on a continuum of traits they share with the other taxa. Thus, according to the alternative cladogram in Figure 5.16, orangutans, gorillas, chimpanzees, *and* humans should all be placed in the same taxon, which is in turn considered monophyletic.

5.9.3 Microevolution

Another way in which evolutionary relationships can be inferred from the fossil record is to trace the succession of different species through time in the sedimentary record. For many decades, the succession of species in the fossil record was viewed as occurring by slow, gradual processes collectively termed **microevolution**. Microevolution was thought to result from the kinds of short-term genetic processes

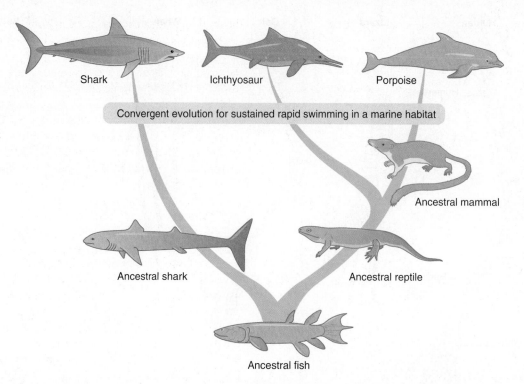

FIGURE 5.13 Analogous structures arise through convergent evolution. The streamlined bodies of fish, ichthyosaurs (extinct marine reptiles) and dolphins (mammals) arose through convergent evolution. If these three taxa were classified into the same taxon based on body shape alone, the taxon would be a grade, not a clade. However, many other features indicate that these taxa are not closely related.

like those documented in genetic experiments and biogeographic studies of the distribution of species.

The pattern of speciation in the fossil record that was thought to result from microevolution was referred to as **phyletic gradualism** (**Figure 5.17**). The gradual transition of one species into another is called **anagenesis** (as opposed to cladogenesis) and results in the **pseudoextinction** of the first species. Although anagenesis was often inferred from the fossil record, it was rarely observed. The lack of observed transitions was therefore dismissed as an artifact of the geologic record due to nonpreservation or erosion.

However, an alternative mode of speciation was developed in the 1970s by the paleontologists Niles Eldredge and Stephen Jay Gould called **punctuated equilibrium**.

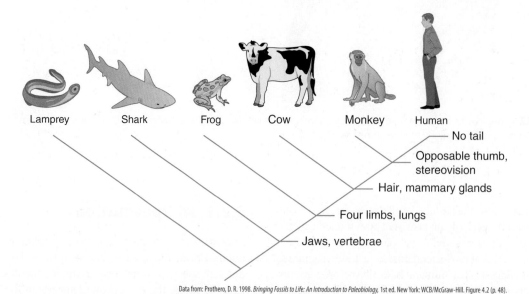

Data from: Prothero, D. R. 1998. *Bringing Fossils to Life: An Introduction to Paleobiology*, 1st ed. New York: WCB/McGraw-Hill. Figure 4.2 (p. 48).

FIGURE 5.14 A cladogram of some vertebrates showing the successive appearance of major features through time at each branching point. Jaws are considered a primitive trait because all of the taxa, except lampreys, possess them, whereas the other features are said to be derived.

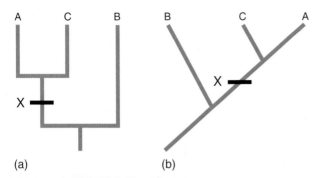

(a)   (b)

FIGURE 5.15 **(a)** Traditional evolutionary tree, in which phylogenetic relationships are depicted between three taxa (A, B, and C). Two taxa (A and C) share a newly derived character (X) and taxon B is ancestral to taxa A and C. Time is vertical (upward). **(b)** Cladogram of relationships between A, B, and C. In this hypothetical example, both diagrams indicate that A and C are more closely related to one another (because they both share "X") than to B.

These workers argued that the fossil record was not nearly as incomplete as most workers had said and that apparent phyletic gradualism really represented long intervals of slow, nondirectional change (what they called "stasis") punctuated by abrupt appearances of new species (Figure 5.17). Punctuated equilibrium is basically a record of allopatric speciation preserved in the fossil record (**Figure 5.18**). According to this theory the main parental population dominates in the fossil record because it is widespread and therefore more likely to be preserved and eventually found. However, occasionally the main population goes extinct and disappears from the fossil record. At this time the peripheral isolates that have been present all along (but not preserved in the fossil record because of their patchy distribution and small population sizes) expand and radiate into the environment left vacant by the extinction of the main population (Figure 5.18). The peripheral isolates therefore appear suddenly in the fossil record as new species. Recent studies of more than 100 animal, plant, and fungal lineages suggest that new species can evolve in these peripheral isolates relatively rapidly, as would be expected given their small population size and the potential for the spread of mutations that might result in reproductive isolation. After they are established, these species settle down to stasis once again. Usually, the taxa established by the peripheral isolates eventually go extinct or are replaced, but on rare occasions lineages might persist to the present essentially unchanged as **living fossils**. One of the most famous living fossils is the coelacanth, which belongs to the same taxon (lobe-finned fishes) that gave rise to amphibians (**Figure 5.19**).

Punctuated equilibrium explains a common misconception of creationists regarding what are called **missing links**. The term "missing link" builds on the concept of the Chain of Being. Missing links are taxa that are intermediate in their

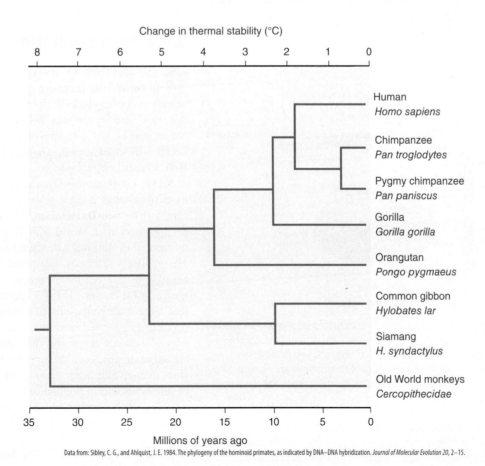

FIGURE 5.16 The evolutionary relationships of humans portrayed. This particular diagram is based on the thermal stability of DNA, which has been used as a measure of evolutionary relationships between different taxa.

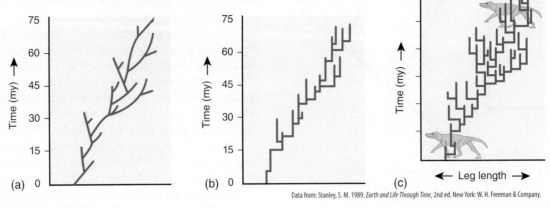

FIGURE 5.17 Modes of evolution suggested by the fossil record. **(a)** Phyletic gradualism, which represents slow, gradual evolution. **(b)** Punctuated equilibrium, which is basically the record of allopatric speciation in the fossil record. Occasionally, following punctuated speciation, lineages persist to the present essentially unchanged as living fossils. **(c)** A hypothetical example of species sorting, in which species lineages are "pruned" sometime after they appear, and which can cause lineages to evolve in new directions.

Data from: Stanley, S. M. 1989. *Earth and Life Through Time*, 2nd ed. New York: W. H. Freeman & Company.

traits between one taxon and another. A prime example is *Archaeopteryx,* which is commonly considered to be the first bird. *Archaeopteryx* possesses the traits of both reptiles (teeth, long tail, claws) and birds (hollow bones, feathers). However, missing links like *Archaeopteryx* are rare. Creationists argue that if evolution occurs, there should be many more examples of missing links in the fossil record. However, evolution in relatively small populations such as peripheral isolates means that it is unlikely that so-called missing links are preserved in the fossil record.

CONCEPT AND REASONING CHECKS

1. How does the use of homologous and analogous structures to infer evolutionary relationships differ from the use of cladograms?
2. Indicate where the processes of allopatric speciation are taking place next to a diagram of punctuated equilibrium.
3. Why is population size important to speciation?

5.9.4 Macroevolution

The fossil record reveals other evolutionary patterns and therefore perhaps other evolutionary processes. Indeed, the fossil record indicates that new species or taxa belonging to categories above that of the category of species may evolve because of rapid genetic change, or **macroevolution**, in small populations.

The coupling of Darwinian evolution with modern genetics is often referred to as the modern synthesis or **neo-Darwinism**. The modern synthesis appeared after World War II and has strongly influenced evolutionary thought to the present day. The modern synthesis dismissed macroevolution because of the results of genetic experiments on human time scales. These experiments indicated that large-scale mutations that presumably had to occur to cause rapid or drastic evolutionary change normally do not occur and that when they do they are lethal. Instead, evolutionary biologists opted for slow, microevolutionary changes in the occurrences of genotypes within populations.

One of the earliest pieces of evidence cited in support of macroevolution was the biogenetic law proposed by the German zoologist, Ernst Haeckel, in the late 1800s (Box 5.3). The **biogenetic law** is usually stated as *ontogeny tends to recapitulate phylogeny.* This means the development of the individual

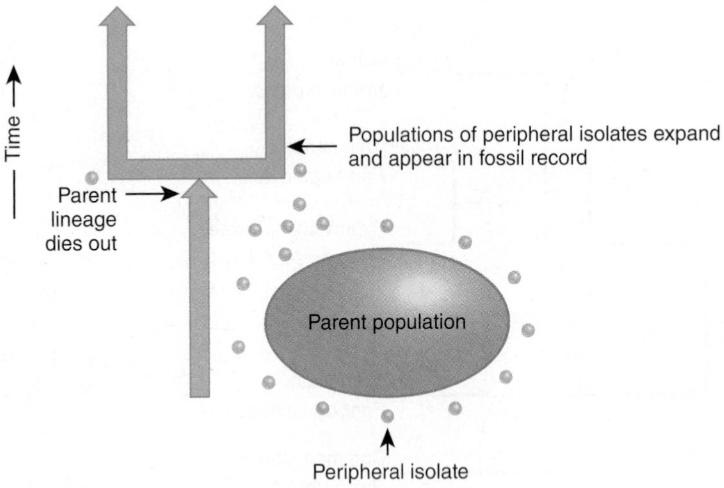

FIGURE 5.18 The relation of speciation to punctuated equilibrium. Allopatric speciation is one process by which new species arise, whereas punctuated equilibrium is the pattern of allopatric speciation preserved in the fossil record. The parent population persists relatively unchanged and is more likely to be found in the fossil record than peripheral isolates because of the parent population's larger numbers and geographic extent. However, if the parent population dies out, one or more of the peripheral isolates will move into the vacated habitat and expand in numbers and geographic area so that new species will be found suddenly appearing in the fossil record. The new species will then remain relatively unchanged because of their large population size. However, they too can eventually go extinct; thus, the pattern is repeated in the fossil record.

Time →

Populations of peripheral isolates expand and appear in fossil record

Parent lineage dies out

Parent population

Peripheral isolate

(a)

© Vova Pomortzeff/Alamy.

(b)

© Michael Pettigrew/Shutterstock, Inc.

(c)

© iStockphoto/Thinkstock.

FIGURE 5.19 **(a)** The coelacanth (*Latimeria chamulnae*) was long thought to be extinct but was accidentally dredged up from deep water off the island of Madagascar by fishermen. **(b)** Leaves of the Gingko tree (*Ginkgo biloba*). **(c)** The Atlantic horseshoe crab (*Limulus polyphemus*).

organism (**ontogeny**) reflects (recapitulates) its evolutionary relationships or phylogeny (evolutionary history). The prime example is the appearance and later disappearance of gill slits (which are used for breathing by fish) during the early development of the human embryo (**Figure 5.20**). The appearance of gill slits during human embryonic development presumably reflected the fact that humans were ultimately descended from fish and was frequently cited in older textbooks as proof of evolution. However, the biogenetic law was later downplayed because it was realized that embryologic development is far, far more complex than such a sequence of figures like that in Figure 5.19 suggests.

In retrospect, however, Haeckel had discovered something of tremendous importance. Recently, scientists have discovered regulatory genes called **homeobox**, or *Hox*, genes. These genes control the early development of certain body regions into particular segments or structures (**Figure 5.21**). *Hox* genes occur along the length of chromosomes, almost like beads on a string, with each bead corresponding to a particular body region or segment (Figure 5.21). Each *Hox* gene or cluster of genes has been given a name. For example, the protein Sonic Hedgehog (named after one worker's favorite

video game character) stimulates the formation and secretion of a growth-stimulating protein during early embryonic development of vertebrate limbs. Increasingly, however, *Hox* genes are referred to as "*Hox 1*," "*Hox 2*," and so on for the sake of clarity. Some *Hox* genes establish the anterior (front) to posterior (tail) axis of the animal or the dorsal (top) to ventral (bottom) orientation of the body. Other *Hox* genes code for the initial subdivision of the body into segments and still others for limb generation, the fate of muscles, and light receptors such as eyes.

Hox genes themselves have been duplicated and modified in different groups through time. As its name implies, the gene *Antennapedia* codes for antennae and legs in the anterior region of the fruit fly. In mice and humans, this gene has been duplicated four times, but it is still involved in the differentiation of the head region. Another example of gene duplication is the long row of vertebrae in extinct marine reptiles such as plesiosaurs and land-dwelling sauropods (see Chapter 13). Because identical or nearly identical *Hox* genes have been found in a variety of creatures ranging from worms to insects to vertebrates (based on sequences of DNA nucleotides), these genes must have been present early

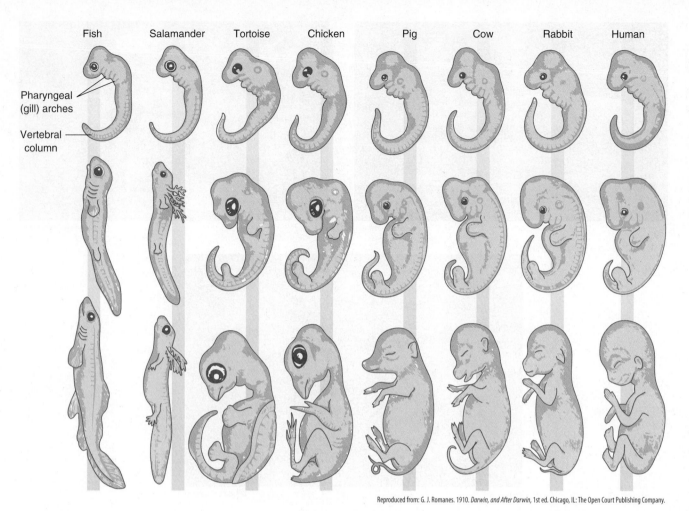

Pharyngeal (gill) arches

Vertebral column

Fish Salamander Tortoise Chicken Pig Cow Rabbit Human

FIGURE 5.20 Early embryonic development of different vertebrates according to Ernst Haeckel. Note the appearance and loss of gill slits, which was used in support of the biogenetic law.

in evolution and passed on in subsequent lineages through time. In fact, approximately 99% of the genes in chimpanzees and humans are identical; the two species differ primarily because of the regulatory genes expressed during development.

Experiments have been conducted in which the *Antennapedia* gene of the fruit fly (an insect widely used in laboratory genetic experiments) was transplanted into worms with a mutant form of the *Antennapedia* gene. After transplantation the worm's offspring developed normally. These sorts of experiments indicate that during embryonic development different *Hox* genes are "turned on" and "turned off" by positive and negative feedback. Proteins coded by certain regions of DNA affect the formation of different types of tissues and organs at specific locations during development. The differentiation of cells into tissues and organs might in turn trigger the activation of other DNA sites for transcription and so on in a kind of developmental cascade (**Figure 5.22**).

Indeed, certain DNA sequences are referred to as master control genes. Master control genes are like the master switch on the electric service panel in a house that controls

all the circuit breakers. The circuit breakers are analogous to genes that code for enzymes in different metabolic pathways. If a circuit breaker is flipped from the "on" to the "off" position, electricity will not flow through the circuits it controls, and the lights will not come on in certain parts of the house or the washing machine won't wash, for example. However, all other circuits to the rest of the house remain unaffected. But if the master switch is thrown, all circuit breakers and circuits are inactivated because the flow of electricity into the entire house has been completely shut off. The same is true of master control genes, which control the DNA transcription of a number of genes and metabolic pathways "downstream."

If a mutation occurs in a *Hox* gene or another gene that is involved in early development, it can cause a cascade of genetic and developmental changes that affect all the other genetic pathways downstream. Normally, mutations that occur early during the embryonic development of an organism are lethal because embryonic development is a tightly controlled process. Early development tends to "lock in" later developmental pathways, such as the differentiation of

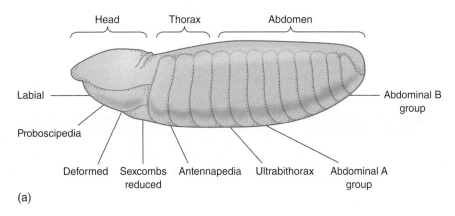

Drosophila embryo

Head — Thorax — Abdomen

Labial

Proboscipedia

Deformed   Sexcombs   Antennapedia   Ultrabithorax   Abdominal A
reduced   group

Abdominal B
group

(a)

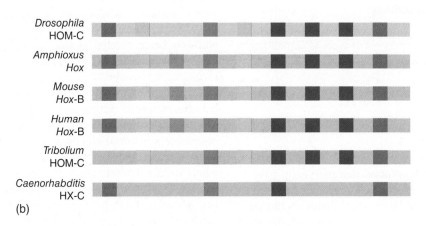

Drosophila
HOM-C

Amphioxus
Hox

Mouse
Hox-B

Human
Hox-B

Tribolium
HOM-C

Caenorhabditis
HX-C

(b)

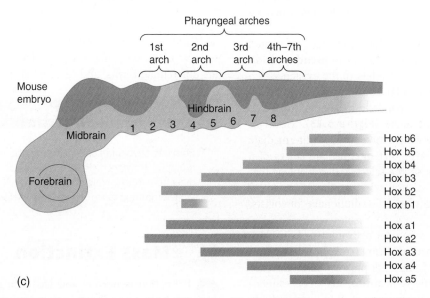

Pharyngeal arches

1st   2nd   3rd   4th–7th
arch   arch   arch   arches

Mouse
embryo

Hindbrain

Midbrain   1 2 3 4 5 6 7 8

Forebrain

Hox b6
Hox b5
Hox b4
Hox b3
Hox b2
Hox b1

Hox a1
Hox a2
Hox a3
Hox a4
Hox a5

(c)

FIGURE 5.21 **(a)** The distribution of *Hox* genes in the fruit fly *Drosophila*. **(b, c)** Note the similarity of the *Hox* genes in *Drosophila* to those of other widely differing taxa, including arthropods (*Tribolium*), worms (*Caenorhabditis*), and the mouse. Forms similar to *Amphioxus* are thought to have given rise to vertebrates.

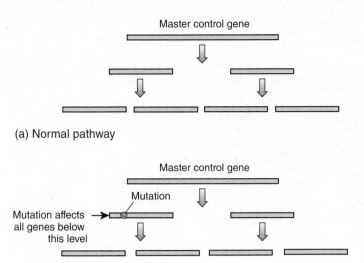

(a) Normal pathway

Master control gene

Mutation

Mutation affects
all genes below
this level

(b) Mutation in developmental pathway

FIGURE 5.22 Effects of mutations in genetic cascades. **(a)** Normal developmental pathway. **(b)** A mutation in a developmental pathway affects later developmental pathways controlled by the DNA in which the mutation occurs.

new tissues and organs and their positions in the embryo. But if important genetic changes occur that affect early development and the changes are nonlethal, they can affect subsequent gene expression and biochemical pathways further along in development (Figure 5.22). These sorts of changes explain anatomic similarities such as homologous structures (Figure 5.12).

A mechanism of rapid evolutionary change also related to changes in developmental pathways is **preadaptation** or *latent potential*. Preadaptation refers to the existence of a structure adapted for a particular function or environmental condition that turns out by chance to be adapted for rapid evolution into a different niche (**Figure 5.23**). How, for example, can an eye or a wing evolve through intermediate stages to its final form? Wouldn't intermediate stages of a structure put creatures at such a selective disadvantage that they would be culled by natural selection almost instantly? This question had long plagued evolutionary biologists, including Darwin.

Since Darwin's time, a number of examples of preadaptation have been documented. Primitive light-sensing and temperature-regulating organs were preadapted to give rise to eyes and wings, respectively, in arthropods. In experimental studies of insects, small wings are used to regulate body temperature; as the wings increase in size, the advantages for flight become dominant just as the benefits of wings for body temperature regulation begin to level off. As demonstrated by the fossil record, the stubby limbs of primitive **lobe-finned fish** (related to the coelacanth) might have helped them scoot along shallow bottoms, but the limbs were preadapted for a rapid transition to land by primitive amphibians called **labyrinthodonts** (now called "temnospondyls"; Figure 5.23). Based on the striking similarity of

bone structure in these two groups, relatively small changes in developmental pathways could easily have transformed the lobe-finned limb into that of a primitive amphibian; as described earlier, the genetic changes could have spread rapidly through relatively small, isolated populations along the shore. In fact, the limb bone structure of both groups has long been considered homologous.

Darwin was also aware of the existence of vestigial structures in a wide range of animals (**Figure 5.24**). As the term indicates, **vestigial structures** are vestiges of organs or structures that previously performed a particular function but that later degenerated and now have little or no use; the appendix of humans and dewclaws of dogs are examples (Figure 5.24). On the surface vestigial structures also posed the problem of intermediate stages in evolution. However, rather than being preadapted for *future* use (as a teleologic view of evolution would demand), vestigial structures were just the opposite: they were a waste of energy to produce and were being selected against, according to Darwin's theory. Thus, vestigial structures argue for natural selection and evolution, not against it.

Another example of macroevolution might be long-term patterns of species' appearances and extinctions in the fossil record called **species sorting**. In theory, species sorting resembles punctuated equilibria, but instead of individuals being selected against, natural selection acts to weed out whole species. Unfortunately, species sorting was originally referred to as "species selection." This upset some evolutionists, who objected to the implication that the mechanism of natural selection, as applied to populations, could be applied to species or higher taxa. In fact, species sorting has nothing to do *directly* with natural selection; rather, it is envisioned to act more like pruning a bush. Some species are more likely to leave more "offspring" (new species) than others so there are differential "births" and "deaths" of species.

CONCEPT AND REASONING CHECKS

1. Both microevolution and macroevolution involve mutation. How, then, do the processes of macroevolution differ from those of microevolution?
2. What is the importance of preadaptation to macroevolution?

5.10 Mass Extinction

Extinction is normal and has occurred through geologic time as a kind of background extinction. Despite all their adaptations produced in response to natural selection, most taxa have become extinct. In fact, the fossil record indicates that more than 99.9% of all species that have ever existed have become extinct. Many taxa died out when the biosphere was decimated during **mass extinctions**. As we will see in later chapters, mass extinctions have resulted from different causes, among them global cooling, massive volcanism, decreased oxygen in the oceans, and meteor impacts.

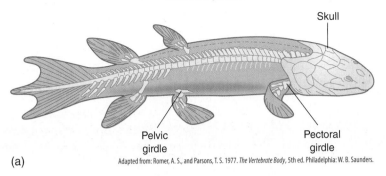

Eusthenopteron

Skull

Pelvic girdle

Pectoral girdle

(a)

Adapted from: Romer, A. S., and Parsons, T. S. 1977. *The Vertebrate Body*, 5th ed. Philadelphia: W. B. Saunders.

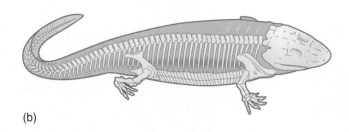

Ichthyostega

(b)

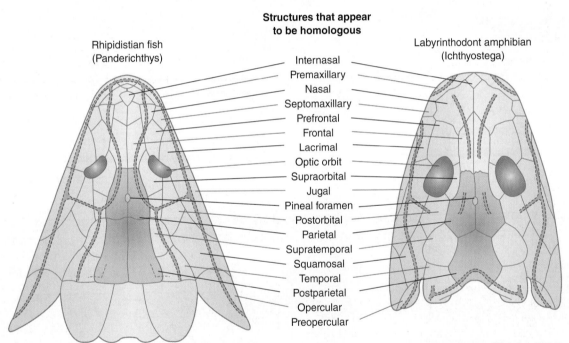

Structures that appear to be homologous

Rhipidistian fish (Panderichthys)

Labyrinthodont amphibian (Ichthyostega)

Internasal
Premaxillary
Nasal
Septomaxillary
Prefrontal
Frontal
Lacrimal
Optic orbit
Supraorbital
Jugal
Pineal foramen
Postorbital
Parietal
Supratemporal
Squamosal
Temporal
Postparietal
Opercular
Preopercular

(c) Data from: Duellman, W. E., and Trueb, L. 1986. *Biology of Amphibians*, 1st ed. New York: McGraw-Hill; Part 2 adapted from: Coates, M. I., and Clack, J. A. 1990. Polydactyly and the earliest known tetrapod limbs. *Nature 347*, 66–69.

FIGURE 5.23 Preadaption in ancient lobe-finned fish and primitive amphibians. **(a)** *Eusthenopteron*, a primitive lobe-finned fish. **(b)** *Ichthyostega*, a primitive labyrinthodont or temnospondyl. **(c)** Comparison of skull structure of a primitive lobe-finned fish and temnospondyl.

Despite the negative consequences for the biosphere, mass extinction is exceedingly important to biologic evolution. Mass extinction might have been necessary to increase Earth's biodiversity, for without extinction far fewer evolutionary opportunities would have occurred because all the habitats and niches would otherwise have been filled long ago. Extinction, then, is a mechanism of macroevolutionary change that might not otherwise occur. Thus, *mass extinction produces history by altering the course of life.*

The phenomenon by which new species fill the niches of previously existing species after mass extinction results from the process of **ecologic replacement**. The process of ecologic replacement can be envisaged as a theater in which the "stage" (planet Earth) has existed through time but on which the "actors" (taxa) have changed through time because of changes to the stage or physical environment of Earth. A prime example is the extinction of dinosaurs at the end of the Cretaceous Period, which allowed the evolutionary

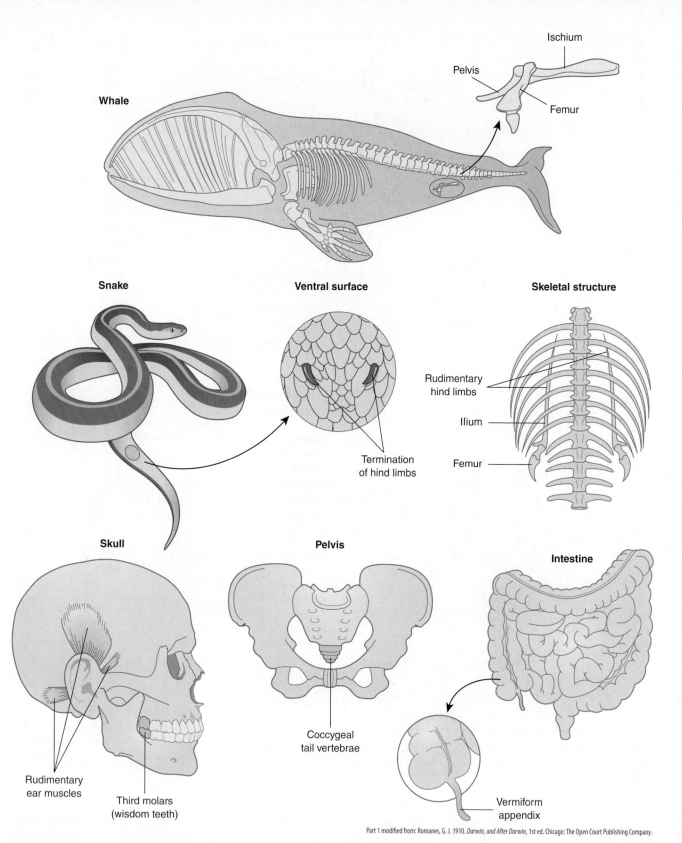

Whale

Ischium

Pelvis

Femur

Snake

Ventral surface

Termination
of hind limbs

Skeletal structure

Rudimentary
hind limbs

Ilium

Femur

Skull

Rudimentary
ear muscles

Third molars
(wisdom teeth)

Pelvis

Coccygeal
tail vertebrae

Intestine

Vermiform
appendix

Part 1 modified from: Romanes, G. J. 1910. *Darwin, and After Darwin*, 1st ed. Chicago: The Open Court Publishing Company.

FIGURE 5.24 Examples of vestigial structures.

diversification of mammals, leading eventually to humans. When massive ecologic replacement (such as that of the dinosaurs by the mammals) or diversification (for whatever reason) occurs among a particular taxon, it is referred to as an **adaptive radiation**.

All mass extinctions seem to have some common features. First, the survival of species depends on their tolerance to environmental change. Eurytopic species are more likely to survive than stenotopic ones because eurytopic taxa are tolerant of environmental change and are more likely to be widespread, increasing their chances of survival in refuges (Chapter 3). Eurytopic taxa are also much more variable genetically than stenotopic taxa; if this were not the case, eurytopic taxa would not be as tolerant of environmental change and would not be as widely distributed. Thus, once extinction has ceased, eurytopic taxa rapidly evolve new adaptations as they radiate into the habitats and niches left vacant by the extinction of previously existing taxa.

Second, after extinction (called the **recovery phase**), any and all mechanisms of microevolution and macroevolution can give rise to new taxa through adaptive radiation. However, extinction does not mean that "anything goes" as the biosphere recovers and new taxa evolve. In fact, the founding taxa are subject to similar natural selection pressures and limitations of genetic constraints as their ancestors. This is evidenced by the phenomenon of convergent evolution *between* taxa belonging to widely different evolutionary lineages to produce analogous structures or features (Figure 5.13). A similar phenomenon known as **iterative evolution** occurs *within* the same taxa that are more closely related by evolution. For example, at different times during Earth's history, both a mass extinction and a smaller minor extinction decimated planktonic organisms called foraminifera. After each of these extinctions, new species of foraminifera evolved that closely resembled the taxa that lived before each extinction.

Thus, the fossil record indicates that the evolution of new taxa after extinction is constrained by the "genetic baggage" of the surviving taxa. Surviving taxa serve as the ancestors of the taxa that diversify during the recovery phase. Thus, certain genetic programming is already "hard-wired" into the founding taxa and constrains the evolution of new traits. In other words the evolution of new taxa during the recovery phase is constrained by a kind of founder effect but on a much more massive scale. Many new classes, orders, families, genera, and species can certainly evolve after an extinction—especially a mass extinction—but no new phyla appear because the basic body plans (**groundplans**) of phyla remained the same. These genetically based groundplans served as the foundation on which the new taxa are reconstructed. In other words, evolution occurred *within* phyla, but wholly new phyla or groundplans did not appear.

CONCEPT AND REASONING CHECKS

1. How does the genetic composition of a species influence its evolutionary potential?

5.11 Biodiversity Through the Phanerozoic

So far, we have examined evolution at the micro and macro levels with regard to the appearance of new species and new higher taxa and their extinction. But how have some of these changes played out during the broad evolution of the biosphere through the Phanerozoic Eon?

The fossil record reflects major changes in the evolution of marine and terrestrial communities through the Phanerozoic (**Figure 5.25**). Fossils first became abundant in the marine geologic record about 540 million years ago at the beginning of the Cambrian Period. These fossils belong to the evolutionary fauna called the **Cambrian Fauna**. The communities of the Cambrian Fauna appear primitive compared with those of later times and were dominated by taxa such as trilobites and jawless fish that fed very close to the sediment surface or just beneath it.

The **Paleozoic Fauna** began to diversify during the Ordovician Period, beginning about 500 million years ago, as the Cambrian Fauna began to wane (Figure 5.25). The Paleozoic Fauna was much more diverse than the Cambrian Fauna and was dominated by *suspension-feeding* taxa that fed on suspended organic matter and plankton above the bottom, including corals and coral-like taxa and strange creatures called crinoids that resemble an upside-down starfish on a popsicle stick. In fact, one of the most significant developments of the Silurian and Devonian was the appearance of widespread reefs. Jawed predators such as sharks that were capable of crushing the hard shells of prey or feeding on fish became prominent by the end of the Devonian and Early Carboniferous periods. It therefore appears that marine food chains were lengthening and food webs becoming more complex during the Paleozoic. The lengthening of the food chains suggests increasing food (energy) availability at the base of food pyramids that could be passed upward to support the predators (see Chapter 3), which likely had elevated metabolism for locomotion and predation. Food webs also became more complex on land. The spread of forests into the interiors of continents led to the widespread coal-forming forests of the Carboniferous Period. These forests provided habitats and niches for the evolutionary diversification of insects, amphibians, and eventually reptiles. The Paleozoic Fauna suffered a number of minor extinctions and two mass extinctions at the end of the Ordovician and Devonian periods that contributed to evolutionary turnover through the Paleozoic. Nevertheless, the Paleozoic Fauna remained recognizable to the end of the Paleozoic Era.

The Paleozoic Fauna was replaced by the **Modern Fauna** in the Triassic Period after the greatest mass extinction in Earth's history about 250 million years ago (Figure 5.25). As its name implies, the Modern Fauna was much more like the faunas we see today. Still, the biotas of the Modern Fauna also changed through time. Marine plankton underwent a tremendous diversification, with coccolithophorids and ultimately diatoms becoming prominent in the fossil record. One of the most significant developments

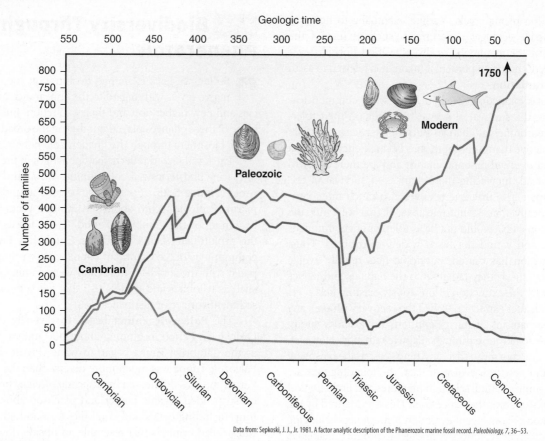

Geologic time

FIGURE 5.25 The so-called "Sepkoski curve" of the major changes in marine faunas of the Phanerozoic. Arrow at time right indicates number of modern families. The actual levels of family and genus biodiversity have been questioned in recent years because of issues of sampling and preservation and remain debated. However, the basic succession of the three faunas has remained the same.

Data from: Sepkoski, J. J., Jr. 1981. A factor analytic description of the Phanerozoic marine fossil record. *Paleobiology, 7,* 36–53.

in the marine realm was the great evolutionary expansion of predatory gastropods (snails), more modern fish, and large marine reptiles, again suggesting increased food availability and metabolism. On land, terrestrial forests continued to expand during the Mesozoic (Figure 5.25). Roaming these forests were the dinosaurs, whereas the seas were prowled by whale-like and dolphin-like reptiles. Flowering plants blossomed during the Cenozoic Era, whereas the dinosaurs were replaced by the mammals on both land and in the seas after yet another mass extinction at the end of the Cretaceous Period.

Thus, biodiversity *appears* to have increased through the Phanerozoic. But questions remain. The actual levels of family and genus biodiversity have been questioned in recent years because of issues of sampling and preservation. However, the basic succession of the three faunas has remained the same. Why, for example, did the Paleozoic Fauna maintain a relatively stable level of biodiversity? No one knows for sure. And, is the steep increase in diversity of

the Modern Fauna real or is it an artifact of the fossil record, or both? These are among the most fundamental questions confronting earth scientists today. Younger rocks and their fossils (like those of the Mesozoic and Cenozoic eras) are more likely to be found at Earth's surface than much older ones (like those of the Paleozoic) because there has been less time for younger rocks to be eroded and for the fossil biotas to be destroyed.

CONCEPT AND REASONING CHECKS

1. Do you suppose all the processes involved in evolution are observable or can be inferred from laboratory experiments? Why or why not?
2. What does the succession of major faunas through time indicate about directionality of Earth's history (see Chapter 1)?

Summary

- Charles Darwin's theory of evolution irrevocably changed the view of Earth from one of equilibrium to one of directionality and history. Darwin's theory has remained largely intact to the present.

- Darwin reasoned that new species evolve based on the following simple observations and inferences: the great diversity of plants and animals and their variation, their geographic distributions, and the fact that natural populations do not grow unchecked.

- Therefore, he reasoned, there must be a struggle for existence that eliminates unfit individuals. Based on breeders' experiments with agricultural stocks, Darwin reasoned the traits of those individuals that survive and reproduce are passed on to their offspring (differential reproduction), so there is descent with modification. Thus, there is a naturally occurring selection for fitter individuals.

- Experiments by Gregor Mendel demonstrated that genetic traits were passed from parents to offspring. Later, these traits were correlated with the appearance and behavior of chromosomes during cell division. Eventually, it was found that genetic variation is produced by mutation, coupled with the formation of gametes and sexual reproduction. Eventually, the genetic material was determined to be DNA.

- New species typically arise through allopatric speciation. In allopatric speciation, natural selection acts on demes or peripheral isolates, which evolve through the accumulation of mutations, genetic drift, and the founder effect.

- In the fossil record, new species can appear gradually through the process known as phyletic gradualism or through the process known as punctuated equilibrium. After they have appeared, new species can be sorted.

- Rapid evolution of new structures and possibly whole new taxa can take place through mutations in master control or *Hox* genes that affect developmental pathways.

- Although the exact causes of extinction vary, extinction can also be considered as an agent of macroevolution because it opens up many habitats and niches simultaneously. Eurytopic species are more likely to survive extinction than stenotopic ones because eurytopic taxa are tolerant of environmental change and are more likely to be widespread, increasing their chances of survival in refuges.

- Eurytopic taxa are also much more variable genetically than stenotopic taxa, so that once extinction has ceased, eurytopic taxa rapidly evolve new adaptations as they radiate into the habitats and niches left vacant by the extinction of previously existing taxa.

Key Terms

active site

adaptive radiation

allopatric speciation

anagenesis

analogous structures

balancing selection

binomial nomenclature

biogenetic law

blending inheritance

Cambrian Fauna

Central Dogma

chromosomes

clades

cladistics

cladogram

convergent evolution

demes

derived traits

descent with modification

differential reproduction

directional selection

divergent evolution

DNA

ecologic replacement

enzymes

eugenics

founder effect

genetic drift

genetic engineering

genetic recombination

groundplans

homeobox

homologous structures

hybrid vigor

industrial melanism

iterative evolution

labyrinthodonts

living fossils

lobe-finned fish

macroevolution

mass extinctions

microevolution

missing links

Modern Fauna

natural selection

neo-Darwinism	proteins	species sorting
ontogeny	pseudoextinction	subspecies
Paleozoic Fauna	punctuated equilibrium	taxonomy
particulate theory of inheritance	races	teleological
peripheral isolates	recovery phase	theory of the inheritance of acquired characteristics
phyletic gradualism	reproductively isolated	
phylogeny	sexual selection	transcription
preadaptation	social Darwinism	translated
primitive	species	vestigial structures

REVIEW QUESTIONS

1. What are the basic observations that led to Darwin's theory of evolution?
2. Give any modern and ancient examples for (a) natural selection and (b) evolution.
3. What was the significance of Mendel's discoveries?
4. Diagram the replication, transcription, and translation of DNA.
5. Diagram the Central Dogma of cell biology. How does a mutation affect the kinds of proteins produced?
6. How does a mutation affect the active site of an enzyme?
7. What is involved in genetic recombination?
8. What is the difference between the terms natural selection, founder effect, and genetic drift?
9. Arrange the following in order (first to last): geographic isolation, allopatric speciation, reproductive isolation. How does one give rise to the next?
10. How can races and subspecies lead to the production of new species?
11. Construct a simple, hypothetical set of taxa with a few characteristics and then draw a cladogram for them.
12. Distinguish between monophyletic, polyphyletic, and paraphyletic. Are paraphyletic taxa polyphyletic?
13. What are the processes involved in (a) microevolution and (b) macroevolution?
14. What is a living fossil, and what does it represent in terms of punctuated equilibrium?
15. Why is preadaptation important for evolution? Give examples.
16. At what taxonomic categories do microevolution and macroevolution act? Where might they overlap?
17. Using diagrams, show how allopatric speciation is related to the pattern of punctuated equilibrium seen in the fossil record.
18. What pattern(s) do microevolution and macroevolution generate in the fossil record?
19. Contrast natural selection with species sorting.
20. Why are "missing links" rare?
21. What is the role of extinction in evolution?
22. List as many examples of natural selection as you can that were discussed in this chapter.

FOOD FOR THOUGHT:
Further Activities In and Outside of Class

1. Do you think evolution is "teleologic," meaning that it has a purpose or goal?
2. Compare the concept of species as defined by biologic criteria with a species as it is recognized in the fossil record. How do they differ? How are they similar? Why are they similar?
3. Exactly what is "success" in terms of natural selection?
4. What is the difference between a preadapted structure and a vestigial structure? How do both argue for evolution? Give examples.
5. Can one oppose the theory of evolution without opposing the scientific method?
6. Does science work in a cultural and political vacuum?
7. What differentiates mass, minor, and background extinction?
8. Comment on the role of causality in extinction (look ahead to Chapters 11, 12, and 13).
9. What is the role of contingency (Chapter 1) in extinction?
10. Why is it important to find out if the increase of fossil biodiversity is real or not?

Sources and Further Reading

Albritton, C. C. 1980. *The abyss of time: Changing conceptions of the earth's antiquity after the sixteenth century.* San Francisco: W. H. Freeman.

Bambach, R. K. 1933. Seafood through time: changes in biomass, energetics, and productivity in the marine ecosystem. *Paleobiology, 19,* 372–397.

Bambach, R. K. 1999. Energetics in the global marine fauna: a connection between terrestrial diversification and change in the marine biosphere. *GeoBios, 32,* 131–144.

Bowler, P. J. 1989. *Evolution: The history of an idea.* Berkeley, CA: University of California Press.

Carroll, S. B. 2005. *Endless forms most beautiful: The new science of evo devo and the making of the animal kingdom.* New York: W. W. Norton.

Desmond, A. and Moore, J. 1991. *Darwin: The life of a tormented evolutionist.* New York: Warner Books.

Eldredge, N. 1995. *Reinventing Darwin: The great debate at the high table of evolutionary theory.* New York: John Wiley & Sons.

Henig, R. M. 2000. *The monk in the garden: The lost and found genius of Gregor Mendel, the father of genetics.* Boston: Houghton Mifflin.

Hannisdal, B., and Peters, S. E. 2011. Phanerozoic Earth system evolution and marine biodiversity. *Science, 334,* 1121–1124.

Katz, M. E., Wright, J. D., Miller, K. G., Cramer, B. S., and Fennel, K. 2005. Biological overprint of the geological carbon cycle. *Marine Geology, 217,* 323–338.

Martin, R. E. 1996. Secular increase in nutrient levels through the Phanerozoic: implications for productivity, biomass and diversity of the marine biosphere. *Palaios, 11,* 209–219.

Henig, R. M. 2000. *The monk in the garden: The lost and found genius of Gregor Mendel, the father of genetics.* Boston: Houghton Mifflin.

Raff, R. 1996. *The shape of life.* Chicago: University of Chicago Press.

Schwartz, J. H. 1999. *Sudden origins: Fossils, genes, and the emergence of species.* New York: John Wiley & Sons.

Schwartz, J. 2008. *In pursuit of the gene,* Cambridge, MA: Harvard University Press.

Shermer, M. 2002. *In Darwin's shadow: The life and science of Alfred Russell Wallace. A biographical study on the psychology of history.* Oxford, UK: Oxford University Press.

Vendetti, C., Meade, A., and Pagel, M. 2010. Phylogenies reveal new interpretation of speciation and the Red Queen. *Nature, 463,* 349–352.

Wallace, B. 1992. *The search for the gene.* Ithaca, NY: Cornell University Press.

Watson, J. 1968. *The double helix.* New York: Signet.

Zimmer, C. 1998. *At the water's edge: Macroevolution and the transformation of life.* New York: The Free Press.

Geologic Time and Stratigraphy

MAJOR CONCEPTS AND QUESTIONS ADDRESSED IN THIS CHAPTER

(A) What is the science of stratigraphy?

(B) What are relative and absolute ages, and how are they established?

(C) How is radiometric dating established?

(D) How are relative and absolute ages integrated to date rocks?

(E) How did the geologic time scale develop?

(F) What is correlation, how is it accomplished, and what are its pitfalls?

(G) Why can't rocks be equated with time?

(H) How are fossils used in correlation?

(I) How does the geographic distribution of organisms affect correlation?

(J) How complete is the geologic record?

(K) Why is sea-level change important?

CHAPTER OUTLINE

The Grand Canyon. There is currently a great deal of debate about when the Grand Canyon formed in response to downcutting by the Colorado River as the Colorado Plateau was being uplifted. One recent estimate places the age at 70 million years ago, but this estimate has been criticized for methodological reasons. Other estimates are much younger: between about 6 and 17 million years ago. The rocks exposed along the sides of the canyon range in age from Precambrian to Permian.

© dibrova/Shutterstock, Inc.

6.1 Introduction

James Hutton is said to have first grasped the enormity of geologic time at Siccar Point, Scotland, in the late 1700s (see Chapter 1). Here, it is said that Hutton realized the juxtaposition of horizontal and vertical rocks could only mean one thing: processes were acting on the planet that were not directly observable to humans because they acted so slowly as to be imperceptible and therefore Earth was incomprehensibly old (recall that Hutton lived well before Charles Darwin, whose theory of evolution also implied that Earth was quite old; see Chapter 5).

The concept of *geologic* durations of time—encompassing hundreds of millions to billions of years—is thought by some to be geology's greatest contribution to human thought. Another important implication of the great durations of passing time is that there has been plenty of time for Earth's systems to evolve. Given enough time the behavior and evolution of Earth's systems are not necessarily due only to the processes we can observe and measure on human time scales. But to unravel Earth's systems and their history, we must know not just *what* happened but *when* it happened. Knowing when events occurred, we can begin to understand cause and effect. If one event follows another in the geologic record, it is highly unlikely that the second event caused the first, but it is possible that the first event might have caused the second event, or at least set the stage for the second through contingency (see Chapter 1).

Age relationships in geology are typically established by the discipline of stratigraphy. **Stratigraphy** is the study of stratified (layered) rocks, or strata, and their age relationships. Stratigraphy studies sedimentary rocks because sedimentary rocks are typically stratified. It is by stratigraphy that we determine when things happened and decipher the history of a region from its geologic record. By establishing the history of events, we might also be able to infer the processes involved. Much of this chapter is concerned with how the ages and relationships of stratified rocks are established.

6.2 Relative Ages

The easiest way to establish the age of something is to establish its age *relative* to something else. Such ages are called **relative ages**, meaning rocks and the fossils in them are older or younger than something else. Establishing relative ages involves recognizing *sequences* of events and therefore history.

In the case of Siccar Point, we can establish a sequence of events using relative ages. The **Principle of Superposition** states that younger sedimentary rocks lie on top of older rocks. The related **Principle of Original Horizontality** states that sedimentary strata are deposited initially as horizontal layers. Based on superposition and original horizontality, at Siccar Point there was first horizontal deposition of sedimentary layers or strata, followed by deformation and tilting of the strata. Then erosion of the tilted strata occurred, followed by more deposition of horizontal strata (see Chapter 1).

6.3 Absolute Ages

Superposition yields relative ages. Although the general sequence of events in the history of a region could still be reconstructed without absolute ages, it is uncertain by exactly how much one layer is older than another.

By contrast, **radiometric dates** ("metric" for measure) provide **absolute ages** of rocks, or the age of rocks in years. Radiometric dates not only provide the ages of rocks and fossils, they also allow scientists to calculate the approximate *rates* of processes such as deposition, erosion, and uplift because they know how much time has elapsed during such processes.

Radiometric dates are derived from the decay of radioactive isotopes (**Figure 6.1**). These are the same types of isotopes that generate Earth's internal heat. A radiometric date is calculated based on the concept of the half-life. A **half-life** is the amount of time it takes for one-half of *any* amount of a radioactive isotope (called the **parent isotope**) to decay to its **daughter** product. For example, it takes 704 million years for one-half of any quantity of the radioactive isotope ^{235}U to decay to the nonradioactive element ^{207}Pb (lead-207; see Chapter 2 for notation of isotopes). The ← age of a rock can be estimated from the ratio of the parent to daughter (Figure 6.1). After the end of one half-life, half of the radioactive parent should be present. Thus, the ratio of daughter to parent is 0.5:0.5 or 1:1. After two half-lives have passed, one-fourth of the radioactive parent remains, and the daughter-to-parent ratio is 3:1, and so on.

Radioactive isotopes vary substantially in the durations of their half-lives. Some, such as uranium-lead and thorium-lead, have half-lives of billions of years. These isotopes have been present since the solar system formed and are commonly used to date igneous intrusive rocks. The same isotopes have also been used to date moon rocks and meteorites. The potassium-argon method is used to date volcanic (extrusive) igneous rocks such as ash layers from which individual crystals cannot be extracted.

One of the most famous radiometric dating methods is that of carbon-14 (^{14}C). Carbon-14 is produced naturally in Earth's atmosphere by cosmic ray bombardment (**Figure 6.2**). Carbon-14 is initially incorporated into the carbon cycle by photosynthesis and is then passed to animals when the plants are eaten. As long as an organism is alive, plant or animal, the ^{14}C that decays is replaced and is absorbed by organisms, along with ^{12}C, in a nearly constant ratio; after the organism has died the ratio begins to change as ^{14}C decays back to ^{14}N. Carbon-14 has a half-life of 5,730 years, which means it can be used back to about 70,000 years. Because the latter portion of this time interval includes the spread of human civilization, the carbon-14 technique has been widely used in archaeology to date charcoal, bones, and shells of organisms such as clams eaten by humans.

Sometimes the ^{14}C technique can be coupled with other means of dating that yield absolute ages. One widely used method is tree rings, which are studied by the science of **dendrochronology** ("dendro" for tree, "chronology"

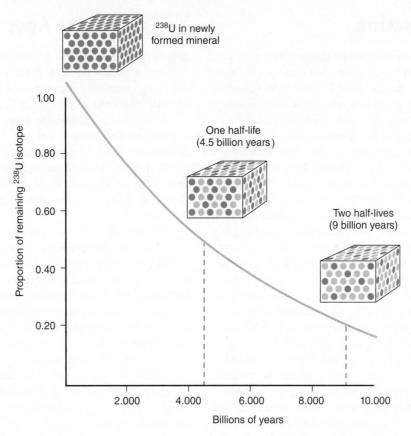

FIGURE 6.1 A general curve for the radioactive decay of an element. After one half-life has passed, one-half of the original radioactive parent is present. After two half-lives, one-fourth of the parent is still present, and so on. By finding the ratio of the daughter product to radioactive parent, one can determine the number of half-lives that have passed and thus the age of the rock.

for time; Figure 6.2). In dendrochronology, tree rings are counted to yield precise dates (in years). However, if the tree is dead the rings do not tell us *when* the tree died but its time of death can be determined by carbon-14 dating. We also can use tree rings to infer climatic conditions. Thick tree rings indicate relatively wet conditions, whereas thin rings indicate dry conditions or even drought. Coupled with carbon-14 dates, dendrochronologic studies can therefore be used to precisely infer the timing of past climatic conditions. Such studies can then be related to archaeological ones to determine the effect of past climatic conditions on humans.

We must keep in mind several factors when using radiometric dates, however. First, the radiometric "clock" only starts from the time when the mineral reached its blocking temperature. The **blocking temperature** of a mineral is typically hundreds of degrees cooler than the melting temperature of the mineral. The blocking temperature is therefore sufficiently cool to prevent the parent and daughter isotopes from breaking chemical bonds with the rest of the mineral and moving into and out of it, which can alter the age. Different minerals have different blocking

temperatures; hornblende's is much higher than that of biotite's, for example.

Second, the mineral must have remained a closed system after the blocking temperature was reached. **Closure** means the loss or addition of either parent or daughter matter, which can alter the age. For example, the potassium-argon method measures the ratio of potassium-40 to argon-40. However, argon is a gas and can leak from the mineral crystal, lowering its apparent age. Leakage can occur in response to heat and pressure associated with metamorphism. If the daughter isotope is completely removed by metamorphism, we are actually dating the time since metamorphism.

Third, weathering and leaching can also affect ratios of different radioactive isotopes. Thus, only fresh, unweathered samples must be used for dating.

One means of determining the accuracy of dates is the method of cross-checks. If dates obtained using two different isotopic pairs agree, or are **concordant**, it is likely that this is the correct age of the rock. If the dates disagree, or are **discordant**, it is possible that one or both methods are incorrect, and we must use other techniques to determine which date, if either, is correct.

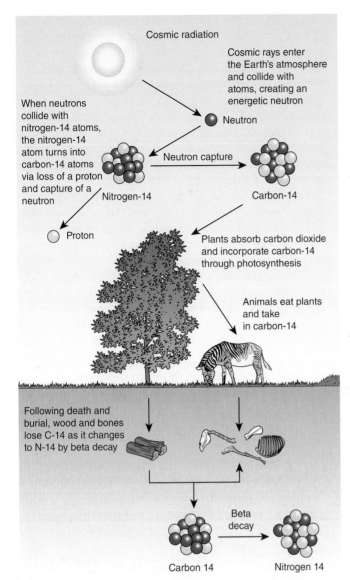

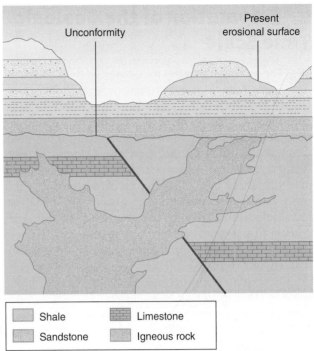

Data from: Levin, H. 2006. *The Earth Through Time.* 8th ed. Hoboken, NJ: John Wiley. Figure 2.11 (p. 20).

FIGURE 6.3 Radioactive isotopes provide absolute ages of rocks using the principle of cross-cutting relationships. Although the general sequence of events in the history of a region can be reconstructed without absolute ages, radiometric dates give scientists greater certainty to their conclusions and allows scientists to calculate rates of processes such as deposition, erosion, and uplift. However, even absolute dates give only approximate ages, as shown in this example. For example, in the figure, the sedimentary rocks penetrated by the igneous dike must have existed before the dike was intruded, but we cannot tell how long the sedimentary rocks existed before the intrusions without other evidence. Similar considerations apply to the fault cut by the dike and the unconformity at the top.

FIGURE 6.2 The carbon-14 dating technique. We can use this technique to date archaeological objects, bones, and tree rings, which yield paleoclimate information as shown. In temperate and higher latitudes, trees add annual layers (rings) that, when coupled with carbon-14 dates, we can use to precisely assess when past changes in precipitation occurred and their possible impact on human settlements.

radiometric date is subject to error, sometimes up to millions of years, so that when an age is reported it must be remembered that the age really could be anywhere within the margin of error.

This means we must seek other data to establish finer subdivisions of time to produce a reliable geologic time scale. These finer subdivisions are based on the appearance and extinction of different fossil species through time.

D Despite the advantage of radiometric dates, absolute ages are not exact. Radiometric dates usually come from igneous rocks. Thus, if one is dealing primarily with a sequence of sedimentary rocks, the radiometric dates come from igneous plutons, dikes, or sills intruded into the sedimentary sequence (**Figure 6.3**). Based on the **Principle of Cross Cutting Relationships**, which states that rock bodies that cut across others must have come after them, the dikes and sills must have been injected after the deposition of the sedimentary rocks into which they were intruded. Still, we don't know exactly when the dikes or sills were intruded after the surrounding strata were deposited. Also, a

CONCEPT AND REASONING CHECKS

1. What are the principles on which relative ages are based?
2. Do radiometric ages by themselves give the precise age of a particular stratum?
3. How many human life spans—or generations—do you believe an individual stratum might normally represent? A few? A lot? How much time might this represent in terms of absolute time?

6.4 Evolution of the Geologic Time Scale

The earliest time scales were based on the Principle of Superposition. Similar sequences of rocks were observed by workers in Europe and Russia about the same time and led them to erroneous conclusions. Because the same kinds of rocks were geographically widespread, these workers concluded they must have all formed at once; therefore, each basic rock type represented a distinct phase in Earth's history. In fact, the doctrine of **Neptunism**, which was a form of catastrophism (see Chapter 1), stated that rocks were successively deposited from a global ocean. Catastrophism arose partly as an attempt to "fit" the slowness of geologic processes to the short time scales derived from biblical scripture (**Box 6.1**).

These early time scales had several major subdivisions based on lithology, or rock type (igneous rocks, shales, sandstones, limestones, and so on). The founders of the earliest time scales equated rock with time because the significance of fossils was unknown at that time, as was the theory of evolution. The earliest rocks at the base of the scale were called "primary" and consisted of crystalline rocks with metal ores (Ganggebirge for "ore mountain"; **Table 6.1**). "Secondary" rocks formed on top of primary rocks, were stratified, and contained fossils (Flötzgebirge or "layered mountain"). On top of secondary rocks came "tertiary" (or "third") rocks, which also contained fossils. These rocks were capped by the poorly consolidated alluvium such as the sands and gravels of stream and river channels. Alluvium was also referred to by wonderful tongue-twisters such as "Angeschwemmtgebirge," "Aufgeschwemmtgebirge," and "Neues Flötzgebirge" (Table 6.1). Although most of these terms have long since been dropped, some of the older terms such as "Tertiary" are still used, at least informally.

During the 18th and early 19th centuries, earth scientists such as Cuvier began to accept the existence of fossils as evidence for prehistoric life. However, it was not until 1815 that William Smith proposed the **Principle of**

BOX 6.1 Age of the Earth

In 1654, the Anglican Archbishop James Ussher of Ireland, a respected biblical scholar, calculated that Earth began on Sunday, October 23, 4004 B.C. According to Ussher, Earth was only 6,000 years old. He based his determination on Middle Eastern and Mediterranean history and biblical accounts of genealogies (family histories of ancestor–descendant relationships). Ussher also calculated that Adam and Eve were expelled from paradise 18 days after creation. Other scholars had earlier arrived at similar ages for Earth. These estimates severely constrained the thinking of many scientists who studied the history of Earth.

Much later the physicist Sir William Thompson (1824–1907), later known as Lord Kelvin, became a major antagonist in debates about the age of Earth and therefore the amount of time available for Darwinian evolution. By the last half of the 19th century, most investigators had accepted the idea that Earth was originally molten, based on phenomena like volcanism and that Earth was therefore losing heat. Kelvin pointed out that Lyell's concept of an equilibrium view of Earth (see Chapter 1) therefore had to be incorrect because Earth was losing heat. Kelvin reasoned that if Earth is hot now, it must have been hotter, and more likely molten, much earlier in its history.

After the initial publication of Darwin's *On the Origin of Species* in 1859, Kelvin published the paper "The 'Doctrine of Uniformity' in Geology Briefly Refuted." He calculated the rate of Earth's heat loss by assuming that (1) heat is lost from Earth at a constant rate, (2) Earth's composition is fairly uniform, and (3) there are no renewable sources of heat in Earth. Kelvin then extrapolated backward in time to Earth's molten state and calculated an age for Earth of about 100 million years. Although this is of course a long time, it was far less than Darwin's estimate of hundreds of millions of years based on the fossil record or Hutton's concept of nearly beginning-less time. In subsequent years, Kelvin continued to revise his age estimates downward, eventually settling on about 20 million years.

Kelvin's calculations for the age of Earth put Darwin in a bind from which he never escaped. To the end of his life Darwin believed Kelvin's calculations were wrong, although he could not explain why. Many scientists accepted Kelvin's calculations, mainly because they were based on physics and mathematical calculations rather than the science of the "higgle-de-piggledy" (as the famous 19th astronomer Sir John Herschel once referred to evolution). Also, Kelvin's stature in the scientific community was such that most other scientists

BOX 6.1 Age of the Earth (Continued)

avoided disagreeing with him. Consequently, during the later 1800s the acceptance of Darwin's views of long, slow rates of evolution began to decline, even among some of Darwin's strongest supporters. Many began to look for processes that could act at much faster rates of evolution to accommodate Kelvin's conclusion, even resurrecting a version of Lamarck's old theory of the inheritance of acquired characteristics (see Chapter 5).

We now know that the age of Earth is approximately 4.5 to 5 billion years based on radiometric dates. However, radioactivity, the source of Earth's internal heat, was not discovered until the late 1800s, late in Kelvin's life, and estimates of the age of Earth based on radioactive dating techniques did not start to become available until decades later. Initially, it was concluded that Earth was at least 3 billion years old based on dates reported by Ernest Rutherford and Arthur Holmes (of continental drift fame; **Box Figure 6.1A**).

However, in 1956 Clair Patterson at Cal Tech found the decay of radioactive isotopes in the Canyon Diablo meteorite (which formed Meteor Crater in northern Arizona) indicated an age for Earth of 4.56 billion years (Box Figure 6.1A). Patterson assumed meteorites were left over from the birth of the solar system (see Chapter 7). Patterson's dates were later corroborated by concordant ages obtained for crustal rocks from Earth using rubidium-strontium (Rb-Sr) and potassium-argon (K-Ar) isotope pairs (see Chapter 2) and by dates on Moon rocks brought back by the Apollo astronauts. Like meteorites, the Moon also formed as our solar system originated.

So, Darwin was basically right. Nevertheless, Kelvin maintained to his dying day that Earth could not be as old as Darwin or later scientists claimed. Today, Kelvin lies buried in Westminster Abbey, next to Sir Isaac Newton—and Charles Darwin.

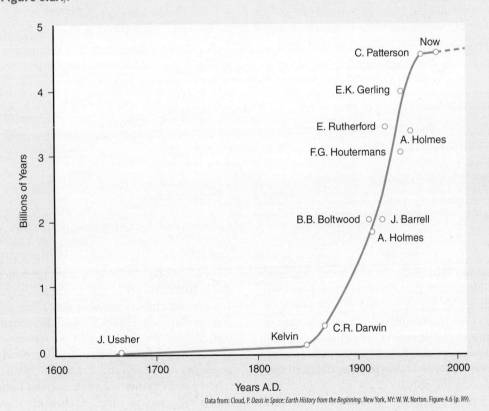

Data from: Cloud, P. *Oasis in Space: Earth History from the Beginning.* New York, NY: W. W. Norton. Figure 4.6 (p. 89).

BOX FIGURE 6.1A Changing estimates of the age of the Earth.

TABLE 6.1

The Evolution of the Geologic Time Scale*

Early Subdivision				Modern Usage			
Arduino 1760	Lehmann 1756 Fochsel 1760–1773	Werner ca. 1800	English Equivalents	Eras	Periods	Epochs	Alternate Periods
Volcanic Alluvium Tertiary	Aufgeschwemmt-gebirge	Aufgeschwemmt-gebirge or Neues flötzgebirge	Alluvium Tertiary	Cenozoic Phillips 1841	Neogene Hoernes 1853	Holocene Pleistocene	Quaternary Desnoyers 1829
						Pliocene Miocene	Tertiary Arduino 1760
					Paleogene Naumann 1866	Oligocene Eocene Paleocene	
Secondary	Flötzgebirge	Flötzgebirge	Secondary	Mesozoic Phillips 1841	Cretaceous d'Halloy 1822		
					Jurassic von Humboldt 1799		
					Triassic von Alberti 1834		
				Paleozoic Sedgwick 1838	Permian Murchison 1841		
					Pennsylvanian Williams 1891	Carboniferous Conybeare and Phillips 1822	
					Mississippian Winchell 1870		
Primitive	Ganggebirge	Übergangs-gebirge	Transition		Devonian Murchison, Sedgwick 1839		
					Silurian Murchison 1835		
					Ordovician Lapworth 1879		
					Cambrian Sedgwick 1835		
		Ürgebirge	Primary	Precambrian			

*Workers and the date of recognition of the most commonly used modern units are indicated.
Reproduced from: Mintz, L. W. 1977. *Historical Geology: The Science of a Dynamic Earth*, 2nd ed. Columbus, Ohio: Charles E. Merrill. 588 pp. (Figure 2.8, p. 14).

Faunal Succession (**Table 6.2**). Smith noticed that fossil assemblages always occurred in the same superpositional sequence, which led to his recognition of this principle. Smith was a surveyor responsible for the construction of canals in England as the Industrial Revolution grew; he therefore had ample opportunity to observe rocks and fossils over large areas and to test his hypothesis of faunal succession by predicting the sequences elsewhere. In reality, the Principle of Faunal Succession is based on the fact that fossil biotas have changed through time because of biologic evolution. Thus, life, unlike rocks, *does not tend to repeat itself.* The directionality of life (sometimes called Dollo's Law) is of course the result of biologic evolution; however, the recognition of Faunal Succession preceded Charles Darwin's *On the Origin of Species,* published in 1859, by nearly half a century.

Like superposition, fossils also give relative ages. However, fossils produce much finer subdivisions of the time scale than rocks alone (Table 6.1). Consequently, the primary is now called the **Precambrian**, which records the origin and development of Earth and its early life during its three main subdivisions or **eons**: the Hadean (from Hades, referring to Earth's early molten state), **Archeozoic** ("ancient life"), or just **Archean**, and **Proterozoic**, or "first life." Each of these eons is characterized by distinctive rock types and climate.

Many paleontologists concentrate their efforts somewhere within what used to be called the "Secondary," the "**Tertiary**," and the "**Quaternary**." Together, these intervals encompass

TABLE 6.2

Geologic Ages and Associated Biotic Events

Time Scale				Millions of Years Before Present (approx.)	
Eon	Era	Period	Epoch		Some Major Biotic Events
Phanerozoic	Cenozoic	Quaternary	Holocene (last 11,700 years)	0.01	Appearance of humans
			Pleistocene		
		Tertiary	Pliocene	2.8	Dominance of mammals and birds
			Miocene	5.3	Proliferation of bony fishes (telosts)
			Oligocene	23	Rise of modern groups of mammals and invertebrates
			Eocene	34	Dominance of flowering plants
			Paleocene	56	Radiation of primitive mammals
	Mesozoic	Cretaceous		65	First flowering plants Extinction of dinosaurs
		Jurassic		145	Rise of giant dinosaurs Appearance of first birds
		Triassic		200	Development of conifer plants
	Paleozoic	Permian		251	Proliferation of reptiles Extinction of many early forms (invertebrates)
		Carboniferous	Pennsylvanian	299	Appearance of early reptiles
			Mississippian	318	Development of amphibians and insects
		Devonian		359	Rise of fishes First land vertebrates
		Silurian		416	First land plants and land invertebrates
		Ordovician		443	Dominance of invertebrates First vertebrates
		Cambrian		488	Sharp increase in fossils of invertebrate phyla
Precambrian	Proterozoic	Neoproterozoic		542	Appearance of multicellular organisms
		Mesoproterozoic		1,000	Appearance of eukaryotic cells
		Paleoproterozoic		1,600	Appearance of planktonic prokaryotes
	Archean			2,500	Appearance of sedimentary rocks, stromatolites, and benthic prokaryotes
	Hadean			4,000	From the formation of Earth until first appearance of sedimentary rocks; no observable fossil organisms

Note: Dates derived mostly from Gradstein et al. *A Geologic Time Scale*. Cambridge University Press, 2004, and from Geologic Time Scale, available from http://www.stratigraphy.org, accessed August 2012.

about the last 540 million years. This interval is also called the **Phanerozoic** Eon, or the "time of apparent life," because it is when fossils are typically abundant. The distinctive changes in fossil assemblages of the Phanerozoic were recognized by Sir John Phillips to consist of three main **eras**, each now recognized to span tens to hundreds of millions of years. Phillips recognized that each era was represented by very distinct groups of fossils: **Paleozoic** ("ancient life"), **Mesozoic** ("middle life"), and **Cenozoic** ("recent life"; Table 6.1).

Many of the boundaries between the units of time scale correspond to major faunal, floral, and climatic change, including mass and minor extinctions (see Chapter 5). Each of the eras has its own distinct fossil assemblages or biotas that were later used to subdivide the eras into smaller units called **periods**. Periods are roughly tens of millions of years in duration. It is this sequence of fossils through the periods that Smith observed.

By the middle of the 19th century most of the geologic periods had been recognized in Europe based on fossil biotas (Table 6.1). Each geologic period's name was assigned based on characteristic features of the rocks, the area in which the fossils were first described, or both. The **Cambrian** is the first period of the Paleozoic and is an ancient name for Wales, whereas the next two periods, the **Ordovician** and **Silurian**, were named after ancient Welsh tribes. **Devonian** refers to Devonshire, England, and **Carboniferous** to coal-bearing units first recognized in England. The Carboniferous is often divided into the Early and Late Carboniferous in Europe, which correspond to the **Mississippian** and **Pennsylvanian** periods of the United States. The last period of the Paleozoic is the **Permian**, named for the Perm province in the Ural Mountains far to the east of Moscow in Russia, where they were described. Next is the Mesozoic Era, comprised of the **Triassic** and named for a distinctive, three-fold ("tri") sequence of rocks in eastern Germany; the **Jurassic**, named for the Jura Mountains of southeast France and Switzerland; and the **Cretaceous** for the abundant chalks ("creta") of this time. The Tertiary ("third") and Quaternary (meaning "fourth") follow.

All geologic periods are further subdivided into **epochs**. The most widely used epochs are those of the Tertiary and Quaternary periods. Epochs were first formally recognized by Lyell in the *Principles of Geology* based on rocks of the Cenozoic Era exposed in the vicinity of London and Paris. Lyell used the ratio of *extant* (or still-living) mollusks (mainly clams and snails) to extinct mollusks present in the rocks to recognize the epochs. From this fraction, he calculated the percentage of extant species also found as fossils. He found that as the rocks became older, the number of species still living today and found as fossils in the rocks decreased. Based on these percentages Lyell recognized the epochs **Eocene** ("dawn recent"), **Miocene** ("middle recent"), and Lower and Upper **Pliocene** ("more recent"). The Upper Pliocene was later renamed **Pleistocene** ("most recent") and represents the time of the "Ice Ages." About the last 10,000 years of the Pleistocene are often recognized as the **Holocene** ("wholly recent") for when human civilization spread over the globe. Other workers recognized the **Oligocene** ("few recent") and

Paleocene ("ancient recent") epochs, which was split off from the lower Eocene based on fossil floras.

The Paleocene through Oligocene periods are often lumped together as the **Paleogene** and the Miocene to Recent into the **Neogene** because these two intervals represent fairly distinctive climate modes: the Paleogene was generally a warm interval with widespread seas until the Late Eocene and Oligocene, whereas the Neogene generally represents a time of widespread glaciers, colder conditions, and lower sea level. Geologic committees have recently recommended the use of the terms Paleogene and Neogene in place of the epochs. It has also recently been suggested that the epochs of the Cenozoic and the term "Tertiary" should be dropped and only the terms Paleogene, Neogene, and Pleistocene should be used. The date for the beginning of the Pleistocene has also been moved back to 2.8 Ma. We will continue to use the epoch terms, however, because they are well embedded in the literature and provide further subdivisions of the Cenozoic that are useful in further discussions in this text.

CONCEPT AND REASONING CHECKS

1. Why is the Principle of Faunal Succession an example of directionality in Earth's history?

6.5 Correlation

As the geologic time scale began to be established, workers attempted to recognize rocks of the same fossil content and therefore age. The procedure that infers the age equivalence of rocks is called **correlation**. We can carry out correlation in a number of different ways.

6.5.1 Lithocorrelation

The study of the stratigraphic relationships of rocks based solely on lithology is called **lithostratigraphy**. Correlation using lithology, or rock type, alone is therefore called **lithocorrelation**. Sometimes we have no choice other than to correlate using lithology because fossils are absent.

Lithocorrelation is usually done in the field by the folksy-sounding process of **walking the outcrop** (or exposure), which means one maps the rock unit(s) in the field visually over the area in which they are exposed. In lithocorrelation, one assumes certain distinctive rock units, or **distinctive sequences** of rock units, are *contemporaneous*, meaning they were formed more or less at the same time (**Figure 6.4**). Such units could be a distinctively colored sandstone or a limestone that is very resistant to weathering or perhaps a distinctive sequence of rocks such as a shale overlain by a coal bed and then a limestone; such distinctive sequences are common in coal-bearing areas of the Carboniferous.

Another method of lithocorrelation uses distinctive **marker beds**, such as volcanic ash falls, which can range in thickness from a few centimeters to perhaps a meter or so in thickness, and are often spread over wide areas. Volcanic ashes are extremely useful because, geologically speaking,

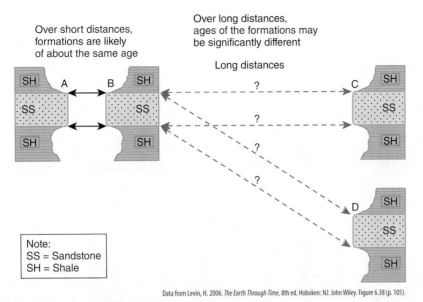

Over short distances, formations are likely of about the same age

Over long distances, ages of the formations may be significantly different

Long distances

Note:
SS = Sandstone
SH = Shale

Data from Levin, H. 2006. *The Earth Through Time*, 8th ed. Hoboken: NJ: John Wiley. Figure 6.38 (p. 105).

FIGURE 6.4 Lithocorrelation correlates rocks over relatively short distances (localities A and B in the diagram) by assuming that the rocks formed more or less simultaneously. Volcanic ashes can also form distinctive beds that can be correlated over large areas. Distinctive sequences of beds can also be used for correlation. However, lithocorrelation becomes unreliable over greater distances (localities C and D) because, unlike fossils, rocks tend to repeat themselves through geologic time. See text for further discussion.

they are instantaneous events that almost resemble the ticks of the second hand of a watch.

When rocks cannot be observed at the surface, another method of lithocorrelation involves matching distinctive shifts in **well-logs**. Well-logs are commonly used to infer the lithologies of subsurface strata and their pore fluids (water, oil, gas). The well-logs result from lowering remote-sensing instruments down wells and measuring the ability of the fluids in the well and the surrounding rocks to conduct electricity, emit radioactivity, or other features (**Figure 6.5**). Well-logging techniques are widely used in petroleum exploration to determine the thicknesses and pore fluids in subsurface rocks and to correlate subsurface rocks that conduct large volumes of water, or **aquifers**.

Another method of lithocorrelation uses the magnetic properties of rocks. As igneous rocks cool from a molten state, iron-rich mineral particles present in the magma align themselves with Earth's magnetic field, which acts much like a bar magnet (**Figure 6.6**). The temperature below which the alignments of the particles become fixed in the rock is called the **Curie point** (named after Marie Curie, one of the codiscoverers of radioactivity).

The Earth's magnetic field is not completely stable, however. The iron-rich particles in solid magma indicate past states of Earth's magnetic field. The modern Earth's magnetic field is said to be normal, and the north magnetic pole (which differs from Earth's geographic North Pole or axis of rotation) is today found in northern Canada (**Figure 6.7**) but is currently moving toward Russia at a rate of about 40 miles per year. Earth's magnetic field, which derives from the core, also flip-flops back and forth between "normal" and "reversed" polarity. In a reversed magnetic field, Earth's magnetic south polarity shifts to the northern hemisphere and the north magnetic polarity shifts to the southern

hemisphere (Figure 6.7). The changes in magnetic polarity, or **magnetic reversals**, occur very quickly (geologically speaking) all over the world. Moreover, this behavior of iron-rich particles can occur not only in intrusive or extrusive rocks, but also sediments, both on land and on the seafloor. Because the reversals are recorded on land and in seafloor crust, both in igneous rocks and sediments, the changes in polarity can potentially be correlated between land and sea. In fact, paleomagnetic data and other types of datums are frequently integrated with what is called the oxygen isotope record of the Pleistocene Epoch to distinguish the different oxygen isotope stages, which reflect ice volume during the Pleistocene but which tend to resemble one another (see Chapter 15).

All these methods of lithocorrelation equate rock with time. Over short distances, equating rock with time is not a problem, and lithocorrelation is usually quite successful. However, over long distances, we cannot be sure that, for example, the sandstone one observes in an outcrop is the exact same sandstone far away (Figure 6.4). These sorts of problems led the originators of the early time scales awry. This is because, unlike fossils, rocks *repeat themselves through time*. Similarly, despite the great precision that oxygen isotope and paleomagnetic records offer for correlation over long distances, oxygen isotope stages and paleomagnetic reversals tend to look alike and they repeat themselves through time, just like rocks. Thus, we must distinguish and correlate different isotope stages or reversals by using other means such as fossils, which do not recur through time.

6.5.2 Formations and facies

We have now recognized a fundamental principle of stratigraphy: *the same kinds of rocks repeat themselves*

Walakpa 1

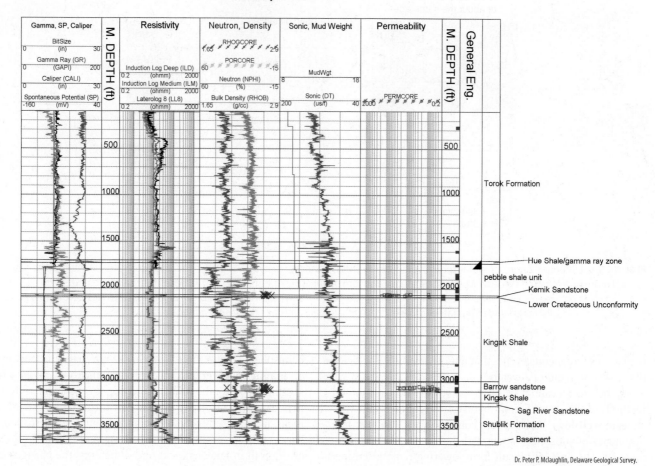

Dr. Peter P. Mclaughlin, Delaware Geological Survey.

FIGURE 6.5 A typical well-log. The squiggles indicate the lithology and kinds of fluids such as water, oil, or gas that fill the pores of the rocks.

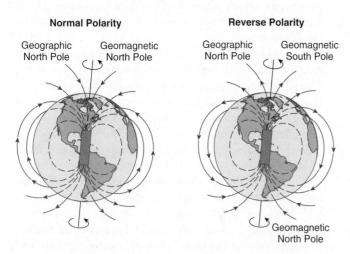

GEOMAGNETIC POLARITY REVERSALS

FIGURE 6.6 Changes in the Earth's magnetic field. The current orientation of the magnetic field is said to be normal, but it has flip-flopped through geologic time between normal and reversed states, with the south magnetic pole located in the northern hemisphere and the north magnetic pole located in the southern hemisphere.

through geologic time. Furthermore, *different types of rocks can occur at the same time.* With regard to sedimentary rocks these statements are sometimes referred to collectively as **Walther's Law**. We also can generalize the statements to include the environments in which the rocks were deposited.

The fact that rocks cannot be considered exactly equal to time is embodied by the concepts of formation and facies. A **formation** is a mappable unit of rock that is recognized based only on its lithology. A formation is therefore descriptive only, and there should be no interpretation of its age. An individual formation might range from only a few to hundreds of meters in thickness and might be traceable over areas of hundreds to thousands of square kilometers. Each formation is given a name based on the approximate location—or type locality—where it was originally recognized and described. Type localities can be towns, crossroads, rivers, streams, or some other geographic feature. Thus, formations bear names like Antietam Sandstone (found in the Appalachian Mountains), Niagara Limestone (of Niagara Falls), and Green River Shale (of Wyoming, which is famous for its fossil fish; see Chapter 14).

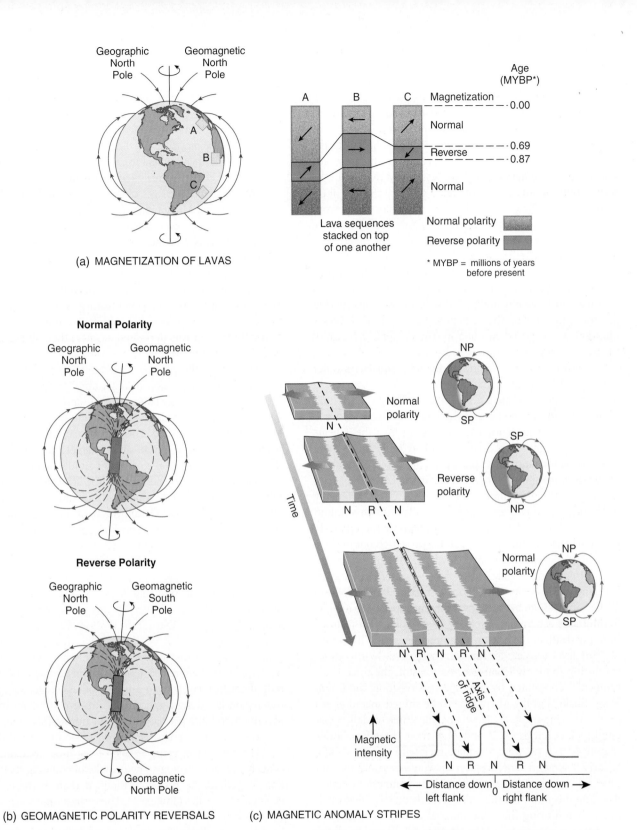

(a) MAGNETIZATION OF LAVAS

Age (MYBP*)

Magnetization — 0.00

Normal

— 0.69
Reverse
— 0.87

Normal

Lava sequences stacked on top of one another

Normal polarity

Reverse polarity

* MYBP = millions of years before present

Normal Polarity

Reverse Polarity

(b) GEOMAGNETIC POLARITY REVERSALS

(c) MAGNETIC ANOMALY STRIPES

FIGURE 6.7 As igneous rocks cool from a molten state, iron-rich mineral particles that are present in the magma align themselves with the Earth's magnetic field as the temperature decreases below the Curie point to produce "magnetic stripes" (see Chapter 2). This behavior can occur in intrusive or extrusive rocks on land or in seafloor crust. Similar behavior is exhibited by iron-rich mineral grains as they are deposited in sediment on the ocean floor.

The concept of **facies** (meaning "aspect") recognizes that (1) different sedimentary rocks and the environments in which they were deposited can exist at the same time and (2) the same rocks (environments) have existed at different times. For example, when the limestones of a reef's crest are forming, biogenic or oolitic calcareous muds and sands might be deposited in the back reef lagoon adjacent to the core (**Figure 6.8**). Although these relationships are easily visualized in a diagram, they are less likely to actually be seen in outcrop. In the case of Figure 6.8, what would likely be observed in the geologic record are muds or sands cropping out in separate exposures over a large area, with the reef core cropping out in other exposures, most likely over a smaller area. The lithologies might be recognized as separate formations because they are relatively distinctive, but the fact that the sediments and their environments coexisted simultaneously would not be immediately obvious. Unless the interfingering of the two lithologies (called a **facies change**) was exposed in outcrop, the contemporaneity of the two lithologies could only be established by correlation.

A classic example of how formations and facies differ is provided by rocks at the base of the Grand Canyon in northern Arizona (**Figure 6.9**). Immediately above the Precambrian rocks at the base of the Grand Canyon are three formations that occur in an upward sequence: the Tapeats Sandstone, Bright Angel Shale, and Muav Limestone. These three formations formed as parallel bands of sediment along the western margin of North America in the Early Cambrian, roughly about 540 million years ago. Because of the energy distribution of waves and currents, sandstones were found closest to shore and shales farther away; only the relatively coarse particles of the Tapeats Sandstone settled out in wave and current-swept environments, whereas the finer particles of the Bright Angel Shale settled in quiet water farther away from shore. The rocks of the Muav Limestone tended to form farthest from shore because the sediment supply was too low to dilute the accumulation of calcareous shells that resulted from the death of the organisms living there (see Chapter 4).

Sea level rose across the western part of North America during the Early Cambrian. As a result, the three environments represented by the Tapeats Sandstone, the Bright Angel Shale, and the Muav Limestone shifted inland as sea level rose (Figure 6.9). Now imagine three widely separated rock outcrops (X, Y, and Z) of these three formations (Figure 6.9). If we were to equate rock with time at sites X, Y, and Z, based on superposition we would have "sandstone-time" followed by "shale-time" followed by "limestone-time." Based on this interpretation, *all* the sandstone was deposited everywhere during the same interval of time, then all the shale everywhere on top of the sandstone, followed by all the limestone. The historical reconstruction of this area would then be as follows: (1) sea level rose and all the sandstone was deposited uniformly throughout the area, (2) then all the shale was deposited on top of the sandstone in deeper water over the entire area as sea level rose further, and (3) then all the limestone was deposited on top of the shale after the terrigenous sediment supply from land was shut off. In this reconstruction *rock has been equated with time* and the ages of each of the Tapeats Sandstone, Bright Angel Shale, and Muav Limestone are inferred to be the same throughout their whole extent.

But this isn't what happened. All three environments coexisted as facies (Figure 6.9). In other words these three types of rocks formed in different environments that existed *at the same time,* not at succeeding times. Thus, the time lines must not be drawn parallel to and separating the formations (Figure 6.9). Rocks normally "cut across time," so to speak. In this interpretation, the ages of the Tapeats Sandstone, Bright Angel Shale, and Muav Limestone at site X are different from their ages at site Z, even though the rock types are identical (Figure 6.9). This is because as sea level rose and fell, the three environments and their associated lithologies gradually shifted landward and seaward, respectively.

Thus, to equate rock and time can completely obliterate any notion of the true geologic history of a region and the processes and the rates involved. In the incorrect reconstruction, all the sand was deposited at once, then all the shale, and then all the limestone.

What does the incorrect interpretation say about sediment supply to the region through time? In this case, there must have been a rather large influx of sand into a very widespread and shallow marine setting while the Tapeats Sandstone was being deposited. This suggests an uplift nearby that resulted in extensive erosion and redeposition. Why, then, did sandstone deposition stop rather suddenly and give way to shale in this interpretation? Did the sea level rise so rapidly that coarse-grained sediments like sand were prevented from reaching the area? Or did the basin in which the sediments were accumulating suddenly deepen, which would also appear as a sudden sea-level rise? Or perhaps the source rocks for the sand were largely eroded by the end of "Tapeats time"?

Note that this last interpretation really doesn't make sense: the time it probably took for the Tapeats to be deposited—millions of years—was undoubtedly much shorter than it would take for the erosion of a mountain range given the slow rates of weathering. So, if we were leaning toward accepting the incorrect interpretation, we ought to have some intuitive realization of our mistake and begin to rethink the history of the area. In fact, other evidence suggests this region lay along a quiescent continental margin where mountain building was largely absent during this time, so it would seem that extensive uplift and erosion was unlikely.

The concept of facies is one of the most fundamental concepts in all of geology. The mistake of equating rock and time is perhaps no better exemplified than in the earliest geologic time scales (Table 6.1). The earliest time scales were essentially based on a kind of lithocorrelation: younger rocks lay on top of older rocks, and, with no evidence to the contrary, it was only natural for some workers to conclude that the succession of lithologies in the areas where they worked were characteristic of certain times in Earth's history. Hence, the subdivisions Primary, Secondary, and Tertiary of the earliest time scales were thought to apply to the whole Earth, like the layers of a cake (Table 6.1). However, to equate rock with time at the scale of the whole Earth is no more satisfactory than, in our example, equating rock with time in

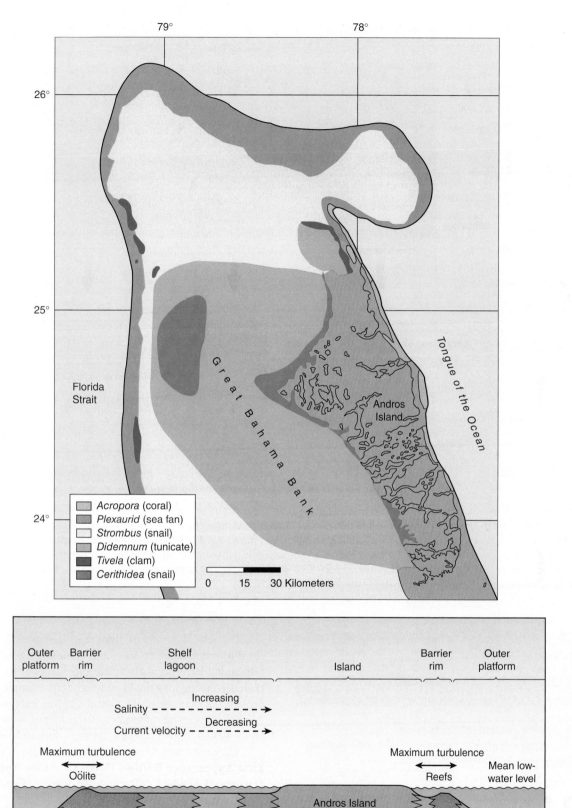

FIGURE 6.8 Map view and cross-section of modern environments on the Great Bahama Bank. A number of different environments and rock types exist at the same time. In the geologic record, these rock types would be recognized and mapped as formations, but the fact that they were deposited simultaneously would not be established until time lines were added.

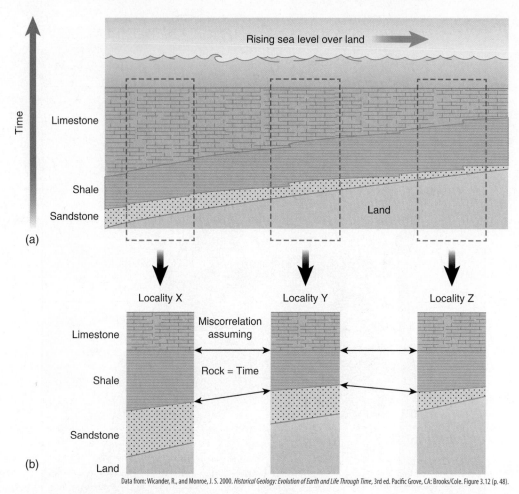

(a)

(b)

Data from: Wicander, R., and Monroe, J. S. 2000. *Historical Geology: Evolution of Earth and Life Through Time*, 3rd ed. Pacific Grove, CA: Brooks/Cole. Figure 3.12 (p. 48).

FIGURE 6.9 The difference between facies and formations, illustrating that rock and time are not the same. **(a)** As sea level rises, the three formations represented by sandstone, shale, and limestone, move landward. All three environments exist through time, but the ages of each of the types of rocks, which are recognized as formations, differs over the area in which the rocks are exposed. **(b)** Miscorrelations result if rock and time are considered equal (arrows). The red and orange dots represent the extinctions of trilobites that are used instead to produce more accurate time lines for correlation (see section "6.5.3 Biostratigraphy" for further discussion). Note how the ages of the rocks are not the same between the localities, even within the same formation.

the field at the scale of a few outcrops at the base of the Grand Canyon.

CONCEPT AND REASONING CHECKS

1. Why doesn't rock normally equal time?
2. Diagram the shift of facies in a transgressing and regressing seaway. Then, indicate the formations.

6.5.3 Biostratigraphy

 Very often, though, lithocorrelation works quite well. In fact, geologists use it all the time to correlate well-logs within relatively small areas. However, rock can only be confidently equated with time over relatively short distances. If we are to correlate successfully over longer distances and avoid making erroneous reconstructions of Earth history, we must use fossils because the succession of fossils can be equated with time.

The use of fossils to study the stratigraphic relationships of sedimentary rocks is called **biostratigraphy**. After formations have been identified and mapped, they can be correlated using their enclosed fossils. For example, the time lines in the example of the Grand Canyon were established using the extinctions of fossils called trilobites, which are distantly related to insects, crabs, lobsters, shrimp, and spiders.

The easiest way to conduct biostratigraphy is to use the **First Appearance Datums (FADs)** and **Last Appearance Datums (LADs)** of different fossil species (**Figure 6.10**). Usually, the FAD of a species represents the first evolutionary appearance of the species, and the LAD its extinction. The use of FADs and LADs was developed over the past few decades by the Deep Sea Drilling Program; its successor, the Ocean Drilling Program (now superseded by the Integrated Ocean Drilling Program), and petroleum companies. These groups needed methods that could be used in rapid biostratigraphic correlation of sediments in deep-sea cores and oil wells. In fact, of all the types of biostratigraphic datums, FADs and LADs are probably the most easily and rapidly

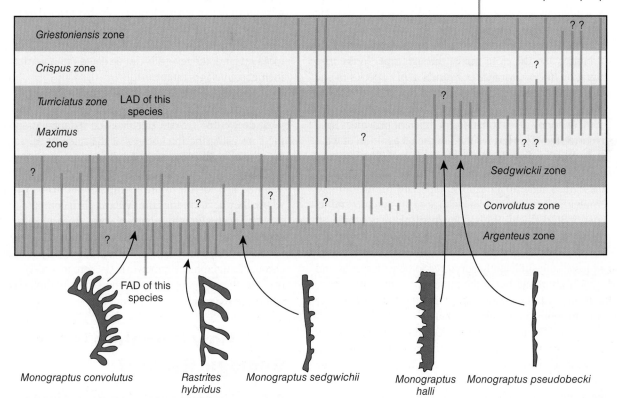

FIGURE 6.10 First and Last Appearance Datums of different species of the extinct planktonic taxon called graptolites (see Chapter 11). The base of the range of each species (orange vertical lines) represents its First Appearance Datum (FAD) and the last (uppermost) occurrence its Last Appearance Datum (LAD). The overlapping ranges are then used to define biostratigraphic units called zones that are used for correlation, as discussed in the text.

determined. This is particularly advantageous when deep-sea cores or oil-well samples suddenly become available, especially in the middle of the night (because drilling continues around the clock). An age determination can be critical to determining whether to continue drilling, which takes time and costs money.

The use of FADs and LADs is really a highly simplified version of a much older process of biostratigraphic correlation. This process was developed by the German paleontologist, Albert Oppel, while he was working in the Jura Mountains in the 1850s. Instead of using just one species, as with FADs and LADs (Figure 6.10), Oppel plotted the stratigraphic ranges of multiple fossil species at each of many different localities. He used extinct fossils called *ammonites*, which are related to the modern octopus, squid, and pearly *Nautilus*. Oppel found that the ranges of the different ammonite species overlapped: some species were short ranging geologically, whereas others had intermediate to long ranges and spanned greater thicknesses of rock.

Oppel termed these intervals of overlap **overlapping** or **concurrent range zones** (Figure 6.10). Concurrent range zones consist of the ranges of individual species, each with their own FAD and LAD, within a biogeographic province. These individual ranges, called "teilzones," are plotted next to one another; an interval of overlapping teilzone ranges comprises a concurrent range zone. Thus, Oppel's method of using

concurrent range zones was really the forerunner of the use of FADs and LADs. Oppel picked the upper and lower boundaries of each concurrent range zone where there were multiple appearances or disappearances of species (Figure 6.10). The technique of concurrent range zones produces zones as short as about 0.5 to 1 million years in duration, which is a substantial refinement of the ages provided by radiometric dates. However, the technique is labor intensive and time consuming, which is why FADs and LADs are now typically used.

An important aspect of biostratigraphy is that species are not uniformly distributed over Earth's surface because of differing tolerances to environmental factors (temperature, salinity, etc.) and barriers to migration and dispersal. An area on Earth's surface characterized by a particular group of marine species is called a biogeographic province (Chapter 3), and each ancient biogeographic province can have its own biostratigraphic zonation.

How, then, are the zonations of different provinces correlated? One method is to use more eurytopic species. Eurytopic species are fairly tolerant of environmental change, unlike stenotopic species (see Chapter 3). Eurytopic species are therefore more wide ranging geographically than stenotopic ones and more likely to occur in more than one province. One can also correlate between provinces by using the migration of a species from one province to another due to the dispersal of larvae, spores, pollen, or adults by atmospheric or oceanic currents.

No matter what the means, though, correlating zonations of different provinces is necessarily imprecise. Eurytopic species tend to have longer geologic ranges than stenotopic ones because they are more likely to survive environmental disturbances and therefore persist through time. In the case of migration, the appearance or extinction of a species in one province is unlikely to occur at exactly the same time as in another province. However, this is the best we can do.

Still, in many cases zonations of different provinces have been correlated. This has resulted in the recognition of what are called **index fossils**. An index fossil is a fossil species that is, ideally, easily identified, widespread, and abundant. All these traits make an index fossil especially useful for rapid determination of the approximate age of an outcrop and its correlation in the field. With index fossils there is a trade-off between eurytopy (being widespread) and rapid evolution and extinction that provide FADs and LADs for correlation. Index fossils are sufficiently eurytopic to be widespread and reasonably abundant but at the same time sufficiently stenotopic to undergo evolution and extinction fairly frequently and produce biostratigraphic markers.

Although Oppel did not originate the concept of index fossils, it is implicit in his work. The geologic range of an index fossil consists of the sum of the ranges of the same species in all the provinces in which it occurs, or its **total range zone**. By establishing the total range zone of a species, one establishes the total geologic range of an index fossil. The ranges of index fossils get down to about the epoch level of temporal resolution, or tens of millions of years. Although index fossils are not nearly as precise as using concurrent range zones, when finding an index fossil, we can be confident that we are dealing with rocks deposited within a particular interval of time. This provides a quick way of assessing the ages of rocks, especially if you are in the field, perhaps far removed from any form of civilization and no other resources are available.

CONCEPT AND REASONING CHECKS

1. What is the difference between the individual range zone of a species, concurrent range zones, and the total range zone of a species?
2. Diagram how the stratigraphic range of an index fossil is derived.

6.5.4 Integrating different stratigraphic datums

The different types of datums—lithologic, biostratigraphic, and so on—are by no means used separately. They are routinely integrated into standardized **chronostratigraphic** ("chrono" for time) frameworks into which absolute ages are incorporated to provide as high a time resolution as possible. Take, for example, a deep-sea core for which both FADs and LADs have been determined (**Figure 6.11**). These biostratigraphic datums yield relative dates. How can we determine the absolute ages for the datums? First, we can use a magnetometer to detect magnetic reversals that formed as the sediment was being deposited and that are recorded in the cores. If these same reversals can then be correlated to, say, igneous rocks on land, the reversals can be dated radiometrically and their absolute ages established.

But what if the levels of the magnetic reversals in the cores don't match those of the biostratigraphic datums? Now what do we do? We can calculate the sedimentation rate in the core using the thicknesses of the relevant sections and the available absolute ages. This gives us an average sedimentation rate that we can use to calculate an approximate absolute age for the biostratigraphic datums at the levels where they occur in the core. We can then compare the ages for the biostratigraphic datums with those in other cores as they become available or when new data for magnetic reversals or absolute ages are determined. In this way the chronostratigraphic framework is progressively refined because the ages obtained for the biostratigraphic datums will begin to converge on a particular age.

6.6 How Complete Is the Geologic Record?

6.6.1 Unconformities and diastems

The geologic record is by no means complete, and it does not record every event faithfully. Assume that a particular spot on Earth's surface started out with an uninterrupted sediment accumulation rate of 1 meter per thousand years or 1 kilometer per million years. This is not an unusual rate for relatively rapid sedimentation adjacent to a large delta depositing large amounts of sediment, for example. This would mean that over the course of the Phanerozoic, or about 540 million years, 540 kilometers of sediment would have accumulated. This thickness is many times greater than the sections of the thickest passive continental margins! Clearly, much of the sediment must have bypassed the site of deposition to deeper sites, was eroded, or the sediment was never deposited in the first place.

In fact, much of the geologic record is not represented by sediment at all but by surfaces of erosion or nondeposition called **unconformities** (**Figure 6.12**). Unconformities are only the most obvious expression of erosion or nondeposition of sediment and typically represent gaps in time on the order of hundreds of thousands to millions of years or more. It was a particular type of unconformity called an **angular unconformity** that Hutton observed at Siccar Point (see Chapter 1); an angular unconformity is also present at the base of parts of the Grand Canyon (refer to this chapter's frontispiece). Other types of unconformities are much more common. One type is a **disconformity** in which the erosional surface is overlain by more horizontal layers of sediment (Figure 6.12). Sometimes, a disconformity is so subtle it is not at first obvious, in which case it might be called a "paraconformity." Igneous and metamorphic rocks can also be eroded and sediments deposited on top to produce a **nonconformity** (Figure 6.12).

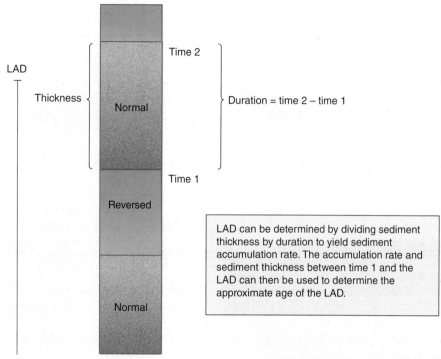

LAD

Thickness {

Time 2 }

Normal

Duration = time 2 − time 1

Time 1

Reversed

Normal

LAD can be determined by dividing sediment thickness by duration to yield sediment accumulation rate. The accumulation rate and sediment thickness between time 1 and the LAD can then be used to determine the approximate age of the LAD.

Data from: Tarbuck, E. J., and Lutgens, F. K. 2000. *Earth Science*, 9th ed. Upper Saddle River, NJ: Prentice-Hall. Figures 7.19 and 7.20 (p. 202).

FIGURE 6.11 Simplified example of how the absolute ages of biostratigraphic and other types of datums are determined and integrated. In this example, the FADs and LADs have been determined for a deep-sea core, yielding relative dates. Magnetic reversals recorded in volcanic rocks while the sediment was being deposited are determined with a magnetometer. If we can date the reversals radiometrically (from, say, intruded volcanic rocks or seafloor stripes), we can determine the sedimentation rate by using the thicknesses of the core section and the absolute ages. This yields an average sedimentation rate that we can use to calculate absolute ages for biostratigraphic datums in the core. The ages of the datums are refined by repeating this process over and over for the same datums in different cores from different areas.

The time represented by the missing sediment is called a **hiatus** (Figure 6.12). The term "hiatus" is distinguished from an unconformity because even though no sediment was deposited during a particular interval of time or was later eroded, *time still passed*. Thus, not all time is represented by sediment in the geologic record.

At the other extreme from unconformities are **diastems**. Diastems are gaps in the stratigraphic record so short they are virtually undetectable. Much of the sedimentary record is missing more because of diastems than unconformities. Diastems are caused by very short-term changes in deposition. Sedimentation within a particular environment is not uniform everywhere and continuous throughout the same short interval of time, even during "normal" times, whereas short-term events such as storms can redistribute previously deposited sediment but leave no record of the storm. Inbetween, unconformities and diastems are stratigraphic gaps and hiatuses of all magnitudes and durations like those of the Grand Canyon (Figure 6.12).

6.6.2 Sequence stratigraphy

Unconformities, especially disconformities, are the basis for the discipline of **sequence stratigraphy**. The basic unit of sequence stratigraphy is the **depositional sequence** that consists of rocks or sediments laid down relatively continuously but are bounded above and below by an unconformity. Each unconformity represents a **sequence boundary**.

The concept of depositional sequences was first recognized by Lawrence Sloss, who related them to major sea-level rises, or **transgressions**, and falls, or **regressions** over the North American continent during the Phanerozoic (**Figure 6.13**). Each of Sloss's sequences is represented by numerous formations and environments, all of which comprise a thick sedimentary section spanning many tens of millions of years. Each of his sequences is bounded above and below by unconformities that can be traced over large portions of North America. The transgressions and regressions responsible for these sequences might have resulted

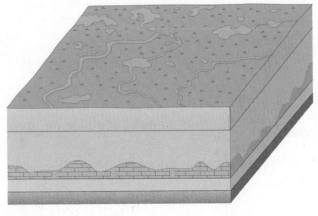

Disconformity

(a)

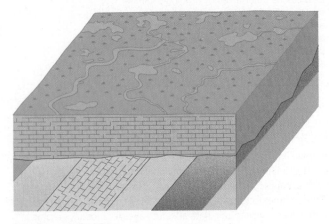

Angular unconformity

(b)

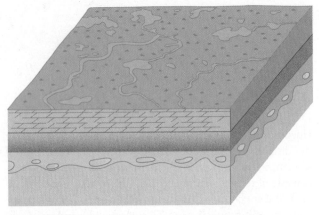

Nonconformity

(c)   Data from: Monroe, J. S., and Wicander, R. 1997. *The Changing Earth: Exploring Geology and Evolution,* 2nd ed. Belmont CA: West/Wadsworth. Figure 17.10 (p. 418).

FIGURE 6.12 Types of unconformities. **(a)** A disconformity separating rocks of different ages. Sometimes, the disconformity is not particularly obvious, in which case it is called a paraconformity. **(b)** Angular unconformity. **(c)** A nonconformity involving erosion of an igneous intrusion, followed by deposition of sediment. Eroded fragments of the igneous rock body can be found in the overlying sediment.

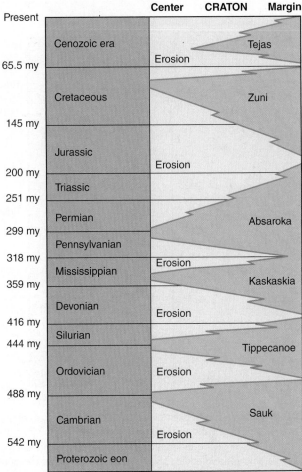

Data from: Monroe, J. S., and Wicander, R. 1997. *The Changing Earth: Exploring Geology and Evolution,* 2nd ed. Belmont CA: West/Wadsworth. Figure 21.5 (p. 533).

FIGURE 6.13 The major depositional sequences recognized by Sloss for North America. The seas (blue) are shown to be moving gradually landward onto the continents (craton) and then retreating as the craton is eroded.

from changes in the size of large polar ice caps or changes in the volume of mid-ocean ridges, which would change the volume of water in the ocean basins by altering the volume of seafloor crust (see Chapter 3). The initial transgression in western North America that laid down the sequence of the Tapeats Sandstone, Bright Angel Shale, and Muav Limestone at the base of the Grand Canyon occurred as the basal part of the Sauk transgression or sequence.

Sloss's approach was further refined by Peter Vail, Robert Mitchum, and colleagues, who were former students of Sloss and had moved to the U.S. Gulf Coast. They developed sequence stratigraphy to better predict the location of oil and gas reservoir sands, but the technique of sequence stratigraphy is now widely used by all stratigraphers (see Chapter 17 for petroleum exploration). Like Sloss's sequences, their depositional sequences are cyclic in nature and caused mainly by sea-level fall and rise. However, their depositional sequences typically represent durations of tens of thousands to several hundred thousand years, depending on environmental setting and the mechanisms causing sea-level change. In the Gulf Coast, for example, the younger sequences are often related to sea-level change caused by

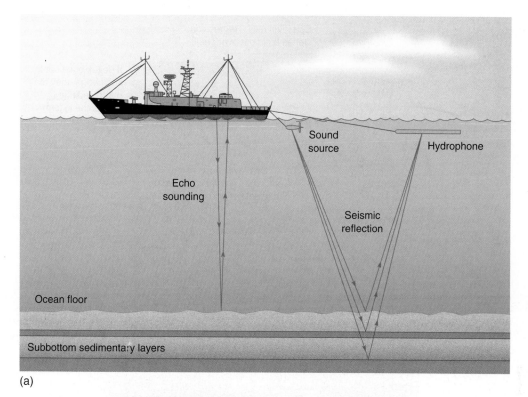

(a)

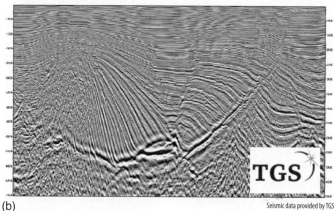

(b)

Seismic data provided by TGS.

FIGURE 6.14 **(a)** The collection of seismic data. Pulses of energy are sent out at specified intervals from the ship. These energy pulses are either reflected back from the strata or travel through the rocks and then return to the surface. Here, the reflected energy is collected by special instruments called hydrophones which are being towed by the ship. A similar procedure is used on land. The data are recorded and then processed to produce a seismic section like that shown here. **(b)** Seismic section. Arrows indicate some of the major faults along which sediments have moved downward. Alternating dark and light layers roughly correspond to systems tracts (see Figure 6.15).

the waxing and waning of glaciers during the Neogene (**Figure 6.14**). Individual depositional sequences and their component systems tracts can be identified on seismic sections, which are produced by remote sensing techniques (Figure 6.14).

These depositional sequences are normally first recognized on seismic sections and later correlated to well-logs and biostratigraphic extinctions (called "tops" in oil industry parlance) seen in oil wells. At the base of each depositional sequence is a sequence boundary formed through erosion during sea level by regression (**Figure 6.15**). As sea level falls, river deltas prograde across the shelf; if sea level falls sufficiently, eventually the deltas dump sediment at or

beyond the continental shelf edge to form a type of deposit called a **lowstand systems tract**, which consists of submarine fans. Very often, oil and gas are trapped in reservoirs formed by sand channels within the submarine fans.

Two other systems tracts occur above the lowstand systems tract in a complete depositional sequence. As sea level begins to rise and the deltas retreat shoreward, a **transgressive systems tract** forms over the shelf once occupied by the deltas (Figure 6.15). As the rate of sea-level rise slows and approaches its maximum height, a **highstand systems tract** develops (Figure 6.15). During this time, if there is enough sediment entering the system, the deltas build back out across the continental shelf. Ideally, all systems tracts are

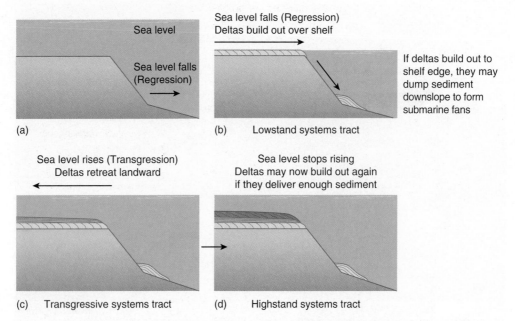

Sea level

Sea level falls
(Regression)

(a)

Sea level falls (Regression)
Deltas build out over shelf

If deltas build out to
shelf edge, they may
dump sediment
downslope to form
submarine fans

(b)   Lowstand systems tract

Sea level rises (Transgression)
Deltas retreat landward

(c)   Transgressive systems tract

Sea level stops rising
Deltas may now build out again
if they deliver enough sediment

(d)   Highstand systems tract

FIGURE 6.15 How a depositional sequence and its systems tracts form in response to sea-level transgression and regression.

preserved in a depositional sequence, but quite often parts of one or more of the system tracts are destroyed by erosion resulting from the next sea-level regression.

Despite the incompleteness of the stratigraphic record, the geologic record still provides a valuable means of understanding past environmental change on scales of time longer than those observable by humans. Because strata and fossil assemblages accumulate over relatively long periods of time, they are more likely to indicate longer-term environmental conditions, including unusual conditions that we might otherwise not detect on human time scales.

CONCEPT AND REASONING CHECKS

1. Diagram the formation of lowstand, transgressive, and highstand systems tracts in response to sea-level change.
2. Indicate the positions of disconformities on your diagram in question 1.
3. What sort of unconformity lies at the base of Siccar Point (see Chapter 1)?

6.7 Why Is Sea Level So Important?

Sea level is important for several reasons, which we've already discussed. First, sea level determines the broad distribution of sedimentary facies and their fossils like those of Figure 6.9. It is the sedimentary rocks and fossils that

record many of the changes in Earth's ancient climates. Second, sea level affects climate by affecting Earth's albedo, or surface reflectivity, which in turn affects global temperatures (see Chapter 3). Third, sea level is a broad indicator of the overall tectonic and climatic conditions on Earth. We have already alluded to this in our discussion of the causes of Sloss's sequences (Figure 6.13).

Indeed, the relative importance of the geologic processes that cause transgressions and regressions vary with the durations—or scales—of time involved (**Figure 6.16**). For example, if we were to take continuous measurements of the elevation of the ocean's surface at monitoring stations for a period of years, sea level would be seen to oscillate according to the tides (which are caused by the gravitational attraction of the moon on the ocean's surface). Such a monitoring network exists through tide gauges and satellite measurements. Other than the tidal fluctuations, however, sea level would appear to be relatively constant at each station from one year to the next.

Sea level also fluctuates on longer time scales, and the processes involved change accordingly. On time scales of decades or centuries, sea level can change by millimeters to centimeters. Although this does not sound like much, it is if your home is located in a low-lying area or along an eroding beach. These sea-level oscillations are most likely due to changes in water temperature and atmospheric circulation that are related to climate. Warmer temperatures cause ocean water to expand and global sea level to rise, whereas changes in the direction and speed of ocean currents, in response to changes in atmospheric circulation patterns, can cause sea level to fluctuate over large areas. On longer time scales of

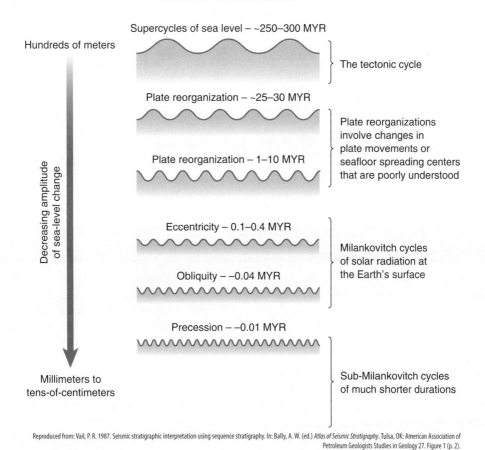

Nature exists as hierarchies

Reproduced from: Vail, P. R. 1987. Seismic stratigraphic interpretation using sequence stratigraphy. In: Bally, A. W. (ed.) *Atlas of Seismic Stratigraphy*. Tulsa, OK: American Association of Petroleum Geologists Studies in Geology 27. Figure 1 (p. 2).

FIGURE 6.16 The hierarchy of processes that affect rates of sea-level change. A hierarchy is like a set of smaller and smaller boxes nested within one another, in this case involving shorter and shorter cyclic changes of sea level as one moves down the diagram. The processes that affect sea level also change with shorter durations of time (down the diagram), as shown. Eccentricity, obliquity, and precession refer to Milankovitch cycles of solar radiation reaching the Earth's surface; these cycles affect the advance and retreat of glaciers, which in turn affect sea level on these time scales (see Chapter 15). Even shorter "sub-Milankovitch cycles" of different durations occur, too (Chapter 16). If we use a "yardstick" (time scale) that is too short, such as observations made over human time scales (years, decades, centuries), the environment might appear relatively constant to us, with regard to longer-term processes like those nearer the top of the diagram. As the yardstick (time scale) lengthens, though, we are more likely to detect significant processes and changes on the Earth that cannot be observed or measured on human time scales and can only be detected in the rock and fossil record.

a few thousand years up to hundreds of thousands of years, sea-level change is more likely due to the advance and retreat of glaciers, like that described earlier for the formation of depositional sequences during the Plio-Pleistocene. In these cases, sea level might have oscillated up to 100 meters or so.

These particular cycles are called Milankovitch cycles and cause changes in the amount of solar radiation reaching the Earth's surface resulting from changes in the elliptical nature of the Earth's orbit around the Sun (eccentricity), changes in the angle of tilt of the Earth's axis of rotation, and shifts in where the seasons occur in the Earth's orbit (precession; see Chapter 15). On even longer time scales of many millions of years, sea-level change like that of the Sloss sequences is thought to also occur in response to changes in seafloor volume or the directions in which Earth's lithospheric plates are moving. As the plates change direction, the kinds of boundaries or the amount of seafloor crust produced at the boundaries can change; this alters the volume

of the ocean basins and therefore the amount of water they can hold, causing sea level to go up or down. Finally, on scales of hundreds of millions of years, sea level fluctuates with the assembly and rifting of supercontinents, which affects mid-ocean ridge volume and the volume of water the ocean basins can hold (see Chapter 3). Changes in plate boundaries or mid-ocean ridge volume can cause sea level to fluctuate on the order of several hundred meters. Thus, the relative importance of the processes that affect sea level varies with time scale.

Conversely, our choice of time scale determines the processes that we observe or infer to affect sea level (Figure 6.16; see also Chapter 1). The choice of time scale is like selecting a measuring stick. If we were trying to measure sea-level change occurring in response to plate tectonic processes by using tide gauges on human time scales (a few years to decades or centuries), we would not see anything happening and sea level would appear constant. In fact, to determine that sea level has changed on long scales of time we would need to examine the sedimentary and fossil records; even the most sophisticated human instruments would be useless in this case because the "yardstick" (time scale) would be too short to detect any significant long-term change. Furthermore, detailed instrumental records like those from tide gauges are available only back to about the mid-19th century. The only records we have before this time, whatever the time scale or processes of interest, are the sedimentary and fossil records.

Again we see that Earth's history is an invaluable record of environmental change. Although not necessarily observable on human time scales, environmental change during

Earth's evolution affects us because it determines the conditions under which we and other life on the planet exist.

 CONCEPT AND REASONING CHECKS

1. Why is sea level important to geologists?
2. Would the rates of sea-level change occurring during a tectonic cycle (see Chapter 3) be discernable on human time scales? (Hint: Think about the rates of seafloor spreading discussed in Chapter 2.)

SUMMARY

- Time is important to Earth scientists for two fundamental reasons. First, having "enough" time gives Earth scientists intellectual room to consider geologic processes and conditions not observed today. Second, time is important to establish the causes of phenomena or "effects" preserved in the geologic record.

- The geologic time scale has evolved. Early time scales equated rock with time; thus, distinctive types of rocks were concluded to have all formed at once. The most extreme version of this view was Neptunism, which stated that rocks were precipitated from a global ocean. However, with the recognition of the Principle of Faunal Succession by William Smith, fossils were used increasingly to subdivide rocks, especially those of the Phanerozoic, into smaller units called eras, periods, and epochs.

- We can determine the ages of rocks and fossils by using two different types of dates: relative dates from stratigraphic sequences of rocks and fossils and absolute ages from radiometric dates. Although absolute dates give ages in years, fossils, in particular the field of biostratigraphy, have been used to "interpolate" between dates and to recognize finer subdivisions of the geologic time scale.

- Time relationships of different events are established by the procedure of correlation. Correlation is done in two basic ways: correlation of rocks (lithocorrelation) and correlation of fossils (biostratigraphy). Lithocorrelation equates rock with time and normally works well over short distances but is prone to error over large distances because (1) similar kinds of sedimentary rocks and the environments in which they formed repeat themselves through time and (2) different kinds of sedimentary rocks (environments) have existed at the same time.

- These two points form the basis of the concept of facies. The concept of facies refers to the fact that different sedimentary rocks and the environments in which they were deposited can exist at the same time and that the same rocks (environments) have existed at different times.

- To distinguish rock from time, distinctive rock units are first recognized as formations, which are units of rock that we can recognize and map in the field based on their particular physical traits, such as color or grain size.

- Time lines are then determined by using fossils. The fossils within the formations are correlated between different areas using biostratigraphy. We can use biostratigraphic correlation over much longer distances because fossils are unique: when different groups of organisms go extinct they are gone forever. However, fossils can evolve and go extinct at different times in different places, so biostratigraphy is not simply a matter of "connecting the dots" represented by the first and last appearances of species. Also, the geographic distribution of organisms affects the types of fossils found in different places and their use in biostratigraphic correlation.

- The geologic record is not a continuous record of sediment deposition through time. Much sediment is lost through erosion or nondeposition to produce surfaces called unconformities. Thus, not all time is represented by the rock record. Unconformity-bounded sequences of rock are called depositional sequences and form the basis of sequence stratigraphy. Despite the incompleteness of the stratigraphic record, we can still use the geologic record to understand the longer-term state of the environment.

- Sea level determines the broad patterns of depositional environments and facies, which record much of the history of Earth in their sediments and entombed fossils.

- Without these records a great deal of Earth's history would be lost and our understanding of this planet's evolution would be far more fragmentary because sea level reflects and interacts with tectonics and climate to shape the Earth. The relative importance of the processes affecting sea-level change varies with the choice of time scale.

KEY TERMS

absolute ages

angular unconformity

aquifers

Archean

Archeozoic

biostratigraphy

blocking temperature

Cambrian

Carboniferous

Cenozoic

chronostratigraphic

closure

concordant

concurrent range zones

correlation

Cretaceous

Curie point

daughter

dendrochronology

depositional sequence

Devonian

diastems

disconformity

discordant

distinctive sequences

Eocene

eons

epochs

eras

facies

facies change

First Appearance Datums (FADs)

formation

half-life

hiatus

highstand systems tract

Holocene

index fossils

Jurassic

Last Appearance Datums (LADs)

lithocorrelation

lithostratigraphy

lowstand systems tract

magnetic reversals

marker beds

Mesozoic

Miocene

Mississippian

Neogene

Neptunism

nonconformity

Oligocene

Ordovician

overlapping

Paleocene

Paleogene

Paleozoic

parent isotope

Pennsylvanian

Permian

Phanerozoic

Pleistocene

Pliocene

Precambrian

Principle of Cross Cutting Relationships

Principle of Faunal Succession

Principle of Original Horizontality

Principle of Superposition

Proterozoic

Quaternary

radiometric dates

regressions

relative ages

sequence boundary

sequence stratigraphy

Silurian

stratigraphy

Tertiary

total range zone

transgressions

transgressive systems tract

Triassic

unconformities

walking the outcrop

Walther's Law

well-logs

REVIEW QUESTIONS

1. Why is time important in establishing cause and effect?

2. What is the difference between relative and absolute ages?

3. What were the important events in the development of the geologic time scale?

4. The half-life of ^{238}U is 4.5 billion years. How old is a rock containing a ratio of the daughter ^{206}Pb/^{238}U of 0.5? How old is a rock with a ratio of 0.75?

5. If the daughter isotope leaks from a mineral, will the leakage cause the age of the mineral to be too old or too young?

6. Why is it often incorrect to equate rock and time? Use labeled diagrams to explain your answer.

7. What is the difference between a formation and a facies? Use labeled diagrams in your answer.

8. Describe the different methods of lithocorrelation.

9. Compare FADs and LADs to overlapping range zones using labeled diagrams.

10. How do the biogeographic distributions of species affect their use in biostratigraphic correlation? Use labeled diagrams in your answer.

11. How do the environmental tolerances of species affect their use in biostratigraphic correlation?

12. What are the traits of an index fossil?

13. How is the geologic range of a fossil species determined so that we can use it as an index fossil?

14. What are the approximate durations of (a) overlapping range zones; (b) index fossils; (c) geologic epochs; (d) geologic periods; (e) geologic eras; (f) geologic eons?

15. Diagram a complete depositional sequence, label all systems tracts, and relate the systems tracts to sealevel cycles and sedimentation.

16. What is the difference between an unconformity and a hiatus?

17. How does sea level vary with process?

1. Why is knowing Earth's age important?
2. Determine the percentage of time each of the following represents: (a) Precambrian; (b) Hadean; (c) Archean; (d) Proterozoic; (e) Phanerozoic.
3. Why are radioactive elements with long half-lives required to date rocks billions of years old?
4. Why hasn't Lyell's method of subdivision of the Cenozoic been applied to the Paleozoic and Mesozoic?
5. How would you distinguish an igneous pluton from a nonconformity?
6. Compare and contrast the terms formation, facies, depositional sequence, and systems tract. Draw a depositional sequence with all the possible systems tracts, and then label what you consider to be formations and facies.
7. Divide into groups of several students. On large poster sheets, construct diagrams illustrating superposition, cross-cutting relationships with igneous intrusions and faults, unconformities, and radiometric dates. Challenge other groups to determine the sequence of events in your poster.

SOURCES AND FURTHER READING

Abreu, V., Neal, J. E., Bohacs, K. M., and Kalbas, J. L. (eds.). 2011. *Sequence stratigraphy of siliciclastic systems—The ExxonMobil methodology, Atlas of exercises.* Tulsa, OK: Society for Sedimentary Geology (SEPM).

Berry, W. B. N. 1987. *Growth of a prehistoric time scale: Based on organic evolution.* Palo Alto, CA: Blackwell.

Coe, A. L. (ed.). 2003. *The sedimentary record of sea-level change.* Cambridge, UK: Cambridge University Press.

Dott, R. H. (ed.). 1992. *Eustasy: The historical ups and downs of a major geological concept.* Memoir 180. Boulder, CO: Geological Society of America.

Eicher, D. L. 1976. *Geologic time.* Englewood Cliffs, NJ: Prentice-Hall.

Gorst, M. 2001. *Measuring eternity: The search for the beginning of time.* New York: Broadway Books.

Greene, M. T. 1982. *Geology in the nineteenth century: Changing views of a changing world.* Ithaca, NY: Cornell University Press.

Lewis, C. 2000. *The dating game: One man's search for the age of the earth.* Cambridge, UK: Cambridge University Press.

Martin, R. E. 1998. *One long experiment: Scale and process in earth history.* New York: Columbia University Press.

Peters, E. K. 1996. *No stone unturned: Reasoning about rocks and fossils.* New York: W. H. Freeman.

Rudwick, M. J. S. 2005. *Bursting the limits of time: The reconstruction of geohistory in the age of revolution.* Chicago: University of Chicago Press.

Toulmin, S., and Goodfield, J. 1965. *The discovery of time.* Chicago: University of Chicago Press.

Zen, E. 2001. What is deep time and why should anyone care? *Journal of Geoscience Education, 49*(1), 5–9.

PART

II

THE PRECAMBRIAN: ORIGIN AND EARLY EVOLUTION OF EARTH'S SYSTEMS

In Part I, we examined the behavior of Earth's systems—solid Earth, hydrosphere, atmosphere, and biosphere—largely in relative isolation from one another and from a uniformitarian perspective. Now it is time to begin examining the evolution of Earth's systems in greater detail. Although we will not always explicitly use systems terminology such as "reservoir," "flux," and so on, as discussed in earlier chapters, we will discuss the history and the evolution of Earth from a systems perspective and give appropriate examples of systems behavior.

We begin with the interval of time informally called the Precambrian ("time before the Cambrian"). The Precambrian represents the first 4 to 4.5 billion years of the planet's history; this alone accounts for the first 90% of Earth's duration! Recall in *Principles of Geology*, Lyell refrained from discussing Earth's cosmogony, or origins, because to do so, he said, would violate the doctrine of uniformitarianism (see Chapter 1). Nevertheless, the Principle of Uniformitarianism remains the guiding principle for the study of the Precambrian.

Still, as we will see in this and succeeding chapters, during its first few billion years of history, Earth was a vastly different planet from that of the Phanerozoic. Moreover, Earth evolved into a planet different from its two nearest neighbors, Mars and Venus. This occurred for several reasons, all of which are contingent on the fact that Earth just happens to exist at a particular distance from the sun. This distance resulted in a relatively narrow temperature range that has permitted the formation of liquid water. Liquid water is critical to plate tectonics (see Chapter 2) and the rock cycle; the biosphere and life as we know it; and, of course, liquid water is an integral component of the atmosphere and hydrosphere.

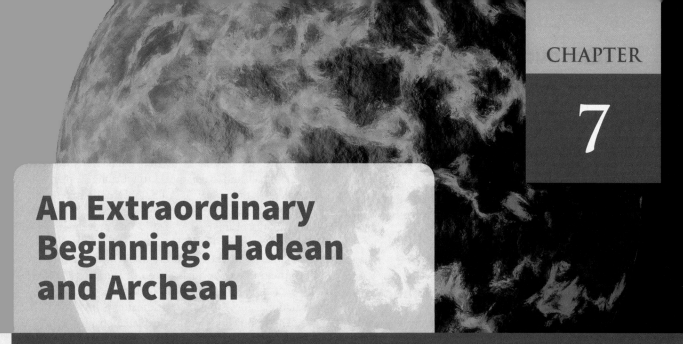

An Extraordinary Beginning: Hadean and Archean

MAJOR CONCEPTS AND QUESTIONS ADDRESSED IN THIS CHAPTER

Ⓐ A seemingly childish question: Why is the night sky dark?

Ⓑ What were early hypotheses for the origin of the universe and our solar system?

Ⓒ Has the universe remained in steady state or has it evolved?

Ⓓ How do we know the universe is expanding?

Ⓔ How did stars and galaxies form?

Ⓕ When and how did matter and the basic forces of nature originate?

Ⓖ Where did the chemical elements of the universe come from?

Ⓗ How did our solar system originate, and what is the evidence?

Ⓘ What role(s) might meteorites and comets have played in the formation of the Earth?

Ⓙ How did Earth's core and its earliest crust and atmosphere form?

Ⓚ What are the hypotheses for the formation of the Moon?

Ⓛ How did early plate tectonics produce the earliest known crust on Earth?

Ⓜ Did Earth's crust form gradually, rapidly, or in pulses?

Ⓝ What is meant by the "Habitable Zone," and how and why did Earth's climate come to differ from that of its neighbors, Mars and Venus?

CHAPTER OUTLINE

The Earth as it might have appeared early in its history after the initial crust had begun to form.

© Iuliia Bycheva / Alamy Stock Photo.

7.1 Introduction

Until the 20th century, most scientists believed the universe was eternal. However, during the 20th century evidence accumulated showing that the universe had a beginning.

Why should we care about how the universe began? There are, of course, obvious philosophical issues. But, on a scientific level, a beginning for the universe again forces us to confront the concept of an equilibrium Earth versus directional change, for the origin of the universe ultimately led to the succession of worlds on an evolving Earth.

7.2 Origin of the Universe

7.2.1 Early observations and theories

Stretching across the night sky is a prominent dusty-white band of stars known as the Milky Way galaxy (**Figure 7.1**). The night glow and dusty appearance of the Milky Way are from the 150 billion suns other than our own, many of which have their own solar system. The visible portion of the Milky Way is about 100,000 light years across, and our solar system is about 30,000 light years from the galaxy's center (distances in cosmology—the study of the structure and origin of the universe—are measured in light years, which is the distance traveled by light in 1 year; the speed of light is about 186,000 miles per second).

Despite its immensity the Milky Way is only 1 of about 100 to 200 billion galaxies in the universe, each of which is estimated to contain hundreds of billions of stars. Why, then, aren't we blinded by the night sky? Why instead is the night sky dark?

A This simple observation, sometimes called **Olbers's paradox** (after the German amateur astronomer, Heinrich Olbers), holds profound implications for the origin of the universe. The paradox was actually first recognized by the German astronomer, Johannes Kepler (1571–1630), who hypothesized that the universe must be finite, or bounded; otherwise, assuming a uniform distribution of stars, there would be so many stars the heavens would glow like a fireball and bake the Earth. Unlike Kepler, Olbers believed the universe was infinite, or unbounded. Olbers therefore proposed an alternate hypothesis: there must be vast amounts of interstellar matter that prevents most starlight from reaching Earth. However, this hypothesis doesn't work either: starlight would eventually heat the clouds and cause them to glow, which of course is not observed. (It is now recognized that so-called "dark matter" exists based on gravitational effects but this dark matter is invisible.)

B The first really plausible explanation for the origin of the universe stems from the solar **nebula hypothesis**, which was proposed by the 18th-century German philosopher, Immanuel Kant, and later revived by the French mathematician, Pierre-Simon Laplace. The nebular hypothesis was initially advanced for the origin of the solar system. Kant and Laplace suggested the origin of the solar system must be considered in the larger context of Sir Isaac Newton's theories of how planets move in their orbits around the sun. Kant and Laplace concluded that because the planets orbit the sun, the solar system evolved from a primordial cloud of dust and gas, and that as the proto-sun condensed, it threw off rings of debris that accreted to form the planets.

Thus, the solar system had a beginning. As a result, the light from distant stars has yet to reach Earth, so that when we look at the night sky we are actually looking back

© iStockphoto/Thinkstock.

FIGURE 7.1 The Milky Way. In this view, you are looking through the plane of the Milky Way.

in time toward the beginning of the universe! The nebular hypothesis also explained several other observations known for hundreds of years: (1) the planets and asteroids all move in the same direction around the sun, (2) most of the planets rotate in the same direction as Earth, and (3) the orbits of the planets are all nearly circular.

However, the nebular hypothesis had one serious flaw: the sun should spin very rapidly, which it does not. According to the nebular hypothesis, the sun should spin rapidly because of angular momentum. **Angular momentum** is the same phenomenon that occurs when ice skaters retract their arms toward their bodies as they spin; as the skater's arms are retracted, the skater spins faster. The momentum of an object *moving in one direction,* such as a train or a car, is defined as its mass (m) times its velocity (v), or $m \times v$. Angular momentum is defined as the mass of all components of a rotating system (m) times the distance of the components from the axis of rotation of the system (d) times the rotational velocity of the components (v), or $m \times d \times v$. Angular momentum is conserved, meaning the total angular momentum of a system remains constant.

Thus, if a spinning mass of matter becomes smaller, such as a primordial planet that is condensing or an ice skater pulling in his or her arms, the distance of the matter from the center will decrease. This should decrease angular momentum because the distance (d) decreases. However, because angular momentum is conserved, the decrease in d is compensated by an increase in velocity (v) of rotation. Thus, if the nebular hypothesis proposed by Kant and Laplace were correct, the sun should spin quite rapidly. Instead, it completes one rotation about once a month. Consequently, the nebular hypothesis came to be viewed skeptically by most scientists.

7.2.2 The Big Bang: from hypothesis to theory

It was not until the 1920s that the idea of a universe with a beginning was revived by Georges-Henri Lemaître, a Belgian priest trained as an astrophysicist. Lemaître's hypothesis required that all the matter in the universe initially be compressed into a single "primeval atom" (in Lemaître's terminology), or **singularity**, as it is known today. Not surprisingly, some prominent physicists of the time, including Albert Einstein, scoffed at Lemaître's idea. Later, in 1950 Lemaître's hypothesis was dubbed the **Big Bang** in a radio broadcast by the famous cosmologist, Fred Hoyle. Although the name was intended to be derogatory, it stuck. The most important objection to the Big Bang hypothesis centered on its implications for physical theory. Namely, the laws of physics (like the law of gravity) must always hold; otherwise, they wouldn't be laws. A body like Lemaître's singularity would be infinitely dense, and it would therefore be impossible to determine what conditions were like before the Big Bang. Thus, Lemaître's hypothesis struck physicists like Hoyle as being absurd.

Hoyle instead favored a **Steady State Theory** of the universe. As its name suggests, the Steady State Theory viewed the universe as being in equilibrium and undergoing no *net* change (just like Charles Lyell's view of Earth; see Chapter 1). The Steady State Theory is therefore a throwback to the static view and for fundamentally the same reasons that Lyell dismissed the cosmogony of Earth: physical laws still apply. Also, if the universe had a beginning, it may well have an end, something that deeply disturbed some physicists and philosophers. Instead, in the steady state view the universe is eternal. Finally, the steady state view meant science would not be confronted with something that was potentially unknowable: what existed before the Big Bang?

Nevertheless, evidence in support of Lemaître's hypothesis had already begun to emerge about the time he proposed it. At this time, the Milky Way galaxy was thought to represent most of the universe. However, the astronomer Edwin P. Hubble was unconvinced because there seemed to be spiral clusters of stars, or **nebulae**, outside the Milky Way. (Hubble initially thought these nebulae were the suns of other solar systems within the universe but these nebulae turned out to be other galaxies with their own solar systems.) Hubble calculated the distance to stars by using a star's brightness and its distance from Earth: the brightness of a star decreases with the square of its distance from Earth. Using this relationship Hubble demonstrated that other nebulae exist far beyond our own galaxy. Thus, the universe was much larger than had previously been imagined.

Hubble's findings are based on the **Doppler effect**, or **red shift**, named after the 19th-century Austrian physicist, Christian Doppler. The Doppler effect is demonstrated by the following example. As a fire truck approaches an observer standing by a street, the observer hears a relatively high-pitched sound from the siren, but as soon as the vehicle passes the sound shifts to a deeper pitch. As the fire truck approaches the observer, the crests of the sound waves are "squeezed" together; thus, the pitch—or frequency of the sound waves (the number of sound waves passing per unit time)—increases as the fire truck approaches. After the fire truck passes, the opposite happens: the sound waves spread out as the distance between the observer and fire truck increases. The siren's pitch deepens and the sound dissipates as the fire truck continues to move away.

The same phenomenon happens with light. Like sound, light can also be thought of as consisting of waves. White (visible) light is actually part of the much larger electromagnetic spectrum. As first demonstrated by Newton, the visible spectrum is composed of a spectrum ranging from blue to red light (**Figure 7.2**). Blue light is composed of closely spaced waves, which are therefore said to have a higher frequency; these waves have more energy because they have a higher frequency. By contrast, the waves of red light are spaced farther apart (lower frequency) and are less energetic. By 1913, it had been found that distant galaxies emitted red light. This finding hinted that the galaxies were moving away from observers on Earth and therefore the universe was expanding. Moreover, the farther away the galaxies were, the greater their redness, or *red shift.* In other words, the farthest galaxies were moving away at faster rates, with speeds up to thousands of kilometers per second! The universe could no

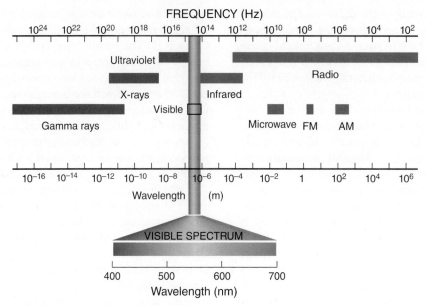

FIGURE 7.2 The electromagnetic spectrum includes visible light. The visible portion of the spectrum can be remembered by the mnemonic: ROY G. BIV (Red Orange Yellow Green Blue Indigo Violet).

longer be viewed as static, with the stars and galaxies fixed in space. The universe seemed to be expanding.

Building on these observations, Hubble developed the law that bears his name. The **Hubble Law** states that there is a relationship between the distance to a galaxy and its red shift: the greater the distance to a galaxy, the greater the red shift (**Figure 7.3**). The Hubble Law allowed calculation of the origin of the universe, based on the farthest detectable galaxies, as occurring sometime between 10 and 20 billion years ago.

Other evidence eventually came to light that supported the Big Bang hypothesis. Supporters of the Big Bang predicted the density of galaxies should increase into deep space because it is closer in time to the Big Bang, when much of the universe's matter was closer together. This prediction was corroborated by the discovery in the early 1950s of *radio galaxies* by radio telescopes (**Figure 7.4**). Radio telescopes detect distant objects by the nonvisible components of the electromagnetic spectrum and also monitor outer space for signs of intelligent life. The greater the distance, the more abundant radio galaxies become, meaning the galaxies had once been much closer together and the universe had once

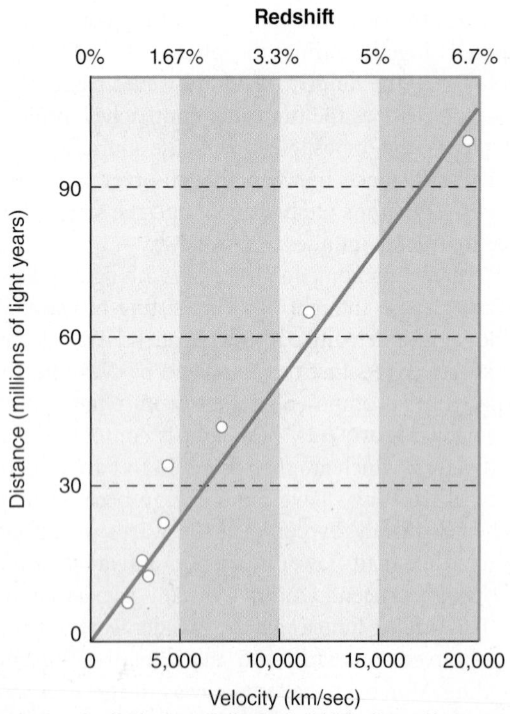

Data from: Smoot, G., and Davidson, K. 1993. *Wrinkles in Time*. New York: William Morrow and Company, p. 49.

FIGURE 7.3 The Hubble law, which states that there is a linear relationship between the distance to a galaxy and its redshift.

© Israel Pabon/Shutterstock, Inc.

FIGURE 7.4 The Arecibo radiotelescope located in Puerto Rico. Radio telescopes detect distant objects by the nonvisible components of the electromagnetic spectrum and monitor outer space for signs of intelligent life.

been much denser. The discovery of quasistellar objects, or *quasars,* in the 1960s also supported the Big Bang Theory. Because of their extremely strong red shifts, quasars are thought to have formed close to the time of the Big Bang.

Finally, and most important, was the discovery of **cosmic background radiation**. This discovery clinched it for the Big Bang Theory (much like that of the discovery of magnetic anomalies in seafloor crust for plate tectonics; see Chapter 2). It was hypothesized in 1948 that the unimaginably intense heat of the Big Bang would have been dissipated as the universe expanded and that only a faint afterglow should remain. This cosmic background radiation is a kind of blackbody radiation. Unlike the electromagnetic spectrum, which consists of regions of radiation with distinct waves, **blackbody radiation** is the randomized energy released when particles collide with one another very rapidly in a warm-to-hot object, such as the surface of a planet warmed by a star or the interior of the star itself. Given the conditions of the Big Bang, such collisions should have been incredibly intense. It was not until 1964, however, that the faint afterglow of the Big Bang was detected, by accident, by Arno Penzias and Robert Wilson of Bell Laboratories in New Jersey. Penzias and Wilson were trying to measure noise levels that might deter communication using radio antennae. They detected **microwave radiation** like that used by home appliances (Figure 7.2), which they initially thought originated for other reasons (such as pigeon droppings on their antenna). However, the signal persisted after they had excluded all other possible sources of the signal (including cleaning the antenna), so they concluded that the signal was real and coming from space. Despite its faintness, total cosmic background radiation is at least 10 times that of all the stars, galaxies, and other radiating objects averaged over the volume of the observable universe.

7.2.3 The inflationary universe

How did the Big Bang lead to the formation of stars, galaxies, and the universe? Based on the aforementioned evidence, cosmologists now believe that about 13.7 billion years ago all matter and energy of the universe was so strongly compressed into a region, it basically did not exist! The universe then burst into existence, giving birth to space, time, matter, and energy. Of course, this seems absolutely ridiculous based on our commonsense notions of everyday occurrences.

You now see why some scientists objected so strongly to the Big Bang hypothesis when it was first proposed. To understand the origin of the universe, we must abandon our everyday, commonsense anthropocentric notions of matter, energy, space, and time. Indeed, when the basic structure of the atom was determined, it was found to consist almost entirely of space. It was eventually realized that given the vast amount of space within even a single atom, it is quite possible for all matter and energy to have condensed into virtually nothing!

Despite its name, however, the Big Bang was *not* the expansion of matter *into* preexisting empty space. It was the expansion of what we know as matter, energy, space, and time into what resembles a foam-like structure (**Figure 7.5**). Imagine two bugs on a balloon (**Figure 7.6**). As the balloon is blown up, the distance between the bugs increases. Similarly, during the Big Bang matter moved *with* space, not *through* it.

Courtesy of NASA/ESA/and The Hubble Heritage Team (STScI).

FIGURE 7.5 Galaxies are not distributed evenly, but are clustered or occur in groups of clusters, and are separated by voids. Some galaxies occur on the surfaces of voids or at the intersections of voids, which occur like bubbles in a foam. Here, two galaxies are shown colliding with each other.

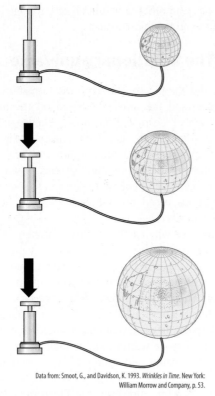

Data from: Smoot, G., and Davidson, K. 1993. *Wrinkles in Time*. New York: William Morrow and Company, p. 53.

FIGURE 7.6 Expansion of galaxies in relation to space. During the Big Bang, stellar masses expanded *with* space, not *into* space. Although the universe started out with a great deal of curvature, it flattened as it enlarged. This is in agreement with cosmologic theory: the universe is actually flat.

CONCEPT AND REASONING CHECKS

1. Why is Olbers's paradox a paradox?
2. How does the Steady State Theory resemble Lyell's original view of uniformitarianism (see Chapter 1)?
3. How does the Big Bang embody secular change (see Chapter 1)?
4. Why did the discovery of cosmic background radiation, and not the red shift, clinch the Big Bang for most people? (Hint: Does a red shift necessarily mean there was a Big Bang? Review the scientific method in Chapter 1 about prediction and hypothesis testing.)
5. How was the discovery of cosmic background radiation like the documentation of seafloor stripes (see Chapter 2)?

7.3 Origin of Matter and Forces of Nature

Exactly what happened during the Big Bang is based on our understanding of the structure of the atom and the forces of nature (**Figure 7.7**). Between the time the Big Bang occurred (time zero) and 10^{-43} seconds (!), the temperature

of the universe was 10^{32} K. (Kelvin [K] equals degrees Celsius [°C] + 273.15.)

During this incredibly short interval, the universe was so hot protons and other subatomic particles existed as even smaller particles called **quarks**. This interval is called the **Grand Unification Epoch**, according to which all known matter, radiation, and three of the four fundamental forces of nature were one and the same. This phase corresponds to Lemaître's "primeval atom." During this time three of the four forces of nature we recognize did not exist: radioactivity, or the weak force; the strong force, which holds atomic nuclei together; and electromagnetic radiation. Cosmologic theory has yet to successfully unite gravity with matter and energy before this time.

By 10^{-35} seconds after the Big Bang began, the universe had cooled to 10^{27} K. At this temperature all matter and radiation of the universe were part of a primal soup of energy confined to an object no larger than the size of a trillionth of a proton (about 10^{-25} cm across). The energy was then so rapidly dispersed the universe experienced an incredibly rapid period of inflation, like a turbocharger accelerating an automobile engine. In effect, a second big bang took place very shortly after the first Big Bang.

Several other stages followed in rapid succession. At 10^{-34} seconds, the strong force became distinct from the weak and electromagnetic forces. By 10^{-32} seconds, the universe had expanded to no larger than about 10 meters square (roughly the size of a dormitory room) but might have been as small as a grapefruit. By this time the rapid inflationary period had ended because gravity began to take effect. It is the faint afterglow from this phase that has been detected as cosmic background radiation. By 10^{-12} seconds, the electromagnetic and weak forces separated as distinct forces, and by 10^{-10} seconds, quarks began to coalesce to form protons and neutrons. The temperature of the universe was now 10^{15} K. At 3 minutes and a temperature of 10^9 K, protons and neutrons began to form atomic nuclei because the universe had cooled to the point at which the strong force now exceeded that of the cosmic background radiation.

By 300,000 years the universe was becoming a little more familiar in appearance. The universe was still a cloud of white-hot (3,000 K) dense gas but had expanded to about the size of the present Milky Way. At 1 billion years, the universe's temperature was only 18 K; stars and galaxies were present but had probably appeared much earlier. According to computer simulations the first stars formed in small proto-galaxies that occupied regions where the density of matter was higher. The first stars are thought to have been 100 to 1,000 times as massive as our sun with temperatures of 10^5 to 10^6 K.

So, we now have the basic structure of the universe. But how did the chemical elements form? There are more than 100 chemical elements now known, but two—hydrogen and helium—are by far the most abundant elements in the universe. In fact, hydrogen accounts for about 75% of all matter in the universe and helium for almost the entire remaining 25%. In 1925, it was discovered that the sun consists almost entirely of hydrogen. Hydrogen is the simplest

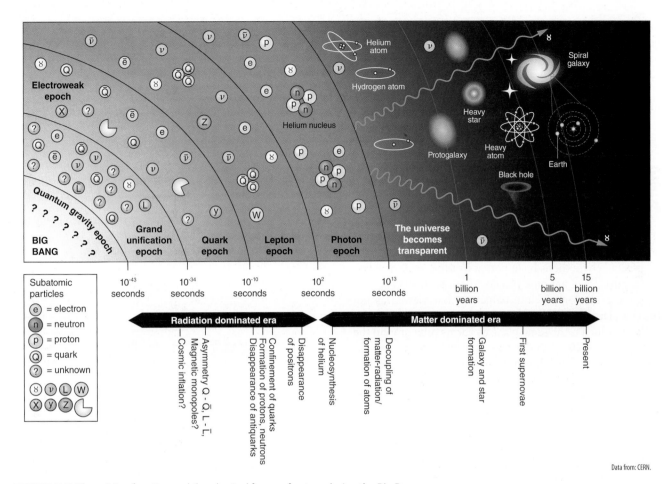

FIGURE 7.7 The origin of matter and the physical forces of nature during the Big Bang.

Data from: CERN.

element; each hydrogen atom has an atomic number of 1, meaning it has one proton in its central nucleus. Because of its simplicity, a hydrogen atom cannot be split by the process known as **fission** into other atoms. Instead, it was proposed that two hydrogen nuclei undergo the process of **fusion** to produce one helium nucleus. Fusion releases vastly greater amounts of energy than that released by the explosion of atomic bombs, which are based on fission.

It was thought that if hydrogen nuclei could fuse to form helium, helium could fuse to form more complex elements with larger atomic numbers. However, experiments demonstrated that although massive amounts of hydrogen and helium could have been formed by the Big Bang, elements heavier than hydrogen and helium could not have formed during the Big Bang because the intense heat would have made them unstable. On the other hand, the stars by themselves probably would not have produced sufficient hydrogen and helium to account for their abundance throughout the universe.

Thus, it is thought that much of the hydrogen and helium originated during the Big Bang. Heavier elements formed later inside the early stars by the fusion of atomic nuclei, or **nucleosynthesis**. Eventually, the stars exploded as **supernovae** and distributed the heavier metallic elements like iron, formed by the fusion of hydrogen and helium throughout their adjacent galaxies and beyond (**Figure 7.8**). Because the

Courtesy of NASA/JPL.

FIGURE 7.8 Supernova 1987A. Stars are thought to have been the sites of formation of the heavier elements of the universe, with supernovae distributing them throughout the universe.

universe was much smaller, the supernovae would have spread heavier elements throughout the universe quite rapidly. As older stars died, new stars formed that contained higher concentrations of metals. The presence of metals in the next generation of stars allowed stars with smaller masses to form and may have accelerated the rate of star formation.

7.4 Formation of the Solar System

By about 10 billion years, our solar system began to form. The more modern version of the nebula hypothesis, the **solar nebula hypothesis**, states that the planets and the sun formed from a gaseous nebula (**Figure 7.9**). The solar nebula hypothesis is supported by some basic observations. First, there are nine planets in our solar system and, with the exception of Pluto (which is now viewed as not being a planet), all lie in or very close to the **plane of the ecliptic**, which is the same plane as Earth's orbit around the sun (**Figure 7.10**). Second, all the planets rotate counterclockwise around the sun in nearly circular orbits. Apparently, as the disk of gas rotated around the proto-sun, regions of it began to condense into solid dust particles. Unlike the Kant-Laplace hypothesis, then, the rings of debris did not originate *from* the sun. Solid dust particles in turn collided and coalesced through gravitational attraction to form larger and larger bodies. As these regions coalesced they began to spin faster because of the conservation of angular momentum, and the proto-planets began to appear.

As the planets formed, they differentiated into two groups: inner or **terrestrial planets** and the outer, gaseous Jovian planets (named after Jupiter; **Figure 7.11**). The inner planets are relatively small in size and include Mercury, Venus, Earth, and Mars. The inner planets consist mostly of silicate minerals and metals and so are much denser than the Jovian planets (**Table 7.1**). In fact, the bulk of the mass of the Jovian planets is composed of gases: hydrogen, helium, methane (CH_4), ammonia (NH_3), and water. Consequently, the Jovian planets dwarf the inner planets in size (Table 7.1). Because of the unusual composition of their atmospheres, the Jovian planets also exhibit unusual atmospheric phenomena, including color banding and gigantic storms, such as Jupiter's gigantic "red spot" that can last for centuries (**Figure 7.12**).

As the planets were coalescing, so too was the sun. The center of the early solar system had the greatest density of matter because of the conservation of angular momentum. The increasing density of matter caused the sun to continue to coalesce through positive feedback. As the sun continued to coalesce, particles began to collide more frequently. In fact, 600 million tons of hydrogen atoms fuse into helium *every second* in the modern sun. This caused temperatures and pressures to increase to tens of millions of degrees and billions of atmospheres pressure based on observations of the planets and computer simulations. At these temperatures and pressures, hydrogen fused to form helium, and the sun reached steady state, or equilibrium, when the outward

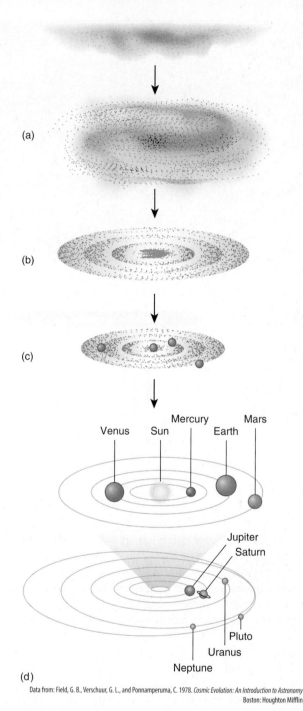

(a)

(b)

(c)

Mercury
Mars
Venus   Sun   Earth

Jupiter
Saturn

Pluto
Uranus
Neptune

(d)

Data from: Field, G. B., Verschuur, G. L., and Ponnamperuma, C. 1978. *Cosmic Evolution: An Introduction to Astronomy.* Boston: Houghton Mifflin.

FIGURE 7.9 Formation of the planets according the solar nebula hypothesis.

pressure from the interior of the sun began to counteract the gravitational collapse of matter toward its center.

The remnants of the formation of the solar system are represented by **asteroids** (from the Greek, *aster* for "star" and *eidos* meaning "form"). Most asteroids orbit the sun in a belt located between Jupiter and Mars and are probably the remains of small planets or **planetesimals**. Why a planet did not form within this region is puzzling. Possibly, the gravitational field of Jupiter caused a large nearby planetesimal to enter the region where the asteroid belt is now located; this might have increased the likelihood of the

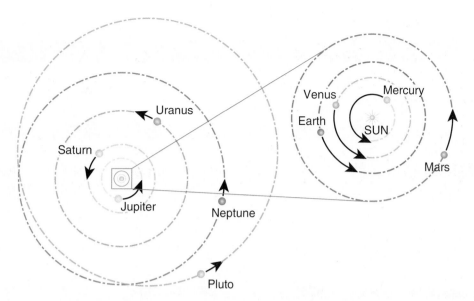

FIGURE 7.10 The solar system. With the exception of Pluto or perhaps a dwarf planet, the planets all occur more or less in the plane of the ecliptic. For this reason, Pluto is now considered to be a large asteroid captured by the solar system after it formed.

Courtesy of NASA/JPL.

FIGURE 7.11 The terrestrial and Jovian planets shown to the same scale in relation to the sun. The mass of the Earth is only 0.000003 that of the sun, which is considered an average star. A massive supernova on the lower right is shown for scale.

TABLE 7.1

Major features of the planets

	Mercury	Venus	Earth	Mars	Jupiter	Saturn	Uranus	Neptune
Diameter (km)	4,878	12,104	12,756	6,794	142,800	120,540	51,200	49,500
Diameter in relation to Earth	38%	95%	X	53%	1,120%	941%	401%	388%
Mass in relation to Earth	5.5%	82%	X	10.7%	31,780%	9,430%	1,460%	1,720%
Density (g/cm³)	5.43	5.24	5.52	3.9	1.3	0.7	1.2	1.6
Rotation period (days)	58.6	−243	0.997	1.026	0.41	0.43	−0.72	0.67
Inclination of axis of rotation to equator (degrees)	0.0	177.4	23.4	25.2	3.1	26.7	97.9	29
Surface gravity in relation to Earth	38%	91%	X	38%	253%	107%	92%	118%

AU = astronomical unit, or the distance between the Earth and sun.

planetesimal's collision with other objects and their fragmentation. Alternatively, because of Jupiter's size, perhaps Jupiter's gravitational field was simply so strong it prevented the formation of a planet from asteroid debris that was already present.

Indeed, asteroids are not thought to have originated from large planets for two main reasons. First, high pressure minerals that typically form within large planets are not present in meteorites. **Meteorites** are fragments found on Earth of extraterrestrial bodies, including asteroids, that collided with Earth (**Figure 7.13**). Some meteorites have been radiometrically dated at approximately 4.5 billion years, indicating these particular meteorites date from about the time of the formation of the solar system. Second, strong

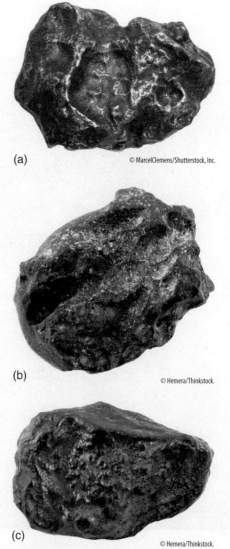

(a)
© MarcelClemens/Shutterstock, Inc.

(b)
© Hemera/Thinkstock.

(c)
© Hemera/Thinkstock.

Courtesy of NASA/JPL.

FIGURE 7.12 Jupiter's red spot might represent a gigantic storm that has persisted for hundreds of years.

FIGURE 7.13 Types of meteorites. From top to bottom are shown: stoney-iron, iron, and stoney meteorites.

chemical differences exist between different meteorites that cannot be explained if the meteorites came from only a few large planets. Based on chemical analyses most meteorites appear to have come from the asteroid belt between Mars and Jupiter.

Meteorites are themselves of several types, depending on the relative amounts of silicate minerals and iron: stony, stony-irons, and irons (Figure 7.13). Stony meteorites are the most abundant and are of two types: **chondrites** and **achondrites**. **Carbonaceous chondrites** contain carbon, whereas **ordinary chondrites** do not (the Murchison meteorite that was dated to determine the age of Earth was an iron meteorite). Possibly, while the Earth was coalescing, ordinary chondrites were heated above about 450 K, vaporizing carbon and water, whereas carbonaceous chondrites were originally located farther from the sun.

Carbonaceous chondrites might have been an important source of water on Earth. This inference is based on the ratio of the hydrogen isotope **deuterium** (which has an extra neutron) to normal **hydrogen** (D/H ratio) in water. The D/H ratio for Earth falls within the range of carbonaceous chondrites. Because deuterium is heavier than normal H_2, it would have remained behind in the water vapor of meteorites as the lighter H_2 was evaporated into space. By contrast, achondrites exhibit a wide range of textures and,

along with irons and stony-irons, are thought to have originated by the melting and recrystallization of primitive planetary material. Iron meteorites appear to represent portions of asteroid cores.

Also thought to be left over from the origin of the solar system are **comets** (**Figure 7.14**). Comets are basically "dirty snowballs." They consist of a nucleus of ice (either water or frozen carbon dioxide) and soot surrounded by a gaseous envelope and a long, trailing tail (hence the name *kometeos*, meaning "long-haired"). Although a comet's nucleus might be only 1 to 10 km in size, the tail can reach tens of millions of kilometers in length. The tail points away from the sun and consists of ionized atoms and dust blown off by the **solar wind**, the immense stream of ionized particles or **plasma** produced by the fusion of hydrogen atoms into helium. The solar wind and ultraviolet radiation observed emanating from young stars is quite intense, and our sun must have been no different.

Comets appear to originate in the **Oort cloud**, which is located far beyond the edge of the solar system at a distance up to 1,000 times the diameter of Pluto's orbit around the sun. When dislodged by gravitational forces of the outer planets or distant stars, comets move into the inner solar system. Then, they assume highly elliptical orbits around the sun at angles to the plane of the ecliptic. Because they contain so much water, comets have been suggested to have played a major role in the formation of Earth's oceans; however, their deuterium-to-hydrogen (D/H) ratios are about twice those of Earth's.

© Datacraft/age fotostock.

FIGURE 7.14 Comet Hale-Bopp. The two tails range from 10 to 15 million miles in length. Comets are "dirty snowballs" and might have brought water to the early Earth.

7.5 The Hadean: Origin of Earth and Moon

7.5.1 Earth's earliest evolution

Calculations suggest it would have taken at least 100 million years for the proto-Earth to grow from about 10 km in diameter to its present size. This phase occurred during the first 0.8 billion years of the development of Earth, stretching to about 3.9 to 3.8 billion years ago. This first phase of Earth's history is known as the **Hadean Eon**. As the name implies, Earth was quite hot. The Hadean represents the beginning of the much longer Precambrian, which stretches from the birth of Earth to about 540 million years ago, when the first good fossils of the Phanerozoic Eon ("time of apparent life") begin to appear consistently.

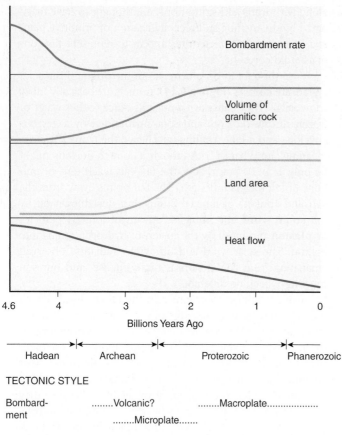

FIGURE 7.15 Generalized heat flux from the Earth through time in relation to tectonic styles. Recent model simulations indicate that although meteor bombardment declined after about 4 Ga, bombardment continued into the Proterozoic Eon as late as 2.0 to 2.5 Ga.

There were several sources of heat during the Hadean and the subsequent **Archean Eon** (**Figure 7.15**). First, there was radioactive decay, which has continued to the present and drives seafloor spreading and plate tectonics. A significant portion of the heat was probably supplied by now-vanished radioactive isotopes. Second, there was the accretion of dust and larger particles through gravitational attraction and the conservation of angular momentum. As particles were drawn closer to the planet as it formed, their potential energy was released as heat energy because of friction. Third, Earth was undoubtedly bombarded by meteors and planetesimals hurtling through the solar system; this bombardment would have also released tremendous amounts of heat. Fourth, accretion of the core released even more heat. Because Earth was molten, heavier elements such as iron and nickel (which is **siderophilic** or "iron-loving") moved rapidly toward Earth's interior to form the core. Calculations suggest the formation of the core was quite rapid and completed within about 100 million years after Earth began to form but before it reached its final size (**Figure 7.16**).

As Earth continued to grow, its mantle started to differentiate (Figure 7.16). A **magma ocean** extending from Earth's surface down to perhaps 400 km is thought to have formed because of the intense heat. Because of rapid heat loss from near Earth's surface, the magma ocean probably crystallized rapidly, geologically speaking. Olivine and pyroxene, mafic minerals rich in iron, magnesium, and calcium, underwent phase transitions to ultramafic minerals such as peridotite. If a crust formed, it was most likely formed of komatiites, which are the volcanic equivalents of peridotites. Ultramafic volcanics such as komatiites crystallize at relatively high temperatures. Thus, for komatiites to have formed, the surface temperature must have exceeded approximately 1,600°C. (In contrast, modern basalt flows have temperatures of about 1,000°C to 1,200°C.) However, no rocks indicative of such an early primary crust have been found from this interval, implying that if such a crust formed, it was melted and destroyed almost as soon as it was produced.

During its existence the magma ocean undoubtedly fed numerous volcanoes that spewed lava onto Earth's surface and gases into the atmosphere. This process is called **outgassing** or *degassing* (refer to this chapter's frontispiece).

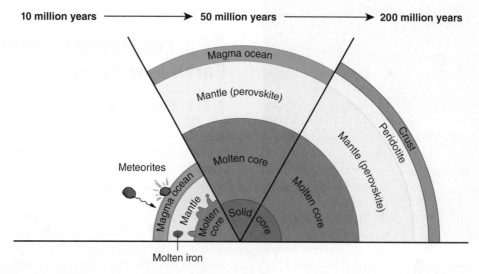

FIGURE 7.16 Formation and differentiation of the early Earth. Ages are in millions of years after the initiation of planetary accretion. Cross-sections of the Earth are not to scale.

Courtesy of SOHO/ESA/NASA.

FIGURE 7.17 Deflection of the solar wind by the Earth's magnetic field permitted the formation of an atmosphere.

These gases would have begun to accumulate in a primitive atmosphere because the magnetic field established by Earth's core had begun to deflect the solar wind around Earth (**Figure 7.17**). The magnetic field is generated by the relative movements of the Earth's inner and outer cores and the rotation of the Earth. The inner core rotates at a faster rate than the outer core which, along with the Earth's rotation, generates currents within the outer core that produce an electric current that in turn generates the magnetic field around the Earth.

The question arises, though, as to whether the atmosphere that developed was primary or secondary in origin. If primary the atmosphere was presumably highly reducing, meaning free oxygen (O_2) was not present. A highly reducing atmosphere would have been composed of gases such as methane (CH_4) and ammonia (NH_3). Hydrogen (H_2) would probably also have been abundant as a result of **photodissociation** or **photolysis** (literally, "breakdown by light") of water (H_2O) vapor at the outer margins of Earth's atmosphere. In this process H_2O was split by the intense ultraviolet light emanating from the sun into the gases hydrogen (H_2) and oxygen (O_2).

There are several problems with the existence of a highly reducing atmosphere during Earth's early existence. First, Earth's magnetic field could not have been established until Earth's iron-nickel–rich core formed. Thus, the solar wind could have easily blown off any primary atmosphere as it formed. Second, it would have been virtually impossible for ammonia to have accumulated in abundance in the atmosphere because, like water, ammonia is easily photolyzed by

ultraviolet radiation. Third, some workers believe if CH_4 had been abundant, carbon-rich organic compounds should also have been deposited in sediments, but such compounds are rare during this time.

Consequently, scientists now believe early Earth's atmosphere was probably secondary in origin, meaning it formed after the solid Earth had accreted. Because the core had already formed, most of the iron, with which O_2 easily combines, might have been sequestered in the core. This would have left relatively little iron in the mantle with which free O_2 could combine. Instead, oxygen would have combined with other elements such as carbon and hydrogen to produce an atmosphere enriched in CO_2 and water vapor (H_2O), along with nitrogen (N_2). This secondary atmosphere would have been only **mildly reducing** because of the increased presence of O_2. The early secondary atmosphere probably resembled the present atmospheres of Mars and Venus (**Table 7.2** and **Box 7.1**).

The formation of a secondary atmosphere by degassing of the mantle is supported by isotopic studies of the inert gas, argon-40 (^{40}Ar). Argon belongs to the same chemical group as helium and forms from the radioactive decay of potassium-40 (^{40}K; see Chapter 2). Today, Earth's atmosphere is enriched in ^{40}Ar, which was probably originally produced in the mantle and then released to the atmosphere. Using the ratios of concentrations of these isotopes in the atmosphere to their ratios in the mantle, scientists have calculated that 80% to 85% of Earth's initial atmosphere was outgassed in the first million years, and the rest of the mantle's gases released slowly during the subsequent 4.5 billion years.

TABLE 7.2

Comparison of the present atmospheres of the planets in Earth's solar system

Planets	Surface Temperature (°C)	Surface Pressure (Bars)	Principal Gases
Mercury	−184 to 465	0	None
Earth	−20 to 40	1	Nitrogen, oxygen
Venus	450 to 500	90	Carbon dioxide
Mars	−130 to 25	0.01	Carbon dioxide
Jupiter	−140	2	Hydrogen, helium
Saturn	−150	2	Hydrogen, helium
Uranus	−150	5	Hydrogen, methane
Neptune	−150	10	Hydrogen
Pluto*	−200	0.005	Methane

*Now considered a dwarf planet or large asteroid because it is only one of several large bodies within the Kuiper asteroid belt located outside the solar system.

7.5.2 Origin of the Moon

The Moon also began to form during the Hadean (**Figure 7.18**). Samples returned by the American Apollo space missions and Russian unmanned landings have been dated at 4.5 to 4.6 billion years old, confirming ages obtained from meteorites. Thus, the Moon must have formed as the early Earth was coalescing and differentiating.

The surface of the Moon exhibits several major features that yield clues to its origin. It is thought that the crust of the lunar highlands is primary crust, which formed from a magma ocean that extended to depths of about 500 km (Figure 7.18). The lunar highlands consist of anorthosite,

a rock enriched in the mineral plagioclase that might have formed as plagioclase floated close to the surface of a magma ocean, whereas denser olivine and pyroxene settled into the upper mantle. The maria of the moon, originally thought to have been seas ("mare," or sea) by the astronomer Galileo, are actually huge basaltic outpourings. Because the maria are so flat, the flows appear to resemble huge flood basalts that comprise about 17% of the Moon's surface. The basins themselves appear to have been formed by impacts before about 3.9 to 3.8 billion years ago and were then flooded repeatedly by lava between about 3.9 and 3.2 billion years before present. After about this time, the incidence of cratering decreased on the Moon and the terrestrial planets.

BOX 7.1 How Do We Know the Atmospheric Composition of Other Planets?

We can determine the composition of other planets' atmospheres by using the technique of *spectroscopy*. Different gases characteristically absorb or emit electromagnetic radiation as visible light or infrared radiation (**Box Figure 7.1A**). For example, the significant downward spikes in the spectra for Earth, Mars, and Venus all indicate the presence of carbon dioxide.

Scientists have also used unique opportunities to study planetary atmospheres. For example, the spectacular collisions of the Shoemaker-Levy comets with Jupiter in 1994 spewed vast amounts of gas into space that was examined spectroscopically from Earth (**Box Figure 7.1B**). The interior of Jupiter's atmosphere is otherwise shielded from view.

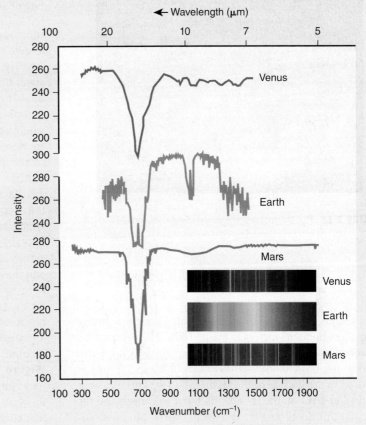

BOX FIGURE 7.1A Infrared spectra of the Earth and Venus indicate the gaseous composition of the respective atmospheres. The spectra indicate that the atmospheres of both Venus and Earth contain carbon dioxide but that only Earth's atmosphere also contains water vapor and ozone (and therefore presumably oxygen).

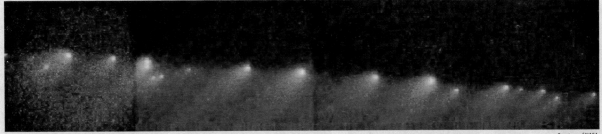

Courtesy of NASA.

BOX FIGURE 7.1B The collision of the Shoemaker-Levy comets with Jupiter, as shown in this sequence of photographs (left to right), ejected vast amounts of gas into space where it could be analyzed using spectroscopy.

Courtesy of NASA/JPL.

FIGURE 7.18 The Moon. Note the lunar highlands, broad flat areas called maria (so-named by Galileo because he thought they were seas), and rays of material spewed beyond the rims of the craters by impacts.

Total thicknesses of the flood basalts are thought to range from a few hundred meters to about 4 km. The mare basins are more prominent on the side of the Moon facing Earth because the crust is thinner there (about 60 to 70 km) as opposed to the far side of the Moon (80 to 90 km). Possibly, the thicker crust on the far side inhibited volcanism and the formation of flood basalts. Volcanism on the Moon is estimated to have ceased at about 2.8 billion years ago.

There are four major hypotheses for how the Moon originated. Each of these hypotheses has its own advantages and drawbacks. The **fission hypothesis** states the Moon was essentially thrown off from Earth because Earth was spinning so rapidly it ejected material that overcame gravitational attraction. This mechanism accounts for, among other features, (1) the absence of a large metallic core in the Moon, because the Moon would have been derived predominantly

from mantle material of early Earth, and (2) the similarity of the chemical composition of lunar samples and Earth. However, one of the drawbacks of the fission hypothesis is that the Moon's orbit occurs at an angle to Earth's orbit, which should not be the case if the two bodies were once close together. Moreover, if the two objects were once extremely close, the Moon would have needed to overcome about four times the angular momentum of the present Earth–Moon system to have moved to its present position.

The **simultaneous accretion** or *double planet hypothesis* states that the Moon formed through accretion just as Earth did. However, this hypothesis does not explain why the Moon's orbit is inclined relative to Earth's. Also, like the fission model, how could a body of the Moon's size have been so close to Earth while resisting the angular momentum of the Earth–Moon system and remaining separate from Earth?

As their name implies, **capture hypotheses** invoke the capture of a planetesimal, either whole or as fragments, which later accreted to form the Moon. Although capture hypotheses can account for the inclined lunar orbit, they cannot account for the similar chemical composition of Earth and the Moon. Moreover, capture models involving an intact body require a very specific approach velocity and trajectory for the planetesimal, whereas the capture of fragments cannot account for the angular momentum of the system.

Currently the **impact hypothesis** seems to best account for the features of both Earth and the Moon (**Figure 7.19**). Computer simulations suggest that early in the evolution of the solar system, there were at least 10 bodies larger than the planet Mercury, and some larger than Mars, that could have sideswiped Earth with a glancing blow. The impact model accounts for chemical similarities between Earth and the Moon, as well as chemical differences; volatile compounds would have been vaporized from Earth materials during impact before the remaining materials coalesced to form the Moon. The impact model also accounts for the inclined orbit of the Moon and the angular momentum of the system. According to the impact hypothesis, the Moon would initially have been relatively close to Earth. In fact, after its formation the Moon would have been close enough

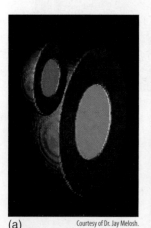

(a)　　Courtesy of Dr. Jay Melosh.

(b)　　Courtesy of Dr. Jay Melosh.

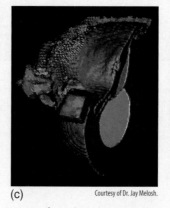

(c)　　Courtesy of Dr. Jay Melosh.

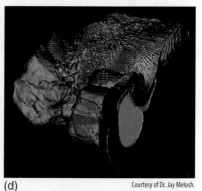

(d)　　Courtesy of Dr. Jay Melosh.

FIGURE 7.19 Simulation of the formation of the Moon by the impact of another planet or Mars-sized object with the Earth.

to Earth to have filled much of the sky! Still, the Moon was far enough from Earth to have gradually shifted away at a rate of about 4 centimeters per year.

Despite their differences, all the hypotheses for the Moon's origin share one thing in common: a fiery origin for the Moon, just as for Earth. Nevertheless, recent studies indicate that the Moon, at least at one time, held water. The recent impact of an empty rocket stage of a NASA satellite with the Moon's surface sent up a plume of dust and water vapor. Traces of water have also been detected in volcanic glass and crystals of the phosphatic mineral apatite have been found in the basalt of the Moon's maria. Unlike the volcanic glass, the maria formed by gentle eruptions would have been less prone to expel water from the mineral's crystal structure. The implication of these findings is that perhaps the Moon did not form from a collision with a Mars-sized object after all and that comets might have been a much more important source of water than previously thought. In a way, then, maybe Galileo was right about the origin of the Moon's maria.

7.6 The Archean: Beginnings of a Permanent Crust

7.6.1 Shields and cratons: cores of continents

Rocks of Archean and Proterozoic age comprise large, exposed areas, or **shields**, of the modern continents. The rocks are predominantly of igneous and metamorphic origin and are frequently deformed, indicating orogenesis. Among the largest of the shields is the Canadian Shield (**Figure 7.20**). Overlying the shields at some distance from their centers are relatively flat-lying sedimentary rocks that have undergone little or no deformation. Together, the shield and its sedimentary cover form the "cores" of the continents, or **cratons**.

The distribution of rocks within the cratons indicates that as time passed, orogenesis generally enlarged the cratons. In the case of North America, the craton enlarged toward its margins, but the pattern of enlargement is less distinctive for other cratons. Unlike the mountains formed during the Phanerozoic Eon, which are still rising (such as the Himalayas) or which have undergone extensive erosion since their completion (such as the Appalachians), those of the Precambrian are old enough to have been beveled by erosion. Indeed, Precambrian rocks underlie much of North America and other continents, but the only Precambrian rocks now largely seen outside of the shields are exposed in mountain ranges later uplifted by the orogenies of the Phanerozoic.

7.6.2 Gneiss terranes

The Archean rocks of the shields are of predominantly two types: **gneiss terranes** that surround and often intrude into **greenstone belts** (**Figure 7.21**). The oldest known crust on Earth is represented by the **Acasta Gneiss**, which has been dated at about 4 Ga ("giga-annums" or billions of years) and is found in the Arctic of northwestern Canada. Slightly younger rocks (about 3.9 Ga) belonging to the **Isua Complex** of southern Greenland and Labrador were until recently the oldest rocks known.

The Acasta Gneiss consists of intrusive **tonalite-trondhjemite-granodiorites (TTGs)**. These rocks are known collectively as **granitoids**. As the name implies, these rocks contain an unusual combination of minerals compared with the sequence of mineral crystallization in granites (see Chapter 3). Unlike true granites, tonalites consist of abundant calcium and sodium-rich plagioclase along with quartz, whereas granodiorites contain potassium feldspar. Granitoids also frequently contain fragments of mafic (basaltic) and ultramafic rocks that might represent older oceanic crust. Because they are neither ultramafic nor mafic rocks, granitoids must have undergone some form of magmatic differentiation that allowed the mineral assemblages to form. Today, TTGs are found only in regions where relatively young, hot oceanic crust is rapidly subducted. One such region occurs along the margin of southern Chile in South America. The conditions for TTG formation today imply that, as expected, the formation and subduction of oceanic crust occurred at much faster rates in the Archean than later because of the greater rates of heat flux (loss) from Earth's interior (Figure 7.15).

Even older silicate minerals called **zircons** have been dated at about 4.4 Ga. These particular minerals are believed to be detrital because they are found in sedimentary rocks; "detrital" means the zircons were eroded and transported from previously existing crust. If this is indeed the case, *liquid water must have been present on Earth by about 4.4 Ga.* Thus, the Acasta Gneiss and other early crustal rocks are probably secondary in origin, meaning that any original crust that might have existed before the Acasta Gneiss has been destroyed, possibly by subduction or erosion.

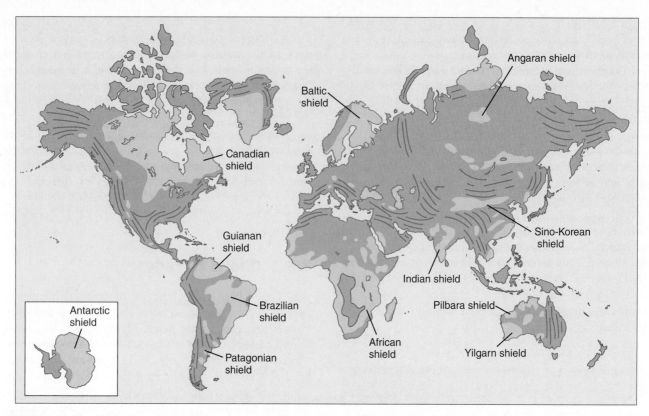

FIGURE 7.20 The distribution of the shields and cratons of the continents. Archean rocks help form the "cores" of continents around which other rocks accreted as the continents grew. Orange lines indicate general trends of today's mountain belts.

7.6.3 Greenstone belts

By contrast to gneiss terranes, greenstone belts typically consist of metamorphosed, trough-like, folded or synclinal sequences of rocks that are intruded by granitic magmas; these sequences might also be cut by thrust faults (**Figure 7.22**). Greenstone belts are variable in size, but range up to about 500 km across. In the Canadian Shield, Archean greenstone belts occur mostly in the Superior and Slave regions (**Figure 7.23**).

Greenstone belts consist of a sequence of mostly extrusive igneous rocks followed upward by sedimentary ones. The oldest rocks in the idealized sequence of an Archean greenstone belt are pillow basalts, which indicate submarine volcanic eruptions. Volcanoes are also indicated by the presence of komatiites. However, komatiites are relatively rare after the Archean. Thus, the occurrence of komatiites again indicates that the Archean Earth's surface was substantially warmer than today (Figure 7.15). Lying directly above these lower units are mostly basalts, tending to grade upward into more andesitic volcanics. The various volcanics were altered by low-grade metamorphism to produce chlorite, which imparts its characteristic greenish color to the greenstone belts.

It is the upper, younger portions of greenstone belts that tend to contain sedimentary rocks (Figure 7.21). Graywackes, which are poorly sorted sandstones containing volcanic rock

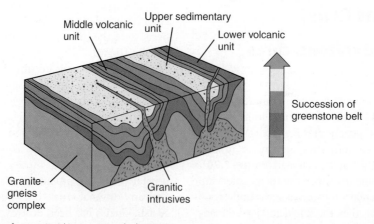

FIGURE 7.21 Reconstruction of a granitoid-greenstone belt setting.

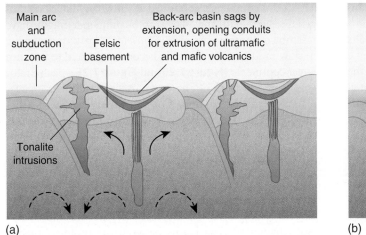

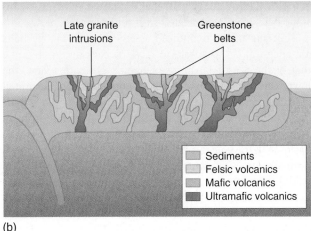

(a)   (b)

FIGURE 7.22 The back-arc basin hypothesis for the formation of greenstone belts. **(a)** Ancient tectonic setting. **(b)** After collision and accretion to form larger continent with greenstone belts and gneiss terranes. Tonalite is an igneous rock rich in quartz and plagioclase.

fragments, occur nearer the base of the sedimentary sequence, suggesting the presence of island arcs nearby. The volcanic fragments tend to become progressively more andesitic or felsic in younger greenstone belts, implying crustal rocks were undergoing some form of differentiation. Also present are turbidites, indicating submarine slumping. Nearer the top of the sedimentary sequence, conglomerates composed of granitoid cobbles hint that the sediments were being eroded from nearby land masses and rapidly redeposited. Also, quartz sandstones, some with ripple marks, indicate shallow-water, nearshore environments above wave base. Besides conglomerates, quartz sandstones, and graywackes, cherts and carbonates have been found in the upper portions of greenstone

belts. Cherts and especially carbonates are often indicative of shelf environments (see Chapter 4).

Where and how could such a complex sequence of rocks have formed? One hypothesis suggests that some greenstone belts formed within intracontinental rift basins. According to this hypothesis, a mantle plume initiated rifting of a preexisting felsic (granitoid) crust and volcanism (Figure 7.22). The plume served as the source of the lower and middle volcanic units. As rifting occurred the sides of the rift valley were eroded and the sediments rapidly shed into the basin. Ultimately, the rift basin closed, causing deformation, uplift, and further erosion so that sedimentary rocks dominated nearer the top of the sequence. However, the submarine volcanism,

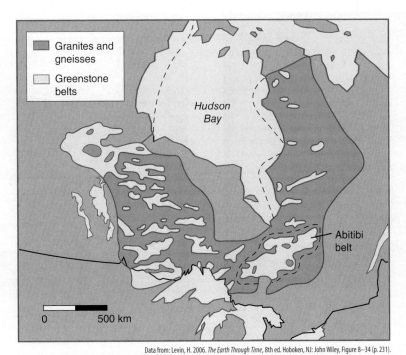

Data from: Levin, H. 2006. *The Earth Through Time*, 8th ed. Hoboken, NJ: John Wiley, Figure 8–34 (p. 231).

FIGURE 7.23 The general occurrence of greenstone belts in the Superior and Slave cratons.

complex folding, and thrust faulting found in many greenstone belts are not associated with rift basins.

An alternative hypothesis asserts that greenstone belts are ancestral, so to speak, to modern back-arc basins, like that of the Sea of Japan (see Chapter 2). In this view, the lower volcanic portion of the greenstone belts suggests initial extension and volcanism, whereas the sedimentation, folding, faulting, later igneous intrusion, and metamorphism are associated with the closure, uplift, and erosion of the basin. Still, the volcanism associated with such subduction is normally andesitic, whereas the rift valley hypothesis more readily accounts for the ultramafic volcanism. Also, the rift hypothesis accounts for the variable size of greenstone belts. Greenstone belts might therefore have formed in two very different types of settings, not just one.

7.6.4 Microplate tectonics and differentiation of the early crust

What else do the gneiss terranes and greenstone belts tell us about the early evolution of Earth? Archean continents probably grew by collision and accretion along convergent plate boundaries. This view implies the application of a uniformitarian view of modern plate tectonics to the Archean. However, the rocks themselves indicate that we cannot apply the Principle of Uniformitarianism too strictly to Archean plate tectonics. Indeed, rather than a dozen or so crustal plates like today, Earth's outer shell might have been composed of 100 or more **microplates** moved by many different spreading centers or mantle plumes trying to rid Earth of the internal heat generated by its initial formation and radioactive decay (Figure 7.15).

Thus, Archean continental crust was probably represented by smaller cratons, or microcontinents (**Figure 7.24**).

The proto-continents continued to grow by collision, orogenesis, and suturing. The proto-continents might have resembled bumper cars at an amusement park, but moving in slow motion and eventually becoming sutured to produce larger continents (**Figure 7.25** and **Figure 7.26**). Each proto-continent consisted of one or more gneiss terranes characterized by distinct groupings of radiometric dates and structural trends; thus, each proto-continent resembled the exotic terranes of later times (see Chapter 2). Radiometric dates generally decrease away from the centers of the shields. This indicates the microcontinents collided and accreted to form larger continents. However, in some cases younger dates occur where they are not expected, suggesting the radiometric "clock" might have been "reset" by later metamorphism (see Chapter 6). The basins between the proto-continents are represented by pillow basalts and are filled by sediments such as graywackes. When the proto-continents collided, the basins were further filled with sediment, deformed, and metamorphosed to produce the greenstone belts.

Taken together, these data suggest that Earth's early crust differentiated gradually through time. This gradual differentiation was punctuated by episodes of somewhat more rapid crustal formation when proto-continents collided. This hypothesis has been tested using ratios of what are called **Rare Earth Elements (REEs)**, of which there are 14 in all. Unlike other elements such as calcium and magnesium, REEs are relatively insoluble in water and are not easily incorporated into the crystals of other minerals. Consequently, their relative abundances change little between the time they are weathered from igneous source rocks and deposited in sedimentary rocks, such as shales. Because shales tend to reflect the average composition of the rocks from which they are eroded, we can use their REEs to infer the relative abundance of different types of igneous source

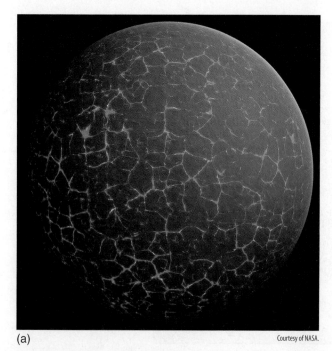

(a) Courtesy of NASA.

(b) Courtesy of USGS.

FIGURE 7.24 **(a)** Hypothetical reconstruction of Earth as it might have appeared during this time. Note the numerous small plates. **(b)** Satellite image of granitic plutons (light) and deformed greenstone-sediment belts (dark) of western Australia.

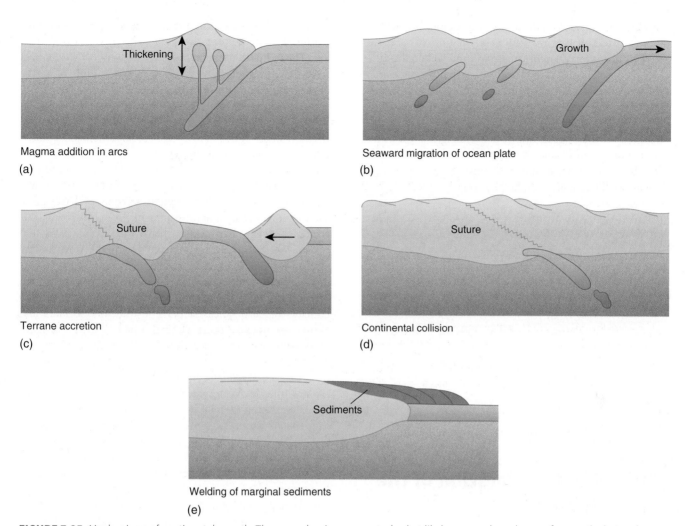

FIGURE 7.25 Mechanisms of continental growth. These mechanisms occur today but likely occurred much more frequently during the formation of greenstone belts, as described in the text. Note how remnants of subducted plates descend into the mantle (see Chapter 2).

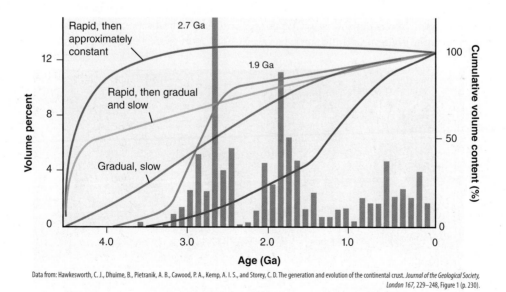

Data from: Hawkesworth, C. J., Dhuime, B., Pietranik, A. B., Cawood, P. A., Kemp, A. I. S., and Storey, C. D. The generation and evolution of the continental crust. *Journal of the Geological Society, London 167*, 229–248, Figure 1 (p. 230).

FIGURE 7.26 Some of the growth rate models of the continents proposed by various investigators. Note the clumps of peak ages. 100% represents modern crust.

crust, even if it has been destroyed. For example, continental crust is enriched in the REE neodymium (Nd), whereas oceanic crust is enriched in samarium (Sm). When the ratios of these two elements are plotted for the Archean and earliest Proterozoic, the ratios show an increasing range of values. This suggests an increasing differentiation of oceanic versus continental crust after the initial crust formed. Crustal differentiation probably occurred by the recycling of crust during subduction, melting, and magmatic differentiation through time. By contrast, lunar samples cluster near 0, indicating little or no differentiation of the Moon's crust since it formed billions of years ago.

7.7 Climatic Evolution of the Inner Planets

7.7.1 Habitable zone

 The region within the solar system where life can exist is known as the **habitable zone (HZ)**. The initial mass of the sun and the positions of the planets relative to the sun set the broad constraints on where the HZ occurs. In effect, the HZ is a "Goldilocks zone." Life is dependent on liquid water, so the HZ represents the zone around the sun where temperature permits condensation of water and the potential existence of life (**Figure 7.27**).

The radius of the solar system is roughly 100,000 **astronomical units**, or **AUs** (1 AU equals the radius of Earth's orbit around the sun, or about 98 million miles). If Earth had accreted in an orbit 5% closer to the sun (or 0.95 AUs), Earth's water would have evaporated into the atmosphere. Because water vapor, like carbon dioxide, is a greenhouse gas, this would have caused Earth to warm, which would have caused more evaporation, and so on. Thus, a **runaway greenhouse** would have ensued through positive feedback. On the other hand, if Earth had accreted at 1.01 AUs from the sun, a **runaway glaciation** might have occurred. In this scenario, because Earth exists farther from the sun, it would receive less solar radiation and cool down. The growth of ice and snow would have increased the albedo of Earth's surface and served as positive feedback on snow and ice cover.

We can now account for why the atmospheres and the climates of Earth's nearest planetary neighbors, Mars and Venus, are not like those of Earth (Table 7.2). First, Earth is the only planet to fall within the HZ for the entire 4.5 to 5 billion years of its existence. Second, besides its distance from the sun, a planet's surface temperature is a function of the greenhouse effect, which is controlled by the amount of water vapor and carbon dioxide present in the atmosphere. The water vapor pressure in each planet's atmosphere is determined by the distance of the planet from the sun. Starting with the estimated surface temperature of each of the three planets, water vapor must have increased in each planet's atmosphere during the Hadean through degassing and perhaps meteor impact (as discussed earlier) until the atmosphere became saturated with water vapor and began to condense (**Figure 7.28**). On Earth, increasing the amount of water vapor resulted in increased precipitation and the formation of the oceans (Figure 7.28).

By contrast, Venus's atmosphere might initially have held much more water vapor before any oceans could have formed. Venus is closer to the sun and therefore warmer, causing more evaporation and allowing the atmosphere to hold more water vapor before saturation and precipitation

FIGURE 7.27 The continuously habitable zone (CHZ) of the Earth. By contrast to Earth, Venus suffered a runaway greenhouse and Mars a runaway icehouse.

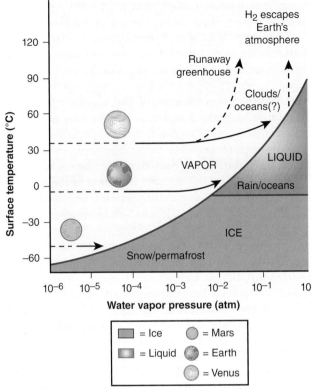

FIGURE 7.28 Atmospheric evolution of Venus, Earth, and Mars in terms of surface temperature and water vapor pressure. Water vapor pressure is shown as fractions of the Earth's current atmospheric pressure at its surface (1 atmosphere). As sufficient water vapor accumulated in the atmosphere on Mars, it presumably began to precipitate as snow and permafrost because of Mars' distance from the Sun. On the other hand, water vapor on Earth eventually began to precipitate as liquid water. Water vapor on Venus accumulated in the atmosphere to produce clouds and a runaway greenhouse because Venus is located closer to the Sun. Water vapor was also photolyzed to produce hydrogen gas that escaped from the atmosphere.

could occur. A runaway greenhouse is therefore thought to have ensued, during which temperature and evaporation increased. The atmosphere became so full of water vapor it prevented heat (in the form of infrared radiation) from escaping and caused the planet to heat further through positive feedback. Eventually, however, so much water is thought to have evaporated into Venus's atmosphere that almost all of it was lost! Water is thought to have split into molecular hydrogen (H_2) and O_2 by photolysis. The hydrogen would have escaped into space, whereas the oxygen might have combined with iron in rocks during weathering. This scenario is supported by the occurrence of deuterium in Venusian water vapor in concentrations over 150 times that of Earth's oceans. Because deuterium is heavier than normal hydrogen (H_2), it would have remained behind as lighter hydrogen was lost to space.

On the other hand, Mars was once thought to have been quite cold throughout much of its existence because of its greater distance from the sun (Figure 7.28). As a result, precipitation was thought to have occurred as snow and ice.

According to this scenario, as cooling continued carbon dioxide froze to produce polar caps composed of dry ice (like that used for temporary refrigeration). As a result Mars became a cold desert.

Nevertheless, recent evidence seems to support the presence of a vast ocean as much as 1.6 km deep early in Mars's history. Indeed, water might still be present beneath the surface of Mars. Features that appear to have been formed by water include apparent conglomerates and river-like channels (**Figure 7.29**). Rounded pebbles in the conglomerates indicate substantial abrasion during transport. Cross-beds, each no more than a few centimeters thick, have also been observed. Some features resembling glacial outwash channels like those on Earth are up to 100 km wide and 2,000 km long. By calculating backward from the size of

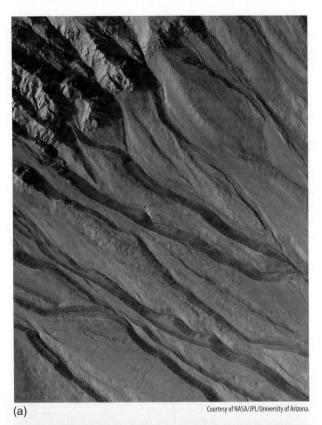

FIGURE 7.29 Evidence for water on Mars. (a) Possible river channels or canyonlands. (b) Possible cross-bedding and mud cracks.

these features, scientists have estimated they were produced by water flows on the order of 300 million cubic meters per second! For comparison, during the great flood of the summer of 1993 in the American Midwest the flow rate of the Mississippi River when it crested at St. Louis, Missouri, was roughly 28,000 cubic meters per second.

How was the water lost? Recent spectroscopic data obtained from satellites (see Box 7.1) indicate relatively large amounts of molecular H_2 in the atmosphere. The molecular H_2 is thought to have been produced by another type of dissociation involving the chemical reaction of water with iron-rich rocks, which might have been abundant early in Mars's history. In the process, water was split, releasing the H_2 to the atmosphere, whereas the O_2 combined with the iron, giving Mars its red color. The cratering record of Mars has been inferred to indicate that much of this early atmosphere was then stripped away during bombardment by asteroids and comets, leaving behind an atmosphere of mostly carbon dioxide. Although much of the water of this phase appears to have been lost, some liquid water might have been present only tens to hundreds of millions of years ago. The relatively recent formation of the sedimentary features is suggested by their fresh, uncratered appearance.

7.7.2 Faint young sun

If the levels of greenhouse gases (whatever their chemical composition) in Earth's atmosphere were so high early in the Archean, why didn't a greenhouse effect cook Earth's surface? Based on computer models that omit the greenhouse warming, Earth should have been frozen until well more than a billion years ago (**Figure 7.30**). If this had occurred, the increased albedo caused by large amounts of snow and ice should have made Earth even colder via positive feedback. However, there is no evidence for a perpetually frozen Earth during its first three billion years of existence based on the near absence of glacial deposits like tillites. Indeed, as we have seen, there is evidence for liquid water going back more than four billon years. It is therefore thought that greenhouse warming occurred as a result of large amounts of greenhouse gases that counteracted the faint young sun.

The cosmologist Carl Sagan proposed what is now called the **Faint Young Sun Paradox**: the existence of a faint young sun during the early evolution of the solar system. Computer models indicate that early in Earth's history the sun was only about 70% to 75% as bright as today's (Figure 7.30). When the Faint Young Sun Paradox was first presented, ammonia (NH_3) was thought to have been the greenhouse gas involved in warming Earth, but it was later realized that NH_3 would have been broken down by photolysis. It was then that the possible role of high CO_2 levels in Earth's early climate was hypothesized. However, more recent examination of ancient soils, or **paleosols**, suggests that CO_2 levels at 3.5 Ga were at least 5 times lower than those necessary for the Faint Young Sun Paradox to occur.

Thus, another greenhouse gas might have been more important, perhaps methane (CH_4). But methane was supposed to characterize Earth's primary atmosphere and was then presumably "blown off" by the solar wind, or perhaps oxidized during the formation of a secondary atmosphere. What, then, could have been the source for this methane, if it was in fact present? Possibly, *methanogenic,* or methane-producing, bacteria are responsible. The recent detection of methane in Mars's atmosphere hints that methanogenic

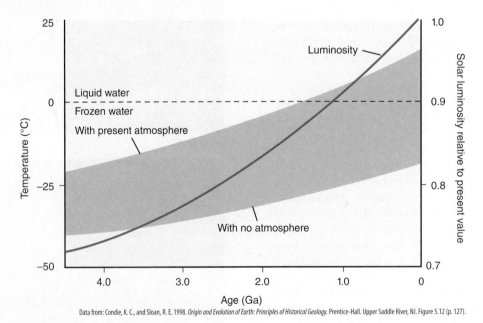

Data from: Condie, K. C., and Sloan, R. E. 1998. *Origin and Evolution of Earth: Principles of Historical Geology.* Prentice-Hall. Upper Saddle River, NJ. Figure 5.12 (p. 127).

FIGURE 7.30 Effect of the sun's increasing brightness on the Earth's surface temperature through time. The region shaded in blue represents the range of temperatures resulting from a greenhouse effect caused by water vapor. The magnitude of the greenhouse effect increased through time because as the sun's intensity increased, the amount of water vapor evaporated into the atmosphere also increased. However, note that in this simple model, the Earth should have remained perpetually frozen for well more than 1 billion years ago. This result is at odds with the geologic record, which indicates the presence of liquid water no later than 3.8 billion years ago and likely much earlier.

bacteria might also occur there. We will explore these topics in greater detail in Chapter 8.

Whatever the chemical composition of the greenhouse gases, the geologic record indicates the sun was unable to generate the runaway greenhouse on Earth that apparently occurred on Venus. Because it is closer to the sun, during its early history Venus received 40% more solar radiation than Earth receives *today*, so a runaway greenhouse probably still occurred on Venus.

7.7.3 Weathering and tectonism on the inner planets

It is unlikely that the weathering of rocks during the geologic cycle of carbon (see Chapter 1) accounts for any sort of climatic feedback early during Earth's history. During the Hadean and early Archean, this cycle, which later provided feedback on atmospheric carbon dioxide levels on Earth, might have been mostly inoperative. Any continents that might have been present were probably small and of low relief, which would have inhibited continental weathering. Also, if Earth's surface temperature were somewhat lower because of lower solar radiation, chemical weathering reactions between carbon dioxide and continental rocks would have been lower. It is also unlikely that a lower albedo could have helped warm Earth: given the amount of water vapor that must have been present in Earth's early atmosphere, there must have been extensive cloud cover and, if anything, *higher* albedo (**BOX 7.2**).

Feedback from rock weathering probably did not operate on the other inner planets, either. Because of their small

BOX 7.2 The Fate of Earth

Our sun's fate will be no different from that of any other star of comparable mass (**Box Figure 7.2A**). The sun is now about halfway through its approximately 10- to 12-billion-year life span. As the sun grows older, it will also become brighter, causing the HZ to migrate outward across the solar system.

As a result of the expansion of the HZ, it has been projected that in a few hundred million years, Earth will begin to revert toward an increasingly harsh state like that of the Archean. As Earth initially warms, animals, which have an upper thermal survival limit of about 45°C (113°F), might migrate toward the poles, where it is cooler. Also, continental weathering will eventually use up carbon dioxide because of the warmer temperatures. As a result land plants will die and terrestrial ecosystems collapse, all of which has been projected to occur roughly 500 million to 1 billion years from now. As continental weathering slows because of lowered carbon dioxide levels and the loss of terrestrial plants, nutrient inputs from land to the seas will slow, decreasing the productivity of marine plankton so that marine ecosystems will also collapse. At that point, Earth will be tolerable to only the most primitive forms of life, and after that, perhaps none at all.

Earth is eventually projected to exist in one of several states, depending on how fast the oceans evaporate. If the oceans evaporate quickly, there will be no severe greenhouse effect from water vapor and Earth's surface temperature will increase by another 30°C in the next 6 billion years. Earth will become relatively dry, perhaps desert-like. However, if the oceans evaporate more slowly, Earth will be left in a Venusian-like state with a runaway greenhouse, in which case Earth's surface would be like an autoclave used to sterilize medical instruments.

With the loss of water, plate tectonics will shut down and orogeny will cease between 750 million and 1.2 billion years from now. Under these conditions, 3.5 billion years into the future and when the sun is 40% brighter, Earth's surface could warm to as much as 1,000°C. This temperature is hot enough to melt Earth's surface and release all the CO_2 stored in limestones, which would only make Earth hotter. On the other hand, plate tectonics could also cease long before the oceans are lost if the heat flux from Earth's interior slows. In this case orogeny would also stop; the continents would be beveled, perhaps disappearing beneath the ocean's surface in a kind of "water world," perhaps resembling that of the Archean 4 billion years ago.

In any case, as the sun's nuclear fuel continues to dwindle, the gravitational attraction of the remaining matter would eventually cause the sun to collapse upon itself like a soufflé. As the sun collapses about 6 to 7 billion years in the future, its nuclear furnaces will be restoked briefly and it will swell into a red giant and engulf our solar system, only to shrink to a white dwarf (Box Figure 7.2).

(continues)

BOX 7.2 The Fate of Earth (Continued)

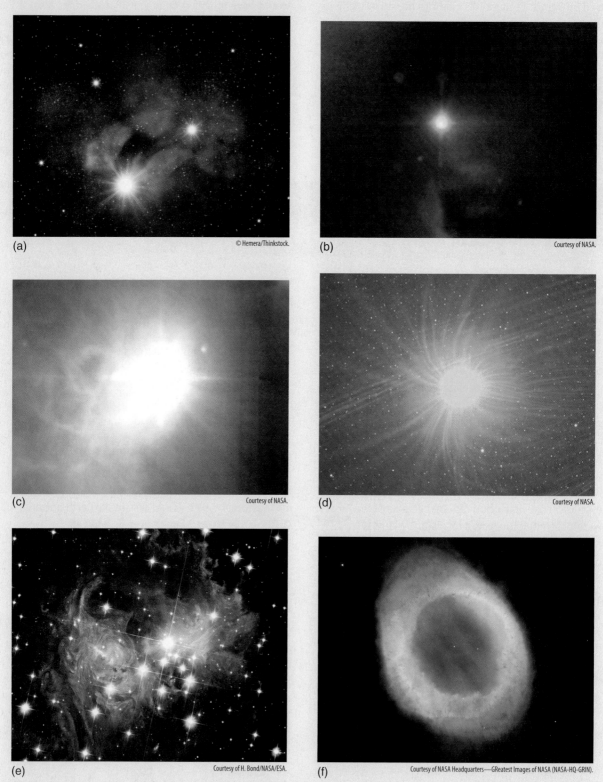

(a)   © Hemera/Thinkstock.

(b)   Courtesy of NASA.

(c)   Courtesy of NASA.

(d)   Courtesy of NASA.

(e)   Courtesy of H. Bond/NASA/ESA.

(f)   Courtesy of NASA Headquarters—GReatest Images of NASA (NASA-HQ-GRIN).

BOX FIGURE 7.2 Star evolution. Photos: **(a)** dust and gases; **(b)** protostar; **(c)** main sequence star; **(d)** star in giant stage; **(e)** star in variable stage; and **(f)** planetary nebula stage.

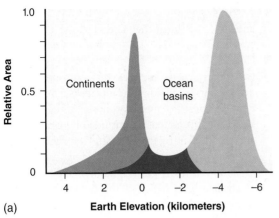

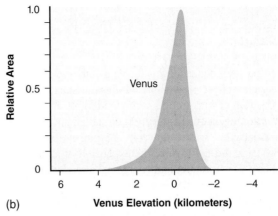

FIGURE 7.31 The relative distribution of surface elevations on Earth and Venus. The distribution on Earth is bimodal, reflecting the presence of ocean basins (negative elevations) and continents (positive elevations). Note that elevations for Venus are clumped at or near 0, suggesting that plate tectonics on Venus ceased billions of years ago because it had lost its water in response to a runaway greenhouse.

size, Mars and Mercury (and also the Moon) probably radiated their internal heat relatively rapidly to form rigid one-plate lithospheres. The larger a sphere (or planet), the smaller its surface area, because the volume of a sphere depends on the cube of its radius, whereas its surface area increases only by the square of the radius. Thus, small spheres or planets lose heat more readily than large ones. Mercury has a global network of faults that might have resulted from thrust faulting of the early crust during cooling and contraction, but it still resembles the Moon in its highly cratered appearance. This implies that once craters were formed they did not disappear by the subduction of crust. Mercury is also so close to the sun that it has no atmosphere that could be involved in chemical weathering and climate regulation.

As evidenced by its huge rift valleys, Mars might have had some plate tectonic activity in the past. However, there is no evidence of large-scale features such as mountain chains, which form as subduction produces volcanic arcs or continental collisions. In fact, Mars harbors the largest-known volcano in the solar system, Olympus Mons, which is the size of the state of Ohio and approximately 23 km (about 76,000 feet!) high (by contrast, Mt. Everest is "only" about 29,000 feet high). The great size of Olympus Mons hints that the crust over the site was stationary, allowing the volcano to grow to a huge size. Volcanism is thought to have occurred on Mars up to about 1 billion years ago. Some of the carbon dioxide emitted by volcanism might have been transferred by rock weathering to *chemical* (not biogenic) limestones;

these limestones might have then been subducted and permanently trapped in the crust when subduction stopped. Because it is farther away from the sun, any carbon dioxide remaining in Mars's atmosphere might have frozen and solidified, which kept it from becoming involved in further rock weathering and long-term climatic feedback cycles (see Chapter 1 on the geologic cycle of carbon). In contrast to Mars, Venus likely lost its heat more slowly because Venus's size is nearly that of Earth's (Table 7.1). Still, Venus has a relatively smooth surface compared with Earth (**Figure 7.31**). This suggests plate tectonics has also largely or entirely ceased on Venus. Because Venus has lost almost all of its water, it is also dry and essentially resembles the frozen state of Mars; thus, there is little or no liquid water into which carbon dioxide can dissolve and weather rocks. The lack of liquid water would have also inhibited the melting of plates during subduction (see Chapter 2). Consequently, carbon dioxide accumulated in the atmosphere during a runaway greenhouse so that Venus is now a hot, dry planet characterized by surface temperatures of 450°C to 500°C.

CONCEPT AND REASONING CHECKS

1. Will the HZ of Earth eventually shift to the point in which Earth is no longer habitable for life?
2. Why was early Earth habitable despite a faint young sun?
3. Why is Venus quite warm and Mars quite cold?

SUMMARY

- Profound questions and startling answers arise from one simple question: why is the night sky dark? Because the light from distant stars and galaxies has yet to reach us. Why are these bodies so distant? Because energy and mass, space and gravity, and all the atomic and subatomic particles and forces that we know today were once clustered so close together that perhaps they did not exist!

- The universe is thought to have exploded into existence during the Big Bang about 13 to 14 billion years ago. The alternative view, that the universe has always existed and been in steady state, allows application of "normal" physical theory to understanding the universe. However, to understand the Big Bang we must suspend our commonsense notions of what is normal. Within time intervals of far, far less than a split second during the Big Bang, subatomic particles and forces began to be established, and the universe began to inflate toward its present size.

- The basic evidence for the Big Bang comes from (1) the Doppler effect (red shift) of distant stars, (2) the detection of background cosmic radiation, and (3) the nonhomogeneous distribution of matter, such as stars and galaxies, in the universe. Despite its name, the Big Bang did *not* involve the expansion of matter *into* space; rather, matter and energy expanded with space and time to form the universe.

- In the billions of years after the Big Bang, heavier elements such as iron were formed from fusion of hydrogen and helium in early stars; heavier elements were then distributed to the surrounding universe by supernovae.

- According to the solar nebula hypothesis, after the formation of matter, gaseous nebulae condensed and the planets of our solar system began to accrete. The inner or terrestrial planets are composed primarily of silicate minerals and iron, whereas the outer or Jovian planets, which dwarf the inner ones in size, are composed mainly of gas. Dates from the oldest rocks on Earth, lunar samples, and meteorites indicate the solar system formed no later than about 4.5 to 4.6 Ga.

- The earliest phase of Earth's evolution occurred during the first 0.8 billion years, referred to as the Hadean. Earth was quite hot during this time because of the release of heat from radioactive decay, gravitational attraction of matter as the planet formed, and meteor bombardment.

- During this time Earth's core, mantle, and earliest crust began to form. The core formed as molten iron and nickel moved toward the center of Earth, which also released heat, whereas ultramafic and mafic minerals underwent phase transitions to form the mantle.

- The formation of Earth's core eventually established a magnetic field around Earth, which prevented Earth's atmosphere from being blown out into space by the solar wind. Earth's earliest permanent atmosphere was probably secondary in origin and resulted from the outgassing of volcanoes; the atmosphere is now thought to have been composed mainly of H_2O, N_2, and CO_2. The atmosphere was probably mildly reducing because no free O_2 was present.

- The Moon also originated during the Hadean, most likely as a result of a collision between a planetesimal and Earth.

- The oldest known rocks on Earth are represented by the Acasta Gneiss (~ 4.0 Ga) of the earliest Archean. These rocks appear to represent secondary crust that formed after destruction of the primary crust. The presence of liquid water and possibly also erosion on Earth earlier than this time is indicated by the occurrence of detrital zircons.

- Archean rocks help comprise large, exposed shields of the modern continents and consist predominantly of igneous and metamorphic rocks. Together the shield and its cover form the "cores" of the continents, or cratons.

- The Archean rocks of the shields are of predominantly two types: gneiss terranes that surround and often intrude into granitoid greenstone belts. Greenstone belts typically consist of metamorphosed, trough-like, folded or synclinal belts that are intruded by granitic magmas and may be cut by thrust faults. The submarine volcanism, metamorphism, complex folding, and thrust faulting found in many greenstone belts suggests that most greenstone belts represent the earliest counterparts of modern back-arc basins. The gneiss terranes represent small proto-continents that collided to form larger continents. The basins between the proto-continents are represented by pillow basalts and are filled by sediments such as graywackes. When the proto-continents collided, the basins were further filled with sediment, deformed, and metamorphosed to produce the greenstone belts.

- Despite their similar (but by no means identical) starting points, the inner or terrestrial planets followed different paths toward their present climates. The climatic evolution of the inner planets is a function of their distance from the sun or habitable zone (HZ). Mercury is far too close to the sun and therefore too hot to harbor life as we know it. Although tectonism might have occurred on Mercury early in its history, its surface resembles the pock-marked surface of the Moon, which has been tectonically dead for billions of years. Any form of plate tectonics on Mars and Venus appears to have also ceased long ago. Although surface features of Mars suggest the planet harbored liquid water at one time, because Mars lies farther from the sun than Earth, Mars cooled and liquid water and CO_2 both froze. The freezing of water made it unavailable to combine with CO_2 to form carbonic acid (H_2CO_3) and weather rocks. On the other hand, Venus lies closer to the sun than Earth. As a result most water vapor was probably lost from the top of the Venus's atmosphere through photodissociation, leaving behind a hot, dry planet with an atmosphere enriched in CO_2 but no H_2O. Consequently, the rock weathering cycle also shut down on Venus.

- Because of its distance from the sun, only on Earth did conditions fall within a HZ suitable for continued planetary evolution. Although Earth's atmosphere was enriched in CO_2 like that of Venus and Mars, Earth was sufficiently far from the sun to prevent loss of water vapor through either photolysis or freezing.

- According to the Faint Young Sun Paradox, the large amount of CO_2 or other greenhouse gases in the atmosphere counteracted the sun's relatively low brightness, which might have been 25% to 30% lower than it is today. Hence, greenhouse conditions were sufficiently warm, but not too warm, to permit both the evaporation of water to warm Earth and the condensation of water to form oceans.

KEY TERMS

Acasta Gneiss	Faint Young Sun Paradox	microplates	quarks
achondrites	fission	microwave radiation	Rare Earth Elements (REEs)
angular momentum	fission hypothesis	mildly reducing	red shift
Archean Eon	fusion	nebulae	runaway glaciation
asteroids	gneiss terranes	nebula hypothesis	runaway greenhouse
astronomical units	Grand Unification Epoch	nucleosynthesis	shields
Big Bang	granitoids	Olbers's paradox	siderophilic
blackbody radiation	greenstone belts	Oort cloud	simultaneous accretion
capture hypotheses	habitable zone (HZ)	ordinary chondrites	singularity
carbonaceous chondrites	Hadean Eon	outgassing	solar nebula hypothesis
chondrites	Hubble Law	paleosols	solar wind
comets	hydrogen	photodissociation	Steady State Theory
cosmic background radiation	impact hypothesis	photolysis	supernovae
cratons	Isua Complex	plane of the ecliptic	terrestrial planets
deuterium	magma ocean	planetesimals	tonalite-trondhjemite-granodiorites (TTGs)
Doppler effect	meteorites	plasma	zircons

REVIEW QUESTIONS

1. Describe the Doppler effect using diagrams.
2. Compare the Big Bang and Steady State theories of the origin of the universe with regard to the following: (a) age, (b) Doppler effect, and (c) cosmic background radiation. Which theory is corroborated by the evidence?
3. Why do scientists like nature to be in steady state?
4. No one ever saw the Big Bang. Why do we believe it happened?
5. No one ever observed the formation of the planets and the solar system. How, then, do we believe we know what happened?
6. Why are hydrogen and helium so abundant in the universe?
7. In table form, list the features of each of the hypotheses for the origin of the Moon. Is the evidence in support of each hypothesis mutually exclusive? Do you believe this often happens with scientific hypotheses and theories? Why?
8. How did Earth's core form? What were the implications of core formation for the formation of Earth's atmosphere?
9. What were the sources of heat that kept early Earth molten? Which of these heat sources was most important for later differentiation of Earth's crust?
10. Why do dates of lunar rocks indicate that the earliest crust formed on Earth has probably been destroyed?
11. Compare the evolution of the atmospheres of the terrestrial planets diagrammatically with respect to (a) the sun's brightness through time, (b) the HZ, (c) plate tectonics, and (d) rock weathering.
12. What are the hypotheses for evolution of the Venusian and Martian atmospheres? What is the evidence for these scenarios?
13. What is the significance of the finding that liquid water existed on Mars?
14. Diagram greenstone belts and gneiss terranes and their relationships to each other in a plate tectonic setting.
15. What is the geologic evidence for high heat flux from Earth during the Hadean and Archean?

FOOD FOR THOUGHT:
Further Activities In and Outside of Class

1. Compare the Steady State Theory of the universe with the Principle of Uniformitarianism described in Chapter 1. What do the two share in common?
2. The Andromeda galaxy shows a blue shift, not red. Explain.
3. The open oceans are deep blue in color. Why? (Hint: Remember that blue light is more energetic than red.)
4. The planets of our solar system are spaced almost equally apart. What might account for the spacing?
5. Compare the Steady State Theory of the universe, the Big Bang Theory, and the theory of plate tectonics (see Chapter 2). What are the traits of a "good" theory?
6. The Moon has undergone little or no erosion. Why?
7. How can scientists assign relative dates to phenomena that happened on other planets?
8. The concentrations of REEs in different types of crustal rocks are plotted relative to their concentration in meteorites. Why do you believe meteorites were chosen for comparison?
9. The Sm/Nd ratios values for lunar samples cluster near 0, indicating little or no differentiation of lunar crust. Why didn't the Moon's crust differentiate like Earth's?
10. Construct an inductive argument for greenstone belts representing back-arc basins.
11. Why are the oceans salty but lakes and rivers are not?
12. Why is the Faint Young Sun Paradox called a paradox?
13. The laws of chemistry and physics were involved in the early evolution of Earth, Mars, and Venus. Describe how these laws were constrained by historical circumstances to produce the different climates of each of these three planets.
14. Do you believe the apparent cross-bedding found on Mars is definite proof that water once existed on that planet? What evidence could you use to infer the cross-bedding was formed by water?
15. In the movie *2001: A Space Odyssey*, astronauts circle the planet Jupiter while they investigate the occurrence of extraterrestrial communication traced there from Earth's moon. How long would it take for a radio signal from Earth to reach them? (Look up the distance on the web; remember: rate × time = distance.)
16. How do scientists determine the atmospheric composition of other planets?
17. Venus is the second planet located from the sun but its average surface temperature (approximately 850–860°C) is greater than that of Mercury, which is located closest to the sun. Explain.

SOURCES AND FURTHER READING

Allégre, C. J., and Schneider, S. H. 1994. The evolution of the Earth. *Scientific American, 271*(10), p. 66–75.

Bottke, W. F., Vokrouhlicky, D., Minton, D., Nesvorný, Morbidelli, A., Brasser, R., Simonson, B., and Levison, H. F. 2012. An Archean heavy bombardment from a destabilized extension of the asteroid belt. *Nature* doi:10.1038/nature10967.

Bowler, P. J. 1989. *Evolution: The history of an idea.* Berkeley, CA: University of California Press.

Bowring, S. A., and Housh, T. 1995. The Earth's early evolution. *Science, 269,* 1535–1540.

Bryson, B. 2004. *A short history of nearly everything.* London: Black Swan.

Bullock, M. A., and Grinspoon, D. H. 1999. Global climate change on Venus. *Scientific American, 280*(3), 50–57.

Cowen, R. 2001. A dark force in the universe: Scientists try to determine what's revving up the cosmos. *Science News, 159,* 218–220.

Delsemme, A. H. 2001. An argument for the cometary origin of the biosphere. *American Scientist, 89*(5), 432–442.

Grotzinger, J., Beaty, D., Dromart, G., Griffes, J., Gupta, S., and Harris, P. 2011. The sedimentary record of Mars. *The Sedimentary Record, 9,* 4–8. doi:10.2110/sedred.2011.2.

Halliday, A. N. The origin of the moon. *Science, 338,* 1040–1041.

Hand, E. 2009. Lunar impact tosses up water and stranger stuff. *Nature News,* 13 November. doi:10.1038/news.2009.1087.

Hand, E. 2010. Old rocks drown dry Moon theory. *Nature, 464,* 150–151.

Hawkesworth, C. J., Cawood, P. A., and Dhuime, B. 2016. Tectonics and crustal evolution. *GSA Today*, 26(9), 4–11. doi:10.1130/GSATG272A.1.

Hawkesworth, C. J., Dhuime, B., Pietranik, A. B., Cawood, P. A., Kemp, A. I. S., and Storey, C. D. 2010. The generation and evolution of the continental crust. *Journal of the Geological Society, London, 167*, 229–248.

Johnson, B. C., and Melosh, H. J. 2012. Impact spherules as a record of an ancient heavy bombardment of Earth. *Nature* doi:10.1038/nature10982.

Kasting, J. F. 2010. *How to find a habitable planet*. Princeton, NJ: Princeton University Press.

Kasting, J. F., and Siefert, J. L. 2002. Life and the evolution of Earth's atmosphere. *Science, 296*, 1066–1068.

Kerr, R. A. 2000. A wetter, younger Mars emerging. *Science, 289*, 714–716.

Khurana, K. K., Jia, X., Kivelson, M. G., Nimmo, F., Schubert, G., and Russell, C. T. 2011. Evidence of a global magma ocean in Io's interior. *Science, 332*, 1186–1189.

Krasnopolsky, V. A., and Feldman, R. D. 2001. Detection of molecular hydrogen in the atmosphere of Mars. *Science, 294*, 1914–1917.

Kyte, F. 2012. Focus on ancient bombardment. *Nature* doi:10.1038/nature11190.

Larson, R. B., and Bromm, V. 2001. The first stars in the universe. *Scientific American, 285*(6), 64–71.

Lemonick, M. D. 2001. How the universe will end. *Time, 157*(25), June 25, pp. 48–56.

Mojzsis, S. J., Harrison, T. M., and Pidgeon, R. T. 2001. Oxygen-isotope evidence from ancient zircons for liquid water at the Earth's surface 4,300 Myr ago. *Nature, 409*, 178–181.

Peebles, P. J. E., Schramm, D. N., Turner, E. L., and Kron, R. G. 1994. The evolution of the universe. *Scientific American, 271*(10), pp. 52–57.

Robert, F. 2001. The origin of water on Earth. *Science, 293*, 1056–1058.

Shirley, S. B., and Richardson, S. H. 2011. Start of the Wilson cycle at 3 Ga shown by diamonds from subcontinental mantle. *Science, 333*, 434–436.

Smoot, G., and Davidson, K. 1993. *Wrinkles in time*. New York: William Morrow and Company.

Taylor, S. R., and McLennan, S. M. 1996. The evolution of continental crust. *Scientific American, 279*(1), 76–81.

Thompson, H. 2012. Ancient asteroids kept on coming. *Nature, 484*, 429.

Trefil, J. S. 1983. How the universe began. *Smithsonian, 14*(2), 32–51.

Ward, P. D., and Brownlee, D. 2002. *The life and death of planet Earth: How the new science of astro-biology charts the ultimate fate of our world*. New York: Times Books.

Wilde, S. A., Valley, J. W., Peck, W. H., and Graham, C. M. 2001. Evidence from detrital zircons for the existence of continental crust and oceans 4.4 Gyr ago. *Nature, 409*, 175–178.

Williams, R. M. E., Grotzinger, J. P., Dietrich, W. E. et al. (there are at least 20 authors). 2013. Martina fluvial congliomerates at Gale Crater. *Science* 340:1068–1072.

Windley, B. F. 1984. *The evolving continents*. New York: John Wiley & Sons.

Origins of Life

MAJOR CONCEPTS AND QUESTIONS ADDRESSED IN THIS CHAPTER

A What were some of the early hypotheses for the origin of life?

B What are the characteristics of life?

C Why is carbon the basis of life as we know it?

D What sorts of experiments demonstrated how life's precursors could have initially formed?

E What are possible sites for the origin of life?

F Why are hydrothermal vents thought to be a likely site for the origin of life?

G What is the pyrite world?

H What is the RNA world?

I How did higher cell forms originate from more primitive types of cells?

J Did life catalyze its own formation?

K What do "molecular clocks" tell us about when life might have originated?

L Is there life on other planets?

CHAPTER OUTLINE

Grand Prismatic thermal spring, Yellowstone National Park, Wyoming. The pool is ~80 meters in diameter and near boiling. As the water flows over the pool's edges, it cools, allowing various types of bacteria to flourish. Certain of these bacteria produce orange-colored pigments called carotenoids, as seen here. (Carotenoids are orange and red pigments most familiar to us because of fall foliage.) These types of bacteria or their ancestors might have been among the first creatures on Earth. Because the Earth was much warmer early in its existence than it is now, hot springs might have been a site for the origin of life.

© Aleix Ventayol Farrés/Shutterstock, Inc.

8.1 Life as a Geologic Force

The surface of earliest Earth must have been sterilized by the enormous heat flux from within the planet and incessant asteroid impacts. Nevertheless, based on the fossil record, life arose no later than about 3.8 billion years ago. Primitive life is surprisingly resourceful and resilient. Certain bacteria live at temperatures ranging from well below freezing to above boiling (100°C; refer to this chapter's frontispiece) and most recently have been found living deep within Earth's crust.

As has been emphasized repeatedly in this book, life, the biosphere, is one of the most fundamental and powerful of all Earth systems. As we will see, after its origin, *life began to interact with other Earth systems and became a geologic force.* One of the most fundamental and vexing questions of the evolution of Earth's systems is this: how did life originate? The question is so complex that we will devote this chapter to the attempts to answer it.

Theories of the origin of life are by no means mutually exclusive. Life might be far too complex to have originated as a single event or in a particular type of environment. Instead, life might have originated in a series of steps, each of which built on the complexity established by previous mechanisms. During this process life might have undergone evolutionary selection for certain metabolic pathways over others. Because life itself consists of open systems, some of these pathways, such as photosynthesis, would eventually begin to act as a geologic force by exchanging matter and energy with physical and chemical systems such as the rock and hydrologic cycles. As we will see beginning in Chapter 9, the interaction and integration of life with other systems profoundly influenced the further evolution of Earth.

8.2 Early Theories of the Origin of Life: Spontaneous Generation and Panspermia

Ⓐ Many philosophers and scientists from the ancient Greeks to the Enlightenment of the 18th century had difficulty in distinguishing nonliving from living entities. Nevertheless, some of the teachings of the ancient Greeks contained the seeds of later ideas about the origin of life. Thales (ca. 625 to ca. 547 B.C.) taught that water is the substance from which all living things derive and into which they are transformed when they die. Somewhat later, Democritus (ca. 460 to ca. 370 B.C.), who later coined the term "atom" ("indivisible"), taught that the first animals originated from warm mud, when soil "atoms" combined with "fire" atoms by "chance and of necessity." In Democritus's teachings, we get a first glimpse of the important roles of energy and matter in maintaining Earth systems such as the biosphere (see Chapter 1). Aristotle (384–322 B.C.) later countered with his views. He believed nonliving matter grades into living plants and animals, which grade into each other. He also believed living creatures were immutable (unchanging through time) and all living things can be arranged on a scale of increasing complexity, called the Chain of Being (see Chapter 5), with humans at the top.

These observations, and many others, led to the theory of **spontaneous generation**. According to this theory living entities are generated by the power of nature, and new living forms are constantly being generated from nonliving. This theory held sway for almost 2,000 years.

By 1809, Jean-Baptiste de Lamarck (1744–1829) had published his theory of evolution (see Chapter 5). Earlier scientists had realized they could not explain the gradual changes in characters between individuals of the same species if all species were fixed. To explain the gradual transitions between members of a species, Lamarck concluded that all organisms are being produced *constantly and spontaneously* from nonliving matter by the addition of heat, moisture, or electricity (which at that time was regarded as a fluid). Lamarck's view stemmed in part from the Chain of Being. For example, new types of unicellular protists were constantly being generated from nonliving matter and were gradually transformed into higher forms of life. Thus, fish, birds, and unicellular protists (such as *Amoeba* or *Paramecium*) represented different stages of a continuous transformation toward "perfection," or humans.

Spontaneous generation was dealt a death blow by a series of experiments, most notably those of the famous French microbiologist, Louis Pasteur (1822–1885). Large amounts of rations were needed to feed France's troops during the Napoleonic wars, and it had been earlier demonstrated that boiling food in glass bottles sterilized and preserved it for long periods of time. Pasteur's experiments were similar. He sterilized culture medium and then left some flasks of culture medium open to the atmosphere while sealing others. The sealed flasks never developed microbes, and the resulting process, pasteurization, is still used today. Unfortunately, the results of Pasteur's experiments inadvertently brought origin of life studies to a halt for more than 50 years because they demonstrated bacteria could only arise from other bacteria (**Figure 8.1**). Consequently, by about the turn of the century some investigators were pointing to a mysterious "vital force" to explain life.

But, the results of Pasteur's experiments merely begged the question: from where did bacteria and cells come? Some, such as the famous physicist Lord Kelvin (who declared Earth to be much younger than did Darwin; see Chapter 6), invoked **panspermia** ("seeds everywhere"). According to panspermia, life on Earth was seeded by chemical compounds or perhaps even bacterial spores found in meteorites or comets. In other words, life came from outer space. Famous scientists, including Fred Hoyle (see Chapter 7) and Francis Crick (the co-discoverer of the structure of DNA; see Chapter 5), invoked the same argument again in the 20th century because they thought it would have otherwise taken too long for life to have originated on Earth by trial and error. Sugars and **amino acids**, the building blocks of **proteins**, have in fact been detected in meteorites (**Figure 8.2**), in comets, and in interplanetary dust. Possibly, sugars and other compounds were manufactured in interstellar space by irradiation by starlight of icy water, ammonia, and carbon monoxide that coated

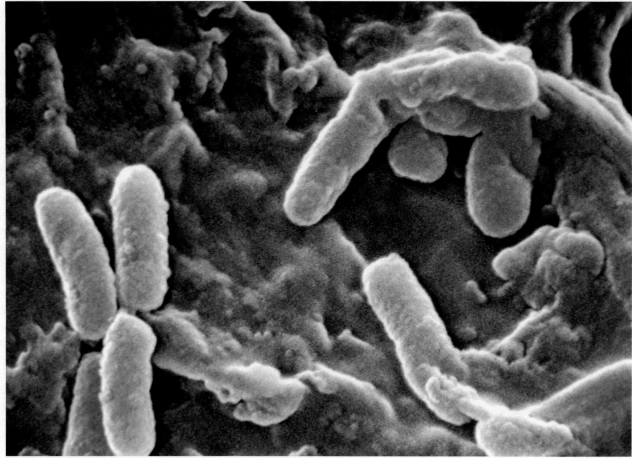

Courtesy of Janice Haney Carr/CDC.

FIGURE 8.1 Electron micrograph of bacteria. The cell membrane is surrounded by a cell wall which maintains the cell's shape. Within the protoplasm, which consists primarily of water and organic compounds, is found a nucleoid that contains the cell's genetic material (deoxyribonucleic acid or DNA) but lacks a membrane. The bacterial chromosome, which is composed of DNA, is a single circular structure. Bacteria also lack distinctive structures called organelles found in higher types of cells.

Courtesy of Argonne National Laboratory.

FIGURE 8.2 The Murchison meteorite is a carbonaceous chondrite that fell near Murchison, Australia, and from which amino acids and sugars have been detected. The presence of such compounds has led some workers to conclude that life on Earth, or at least its precursors, were seeded from outer space. However, even if this did happen, such a hypothesis still begs the question of how life originated in the first place.

the surfaces of small dust particles. However, if life originated on other extraterrestrial bodies, how did life originate there? Like the Steady State, or equilibrium, Theory of the universe (see Chapter 7), panspermia is a kind of doctrine promoting the eternity of life and the universe.

CONCEPT AND REASONING CHECKS

1. How does panspermia "beg the question" with respect to the origins of life?
2. How does panspermia resemble the Steady State (equilibrium) view of Earth and the universe (see Chapters 1 and 7)?
3. Do you believe spontaneous generation is occurring now? Why or why not?

8.3 What Is Life?

8.3.1 Basic traits of life

B Given that life evolved from nonliving matter, there seemingly have been as many definitions of life as investigators defining life. This is not surprising when you consider that living and nonliving entities share many important features, just as the ancient Greeks had noted.

First, a living organism is an open system that exchanges matter and energy with the surrounding environment (see Chapter 1). However, by this criterion alone life is no different from mantle convection cells or early Earth as it radiated heat into space and its crust solidified to form the continents. These latter two systems are also open systems, but they are not considered "living."

Another criterion of life is that living organisms are surrounded by a cell membrane, which acts as a barrier to separate the cell contents from the surrounding environment. Still, this by itself is not enough to invoke the status of life. Earth itself is surrounded by the envelope of the atmosphere that separates it from space and protects its surface. Living systems are also able to reproduce themselves from nonliving matter and energy taken from their surroundings. However, given enough matter, energy, or both some artificial systems can also reproduce themselves; an example is a pot of boiling water that produces multiple convection cells.

Living systems are ultimately distinguished from other open (but nonliving) systems by not just one but *several* key features. First, living systems are more *complex* than nonliving systems, meaning living systems consist of many more compartments than do nonliving systems. The parts are associated with different processes, all of which act to maintain homeostasis via feedback. Most important, living systems *store and process information and transmit it to their offspring*. We now know these functions are associated with the cell nucleus, in particular the structures called chromosomes composed of DNA (deoxyribonucleic acid). The DNA is arranged in units called genes that code for other cell components; DNA is also used as a template or "mold" to make more copies of DNA for offspring during reproduction. In addition, the hereditary material undergoes changes or mutations that can be passed on to and expressed by the offspring (see Chapter 5). Thus, living systems can *evolve*.

8.3.2 Composition of life

C Life as we know it is based on the element carbon. Many molecules of living organisms are based on long chains of carbon atoms that serve as the "backbone" or support of these structures. The elements nitrogen, sulfur, phosphorus, and hydrogen are also prominent components of living systems. Why were these elements involved in the origin of life? And why, especially, did carbon become the "backbone" of life? Why not silicon? Although it has been suggested that silicon could play a role similar to carbon, silicon tends to form bonds only with oxygen, which is why it is mainly locked up in silicate minerals in Earth's crust (see Chapter 3).

Carbon is prominent in life for several reasons. Carbon formed early during the nucleosynthesis of heavier elements from hydrogen and helium (see Chapter 7). Consequently, it is one of the most abundant elements in the universe. Carbon is also quite versatile, chemically speaking, because it

is able to form one or more covalent bonds with hydrogen, nitrogen, and sulfur. This allows carbon to form the complex chain and ring structures used in biochemical pathways like photosynthesis and respiration.

Nitrogen resembles carbon in certain respects. First, it is the most abundant gas in Earth's atmosphere. Second, nitrogen forms multiple covalent bonds with other elements. Thus, like carbon, nitrogen can also form a variety of complex compounds, especially amino acids and proteins.

So, too, do sulfur and phosphorus. Sulfur might have been involved in **prebiotic** ("before life") energy storage and transfer in nonliving systems as they evolved to living entities. Phosphorus was probably also a prominent component of energy storage and transfer molecules during Earth's prebiotic history. Phosphorus eventually became associated with the energy storage and transfer molecule called **adenosine triphosphate (ATP)**, which is also a building block of the nucleotide adenine in DNA.

Hydrogen, of course, was the first element to be synthesized in the universe and is also the most abundant element (see Chapter 7). It also has the ability to form very weak hydrogen bonds with other molecules. These bonds can be easily formed and broken, with little energy requirements. This makes hydrogen bonds ideal for many biochemical reactions that must take place quickly or for bridging longer molecules together. For example, hydrogen bridges the two strands of DNA. The hydrogen bonds are easily broken and reformed during the replication and transcription of the DNA molecule.

CONCEPT AND REASONING CHECKS

1. What are the basic traits of life?
2. Why is carbon more suited for life than silicon? What about nitrogen and phosphorus?

8.4 Chemical Evolution

8.4.1 Early theories

While Kelvin was promoting panspermia, Charles Darwin wrote a passage imagining a "warm little pond" with abundant nutrients in which life could have originated. As early as 1868, only 9 years after the first publication of the *Origin,* Thomas Huxley, one of Darwin's staunchest supporters, stated, "vital forces [meaning biological forces] are molecular forces." Thus, early on Darwin and Huxley recognized the importance of **chemical evolution**: the production of simpler precursors that could be assembled to produce life.

In fact, by the time Darwin wrote the passage regarding the "pond" (1871), there was already plenty of evidence demonstrating the production of carbon-based compounds, like those found in living organisms, from inorganic nonliving precursors. Friedrich Wöhler (1800–1882) was the first to do this when he synthesized urea, a key component of the nitrogenous (nitrogen-bearing) wastes of animals. Baron

Jöns Jakob Berzelius postulated substances called **enzymes**, which are biological catalysts that speed up biochemical reactions (see Chapter 5). The occurrence of enzymes and their ability to act *outside* living cells were later demonstrated when nonliving extracts of yeast cells were used to ferment alcohol and produce wine.

Thus, by the early 20th century, the gap between the nonliving and living worlds had narrowed significantly. The scientific approach in studies of chemical evolution is based on the **Principle of Biological Continuity**. Because present cell **metabolism** (Greek *metabole*, for change), such as respiration and photosynthesis, originated in the past, it must be continuous with ancient metabolism: metabolism recapitulates (or reflects) biogenesis.

Under the umbrella of chemical evolution, there are two basic approaches to the study of the origin of life: (1) **bottom-up** approaches, in which the fundamental building blocks of life are synthesized from simpler precursors and the building blocks then assembled to form living systems, and (2) **top-down** strategies, in which existing biological systems and biochemical pathways are extrapolated backward to earlier, simpler components and systems. The bottom-up approach is a reductionist approach (see Chapter 1) that tries to build up systems from simpler components, whereas the top-down approach examines the role of various factors that affect and constrain systems consisting of simpler components like prebiotic precursors. Both are valid approaches to studying the origin of life.

Studies of chemical evolution did not really begin to blossom, though, until the translation of Aleksandr I. Oparin's landmark book, *The Origin of Life,* which was originally published in Russian in 1924. In it, Oparin integrated organic chemistry, biochemistry, geochemistry, and cosmology into a coherent whole that generated testable hypotheses. Oparin suggested that Earth's early atmosphere was reducing. This would have allowed the synthesis and preservation of more and more complex organic molecules that formed aggregates, called coacervates, that were able to absorb other organic compounds. The English biochemist and geneticist, J. B. S. Haldane, had published similar views in 1929, shortly before publication of the translation of Oparin's book. According to Haldane, the presence of abundant dissolved carbon dioxide (CO_2) and ammonia (NH_3) dissolved in the surface layer of the oceans in the presence of ultraviolet radiation resulted in a primordial, or prebiotic, "soup." Eventually, a cell membrane formed and the first cells appeared. After many failures, a single cell survived and became the common ancestor of all subsequent life.

This hypothesis was later supported in 1953 by the classic experiment of Stanley Miller. As a graduate student, Miller worked under the supervision of Harold Urey (1893–1981) at the University of Chicago (among other discoveries, Urey helped establish the use of stable isotopes in studies of ancient climate, as discussed in later chapters). Miller's experiment was run only 3 weeks before publication of the structure of DNA by James Watson and Francis Crick. The **Miller-Urey experiment** is the prime example of the bottom-up approach. Miller's laboratory apparatus consisted of a flask filled with water, methane (CH_4), ammonia (NH_3), and hydrogen (H_2) through which sparks of electricity measuring 60,000 volts were discharged (**Figure 8.3**). After only a few days the sides of the flask had changed color. When analyzed the solution contained, among other compounds, about two dozen different amino acids. The resulting solution also contained fatty acids which are the building blocks of lipids (fats) that form cell membranes (**Figure 8.4**). Presumably, during the actual origin of life these precursors combined and changed into more complex living entities that became heterotrophic, meaning they obtained food from preformed molecules.

However, it was later realized that the Miller-Urey experiment had certain drawbacks. First, the amino acids produced by the experiment were both left- and right-"handed." Second, and more fundamentally, Miller and Urey had assumed Earth's atmosphere contained little or no oxygen and was therefore highly reducing. This is indicated by the chemical compounds introduced into the flask. However, if the core and mantle had already more or less formed within the first 10 to 100 million years of Earth's history, Earth's atmosphere must have been only mildly reducing (see Chapter 7). Under these conditions CH_4 and NH_3 are largely excluded, and the atmosphere is dominated by CO_2, N_2, and H_2O, with lesser amounts of H_2, carbon monoxide (CO), sulfur dioxide (SO_2), and hydrogen sulfide (H_2S or "rotten egg gas").

When the Miller-Urey experiment is run again under mildly reducing conditions, the results change. The formation of formaldehyde (H_2CO), which is a precursor of sugars, still proceeds via *photolysis* of water (the splitting of H_2O) and CO_2. However, amino acids are not synthesized. Moreover, hydrogen cyanide (HCN) is much less likely to be synthesized. Despite its highly poisonous nature (it rapidly blocks aerobic respiration), HCN is considered an important precursor in the formation of amino acids and DNA and RNA.

In this regard the panspermia hypothesis presents an advantage: amino acids and other compounds could have been synthesized under strongly reducing conditions elsewhere and then brought to Earth. Mars, for example, might at one time have been capable of harboring life. The report of bacteria-like structures from Martian meteorites (**Figure 8.5**) also lends credibility to the panspermia hypothesis, but for one fact: the Martian structures are not believed to be bacteria. It is quite possible the Martian structures are inorganic chemical precipitates. However, even if the Martian structures were to be confirmed as life forms, the same questions about the origin of life are still left unanswered.

The Miller-Urey experiment has also been questioned for another important reason. Some workers suggested the prebiotic soup resembled a giant oil slick. Upon reflection, however, it seems highly unlikely the chemical constituents of a prebiotic soup would be sufficiently concentrated to react with one another to produce more complex molecules. For example, given its atmospheric composition, the gaseous planet, Jupiter, is one enormous Miller-Urey experiment that has been running for billions of years. Nevertheless, life as we know it does not exist on Jupiter.

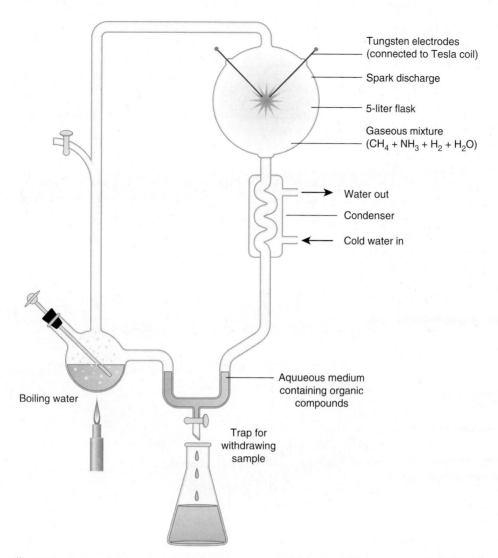

FIGURE 8.3 The Miller-Urey experiment. Apparatus used to simulate a prebiotic soup and the production of organic compounds from simple precursors. The original experiment attempted to simulate an ancient sea with a highly reducing atmosphere experiencing electrical discharges made by lightning.

More recent research has focused on other mechanisms or sites where prebiotic compounds might have been concentrated and life evolved. One of the first suggestions by John D. Bernal was that clay minerals were involved in both concentrating substances and catalyzing their reaction. Individual clay particles display both positive and negative charges at their surfaces that could adsorb simple molecules like amino acids or sugars of complementary charge. The adsorbed molecules could then be catalyzed to form more complex molecules before their release into water. Recent studies have manufactured the precursors of RNA on the surfaces of clays. The surfaces of the clay minerals serve as a scaffolding, like that used during building construction or painting.

This **clay world** (here, "world" is used in the sense of "hypothesis") was suggested in the 1960s by A. G. Cairns-Smith to have been the site of origin of what he called **clay life**. Because clays are layered, their crystal structure is essentially repeated (or replicated) and could therefore have

acted as a template that served as a source of information coding for the absorption and reaction of prebiotic precursors. Defects in the crystal structure could have served as "mutations," producing new compounds when precursors were aligned in new arrangements along the defects (see BOX 8.1).

Because clay minerals can occur almost anywhere, life could have developed in one of a number of different settings (**Figure 8.6**). One possibility is hot springs, like those of Yellowstone National Park, which can be associated with a hotspot (refer to this chapter's frontispiece). Here, spectacularly colored films of **thermophilic** ("heat-loving") bacteria often develop. Thermophilic bacteria are among the most primitive known on Earth. However, early Earth's surface temperature was undoubtedly much warmer than today's hot springs. Recent climate models suggest an *average* surface temperature of 358 K (85°C), as opposed to 288 K (15°C) today. Thus, if the earliest cells did in fact evolve in such environments, they were likely **hyperthermophilic**

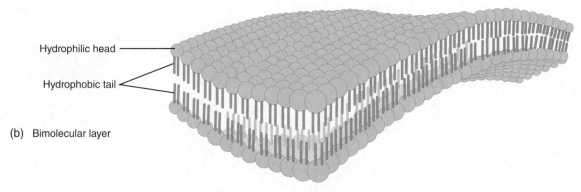

(a) A simplified fatty acid

Hydrophilic head

Hydrophobic tail

(b) Bimolecular layer

FIGURE 8.4 The structure of cell membranes. Fatty acids comprise cell membranes. Fatty acids consist of a slightly charged, hydrophilic "heads" and long, hydrophobic "tails" of carbon atoms. The long "tails" of fatty acids are hydrophobic ("avoid water"), whereas the "heads" are hydrophilic ("water loving"). Thus, in solution, fatty acids tend to group together into two layers (bilayer), in which the hydrophobic tails point inward to avoid water and the heads are either in contact with the aqueous environments of the cell's surroundings or its protoplasm. The bilayer structure is the fundamental structure of all cell membranes, from bacteria through the most complex life forms.

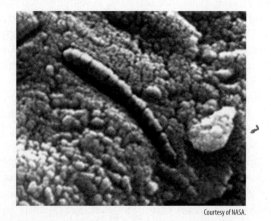

Courtesy of NASA.

FIGURE 8.5 Martian microbes? The presumed microbe recovered from a Martian meteorite found in Antarctica. The structure is composed of $CaCO_3$ and is ~0.5 μm long. Although it resembles a bacterium, this structure might be an inorganic chemical precipitate.

("extreme heat loving") bacteria, which today generally range from 70°C to 115°C and perhaps as high as 150°C (organisms such as these are now frequently lumped under the broader category of **extremophiles**; refer to this chapter's frontispiece). Volcanic hot springs would have also provided dissolved ions of elements such as iron and sulfur that were needed by early life for biochemical reactions. Iron and sulfur are components of various enzymes, and iron is a nutrient essential to photosynthetic and respiratory pathways (see Chapter 3).

These early photosynthetic life forms, such as **green** and **purple sulfur bacteria**, might have been **chemoautotrophic** or **chemosynthetic**. Rather than using energy from light ("photo") to manufacture carbon compounds ("autotrophy"), chemoautotrophs obtain energy from chemical reactions ("chemo") to manufacture their own food. In the case of purple sulfur bacteria, hydrogen sulfide (H_2S) is used

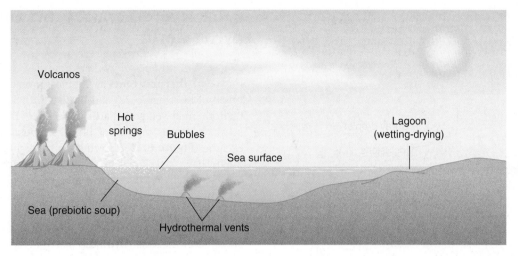

FIGURE 8.6 Possible sites for the origin of life. Note the wide disparity of environments that have been proposed.

instead of water, and elemental sulfur (S) is released instead of oxygen (O_2):

photoautotrophy (photosynthesis):

$$CO_2 + H_2O \rightarrow (CH_2O)_n + O_2 \uparrow \qquad (8.1a)$$

chemoautotrophy:

$$CO_2 + 2H_2S \rightarrow (CH_2O)_n + 2S \downarrow + H_2O \qquad (8.1b)$$

The elemental sulfur (S) generated by chemoautotrophy is deposited as yellowish granules inside or outside the bacteria (as indicated by the downward-pointing arrow in the equation).

Volcanic hot springs would also have been ideal settings for the synthesis of prebiotic precursors by repeated cycles of wetting and drying, or **dehydration-rehydration reactions**. In fact, small microspheres were synthesized by Sidney Fox by heating mixtures of amino acids to 150°C to 180°C. The resulting structures were called **proteinoids** ("protein-like"). The microspheres produced buds like those seen in dividing yeast cells (**Figure 8.7**) and were therefore thought to be precursors of cell membranes. It was suggested that the microspheres were washed from their site of formation into the sea, where they encapsulated and concentrated other substances, leading to the origin of primitive cells. However, Fox's hypothesis has been largely dismissed because cell membranes are now known to consist of fatty acids oriented into bilayers, not proteins (Figure 8.4). As their name implies, **lipid bilayers** consist of two layers: the "tails" of the fatty acids are oriented toward the interior of the bilayer because they are **hydrophobic** ("water-hating"), whereas the "heads," which are **hydrophilic** ("water-loving"), are oriented toward the surroundings. Bilayers could initially have formed as surface films on bubbles or the surfaces of water in open ocean, lagoons, and mudflats along the seashore.

Tidal pools are also seemingly favorable for the production of prebiotic precursors because of dehydration-rehydration reactions and the concentration of substances in evaporating waters. During dehydration, lipid bilayers

(a)
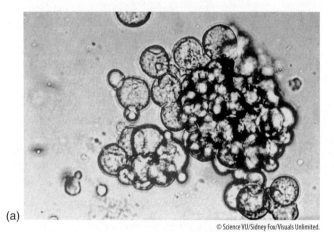

(b)
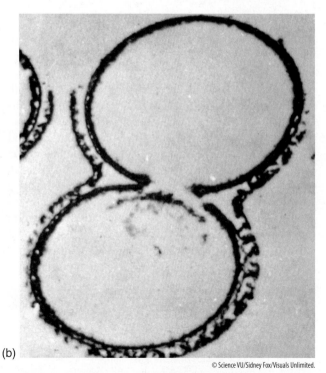

FIGURE 8.7 Proteinoids produced by heating mixtures of amino acids sometimes resemble bacteria dividing by binary fission.

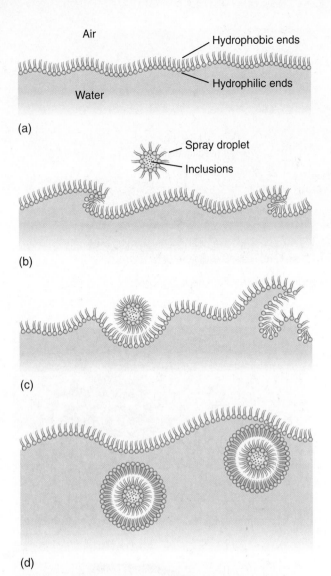

Air

Hydrophobic ends

Hydrophilic ends

Water

(a)

Spray droplet

Inclusions

(b)

(c)

(d)

FIGURE 8.8 Encapsulation of molecules and salts by lipid bilayers during cycles of dehydration-rehydration.

could have collapsed and become sandwiched together with other macromolecules and dissolved substances such as salts (**Figure 8.8**). As water was added back to the system, the bilayers could have trapped the macromolecules and salts in tiny vesicles surrounded by a bilayer membrane (Figure 8.8). Similarly, each individual vesicle of a piece of wet lava or other rocks could have served as a miniature chemical factory. There must have been countless such sites on early Earth, especially after the formation of the moon (see Chapter 7), which would have generated tides and tidal currents.

CONCEPT AND REASONING CHECKS

1. What is the significance of the Principle of Biological Continuity for studies of the origin of life?
2. What is the significance of chirality with regard to question 1?

8.4.2 Hydrothermal vents and the pyrite world

F G A number of workers now view deep-sea **hydrothermal vents** as the most promising site for the origin of life (Figure 8.6 and **Figure 8.9**). The existence of hydrothermal vents on Earth is also why the discovery of active volcanoes on one of Jupiter's moons, Europa, is so exciting (**Figure 8.10**). If vents are present on Europa, life might have developed beneath the surface of Europa's oceans, which are covered in ice.

Hydrothermal vents on Earth were discovered along mid-ocean ridges in 1977 near the Galápagos Islands in the eastern Pacific Ocean and have since been found at many other deep-sea sites. Hydrothermal vents are very appealing as a site for the origin of life because they incorporate a number of different aspects of early evolution into a single hypothesis. First, early Earth was quite warm and seafloor spreading centers (mid-ocean ridges) must have been very abundant. It is even possible that hydrothermal vents or volcanic islands frequently broke the oceans' surface and contributed to some form of a localized prebiotic soup there. Second, because of their depth, many vents would have been less likely to be affected by impacts during the Hadean and Archean. Even if a vent region were destroyed by an impact, there would have been many vents left to continue the synthesis of prebiotic compounds. Third, the most primitive living creatures known are hyperthermophiles, and this group might therefore be related to the ancestors of later groups. Fourth, and perhaps most important, hydrothermal vents are characterized by reducing conditions.

One objection raised by Stanley Miller to the hydrothermal vent hypothesis for the origin of life is that under experimental conditions amino acids break down under high temperatures (about 200°C) and pressures like those that exist at hydrothermal vents. However, hydrothermal vents often appear as **black smokers**, in which columns of sulfide-laden water rise upward and at the base of which large mounds of iron sulfide minerals like pyrite are deposited (see **Box 8.1**). As iron sulfides are added to experimental systems, amino acids remain intact for days, which leaves time for them to react with other molecules. Consequently, Günter Wächtershäuser, a German organic chemist turned patent lawyer, has promoted the origin of life in a **pyrite world** (or hypothesis) associated with hydrothermal vents. The "beauty" (meaning the simplicity) of the pyrite world is that it explains the appearance of early metabolic pathways involving enzymes (proteins) composed of amino acids, the early synthesis of the building blocks, or **nucleotides**, of DNA and RNA, and the formation of membranes for compartmentalization, all in close proximity in the same system. In this scenario molecular evolution started with the fixation of CO_2 on the surfaces of iron sulfides (FeS) as hydrogen sulfide (H_2S)-rich waters percolated through the vents. This resulted in the production of more complex organic compounds:

$$4CO_2 + 7FeS + 7H_2S \rightarrow$$
$$(CH_2COOH)_2 + 7FeS_2 + 4H_2O \quad (8.2)$$

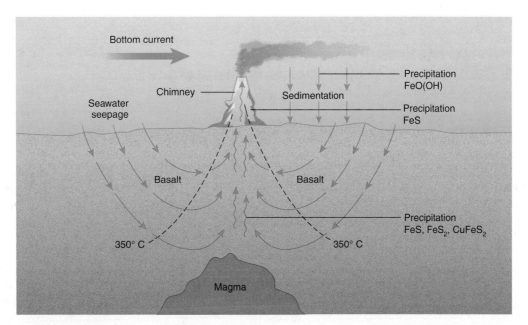

FIGURE 8.9 How life might have originated at hydrothermal vent communities. These processes only indicate the succession of processes that might have occurred on *geologic* scales of time, not on much shorter time scales, as might be implied by the diagram. Seawater percolates through hot ocean crust to produce the plumes of sulfide-rich water.

In this reaction, FeS (iron sulfide) is oxidized to produce Fe^{+3} from Fe^{+2}. Because of the loss of an electron (higher valence state) of iron, sulfur (S) from the H_2S combines with FeS to produce FeS_2, also known as the mineral pyrite. The hydrogen from the H_2S then combines with multiple molecules of CO_2 to produce organic compounds. In this particular case, a carboxyl (–COOH) group is formed. Carboxyl groups are involved in the linking of amino acids into proteins (see Chapter 5). In the pyrite world, the carboxyl group is ionized (–COO⁻) and adsorbed onto the positively charged surface of pyrite, where it reacts with other molecules.

Pyrite surfaces would have catalyzed a variety of other reactions. Some of these reactions are well known from industrial chemistry, such as the reaction of N_2 and H_2 to yield ammonia (NH_3). The simple products of these reactions could in turn have been used in other chemical reactions to produce more complex molecules. Moreover, the proximity of Fe and S to the reactions occurring on the surfaces of iron sulfides would have also made it relatively easy to incorporate both elements into early enzymes and biochemical pathways. Because the vents were located so far below the photic zone, some of the earliest life forms might have also been chemoautotrophic.

8.4.3 The RNA world

The hypothesis of an **RNA world** is the prime example of a top-down approach. The RNA world is based on the **Central Dogma of Cell Biology**: DNA → RNA → protein (see Chapter 5). Genetic information is stored in DNA, which is then read during the process of transcription to produce **messenger ribonucleic acid**, or **mRNA**. The mRNA is then read by the process of translation to produce proteins such as enzymes. We can extrapolate backward from the Central Dogma based on the Principle of Biological Continuity. Based on our reversing of the Central Dogma, it is likely that DNA evolved *last*. Thus, DNA might have evolved as a means of separating information storage and retrieval from other metabolic processes that might otherwise have degraded the DNA.

In other words, *RNA preceded DNA* according to the RNA world. In the RNA world RNA served as both the information repository and as an enzyme for protein synthesis (**Figure 8.11**). This originally seemed highly unlikely

FIGURE 8.10 Volcanic activity on one of Jupiter's moons, Io.

BOX 8.1 The Unusual Communities of Hydrothermal Vents

In the late 1970s, deep-sea submersibles began finding biologic communities comprising worms, crabs, and other organisms at hydrothermal ("hot water") vents. These sites are frequently associated with seafloor spreading centers. The deep-sea communities (**Box Figure 8.1A**) of the vents are too deep to depend on sunlight for photosynthesis. Instead, specialized bacteria produce food by using the process called chemosynthesis. Chemosynthesis is like photosynthesis in that organic matter (food) is produced. However, instead of water, the bacteria use dissolved sulfides produced by hydrothermal exchange between seafloor crust and seawater at the vents to produce elemental sulfur instead of oxygen:

Photosynthesis:

$$CO_2 + H_2O \rightarrow (CH_2O)_n + O_2 \uparrow$$

Chemosynthesis:

$$CO_2 + 2H_2S \rightarrow (CH_2O)_n + 2S \downarrow + H_2O.$$

Modern hydrothermal vents are significant for several reasons. Hydrothermal vents were presumably much more common after the formation of the early Earth billions of years ago, when the planet was much hotter, and might have provided sites for the production of complex compounds that eventually led to the origin of life. The vents are also of interest because metal-bearing ores found on land might have formed at such sites in the past. As seawater is pumped through modern vents, elements such as iron, copper, and zinc dissolve into seawater at "black smokers" (**Box Figure 8.1B**) and then later precipitate in the form-rich deposits of sulfide minerals. These modern vents could conceivably be mined someday.

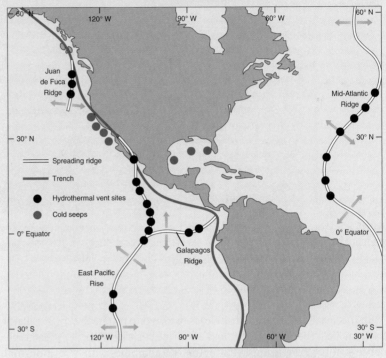

BOX FIGURE 8.1A Location of deep-sea vent communities.

Courtesy of OAR/NURP/NOAA.

BOX FIGURE 8.1B A "black smoker" associated with a hydrothermal vent.

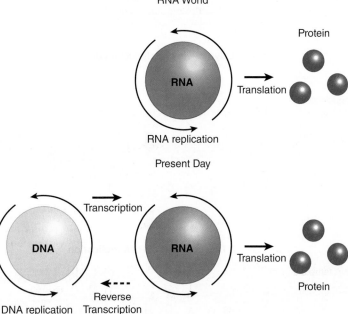

RNA World

Protein

RNA

Translation

RNA replication

Present Day

Transcription

DNA

RNA

Translation

Protein

Reverse
DNA replication   Transcription

FIGURE 8.11 The RNA world led to the world in which DNA is preeminent. In the RNA world (top), RNA was originally translated directly into protein. However, at some point it became advantageous for information to be encoded separately from metabolic processes so as to avoid degradation of encoded genetic information. At this point, ribozymes might have produced DNA by the process of reverse transcription (bottom), leading to the origins of the Central Dogma.

because it would require RNA to be able to catalyze *and* synthesize itself. However, in the early 1980s, specialized RNA molecules called **ribozymes** were discovered that both replicate RNA and synthesize proteins and catalyze other reactions. Presumably, at some point during the production of prebiotic precursors these ribozymes produced DNA, which then began to serve for information storage. This is by no means implausible, as indicated by certain viruses. Polio viruses are **RNA viruses**, in which the RNA is read directly to produce protein, whereas the acquired immunodeficiency (AIDS) virus undergoes a process called **reverse transcription**, in which the RNA is first transcribed into DNA before protein synthesis.

Unlike the pyrite world, however, the RNA world would have required some form of **protometabolism** to produce the nucleotides that form RNA and DNA. Possibly, nucleotides and strings of nucleotides were formed by catalysis on clays or pyrite. Another possible mechanism for the origin of strings of nucleotides is based on the Central Dogma. According to the Central Dogma, proteins result from the translation of RNA. Thus, in the distant past, short strands of proteins called **polypeptides** might have formed a backbone on which nucleotides were assembled. Then, the polypeptide backbone was replaced by the sugar-phosphate backbone now found in DNA and RNA.

A third possibility is that sulfur-based high-energy covalent bonds called **thioester** bonds were involved in early metabolism. Today, thioesters are involved in certain biochemical reactions that require high-energy transfer. Because life might have begun at hydrothermal vents enriched in sulfur and iron sulfides (see the previous section), thioester

bonds might have been much more widespread during prebiotic evolution. Perhaps, then, sometime later during the protometabolism phase leading to life, high-energy bond formation and energy transfer were taken over from thioester bonds by ATP. This might in turn have led to the **ATP world**. Because ATP is also involved in the formation of certain nucleotides in DNA and RNA, the ATP world might in turn have led to the RNA world. Sometime after the RNA world began to evolve, RNA and its associated systems were then encapsulated by a membrane (**Figure 8.12**).

8.4.4 Autocatalysis

In each of these "worlds"—which are really hypotheses—life might well have driven its own formation through the process of **autocatalysis**, in which reactants and products organize themselves into a more complex system. In autocatalysis the product of a chemical reaction catalyzes the production of itself and other products (hence the term auto, or self, catalysis (**Figure 8.13**):

$$A + B \rightarrow C + D \qquad (8.3)$$

Here, *A* and *B* are the reactants and *C* and *D* the products of the reaction, in which *C* catalyzes the reaction to produce more of itself and product *D*. The amount of *C* and *D* produced by the reaction depends on how much of the products were present at the outset. As long as reactants *A* and *B* are able to enter the system—which is an open system—*C* and *D* will continue to be produced. Thus, autocatalysis is a kind of positive feedback that promotes the ability of the product to make more of itself.

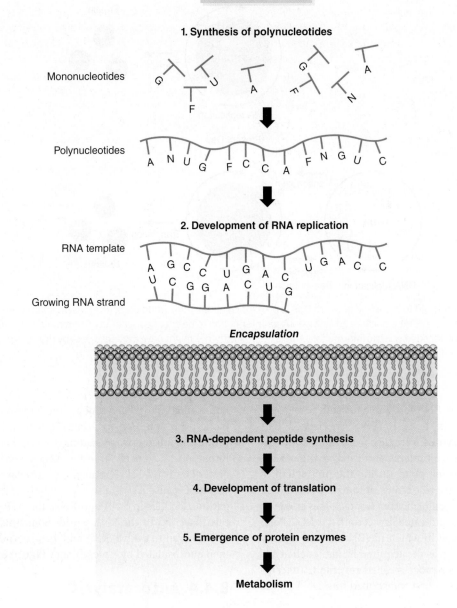

Protometabolism

1. Synthesis of polynucleotides

Mononucleotides

Polynucleotides

A N U G F C C A F N G U C

2. Development of RNA replication

RNA template

A G C C U G A C U G A C C

Growing RNA strand

U C G G A C U G

Encapsulation

↓

3. RNA-dependent peptide synthesis

↓

4. Development of translation

↓

5. Emergence of protein enzymes

↓

Metabolism

FIGURE 8.12 The transitions from molecular precursors to cells and metabolism are thought to have occurred. Sometime after the RNA world began to evolve, RNA and its associated systems were encapsulated by a membrane.

Autocatalysis is also a type of **self-organizing system**, in which simpler components organize themselves into more complex systems. This kind of behavior has been observed repeatedly in laboratory settings, so there is no reason to believe it does not occur in nature. Both prebiotic precursors and primordial metabolic pathways could have appeared and been concentrated on the surfaces of rocks or minerals by autocatalysis as seawater percolated through the pores of seafloor crust at hydrothermal vents, through the sediments of hot springs, or in tide pools, for example.

CONCEPT AND REASONING CHECKS

1. How might the clay hypothesis and pyrite world explain the left-handedness of life?
2. What are the characteristics of a ribozyme, and why are they important to the understanding of the origin of life?
3. What is the significance of autocatalysis and self-organization to the origins of life?
4. Are the hypotheses for the origin of life necessarily mutually exclusive?

C catalyzes Reaction

A + B ⟶ C + D

FIGURE 8.13 Pathways in an autocatalytic system, in which the products of a reaction catalyze their further production.

8.5 Origin of Eukaryotic Cells

J Whatever the exact site(s) and mechanism(s) of the origin of life, prokaryotic cells were the initial result. **Prokaryotic cells** ("first cells") are the simplest cells and include the bacteria (Figure 8.1). Prokaryotes contain a single circular **chromosome** found in a region called the **nucleoid**; this nucleoid also lacks a membrane. By contrast, **eukaryotic cells** ("true cells") such as protists and higher organisms possess a distinct membrane-bound **nucleus** (**Figure 8.14**). Prokaryotes are smaller than eukaryotes and lack the membrane-bound **organelles** of eukaryotes like **mitochondria** (where aerobic respiration takes place) or **chloroplasts** (where photosynthesis occurs) (**Figure 8.15** and **Table 8.1**).

A great deal of evidence exists for the appearance of prokaryotic cells by as early as approximately 3.5 billion years ago. These early prokaryotes are called **cyanobacteria** (in older texts they are referred to as blue-green algae; "cyanin" refers to their blue-green color). Modern cyanobacteria often form slimy films on rocky or muddy surfaces because they tolerate a wide range of harsh environmental conditions. Although the earliest apparent cyanobacteria are microscopic filaments (**Figure 8.16**), they would come to form much larger and widespread structures, called **stromatolites**, in sunlit waters during the Proterozoic Eon (see Chapter 9).

For many years, however, Western biologists were in a quandary as to how eukaryotic cells originated. Unknown to them, around the turn of the 20th century, several Russian

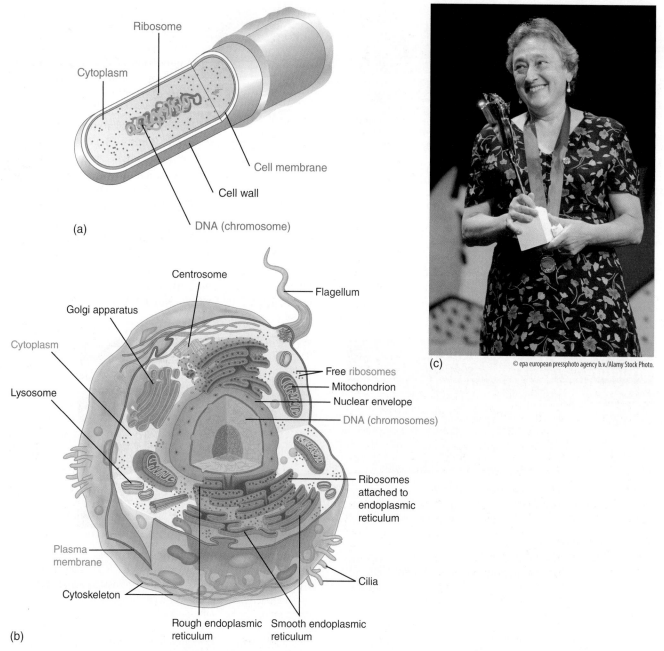

(a)

(b)

(c) © epa european pressphoto agency b.v./Alamy Stock Photo.

FIGURE 8.14 **(a & b)** Prokaryotes and eukaryotes compared. **(c)** The origin of eukaryotic cells via endosymbiosis was initially proposed by Russian scientists but proposed in Western science independently by Lynn Margulis.

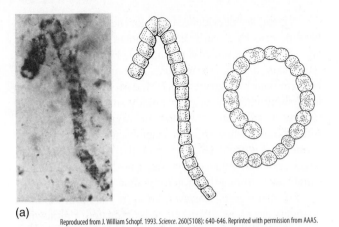

(a)

Reproduced from J. William Schopf. 1993. *Science.* 260(5108): 640-646. Reprinted with permission from AAAS.

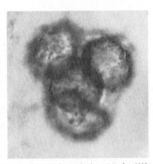

(b)

Reproduced from J. William Schopf. 1993. *Science.* 260(5108): 640-646. Reprinted with permission from AAAS.

FIGURE 8.15 Examples of fossil prokaryotes **(a)** and eukaryotes **(b)**. The larger size of the presumed eukaryotic fossils is thought to indicate that they are eukaryotes.

botanists had proposed that eukaryotic cells originated through a process called **endosymbiosis** (loosely translated as "living together internally"). As has so often been the case, however, Russian and Western scientists worked at cross-purposes because of cultural and intellectual isolation. Thus, it was not until the 1960s that the American scientist,

Lynn Margulis, began to champion the **endosymbiotic theory of cell evolution**. The theory was not well received initially, but it later won wide acceptance among many Western workers.

Symbiosis is everywhere. A prime example is lichens (which consist of algae and fungi), which are especially noticeable encrusting and weathering the surfaces of rocks. Other examples include algal symbionts that live in the tissues of corals and aid the corals as they calcify and produce their skeletons, while the corals supply waste products and nutrients for photosynthesis (see Chapter 4). Other examples are protists that live in the guts of termites and ruminants such as cows, horses, and elephants; without the protists, the termites and ruminants would be unable to digest the complex carbohydrate cellulose that forms plant cell walls.

In the case of eukaryotic cells, there are two very strong hints of endosymbiosis: the presence of mitochondria and, in plants, chloroplasts (Figure 8.14). The structure of both these organelles hints that they originated by the engulfment of prokaryotes by another cell (**Figure 8.17**). First, both mitochondria and chloroplasts have their own DNA nucleoid, like those of bacteria, suggesting their ancestors once lived independently. Although much of their genome has been incorporated into that of the host cell by genetic recombination (see Chapter 5), both mitochondria and chloroplasts divide and reproduce independent of their host cells. Mitochondria and chloroplasts also possess specialized organelles called **ribosomes** composed of a special type of RNA called **ribosomal RNA**, or rRNA; ribosomes

TABLE 8.1

Comparison of prokaryote and eukaryote structure		
	Prokaryotes	Eukaryotes
Organisms	Bacteria and cyanobacteria	Protists, fungi, plants, and animals
Cell size	1 to 10 micrometers*	10 to 100 micrometers
Genetic organization	Loop of DNA in cytoplasm	DNA in chromosomes in membrane-bound nucleus
Organelles	No membrane-bound organelles	Membrane-bound organelles (chloroplasts and mitochondria)
Reproduction	Binary fission, dominantly asexual	Mitosis or meiosis, dominantly sexual

*A micrometer is 1/1,000 of a millimeter.

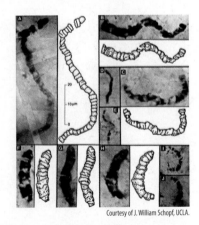

Courtesy of J. William Schopf, UCLA.

FIGURE 8.16 Possible cyanobacteria from the Apex cherts of western Australia, which have been dated at ~3.5 (giga-annums, or billions of years). These apparent microfossils might not be cyanobacteria at all, but might still be biogenic, perhaps representing chemosynthetic bacteria. This is consistent with their presence in chert, which might have been precipitated from hot fluids near hydrothermal vents. Recent studies have found both microfossils and isotopic evidence for the existence of sulfur-metabolizing bacteria by 3.4 billion years ago.

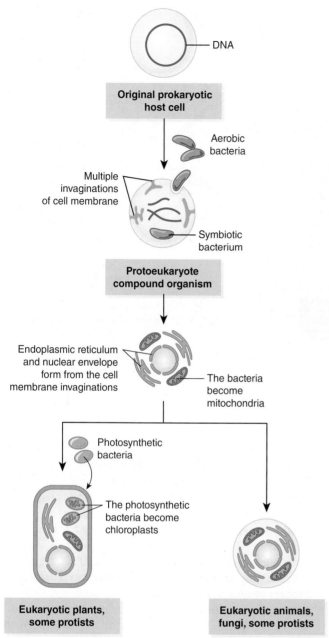

FIGURE 8.17 The origin of mitochondria (or chloroplasts) via engulfment of a prokaryotic cell by an ancestral eukaryote.

Labels in figure:
- DNA
- Original prokaryotic host cell
- Aerobic bacteria
- Multiple invaginations of cell membrane
- Symbiotic bacterium
- Protoeukaryote compound organism
- Endoplasmic reticulum and nuclear envelope form from the cell membrane invaginations
- The bacteria become mitochondria
- Photosynthetic bacteria
- The photosynthetic bacteria become chloroplasts
- Eukaryotic plants, some protists
- Eukaryotic animals, fungi, some protists

Data from: Margulis, L. 1993. *Symbiosis in Cell Evolution*, 2nd ed. New York: W. H. Freeman and Company.

the prokaryote survived and began to propagate itself within the host cell. Moreover, many of the genes of the prokaryote were incorporated into the DNA of the host!

Was this merely chance? Perhaps not. Despite the fact that most life on Earth requires oxygen for survival, oxygen is actually quite toxic. Free oxygen (O_2) forms compounds such as superoxides (O_2^-), hydroxides (OH^-), and peroxides (H_2O_2), all of which readily attack cell constituents. In eukaryotic cells mitochondria are the "powerhouses" where carbon-rich foodstuffs are broken down through aerobic respiration. It is therefore thought that the acquisition of mitochondria permitted survival of the host cell in response to increasing oxygen levels in the atmosphere and oceans. The evolution of new biochemical pathways for survival at higher oxygen concentrations also allowed the host cells to extract more energy from food. The biochemical reactions of aerobic respiration that take place in mitochondria produce far more energy in the form of high-energy ATP from the same amount of food than the breakdown of foodstuffs in the absence of oxygen.

This series of steps is thought to have given rise to an "animal"-like line of eukaryotic cells. Primitive eukaryotic plant cells then branched off from these cells by the engulfment of photosynthetic bacteria. These bacteria were therefore the forerunners of chloroplasts.

CONCEPT AND REASONING CHECKS

1. Make a table characterizing prokaryotic versus eukaryotic cells.
2. Diagram how a prokaryotic organism might have established symbiosis with its ancestral host cell.

8.6 Precambrian Fossil Record and Molecular Clocks

So when, exactly, did the first prokaryotes and eukaryotes arise? Not surprisingly, answers from the fossil record to this question are rather vague because the fossil record is rather poor in the Precambrian. Early Precambrian fossils are soft-bodied and microscopic and therefore easily destroyed by metamorphism or erosion. Moreover, billions of years have passed during which much of the Precambrian fossil record could have been destroyed. It is also quite possible that the ancestors of some important fossil groups were never preserved or will never be found because rapid evolution that produced missing links (transitional forms) occurred in relatively small, isolated populations (see Chapter 5). To complicate matters further, some of the earliest apparent fossils are not really fossils at all but likely chemical precipitates of some type (see the discussion of Martian meteorites earlier in the chapter).

On the other hand, some enigmatic fossilized structures might not represent the actual organisms but biogenic deposits produced by the organisms. Some of these very early fossils might have been chemosynthetic bacteria. This inference

read mRNA during translation and protein synthesis. The host cell also has its own ribosomes, but their subunits are larger in size and resemble those of other eukaryotes. Third, both mitochondria and chloroplasts are surrounded by a double membrane, each of which consists of a lipid bilayer.

According to the endosymbiotic theory, the inner membrane of each mitochondrion is descended from the ancestral prokaryote's membrane, whereas the outer membrane is descended from the time of engulfment of the prokaryotes. Normally, engulfment of one cell by another results in digestion of the prey within an organelle called a food vacuole. Within the food vacuole the prey cell is broken down into simpler constituents, which are then absorbed into the predator. However, according to the endosymbiotic theory

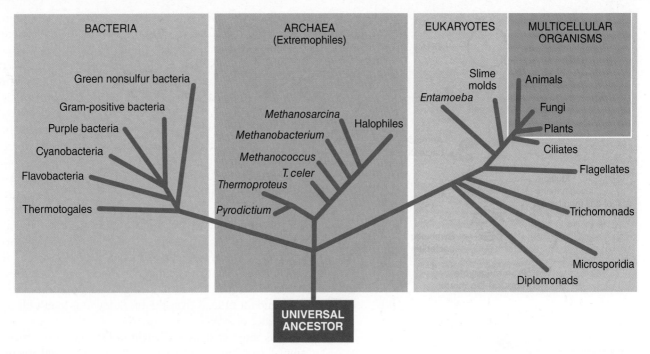

BACTERIA

Green nonsulfur bacteria
Gram-positive bacteria
Purple bacteria
Cyanobacteria
Flavobacteria
Thermotogales

ARCHAEA
(Extremophiles)

Methanosarcina
Methanobacterium
Methanococcus
T. celer
Thermoproteus
Pyrodictium
Halophiles

EUKARYOTES

Slime
molds
Entamoeba

MULTICELLULAR
ORGANISMS

Animals
Fungi
Plants
Ciliates
Flagellates
Trichomonads
Microsporidia
Diplomonads

UNIVERSAL
ANCESTOR

FIGURE 8.18 The tree of life. Each branch, or domain, includes taxa that are considered to belong to separate kingdoms, such as plants and animals. Archaea is a primitive type of bacteria that today include thermophiles, halophiles, and methanogens. The Eubacteria include the cyanobacteria and bacteria that use H_2S instead of water during photosynthesis. The Eukarya include the higher plants, animals, fungi, and unicellular groups such as ciliated unicellular groups.

agrees with their presence in cherts that appear to have been precipitated from fluids surrounding hydrothermal vents (Figure 8.16). Also, recently, substances with carbon isotope ratios indicative of photosynthesis by eukaryotes have been found in rocks as old as 2.7 giga-annums (billion years; [Ga]). This suggests eukaryotes appeared at least by the late Archean (Table 8.2). In fact, recent studies have found both microfossils and isotopic evidence for the existence of sulfur-metabolizing bacteria by 3.4 billion years ago.

Scientists are further attempting to find out when prokaryotes and eukaryotes originated using the technique of molecular clocks. **Molecular clocks** compare the sequences of RNA and proteins and are based on the Principle of Biological Continuity, which states that metabolic and biochemical pathways reflect their evolutionary history. Despite their sophistication, however, molecular clocks have only added fuel to the fire of debate about the origins of the first cells.

Carl Woese and colleagues have used the technique of RNA sequencing to determine the nucleotide sequences of rRNA. rRNA comprises the ribosomes used to translate messenger RNA (mRNA) during protein synthesis. rRNA is thought to be so important to the synthesis of proteins that portions of it should exhibit very little difference between the major groups of organisms; these groups are called **domains** in the terminology of sequencing. Certain stretches of DNA that code for the domains should therefore be found in all known organisms, no matter what the domain. These domains are thought to have been present in the last common ancestors of the domains as they appeared. Other portions of the DNA that code for stretches of rRNA changed because of mutations, which altered the sequences of nucleotides in

the rRNA. By extrapolating backward in time from the rRNA of living organisms, a point should be reached when no differences exist in the rRNA. This point in time presumably represents the time at which the domains first diverged.

Based on RNA sequencing, several domains of organisms are now recognized in addition to the kingdoms Animalia and Plantae: Fungi (mushrooms, etc.), Protoctista or Protista (single-celled plants and animals like *Amoeba* and *Paramecium*), the **Archaea**, and the **Eubacteria** (**Figure 8.18**). Archaea include a primitive type of bacteria found today in extreme, anoxic environments such as hot springs or high-salinity environments; these and other habitats are thought to be more typical of the primitive Earth billions of years ago, as described above. Eubacteria include all other bacteria and are distinguished from the Archaea by various features of their metabolism. Archaea and Eubacteria are sufficiently distinct that they are placed in their own domains, whereas the higher plants, animals, fungi, and protoctists are all included in the domain **Eukarya**. It is thought that the ancestral eukaryote was derived from the Archaea. Based on RNA sequencing, it was originally thought that the domains diverged quite early in Earth's history, at least 3 to 4 billion years ago.

However, another type of molecular clock, called **protein sequencing**, indicates otherwise. Protein sequencing works in basically the same manner as RNA sequencing, but instead of differences in nucleotides, differences in amino acid sequences of selected proteins are used. Based on the Central Dogma, differences in proteins should reflect differences in mRNA and therefore the DNA that serves as the template for mRNA. Based on this technique, R. F. Doolittle and colleagues determined that the Archaea, Eubacteria, and Eukarya

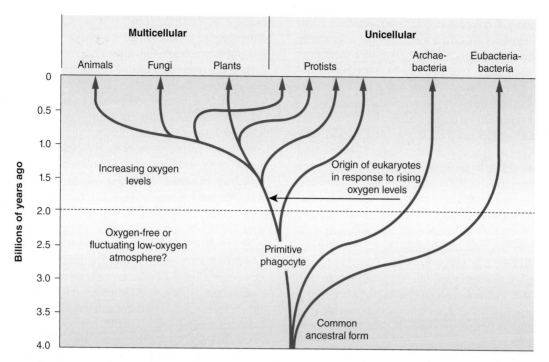

FIGURE 8.19 Relationships and estimates of the divergence times of the major groups of life. Note that in this diagram prokaryotes and eukaryotes are thought to have diverged approximately 2 billion years ago, when atmospheric oxygen levels were rising and the endosymbioses described in the text were presumably established.

all diverged *no earlier* than about 2.0 to 2.5 Ga, which is at least a billion years later than what the fossil record and RNA sequencing indicate (**Figure 8.19**; **Figure 8.20**).

What could account for the seemingly contradictory evidence between the molecular clocks and between the clocks and the fossil record? Recently, the results of the protein clock have been challenged. Protein clocks, like other molecular clocks, make one basic assumption: the rate of mutation in DNA, and therefore the rates of change in the proteins, has been constant through time.

This might not be the case. Oxygen is thought to have begun accumulating in Earth's atmosphere by about 2.0 Ga

(see Chapter 9), about when protein clocks indicate that the Eubacteria, Archaea, and Eukarya diverged. Possibly, as free O_2 began to accumulate in the atmosphere, anaerobic bacteria and Archaea were driven out of their ecologic niches. Some of these might have adopted symbiotic relationships to survive. But, the development of symbiotic relationships would have resulted in massive genetic recombination like that described earlier between mitochondria and chloroplasts and their host cells. Such massive genetic shuffling would in turn have affected protein synthesis. This genetic reorganization might have been so massive it could be what is really being reflected by the protein clock and not the

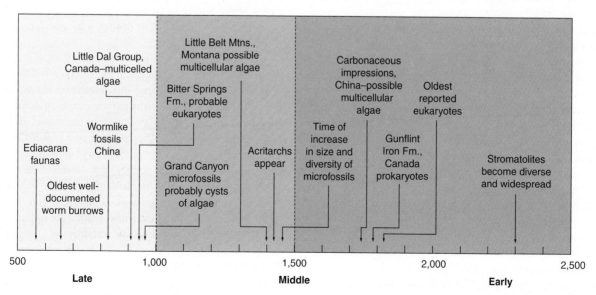

FIGURE 8.20 Occurrence of microfossils through the Precambrian, leading to the eventual appearance of multicellular animals (see Chapters 9 and 10).

actual evolutionary divergence of these taxa. Another possibility is that eukaryotes only diversified after the appearance of sexual reproduction, which would have increased rates of genetic recombination (see Chapter 5). Whether or not similar sorts of changes have led to the origin of life on other planets is open to question.

Seemingly polarized results and the polarized personal viewpoints that, unfortunately, all too often accompany them are not atypical in scientific investigation (see **method of multiple working hypotheses** in Chapter 1). Are the different types of data definitely contradictory? Is one interpretation of the data completely right and the other completely wrong? Not necessarily. The molecular clocks and the fossil record are looking at the same phenomenon—the origin of prokaryotes and eukaryotes—but from different perspectives. Different techniques that measure or detect different objects or the results of different processes are also involved. The study of molecular clocks and the fossil record have different approaches and assumptions upon which the data were both collected and interpreted. This is quite common in science.

CONCEPT AND REASONING CHECKS

1. What is the underlying principle of molecular clocks?
2. What is the fundamental assumption of molecular clocks?

8.7 Is There Life on Other Planets?

The occurrence of simple organic compounds like amino acids in meteorites and comets begs another question: is there life elsewhere? The structures found recently in a Martian meteorite are certainly tantalizing, but their origin is questionable (Figure 8.5). Given that our habitable zone (HZ) migrates through the solar system (see Chapter 7), it is certainly possible that at least primitive microbes once existed on Venus or Mars. However, space probes to these planets have yet to confirm that microbial life still exists—or ever existed—on these planets.

Is there *intelligent* life (meaning those with large brains, like humans and their relatives) on planets outside our solar system? Gigantic radio telescopes monitor outer space for clues (see Chapter 7), but to date no signs have been detected. What are the chances of civilizations occurring on other planets? The probability of intelligent life elsewhere in the universe has been calculated using the *Drake equation,* named after the astronomer Frank Drake:

$$N = N_g f_p n_e f_l f_i f_c f_L \qquad (8.4)$$

where N is the number of civilizations *in our galaxy* sufficiently advanced to transmit detectable signals through space.

Despite its appearance, the Drake equation is far from formidable. The Drake equation assigns probabilities to each factor in the equation and then multiplies them together to estimate N. The Drake equation is therefore basically similar to estimating the chances of getting one or more heads (or tails) in a row in a series of coin flips. The chances of getting a head (or tail) in a single coin flip are 50:50, or one half. The chances of getting two heads in a row are therefore $\frac{1}{2} \times \frac{1}{2}$ or $\frac{1}{4}$ (=0.25). The chances of obtaining 3 heads in a row are 1 in 8, and so on.

The real problem is estimating the probabilities of the different factors for the Drake equation. N_g, the number of stars in our galaxy, is probably the easiest to determine. N_g equals about 400 billion for the Milky Way. Similar estimates have been obtained for distant galaxies. f_p is the fraction of stars that have planets, which is estimated to be about 0.1. This number is used because up to 90% (or 0.9) of the stars might be binary stars, or two-star systems. Many binary stars might not coexist with planets, and if they do, the planets probably do not occur within the habitable zone. n_e is the number of Earth-like planets per solar system. Based on our own circumstances, n_e is estimated at 0.5, or 50:50; that is, either a planet exists within the habitable zone or it doesn't.

The next factors are much more difficult to assess. f_l is the fraction of planets on which life evolves. One way to estimate f_l is to determine the presence or absence of O_2 in a planet's atmosphere by using spectroscopy (see Chapter 7). The presence of O_2 or ozone in another planet's atmosphere hints at the possible existence of life *as we know it*, although the life forms may be no more complex than bacteria. f_i is the probability that life will evolve to an intelligent state. This is probably also a function of O_2 because O_2 is necessary for advanced metabolism, as we understand it, in eukaryotic cells. Furthermore, the early appearance of cyanobacteria on Earth implies that given suitable conditions, the evolution of photosynthesis and oxygen production is relatively high. However, based on the evolution of life on Earth, it might take billions of years for this to happen! f_c is the probability that, *once established,* intelligent life will develop the capability to communicate. f_L is the fraction of a planet's lifetime during which technology adequate for extraterrestrial communication exists. The very approximate lifetime of a species is about 1 million years. Assuming that humans do not make Earth uninhabitable or destroy it by war, f_L equals very roughly $10^6/10^{10}$, or 10^{-4}.

Destruction of Earth raises another point not explicitly included in the Drake equation. A planet could not be too susceptible to impacts like those that devastated life on Earth in the past. At the same time the demise of the dinosaurs might have been essential to the evolution of mammals, and therefore eventually humans. The presence of a planet the size of Jupiter might also affect the probability of an impact on Earth. Because of its gigantic size, Jupiter's gravitational field can affect the rate at which asteroids are dislodged from the belt between Mars and Jupiter (see Chapter 7). At the same time, however, Jupiter's gravity can also help shield Earth from impacts.

Given the tremendous uncertainty of the individual factors, the final "answer" for N ranges from 100 to 100,000,000 civilizations! Thus, depending on whether one is a pessimist or an optimist, the probability of existence of advanced civilizations on other planets is either quite small (100) or seemingly quite large (100,000,000). However, in either case the *possibility* of intelligent life does not appear to be restricted to Earth.

The disciplines of exobiology and astrobiology have in fact begun to search for clues to prebiotic and microbial evolution on other planets by examining the occurrence of microbes called extremophiles in unusual environments on Earth. Kinds of extremophiles include hyperthermophiles (organisms capable of reproducing at temperatures of 80 degrees or more), halophiles (high-salinity tolerant forms), and acidophiles (those tolerating acidic conditions), among others. One such site is Yellowstone National Park, with its abundant hot springs (refer to this chapter's frontispiece). Another is Mono Lake, located in the Sierra Nevada of eastern California (**Figure 8.21**). Mono Lake is the oldest

(a) © Geir Olav Lyngfjell/ShutterStock, Inc.

(b) © BioPhoto Associates/Getty Images.

(c) © NASA.

(d) © Juan Carlos Muñoz / Science Source.

(e) © Thierry Berrod, Mona Lisa Production/Science Source.

FIGURE 8.21 Modern Earth environments that serve as analogs for possible sites for the origin of life on other planets. **(a)** Mono Lake in eastern California is hypersaline and hyperalkaline. **(b)** The deep channel leading from Gusev Crater on Mars suggests the former presence and outflow of water. **(c)** *Spirochaeta americana*, a new species of alkaliphilic bacterium discovered at Mono Lake. The red spiral-shaped figures are dead cells; green ones are living. Such a discovery provides further evidence of how life could originate and evolve on other planets. **(d)** The Rio Tinto River in Spain is heavily polluted by acid mine drainage and can serve as an analog for ancient conditions on Mars. **(e)** Water sample from the river in which different eukaryotic cells and prokaryotes (much smaller) can be seen. See more at: http://www.astrobio.net/topic/origins/extreme-life/living-on-fools-gold/#sthash.e8LBBImF.dpuf.

lake in North America and is both naturally hypersaline and alkaline because it has no outlet; water leaves only by evaporation and by aqueducts to Los Angeles! This results in the concentration of salts and their precipitation to form unusual columns of a particular kind of limestone called "tufa." Surviving microbes include halophiles and *alkaliphiles* (those tolerant of nonacidic conditions), such as a new species of bacterium *Spirochaeta americana*, which are entombed in the tufa and can also be studied as fossils (compare Figures 8.15 and 8.16). Such discoveries provide further evidence of how life could originate and evolve on other planets. Meteorites from Mars have been found to contain carbonates like those of Mono Lake, and Gusev Crater on Mars resembles Mono Lake in photographs.

Some sites such as the Rio Tinto River ("Red River" or "River of Fire") in southern Spain are heavily polluted. The Rio Tinto is the birthplace of the Copper and Bronze Ages,

with the first mines having been excavated about 3000 B.C.; some of the first coins were made about 2000 B.C. by the Romans from gold and silver from these mines. It is also where Columbus set sail for the Americas. The Rio Tinto now serves as an important source of iron, copper, and sulfur and is polluted by acid mine drainage (Figure 8.20). Because of these conditions, the Rio Tinto serves as an analog for ancient conditions at the Meridiani Planum on Mars, given the evidence for water and the detection of iron-rich minerals, and the possibility of sulfur-based life on another of Jupiter's moons, Europa. Despite the heavy pollution, the river exhibits a remarkable diversity of prokaryotic organisms such as iron *stromatolites* formed by cyanobacteria and eukaryotic taxa, including flagellates, amoebae, and ciliated protozoans. The Rio Tinto is now an expedition target for the future unmanned robotic Mars Analog Research and Technology Experiment (MARTE).

SUMMARY

- Life is characterized by a number of traits: (1) it must partially, but not completely, insulate itself from its environment; (2) it must synthesize its own constituents from the matter in its surroundings; (3) it must extract energy from the environment and convert it into high-energy bonds that are used to perform the work of metabolism; (4) it must catalyze the various biochemical reactions of metabolism; (5) it must reproduce itself accurately; and (6) it must regulate each of its activities.

- The multiplicity of potential sites and mechanisms for the origin(s) of life are a prime example of the method of multiple working hypotheses (see Chapter 1).

- The oldest hypotheses for the origins of life basically invoked some form of spontaneous generation because scientists were ignorant of biochemistry and cell structure.

- Pasteur's experiments demonstrated that spontaneous generation did not occur, so scientists eventually began investigating the chemical and biochemical events of the prebiotic world.

- These investigations involve one of two basic approaches: bottom up and top down. The earliest bottom-up approach invoked the formation of a prebiotic soup, as demonstrated by the classic Miller-Urey experiment. According to this scenario, as sufficient prebiotic building blocks accumulated in the oceans, they were eventually surrounded and compartmentalized by a cell membrane. The genetic code and protein synthesis evolved after protometabolism had been established. However, the Miller-Urey experiment was based on the assumption that Earth's early atmosphere was fully reducing rather than mildly reducing.

- Consequently, other bottom-up scenarios for the origin of life have evolved. These include the clay world, in

which clay particles served as scaffoldings or perhaps even catalysts for the production of more complex compounds. In the pyrite world metabolism evolved in association with the surfaces of the mineral pyrite, especially at hydrothermal vents. Because early Earth was undoubtedly quite hot, vent systems would have been abundant and widespread. According to this scenario early metabolism, the genetic code, and cell membranes all evolved more or less simultaneously on the surface of pyrite or other sulfide minerals.

- The most popular top-down scenario for the origin of life is the RNA world. The RNA world is based on the Central Dogma of biology: DNA → RNA → protein. According to this scenario RNA and protein synthesis appeared first, and DNA evolved later as a means of storing and protecting information, with RNA remaining as the intermediary between DNA and protein. However, the RNA world requires the evolution of some form of primitive metabolism before the RNA world could have appeared.

- After the earliest cells became established, they underwent profound changes. The most important change was the appearance of eukaryotic cells, which seem to have evolved by engulfment of members of an ancient group of bacteria (Archaea) by other prokaryotes in response to rising oxygen levels in the atmosphere and oceans.

- Although the chances of encountering life, especially intelligent life, on other planets in our galaxy are relatively slim, the possibility exists. Of course, the chances of finding life on other planets increase substantially if the probabilities associated with finding intelligent life are excluded.

KEY TERMS

adenosine triphosphate (ATP)

amino acids

Archaea

ATP world

autocatalysis

black smokers

bottom-up

Central Dogma of Cell Biology

chemical evolution

chemoautotrophic

chemosynthetic

chloroplasts

chromosome

clay life

clay world

cyanobacteria

dehydration-rehydration reactions

domains

endosymbiosis

endosymbiotic theory of cell evolution

enzymes

Eubacteria

Eukarya

eukaryotic cells

extremophiles

green sulfur bacteria

hydrophilic

hydrophobic

hydrothermal vents

hyperthermophilic

lipid bilayers

messenger ribonucleic acid (mRNA)

metabolism

method of multiple working hypotheses

Miller-Urey experiment

mitochondria

molecular clocks

nucleoid

nucleotides

nucleus

organelles

panspermia

polypeptides

prebiotic

Principle of Biological Continuity

Prokaryotic cells

proteinoids

proteins

protein sequencing

protometabolism

purple sulfur bacteria

pyrite world

reverse transcription

ribosomal RNA

ribosomes

ribozymes

RNA viruses

RNA world

self-organizing system

spontaneous generation

stromatolites

thermophilic

thioester

top-down

REVIEW QUESTIONS

1. In terms of scientific methodology, were the hypotheses advanced by the ancient Greeks about the origin of life as ridiculous as they might seem? Why or why not?

2. What were the origins of the theory of spontaneous generation? What led to its demise?

3. What is the methodological drawback of panspermia as a hypothesis for the origin of life?

4. Why are the elements carbon, hydrogen, nitrogen, oxygen, phosphorus, and sulfur so prominent in living organisms?

5. Discuss the significance of each of the following to prebiotic evolution: amino acid, adenosine triphosphate, autocatalysis, cell membrane (lipid bilayer), enzyme, ribozymes, and thioester bond.

6. What is chirality, and what is its significance for the origin of life?

7. Describe the fundamental experiment that supported the basic "bottom-up" approach. What were its drawbacks?

8. What are the scientific advantages of a hydrothermal origin of life?

9. Describe the possible role of pyrite in the origin of life.
10. Describe the steps in the origin of the RNA world. What are the advantages and disadvantages of the RNA world?
11. How are the sequences of biochemical events invoked in each origin of life scenario related to the proposed *site* of origin?
12. How are autocatalysis and self-organizing systems related?
13. What is the evidence for the origin of eukaryotic cells from prokaryotes?
14. What are the differences between Eukarya, Archaea, and Eubacteria?
15. Diagram the steps leading to eukaryotic cells from prokaryotic cells.
16. What is the relevance of rRNA and protein sequencing to studies of the origin of life? How do the two techniques differ? What information does each technique yield?
17. How might contingency have been involved in making life "left-handed"?

FOOD FOR THOUGHT:
Further Activities In and Outside of Class

1. What are the conditions necessary for any type of life to evolve on a planet? What are the conditions necessary for intelligent life?
2. Discuss the various hypotheses or "worlds" for the origin of life in terms of the method of multiple working hypotheses (see Chapter 1). Are the hypotheses for the origin of life mutually exclusive?
3. Why is spontaneous generation called a theory rather than a hypothesis? Was spontaneous generation a "good theory"?
4. Is a virus living or nonliving? Explain.
5. Besides making sure that their apparatus contained no O_2, what else would Miller and Urey have had to do to the contents? (Hint: Think of Pasteur's experiments.)
6. What is meant by a phrase such as "*the beauty* of the pyrite world …"? Is beauty a desirable trait of a scientific theory? Why or why not?
7. The atmosphere of Archean Earth is frequently depicted as overcast and smoggy rather than sunny. Why do suppose this is?
8. *Assume* that the amino acids of the Murchison meteorite were found to be enriched in the L-form. Comment. What if they were found to be enriched in the D-form?

9. Why are we attempting to detect signals from other civilizations only within our own galaxy?
10. Instead of N_g in the Drake equation, the *rate* of star formation is sometimes used. How could you calculate this? What assumptions do you have to make about the basic state of the universe (see Chapter 7)? Are these assumptions legitimate? (Hint: What is a steady state?)
11. Discuss the contradictory results of sequencing versus the fossil record with regard to the divergence time of the Eukarya, Eubacteria, and Archaea. What are the strengths and weaknesses of the different approaches? Which set of results do you find most acceptable?
12. How might contingency have been involved in making life "left-handed"?
13. Symbiosis is thought to have been intimately involved in the appearance of eukaryotic cells. Symbiosis is also quite common in the modern world. Surprise yourself by looking up examples on the web.
14. Estimate N in the Drake equation using your own estimates of f_l, f_i, and f_c. Justify your estimates of f_l, f_i, and f_c. Compare your estimates with the values given for N in the text.

SOURCES AND FURTHER READING

Bada, J. L., and Lazcano, A. 2003. Prebiotic soup—revisiting the Miller experiment. *Science, 300,* 745–746.

Baltscheffsky, H., and Baltscheffsky, M. 1994. Molecular origin and evolution of early biological energy conversion. In Bengston, S. (ed.), *Early life on earth* (pp. 81–90). Nobel Symposium No. 84. New York: Columbia University Press.

Bordenstein, S. http://serc.carleton.edu/microbelife/topics/monolake/index.html

Bowler, P. J. 1989. *Evolution: The history of an idea.* Berkeley, CA: University of California Press.

Brasier, M. D. et al. 2002. Questioning the evidence for Earth's earliest fossils. *Nature, 416,* 76–81.

Brocks, J. J. et al. 1999. Archean molecular fossils and the early rise of eukaryotes. *Science, 285,* 1033–1036.

Cady, S. L., and Noffke, N. 2009. Geobiology: evidence for early life on Earth and the search for life on other planets. *GSA Today, 19,* 4–10.

Cantu-Guzman, J. C., Oliviera-Bonavilla, A., and Sanchez-Saldana, M. E. 2015. A history (1990–2015) of mismanaging the vaquita into extinction—A Mexican NGO's perspective. *Journal of Marine Animals and Their Ecology 8,* 15–25.

DeDuve, C. 1995. *Vital dust: Life as a cosmic imperative.* New York: Basic Books.

DeDuve, C. 1995. The beginnings of life on earth. *American Scientist, 83*(5), 428–437.

DeDuve, C. 1996. The birth of complex cells. *Scientific American, 274*(4), 50–57.

Delsemme, A. H. 2001. An argument for the cometary origin of the biosphere. *American Scientist, 89*(5), 432–442.

Dyer, B. D., and Obar, R. A. 1994. *Tracing the history of eukaryotic cells: The enigmatic smile.* New York: Columbia University Press.

Gee, H. 2002. That's life? *Nature, 416,* 28.

Hankins, T. L. 1985. *Science and the enlightenment.* Cambridge, UK: Cambridge University Press.

Hazen, R. M. 2001. Life's rocky start. *Scientific American, 284*(4), 76–85.

Jakosky, B. 1998. *The search for life on other planets.* Cambridge, UK: Cambridge University Press.

Kasting, J. F. 2010. *How to find a habitable planet.* Princeton: Princeton University Press.

Kaufman, M. 2011. *First contact: Breakthroughs in the hunt for life beyond Earth.* New York: Simon and Schuster.

Lahav, N. 1999. *Biogenesis: Theories of life's origins.* Oxford, UK: Oxford University Press.

Margulis, L. 1970. *Origin of eukaryotic cells.* New Haven, CT: Yale University Press.

Margulis, L., and Sagan, D. 1986. *Microcosmos: Four billion years of microbial evolution.* New York: Simon and Schuster.

Margulis, L., and Sagan, D. 1995. *What is life?* New York: Nevraumont Publishing.

Marshall, C. P., Emry, J. R., and Marshall, A. O. 2011. Haematite pseudomicrofossils present in the 3.5-billion-year-old Apex Chert. *Nature Geoscience, 4,* 240–243.

Morgan, K. 2003. A rocky start: fresh take on life's oldest story. *Science News, 163*(17), 264–266.

Morowitz, H. J. 1992. *Beginnings of cellular life: Metabolism recapitulates biogenesis.* New Haven, CT: Yale University Press. http://science.nasa.gov/science-news/science-at-nasa/2003/30jul_monolake/ A New Form of Life http://science.nasa.gov/headlines/y2003/30jul_monolake.htm

Pattee, H. 1978. The complementarity principle in biological and social structures. *Journal of Social and Biological Structures, 1,* 191–200.

Schopf, J. W. et al. 2002. Laser-raman imagery of earth's earliest fossils. *Nature, 416,* 73–76.

Schopf, J. W., and Kudryavtsev, A. B. 2011. Biogenicity of Apex chert microstructures. *Nature Geoscience, 4,* 346–347.

Schrödinger, E. 1944. *What is life?* [Reprint] Cambridge, UK: Cambridge University Press.

Sephton, M. A. 2001. Life's sweet beginnings? *Nature, 414*, 857–858.

Simpson, S. 2003. Questioning the oldest signs of life. *Scientific American, 288*(4), 70–77.

Ulmschneider, P. 2006. *Intelligent life in the universe: Principles and requirements behind its emergence*. Berlin: Springer.

Wacey, D., Kilburn, M. R., Saunders, M., Cliff, J., and Brasier, M. D. 2011. Microfossils of sulphur-metabolizing cells in 3.4 billion-year-old rocks of Western Australia. *Nature Geoscience* doi:10. 1038/ NGEO1238.

Wallace, B. 1992. *The search for the gene*. Ithaca, NY: Cornell University Press.

Walter, M. 1999. *The search for life on Mars*. Cambridge, MA: Perseus Books.

Ward, P. D. 2000. *Rare Earth: Why complex life is uncommon in the universe*. New York: Copernicus.

Ward, P. D. 2005. *Life as we do not know it: The NASA search for (and synthesis of) alien life*. New York: Viking Press.

Ward, P. D. and Brownlee, D. 2002. *The life and death of planet Earth: How the new science of astrobiology charts the ultimate fate of our world*. New York: Time Books (Henry Holt and Company).

Wills, C., and Bada, J. 2000. *The spark of life: Darwin and the primeval soup*. Cambridge, MA: Perseus Books.

The Proterozoic: Life Becomes a Geologic Force

MAJOR CONCEPTS AND QUESTIONS ADDRESSED IN THIS CHAPTER

A What is the significance of the Proterozoic?

B What features of the rocks show how the Proterozoic differed from the Archean?

C How did North America form?

D When and how did modern plate tectonics appear?

E What is the evidence for the origin of continental shelves?

F How did continental shelves influence oxygenation of the atmosphere?

G Did oxygenation of Earth's atmosphere occur all at once or in stages? What is the evidence?

H When did ozone form in the earth's atmosphere, and how did it influence the origin of life?

I Was the Earth frozen and fried multiple times during the late Precambrian? If so, how, and what is the evidence?

J How did the behavior of Venus and Mars differ from that of Earth during the Proterozoic Eon?

K Why didn't the Earth continue to be frozen and fried during the Phanerozoic?

CHAPTER OUTLINE

The stromatolite-covered world of the Proterozoic Eon may have looked like this modern example at Shark Bay, Australia. During the early Precambrian, the continents continued to grow beyond the protocontinent stage to near their present size and developed extensive sediment-covered continental shelves, where life began to flourish.

© Rob Bayer/Shutterstock, Inc.

9.1 Significance of the Proterozoic

A During the early Archean, Earth must have seemed like a cauldron, with continents moving about in a prebiotic broth simmering at the surface. However, by the end of the Archean, about 2.5 billion years ago, Earth's surface had cooled sufficiently to enter the final phase of the Precambrian: the **Proterozoic Eon**.

The Proterozoic Eon was of almost 2 billion years' duration. During this time Earth continued to evolve farther away from its much earlier hellish state as the continents grew in size and plate tectonics and the tectonic cycle became more modern in aspect (**Table 9.1**). Indeed, many Archean rocks are metamorphosed, whereas metamorphism is less prominent in the Proterozoic. Sedimentary rocks also became more prominent during the Proterozoic, implying uplift, weathering, and erosion of the continents and the deposition of sediment on widening continental shelves. The term Proterozoic ("first life") implies that life was also present. In fact, it was during the Proterozoic that life increasingly became a major geologic force by oxygenating Earth's atmosphere. It is thought that rising oxygen levels in turn contributed to the appearance of multicellular organisms near the end of the Proterozoic.

As the lithosphere, biosphere, hydrosphere, and atmosphere grew in size, they eventually began to interact with one another and to evolve toward their modern states. These interactions appear to have had drastic consequences for Earth's climate during the later Proterozoic.

9.2 Appearance of Modern Plate Tectonics

B As Earth continued to cool, plate tectonics and crustal composition became more and more modern in aspect during the Proterozoic. Radiometric dates of Proterozoic crust from around the world cluster at about 2.8 to 2.6, 1.9 to 1.7, and 1.0 billion years ago (**Figure 9.1**). The clustering of the dates and the tectonic settings inferred from the rocks both hint that the microplate tectonics of the Archean had begun to give way to a more modern tectonic cycle of continental assembly and rifting involving larger plates and continents.

The initial development of widespread permanent crust appears to have been rapid. As much as 50% to 70% of continental crust formed during a 500-million-year interval from approximately 3 to about 2.5 billion years ago. After this time the relative heights of the continents and depths of

TABLE 9.1

Comparison of the major tectonic features of the Archean and Proterozoic		
	Archean	Proterozoic
Regional setting	Greenstone belts and extensive granitoid complexes.	Cratonic sequences little deformed or metamorphosed; mobile bilts strongly folded, faulted, and metamorphosed (low to high pressure).
Main structures and metamorphism	Synclinal or nappe-like greenstone belts; strike–slip faults common; mainly low to moderate metamorphism (low pressure). Complex structure and moderate to high grade metamorphism (low to high pressure) in gneissic terrains.	Cratonic sequences little deformed or metamorphosed; mobile belts strongly folded, faulted, and metamorphosed.
Volcanic rocks	Ultramafic–mafic komatiites.	Basalts, minor Komatiite-like rocks.
Intrusive rocks	In some terranes, early granitoids mainly tonalite, trondjhemite, and granodiorite, whereas late granitoids mainly granodiorite and granite; temporal overlap between subvolcanic ultramafic, mafic, and felsic intrusives in greenstone belts. Dike swarms.	Large, layered mafic–ultramafic intrusives, dike swarms. Granodiorite, granite.
Sediments	Chert, Banded Iron Formations, conglomerate, volcaniclastic graywacke and siltstone; minor carbonate. Quartz-rich sandstone, aluminous shale are uncommon.	Quartz-rich sandstone (quartzite) arkose, conglomerate, carbonate (mainly dolomite), chert, Banded Iron Formations, shale, graywacke.
Biologic activity	Putative microfossils from ~3.4 Gyr; sporadic stromatolite occurrences.	Proliferation of microorganisms: common stromatolites. First eukaryotic microorganisms and megascopic algae.

Modified from Lambert and Groves. 1981. Early earth evolution and metallogeny in K. H. Wolf (ed.), *Handbook of Strata Bound and Stratiform Ore Deposits*. Amsterdam: Elsevier, p. 352.

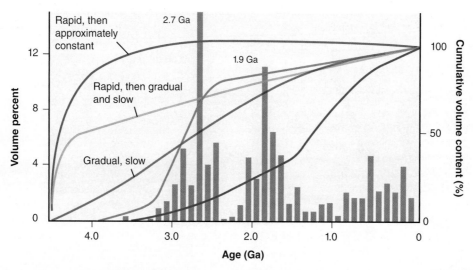

FIGURE 9.1 Growth of continental crust through geologic time. The histogram of the volume distribution of juvenile continental crust is based in part on a compilation of U–Pb zircon ages. Note the significant growth spurt of continental crust about 3 to 2.5 billion years during the transition from the Archean to the Proterozoic, and how the other peaks occur in a cyclic manner, suggesting the production of crust in response to the tectonic cycle (Chapter 2).

the ocean basins probably began to approach modern values more slowly.

The continued differentiation of Earth's crust through the Archean and Proterozoic was crucially dependent on one simple compound: H_2O. Much of the water in Earth's interior was probably released by volcanic degassing, some of which condensed as oceans. Even so, apparently a lot of water still is present in Earth's interior. Experiments indicate that even today minerals that make up Earth's lower mantle hold small amounts of water in their crystal structure. Given the total volume of these rocks, Earth's lower mantle alone might still hold up to about five times the amount of water present in the modern oceans!

The presence of water in Earth's early crust lowered the melting point of that crust. This allowed minerals that form more granitic rocks to melt first and separate from more mafic crust as it melted. Indeed, during the transition from the Archean to the Proterozoic, continental crust began to shift from mafic compositions toward more sodium-rich granites, which crystallize at lower temperatures (see Chapter 3).

It is not surprising, then, to find that the ages of crustal rocks on Earth tend to be much younger than those on Venus, Mars, and the Moon. Plate tectonics presumably ceased on the other inner planets and the Moon for one or both of two reasons: (1) the rapid radiation of most or all their internal heat and (2) the loss of liquid water due to a runaway greenhouse effect in the case of Venus (see Chapter 7). In contrast, there has always been sufficient liquid water, along with heat energy (through radioactive decay), on Earth to drive plate tectonics, the differentiation of magmas, and the formation and growth of continents.

Earth's heat flow has also diminished through time based on geologic record. As Earth continued to cool, the number of plates probably decreased well below the much larger number of microplates in the Archean. The subduction of oceanic crust also became more localized along continental margins and island arcs. Subduction along continental margins would have allowed more granitic magmas to form. This inference is supported by the definite appearance of ophiolites by about 2.0 billion years ago. Ophiolites are normally used to identify ancient convergent plate boundaries (see Chapter 2).

Proterozoic crust was added on, or accreted, around Archean rocks by the process of **continental accretion** (**Figure 9.2**). Like the earlier protocontinents of the Archean (see Chapter 7), each accreted region probably represents a separate piece of crust or terrane that had its own plate and its own geologic history. As these terranes were accreted to form larger continents, preexisting Archean rocks were reworked, deformed, and metamorphosed in **mobile (orogenic) belts** during the collisions. The **Trans-Hudson Orogen**, for example, which runs through what is now Canada, represents the suturing of the Superior, Wyoming, and Hearne Cratons from about 2.0 to 1.8 billion years ago (see Figure 9.2). This particular orogen indicates rifting and volcanism, followed by sedimentation, the formation of ocean crust, and then closure of the basin. After the basin closed, subduction began, leading to island arcs, intrusion of granitic plutons, and later deformation and metamorphism associated with suturing of the three cratons.

Perhaps the best-known collision involving an island arc during the Proterozoic is the **Wopmay Orogen** (named after a famous bush pilot) about 1.9 to 1.8 billion years ago. The Wopmay Orogen is located in what is now northwest Canada near the Arctic Circle (**Figure 9.3**). About 2.1 billion

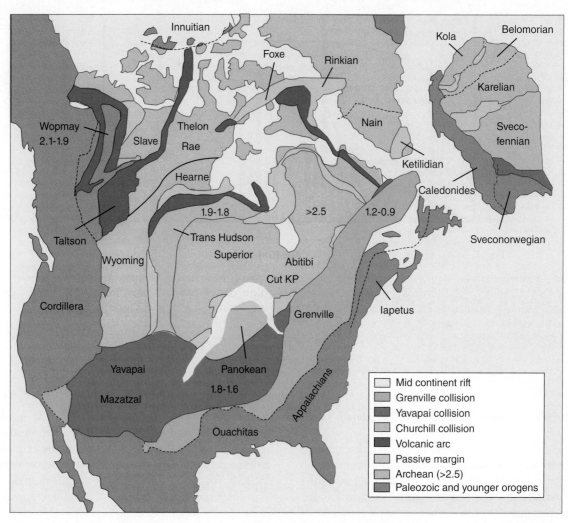

FIGURE 9.2 Growth of the North American craton by continental accretion of mobile (orogenic) belts. Note that most of these rocks are of Proterozoic age and tend to become progressively younger toward the continental margins. Each of the mobile belts probably represents a separate terrane with its own geologic history and were mostly added on ("accreted") to the preexisting continent in order of their ages, so that the rocks tend to become progressively younger as one approaches the margins of the continent.

years ago old Archean crust began to rift apart to produce basins. Then, what is called the Slave plate (named after the lake in northwestern Canada) began to be subducted, eventually resulting in collision and suturing (Figure 9.3). Later, the direction of subduction reversed, producing yet another island arc and collision about 1.8 billion years ago. The Wopmay Orogen is also of interest because of its well-developed sequence of sedimentary rocks; this sequence indicates the appearance of a very important feature—continental shelves—the significance of which is discussed shortly.

A similar style of more modern tectonism also occurred south and west of the Superior Craton from about 1.8 to 1.65 billion years ago. This activity involved extensional and compressional phases, island arc formation and accretion, granitic intrusion, metamorphism, and deformation. In the process, the **Yavapai** and **Mazatzal-Pecos Orogens** were accreted along the southern margin of North America into the western, southwestern, and central United States (Figure 9.2). Both orogens contain greenstone belts (see Chapter 7) and sedimentary rocks. Although the greenstone

belts probably formed in back-arc marginal basins, they contain few if any ultramafic igneous rocks. All told, the Yavapai and Mazatzal-Pecos collisions added about 1,000 km of continental crust along the southern margin of an evolving North America. Other collisions occurred about the same time in what are now Scandinavia, western Africa, and Brazil.

Later deformation and continental accretion occurred in North America during the **Grenville Orogeny**, which spanned from about 1.2–0.9 billion years ago (Figure 9.2). Rocks of the Grenville Orogen consist of sandstones, limestones, and shales along with metamorphic and igneous intrusive rocks. These rocks form most of the deep crustal, or basement, rock of the eastern and southern portion of North America and extend north into Canada, adjacent to the Superior Craton. Rocks of Grenville age are also exposed in the Adirondack Mountains of northern New York state, the Blue Ridge and Great Smoky Mountains of the central Appalachians, the Llano Uplift of central Texas, and what is now Scandinavia. The Grenville Orogeny is also noteworthy because it appears to mark the collision of what was then eastern North America

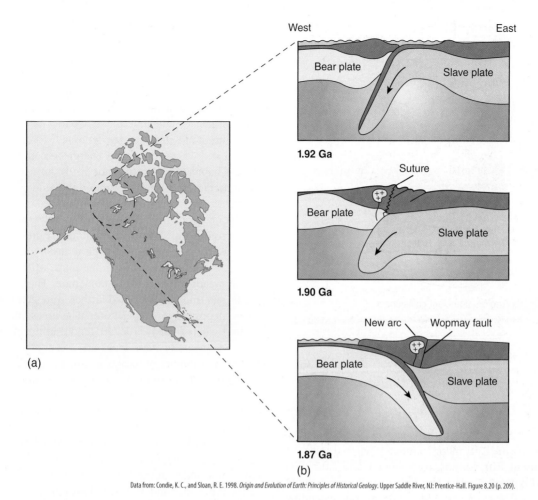

West   East

Bear plate   Slave plate

1.92 Ga

Suture

Bear plate   Slave plate

1.90 Ga

New arc   Wopmay fault

Bear plate   Slave plate

1.87 Ga

(b)

(a)

Data from: Condie, K. C., and Sloan, R. E. 1998. *Origin and Evolution of Earth: Principles of Historical Geology*. Upper Saddle River, NJ: Prentice-Hall. Figure 8.20 (p. 209).

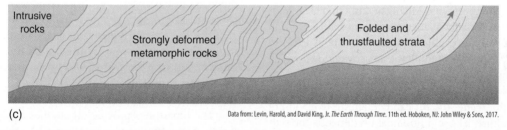

Intrusive rocks

Strongly deformed metamorphic rocks

Folded and thrustfaulted strata

(c)   Data from: Levin, Harold, and David King, Jr. *The Earth Through Time*. 11th ed. Hoboken, NJ: John Wiley & Sons, 2017.

FIGURE 9.3 (a-b) Location and tectonic history and sediments of the Wopmay orogenic belt in northern Canada. The name of the Wopmay belt comes from the name of a famous bush pilot and World War I Ace named Wop May. **(c)** Reconstruction of the shelf margin of the Slave Province.

with what is now northern South America. In South America rocks and structural trends of approximately the same age occur as those of the Grenville province.

Mountain building also occurred around much of the rest of the margin of North America to produce the supercontinent **Rodinia** (Russian for "motherland"). North America (called Laurentia) lay more or less in the center of this supercontinent (**Figure 9.4**). At this time western North America is also thought by some workers to have been sutured to a smaller supercontinent consisting of Antarctica, Australia, and India.

Then, about 750 to 800 million years ago Rodinia began to rift apart. At the same time, the smaller continent composed of Antarctica, Australia, and India rifted away to

produce the ancestral Pacific Ocean, leaving behind a passive margin. Evidence for this so-called **SWEAT hypothesis** (for *South West U.S./East Antarctica*; see Chapter 11) includes similar sequences of rocks in the **Transantarctic Mountains** of Antarctica (now buried, except for their peaks, by many kilometers of ice) deposited in passive margin settings. Rocks similar to those of the Grenville province in eastern North America are also found in eastern Antarctica according to this hypothesis (see Chapter 11). Whatever happened, after this phase of rifting, the western margin of North America would never again collide with another large continent, although, as we will see in succeeding chapters of Part III, smaller microcontinents or terranes frequently accreted along its western margin, resulting in a very complex geologic history.

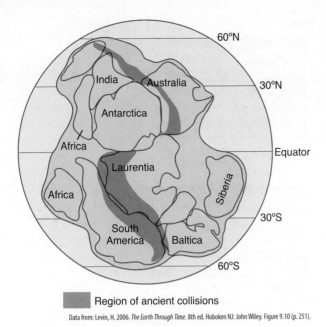

FIGURE 9.4 The supercontinent Rodinia as it appeared about 1.1 billion years ago. The dark brown band shows the location of collisional orogens where continents collided in constructing the supercontinent.

Region of ancient collisions

Data from: Levin, H. 2006. *The Earth Through Time.* 8th ed. Hoboken NJ: John Wiley. Figure 9.10 (p. 251).

Rifting also occurred in the *center* of North America. The **Keweenawan Rift**, more commonly referred to as the **Mid-Continent Rift**, formed from about 1.2 to 1.0 billion years ago (**Figure 9.5**). The Mid-Continent Rift consists of two branches that cut across older orogens: one extending from the Great Lakes region southwest to Kansas and the other trending southeast into Michigan and Ohio. The Mid-Continent Rift is actually a failed rift basin or aulacogen; in fact, failed rift basins are quite common from about 1,000 to 700 million years ago (including the aulacogen down which the Mississippi River flows). The Mid-Continent Rift basin is mostly buried by younger sediments except in the vicinity of Lake Superior. Here, the rift can contain up to 15 km of basalts and reddish lake and floodplain sediments; the sediments are red due to the oxidation of iron in the sediment.

CONCEPT AND REASONING CHECK

1. What evidence is available to reconstruct the sequence of orogenies described above?

9.3 Sedimentary Rocks and Continental Shelves

D **E** We have already noted two of the most significant changes that occurred on Earth during the Proterozoic: the appearance of continental shelves and well-developed sequences of sedimentary rocks characteristic of shelves. The co-occurrence of continental shelves and

well-developed sedimentary sequences is not accidental. As the Proterozoic passed, the continents collided and grew in size and height, becoming more susceptible to erosion. Erosion, followed by deposition along the margins of the early continents, produced the shelves (**Figure 9.6**).

Most notable among these sediments are quartz sandstones, shales, and limestones. Quartz sandstones, shales, and limestones tend to be deposited according to water depth and wave and current energy on continental shelves (see Chapter 4). Quartz sands likely formed by chemical weathering of the early granitic crust and gneisses. Many of these quartz sandstones were deposited close to shore in high-energy environments. Consequently, they tend to be compositionally and texturally more mature, with rounded and well-sorted grains and little if any mud. Granites and gneisses also contain feldspar and mica, which are relatively easily weathered to clays, forming mud. These muds tended to be suspended and carried farther from shore, where they were deposited in deeper water. The limestones formed in waters where there is little terrigenous influx of sediment.

Sediments of these types are exceptionally well developed in a number of localities. Two examples are the Early Proterozoic rocks of the Wopmay Orogen and in the vicinity of the modern Great Lakes. Thick sequences (up to about 50,000 feet!) of shelfal sedimentary rocks also developed from about 1.45 billion to 850 million years ago in basins located in what is now the western United States and Canada. Some of the most spectacular exposures of these rocks are found today in the Belt Supergroup of Glacier National Park (Montana) and Waterton Lakes National Park (southern Alberta, Canada; **Figure 9.7**). Some sediments were laid down in an aulacogen and display features of intermittent exposure to air, such as mud cracks. Among other spectacular exposures of Proterozoic rocks are those belonging to the Grand Canyon Supergroup at the base of the Grand Canyon and which are separated from the Early Cambrian Tapeats Sandstone by a pronounced angular unconformity (Figure 9.7).

The limestones provide other clues about Earth's evolution. Many Proterozoic limestones like those of the Belt Supergroup resulted from the activity of microscopic, prokaryotic cyanobacteria. During the Proterozoic, definite cyanobacteria constructed mound and columnar-shaped structures called stromatolites that extended over large areas (**Figure 9.8**; refer to this chapter's frontispiece). The stromatolites grew as the cyanobacteria secreted a sticky, mucous-like substance that trapped fine-grained sediment into layers. The algae simply continued to grow through the sediment to the surface as the sediment thickness increased, giving the structures a layered appearance in cross-section.

Some limestones might have also been chemically precipitated from seawater. Normally, limestone is composed of the calcitic shells of dead organisms such as clams and corals, which evolved during the Phanerozoic. However, in the Proterozoic, none of these creatures are thought to have existed, hinting that perhaps some limestones, other than those formed by stromatolites, were precipitated directly from seawater.

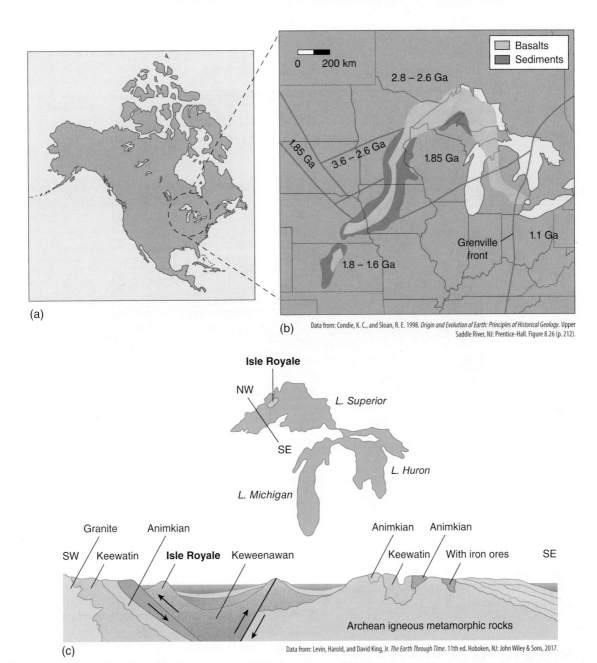

FIGURE 9.5 (a-b) Location and rocks of the Keweenawan, or Mid-Continent, Rift. **(c)** Generalized cross-section showing major rock units.

Dolomites are also present during the Proterozoic, indicating post-depositional alteration of the limestones by highly saline seawater formed in evaporitic basins. In fact, the geologic record indicates that evaporites became more frequent during the Proterozoic. To concentrate seawater sufficiently to cause precipitation, evaporation most likely occurred in restricted marine basins, such as rift basins located in warm climates and flooded by the oceans. As already described, modern plate tectonics developed during the Proterozoic, so rift basins were undoubtedly present. Given the extent of evaporites in the Proterozoic, rift basins might have been very widespread.

The presence of evaporites also indicates the presence of sulfate minerals such as gypsum and anhydrite. Sulfur can exist in a variety of chemical combinations, but the presence of dissolved sulfate ions (SO_4^{2-}) indicates something very important; namely, that oxygen was finally available in sufficient quantities to combine with other elements such as sulfur. When and how did oxygen become abundant in Earth's atmosphere? As we will see, cyanobacteria and oxygenation of Earth's atmosphere are linked.

CONCEPT AND REASONING CHECKS

1. What is the evidence for the development of continental shelves during the Proterozoic?
2. How do continental shelves influence sedimentation (see Chapters 3 and 5)?

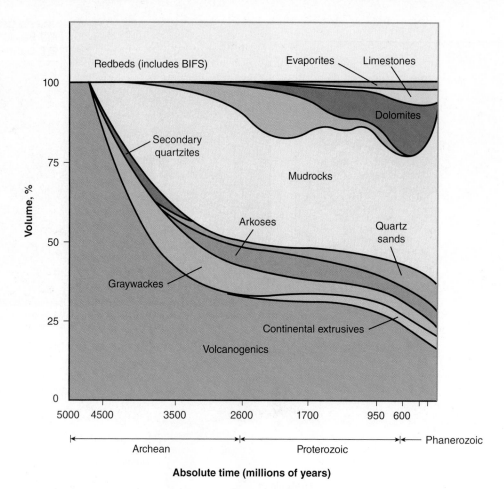

FIGURE 9.6 Occurrence of sediments through geologic time. Note how sandstones like arkoses and graywackes give way to quartz sandstones, shales, and limestones through the Archean into the Proterozoic. Arkoses are texturally and compositionally immature and signal uplift, erosion, and rapid deposition of sediments derived from granitic rocks, whereas graywackes indicate deposition in deep basins associated with volcanic island arcs. By contrast, quartz sandstones, shales, and limestones suggest the presence of well-developed continental shelves (see Chapter 4 and Table 9.1).

9.4 Oxygenation of Earth's Atmosphere

9.4.1 Appearance of oxygen

F If oxygen was not present in early Earth's atmosphere, from where did it come? And why didn't Earth's atmosphere immediately become oxygenated?

Very early in Earth's history, free O_2 was likely produced by photolysis of water being degassed from volcanoes. Much of the hydrogen produced would presumably have escaped to space, leaving behind free O_2. However, another hypothesis suggests that Earth's atmosphere was initially enriched more in methane (CH_4) than CO_2. According to this hypothesis methane was generated by **methanogenic**, or methane-producing, bacteria, which decompose organic matter while releasing methane and O_2. As the atmosphere became enriched in methane, some methane would have been photolyzed, with the resulting H_2 escaping into space.

Any free O_2 present then oxidized (reacted with) carbon to produce carbon dioxide (CO_2) which, like methane, is a greenhouse gas.

Whatever the case, the geologic record indicates that free O_2 also combined with iron-rich minerals, which were abundant in the mafic-to-ultramafic Archean crust. The reaction of iron with oxygen was of profound importance to the evolution of Earth's atmosphere. A peculiar type of sedimentary rock called **Banded Iron Formations (BIFs)**, which appeared during the Archean, became especially prominent during the early Proterozoic (**Figure 9.9**). BIFs consist of alternating thin (millimeter) to thick (>1 m) layers of iron and chert (sometimes in its red variety called jasper). How could these unusual configurations have formed? Some clues are provided by modern oceans. Today, deep ocean waters become oxygenated when they upwell to the surface and come into contact with the atmosphere. Oxygen then diffuses from the atmosphere into the surface waters before they descend for another circuit through the oceans. Because there was relatively little free O_2 in early Earth's

(b)

(a) Courtesy of Dr. Allen Thompson, University of Delaware.

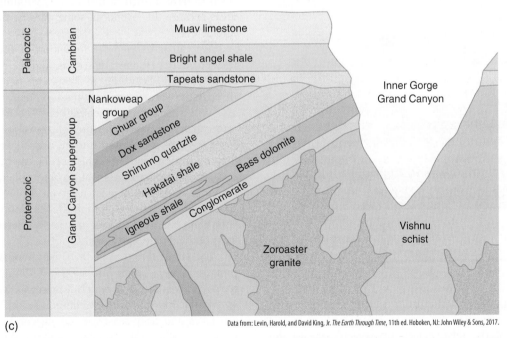

(c) Data from: Levin, Harold, and David King, Jr. *The Earth Through Time*, 11th ed. Hoboken, NJ: John Wiley & Sons, 2017.

FIGURE 9.7 Spectacular exposures of Proterozoic rocks in the western United States. **(a)** Chief Mountain, Glacier National Park, Montana. The lower two-thirds of the cliffs are Cretaceous rocks, over which rocks of the Neoproterozoic Belt Supergroup were overthrust during the Sevier Orogeny of the Cretaceous Period (see Chapter 13). The Lewis thrust fault separates the two rock groups and is indicated by the arrow. **(b)** Proterozoic rocks at the base of the Grand Canyon belong to the Grand Supergroup and are separated from the Early Cambrian Tapeats Sandstone by a pronounced angular unconformity and nonconformity. **(c)** Stratigraphy of the base of the Grand Canyon. Note the intrusion of the metamorphosed Vishnu Schist by the Zoroaster Granite, which formed in a subduction zone assocated with volcanic island arcs. The Vishnu Schist has been dated at about 1.7 billion years, whereas the Tapeats Sandstone is roughly 540 million years old. So at one time, the Grand Canyon was a mountainous region, which was then eroded and deformed, and eventually followed by deposition of the Tapeats. This means that the hiatus corresponding to the angular unconformity at the base of the Tapeats may represent over a billion years!

(a)

© Monica Johansen/Shutterstock, Inc.

(b)

Courtesy of Ron Martin.

FIGURE 9.8 Stromatolites. **(a)** Modern stromatolites, Shark Bay, Australia. **(b)** Ancient stromatolites in cross-section.

© Stefanie van der Vinden/Shutterstock, Inc.

FIGURE 9.9 Banded Iron Formations (BIFs). BIFs contain huge amounts of iron and once formed the basis of the U.S. steel industry. Iron ore was transported from the Great Lakes by ship and rail to the steel mills of Pennsylvania, where abundant coal deposits were used to melt the ore and extract iron.

atmosphere, the deep Proterozoic oceans must have also largely lacked oxygen.

Because of the relative lack of oxygen beneath the ocean's surface, iron was undoubtedly present in dissolved form because of hydrothermal exchange between seawater and the widespread, warm mafic-to-ultramafic ocean crust. Because of the low oxygen conditions, iron would have existed in the +2 valence state (Fe^{2+}). In this valence state, iron remains easily dissolved in seawater. However, if Fe^{2+} comes into contact with O_2, such as might have happened at the ocean's surface during upwelling, the Fe^{2+} could have

been further oxidized to produce Fe^{3+}, which is highly insoluble and precipitates as iron oxide. It was therefore long thought that BIFs formed when Fe^{2+}-rich anoxic waters upwelled in the vicinity of ancient continental shelves or shallow seaways covering the cratons (**Figure 9.10**). BIFs seem to have formed in just such environments over large areas because the layers can sometimes be traced for many kilometers.

We can redescribe the formation of BIFs in terms of the behavior of systems, reservoirs, and their interactions in such a setting. Seawater (hydrosphere) percolates (fluxes) through the warm ocean crust reservoir of iron (lithosphere), and iron dissolves into the seawater. The iron then comes into contact with oxygen (through upwelling) that has diffused from the atmosphere into the upper layers of the water column (hydrosphere). As this occurs, the iron precipitates back to the ocean bottom of the lithosphere, in this particular case the sedimentary reservoir of the lithosphere, to produce BIFs. Notice how, in this redescription, both matter (iron) and energy (heat) are transferred between the reservoirs of the different systems.

The laminated (thin-bedded) nature of many BIFs also provides clues to their formation. The laminations resemble the varves of lakes, which represent seasonal cycles of sedimentation. Seasonal upwelling is found in many modern lakes and deep-sea basins along continental margins. Iron is one of several nutrients that limits photosynthesis by phytoplankton. Thus, seasonal upwelling of dissolved iron might have stimulated seasonal pulses of photosynthesis in surface waters, after which the precipitated iron settled to the bottom. When upwelling was not occurring, normal background formation of chert occurred. Whether or not the chert was biogenic or abiogenic in origin remains unclear, however. Thick-bedded BIFs are also thought by some workers to have formed by the alteration of iron-rich

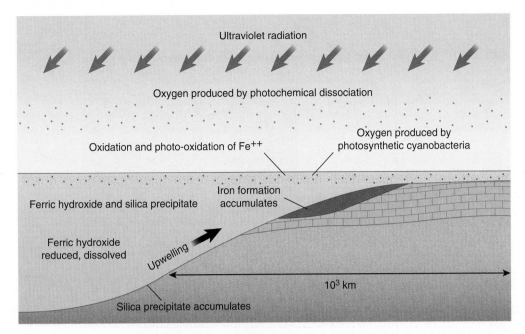

FIGURE 9.10 Hypothesis for the formation of Banded Iron Formations described in terms of the behavior of systems and their interactions. This particular hypothesis, long-favored by scientists, involves upwelling of iron-rich waters into the photic zone, where iron precipitates out as distinct bands. However, some BIFs might have formed from turbidity currents, as described in the text.

turbidity currents after they had slumped from the sides of hydrothermal vents or submarine volcanoes. These sorts of BIFs are therefore thought to have formed in deeper water, perhaps much farther from the continental shelf or slope than the laminated types just described.

Whatever their mechanism of formation (which might have been multiple), the production of BIFs continued until most of the iron was used up, which took billions of years based on the persistence of BIFs in the geologic record. After this time BIFs tapered off and eventually mostly disappeared (Figure 9.6). Given the volume of BIFs deposited during this time, the BIFs must have been a huge oxygen sink. But we have already said that Earth's early atmosphere was oxygen poor. Where did so much oxygen come from to produce the BIFs?

Ⓖ Perhaps some O_2 came from photolysis of H_2O, as described earlier. Another likely source was photosynthesis. But photosynthesis requires plants; there were no higher plants in the Archean and Proterozoic, but there were cyanobacteria. The effect of cyanobacteria on Earth's atmosphere is a prime example of the interaction of the physical Earth and life. Because cyanobacteria require sunlight for photosynthesis, they needed to have lived in the photic zone of the seas. It is perhaps no coincidence, then, that stromatolites and the oxygenation of the atmosphere occurred as continental shelves became prominent. The shelves would have provided broad expanses of sunlit water while also facilitating the burial of organic carbon and the release of O_2 (see this chapter's frontispiece). The early success of cyanobacteria might have also been due to their ability to use free nitrogen (N_2). Nitrogen is an important nutrient because it is a prominent component of amino acids and proteins

(see Chapter 4). Nitrogen was probably very abundant in the Precambrian Earth's atmosphere; even today, the atmosphere is 78% nitrogen.

Despite their success, by the end of the Proterozoic, stromatolites were on the wane, possibly as a result of grazing by newly evolving herbivores (see Chapter 10). Although stromatolites continued into the early Phanerozoic, they were never again as prominent as they were in the Proterozoic. Today, stromatolites resembling those of ancient seas are sometimes found in hypersaline lagoons in places such as the Bahamas. The high salinity prevents grazing animals from living in such habitats, allowing the stromatolites to flourish.

9.4.2 Stages in the oxygenation of Earth's atmosphere

Ⓗ Based on the rock record, the oxygenation of Earth's atmosphere probably occurred in a sequence of stages (**Figure 9.11**). BIFs did not become prominent components of the marine sedimentary record until the early Proterozoic, which began about 2.5 billion years ago. Terrestrial **redbeds**, which like BIFs are reddish due to the oxidation of iron, formed on continents in streams and rivers. Redbeds did not appear until the early Proterozoic at about 2.4 billion years ago and did not become prominent until after about 1.5 billion years before present. Similarly, gypsum and anhydrite, the sulfur of which has been oxidized to sulfate, did not become prominent until about 2 billion years ago.

Conversely, the mineral **uraninite** (UO_2) was prominent before about 2.2 to 2.3 billion years ago, well before redbeds and evaporites appeared. Uraninite occurred in detrital form, as sedimentary grains. Like iron, uranium

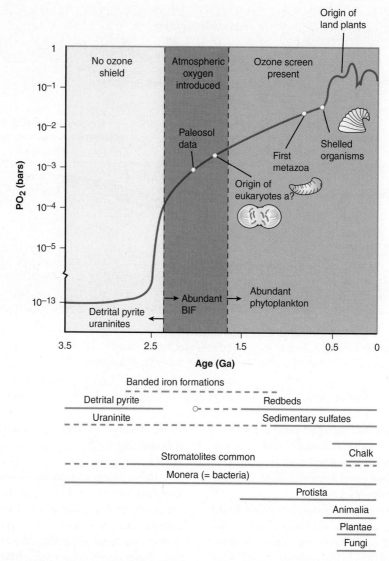

FIGURE 9.11 Generalized stages in the evolution of the oxygenation of the atmosphere and oceans. Geologic and chemical evidence discussed in the text is shown at the bottom. Note that the stages are not abrupt but transitional to one another.

exists in two valence states: U^{4+} and U^{6+}. Unlike iron, however, the most oxidized state of uranium (U^{6+}) forms soluble compounds, whereas the less-oxidized (or more reduced) U^{4+} forms insoluble compounds. Thus, if oxygen levels increased in the atmosphere through time, any U^{4+} should have been oxidized to U^{6+} and dissolved away. This is apparently what happened, given the decreased abundance of uraninite in the Proterozoic.

These changes are corroborated by other indicators. Detrital pyrite (iron sulfide) was also abundant in the Archean and early Proterozoic, but not after about 2.2 billion years ago; apparently, the detrital pyrite was oxidized to produce sulfates. Paleosols, ancient soils or weathering profiles, also indicate the oxidation state of Earth. Paleosols older than about 2.2 billion years ago have lost much of their iron, presumably because it was dissolved away. By contrast, paleosols younger than this time have retained much of their iron because it was precipitated in the presence of oxygen.

Based on these data, we can infer that oxygen levels in the Earth's ancient atmospheres relative to the present atmospheric level (PAL):

- **In stage I**, before about 2.2 to 2.4 billion years ago, there was little or no free oxygen in Earth's atmosphere (Figure 9.11). Although photolysis probably split water to produce oxygen, only relatively small amounts of O_2 were likely produced and were quickly consumed by chemical weathering. Oxygen levels during this stage were less than 0.005 PAL and perhaps ranged down to as low as 10^{-8} to 10^{-14} PAL, as indicated by the presence of detrital uraninite and pyrite and iron-poor paleosols.

- **During stage II**, which was under way by about 2.2 to 2 billion years ago, atmospheric levels had increased to about 0.01 PAL. During this stage, the uppermost layers of the oceans in contact with the atmosphere must have been partially oxygenated for BIFs to precipitate, while detrital uraninite and pyrite disappeared, and iron-enriched paleosols developed (Figure 9.11). Based on the geologic record of sedimentary rocks (Figure 9.6), by this time continental shelves had become widespread and would have provided habitats for stromatolites. Increasing erosion of continents and sedimentation on shelves would have also enhanced the burial of dead organic matter that is enriched in carbon. When carbon is removed from surface systems through burial, oxygen is left behind. Thus, expanded continental shelves might have promoted oxygenation of the atmosphere and oceans in two ways: the growth of stromatolites and the burial of carbon.

■ **In stage III**, beginning about 1.8 billion years ago, BIFs largely disappeared from the rock record. With this oxygen sink gone, oxygen began to accumulate in abundance in Earth's atmosphere and increased to somewhere between 0.01 and 0.1 PAL.

■ **By stage IV**, the deep oceans also began to become better oxygenated. However, exactly when this occurred is unclear (see Chapter 11).

9.4.3 Rise of ozone

After the production of BIFs had ceased, photosynthesis not only oxygenated the atmosphere but set in motion *negative feedback* critical to the further evolution of life. Negative feedback came in the form of **ozone** (O_3) and its effect on ultraviolet light. Ozone is best known as the peculiar odor one smells in the vicinity of faulty electrical machinery, such as vacuum cleaners. Short circuits in the wiring cause electrical discharges into the surrounding air, producing ozone from the oxygen. In the atmosphere, ozone is produced by ultraviolet radiation, which splits diatomic oxygen to produce atomic oxygen:

$$O_2 \rightarrow O + O^- \tag{9.1}$$

Atomic oxygen then quickly combines with diatomic oxygen to form ozone:

$$O^- + O_2 \rightarrow O_3 \tag{9.2}$$

Earth's ozone layer probably began to develop during stage III, as free O_2 began to accumulate in the atmosphere. The formation of ozone shields Earth from over 90% of the ultraviolet radiation emanating from outer space, especially from our sun. Ultraviolet light is highly energetic and destructive to higher living organisms. For example, an increase in the incidence of skin cancers in the southern hemisphere has been attributed to the destruction of ozone and the formation of a hole in the ozone layer centered over Antarctica. Without the ozone shield, then, higher life forms other than primitive groups like cyanobacteria might never have evolved on Earth.

As we have already seen, the abundance of oxygen also presented problems for life (see Chapter 8). Although many heterotrophic organisms require O_2 for respiration, oxygen is toxic to many living systems because it is highly reactive and attempts to form compounds with virtually any other substances available. As oxygen accumulated in the atmosphere, life might have responded by evolving eukaryotic cells (see Chapter 8).

CONCEPT AND REASONING CHECKS

1. Describe how BIFs served as negative feedback on the oxygenation of the atmosphere.
2. What directional changes occurred during the oxygenation of the atmosphere?
3. What had to have happened during early Earth's formation for an atmosphere to appear and evolve (see Chapter 7)?

9.5 Snowball Earth: Earth Out of Balance?

9.5.1 Previous hypotheses for Snowball Earths

The last phase of the Precambrian is called the **Neoproterozoic**, or "new Proterozoic," and stretches from about 1.3 billion to about 540 million years ago. The time is indeed new because Earth's seemingly slow, quiet, almost magisterial evolution was, according to some workers, interrupted by a series of abrupt, radical climate changes involving runaway glaciations of most or all of Earth. Each of these glaciations is thought to have lasted millions of years. The last glaciation occurred about 600 million years ago and is evidenced by widespread tillites on several continents. Each glaciation was terminated by severe warmings. Thus, Earth would have been both frozen and fried repeatedly during the Neoproterozoic.

Large-scale glaciation has long been suspected to have occurred on Earth during the Neoproterozoic. In 1964, the geologist Brian Harland pointed out the occurrence of glacial deposits on many continents, including North America, Australia, and Africa during the Neoproterozoic. When the ancient positions were reconstructed according to polar wandering curves, both the continents and the glaciers occurred in the tropics! According to Harland, when the glaciers began to grow from about 30 degrees latitude toward the equator, Earth's albedo increased so much that runaway glaciation ensued via positive feedback. Glaciers in tropical latitudes do not, of course, agree with observations *today*. Thus, Harland's hypothesis was largely ignored, and only recently have his ideas come to be reexamined in earnest.

The glacial deposits themselves are of a special kind called **diamictites**, which consist of poorly sorted ice-rafted debris, including dropstones, deposited in seaways at the margins of melting sea ice (**Figure 9.12**). Mountain glaciers could certainly occur at sufficiently high elevations in the tropics, as they do today, but the environments of deposition indicate the glaciers were not mountain glaciers. Thus, the Neoproterozoic glaciers were apparently so huge they must have extended over continents and large amounts of

(a)
© Marli Miller/Visuals Unlimited, Inc.

(b)
Courtesy of iStockphoto/Thinkstock.

FIGURE 9.12 **(a)** Diamictite. **(b)** Cliff of carbonate capstones in Namibia (Africa).

the surrounding ocean's surface as sea ice, so that at times Earth must have resembled a gigantic snowball.

This inference has given rise to various hypotheses for a **Snowball Earth**. But, how could glaciation have happened in the tropics?!? One hypothesis that might account for each of the glaciations involves the sun's brightness. By about 750 million years ago the sun's brightness is calculated to have increased to 93% to 94% of modern values (see Chapter 7). Nevertheless, the somewhat lower intensity of the sun at this time might still have been sufficient to prevent runaway glaciation. Furthermore, if only solar radiation were involved, there should have been glaciations long before this time during the Proterozoic, when the sun was fainter. Instead, the only other time of widespread glaciation was near the beginning of the Proterozoic, about 2.5 billion years ago (**Figure 9.13**). So, the hypothesis based on the sun's brightness alone, simple and appealing as it is, does not seem to explain the occurrence of the glacial deposits through time.

Another hypothesis stems from the distinctive sequence of rocks. In this hypothesis, large amounts of carbon were initially buried in the oceans, drawing down atmospheric carbon dioxide levels and cooling the Earth sufficiently to lead to massive glaciation and the widespread deposition of diamictites. The diamictites, which can be up to thousands of feet in thickness, suddenly give way upward to thick limestones that were presumably laid down in warm, tropical seaways like modern limestones. The limestones are referred to as **cap limestones** because they lie on top of, or cap, the glacial deposits; these cap limestones can be hundreds of meters or more thick, suggesting it took millions of years for them to be deposited (Figure 9.12). According to this hypothesis, as carbon dioxide levels declined in the atmosphere, Earth cooled and glaciers advanced. Eventually, the oceans reached a point at which they began to circulate more rapidly (because of the presence of large glaciers), resulting in the upwelling of massive amounts of carbon dioxide

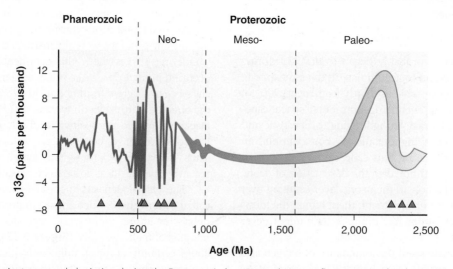

FIGURE 9.13 Carbon isotopes and glaciation during the Proterozoic (see Box 9.1). Strong fluctuations of carbon isotope ratios suggest significant fluctuations of primary productivity during multiple glaciations (triangles) of the Neoproterozoic associated with Snowball Earth. The strong rise in carbon isotope ratios at about 2.2 billion years ago is close to the first well-documented glaciations on Earth and might have resulted from photosynthesis by stromatolites. Triangles indicate times of glaciation.

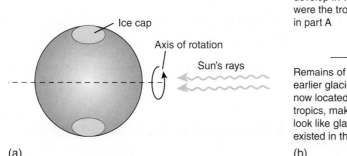

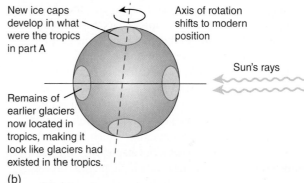

(a)   (b)

FIGURE 9.14 The "high-tilt" hypothesis for Snowball Earth. This hypothesis suggests that **(a)** the Earth very nearly laid on its side until near the end of the Proterozoic. This orientation would have put what are now the poles closer to the sun and warmed them, whereas what are now the tropics would have received less solar insolation. **(b)** The normal orientation of the Earth, in which the tropics receive most solar insolation and the poles relatively little.

already stored in the oceans that in turn led to intense weathering (see Chapter 1) and the widespread deposition of limestones. However, such events would have happened relatively quickly; thus, this hypothesis cannot account for the prolonged deposition of both diamictites and then the sudden deposition of tremendously thick limestones that were probably laid down for millions of years.

A third hypothesis suggests that glaciation was not in fact global at all but only appears so. This hypothesis suggests that Earth's axis of rotation, which is currently about 23.5 degrees from the vertical, was much greater in the past and exceeded 54 degrees until the end of the Proterozoic (**Figure 9.14**). Thus, Earth's climatic zonation (see Chapter 3) would have been reversed in the Neoproterozoic. What later became the high latitudes initially lay in the tropics, and what later became the low latitudes first occurred nearer the poles. Because what would become the tropics during the Phanerozoic were actually closer to the poles during the Neoproterozoic, glacial deposits came to be located in the tropics when Earth's tilt moved toward more normal conditions in the Phanerozoic. If this sounds far-fetched, some scientists believe a similar mechanism might account for the formation of the apparent channels by water found on the surface of Mars (see Chapter 7). After the angle of tilt decreased, it should not have increased again. However, Snowball Earths, if they did occur, happened more than once. Thus, this hypothesis does not explain how *multiple* low-latitude glaciations could occur or how presumably tropical carbonates could come to lie directly over each of the glacial deposits. For this to happen, the angle of tilt would have needed to increase and decrease repeatedly.

A fourth hypothesis that attempts to explain Snowball Earths involves **true polar wander**. In contrast to apparent polar wander (see Chapter 2), true polar wander involves the rotation of the entire crust and mantle as a single unit around Earth's core. According to the true polar wander hypothesis, small changes in the distribution of continental masses caused the entire crust/mantle shell to rotate rapidly from the poles to the tropics because of the conservation of angular momentum (see Chapter 7); the crust/mantle shell

would have carried the continents from the poles to the tropics in the process (**Figure 9.15**). These continental movements were completed in as short a time as a few million years, which is much, much faster than ordinary rates of seafloor spreading could have done. A rapid shift in the crust/mantle shell could have resulted if Earth's longest axis were originally oriented in the equatorial plane rather than more vertically, as it is now. Thus, the true polar wander hypothesis explains the occurrence of tropical carbonates on top of glacial deposits as follows: (1) glaciers formed on continents at higher latitudes, (2) the continents then moved rapidly to the tropics due to true polar wander, and (3) limestones then formed. The true polar wander hypothesis has its drawbacks, too. Like the third hypothesis for the cause of Snowball Earths, the true polar wander hypothesis has glaciers forming at higher latitudes, with the diamictites only *appearing* to form at low latitudes. But, the angle of inclination of magnetized particles in the diamictites is low, indicating the diamictites formed at relatively low latitudes, not high. This hypothesis also cannot explain multiple glaciations.

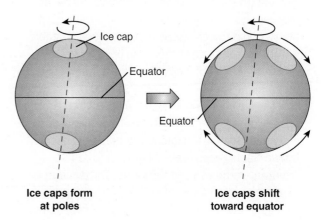

FIGURE 9.15 Formation of glacial deposits and cap limestones according to the true polar wander hypothesis. Glaciers formed on continents at high latitudes. Then, the continents shifted rapidly to low latitudes where limestones formed on top of the glacial deposits.

Yet another hypothesis invoked to explain the diamictites suggests they are actually impact ejecta formed by asteroid impacts. According to this hypothesis, the grooved and striated rock fragments are indicative of impact ejecta, not glaciers. Based on studies of Late Cretaceous extinctions, if an impact of sufficient size were to cause drastic climate change, plankton should have been devastated and photosynthesis shut down immediately. These abrupt shifts can be measured by using carbon isotope ratios, which are indicative of changes in photosynthesis (see Chapter 3). If the carbon isotope ratio values were relatively high, photosynthesis was presumably active, whereas if the ratios were low, photosynthesis decreased. The impact hypothesis therefore predicts that strong decreases in carbon isotope ratios should be found at the top of the presumed impact ejecta identified by others as diamictites. However, the negative spikes occur below the tops of the presumed impact ejecta (diamictites),

thus predating any impact and contradicting a key prediction of the impact scenario.

CONCEPT AND REASONING CHECK

1. What fundamental principle must one largely ignore to entertain the different Snowball Earth hypotheses?

9.5.2 Snowball Earth reexamined

The most recent hypothesis that attempts to explain Snowball Earth has, as might be expected, been received with cautious acceptance by some scientists and extreme skepticism by others. The most recent Snowball Earth hypothesis incorporates tectonics, climate change, feedback, photosynthesis, and carbon isotopes into a cycle (**Figure 9.16**).

BOX 9.1 Stable Isotopes and the Cycling of Elements

We have previously discussed radioactive isotopes, which decay spontaneously to produce the Earth's internal heat and which are used in radiometric dating (see Chapter 6).

But there is another form of isotope, called **stable isotopes**, which are of tremendous value in studying the behavior of the cycling of compounds and chemical elements through time. As their name implies, stable isotopes do not undergo radioactive decay. However, stable isotopes of the same element behave somewhat differently in chemical reactions; because it is less heavy, the "lighter" isotope of an element (the one with fewer neutrons) participates more easily in chemical reactions and is more easily transferred between reservoirs. Consequently, stable isotopes of the same element can be separated by chemical analysis of their atomic mass (number of protons + neutrons).

The element carbon is, of course, central to photosynthesis. Besides carbon-14, which is used in radiometric dating (Chapter 6), carbon is represented by two stable isotopes also of interest to earth scientists: carbon-13 and carbon-12. The different mass numbers (number of protons + neutrons) are indicated next to the symbol for the element, in this case ^{13}C and ^{12}C. Carbon-13 (^{13}C) and carbon-12 (^{12}C) both have the same number of protons (atomic number 12), but ^{13}C has one more neutron.

The flux of carbon on geologic time scales between the different reservoirs of organic carbon is studied determining the ratios of the two isotopes, $^{13}C/^{12}C$, and is measured in parts per thousand (‰). By determining the $^{13}C/^{12}C$ ratio (as recorded in fossils, for example), scientists can determine the extent of past changes in photosynthesis and carbon burial. Carbon dioxide containing ^{12}C is more easily incorporated into organic matter during photosynthesis because ^{12}C has one less neutron than ^{13}C and is therefore lighter and slightly more chemically reactive. This means that if photosynthesis increases, more carbon dioxide enriched in ^{13}C is left behind in the atmosphere. Thus, if photosynthesis increases, the ratio of the two isotopes also increases in the atmosphere and there is an increase in the ratio $^{13}C/^{12}C$. Conversely, if photosynthesis decreases, the isotope ratio goes down, because less ^{12}C-enriched carbon dioxide has been incorporated into organic matter. This indicates that photosynthesis might have decreased. Another possibility is that more carbon dioxide (which is enriched in ^{12}C) has been injected into the atmosphere such as by volcanism or other processes that sometimes coincide with mass extinction.

Besides carbon isotopes, we will discuss stable isotopes of strontium and oxygen later in the text.

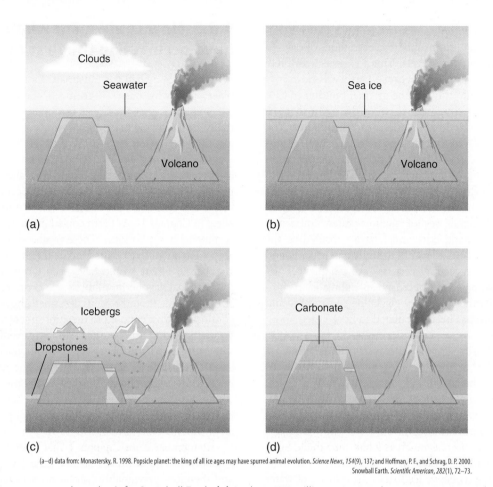

(a–d) data from: Monastersky, R. 1998. Popsicle planet: the king of all ice ages may have spurred animal evolution. *Science News*, *154*(9), 137; and Hoffman, P. F., and Schrag, D. P. 2000. Snowball Earth. *Scientific American*, *282*(1), 72–73.

FIGURE 9.16 The most recent hypothesis for Snowball Earth. **(a)** At about 750 million years ago, the supercontinent Rodinia began to rift apart and its shallow continental shelves became prime sites for photosynthesis by stromatolites. Photosynthesis extracted large amounts of carbon dioxide from the atmosphere, and buried it as dead organic carbon in sediment. **(b)** Because so much carbon dioxide was extracted from the atmosphere, and the sun's brightness was still less than modern values, the Earth became quite cold, and the ocean surface froze. This shut down photosynthesis **(c)** However, volcanism continued and carbon dioxide slowly built up in the atmosphere over millions of years. Eventually, the sea ice melted, producing widespread diamictites. **(d)** The large amount of carbon dioxide in the atmosphere caused widespread precipitation of limestones on top of diamictites in shallow seaways. In order for such great thicknesses of shallow-water limestones to form, the underlying crust must have subsided as the carbonates accumulated. By assuming rates of crustal spreading, it has been calculated that each Snowball Earth cycle lasted about 10 million years.

We begin with the formation and rifting of a supercontinent. Polar wandering curves yield different results for the positions of the continents during the Middle to Late Proterozoic. Nevertheless, it is widely accepted that the supercontinent Rodinia probably formed about 1.3 to 1 billion years ago (Figure 9.4). At about 750 to 800 million years ago it is thought that Rodinia rifted apart, sending some continents into lower latitudes near the equator. The volcanism associated with rifting would have increased atmospheric CO_2 levels, thereby promoting continental weathering that would have eventually drawn down CO_2, cooling Earth. Widespread continental shelves also might have acted as positive feedback on photosynthesis by marine plankton; epeiric seas would have provided an extensive photic zone and been subject to nutrient runoff derived from continental weathering. Increased photosynthesis and carbon burial account for the high carbon isotope ratios that occurred in association with glaciation (Figure 9.13). Because the sun's brightness was still somewhat lower than today's, Earth might have still been sufficiently susceptible to the extraction of CO_2 from the atmosphere. Thus, it is thought that at times Earth became so cold that the surfaces of the oceans froze. According to the recent scenario, the freezing over of the ocean's surface then shut down photosynthesis, and carbon isotope ratios plummeted.

This portion of the hypothesis has been tested by using climate models. The model successfully reproduced the glaciation at about 750 million years ago. As predicted, continental breakup of the simulated supercontinent led to an increase in CO_2 and runoff. The increased CO_2 in turn promoted increased weathering, feeding back negatively on CO_2 levels, which declined by about 1,300 parts per million (by comparison, modern CO_2 levels are now roughly 400 parts per million and rising; see Chapter 1).

Despite the massive glaciation, the recent Snowball Earth hypothesis also maintains that some volcanism continued. Continued volcanism would have allowed CO_2 to slowly build up in the atmosphere over millions of years to 300 to 400 times modern levels. This CO_2 was less likely to weather continental rocks because Earth was now covered in ice. Eventually, though, the gradual buildup of CO_2 in the atmosphere warmed the Earth to the point that sea ice rapidly melted, producing widespread diamictites. The large amount of CO_2 in the atmosphere then caused extensive continental weathering and widespread precipitation of limestones on top of the diamictites in shallow seaways. The warming appears to have been quite rapid, given the abrupt change from diamictites to overlying limestones. For such great thicknesses of shallow-water limestones to form, the underlying crust must have subsided as the carbonates accumulated. Seafloor subsides as it cools and moves away from spreading centers (see Chapter 2). By assuming approximate rates of crustal spreading, each complete Snowball Earth cycle (from glaciation to warming and back to glaciation) has been calculated to have lasted about 10 million years.

As you might suspect, the most recent hypothesis for a Snowball Earth has been criticized. One criticism is that for the diamictites to be so thick, it seems the glaciers would need to have grown for a long, long time near the equator. But the hypothesis indicates the glaciers advanced and retreated rapidly. Furthermore, the sea ice of Snowball Earth is estimated to have been as much as a half-mile thick. Such a great thickness of ice would have slowed the hydrologic cycle by restricting (negative feedback) the transfer of water vapor from the oceans to the atmosphere and then back to ice by precipitation. This would, seemingly, have prevented the growth of ice sheets.

For these reasons some workers prefer a milder version of the Snowball Earth hypothesis, called the **Slushball Earth**. Much thinner sea ice (like that of the Arctic today) is envisioned to have existed on Slushball Earth. Such ice sheets would have been subject to breakup. Breaks in the ice would have allowed the exchange of water vapor between the oceans and atmosphere, permitting the continuation of the hydrologic cycle and the growth of ice. Recent climate models indicate that just such a Slushball Earth could have formed, possibly with ice reaching no further toward the equator than 45 degrees latitude.

The most recent Snowball Earth hypothesis also implies that carbon isotope ratios remained low during glaciation because photosynthesis shut down. By contrast, a Slushball Earth would have allowed photosynthesis—and thus a greater exchange of CO_2—between the oceans and atmosphere. Uptake of CO_2 by photosynthesis would have potentially decreased the buildup of CO_2 in the atmosphere so that CO_2 levels would have only increased to four to five times modern levels. Lowered CO_2 levels would have increased the possibility of the repeated advance and retreat of the ice sheets because Earth would not have been so warm as to prevent the formation of ice. As it turns out, the carbon isotope ratios for rocks deposited during the glaciation are relatively normal, indicating that at least some photosynthesis would have occurred during glaciation.

9.5.3 How could a Snowball Earth have occurred in the first place?

Why did a Snowball (or Slushball) Earth occur in the first place, and why hasn't it recurred? Perhaps the simplest explanation is that the lower brightness of the sun during the Neoproterozoic made Earth more susceptible to changes in CO_2.

The most recent scenario for a Snowball Earth also implicates tectonics, continent and continental shelf formation, and photosynthesis and carbon burial in the occurrence of Snowball Earth (compare the discussion of "normal" cause and effect in Chapter 1). As Earth cooled in the Archean, incipient continents began to form. These protocontinents were probably unable to affect Earth's climate much before the Proterozoic because there would have been too little rock available for chemical weathering. But, as the continents increased in size during the Proterozoic, as described above, the amount of continental crust exposed to the atmosphere and available for weathering and carbon burial might have eventually surpassed some minimum size necessary to promote rock weathering and CO_2 levels in the atmosphere. The concentration of continents in the warm, humid tropics during the Neoproterozoic would have also promoted chemical weathering while releasing nutrients into seaways and accentuating photosynthesis, all of which would have drawn down atmospheric CO_2.

Remember from Chapter 1 that open systems are presumably in equilibrium ("steady state") with one another and therefore presumably exhibit no net change, *at least on human time scales*. Such systems are relatively well adjusted to one another and interact through negative feedback to damp severe disturbances and keep the system from going out of control. However, the components of open systems that *are just beginning to interact with one another* might exhibit **nonlinear behavior**, meaning that relatively small changes in a system produce an effect that is out of all proportion to the change (**Figure 9.17**). Nonlinear behavior can occur in the system as **undershoots** below or **overshoots** above the equilibrium state.

What, then, about the interaction of the lithosphere (continents), atmosphere, hydrosphere (oceans), and biosphere (life) during the Neoproterozoic? Perhaps as the reservoirs of carbon associated with continental weathering, photosynthesis, and carbon burial all increased, they finally reached sufficient size during the Neoproterozoic to begin to significantly interact with, and affect, one another. In the case of the Precambrian Earth, extreme undershoots and overshoots might have been more likely because the systems had not yet had enough time to establish an equilibrium. In this case, undershoots would have consisted of plummeting temperatures (Snowball Earth), whereas the overshoots consisted of volcanism, CO_2 production, and warming. Given that the rock weathering cycle takes millions of years or longer to produce a noticeable effect on CO_2, undershoots and overshoots would have been prolonged, as is indicated by the geologic record.

Also, one main component of the modern global carbon cycle was missing during the Precambrian: calcareous

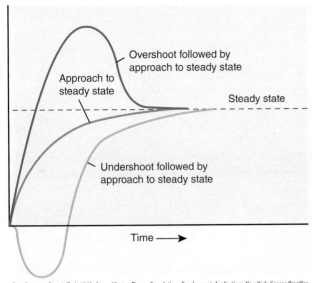

Data from: von Bertainffy, L. 1968. *General System Theory: Foundations, Development, Applications*. New York: George Braziller. Figure 6.2 (p.143).

FIGURE 9.17 The behavior of an idealized open system can vary as it approaches equilibrium ("steady state").

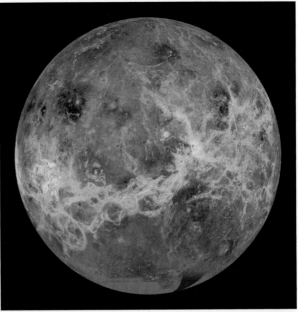

Courtesy of NASA/JPL.

FIGURE 9.18 Global view of Venus centered at 180 degrees east longitude. Most of Venus is covered by smooth plains pockmarked by more than 150 massive volcanoes that are more than 100 km in diameter. The only volcanic chain on Earth of this size is found on the island of Hawaii (the "Big Island") of the Hawaiian chain. Two large "continental" masses occur several kilometers above the average elevation of Venus. Ishtar Terra occurs high in the northern polar region and is about the size of Australia. Ishtar Terra contains the highest mountain on Venus, Maxwell Montes, which is about 11 kilometers high, two kilometers higher than Mount Everest. Aphrodite Terra (the area of light-colored irregular, branching lines) occurs in the equatorial region and is about the size of South America. As can be seen, Aphrodite Terra is crisscrossed by numerous faults. The scattered dark patches surround some of the younger impact craters. Composited from *Magellan* and *Venus Orbiter* data. Colors based on images from Soviet *Venera 13* and *14* spacecraft.

plankton. Calcareous oozes link the weathering of continents to the deposition of calcareous oozes in the deep sea because atmospheric CO_2 is locked up in $CaCO_3$ (see Chapters 1 and 3). The deep-sea oozes are eventually subducted, releasing the CO_2 back into the atmosphere through volcanism. The release of CO_2 by such a mechanism after its drastic drawdown might have otherwise spared Earth from such extreme freeze–thaw conditions. But in the Precambrian this crucial link in the geologic carbon cycle had not yet evolved and would not appear for at least several hundred million more years.

9.5.4 Why were there no Snowball Venuses?

The main problem in deciphering Earth's Precambrian history is that much of Earth has been resurfaced since that time. Enough time has passed for many Precambrian mountain chains to have been uplifted and eroded and much ocean crust to have been subducted back into Earth's mantle.

Neighboring planets provide us with an inkling of Neoproterozoic conditions on Earth because plate tectonics has largely if not entirely ceased on these bodies. For example, tantalizing surface features on Venus have been revealed by recent *Magellan* space probes. Most notable with regard to the Snowball Earth hypothesis is the concentration of continent-like structures in an Equatorial band at Venus's surface (**Figure 9.18** and **Figure 9.19**). But if these features are indeed ancient continents, why didn't they disperse like those on Earth? Perhaps because plate tectonics ceased on Venus. Although much of Venus is covered by huge plains of basaltic lava, plate tectonics does not appear to be occurring on Venus because volcanoes are not concentrated in chains and island arcs as they are on Earth (see Figure 9.19). The

cessation of plate tectonics would have prevented some of the critical steps, such as long intervals of volcanism, in the Snowball Earth cycle to occur.

Several lines of evidence suggest that the tectonic and climatic histories of Earth and Venus began to diverge about 800 million years ago, about when the first Snowball Earth phase began. Based on the number of asteroids present in the inner solar system and crater counts for the Moon, it has been calculated that Venus should, on average, have been impacted 1.2 times every million years. The *Magellan* satellite photographed 963 craters spread randomly over the Venusian surface (see Figure 9.19). Thus, the craters now seen on Venus appear to extend back in time to about 800 million years before present (963 craters/1.2 craters per million years). On Earth most craters have been eradicated by erosion or are buried. However, on Venus these craters appear fresh. In contrast to Earth, Venus is far too hot for liquid water to exist, so little chemical weathering takes place (see Chapter 7). The Venusian atmosphere is also quite thick, which causes most asteroids less than a kilometer in

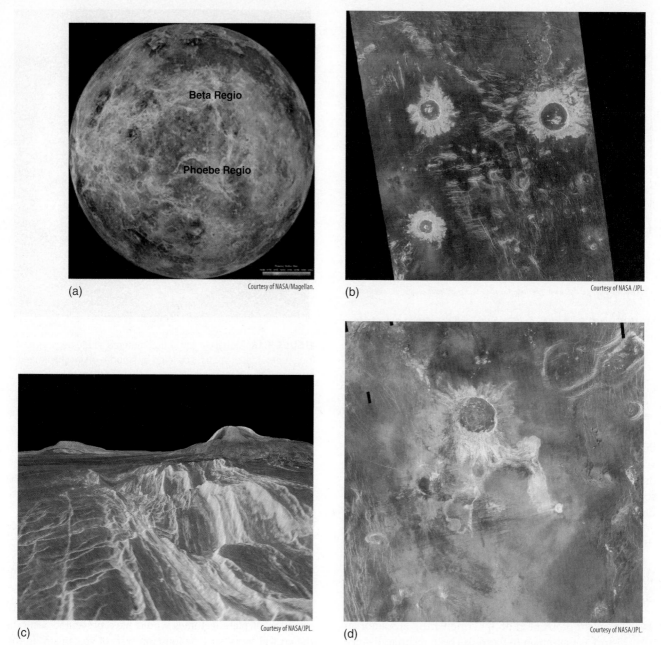

(a)

Beta Regio

Phoebe Regio

Courtesy of NASA/Magellan.

(b)

Courtesy of NASA /JPL.

(c)

Courtesy of NASA/JPL.

(d)

Courtesy of NASA/JPL.

FIGURE 9.19 Surface features of Venus. **(a)** Color-coded map of Venus's surface based on images of *Magellan, Pioneer*, and *Venera* spacecraft and the Earth-based Arecibo telescope. View is centered on 270 degrees east longitude. The smooth plains of Venus are colored blue. Ishtar Terra is seen at the top and the eastern edge of Aphrodite Terra at far left along the equator. The highland regions Beta Regio and Phoebe Regio are also evident. **(b)** Impact craters on Venus. Three large impact craters, with diameters that range from 37 to 50 kilometers (23 to 31 miles), are seen in the Lavinia region, which covers an area 550 kilometers (342 miles) wide by about 500 kilometers (311 miles) long. The craters show many features typical of meteorite impact craters, including rough (bright) material around the rim, terraced inner walls, and central peaks. Numerous domes, probably caused by volcanic activity, are seen in the southeastern corner of the mosaic. The domes range in diameter from 1 to 12 kilometers (0.6 to 7 miles). Some of the domes have central pits that are typical of some types of volcanoes. **(c)** Rift valley on Venus. The volcanoes Sif Mons (left) and Gula Mons (right) are seen in the background. Recall that the East African Rift Valley on Earth is associated with volcanoes such as Mount Kilimanjaro (Chapter 2). **(d)** Relative ages of Venusian terranes can be inferred by a form of "cross-cutting relationships." Crater Isabella, with a diameter of 175 kilometers (108 miles), is the second largest impact crater on Venus. Two large flow-like structures extend to the south and to the southeast. The end of the southern flow partially surrounds a preexisting 40-kilometer (25 mile) circular volcanic area. The southeastern flow shows a complex pattern of channels and flow lobes and is overlain at its southeastern tip by deposits from a later 20-kilometer (12 mile) diameter impact crater. The extensive flows are unique to Venusian impact craters and may represent "impact melt," or rock melted by the intense heat released in the impact explosion. An alternate hypothesis invokes "debris flows," which consist of clouds of hot gases and melted and solid rock fragments that race across the landscape during the impact event. The latter process resembles that which sometimes occurs during violent eruptions on Earth, such as the 1991 Mount Pinatubo eruption in the Philippines.

diameter to burn up before reaching the surface. Otherwise, asteroids of this size would produce craters up to 15 km across! Still, the *Magellan* spacecraft mission about 20 years ago indicated fewer large craters on Venus than predicted.

The question of whether Venus has been geologically active over the past billion years has major implications for the interior dynamics and climate change not only of Venus but also Earth and other planets as they evolve and age. The "catastrophic resurfacing" hypothesis attempts to explain the fresher-than-expected appearance of the Venusian surface by proposing a global pulse of volcanism and lava production from about 1 billion to 300 million years ago. However, if this occurred, a rather complicated sequence of feedbacks might have ensued that led to the eventual cessation of plate tectonics. A global outpouring of lava should have pumped enormous amounts of CO_2 and water vapor into the Venusian atmosphere, warming the planet and allowing the atmosphere to concentrate 10 to 100 times more water vapor than before. Based on computer models, a rapid injection of such large amounts of water vapor would have eventually cooled the planet (via negative feedback) by the formation of clouds, thereby increasing the planet's albedo. But, the H_2O would probably also have begun to dissociate via photolysis. If so, negative feedback on the clouds would have then ensued; as the clouds evaporated Venus's surface would have warmed, causing the clouds to evaporate further and causing Venus to become hot and dry. This should have stopped all plate subduction, which requires liquid water (see Chapter 2) and associated volcanism.

More recent (2010) measurements by the European Space Agency's Venus Express spacecraft suggest that up to nine areas of volcanism and uplift still exist on Venus. The spacecraft detected infrared radiation, which indicates thermal (heat) loss (see Chapter 7) from the planet's surface. These areas resemble the hotspots of Hawaii, which are presumably associated with mantle plumes (see Chapter 2).

Based on the measured heat loss, these lavas are estimated to have formed as recently as 2.5 million years ago and might be as young as 250,000 years or less, indicating Venus is still actively resurfacing, with the lavas being fed by mantle plumes rather than by subducting plates. Thus, accounts of Venus's tectonic death (like those of Mark Twain when he was still alive) may be exaggerated.

CONCEPT AND REASONING CHECKS

1. How are positive and negative feedback presumably involved in the most recent Snowball Earth hypothesis?
2. How was runaway positive feedback involved in the climate history of Venus?

9.5.5 Why didn't Snowball Earths recur after the Precambrian?

 Continental glaciers recurred again on Earth during the Phanerozoic (such as those on Antarctica) but never to

the extent as during the Proterozoic and never at low latitudes. Why?

Perhaps there are several reasons. First, the sun's brightness has increased since the Neoproterozoic. Second, plate tectonics has definitely continued on Earth because of abundant liquid water. Thus, third, rather than being concentrated in the tropics during the Phanerozoic, some continents moved to higher latitudes, if not over the poles, increasing the chances of glaciation there. Also, continental weathering and CO_2 drawdown are not as extensive at high latitudes as in low ones because of the lower temperatures (which slow the chemical reactions of weathering). Fourth, the presence of liquid water allowed Earth to develop the biogeochemical cycles of carbon and other elements involved in the feedbacks that regulate Earth's climate. The fifth reason goes back to the earlier discussion of nonlinear interactions of Earth's systems during the Proterozoic. As the biosphere expanded during the Proterozoic, it might have not only reached a size at which it could influence the other Earth systems, but it might have also started to feed back on itself. Recall that nutrient inputs from land occur relatively slowly, that biotas are typically nutrient-limited, and that biotas therefore depend on the recycling of nutrients from dead organic matter to maintain ecosystem functions (see Chapter 3). With the continued expansion of the biosphere, biomass might have reached the point near the end of the Proterozoic at which most nutrients were sequestered in biomass. As a result, the remineralization of dead organic matter and the recycling of nutrients might have become critical to the continued expansion and maintenance of the biosphere during the Phanerozoic because of nutrient limitation.

In fact, it was during the Neoproterozoic that definite body fossils and trace fossils such as burrows and tracks of multicellular animals first appeared in marine sediments (see Chapter 10). Based on the fossil record, as they evolved multicellular animals increasingly burrowed into the sediments in which they lived. Without burrowing and the breakdown of dead organic matter, the organic carbon would have otherwise remained buried, drawing down CO_2 levels and further promoting potential Snowball Earths. Instead, CO_2 was more likely to be returned to the atmosphere by the remineralization of dead organic matter. If this, too, seems far-fetched, Charles Darwin conducted a simple experiment from which he calculated the effects of the lowly earthworm on soil formation. He determined there were up to about 50,000 worms per acre and they could transport roughly 18 tons of sediment per acre per year!

CONCEPT AND REASONING CHECKS

1. How are directionality and contingency involved in the lack of Snowball Earths after the Precambrian?
2. How might life have acted as a geologic force to help inhibit the occurrence of further Snowball Earths?

SUMMARY

- By about 2.5 billion years ago Earth entered the third main phase of its evolution, the Proterozoic Eon. During this 2-billion-year-long eon, continents became progressively larger, the atmosphere more oxygenated, and life became a geologic force.

- Continents grew to near their present size by the accretion of orogens around preexisting Archean cratons in mobile belts. Rocks became more modern in aspect: ophiolites and rift basins (aulacogens) appeared, and the rocks themselves became less mafic and more granitic in composition. As Earth cooled and volcanism declined, continental shelves became more extensive, as indicated by the presence of quartz sandstones, shales, and limestones. Rift basins invaded by the oceans accumulated evaporites such as gypsum and anhydrite.

- Some of the most significant of all sedimentary deposits were stromatolites, which lived in shallow sunlit waters of continental shelves. Stromatolites are cyanobacteria that photosynthesize and release O_2 as a byproduct.

- Seasonal upwelling brought dissolved iron to the surface, where the iron combined with available O_2 to precipitate varve-like BIFs, which formed on the margins of continents.

- When the iron was exhausted, oxygen began to accumulate in Earth's atmosphere, and the ozone layer, which protects Earth from deadly ultraviolet light, began to form. Both the presence of oxygen and the protective ozone layer resulted in the evolution of more complex life.

- As the biosphere expanded, it began to interact more strongly with Earth's other systems. One possible result was that prolonged runaway icehouses called Snowball Earths might have covered Earth during the Neoproterozoic. Each Snowball Earth was terminated by a pronounced warming. Because the continents appear to have been concentrated in equatorial regions, they were probably subject to extensive chemical weathering that appears to have exceeded climate feedback mechanisms. And because the sun's intensity was no more than about 95% that of today's, Earth may have been especially susceptible to CO_2 drawdown and cooling.

- After the Neoproterozoic, many of Earth's continents moved to higher latitudes, where extreme rates of chemical weathering like that in the tropics and drastic drawdown of atmospheric CO_2 do not occur. At the same time, the continued presence of H_2O in Earth's atmosphere and oceans permitted the development of biogeochemical cycles and feedbacks that helped stabilize Earth's climate.

- Possible proto-continents on Venus might have also concentrated in equatorial regions at about 800 million years ago but moved no further because of the shutdown of plate tectonics. Possibly, the absence of water on Venus doomed the planet to a hot, dry existence in which glaciers and biogeochemical cycles like those on Earth never have a chance to form.

KEY TERMS

Banded Iron Formations (BIFs)

cap limestones

continental accretion

diamictites

Grenville Orogeny

Keweenawan Rift

Mazatzal-Pecos Orogens

methanogenic

Mid-Continent Rift

mobile (orogenic) belts

Neoproterozoic

nonlinear behavior

overshoots

ozone

Proterozoic Eon

redbeds

Rodinia

Slushball Earth

Snowball Earth

stable isotopes

SWEAT hypothesis

Transantarctic Mountains

Trans-Hudson Orogen

true polar wander

undershoots

uraninite

Wopmay Orogen

Yavapai

REVIEW QUESTIONS

1. What is the evidence from the rocks for the appearance of modern plate tectonics?
2. What do ophiolites tell us about the tectonic evolution of Earth?
3. What do sedimentary rocks tell us about the appearance of continental shelves and the oxygenation of the atmosphere?
4. Was there an ozone layer present during Earth's earliest history? Why or why not?
5. What were the different potential sources of O_2 in Earth's atmosphere during the Precambrian?
6. Earth's modern atmospheric CO_2 levels are far lower than those of Venus (see Chapter 7). Where did all the CO_2 go?
7. Why were there apparently no Snowball Venuses?
8. Discuss how photosynthesis and carbon production and decay affect the oxygenation of the oceans.
9. What are some possible reasons for the lack of Snowball Earth episodes during the Phanerozoic?

FOOD FOR THOUGHT:
Further Activities In and Outside of Class

1. What does the near absence of ultramafic rocks in Proterozoic greenstone belts suggest? Explain.
2. Comment on cause and effect (see Chapter 1) with regard to the oxygenation of Earth's atmosphere and ocean. Under what scientific conditions is one most likely to pinpoint a single cause for a phenomenon? Why?
3. The same chemical phenomena that occurred in the Proterozoic also occurred in the Archean, but the results were vastly different. Why?
4. What was your initial reaction to the Snowball Earth scenario? What do you suppose most scientists' reactions were? What are the strengths and weaknesses of the hypothesis?
5. Are the multiple hypotheses for Snowball Earth mutually exclusive? Explain. Do you believe this is typical of science (see Chapter 1)?
6. How did life become a geologic force in the Proterozoic?

SOURCES AND FURTHER READING

Allégre, C. J., and S. H. Schneider. 1994. The evolution of the Earth. *Scientific American*, 271(10), 66–75.

Bullock, M. A., and Grinspoon, D. H. 1999. Global climate change on Venus. *Scientific American*, 280(3), 50–57.

Catling, D. C., Zahnle, K. J., and McKay, C. P. 2001. Biogenic methane, hydrogen escape, and the irreversible oxidation of the atmosphere. *Science*, 293(5531), 839–843.

Donnadieu, Y., et al. 2004. A "snowball earth" climate triggered by continental break-up through changes in runoff. *Nature*, 428, 303–306.

Hawkesworth, C. J., Dhuime, B., Pietranik, A. B., Cawood, P. A., Kemp, A. I. S., and Storey, C. D. 2010. The generation and evolution of the continental crust. *Journal of the Geological Society, London*, 167, 229–248.

Hoffman, P. F., Kaufman, A. J., and Halverson, G. 1998. Comings and goings of global glaciations on a Neoproterozoic tropical platform in Namibia. *GSA Today*, 8(5), 1–9.

Hoffman, P. F., Kaufman, A. J., Halverson, G. P., and Schrag, D. 1998. A Neoproterozoic Snowball Earth. *Science*, 281, 1342–1346.

Hoffman, P. F., and Schrag, D. 2000. Snowball Earth. *Scientific American*, 282(1), 68–75.

Kasting, J. F. 1993. Earth's early atmosphere. *Science*, 259, 920–926.

Kasting, J. F. 2001. The rise of atmospheric oxygen. *Science*, 293(5531), 819–820.

Kasting, J. F., and Siefert, J. L. 2001. The nitrogen fix. *Nature*, 412, 26–27.

Kasting, J. F., and Siefert, J. L. 2002. Life and the evolution of Earth's atmosphere. *Science*, 296, 1066–1068.

Kerr, R. A. 2003. Iceball Mars? *Science*, 300, 234–236.

Kirschvink, J. L., Ripperdan, R. L., and Evans, D. A. 1997. Evidence for a large-scale reorganization of Early Cambrian continental masses by inertial interchange true polar wander. *Science*, 277, 541–545.

Knoll, A. H. 2003. *Life on a young planet: The first three billion years of evolution on Earth*. Princeton, NJ: Princeton University Press.

Martin, R. E., Quigg, A., and Podkovyrov, V. 2008. Marine biodiversification in response to evolving phytoplankton stoichiometry. *Palaeogeography, Palaeoclimatology, Palaeoecology*, 258, 277–291.

Monastersky, R. 1998. Popsicle planet: the king of all ice ages may have spurred animal evolution. *Science News*, 154, 137–139.

Murakami, M., Hirose, K., Yurimoto, H., Nakashima, S., and Takafuji, N. 2002. Water in the Earth's lower mantle. *Science*, 295, 1885–1887.

Navarro-González, R., McKay, C. P., and Mvondo, D. N. 2001. A possible nitrogen crisis for Archaean life due to reduced nitrogen fixation by lightning. *Nature*, 412, 61–64.

Planavsky, N. J., Rouxel, O. J., Bekker, A., Lalonde, S. V., Konhauser, K. O., Reinhard, C. T., and Lyons, T. W. 2010. The evolution of the marine phosphate reservoir. *Nature*, doi:10.1038/nature09485.

Poulsen, C. J. 2003. Absence of a runaway ice-albedo feedback in the Neoproterozoic. *Geology*, 31, 473–476.

Ridgwell, A. J., Kennedy, M. J., and Caldeira, K. 2003. Carbonate deposition, climate stability, and Neoproterozoic ice ages. *Science*, 302, 859–862.

Smrekar, S. E., et al. 2010. Recent hotspot volcanism on Venus from VIRTIS emissivity data. *Science*, 328, 605–608.

Taylor, S. R., and McLennan, S. M. 1996. The evolution of the continental crust. *Scientific American*, 279(1), 76–81.

Windley, B. F. 1990. *The evolving continents*. New York: John Wiley & Sons.

Life's "Big Bang": The Origins and Early Diversification of Multicellular Animals

Major Concepts and Questions Addressed in This Chapter

A What are the earliest known multicellular animals known from the fossil record?

B What sorts of creatures were these earliest known animals, and what was their possible relationship to later animals?

C What other sorts of fossils or traces suggest the appearance of multicellular animals?

D What other sorts of animals began to appear?

E What is the Burgess Shale Fauna, how did it form, and what is its significance?

F What do these early multicellular creatures tell us about rates of evolutionary processes?

G What sorts of processes were involved in the evolution of multicellular creatures?

H How did Earth's earliest biota and the physical environment interact?

I Why were there no more "Big Bangs" of multicellular animals in Earth's history?

J How do molecular clocks compare with the fossil record?

An *Anomalocaris* ("anomalo" or odd, "caris" for shrimp; hence, "odd shrimp"), reconstructed from fossils found in the Burgess Shale. Some species of *Anomalocaris* are thought to have been predatory while others grubbed around in sediment for food like some species of trilobites. The photograph on the right shows an antenna, whereas the photograph on the left is that of a mouth. These structures became disarticulated from the body after death and were preserved as carbon films.

10.1 Enigma of Multicellular Organisms

This chapter reviews the evidence for the origins and early diversification of multicellular animals. As such, it lays the groundwork for the discussion of the major taxa of fossils to be covered in succeeding chapters because the majority of animal phyla that are found today originated shortly before or during the Cambrian period or shortly thereafter.

The origins of multicellular animals, ranks right up there with some of the most remarkable historical milestones in the evolution of Earth's systems discussed in previous chapters: the origin of the Earth and its Moon; the formation of the continents and oceans; the origin and oxygenation of the atmosphere, and the origins of life.

The transition between the Proterozoic and Phanerozoic eons represents a remarkable shift in Earth's biota. The microbial world of the Precambrian was dominated by bacteria (including cyanobacteria) and protists (unicellular eukaryotes). Molecular clocks indicate that eukaryotic cells arose no later than about 2 billion years ago and perhaps much earlier (see Chapters 8 and 9). Organic films appearing by 1 to 1.5 billion years ago have been interpreted as eukaryotic multicellular seaweeds (plants). Still, it took roughly another 0.5 to 1 billion years for the traces of the first definite multicellular animals to appear. Many of the major phyla of organisms present today originated sometime during the transition from the late Proterozoic to the Phanerozoic Eon. The Phanerozoic is characterized by the further diversification of multicellular animals, or **metazoans**, and

higher plants (although bacteria and protists were of course still present). Abundant fossil hard parts eventually began to be preserved in the fossil record, giving the Phanerozoic Eon its name: "time of apparent life." Thus, before we begin discussion of Earth's diverse fossil record in succeeding chapters, we must confront two fundamental questions: Why did it take so long for multicellular animals to appear, and what factor or factors finally triggered their appearance?

The appearance of metazoans is puzzling enough, but many of these creatures also leave us scratching our heads as to what they really were. Were many taxa evolutionary "dead ends" that died out, never to appear again? How did these creatures evolve? Did these taxa pave the way for later biologic evolution? What were the effects of Earth's physical systems on the evolution of these new biotas and biogeochemical cycles? And what was the effect of the new biotas on Earth's physical systems? As we will see, the answers to these questions may well involve the continuing evolution and integration of Earth's systems that began earlier during the Proterozoic (see Chapter 9).

10.2 Stages of Life's Big Bang

Hard parts derived from metazoans first become diverse and abundant during the Cambrian Period, the first major interval of the Phanerozoic. The initial appearance of diverse and abundant fossils has often been coined the "**Cambrian explosion**." Like the origin of the universe, though, the Cambrian explosion occurred in stages. These stages were preceded by recognizable developments in the fossil record spread out over roughly 50 to 70 million years (**Figure 10.1**).

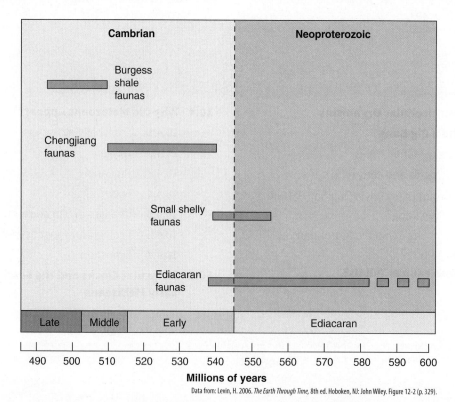

Data from: Levin, H. 2006. *The Earth Through Time*, 8th ed. Hoboken, NJ: John Wiley. Figure 12-2 (p. 329).

FIGURE 10.1 Sequence of appearance of metazoan faunas and major environmental changes during the late Neoproterozoic and Cambrian. The latest Neoproterozoic has been named the Ediacaran after the unusual fauna of the same name (see text for further discussion).

10.2.1 Ediacara fauna

Ⓐ The Ediacara fauna (**Figure 10.2** and **Figure 10.3**) was discovered by R. C. Sprigg in the Pound Quartzite, a fine-grained, hard quartz sandstone found in the Ediacara Hills of South Australia. The Ediacara fauna is but one example of a fossil Lagerstätte, in which soft-bodied taxa that are normally destroyed are preserved (see Chapter 4).

Unfortunately, Sprigg described the Ediacara fauna in a journal that did not receive wide circulation, and his discovery went largely unnoticed. Sprigg believed the Ediacarans occurred above the base of the Cambrian, from which other fossils were already long known, so the Ediacarans did not attract much attention for this reason, either. In fact, for many decades the Ediacara fauna was widely treated as a geologic curiosity more than anything else. It was only much later, after similar faunas had been found on different continents, that it was realized Ediacaran fossils represented a global fauna that predated other known fossils. Ediacarans are now known from rocks of the same age in Siberia, China, Newfoundland, Namibia (West Africa), and California.

Because of their enigmatic nature the evolutionary affinity of many Ediacarans to other metazoans remains a mystery. Some Ediacarans resemble jellyfish (see Figure 10.2d), which belong to the **Phylum Cnidaria**, which also includes corals (see Chapter 5 for biologic classification). Other Ediacarans resemble medusae and sea pens, which are also cnidarians (**Figure 10.4**). The occurrence of cnidarians early in the fossil record comes as no surprise; cnidarians are among the simplest, and therefore relatively primitive, modern metazoans, so they would be expected to appear early in metazoan evolution.

But other Ediacarans are much more problematic. *Dickinsonia* is certainly one of the most unusual (Figure 10.2a). One well-known biologist reported that upon asking seven paleontologists what they considered *Dickinsonia* to be, he

(a) © Ken Lucas/Visuals Unlimited, Inc.

(b) © Ken Lucas/Visuals Unlimited, Inc.

(c) © Sinclair Stammers/Science Source.

(d) © Ken Lucas/Visuals Unlimited, Inc.

FIGURE 10.2 Members of the Ediacaran fauna and other metazoans. The Ediacarans are preserved as impressions in the sand and resemble jellyfish, worms, and sea pens. **(a)** *Dickinsonia,* a flat, headless organism which ranged up to 4 to 5 feet in size. **(b)** *Spriggina* (named after R. C. Sprigg, who discovered the Ediacara fauna), which resembles a modern annelid (segmented) worm. Annelids include many marine worms and the familiar earthworm. *Spriggina* might also have been a frond-like creature with its "head" stuck in the sand. **(c)** *Charnia,* a sea-pen like form. **(d)** *Mawsonites,* a jellyfish-like form that might have actually been a sea anemone.

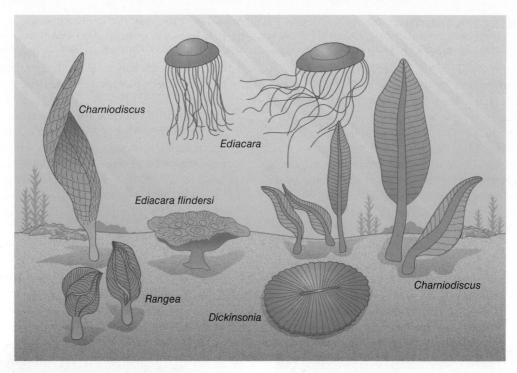

FIGURE 10.3 Reconstruction of the Ediacara fauna. Compare to some of the actual fossils in Figure 10.2.

received seven different answers, including a cnidarian, a flatworm (a group that includes the modern planarians), and a "fungal grade," perhaps like a mushroom! Although *Dickinsonia* ranged up to 1 meter in length, it was no more than 3 mm thick. It was obviously segmented and bilaterally symmetric, with left and right halves that are mirror images of each other. Its anterior (front) and posterior (tail) ends also differed from each other, although nothing like a head, with concentrations of nervous tissue, sense organs, or a "brain," was present.

B What is the relationship of Ediacarans to later metazoans? The simplest hypothesis is that Ediacarans were ancestral to all subsequent metazoans, which then diversified later during the Cambrian Period (**Figure 10.5**). But the Ediacarans just seem too dissimilar from later metazoans to be ancestral to them. Other workers believe the Ediacarans were a distinct offshoot that led only to the cnidarians. Others suggest that Ediacarans harbored algal symbionts, like many modern corals (see Chapter 4), and sunned themselves in a "Garden of Ediacara." In this scenario, the organisms were described as projecting up into the water column like modern sea fans or sea whips or laying on the bottom like "photosynthetic steaks."

Still other investigators have suggested that a part of the Ediacara fauna actually represented a completely different phylum, or perhaps even kingdom, of metazoans called the Vendobionta or **Vendozoa** (after the Vendian interval of the latest Neoproterozoic; Figure 10.5). According to this hypothesis Vendozoans became extinct and were replaced by later, more familiar, metazoans like those found today. In this view, Vendozoans might have been chemotrophic, making foodstuffs from simpler dissolved substances (see Chapter 3). Finally, some workers have suggested that the Ediacarans were lichens—creatures that consist of a symbiotic relationship between algae and fungi. Microscopic objects identified by some workers as phosphatized embryos in the earliest stages of cell division were recently discovered in China in rocks dated at about 570 million years old.

Perhaps even more surprising is the recent suggestion that Ediacarans such as *Dickinsonia* were not marine but *terrestrial* organisms! This conclusion is based on the interpretation of the entombing sediments as paleosols (ancient soils) and the Ediacaran fossils not as marine organisms but as soil lichens or colonies of soil microbes. The evidence cited for paleosols includes their reddish color, geochemical

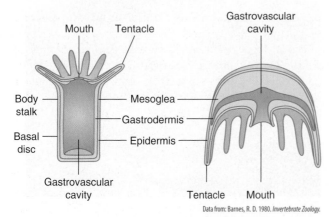

Data from: Barnes, R. D. 1980. *Invertebrate Zoology.*

FIGURE 10.4 Comparison of body structure of polyps ("anemones") and medusae ("jellyfish") belonging to the Phylum Cnidaria. Note how if the polyp is turned upside down, it resembles the medusa, and vice versa. Both forms have a mouth surrounded by stinging tentacles and an incomplete digestive tract.

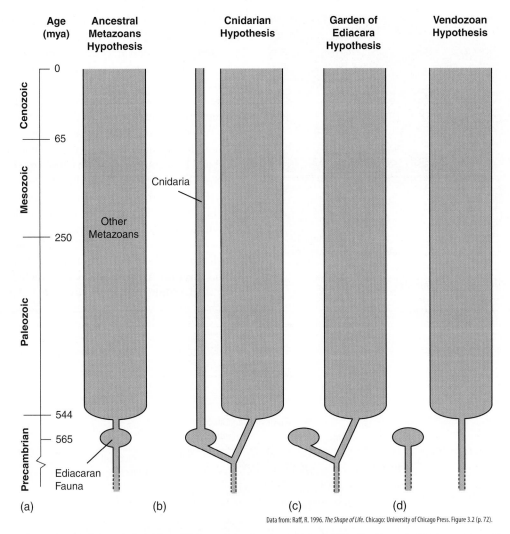

FIGURE 10.5 Four hypotheses of the relationships of Ediacarans to later metazoans. **(a)** In the "Ancestral Metazoan hypothesis," many Ediacarans are ancestral to many later metazoans. **(b)** In the "Cnidarian hypothesis," the Ediacarans are ancestral only to cnidarians (sea anemones, jellyfish, and corals), with other metazoans appearing later as distinct offshoots. **(c)** The "Garden of Ediacara hypothesis" suggests that Ediacarans lived in symbiotic associations with algae and eventually died out with the appearance of grazers and predators. **(d)** The "Vendozoan hypothesis" suggests that the Ediacarans were a distinct phylum or even kingdom, the Vendozoa or Vendobionta, that eventually became extinct and were replaced by more modern metazoans.

composition (including stable isotopes), and the presence of gypsum crystals and carbonate, suggesting caliché-type deposits involving evaporation of temporary pools of water, perhaps among lagoonal or aeolian (dune) deposits. However, the presumed soils would predate well-developed soil-forming terrestrial floras by tens if not hundreds of millions of years (see Chapters 11 and 12). Furthermore, one critic has suggested that the reddish soil color was formed much later by weathering during the Cenozoic and that Ediacaran fossils have been found in what have been concluded to be definite marine deposits. Could the fossils have been washed up onto shore by storms? Possibly, but if this were the case, we might expect the fossils to overlap one another after being washed up and then buried. Still, this interpretation is not out of the question. Also, the stable isotope signatures of the carbonate nodules are consistent with freshwater recharge into coastal aquifers that commonly extend far offshore beneath modern

continental shelves where marine organisms otherwise dominate. Even if this hypothesis turns out to be negated, it at least makes us think about terrestrial life forms long before their appearance in the fossil record. Nevertheless, despite the questions surrounding the evolutionary affinities of the Ediacarans, the presence of metazoans at this time is undisputed.

10.2.2 Trace fossils and early hard parts

Simple burrows and the tracks and trails of undisputed metazoans—many of them deposit feeders—also started to appear in the fossil record about the time of (or perhaps shortly before) the Ediacara fauna (**Figure 10.6**). These ichnofossils (see Chapter 4) are not associated with body fossils like those of the Ediacara that showed what the animals actually looked like. However, the ichnofossils are

Courtesy of G. Mángano.

FIGURE 10.6 Ichnofossils of the Early Cambrian.

still very important because they reveal how the animals that made them moved and behaved.

D The earliest tracks and trails are thought to have been made by a bilaterally symmetric animal. This animal was most likely some kind of simple worm, perhaps resembling a flatworm like a planarian (**Figure 10.7**). The earliest traces were formed at the surface of the sediment, suggesting the burrowers, whatever they were, could not burrow deeply into sediment. Deeper burrowing, like that exhibited by members of the **Phylum Annelida** (segmented worms like those related to the modern earthworm; see Box 10.1), occurred by the Early Cambrian, after the Ediacara Fauna (see Figure 10.7). To burrow deeply requires a certain degree of body coordination controlled by a fairly sophisticated musculature and nervous system.

It is about the Early Cambrian, too, that the first traces of the **Phylum Arthropoda**, which includes insects, spiders, and **trilobites**, occur (**Box 10.1**; Figure 10.1 and

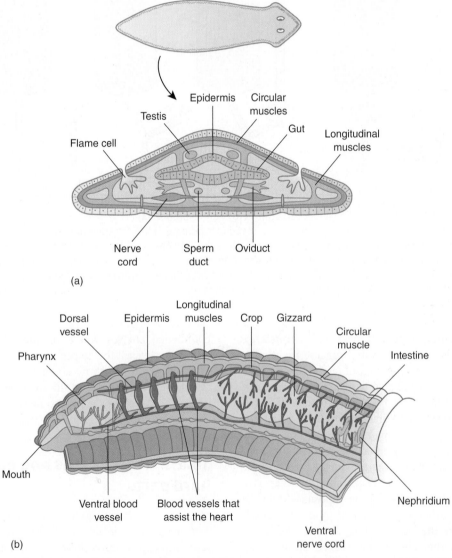

FIGURE 10.7 (a) Internal structure of a planarian. Like cnidarians, planarians have an incomplete gut, in which a single opening serves as both mouth and anus. Planarians move by gliding over the substrate on cilia and a mucus film which they secrete. **(b)** Internal structure of an earthworm (Phylum Annelida, or segmented worms). Note the much more complex musculature and nervous system of the earthworm. The musculature and nervous system are used in deeper burrowing, whereas the flatworm forms shallow burrows at the sediment surface.

BOX 10.1 Important Fossil Phyla Originating During Life's "Big Bang"

Important new taxa or their ancestral forms originating during the late Proterozoic-Cambrian transition are briefly described here in approximate order of increasing complexity, along with references to figures elsewhere in this chapter.

Kingdom Protista. Comprises unicellular algae and amoeba-like forms (Figure 10.22). Includes major benthic and planktonic taxa to be discussed in subsequent chapters.

Phylum Porifera Sponges or "pore-bearing" animals. The most primitive of multicellular animals, these metazoans lack true tissues or organs (Figure 10.12). Pores are connected to branching canals and chambers lined by specialized flagellated cells (choanocytes). The body is radial to irregular, cylindrical to globose. Skeleton is internal and consists of specialized elements called spicules, which consist of calcium carbonate, silica, or an organic compound called spongin. Porifera can form reefs or reef-like structures, especially the sclerosponges, which were the dominant frame-builders of reefs (see Chapter 4) during the first half of the Paleozoic Era (Chapter 11).

Phylum Cnidaria Includes "Jellyfish," sea anemones, corals, sea fans, and sea whips. Possibly some of these forms are represented by the Ediacara fauna (see Figure 10.4). Radially symmetric animals that possess tissues but no true organs. Can exist as solitary or colonial sessile (attached) polyps or as bell or umbrella-like free-floating medusae with the mouth surrounded by tentacles possessing stinging cells called cnidocysts. Digestive system is incomplete (mouth also serves as the anus). Corals belong to the Class Anthozoa. Paleozoic corals typically did not produce reef frames but new taxa of corals (Scleractinia) became dominant reef builders in the Triassic Period (Chapter 13).

Phylum Brachiopoda "Lamp shells," so-called because certain species resemble the oil lamps of the "Arabian nights" tales (Figure 10.13). Clam-like in appearance, consisting of two shells or valves typically composed of calcium carbonate, typically articulated by tooth-and-socket structures along a hinge or beak-like area (Articulates) but which is lacking in earlier primitive forms (Inarticulates). The shell in both groups is anchored to the substrate by a fleshy stalk called a pedicle. Like bryozoans (see below), they possess a lophophore and are therefore often lumped with bryozoans as "lophophorates." Like bryozoans, brachiopods were often prominent components of Paleozoic limestones (Chapter 11) but declined in diversity and abundance in the Modern Fauna. Marine only. Includes the living fossil *Lingula*.

Phylum Bryozoa These are the so-called "moss animals," which consist of stick to fan-like, to encrusting colonies of tiny polyps. Like the brachiopods, the mouth is surrounded by a lophophore, so both phyla are sometimes grouped informally under the "lophophorates." Although not prominent in the Cambrian, trepostome or "stony" bryozoans, which ranged from stick-like to encrusting forms, they became sufficiently diverse and abundant to become important rock formers in the Ordovician. These were followed by the fan-like fenestrate bryozans later in the Paleozoic. Both taxa died out by the end of the Paleozoic and were replaced by the cheilostome bryozoans in the Modern Fauna during the Mesozoic Era and are still prominent today. All three taxa will be discussed in greater detail in Chapters 11, 12, and 13.

Phylum Mollusca Comprises clams (Class Bivalvia), snails (Class Gastropoda), octopuses, squid, and modern pearly *Nautilus* (Class Cephalopoda; Figure 10.11). Bilaterally symmetric animals characterized by the possession of a muscular foot variously modified for burrowing (clams), crawling or attachment (gastropods), or a head (cephalopods, for "head foot"). This phylum is thought to have originated from an ancestral form resembling modern chitons, which possess a muscular foot used for attachment to rocky surfaces and protected by a series of overlapping calcareous plates. Although present during the Paleozoic, bivalves became especially diverse and abundant in the Modern Fauna because of the development of siphons, which allow the animals to burrow deep beneath the sediment surface by piping oxygenated water and food

(continues)

downward while removing wastes to the surface. One group of bivalves, the Rudists, became the dominant reef frame builders during the middle Cretaceous. Predatory gastropods also diversified greatly in the Modern Fauna during the Mesozoic Era. Straight-shelled and coiled cephalopods were prominent during the Paleozoic but gave way to the so-called ammonites of the Modern Fauna (see Chapter 5 and Chapter 13). Cephalopods declined after the Mesozoic as they reduced their shell (belemnites) in response to increasing predation by fish and other vertebrates.

Phylum Annelida Segmented worms typified by the earthworm but including diverse species of marine forms, many of which are benthic suspension feeders or are mobile predators (Figure 10.7). Also includes parasitic forms such as tapeworms and leeches but which have no fossil record.

Phylum Arthropoda By far the most diverse phylum of *invertebrate* (lacking backbones) animals. The "joint-footed" animals represented by crayfish, lobsters, crabs, shrimp, copepods, barnacles, centipedes, millipedes, spiders, mites, ticks, and, especially, insects. The body is covered by a chitinous exoskeleton and typically consists of a head, thorax, and abdomen, each consisting of separate segments or fused together. Extinct members include the trilobites (Subphylum Trilobitomorpha; Figure 10.11), which lived throughout the Paleozoic Era but finally became extinct during a mass extinction at the end of the Permian Period. Also includes the living fossil horseshoe crab.

Phylum Onychophora "Velvet worms." Elongate segmented worms originally inhabiting the marine realm but now found living only in terrestrial leaf litter. These are thought to be related to both annelids and arthropods (Figure 10.11).

Phylum Echinodermata The "spiny skinned" animals represented today primarily by the classes Asteroidea (starfish), Echinoidea (sea urchins and "sea biscuits"), Ophiuroidea (brittle stars), and Holothuroidea (sea cucumbers), all of which are marine. Skeleton consists of a series of calcareous plates embedded in a skin that articulate with each other. Superficially with radial symmetry but actually bilaterally symmetric. All members possess a curious water vascular system that is used for locomotion and attachment. "Weird," enigmatic classes existed at times during the early Paleozoic but became extinct (Edrioasteroids, Carpoids, Cystoids, Eocrinoids, Blastoids; Figure 10.20). Among the taxa that were prominent during much of the Paleozoic was the Class Crinoidea ("sea lilies"), which were mostly shallow-water sessile suspension feeders, resembling an "upside-down starfish on a stick" (Chapters 11, 12). Crinoids today are found only in deeper water, apparently to escape predation.

Phylum Hemichordata This group is made up of tongue worms and pterobranchs and includes an extinct group called the graptolites. Hemichordata are colonial animals that lived either attached to the bottom or as plankton during the Paleozoic but became extinct before the end of the era. These forms are especially useful as index fossils during the early-to-middle Paleozoic Era (see Chapter 11).

Phylum Chordata Comprises animals possessing at some stage in their life cycle: (1) an axial rod-like notocord for body support, (2) a single dorsal tubular nerve cord, and (3) paired gill slits. Several important subphyla exist:

Subphylum Cephalochordata The fish-like lancets (Figure 10.19).

Subphylum Vertebrata All vertebrates: fish, amphibians, reptiles, dinosaurs, and mammals. Those animals possessing a cranium (skull) housing a brain and spinal column consisting of cartilaginous (e.g., sharks, skates, and rays) to bony vertebrae. Paired special sense organs and a two- to four-chambered heart with red blood cells. This subphylum would eventually arise from chordate ancestors that appeared during the Cambrian explosion. Vertebrates will be discussed in greater detail as they appeared and evolved during the Phanerozoic Eon (Chapters 11 through 16).

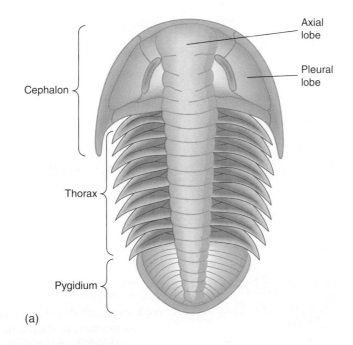

Cephalon

Axial lobe

Pleural lobe

Thorax

Pygidium

(a)

(b)

Courtesy of Ronald Martin, University of Delaware.

(c)

© Marques/Shutterstock, Inc.

FIGURE 10.8 (a) The basic anatomy of a trilobite. Early trilobites were characterized by distinct body regions and segmentation. However, as we will see (see Chapter 11), in later periods trilobites underwent various modifications, such as the fusion of segments. **(b)** *Paradoxides* from the Cambrian Period of Morocco. Some of the earliest trilobites were characterized by distinct segmentation and numerous spines. **(c)** Trilobite tracks.

Figure 10.8). Indeed, the Cambrian Period is often referred to as the "age of trilobites," (although, as we will see, many other new taxa appeared). Many trilobites had complex eyes consisting of multiple facets, similar to modern insects. More complex burrowing as well as sensory organs like eyes would have required even greater sophistication of nervous and muscle systems. Many of the earliest trilobites also had distinct segments and multiple spines (see Figure 10.8).

Another important event occurred by the Early Cambrian: the appearance of **small shelly fossils**. Small shelly fossils are represented by tubes and scale, shield-like, or spiny structures composed of calcium phosphate or calcium carbonate. Small shelly fossils have been found all over the world during this time, including Australia, Newfoundland, Russia, and China. Thus, whatever the animals were that secreted them, they lived far and wide. But the origins of these fossils are not easily explained because the

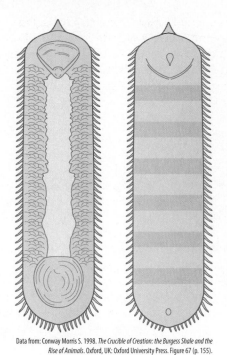

Data from: Conway Morris S. 1998. *The Crucible of Creation: the Burgess Shale and the Rise of Animals*. Oxford, UK: Oxford University Press. Figure 67 (p. 155).

FIGURE 10.9 Reconstruction of *Halkieria*, an animal with chainmail-like armor from the Cambrian.

shapes of the fossils are by no means indicative of what lived in them. One of the most famous of these fossils is the tube-like *Cloudina*, named after Preston Cloud, who pioneered studies of Precambrian paleontology, geology, and climate. Despite its namesake, the affinities of *Cloudina* are clouded. The fossil looks like the stack of paper cups next to a water cooler but is no more than 1 to 2 mm in length.

Despite the temptation to view each type of small shelly fossil as being occupied by an individual animal, recent finds

of the unusual *Halkieria* indicate otherwise (**Figure 10.9**). This fossil was first found in 1990 from Early Cambrian rocks of Greenland and is several centimeters long. Small shelly fossils, some of which had already been found separately and given their own names, are embedded in its surface like the overlapping chain-mail armor of medieval times. In addition, two large plates are located at either end of *Halkieria*. Possibly, *Halkieria* represented an early attempt to secrete a shell.

Halkieria might even be a distant relative of the modern **chiton** (**Figure 10.10**). Chitons (Class Amphineura) are primitive members of the **Phylum Mollusca**, which includes modern clams, snails, and cephalopods like the modern octopus. Like other mollusks, modern chitons have a muscular foot. Chitons use the foot to clamp down onto rocks and have a row of large plates over their backs arranged in such a way that the animal can roll up when disturbed. Chitons also exhibit repeated sets of organs. Mollusks are therefore thought to have evolved from ancestors called **monoplacophorans**, which possess repeated sets of organs. Because chitons and annelid worms, like the earthworm, and arthropods also exhibit segmented body plans, it is thought that the annelids, arthropods, and mollusks (represented by ancient monoplacophorans) all evolved from a common ancestor during the Late Neoproterozoic (**Figure 10.11**). Monoplacophorans were known from the fossil record but were long thought to be extinct until they were dredged from a seafloor trench off the Pacific coast of Costa Rica. Modern monoplacophorans seem little changed from the ancient ancestors and so are regarded as living fossils (see Chapter 5).

Other spiny fossils are likely the *spicules*—or skeletal elements—of sponges, which belong to the **Phylum Porifera** (**Figure 10.12**). Sponges are considered to be among the most primitive of all metazoans because they do not possess true tissues or even a primitive nervous system. The body walls of these animals are penetrated by countless numbers

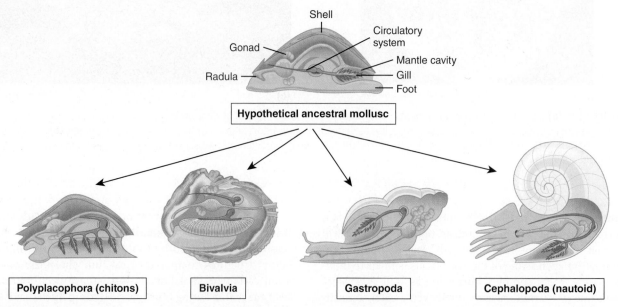

FIGURE 10.10 Modern classes of the Phylum Mollusca. The Hypothetical Ancestral Mollusc is thought to have possessed repeated pairs of gills; because this repetition resembles that of the repeated structures of annelid worms (like the earthworm, Figure 10.7), molluscs and annelids are thought to have diverged from a common ancestor during the initial evolution of metazoans.

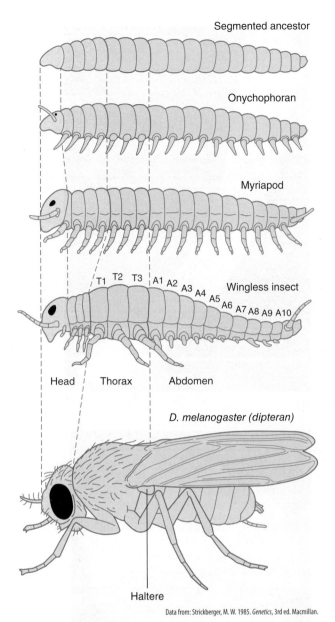

Segmented ancestor

Onychophoran

Myriapod

T1 T2 T3 | A1 A2 A3 A4 A5 A6 A7 A8 A9 A10   Wingless insect

Head / Thorax / Abdomen

D. melanogaster (dipteran)

Haltere

Data from: Strickberger, M. W. 1985. *Genetics*, 3rd ed. Macmillan.

FIGURE 10.11 How arthropod body organization might have initially evolved from a primitive segmented ancestor through a new phylum (Onychophora) to centipedes (myriapods) and then to wingless and eventually winged insects such as flies (Diptera).

of pores through which water is pumped by specialized cells to obtain food and oxygen and to rid the tissues of waste. In life, the spicules embedded in the body wall are cemented together to produce a skeleton that supports the soft parts. Upon death, the soft parts decay and the spicules are incorporated into the sediment. Also appearing during the Cambrian were inarticulate **brachiopods** which, despite their clam-like appearance, belong to their own distinctive phylum. Brachiopods produce a shell consisting of two halves called valves and are anchored to the bottom by a fleshy stalk called a *pedicle* (**Figure 10.13**). In the inarticulate forms, the shells glide past one another using muscles, but in the articulate forms, the valves articulate along a hinge using tooth and socket-like structures. The inarticulate forms never really became prominent (although the modern

inarticulate form *Lingula* represents a living fossil), but the articulate forms would greatly diversify in the Paleozoic Era and often became prominent rock formers (see Chapter 11). The inarticulate brachiopods would eventually give way to their articulate cousins (see Figure 10.13).

10.2.3 Burgess Shale and the "Cambrian explosion"

The only fossil deposit that rivals or surpasses the Edi-acara fauna in fame is the spectacular fauna of the **Burgess Shale** (**Figure 10.14** and **Figure 10.15**). The Burgess Shale is of Middle Cambrian age and was first discovered in the Canadian Rockies of British Columbia in the latter part of the 19th century (**Figure 10.16**). The Burgess Shale fauna was made famous by the dogged efforts of Charles Doolittle Walcott, who became director of the U.S. Geological Survey and later head of the Smithsonian Institution. When Walcott learned of the fauna, he returned every summer with pack mules, camping by the quarry and searching for specimens to be shipped back to the Smithsonian for description.

The Burgess Shale is another example of a Lagerstätte (see Chapter 4). The fossils are preserved as carbon films on the rocks and are found by patiently sitting and splitting open slab after slab of rock. Based on recent interpretations of the ancient depositional environments, the animals of the Burgess Shale lived in the vicinity of a limestone platform in a shallow marine seaway (**Figure 10.17**). During the Cambrian, limestone platforms were quite common along this part of North America. The long-standing interpretation of the preservation of the Burgess Shale Fauna is that it was rapidly buried by a turbidity current at the base of the platform; rapid burial prevented the complete decay of the soft parts. A more recent detailed stratigraphic study concluded instead that the Burgess Shale fauna was associated with deep-water limestone mud mounds in front of the escarpment; geochemical evidence suggests that brine seeps provided food via chemosynthesis to at least some of the fauna.

The Burgess Shale provides us with some other significant glimpses of fossil life that are otherwise not normally preserved. A number of the fossils appear to be combinations of animals that are considered as separate, distinct groups later during the Phanerozoic, including trilobites and other arthropods and various types of worms (compare Figure 10.14 and Figure 10.15). The Burgess Shale also contains the spectacular *Anomalocaris* (refer to this chapter's frontispiece). *Anomalocaris* grew to a meter or more in length, and its pineapple-shaped mouth matches the crescent-shaped wounds of trilobites found in the Burgess Shale and related deposits. *Anomalocaris* has therefore long been presumed to have been a predator, especially given its large claws, but recent studies of its jaw structure suggest that the jaws of some forms might have been too weak for predation and that the animal might instead have grubbed about in the sediment, much like many trilobites are thought to have lived, perhaps eating worms or jellyfish. Possibly, there were different species of *Anomalocaris* that lived in one way or the other.

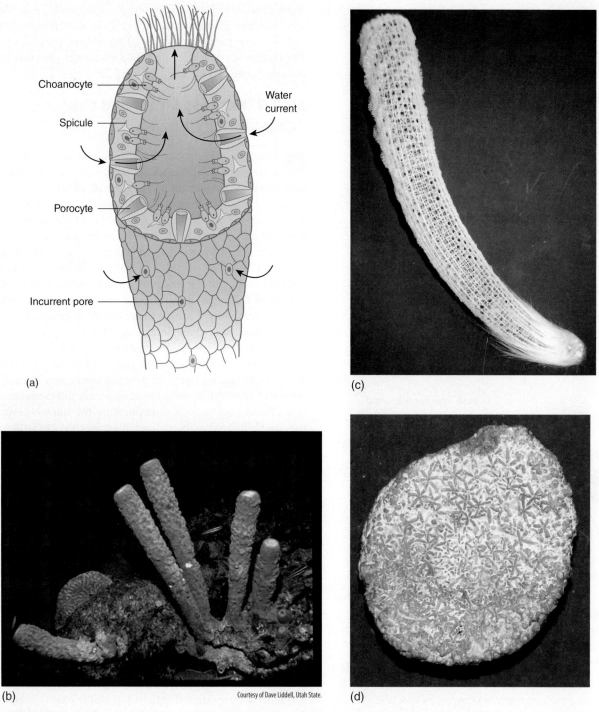

(a)

(c)

(b)

Courtesy of Dave Liddell, Utah State.

(d)

FIGURE 10.12 (a) Basic anatomy of a sponge. **(b)** Modern sponges living on a coral reef. **(c)** The internal skeleton of the modern sponge *Euplectella*, or Venus's Flower Basket. The skeleton consists of siliceous spicules fused together. **(d)** Fossil sponge, which has likely been compressed by burial because of its soft body. Note the star-like spicules that have been preserved, though.

Some forms were even more bizarre. For example, *Opabinia* had five eyes at the end of a stalk (Figure 10.13 and Figure 10.14), and what were thought to be spine-like legs of the aptly named *Hallucigenia* were later realized to be spines down its back that probably served for protection. The specimens were so bizarre, they had originally been reconstructed upside down.

Also present were **hyoliths**, which secreted a cone-shaped shell with a "lid" (or operculum) composed of calcium carbonate (**Figure 10.18**). Hyoliths were worm-like animals, perhaps distantly related to mollusks, but are placed by some workers in their own phylum. They are thought to have been epifaunal benthic deposit feeders and might have been sessile.

One of the most significant finds in the Burgess Shale is the fossil *Pikaia*. This genus appears to have belonged to the **Phylum Chordata** (**Figure 10.19**). Chordates have also been reported from the Early Cambrian shales of southern

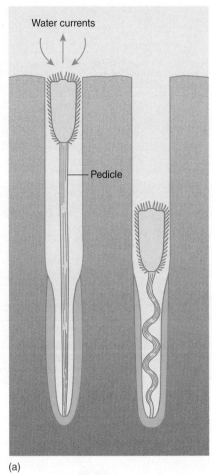

Water currents

Pedicle

Data from: Simpson, G. G., Pittendrigh, C. S., Tiffany, L. H. 1957. *Life: An Introduction to Biology*. New York: Harcourt, Brace, and World. Figure 22-10, Part 1 (p. 534).

(a)

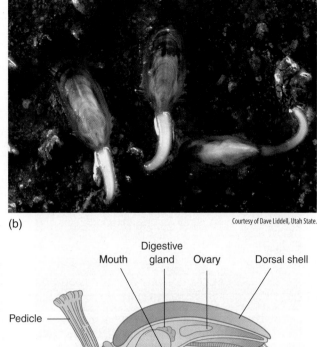

Courtesy of Dave Liddell, Utah State.

(b)

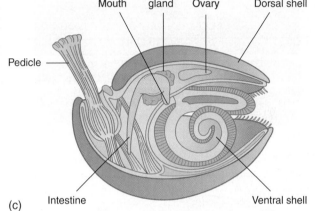

Mouth   Digestive gland   Ovary   Dorsal shell

Pedicle

Intestine   Ventral shell

(c)

Data from: Simpson, G. G., Pittendrigh, C. S., Tiffany, L. H. 1957. *Life: An Introduction to Biology*. New York: Harcourt, Brace, and World. Figure 22-10, left photo and bottom-right figure (p. 534).

FIGURE 10.13 Brachiopod structure and function. Brachiopods feed on suspended organic matter and plankton close to the bottom. Inarticulate brachiopods are primitive compared to later articulate brachiopods, the shells of which are articulated by a series of tooth-and-socket like structures. **(a)** Inarticulate brachiopods. **(b)** Modern inarticulate lingulid brachiopods. This particular form is a living fossil (see Chapter 5). **(c)** Articulate brachiopod structure.

China and therefore appeared before the Burgess Shale. Members of the Phylum Chordata are characterized by the presence at some time during the lives of a stiff cartilaginous rod called a *notochord,* a dorsal (or upper) nerve cord, and paired gill slits like those found in fish. Chordates would later evolve to produce *vertebrates* (animals characterized by a bony vertebral—or spinal—column and a distinct cranium): amphibians, reptiles, birds, and mammals.

But it is not just the creatures of the Burgess Shale that often appeared to be odd. Elsewhere during the Cambrian, bizarre representatives of the **Phylum Echinodermata** (which includes starfish) appear and eventually disappear, never to be seen again (**Figure 10.20**). Some of these forms probably burrowed in sediment, whereas others fed on suspended organic matter and plankton in the water just above bottom.

Microscopic fossils, or microfossils, called **acritarchs** were not only prominent in the Cambrian but also much earlier during the Proterozoic (**Figure 10.21**). Acritarchs are believed to be the cysts, or resting stages, of eukaryotic phytoplankton; however, some acritarchs are so large (a few millimeters) they might represent the reproductive stages of larger benthic eukaryotic algae such as seaweeds. Still, the

smaller acritarchs suggest the presence of eukaryotic phytoplankton. Cysts are produced by many modern unicellular protists in response to adverse temperature, salinity, oxygen, and nutrient levels.

Burgess Shale-type faunas were long thought to have disappeared from the fossil record by the end of the Cambrian. Increasing rates of burrowing had presumably "burrowed away" any soft-bodied faunas like the Burgess Shale after the Cambrian. However, recent studies have found similar faunas extending at least into the Early Ordovician. These faunas appear to have persisted in oxygen-poor environments where burrowers would have been less prevalent.

CONCEPT AND REASONING CHECKS

1. What are the stages of metazoan evolution leading up to the Cambrian?
2. What is the significance of the appearance of sponges and apparent cnidarians before more complex animals like trilobites?

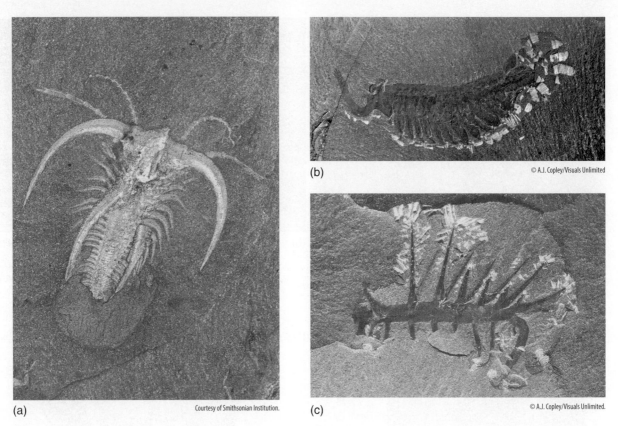

(a)   Courtesy of Smithsonian Institution.
(b)   © A.J. Copley/Visuals Unlimited
(c)   © A.J. Copley/Visuals Unlimited.

FIGURE 10.14 Fossils of the Burgess Shale, preserved as carbon films. **(a)** *Marrella.* **(b)** *Opabinia.* **(c)** *Hallucigenia.*

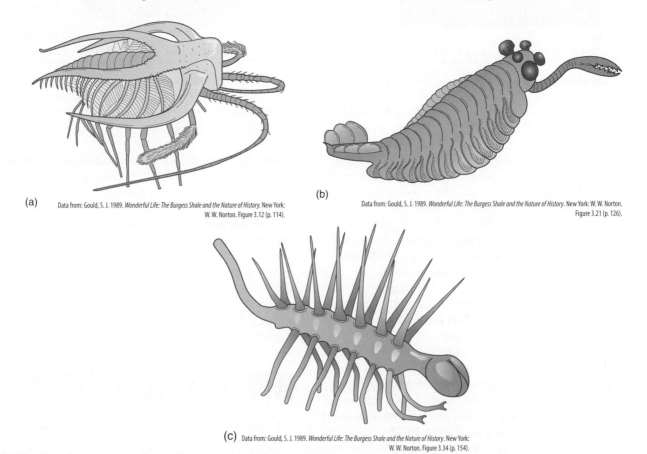

(a)   Data from: Gould, S. J. 1989. *Wonderful Life: The Burgess Shale and the Nature of History.* New York: W. W. Norton. Figure 3.12 (p. 114).

(b)   Data from: Gould, S. J. 1989. *Wonderful Life: The Burgess Shale and the Nature of History.* New York: W. W. Norton. Figure 3.21 (p. 126).

(c)   Data from: Gould, S. J. 1989. *Wonderful Life: The Burgess Shale and the Nature of History.* New York: W. W. Norton. Figure 3.34 (p. 154).

FIGURE 10.15 Three-dimensional reconstructions of some of the bizarre animals of the Burgess Shale. Compare to Figure 10.15. **(a)** *Marrella*, a trilobite-like form. **(b)** *Opabinia*. Note the front nozzle with claw and five eyes. **(c)** *Hallucigenia*, which was originally figured upside down.

© Alan Sirulnikoff/Photo Researchers, Inc.

FIGURE 10.16 Burgess Shale quarry on Mount Stephen.

Courtesy of Smithsonian Institution.

FIGURE 10.18 A hyolith from the Burgess Shale.

10.3 What Do These Faunas Tell Us?

F What do the Ediacara and Burgess Shale faunas and other fossils tell us about the evolution of Earth and metazoan life? First, spectacularly preserved fossil assemblages like the Ediacara and Burgess Shale faunas indicate the fossil record is not quite so incomplete after all (see Chapters 5 and 6). In Darwin's day some workers—including Charles Lyell—believed all fossils appeared abruptly in the geologic record at the same time, giving Darwin's critics plenty of ammunition. If the evolution of new organisms was slow and gradual, as Darwin advocated, why

did the first fossils seem to appear suddenly? Why were there presumably no fossils (such as "missing links"), or even traces of creatures, before the first "good" fossils? Something seemed terribly wrong, and to Darwin's detractors it was his theory of evolution.

In his defense, Darwin could only suggest that there had been a long period of evolution before the first fossils

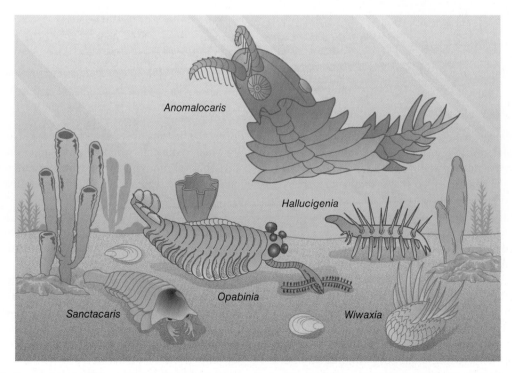

FIGURE 10.17 Reconstruction of the ancient setting of the Burgess Shale. This fossil fauna formed in one of two ways: either the animals lived in the shallow water of the limestone platform and were swept into deeper water and buried by a turbidity current, or the animals actually lived at the base of the platform and were buried by a turbidity current originating on the shallow platform.

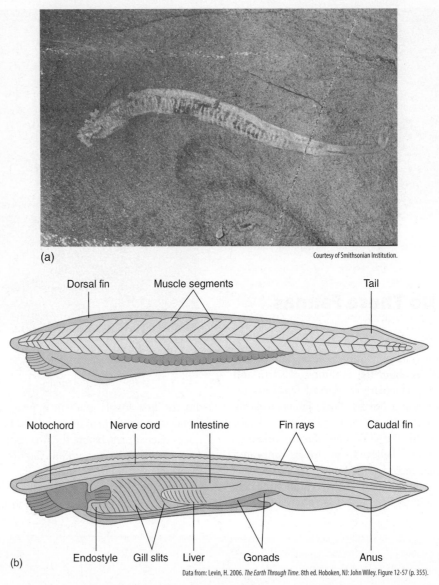

(a)

Courtesy of Smithsonian Institution.

Dorsal fin   Muscle segments   Tail

Notochord   Nerve cord   Intestine   Fin rays   Caudal fin

(b)

Endostyle   Gill slits   Liver   Gonads   Anus

Data from: Levin, H. 2006. *The Earth Through Time*. 8th ed. Hoboken, NJ: John Wiley. Figure 12-57 (p. 355).

FIGURE 10.19 **(a)** *Pikaia*, a primitive chordate from the Burgess Shale that resembles the **(b)** modern chordate *Branchiostoma*.

but that the geologic record was incomplete and the earliest portion of the fossil record lost. Possibly, Darwin suggested, early evolution had taken place in open seaways far removed from the present continents, so that a sedimentary record had not formed on the continents.

The view that the fossil record was terribly incomplete held sway up to and well past the time of Charles Walcott. Walcott conceded that there was what he termed a Lipalian ("lost") interval separating the Precambrian from the Cambrian. But Walcott based his conclusion on work that had been done in the United States. In North America, there is such a Lipalian interval. In fact, geologists had long used the abrupt appearance of trilobites to demarcate the end of the Proterozoic and the beginning of the Cambrian. Sometime during the late Neoproterozoic, the seas withdrew from the continents during a sea-level regression, leaving the continents exposed to erosion and nondeposition. Later, during the Cambrian Period the seas reflooded the continents

during the first major sea-level rise of the Phanerozoic, during which sedimentary deposition resumed and fossils such as trilobites abruptly appeared. However, outcrops were later found in other places that contain fossil faunas transitional to trilobites and the Burgess Shale (see Figure 10.1). In fact, as already noted, the Burgess Shale does not lie at the base of the Cambrian. The formation actually occurs somewhat later in time in the earliest portion of the Middle Cambrian (see Figure 10.1).

Second, and perhaps most important, it appears that representatives of all the major phyla found later through the Phanerozoic were already present by the *beginning* of the Cambrian and perhaps even earlier. Thus, there must have been a great deal of evolutionary experimentation occurring not only in the Cambrian but also well before it. Furthermore, no new major phyla characterized by distinctive **body plans** appear to have evolved after the Cambrian (see Chapter 5). For example, all mollusks such as clams and snails

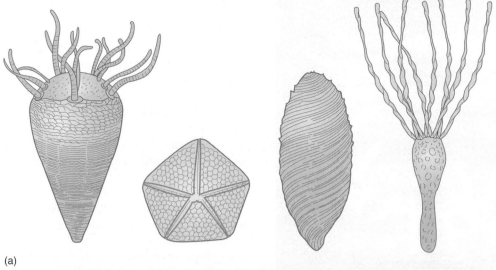

(a)

Data from: Stanley, S. M. 1989. *Earth and Life Through Time*. New York: W. H. Freeman and Company. Figure 12-5 (p. 316).

(b)

© cphoto/Fotolia.com.

FIGURE 10.20 **(a)** Strange echinoderms, which are distant relatives of the modern starfish, from the Cambrian. **(b)** The structure of a modern starfish, showing tube feet which are part of the water vascular system.

have the same basic feature that makes them mollusks: a muscular foot. In clams, the foot is used for burrowing, in snails for crawling, and in cephalopods such as the octopus, for swimming. Similarly, despite their tremendous diversity, all arthropods such as insects, crabs, and trilobites share arthropod features, such as a jointed external skeleton.

Walcott originally classified, and in some cases seemed to "shoe-horn," the Burgess Shale creatures into groups that were already well known from the fossil record or modern seas. Why should he have suspected the creatures were anything different from what were already known? Later, careful examination of the Burgess Shale fossils led to the suspicion that some of the fossils were not what they seemed. Some appeared to represent groups never seen again in the fossil record.

This led to an extreme interpretation of the evolution of Cambrian life by the American paleontologist, Stephen Jay Gould (**Figure 10.22**). In this view, when metazoans began to appear, there were few if any environmental or biologic constraints to prevent them from experimenting with different ways to fill the niches and habitats that lay wide open to them. Life was radiating into an unoccupied frontier. Because many of these new groups apparently did not belong to the phyla that were prominent later in the Phanerozoic, Gould concluded they must represent altogether different phyla with highly unusual features. According to Gould, many of these biologic "experiments" failed, leaving behind the phyla that diversified into the taxa recognized through the rest of the Phanerozoic (see Figure 10.22). As a result, the evolutionary "tree" of these phyla did not resemble a tree

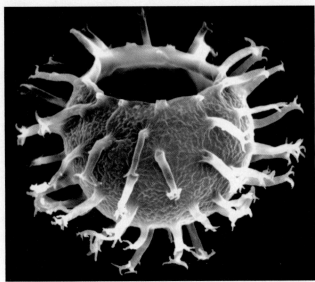
Courtesy of Ronald Martin, University of Delaware.

FIGURE 10.21 Acritarchs are considered to be fossilized cysts of eukaryotic phytoplankton.

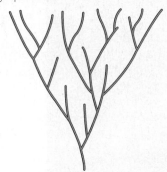

The cone of increasing diversity

(a)

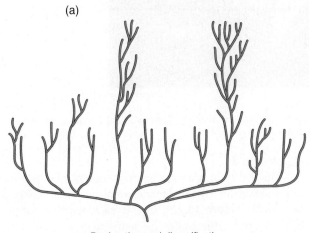

Decimation and diversification

(b)

Data from: Gould, S. J. 1989. *Wonderful Life: The Burgess Shale and the Nature of History.* New York: W. W. Norton. Figure 1.17 (p. 46).

FIGURE 10.22 Divergent interpretations of the Cambrian explosion of metazoan life. **(a)** According to the tree of life, life evolved and progressively diversified from a common ancestor through time. **(b)** Gould's bush of life also sees life as originating from a common ancestor, but instead of progressive diversification, there were many early experiments by rapidly evolving animal groups. According to this hypothesis, most of these experiments failed, but the successful ones led to the establishment of groups found later in the fossil record.

so much as a bush. Moreover, many branches at the base of the bush were progressively pruned after the taxa originated. In Gould's view, this pruning process was an example of the macroevolutionary process of sorting (see Chapter 5) but at a much higher level than species, namely at the level of phyla! However, Gould's interpretations are now widely regarded as an overreaction to early interpretations of the Burgess Shale fossils. Some fossils, such as *Hallucigenia*, are not quite so bizarre after they are correctly oriented (see Figure 10.14 and Figure 10.15). Also, specimens from other localities, such as those in China, indicate that many of the Burgess Shale creatures are in fact allied more closely to later groups than previously thought.

Nevertheless, there was a great deal of biologic evolution and experimentation occurring during the latest Neoproterozoic and Cambrian. Why? For one thing the physical and biologic environments of this time were far from benign (for example, the Snowball Earth episodes of the Neoproterozoic described in Chapter 9). Climate appears to have been fluctuating tremendously, continents were shifting positions, and sea level was going up and down. It is possible, then, that strongly fluctuating environments might have stimulated the rapid evolution of living systems. Just as important, the fossil record indicates that new, *biologic* forces were coming into play: predation, extinction, and macroevolution.

10.4 Why Did Metazoans Appear?

G **H** The most profound implications that the Ediacara and Burgess Shale faunas hold for us involve how the creatures of this time originated and how they began to interact with the environment and one another. Like the origins of life, a diversity of hypotheses has been proposed for the appearance of metazoans (see method of multiple working hypotheses in Chapter 1). Some invoke physical or chemical forces, whereas others resort to biologic mechanisms. Because the hypotheses overlap, the real answer likely involves **multiple causation** (see Chapter 1): a combination of causes that reflect the continuing integration and evolution of Earth's systems and biogeochemical cycles (see Chapter 9). However, to what extent biologic innovation was directly triggered by environmental change, or vice versa, is unclear.

10.4.1 Snowball Earths

It has been suggested that one or more *Snowball Earths* (see Chapter 9) selected for more "fit" metazoans, causing them to evolve. According to this hypothesis, hydrothermal vents provided refugia for creatures during Snowball Earths, from which survivors could have dispersed and evolved when conditions were again favorable. Modern hydrothermal vents are populated by animals that belong to taxa characteristic of continental shelves (annelids and mollusks, for example). However, the taxa that evolved at the vents have not returned to the shelves, so it seems unlikely (although certainly not impossible) that vents could have served as refuges in the past. Moreover, like the hypothesis of panspermia for the origin of life (see Chapter 8), the Snowball Earth origin begs the question: if hydrothermal vents provided refuge, where did the metazoans originate in the first place? Still, as we will see below, we cannot completely dismiss the possibility that severe environmental crises might have had something to do with metazoan evolution.

10.4.2 Oxygen

One of the long-standing hypotheses—first suggested by the American paleontologist Preston Cloud—for the appearance of metazoans is that the concentration of oxygen in the atmosphere and oceans increased during the Neoproterozoic to the point that it allowed metazoans to evolve. The decline of Banded Iron Formations (BIFs) allowed oxygen to begin to accumulate in the atmosphere (see Chapter 9). Based on a variety of indicators, oxygen levels in the atmosphere rose to about 10% of the present atmospheric level by the late Neoproterozoic. Some of the oxygen undoubtedly began to diffuse into the surface oceans and perhaps also into deeper waters via deep ocean circulation. These waters would have been in contact with the sea bottom, where many metazoans presumably originated.

Increasing oxygen availability would have allowed organisms to grow larger. According to this hypothesis, the reason creatures such as *Dickinsonia* were so flat was because the flatness increased the creature's surface area and made its interior closer to the environment for oxygen and carbon dioxide exchange with the environment during respiration. Primitive organisms, like cnidarians or flatworms (such as planarians), have no special respiratory or circulatory systems and can only exchange oxygen and waste carbon dioxide through their body surfaces. Assuming to a first approximation that organisms are spherical, as an organism increases in size its volume increases by the cube of its radius, whereas its surface area increases only by the square of its radius. Thus, given a constant oxygen level and assuming no respiratory or circulatory systems, the maximum size of an organism is *diffusion-limited*. The only way to get more oxygen into such primitive organisms is to increase the concentration of oxygen in the surrounding environment.

Increasing oxygen levels explains some other observations. First, the appearance of skeletons in the Cambrian, which increased the chances of fossilization (see Figure 10.1), might be related to the increase in size of creatures. Skeletons would have provided support and muscle attachment for the soft parts of enlarging metazoans; otherwise, the organisms would have collapsed on themselves and been unable to move or feed. Second, the complexity of burrowing increased as metazoans were evolving. Increasingly sophisticated burrowing behavior would have required increasingly sophisticated musculatures and nervous systems, all of which would have demanded more and more energy, and therefore oxygen, to break down foodstuffs by aerobic respiration.

10.4.3 Predation

Another possible cause of the appearance of metazoans and their diversification might have been predation. As we have

seen, many Ediacaran fossils resembled (or perhaps even were) cnidarians. All cnidarians possess specialized stinging cells called cnidocysts used to capture prey. Although the term "Garden of Ediacara" used above conjures up a paradise, the time of the Ediacara fauna might instead have been characterized by a "rain of cnidocysts" from cnidarians. Elements of the Ediacara fauna might have survived only by retreating to deep water to avoid predators or burrowers.

There is other more definite evidence for increasing predation. Stromatolites were waning by this time, presumably as a result of increased grazing. Also, specimens of *Cloudina* sometimes bear drill holes produced by some as yet unidentified predator, as do other invertebrates such as brachiopods and, later on, (**Figure 10.23**) bivalves; ichnofossils have also been found of trilobites converging on worm trails from which the worm does not emerge. The appearance of *Anomalocaris* in the Cambrian also suggests increasing predation on trilobites (but see discussion above), although this occurred long after the initial appearance of metazoans in the fossil record.

Increasing predation presumably stimulated some groups to develop skeletons. As indicated by the chainmail armor composed of small shelly fossils embedded in *Halkieria,* skeletons appeared in stages, hinting that a **biologic "arms race"** between predators and prey might have begun. In this view, when predators appeared, skeletons followed. Predators responded by becoming more adept at obtaining prey, whereupon more sophisticated skeletons were evolved by prey, and so on. However, most marine animals do not have skeletons and escape from predators through behavioral responses, such as swimming or burrowing. The cloud

Courtesy of Ronald Martin, University of Delaware.

FIGURE 10.23 Drill holes in the bivalve *Chesapecten* were likely produced by predatory gastropods such as *Polinices*. A small specimen of *Polinices* is shown at the lower right, but this genus can grow to a much larger size. These specimens are from the Miocene of Maryland along Chesapeake Bay, but similar drill holes have been found on ancient bivalves, indicating predation.

of ink emanating from a squid when it is frightened is a prime example; the ink distracts the predator as the squid escapes. The occurrence of increasingly complex burrows toward the Cambrian has also been hypothesized to indicate escape from predation.

10.4.4 Food

Swimming, deep burrowing, and escape all take fairly sophisticated musculatures and nervous systems, both of which require energy. Thus, the increasing complexity of burrows, the increase in body size, the formation of skeletons, and predation might all be related to increasing amounts of food available to animals. Animals are open systems that require inputs of matter and energy to maintain themselves (see Chapter 1). Experimentation with increasingly sophisticated musculatures and nervous and sensory systems along with increasing body size would presumably have required more energy—that is, food.

Until the Neoproterozoic, however, food availability might have been relatively limited. Possibly, photosynthesis was limited before this time because of low nutrient availability. Low nutrient availability may have resulted for several reasons. The increasing volumes of sedimentary rocks indicate that continental shelves were expanding during the Proterozoic (see Chapter 9). Still, continents might have been too small until the late Proterozoic for extensive weathering and runoff of nutrients to the oceans to occur that could have supported plankton and benthic ecosystems. Furthermore, up to the Neoproterozoic, much of the phosphorus in the oceans might have precipitated during the formation of BIFs (see Chapter 9). Dissolved phosphorus, which is necessary for marine photosynthesis (see Chapter 3), precipitates in the presence of dissolved iron. However, after most of the iron had precipitated, the levels of dissolved phosphorus and perhaps other nutrients in the oceans might have increased. By the end of the Proterozoic, phosphorus-rich sedimentary rocks called **phosphorites** were becoming very abundant on continental shelves. The presence of phosphorites hints that phosphorus scavenging by iron had declined sufficiently to make phosphorus increasingly available to plankton instead of precipitating on the ocean bottom. Upwelling of phosphorus from deep ocean waters and its deposition in shallower waters may have been promoted by more widespread continental shelves by this time.

Besides changes in ocean circulation and upwelling, there is evidence for increasing nutrient input from land during the Neoproterozoic and Cambrian. During this interval, there is a dramatic and prolonged rise in the stable isotope ratios of the element strontium (see Box 9.1). Continental rocks like granites tend to be enriched in the heavier isotope of strontium (strontium-87, symbolized as ^{87}Sr), whereas seafloor crust is enriched in a lighter isotope (^{86}Sr). Increased strontium isotope ratios ($^{87}Sr/^{86}Sr$) therefore suggest increased continental uplift and exposure of rocks to weathering, resulting in the increased runoff of dissolved ions and nutrients to the oceans. Conversely, decreased ratios are associated with increased rates of seafloor spreading and hydrothermal exchange between seafloor crust and seawater. The increase in ratios suggests increased continental weathering and nutrient runoff into the oceans, stimulating photosynthesis. In fact, the rise in strontium isotope ratios is so high, it even surpasses that associated with all the widespread orogenies during the Cenozoic Era, including the rise of the Rockies, Andes, and the Himalayas!

Increased weathering could have resulted from orogeny associated with tectonic activity. During the Neoproterozoic, assembly of the supercontinent **Gondwana** occurred. The assembly of Gondwana involved the collision of East Gondwana (mainly India, Antarctica, and Australia) and West Gondwana (mainly Africa and South America), along with parts of what are now Florida, the Middle East, southern Europe, southeast Asia, and Tibet. The formation of this supercontinent was associated with continental collision and the widespread **Pan-African Orogeny** that extended over many continents. Overlapping with the Pan-African Orogeny was the rifting of the supercontinent Rodinia and perhaps another called Pannotia (see Chapter 11), which started about 750 million years ago and continued into the Late Cambrian to about 530 million years ago (see Figure 10.1). The rifting of Rodinia is thought to be associated with Snowball Earth episodes (see Chapter 9). Some continents might have been located in low-latitude regions. Here, the continents would have been weathered by warm temperatures and abundant rainfall, setting the stage for Snowball Earths (see Chapter 9). Intense chemical weathering in the tropics would have shed nutrients onto the shelves and into the seaways between the continents.

The net effect of increasing nutrient availability would have been to stimulate marine photosynthesis on continental shelves. Carbon isotope ratios (see Chapter 9) remained relatively low during much of the Proterozoic but then rose and fluctuated significantly during the Neoproterozoic. The overall rise in the carbon isotope ratios suggests that marine photosynthesis was generally increasing, whereas the strong fluctuations in the ratios during the Neoproterozoic and Cambrian might represent the response of the phytoplankton and marine photosynthesis to strongly fluctuating physical conditions associated with orogeny and perhaps glaciation during the presumed Snowball Earth episodes.

Enhanced photosynthesis is also suggested by the fossil record of marine phytoplankton. Acritarchs largely disappeared from the fossil record during the Neoproterozoic. Before this time, however, they were prominent components of the fossil record. The disappearance of acritarchs during the Neoproterozoic has been suggested to have resulted from extinction. However, their disappearance might also have been related to less severe nutrient limitation, thereby decreasing the necessity of cyst formation. An increasing abundance of phytoplankton in the water column during the Neoproterozoic would have provided food for the expanding zooplankton and nekton. When the plankton died, the dead organic matter would have rained down to the sediment, providing food for the evolving benthos.

Increasing marine photosynthesis would have had another important effect on metazoan evolution: increasing levels of oxygen in the atmosphere and oceans. Increasing levels of oxygen would have contributed to the evolution of metazoans, as previously described. Because oxygen is used in aerobic respiration and energy release, however, increasing oxygen levels by themselves might have had little effect on metazoan evolution unless sufficient food was available as an energy source.

10.4.5 Changes in biogeochemical cycles

In effect, then, metazoans might have diversified because of changes in Earth's biogeochemical cycles. Both cnidarians and flatworms, which were likely present in the late Neoproterozoic, have an incomplete gut; also, flatworms do not burrow deeply. An incomplete gut is a relatively inefficient means of digesting food because the gut lacks highly specialized tissues for more efficient digestion.

By contrast, a complete gut, with both separate mouth and anus, is a much more efficient means of digesting food. Complete guts are found in other metazoans such as annelid worms, arthropods, mollusks, echinoderms, and chordates like those already described. These taxa all appeared during the late Neoproterozoic and Cambrian. The tissues along a complete gut can become specialized for different stages of digestion as food passes through the intestine. A one-way gut also means that deposit feeders can eat as they burrow by ingesting sediment. Thus, a complete gut would have effectively increased food availability to early metazoan burrowers because deposit feeding would have allowed locomotion and feeding to be combined into the same activity, thereby economizing on energy use. The energy that was saved could in turn have been devoted to reproduction, population expansion, and thus perhaps biotic diversification. Greater food availability in sediment would have permitted, and perhaps even promoted, deeper and more sophisticated burrowing. Greater sophistication of burrowing during the Neoproterozoic and Cambrian would in turn have required greater amounts of food, and so on. In other words, *positive feedback* might have begun between food availability, deposit feeding, and burrowing.

Put another way, *positive feedback might have occurred between evolution and food (energy) availability.* The appearance of a more efficiently burrowing fauna would have increased the rate of remineralization of dead organic matter and the recycling of nutrients that would otherwise have been trapped in the organic matter. The recycling of nutrients back to the water column would have allowed the biosphere to continue its explosive radiation. Thus, if it had not been for bioturbation, available nutrients might have been locked up so rapidly in living and dead biomass that there would have been few nutrients left to continue expansion of the biosphere!

The evolution of a complete gut could have also affected oxygen availability in the oceans. Before the Neoproterozoic organic matter presumably settled slowly through the water column and was subject to extensive bacterial decay (**Figure 10.24**). The decay of organic matter would have used up much of the oxygen and released carbon dioxide back to the water column causing extensive anoxia. Thus, one way for oxygen levels to have increased in the oceans would have been to make organic matter settle rapidly so that it did not have time to use up all the oxygen in the water column as it settled to the bottom. This might have been accomplished by the evolution of a complete gut and the production of fecal pellets. Today, fecal pellets are produced by many metazoans, both in the benthos (by worms and so on) and by plankton such as microscopic crustaceans called copepods.

At first glance these sorts of mechanisms do not seem very plausible. After all, how much effect can lowly creatures such as plankton or worms really have on the environments? The answer is a lot. Recall, for example, the long-term effects of minute calcareous plankton on the carbon cycle (see Chapter 1), and the potential effect of enormous numbers of small and seemingly unimportant creatures becomes evident. The collective activity of earthworms in bioturbating soils is no less impressive. Recall, too, that the destruction of organic matter by increasing bioturbation might have released increasing amounts of carbon dioxide back to the atmosphere, which might be one of the reasons why Snowball Earths never recurred after the Neoproterozoic (see Chapter 9).

10.4.6 Ecologic and genetic mechanisms

The hypotheses so far described view physical environmental change and the biologic *response* to physical environmental change as the prime factors involved in the appearance of metazoans. In this view, primitive animals radiated into vacant ecologic space made suitable by changing physical environmental conditions. This view is referred to broadly under the umbrella as the **ecologic hypothesis**.

However, as we have already seen, there was also a great deal of genetic "experimentation" occurring during life's Big Bang. The explanation for the origin(s) of metazoans based on macroevolutionary innovation is referred to as the **genomic hypothesis**. Changes in gene transcription and regulation during development were undoubtedly an important force in macroevolution during the Neoproterozoic and Cambrian. Recently, scientists have discovered that regulatory genes called *Homeobox* or *Hox* genes control the early development of certain body regions into particular segments or structures (see Chapter 5). Mutations in the DNA of regulatory genes can cause changes in the regulatory proteins and enzymes that are produced, which in turn affect the activity of other genes. During the Neoproterozoic and Cambrian, any sorts of ecologic controls that might have weeded out such mutants might have been largely missing. Thus, a combination of wide open habitats ("ecospace") coupled with the changing, and perhaps highly stressful, environmental conditions of the Neoproterozoic might have triggered the appearance and rapid diversification of metazoans.

Unfortunately, there appears to have been no time later in the Phanerozoic when the ecologic world was quite as

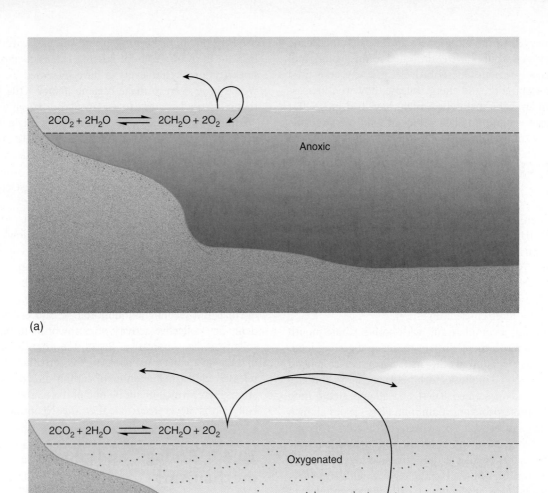

(a)

(b)

FIGURE 10.24 Effects of the evolution of a complete gut on oxygenation of the oceans and food availability at the sea bottom. **(a)** Prior to the evolution of a complete gut, decay of dead organic matter settling through the water column uses up the oxygen in the water column. **(b)** After evolution of a complete gut, fecal pellets (dots) rapidly send organic matter to the sea bottom, preventing decay in the water column and allowing the water column to become oxygenated.

vacant as the late Neoproterozoic and Cambrian. Otherwise, the ecologic and genomic hypotheses could be tested by examining the fossil record of how the biota responded to a similar situation. As we will see in Chapter 13, the only other time that comes close to being completely ecologically vacant followed the massive Late Permian extinctions at about 250 million years ago. Although new taxa did in fact originate after the Permian extinctions, many of whose descendants are with us today (such as corals, clams, snails, and fish), these groups apparently had too much genetic "baggage" to carry around from their ancestors to develop wholly new ground plans and phyla. Consequently, Earth never again witnessed anything like the evolutionary explosions of the late Neoproterozoic and Cambrian.

10.4.7 Extinction

Another possibility that explains changes in animal biodiversity in the late Precambrian and Cambrian is the extinction of the Ediacaran and perhaps other faunas. Several sharp excursions to very low values of carbon isotope ratios suggest photosynthesis suddenly shut down a number of times (see Box 9.1). However, given the amount of environmental disturbance that occurred during the Neoproterozoic, the exact cause of the spikes to lower ratios is unknown. But if extinction had occurred at, say, the Precambrian–Cambrian boundary, which is marked by a strong spike to lower carbon isotope values, it might have allowed some metazoans to rapidly replace the Ediacarans and expand into vacant habitats and ecologic niches in the Cambrian.

This possible Late Precambrian extinction has been likened to that which occurred at the end of the Cretaceous. At the end of the Cretaceous, extinction finished off the dinosaurs, permitting mammals, which had been present but much less prominent than dinosaurs, to diversify, like the Cambrian faunas, *almost explosively* (see Chapter 14).

10.5 Molecular Clocks and the Fossil Record of Early Metazoans

Although relatively primitive, Ediacarans were multicellular and had tissues that were likely specialized to perform a particular function. The presence of tissues therefore implies a period of evolution before the appearance of Ediacarans in the fossil record. The question is how much before?

Sinuous shallow grooves about half a centimeter in width have been found recently in India in rocks reported to be about 1.6 billion years old (**Figure 10.25**). Although their sinuosity suggests worms or worm-like creatures, the grooves also branch, suggesting they perhaps instead represent the impressions of branching algal fronds, not animals. Bead-like structures thought to be of biogenic origin have also been found in rocks about 1.5 billion years old in Glacier National Park. Still, not everyone is convinced these structures are of animal or even of biogenic origin. Then again, if they are not biogenic, what are they?

Molecular clocks also indicate an earlier origin for the metazoans than does the fossil record (**Figure 10.26**), while a recent discovery in Greenland suggests the presence of stromatolites as early as 3.7 Ga, although this report has met, as you might expect, with extreme skepticism in some quarters because of the possibility that the presumed fossils are actually chemical precipitates. Depending

Reproduced from Seilacher, A., et al. 1998. Triploblastic animals more than 1 billion years ago: trace fossil evidence from India. *Science, 282,* 80-8F. Reprinted with permission from AAAS.

FIGURE 10.25 Presumed worm burrows ~1.6 billion years old from sandstones of central India. Given the shallow nature of these structures, these presumed ichnofossils, if they are in fact burrows, might have been made by some form of flatworm. If these structures are worm burrows and the age assignment is correct, the fossils support molecular clocks that indicate a much earlier origin of metazoans.

on the assumptions of the clocks and the choice of genes used in sequencing, estimates of the time of appearance of metazoans range from 1 to 1.5 billion years to 700 million years ago. However, critics have pointed out that the clocks could have been "speeded up" during biologic "explosions" in response to environmental change, like the aforementioned "explosions." Given the apparently radical environmental changes occurring on Earth during the Neoproterozoic, this is certainly a distinct possibility.

But, if recent estimates from molecular clocks and the sporadic occurrences of true fossils are verified by further evidence, there must have been an interval of metazoan evolution long before the appearance of the first undisputed fossils. Thus, the Cambrian "explosion" would have had a long fuse that was lit much earlier in the Proterozoic. If this turns out to be the case, much early metazoan evolution would likely have occurred among very small soft-bodied creatures, so what we see as fossils later in the Proterozoic are the somewhat larger and more preservable result.

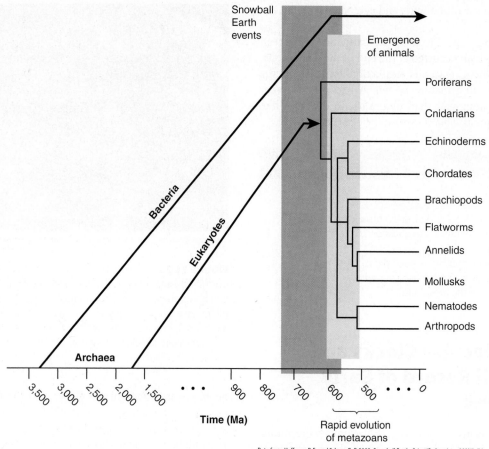

FIGURE 10.26 Timing of the origin of metazoans based on the fossil record versus molecular clocks. Note the relatively sudden appearance of different groups in the fossil record (as discussed in the text), as opposed to the molecular clocks, which indicate much earlier origins. Molecular clocks suggest that the first metazoans may have originated by a billion years or more before the appearance of the Ediacara Fauna.

SUMMARY

- During the late Neoproterozoic, animal life appears to have crossed a threshold from a unicellular to multicellular state. Because of the abrupt appearance of metazoans in the fossil record, their appearance was often referred to as an "explosion." We now know that the abruptness of appearance of metazoans was partly a function of the incompleteness of the fossil record.

- Given the changes that occurred during the Neoproterozoic and Cambrian, there must have been a great deal of evolutionary experimentation occurring. Much more complete stratigraphic sections now reveal a general sequence of events beginning in the Vendian:
  - The appearance of the Ediacara fauna, which appears to be represented by creatures of cnidarian ("jellyfish") affinities; however, the exact relationship, if any, of the Ediacara fauna to later creatures remains a mystery.
  - The increasing complexity of trace fossils (tracks, trails, burrows) in sediment.

  - The appearance of small shelly fossils, which are followed by full-fledged hard parts, such as those of brachiopods and trilobites.
- The evolutionary experimentation of the Ediacaran and Cambrian might have occurred for several reasons. One possibility is that one or more environmental "triggers" opened up new evolutionary opportunities:
  - Increasing oxygen levels in the atmosphere and oceans might have allowed smaller, diffusion-limited creatures to increase in size.
  - The increase in size of soft parts would have required skeletons for support and muscle attachment and perhaps also protection from predators.
  - The evolution of shells would in turn have increased the chances of fossilization.
  - It's also possible that early animals were responding to the reorganization of the biogeochemical cycles of carbon and nutrients such as phosphorus, which might have increased both food and oxygen at the sea bottom.

- Early metazoan evolution also appears to have involved macroevolution, as exemplified by the sometimes enigmatic creatures of the Burgess Shale. Different body plans, representing the major phyla of the Phanerozoic such as worms, arthropods (trilobites), strange echinoderms, and chordates (which include the vertebrates), appeared in response to modification of *Homeobox* genes. After the basic body plans were established, further evolutionary experimentation during the Phanerozoic appears to have been constrained by the basic body plans of the major phyla.

- Molecular clocks hint that metazoan evolution might have been occurring well before the appearance of the definite body and trace fossils in the Vendian, and several fossil occurrences dating to about 1 to 1.5 billion years ago have been reported. If the fossil record and molecular clocks turn out to agree, possibly much early metazoan evolution occurred among quite small soft-bodied creatures that have not as yet been found in the fossil record.

KEY TERMS

acritarch	ecologic hypothesis	multiple causation	Phylum Cnidaria
biologic "arms race"	genomic hypothesis	Pan-African Orogeny	Phylum Echinodermata
body plan	Gondwana	phosphorites	Phylum Mollusca
brachiopods	hyoliths	Phylum Annelida	Phylum Porifera
Burgess Shale	Kingdom Protista	Phylum Arthropoda	small shelly fossil
Cambrian explosion	metazoans	Phylum Brachiopoda	trilobites
chiton	monoplacophorans	Phylum Chordata	Vendozoa

REVIEW QUESTIONS

1. What is the sequence of developments as metazoans appeared in the fossil record?
2. What is the significance of the Ediacara and Burgess Shale faunas?
3. What do small shelly fossils represent?
4. What is the importance of ichnofossils? What do they tell us about early metazoans?
5. Why are Lagerstätten important to the study of Earth history?
6. What were the potential roles of oxygen, predation, extinction, food, and macroevolution in the origins of early metazoans?
7. What do strontium isotope ratios tell us about ancient ocean chemistry, the appearance of metazoans, and the evolution of biogeochemical cycles?
8. Compare the ecologic and genomic hypotheses of Metazoan evolution. Are the two hypotheses mutually exclusive? Why or why not?
9. What is the significance of each of the following fossils: *Anomalocaris*, *Cloudina*, *Dickinsonia*, trilobites. Place them in order of their appearance in the fossil record on a diagram like Figure 10.1.

FOOD FOR THOUGHT:
Further Activities In and Outside of Class

1. Give an example of how an observer's biases or preconceptions can influence the interpretation of observations or data.
2. Did Walcott's classification of the Burgess Shale fossils more likely reflect an equilibrium or an evolving world (see Chapter 1)?
3. Compare cause and effect as determined in a laboratory experiment with the causes for the origins (the effect) of metazoans and the method of multiple working hypotheses described in Chapter 1.

Anbar, A., and Knoll, A. H. 2002. Proterozoic ocean chemistry and evolution: A bioinorganic bridge? *Science*, 297, 1137–1142.

Brasier, M. D., and Lindsay, J. F. 1998. A billion years of environmental stability and the emergence of eukaryotes: new data from northern Australia. *Geology*, 26, 555–558.

Chen, J. Y., Huang, D. Y., and Li, C. W. 1999. An Early Cambrian craniate-like chordate. *Nature*, 402(6761):518–521.

Conway Morris, S. 1998. *The crucible of creation: The Burgess Shale and the rise of animals*. Oxford, UK: Oxford University Press.

Darwin, C. 1896. *The formation of vegetable mould, through the action of worms, with observations on their habits*. New York: D. Appleton and Company (reprinted from original).

Erwin, D. H., Laflamme, M., Tweedt, S. M., Sperling, E. A., Pisani, D., and Peterson, K. J. 2011. The Cambrian conundrum: early divergence and later ecological success in the early history of animals. *Science*, 334, 1091–1097.

Erwin, D. H., and Valentine, J. W. 2013. *The Cambrian Explosion: The Construction of Animal Biodiversity*. Greenwood Village, CO: Roberts and Company.

Fortey, R. 2001. The Cambrian explosion exploded? *Science*, 293(5529):438–439.

Gaines, R. R., Droser, M. L., Orr, P. J., Garson, D., Hammarlund, E., Qi, C., and Canfield, D. E. 2012. Burgess shale-type biotas were not entirely burrowed away. *Geology*, 40(3):283–286.

Gould, S. J. 1989. *Wonderful life: The Burgess Shale and the nature of history*. New York: W. W. Norton.

Johnston, P. A., Johnston, K. J., Collom, C. J., Powell, W. G., and Pollock, R. J. 2009. Palaeontology and depositional environments of ancient brine seeps in the Middle Cambrian Burgess Shale at the Monarch, British Columbia, Canada. *Palaeogeography, Palaeoclimatology, Palaeoecology*, 277, 86–105.

Jones, N. 2010. Cambrian's fiercest hunter defanged. *Nature News* (7 August 2009), doi:10.1038/news.2009.811 News.

Knauth, L. P. 2012. Not all at sea. *Nature* doi:10.1038/nature11765.

Knoll, A. H. 2003. *Life on a young planet: The first three billion years of evolution on earth*. Princeton, NJ: Princeton University Press.

Knoll, A. H., and Carroll, S. 1999. Early animal evolution: Emerging views from comparative biology and geology. *Science*, 284, 2129–2137.

Logan, G. A., Hayes, J. M., Hieshima, G. B., and Summons, R. E. 1995. Terminal Proterozoic reorganization of biogeochemical cycles. *Nature*, 376, 53–56.

Martin, R. E. 1999. *Taphonomy: A process approach*. Cambridge, UK: Cambridge University Press.

Martin, R. E., Quigg, A., and Podkovyrov, V. 2008. Marine biodiversification in response to evolving phytoplankton stoichiometry. *Palaeogeography, Palaeoclimatology, Palaeoecology*, 258, 277–291.

McMenamin, M. A. S. 1998. *The Garden of Ediacara: Discovering the first complex life*. New York: Columbia University Press.

Planavsky, N. J., Rouxel, O. J., Bekker, A., Lalonde, S. V., Konhauser, K. O., Reinhard, C. T., and Lyons, T. W. 2010. The evolution of the marine phosphate reservoir. *Nature*, doi:10.1038/nature09485.

Raff, R. A. 1996. *The shape of life: Genes, development, and the evolution of animal form*. Chicago: University of Chicago Press.

Retallack, G. J. 2012. Ediacaran life on land. *Nature*, 493, 89–92. doi:10.1038/nature11777.

Schwartz, J. H. 1999. *Sudden origins: Fossils, genes, and the emergence of species*. New York: John Wiley and Sons.

Van Roy, P., Orr, P. J., Botting, J. P., Muir, L. A., Vinther, J., Lefebvre, K., el Hariri, K., and Briggs, D. E. G. 2010. Ordovician faunas of Burgess Shale type. *Nature*, 465, 215–218.

Xiao, S. 2012. Muddying the waters. *Nature*, doi:10.1038/nature11765.

THE PHANEROZOIC: TOWARD THE MODERN WORLD

The Phanerozoic Eon ("time of apparent life") represents approximately the last 540 million years of Earth's history. Our understanding of the timing of events during the Phanerozoic is much better than for that of the Precambrian because of the extensive fossil record that begins with the Cambrian Period at the base of the Phanerozoic. In fact, changes in fossil biotas were used as early as the 19th century to recognize the three main eras of the Phanerozoic: Paleozoic ("ancient life"), Mesozoic ("middle life"), and Cenozoic ("recent life"). Each era is marked by shifting continent–ocean configurations; multiple episodes of orogeny; significant changes in sea level, climate, and lithology; and the origin and extinction of different biotas in the oceans and on land. As a result, Earth's systems and their interactions continued to evolve.

We examine the evolution and interactions of Earth's systems in detail in each of the chapters of Part III using the tectonic cycle as a unifying theme. For each interval we examine (in approximate order):

1. The general geologic and climatic setting resulting from the impact of the solid Earth system—especially the tectonic cycle—on continent–ocean configurations, sea level, sedimentary facies, climate, and ocean chemistry (see Chapter 3)
2. The impact of orogeny on these conditions
3. The evolution and integration of the biosphere with other Earth systems in the context of the evolving biogeochemical cycle of carbon
4. The causes of biodiversification and the causes and consequences of mass extinction involving Earth's systems

As we proceed in subsequent chapters, you should be attempting to describe and explain the interactions of Earth's systems in your own words and diagrams.

The Early-to-Middle Paleozoic World

MAJOR CONCEPTS AND QUESTIONS ADDRESSED IN THIS CHAPTER

Ⓐ What were the positions of the continents during the first half of the Paleozoic Era?

Ⓑ How did the movements of the continents affect orogeny, sea level, and climate?

Ⓒ Are there alternative continental configurations for this time? If so, what is the evidence?

Ⓓ What were the effects of continental movements on ocean circulation and chemistry? How did these factors differ from today?

Ⓔ Why were black shales so widespread at times, and why was the calcite compensation depth (CCD) shallower?

Ⓕ How do the major physiographic provinces of eastern North America reflect its geologic history?

Ⓖ What happened during the first two orogenies to form the Appalachian Mountain chain, and where did they occur?

Ⓗ How did western North America and northern Europe behave during this time?

Ⓘ How did orogeny affect the other Earth systems?

Ⓙ What were the dominant plankton of this time?

Ⓚ What were the dominant benthic marine groups, and how did they affect nutrient recycling?

Ⓛ Why were reefs so widespread, and what were the major groups of reef builders?

Ⓜ When did life invade freshwater and land, and what were the first groups to do so?

Ⓝ When did fish originate, and how did they compare to modern fish?

Ⓞ When and how did jaws originate, and what was the effect on ecosystems?

Ⓟ When did amphibians appear, how and from which groups did they originate, and what did they look like?

Ⓠ What is the evidence for the earliest land plants, and where did they live?

Ⓡ What major evolutionary innovations allowed land plants to grow larger and colonize continental interiors far from water?

Ⓢ When did forests appear and why? What were the effects on ecosystems?

Ⓣ When did mass extinction occur? Were there multiple causes?

CHAPTER OUTLINE

A reconstruction of an Early Paleozoic marine benthic community. *Anomalocaris* is especiailly prominent in this reconstruction. During the Early-to-Middle Paleozoic periods of the Cambrian, Ordovician, Silurian, and Devonian periods, benthic marine faunas changed from those dominated by taxa that fed close to or just beneath the sediment surface in the Cambrian, such as trilobites and inarticulate brachiopods, to widespread and well-developed reefs in the Devonian. Other taxa, such as crinoids ("sea lilies") began to feed higher above sea bottom as plankton expanded. Benthic faunas were accompanied by primitive fish called ostracoderms. The dominant predators in the seas changed from nautiloid and ammonoid cephalopods beginning in the Ordovician to jawed fish, including large extinct placoderms and sharks. On land, primitive amphibians and forests were present by the end of the Devonian.

© Corey Ford/Alamy Stock Photo.

11.1 Introduction: Beginnings of the Phanerozoic Eon

The Early-to-Middle Paleozoic is composed of the Cambrian, Ordovician, Silurian, and Devonian periods. These periods span the first half of the Paleozoic tectonic cycle (**Figure 11.1**). During this time a number of the continents—particularly those we associate with the northern hemisphere such as North America and Europe—drifted into warm, tropical latitudes, with North America (Laurentia) laying on its eastern side (**Figure 11.2**). Also during much of this time, sea level was typically much higher than today, sometimes by hundreds of meters, spreading across the continents in warm, shallow epicontinental seaways that resulted from the rifting.

These seaways began to teem with life after the early radiation of metazoans in the Neoproterozoic and the Cambrian explosion (see Chapter 10). Ancient reefs eventually thrived in the sunlit waters of the shallow seaways, far into the interiors of the continents (refer to this chapter's frontispiece). Small jawless fish appeared early but eventually gave way to sharks and extinct groups of jawed fishes that undoubtedly began to terrorize the oceans' other inhabitants. In contrast, vertebrates did not invade the land until much later with the spread of forests in the Devonian.

As the continents moved about, some of them eventually collided, producing mountain chains and destroying the seaways. These mountain chains or their remnants are still with us today, such as portions of the Appalachian Mountains. These particular orogenies involved collisions between what are now eastern North America and Europe during the Late Ordovician and Late Devonian. Mountain building also took place in western North America with the accumulation of different terranes.

Despite the overall warmth of the first half of the Paleozoic Era, plate movements led to the establishment of a continental glacier in the southern hemisphere during the Late Ordovician and perhaps also the Late Devonian. The biosphere was decimated by mass extinctions near the end of the Ordovician and Devonian periods. Global cooling (as a result of glaciation) has been primarily implicated in the Late Ordovician extinction but the Late Devonian extinction might have involved multiple factors, among them orogeny, sea-level change, and anoxia.

11.2 Tectonic Cycle: Impacts on the Hydrosphere, Atmosphere, and Rock Cycle

11.2.1 Sea level, CO_2, and sedimentary facies

The tectonic cycle of the Paleozoic begins with the presence of one or more supercontinents. As we saw in Chapter 10, the supercontinent Gondwana was formed by the final phase of the Pan-African Orogeny during the late Proterozoic. The assembly of Gondwana involved the collision of East Gondwana (mainly India, Antarctica, and Australia) and West Gondwana (mainly Africa and South America) as well as parts of Florida, the Middle East, southern Europe, southeast Asia, and Tibet. During the Cambrian and Early-to-Middle Ordovician periods, various continents began to rift away from Gondwana, which largely remained in the southern hemisphere. The continents that rifted away were **Laurentia** (basically North America and Greenland), **Baltica** (mostly northern Europe west of the Ural Mountains of Russia), **Siberia** and China (including southeast Asia), and **Kazakhstania** (centered on Kazakhstan). As the continents rifted apart, the **Iapetus Ocean** was produced (see Figure 11.2 and **Box 11.1**).

The rifting and continued movement of these continents away from Gondwana had several important effects. First, the level of the Iapetus and other oceans was typically much higher than today, often by as much as several hundred meters (see Figure 11.1). As rifting occurred, seafloor spreading centers are thought to have displaced large volumes of water out of the ocean basins onto the continents and into their interiors to produce epicontinental seas (refer to the Tectonic Cycle in Chapter 3). Indeed, large amounts of igneous rocks were emplaced during this time, indicating

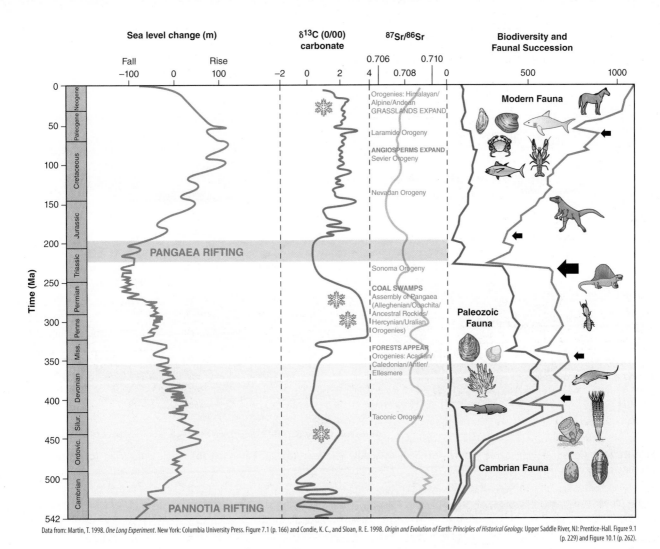

FIGURE 11.1 The physical and biologic evolution of the Earth through the Phanerozoic Eon. The lightly shaded area indicates the Cambrian through Devonian periods of the Phanerozoic Eon to be discussed in this chapter. Periods to be discussed in succeeding chapters will be similarly shaded. Snowflakes indicate times of major glaciation. The behavior of the stable isotopes of carbon and strontium is also indicated (see Chapter 9 for elementary discussion). Arrows indicate the Big Five mass extinctions that we will also begin to discuss in this chapter.

Data from: Martin, T. 1998. *One Long Experiment*. New York: Columbia University Press. Figure 7.1 (p. 166) and Condie, K. C., and Sloan, R. E. 1998. *Origin and Evolution of Earth: Principles of Historical Geology*. Upper Saddle River, NJ: Prentice-Hall. Figure 9.1 (p. 229) and Figure 10.1 (p. 262).

widespread igneous activity associated with rifting, subduction, and volcanism (see Figure 11.1). The planet also began to warm as rifting occurred. Rifting, subduction, and volcanism would have produced large amounts of carbon dioxide (CO_2; see Figure 11.1). Because terrestrial forests did not become widespread until after the Devonian (see Chapter 12), much of the CO_2 accumulated in the atmosphere and was drawn down only very slowly by continental weathering. The extensive seaways would have reduced Earth's albedo, also warming the planet. In fact, so much of Laurentia was covered by water that only part of the Canadian Shield and a few islands remained emergent along a structural high. This feature, known as the Transcontinental Arch, stretched all the way from the Great Lakes to New Mexico (**Figure 11.3**).

In addition to the Transcontinental Arch, other broad structural highs and basins appeared at various times during the Phanerozoic, presumably in response to tectonic activity or plate movements, although the exact mechanisms of

formation are uncertain. The highs or lows are recognizable by the gentle dip of the strata away from the high or into the basin, respectively, but the angle of dip is so gentle that it is not recognizable on a local basis and must be traced over a much larger region. Highs like the Cincinnati Arch (or dome; see Figure 11.3) are also recognizable by the thinning of strata over the highs or the complete erosion of the strata and their replacement by unconformities; such relationships suggest the highs were present as the strata were being deposited. Conversely, basins are recognizable by the thickening and gentle regional dip of the strata toward the center of the basin.

The first of the major sea-level transgressions onto Laurentia occurred during the Cambrian and Early Ordovician on either side of the Transcontinental Arch. This initial broad sea-level rise, called the Sauk Sequence, was so pronounced it was really only significantly interrupted by mild regressions during the Middle and Late Cambrian (see Figure 11.1). In

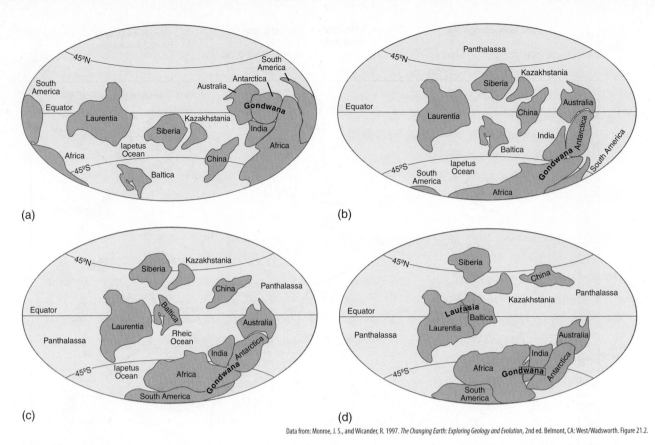

FIGURE 11.2 General paleogeography during the Early-to-Middle Paleozoic. **(a)** Late Cambrian. **(b)** Middle Ordovician. **(c)** Late Silurian. **(d)** Middle Devonian.

Data from: Monroe, J. S., and Wicander, R. 1997. *The Changing Earth: Exploring Geology and Evolution*, 2nd ed. Belmont, CA: West/Wadsworth. Figure 21.2.

BOX 11.1 SWEAT Hypothesis: An Alternative Hypothesis for Early-to-Middle Paleozoic Continent–Ocean Configurations

B Reconstructing ancient continental positions is much like piecing a jigsaw puzzle together. However, unlike a jigsaw puzzle, which has a single solution, different workers can arrive at different reconstructions of continent–ocean configurations. Indeed, not everyone agrees with the reconstructions of continental movements presented in Figure 11.2, and a number of alternative hypotheses for Late Proterozoic and Early Paleozoic continent–ocean configurations have been constructed.

One of these is by Ian Dalziel from the Institute for Geophysics of the University of Texas. Dalziel and colleagues examined similarities in lithology, structural trends (patterns of folding and faulting) in mountain ranges, and fossil distributions in South America, Antarctica, and Australia (much as Alfred Wegener did in his reconstructions; see Chapter 2). During the Late Proterozoic about 1 billion years ago, the supercontinent Rodinia was assembled and then began to rift apart by about 750 to 800 million

years before present (see Chapter 9). According to the *SWEAT hypothesis* (for *S*outh *W*est United States/*E*ast *An*Tarctica), a smaller continent consisting of Australia, Antarctica, and India is thought to have rifted away from western North America to produce the ancestral Pacific Ocean. Dalziel and colleagues believe Laurentia (North America) collided with South America to produce the continent **Pannotia** (**Box Figure 11.1A**). According to this reconstruction, Laurentia came to its equatorial position after colliding and then rifting away from South America, not Baltica. Eventually, South America, Australia, Antarctica, India, and Africa are thought to have collided with a number of smaller continents or terranes to produce Gondwanaland during the Pan-African Orogeny.

C How could such different interpretations of ancient continental positions be drawn from the same data? Continent–ocean configurations are known with much greater certainty for the Mesozoic and Cenozoic because magnetic "stripes" on the

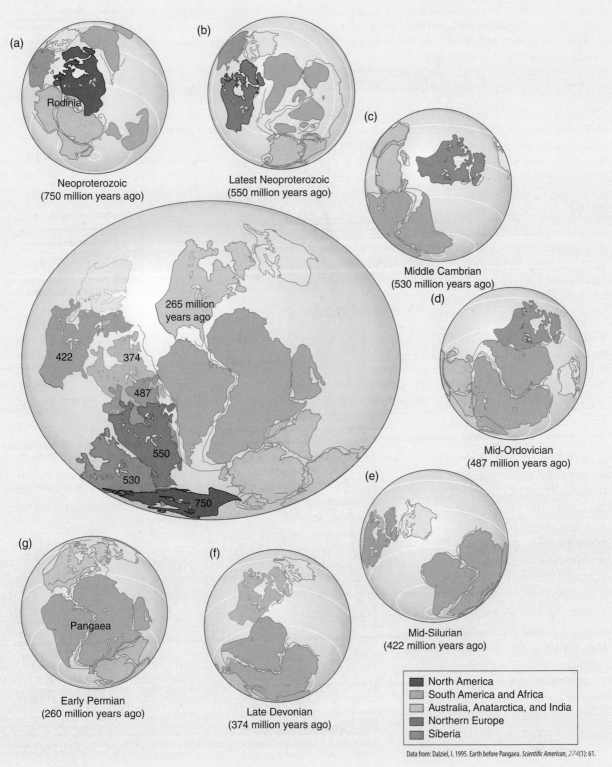

(a) Rodinia

Neoproterozoic
(750 million years ago)

(b)

Latest Neoproterozoic
(550 million years ago)

(c)

Middle Cambrian
(530 million years ago)

(d)

Mid-Ordovician
(487 million years ago)

265 million years ago

422   374

487

550

530

750

(e)

Mid-Silurian
(422 million years ago)

(g) Pangaea

Early Permian
(260 million years ago)

(f)

Late Devonian
(374 million years ago)

Legend:
- North America
- South America and Africa
- Australia, Antarctica, and India
- Northern Europe
- Siberia

Data from: Dalziel, I. 1995. Earth before Pangaea. *Scientific American, 274*(1): 61.

BOX FIGURE 11.1A Movement of the continents during the Late Proterozoic and Early Paleozoic according to Dalziel and coworkers. Dates on the continents indicate time of position.

(continues)

ocean floor can be dated back to the Jurassic and used to reconstruct ancient positions. Very simply, one can begin with the stripes closest to the continents and move the continents back toward their original positions in Pangaea one stripe at a time (see Chapter 2). However, other than occasional ophiolites (see Chapter 2), there is no ocean crust older than the Jurassic because older crust has been subducted. So, before the Jurassic, we are left only with the evidence recorded on the continents to reconstruct their positions. Unfortunately, the paleomagnetic data from continents can tell us about the latitudes of continents or terranes, but not their longitudes (see Chapter 2). Thus, hypotheses about the ancient positions of continents must be tested against lithologic, structural, and fossil data. Even so, the rock record is not complete, and its data are subject to different interpretations. For example, the same or superficially similar rocks and structural trends (such as large-scale patterns of folds) that might be used to match up different continents can occur at very different times. Pieces of what was once the same biogeographic province (see Chapter 3) and their characteristic fossils that now lie on widely separated continents or terranes also allow the reconstruction of the positions of ancient land masses (**Box Figure 11.1B**). Because fossils do not repeat themselves through time (Principle of Faunal Succession; see Chapter 6), they might be more reliable indicators of the juxtaposition of ancient lands; unfortunately, the fossils aren't always present.

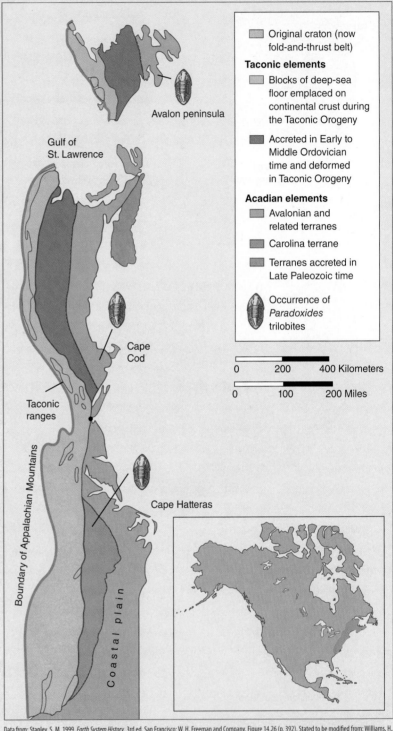

Original craton (now fold-and-thrust belt)

Taconic elements

Blocks of deep-sea floor emplaced on continental crust during the Taconic Orogeny

Accreted in Early to Middle Ordovician time and deformed in Taconic Orogeny

Acadian elements

Avalonian and related terranes

Carolina terrane

Terranes accreted in Late Paleozoic time

Occurrence of *Paradoxides* trilobites

Avalon peninsula

Gulf of St. Lawrence

Cape Cod

Taconic ranges

Boundary of Appalachian Mountains

Coastal plain

Cape Hatteras

Data from: Stanley, S. M. 1999. *Earth System History*, 3rd ed. San Francisco: W. H. Freeman and Company. Figure 14.26 (p. 392). Stated to be modified from: Williams, H., and Hatcher, R. D. *Geology*, *10*, 530–536. Boulder, CO: Geological Society of America.

BOX FIGURE 11.1B Trilobite provinces associated with different terranes in a standard reconstruction of Cambrian paleogeography. See text for discussion of Taconic and Acadian orogenies.

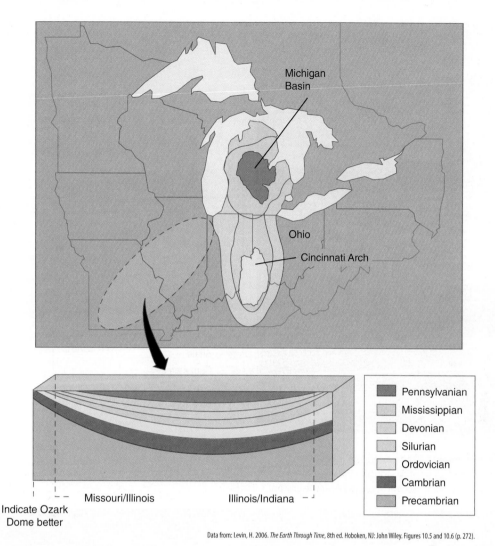

FIGURE 11.3 Location of the Transcontinental Arch and some other major structural features of the midwestern portion of the North American craton.

western North America, the Sauk transgression was embodied by the succession of the Tapeats Sandstone, Bright Angel Shale, and Muav Limestone of the Cambrian (see Chapter 6; **Figure 11.4**); this succession indicates initial high energy conditions on wave-swept shelves, followed by deepening and mud deposition, and then limestone deposition as terrigenous input was shut off (see Chapter 4). The limestones at the tops of the sequences were variously of biogenic or oölitic origin (**Figure 11.5** and **Figure 11.6**). By analogy with environments in which limestones are forming today, these limestones formed in relatively shallow waters where there was little terrigenous input. A similar sequence of rocks is found in what is now eastern North America, indicating the seas were transgressing over this portion of the continent, as well.

The **Sauk Sea** eventually began to retreat, exposing sediments to weathering and erosion to produce a more-or-less craton-wide disconformity. This unconformity marks the boundary between the Sauk Sequence and the subsequent Tippecanoe Sequence (see Figure 11.1; see Chapter 6). Like the Sauk Sequence, the initial sediments laid down in the Tippecanoe Sea were sandstones. However, the sands

were very clean, well-sorted, mature sandstones such as the St. Peter Sandstone of the midwestern United States. The highly mature state of these sandstones indicates that they formed by numerous cycles of weathering and reworking that winnowed out finer sediments while rounding and sorting the residual quartz grains. The sandstones were followed by extensive, and often highly fossiliferous, limestones, including reefs and dolomites.

The **Tippecanoe Sea** itself began to retreat by about Early Devonian, only to be followed by another transgression over Laurentia. This transgression is represented by deposition in the **Kaskaskia Sequence**. The Kaskaskia transgression continued into the Mississippian Period (see Chapter 12). Like the Sauk and Tippecanoe seas, basal sediments of the Kaskaskia seaway often consist of clean, well-sorted sandstones that give way upward to limestones, including reefs.

Other lithologies also indicate high sea level and warmth. Black shales were extensive at times during the Early-to-Middle Paleozoic (**Figure 11.7** and **Figure 11.8**). Black shales, the color of which results from high levels of organic matter, hint that oxygen levels in the oceans were low.

© iStockphoto/Thinkstock.

FIGURE 11.4 Outcrop of cross-bedded sandstone of Cambrian age near the Dells of Wisconsin. These sediments were deposited as the Sauk Sea transgressed over the craton.

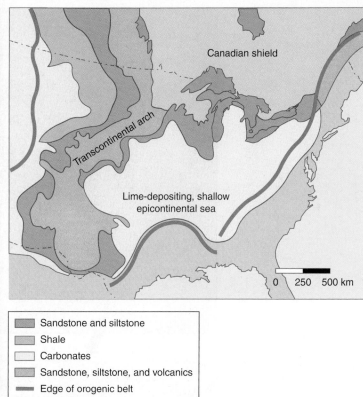

Data from: Levin, H. 2006. *The Earth Through Time*, 8th ed. Hoboken, NJ: John Wiley. Figure 10.9 (p. 275).

Sandstone and siltstone
Shale
Carbonates
Sandstone, siltstone, and volcanics
Edge of orogenic belt

FIGURE 11.5 The widespread deposition of limestones in the interiors of modern continents could only have occurred in shallow, epicontinental seas.

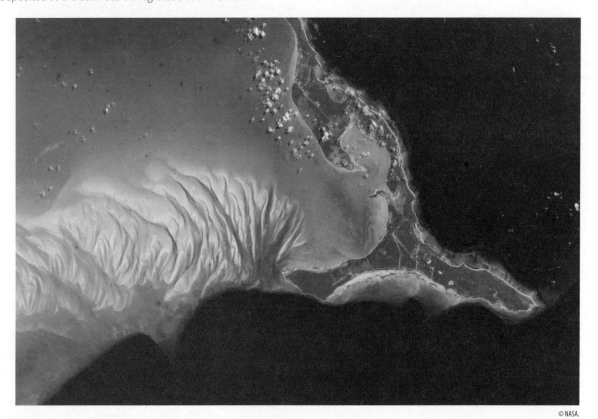

© NASA.

FIGURE 11.6 Aerial photo of Bahamas, with their widespread carbonate environments (compare Figure 6.8). The enormous sand waves mainly consist of oöids. The fluffy white blobs near the island are clouds.

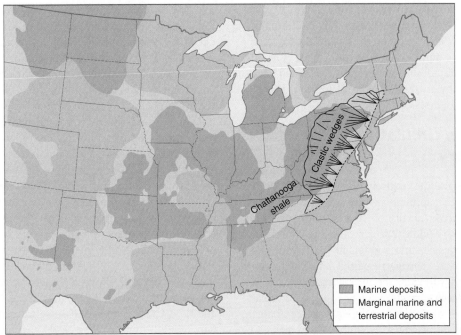

Data from: Stanley, S. M. 1989. *Earth and Life Through Time*. New York: W. H. Freeman and Company. Figure 13.40 (p. 377).

FIGURE 11.7 Map of the general distribution of outcrops of the Chattanooga Shale and contemporaneous deposits.

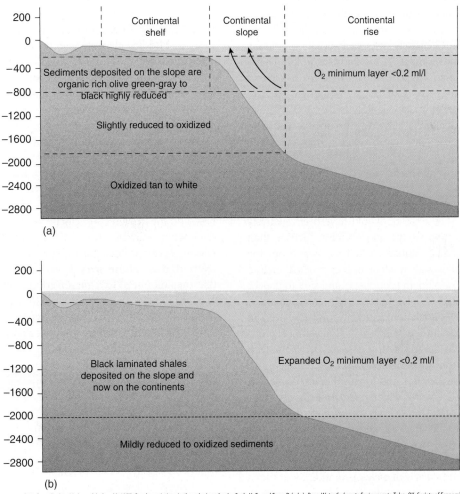

Data from: Fischer, M. A., and Arthur, M. 1977. Secular variations in the pelagic realm. In: Cook, H. E., and Enos, P. (eds.). *Deep-Water Carbonate Environments*. Tulsa, OK: Society of Economic Paleontologists and Mineralogists (Renamed SEPM Society for Sedimentary Geology). Special Publication No. 25, Figure 8 (p. 39).

FIGURE 11.8 How the oxygen minimum zone tracks sea level as it rises over the continents.

The presence of extensive evaporites in what is now the state of Michigan also suggests extremely warm climates during the Silurian (see Figure 11.3). Based on the geologic distribution of rocks beneath the surface, it appears Michigan was surrounded by a gigantic atoll-like structure that restricted ocean circulation sufficiently to result in the deposition of thick evaporite deposits.

11.2.2 Ocean circulation and chemistry

Ⓔ High sea level and the relatively dispersed nature of the continents in the northern hemisphere had important consequences for ocean circulation and chemistry. During much of the Early-to-Middle Paleozoic there was probably plenty of time for tropical surface waters to warm from solar insolation because of the generally dispersed nature of the continents. As compared with today, the ancient trade winds probably drove surface ocean currents over long distances from east to west in relatively low latitudes near the equator (see Chapter 3). Along with the widespread seas, low albedo, and high CO_2 levels, the circulation of surface waters in low latitudes around Earth more than once probably contributed to the warming of Earth. If any portions of these equatorial currents were deflected by land masses to the north and south, they would have carried heat to higher latitudes, as well. Because the ocean waters were warmer, they also lost more water vapor to the atmosphere. Computer models of ancient ocean circulation also indicate that during much of the Early-to-Middle Paleozoic, cool waters might have penetrated toward the tropics.

Circulation within the deep oceans is thought to have been relatively sluggish. Today, ice is found over both poles and ocean circulation is relatively fast; the oceans turn over about once every 1,000 years (see Chapter 3). However, with the exception of the Late Ordovician (and possibly the Late Devonian), glaciers were absent over the poles during the Early-to-Middle Paleozoic. This suggests deep water masses initially formed by evaporation in shallow seaways or perhaps in restricted basins. These waters would have been relatively saline (because of evaporation) and therefore denser than more open ocean surface waters located far away from the continents. The shallow saline water masses might therefore have been able to flow off of the continents and sink to greater depths, driving ocean circulation but at slower rates than today.

Sluggish ocean circulation in turn implies that the oceans held relatively little oxygen (O_2). Sluggish ocean circulation would have decreased the contact between deep ocean waters and the atmosphere, when O_2 diffuses into surface waters before they descend for another circuit within the ocean basins. Also, warm waters carry less dissolved O_2 than cold ones. Indeed, the oceans of the Early-to-Middle Paleozoic tended to become anoxic, as indicated by the frequent occurrence of black shales (see Figure 11.7 and Figure 11.8). These deposits suggest in turn that the oxygen minimum zone (OMZ) was well developed and that at times it might have shallowed sufficiently ("tracking" sea level as it rose over the continents) to spread across the continents on which the black shales were deposited (see Figure 11.8).

The extensive deposition of limestones ($CaCO_3$) in shallow waters also had important implications for seawater chemistry. Widespread limestone deposition suggests plenty of dissolved calcium (Ca^{2+}) and bicarbonate (HCO_3^-) in epeiric seas. Most likely the abundance of these ions resulted from continental weathering resulting from high levels of CO_2 in the atmosphere (a result of rifting, seafloor spreading, subduction, and volcanism associated with the tectonic cycle); weathering itself therefore appears to have been too slow to have drawn down CO_2 levels significantly (see Figure 11.1).

Widespread limestone deposition in the extensive epeiric seaways of the Early-to-Middle Paleozoic might have caused the CCD to shallow well up onto the continental slope. Recall from Chapter 4 that the CCD helps maintain equilibrium between the deposition and the dissolution of limestone ($CaCO_3$) in the deep oceans. When enough $CaCO_3$ is deposited in the deep oceans (by calcareous ooze), the CCD deepens; however, when insufficient $CaCO_3$ is deposited in the deep ocean, the CCD is "starved" of $CaCO_3$ and the CCD shallows to dissolve calcareous ooze to maintain equilibrium. Widespread shallow-water limestone deposition during the Early-to-Middle Paleozoic would presumably have "starved" the deep sea of $CaCO_3$. To dissolve more calcium carbonate and maintain equilibrium, the CCD would have shallowed onto the continental slope. Red clays like those now found below the CCD in the abyssal ocean might therefore have formed at shallower depths of the Early-to-Middle Paleozoic than they do today (**Figure 11.9**).

It also appears that during the Early-to-Middle Paleozoic calcareous hard parts were more prone to consist of the mineral calcite. The most widely accepted hypothesis for the predominance of calcite skeletons is related to rates of seafloor spreading and the presence of mid-ocean ridges. Greater seafloor spreading increases hydrothermal weathering (see Chapter 3) at mid-ocean ridges. Hydrothermal weathering at these sites promotes the movement of magnesium ions (Mg^{2+}) from seawater into ocean crust, whereas calcium ions (Ca^{2+}) move into the seawater (see Chapter 3). Based on the fossil record, calcite skeletons were more abundant when seawater was enriched in Ca^{2+} ions. For this reason, ocean waters during such intervals are sometimes referred to as **calcite seas**. However, recent studies indicate that the amount of CO_2 dissolved in seawater and the ratios of Mg/Ca ions in seawater might also influence the mineralogy of the shell material, so the hypothesis might be taken, for the time being, with a grain of salt.

CONCEPT AND REASONING CHECKS

1. What evidence do we use to reconstruct ancient continent–ocean configurations?
2. Can we interpret this evidence in different ways?
3. Why was Earth presumably so warm during much of the Early-to-Middle Paleozoic Era?
4. Why was limestone deposition so widespread during this time, and how did it affect the CCD?

(a)

Courtesy of Ron Martin.

(b)

Courtesy of Ron Martin.

(c)

Courtesy of Ron Martin.

FIGURE 11.9 **(a)** Flat-lying, interbedded limestones and shales were quite common in the interior of North America during the first half of the Paleozoic Era. **(b)** Laminated shallow-water carbonates and sands were common along the margins of Laurentia during the Ordovician. Widespread limestone deposition may have resulted in a very shallow CCD during the early Paleozoic. **(c)** Ordovician red and green shales outcropping near Reading, Pennsylvania. These shales might have been deposited on the continental slope but below the CCD. Compare to the distribution of red clays in the modern ocean shown in Chapter 4.

11.3 Tectonic Cycle and Orogeny

The seeming calm of the Early-to-Middle Paleozoic Earth was at times disrupted by significant episodes of orogeny. These orogenic episodes affected sea level, climate, and ocean circulation and chemistry as the passive tectonic margins that initially existed along North America (Laurentia) became active margins.

11.3.1 Physiographic provinces of the Appalachian Mountains

The orogenic episodes along the southern margin of Laurentia are primarily responsible for the modern physiographic provinces—including the Appalachian Mountains—of the eastern portion of North America. We must briefly review these provinces to better understand the orogenies of the Early-to-Middle Paleozoic. These particular provinces run from Newfoundland (Canada) through the northeastern United States to the Carolinas and westward as far as Oklahoma (**Figure 11.10** and **Table 11.1**).

The **Valley and Ridge Province** is the westernmost province of the **Appalachian Mountains**. Here, the rocks consist of sandstones, shales, and limestones which were deposited on the shallow continental shelves of epeiric seas. For this reason the Valley and Ridge is sometimes referred to as the "sedimentary Appalachians." The rocks are often folded, indicating that they were deformed during orogenesis as

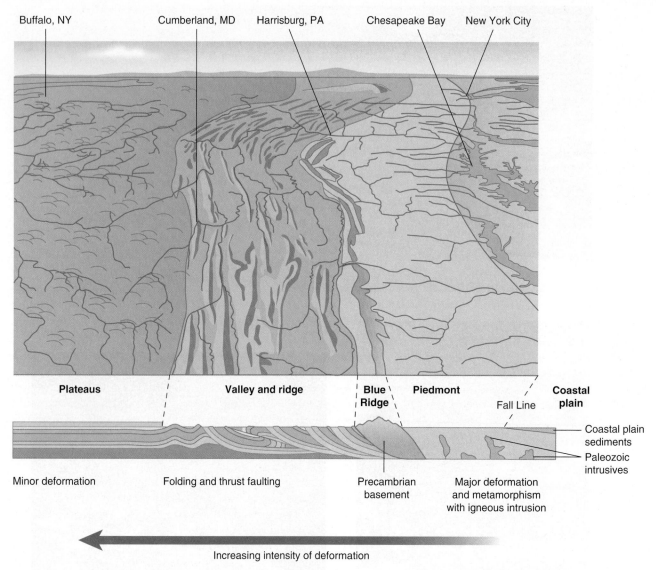

FIGURE 11.10 Major geologic and geographic features of the eastern portion of the United States. Note the occurrence of major eastern cities along the Fall Line or Zone, which marks the boundary between the Piedmont and the Coastal Plain.

TABLE 11.1

Comparison of different regional names for the physiographic and geologic provinces of the Appalachian region				
Geomorphic province	Allegheny and Cumberland Plateaus	Valley and Ridge province	Blue Ridge province (including Green Mountains in north)	New England Upland and Piedmont province	Atlantic Coastal Plain and continental shelf
Structural unit	Allegheny Synclinorium	Sedimentary Appalachians (folded and faulted sedimentary rocks)	Anticlinorium in some places, faulted uplift in others	Crystalline Appalachians (metamorphic and plutonic rocks)	Cretaceous and Tertiary sediments
Rock facies	Foreland	"Shelf"	Exposures of basal Paleozoic and of Precambrian basement	"Trench"	Relatively flat-lying Cretaceous and Cenozoic sediments

(a)

Courtesy of Ronald Martin, University of Delaware.

(b)

Courtesy of Ronald Martin, University of Delaware.

FIGURE 11.11 Landscape of the Valley and Ridge Province of the Appalachian Mountains. **(a)** Typical view from a ridge crest in Pennsylvania. **(b)** The band of hard, white sandstone known as the Tuscarora Sandstone, comprises the crests of some ridges in the eastern half of the Appalachians because it is resistant to erosion. The Tuscarora ranges from a quartz sandstone to conglomerate and was deposited in marginal marine to terrestrial conditions.

thrust sheets during continental collision (see Figure 11.10). A number of the mountain ridges (anticlines) are "held up" by hard sandstones resistant to erosion that are sometimes visible as white bands along the crests of ridges (**Figure 11.11**). Less resistant rocks like shales and limestones have been eroded between the ridges to produce the valleys.

Thrusting of rocks to produce the Valley and Ridge occurred from what is now the southeast toward the northwest. The rocks appear to have been displaced as much as 300 km (~180 miles). As a result of the thrusting, highly metamorphosed and deformed rocks that formed during the Precambrian Grenville Orogeny (see Chapter 9) were brought to the surface and slid over Early Paleozoic sediments. These rocks are now recognized as the **Blue Ridge Province**. The Blue Ridge extends from Alabama and Georgia through the western Carolinas and eastern Tennessee (including the Great Smoky Mountains) into southern Pennsylvania; the Green Mountains of Vermont appear to represent an extension of the Blue Ridge (see Table 11.1).

To the east and southeast of the Blue Ridge in New England lie the "crystalline Appalachians," so named for the predominance of igneous and metamorphic rocks. These rocks once formed the cores of high mountain ranges, but the cores are now exposed because of erosion and isostatic uplift. In New Hampshire, the crystalline Appalachians are represented by the plutonic igneous rocks of the White Mountains, which include Mt. Washington, the highest mountain in the northeastern United States (elevation 1,950 m [6,300 feet]).

South of New England, the crystalline Appalachians are referred to as the **Piedmont**. Here, the climate is more humid so rocks found as uplands and mountains in New England occur at lower elevations because of greater weathering. The Piedmont is represented by schists, gneisses, marbles, and pegmatites, along with the **slate belts** of eastern New York State and the Carolinas (**Figure 11.12**). Piedmont rocks also include graywackes, volcanics, and turbidites that were deposited in deep-water environments adjacent to

(a)

Courtesy of Ronald Martin, University of Delaware.

(b)

Courtesy of Ronald Martin, University of Delaware.

FIGURE 11.12 The Appalachian Piedmont. **(a)** Typical view of rolling hills underlain by schists and gneisses in Virginia. These rocks were likely originally seafloor crust and sediments of a Paleozoic trench. **(b)** View looking northward along Interstate 95 in Wilmington, Delaware. The dark, ugly looking rocks are part of an ancient volcanic arc.

ancient seafloor trenches. These rocks were deformed and metamorphosed during orogeny and suturing of the ancient continents.

The western boundary of the Piedmont is marked by the Fall Line or Fall Zone. Many eastern American cities such as Philadelphia, Baltimore, and Washington are located along the Fall Line because rivers above this boundary are not navigable due to the occurrence of waterfalls and rapids (see Figure 11.12). Along the Fall Line the rocks of the Piedmont disappear beneath the relatively flat-lying sediments of the **Atlantic Coastal Plain**. Coastal plain sediments are of Mesozoic and Cenozoic age and were eroded much later from the Appalachian Mountains that lie to the west and north. Coastal plain sediments in turn grade almost imperceptibly into the continental shelf offshore and can reach about 10 miles in thickness in places.

As the rocks of the Appalachians were uplifted during different orogenic episodes, the sediments were also shed westward into foreland basins (**Figure 11.13**). Many of these rocks range from Pennsylvanian to Permian in age and consist mostly of molasse such as sandstones, conglomerates, and coals deposited in terrestrial to marginal marine environments. The sediments also frequently include redbeds because of the oxidation of iron. The rocks of this ancient foreland basin now comprise the **Allegheny Plateau** of western Pennsylvania and New York State. Further south, such as in Kentucky, the same deposits are referred to as the **Cumberland Plateau** (see Figure 11.10). The deposits of these plateaus are relatively horizontal, indicating that they were much less affected by orogeny. However, the rocks are also deeply cut by streams and rivers so the landscape often consists of large rolling hills. In Pennsylvania, the eastern edge of the plateau forms a steep escarpment called the Allegheny Front adjacent to the Valley and Ridge Province. For these reasons the plateaus were often more of a barrier to early settlers than were the folds of the Valley and Ridge.

11.3.2 Orogenic episodes

The Early-to-Middle Paleozoic orogenies affecting North America and Europe were quite complex. These orogenic phases involved not only the continents of Laurentia and Baltica, but also terranes such as volcanic island arcs and microcontinents and changes in the types of plate boundaries through time.

Taconic orogeny

Ⓖ The first of the major orogenic episodes of the Paleozoic Era was the **Taconic Orogeny**, named after the Taconic Mountains of eastern New York, Massachusetts, and Vermont. As its name implies, the Taconic Orogeny primarily affected what is now New England, but its effects in North America extend as far north as Newfoundland and as far south as the Carolinas, and it is responsible for much of the Valley and Ridge province.

The Taconic Orogeny did not involve the *suturing* of continents, however. Rather, it appears to have primarily involved

the uplift of a volcanic island arc sandwiched between Laurentia and Baltica. As Laurentia and Baltica separated from Gondwana and from each other during the Cambrian and Early Ordovician, several island arcs and microcontinents rifted away from Gondwana and moved toward Baltica and Laurentia. One of these arc-microcontinent complexes, **Avalonia** (basically modern Newfoundland), was accompanied by smaller islands (**Figure 11.14** and **Figure 11.15**; compare Box Figure 11.B1). It was these islands that collided with the North American craton during the Taconic Orogeny in the Middle-to-Late Ordovician as the Iapetus Ocean began to close. Today, these former islands are represented by terranes in Newfoundland, New Brunswick, and Maine. The previous existence of the islands is inferred from their lithologies (which are not found on the craton) and the presence of very different taxa of trilobites that have restricted biogeographic ranges. Apparently, different populations of trilobites were restricted to certain islands so that distinct taxa possibly evolved on the islands because of geographic isolation (see Chapter 5 and Box Figure 11.B2).

The onset of the Taconic Orogeny was signaled by the cessation of limestone deposition in the Sauk Sea and the onset of flysch deposition. These deposits included black shales, graywackes, turbidites, and volcanics (**Figure 11.16** and **Figure 11.17**). Uplift associated with the Taconic Orogeny deformed the flysch to produce the Taconic Mountains. After their uplift the Taconic "Alps" might have been as high as 4,000 meters, based on calculations of the volume of sediment that was shed by erosion. This is comparable with the height of the much younger mountains of western North America, which have been eroded for far less time. However, today the mountains of New England mostly range from a few hundred to a thousand meters or so in height as a result of hundreds of millions of years of erosion.

The molasse shed from the Taconic Mountains formed a clastic wedge called the **Queenston Delta** that replaced the flysch. The volume of molasse was so enormous that it was shed into the foreland basin as far west as the upper Midwest (see Figure 11.13). The molasse included hard, coarse sandstones and conglomerates like the Juniata and Tuscarora formations that are resistant to weathering and often form the ridges of the Valley and Ridge province, described previously (Figure 11.11).

The Taconic Orogeny has its counterpart in northern Europe, particularly Great Britain, Norway, and eastern Greenland. This orogeny was originally named as a separate orogeny, the **Caledonian Orogeny**, before the significance of plate tectonics was recognized (see Figure 11.15). With the acceptance of plate tectonics, the Caledonian orogenic belt is now recognized as the continuation of the Taconic belt, so that the northern portion of Avalonia included what is now the British Isles (see Figure 11.15). The Caledonian Orogeny involved the collision and suturing of Laurentia and Baltica as the northern Iapetus Ocean closed. However, the main phase of the Caledonian Orogeny occurred somewhat later than the Taconic, during the Late Silurian and Early Devonian periods.

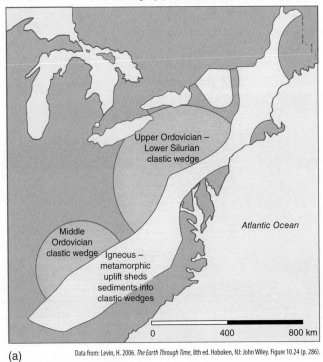

Taconic Orogeny (Late Ordovician)

Upper Ordovician – Lower Silurian clastic wedge

Atlantic Ocean

Middle Ordovician clastic wedge

Igneous – metamorphic uplift sheds sediments into clastic wedges

0   400   800 km

(a)

Data from: Levin, H. 2006. *The Earth Through Time*, 8th ed. Hoboken, NJ: John Wiley. Figure 10.24 (p. 286).

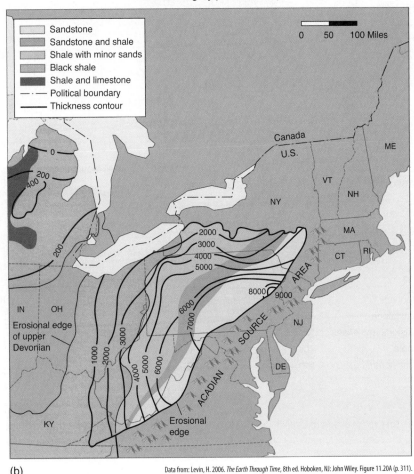

Acadian Orogeny (Late Devonian)

Sandstone
Sandstone and shale
Shale with minor sands
Black shale
Shale and limestone
Political boundary
Thickness contour

0   50   100 Miles

Canada
U.S.

ME
VT
NH
NY
MA
CT   RI

IN   OH
Erosional edge of upper Devonian

NJ

DE

KY   Erosional edge

SOURCE AREA

ACADIAN

(b)

Data from: Levin, H. 2006. *The Earth Through Time*, 8th ed. Hoboken, NJ: John Wiley. Figure 11.20A (p. 311).

FIGURE 11.13 The Taconic and Acadian orogenies occurred in what is now eastern North America. **(a)** Location of Middle and Late Ordovician (Queenston) deltas that resulted from the Late Ordovician Taconic Orogeny. **(b)** Location of Catskill delta that resulted from the Late Devonian Acadian Orogeny.

(continues)

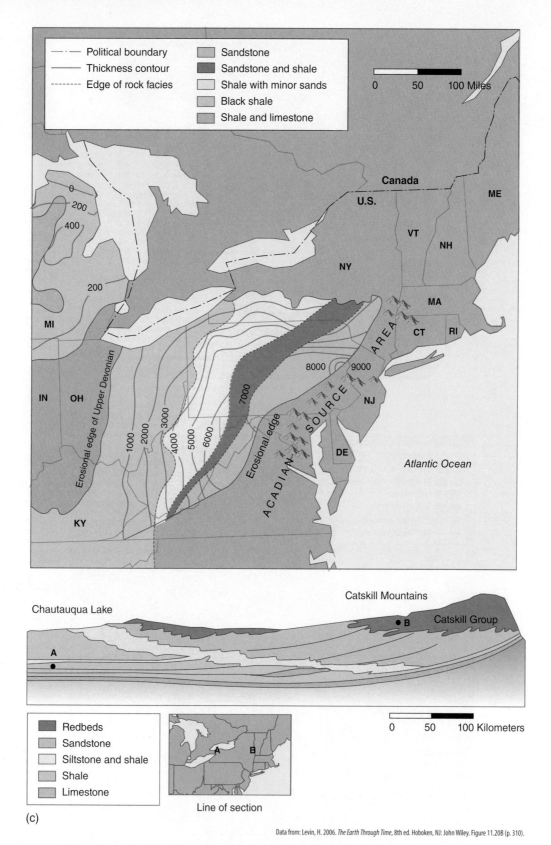

Legend (map):
- — · — Political boundary
- —— Thickness contour
- ----- Edge of rock facies
- Sandstone
- Sandstone and shale
- Shale with minor sands
- Black shale
- Shale and limestone

Legend (cross-section c):
- Redbeds
- Sandstone
- Siltstone and shale
- Shale
- Limestone

Line of section

Data from: Levin, H. 2006. *The Earth Through Time*, 8th ed. Hoboken, NJ: John Wiley. Figure 11.20B (p. 310).

FIGURE 11.13 (Continued) **(c)** Cross-section of Catskill Delta. Notice the downwarping of sediments into what was a foreland basin. Notice how in parts b and c, the sediments tend to become more fine-grained (shale-rich) to the west and northwest of the mountain range into the foreland basin.

© Marli Miller/Visuals Unlimited.

FIGURE 11.14 Strata on the west coast of Newfoundland that were deformed by the Taconic Orogeny.

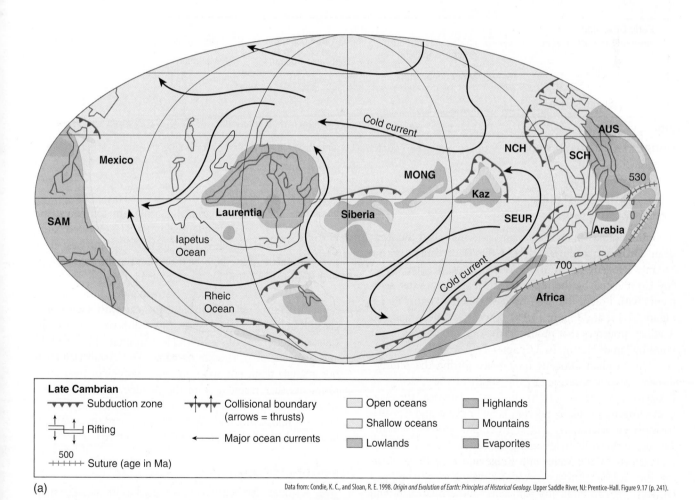

Data from: Condie, K. C., and Sloan, R. E. 1998. *Origin and Evolution of Earth: Principles of Historical Geology.* Upper Saddle River, NJ: Prentice-Hall. Figure 9.17 (p. 241).

(a)

FIGURE 11.15 Plate movements and changing plate boundaries involved in the Early-to-Middle Paleozoic orogenies of North America and Europe (see Chapter 3). **(a)** Late Ordovician.

(continues)

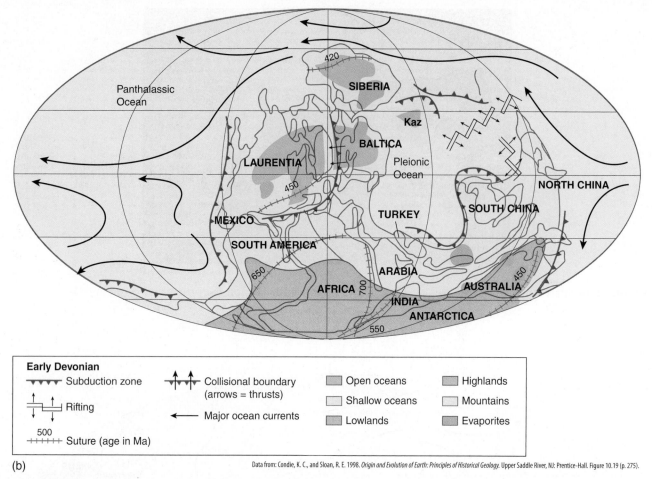

Early Devonian

- ⌄⌄⌄⌄ Subduction zone
- ⊥ Rifting
- 500 +++++ Suture (age in Ma)
- ↑⊤↑ Collisional boundary (arrows = thrusts)
- ← Major ocean currents
- ▢ Open oceans
- ▢ Shallow oceans
- ▢ Lowlands
- ▢ Highlands
- ▢ Mountains
- ▢ Evaporites

(b)

Data from: Condie, K. C., and Sloan, R. E. 1998. *Origin and Evolution of Earth: Principles of Historical Geology.* Upper Saddle River, NJ: Prentice-Hall. Figure 10.19 (p. 275).

FIGURE 11.15 (Continued) **(b)** Early Devonian. Note how equatorial currents are being deflected by an enlarging land mass that straddles the equator.

Acadian orogeny

A continuation of the Caledonian Orogeny, called the **Acadian Orogeny**, occurred during the Devonian. Like the Taconic Orogeny before it, the onset of the Acadian Orogeny is indicated by the replacement of shallow-water sediments with flysch, followed by molasse (see Figure 11.13, Figure 11.15, and Figure 11.16). It was the Caledonian and Acadian orogenies that produced the outcrop seen at Siccar Point by James Hutton (see Chapter 1).

The Acadian Orogeny took place during the Middle-to-Late Devonian further to the south of the Caledonian Orogeny (see Figure 11.15). In North America, the Acadian Orogeny primarily affected New England and Canada, basically emplacing the central geographic provinces of the Appalachians while redeforming the earlier Taconic terranes represented by the Valley and Ridge and Blue Ridge. During the Acadian Orogeny, Avalonia was basically trapped between the Laurentia and Baltica as southern Great Britain was sutured to Scotland and Ireland (**Figure 11.18**).

The Caledonian and Acadian orogenies finally closed the Iapetus Ocean and produced a new continent, the **Old Red Sandstone Continent**. The Old Red Sandstone Continent derives its name from the presence of thick, widespread redbeds of the Old Red Sandstone that were deposited in lacustrine (lake) and marginal marine environments and continental interiors (see Figure 11.18). The Old Red Sandstone is the molasse shed both to the east and west from the mountains uplifted by the orogenies that formed the Old Red Sandstone Continent (see Figure 11.18).

A huge wedge of molasse also extended from the Carolinas to New England, forming the **Catskill Delta**. The Catskill Delta is so named because its sediments comprise the Catskill Mountains of southern New York (see Figure 11.13 and Figure 11.16). Despite its name the Catskill Delta was not a single delta but an enormous complex of river environments that resulted from the Acadian Orogeny. These sediments were shed as far west and northwest as western Pennsylvania, Ohio, and West Virginia toward the continental interior (see Figure 11.13 and Figure 11.16).

Antler orogeny

Ⓗ Terranes also began to be accreted to the western margin of North America during the **Antler Orogeny** (**Figure 11.19**). Early during the Paleozoic, western North America, like eastern North America, consisted of a passive margin. However, by about the early-to-middle Paleozoic an island arc appeared off the western margin of the craton, indicating a change to an active margin. This volcanic arc

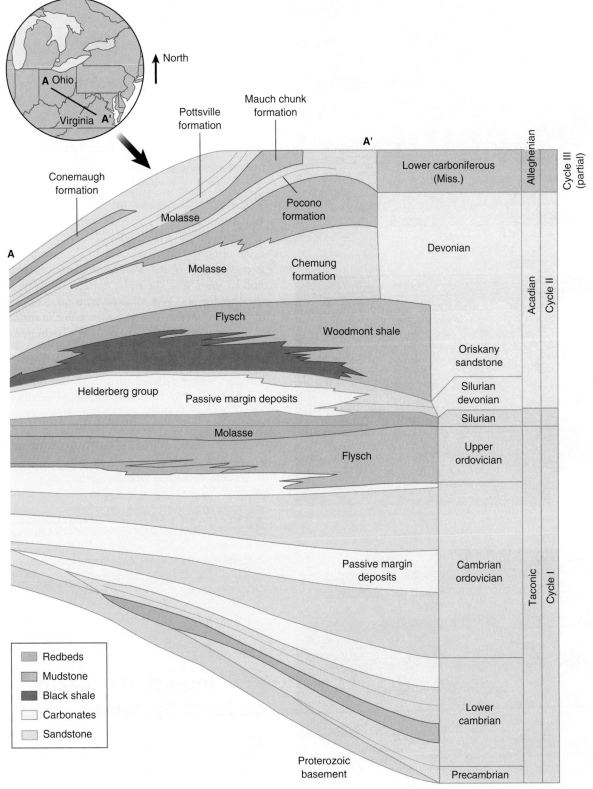

Data from: Stanley, S. M. 1999. *Earth System History*, 3rd ed. San Francisco: W. H. Freeman and Company. Figure 15.24 (p. 423). Stated to be adapted from: Colton, G. W., and Fisher, G. W. (eds.) 1970. *Studies of Appalchian Geology: Central and Southern.* New York: John Wiley and Sons.

FIGURE 11.16 Orogeny and sedimentation during the Early-to-Middle Paleozoic of North America with the Taconic and Acadian orogenies. Note that each orogenic episode was preceded by clean, shelfal sandstones, shales, and limestones, followed by flysch, and was terminated by a huge mass of molasse that thins to the west for hundreds of miles. This pattern would cease with the final phase of the formation of the Appalachians, as shown here (see also Chapter 12).

(a)   Courtesy of Ron Martin.

(b)   Courtesy of Ron Martin.

(c)   Courtesy of Ron Martin.

FIGURE 11.17 The transition from flysch to the molasse of the Taconic clastic wedge (Queenston Delta). **(a)** Folded flysch consisting of black shales and turbiditic sands. **(b)** Volcanic ash (shaly layer directly above hammer) derived from the Taconic volcanic arc. **(c)** Nearby, these rocks give way to thick deposits of shallow marine to nonmarine sediments shed from the rising Taconic Mountains to the east. The beds are standing nearly on end because of later folding.

moved eastward to initiate the Antler Orogeny by the Late Devonian Period and which continued into the later Paleozoic Era. Eventually, a mélange consisting of ophiolites, graywackes, and other submarine rocks and sediments are thought to have been moved over the Roberts Mountain Thrust Fault to lie on top of older shallow-water limestones. In the process the Roberts Mountain terrane was accreted to what was then the western margin of North America. Today, remnants of the Antler Orogeny extend through northern and central Nevada into southern California.

Ellesmere orogeny

Evidence also exists for the **Ellesmere Orogeny** (see Figure 11.18). The Ellesmere Orogeny is named for Ellesmere Island, which today lies on the Arctic Circle off the coast of northwest Greenland. In the Late Devonian, however, this part of the world represented eastern Laurentia and lay on or near the equator. Here, thrust-faulting during the Caledonian Orogeny of (what is now) northern Europe appears to have caused subsidence in the region, allowing westward erosion and accumulation of sediment as deep-sea fans that were later folded and uplifted.

These fans occurred along an elongate trough that has sometimes been referred to as the **Franklin orogenic belt** (see Figure 11.18). The Caledonian and Franklin belts appear to merge in northern Greenland. However, rather than continent–continent collision, the Ellesmere Orogeny might have involved the collision of Laurentia with a volcanic island arc or microcontinent because of the presence of ultramafic or volcanic rocks that were also deformed.

> ### CONCEPT AND REASONING CHECKS
>
> 1. Why does the cyclic nature of lithologies found in eastern North America during the Early-to-Middle Paleozoic indicate orogeny? (Hint: Think about what different lithologies indicate about depositional environments.)
> 2. Why were the mountain belts in northern Europe and North America originally given different names?

11.4 Impact of Orogeny on Earth Systems

The continental movements and orogenies of the Early-to-Middle Paleozoic all affected the general trend of rising sea level during the first half of the Paleozoic tectonic cycle. The orogenic phases involving Laurentia, Baltica, and terranes account in part for sea-level regression during the Middle and Late Ordovician and Late Devonian (see Figure 11.1). Shallow continental shelves were all caught between colliding continents and uplifted relative to sea level during these orogenic episodes. The sediments that were shed into foreland and other basins no doubt contributed to sea-level fall by filling the seaways and effectively driving the seas off the continents.

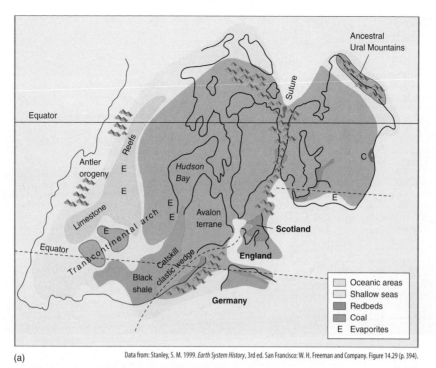

(a)

Data from: Stanley, S. M. 1999. *Earth System History*, 3rd ed. San Francisco: W. H. Freeman and Company. Figure 14.29 (p. 394).

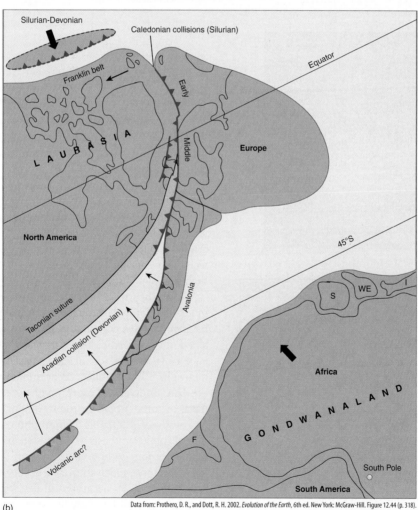

(b)

Data from: Prothero, D. R., and Dott, R. H. 2002. *Evolution of the Earth*, 6th ed. New York: McGraw-Hill. Figure 12.44 (p. 318).

FIGURE 11.18 The Acadian Orogeny and formation of the Old Red Sandstone Continent. **(a)** Map of the the Old Red Sandstone Continent. Note the Transcontinental Arch and Catskill Delta. Western North America as we know it had not yet begun to form. **(b)** Suturing of the British Isles. The Franklin orogenic belt was involved in the Ellesmere Orogeny, as discussed in the text. Note the location of the South Pole.

(continues)

© David Woods/ShutterStock, Inc.

(c)

FIGURE 11.18 (Continued) **(c)** Spectacular outcrop of the Old Red Sandstone in the British Isles.

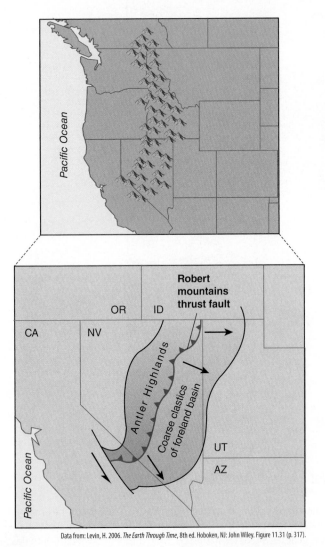

FIGURE 11.19 Location of the Late Devonian Antler Orogeny.

Changes in ocean circulation also occurred as a result of shifting continent–ocean configurations. With the formation of the Old Red Sandstone Continent, the east-to-west flow of surface ocean currents at low latitudes began to be blocked. This diverted the warm, low latitude currents to flow north along the eastern margins of the continent (see Figure 11.15; see Chapter 3). Continental movements also shifted what is now North Africa over the South Pole by the Late Ordovician, resulting in the growth of large glaciers in the southern hemisphere (see Figure 11.2). Despite the overall warmth of Earth, the occurrence of continental glaciers at this time is evidenced by the presence of tillites of Late Ordovician age today in the Sahara Desert. Possibly, the growth of southern hemisphere glaciers was aided by the diversion of warm water currents and their associated moisture-laden air to higher, cooler latitudes.

After they were initiated, the glaciers might have promoted their growth through positive feedback. Positive feedback might have occurred in several ways. First, the overall albedo of Earth would have increased because of the presence of snow and ice. Second, glaciation in the southern hemisphere likely enhanced deep-ocean circulation during the Late Ordovician. More rapid ocean circulation would have, in turn, promoted the upwelling of nutrients to surface waters, stimulating marine photosynthesis. Enhanced marine photosynthesis is suggested by a marked shift to higher carbon isotope values during the Late Ordovician, and computer models indicate a decline in atmospheric CO_2 (see Figure 11.1). Because well-developed terrestrial floras did not occur until the Late Devonian, a possible cause of the carbon isotope shift was increased photosynthesis by marine phytoplankton. Third, marine photosynthesis might have been stimulated by enhanced weathering and erosion due to glaciation which could have promoted nutrient runoff from land. Indeed, there was a strong shift in strontium isotopes (^{87}Sr/^{86}Sr) toward higher ratios during the Late Ordovician when the Taconic Orogeny was occurring. This might suggest increased weathering and erosion of the land relative to hydrothermal exchange at seafloor spreading centers (see Figure 11.1; see Chapter 10). However, the question as to why the Late Ordovician glaciation ended remains unanswered.

Glaciation might have occurred again in the southern hemisphere during the Late Devonian about the time of the Acadian Orogeny. Like the Late Ordovician, stable isotope values for carbon and strontium both increase, suggesting increased nutrient inputs to the oceans and enhanced photosynthesis (see Figure 11.1). However, tillites appear to have been more limited in extent during the Late Devonian compared with the Late Ordovician.

CONCEPT AND REASONING CHECKS

1. How might have positive feedback contributed to the growth of glaciers in the southern hemisphere during the Late Ordovician?
2. Today, the warm Gulf Stream is deflected by eastern North America to high latitudes, where it supplies moisture for the growth of northern hemisphere ice sheets (see Chapter 15). What does this suggest during the Late Ordovician?

11.5 Diversification of the Marine Biosphere

The diversity of the marine biosphere began to increase tremendously during the Late Proterozoic and Cambrian. This tremendous diversification likely occurred in response to a combination of environmental change and evolutionary innovation beginning in the latest Proterozoic and Cambrian (see Chapter 10). It was this initial biotic diversification that gave rise to the relatively primitive taxa of the so-called **Cambrian Fauna** (see Figure 11.1; Chapter 5). The Cambrian Fauna quickly gave way, in turn, to the **Paleozoic Fauna** that dominated the marine benthos during the rest of the Paleozoic Era. During this time benthos, plankton, and nekton all diversified, and food webs became more complex, including the eventual appearance of new top predators. The initial post-Cambrian diversification of the Paleozoic Fauna is termed the **Great Ordovician Biodiversification Event (GOBE)**; if the fossil record of marine biodiversity is taken at face value (see Figure 11.1 and refer to Chapter 5), biodiversification during the GOBE appears to have surpassed even that of the Cambrian. What accounts for the tremendous diversification of the marine biosphere during the GOBE?

The GOBE can be viewed as a follow-up to the Cambrian explosion, but the significant time lag of ~100 million years between the two phases of biodiversification requires explanation. For one thing, body plans of the major phyla (see Chapter 10) had to be in place before diversification at lower taxonomic levels within the phyla could occur.

On the global scale, some of the terrestrial processes that might have promoted the GOBE were part of a continuum from the Cambrian into the Middle Ordovician: continental rifting and the development of new terranes with their own biogeographic provinces, the increase in shelf area resulting from the related sea-level rise, and climate change. Volcanic activity and tectonism might have also contributed by shedding nutrients into the seaways. Many of these processes were likely interrelated but all impinged on both the benthos and the plankton, and the continued evolution of plankton probably had a major effect on the evolution of benthic ecosystems. Thus, The GOBE probably had more to do with positive feedbacks and the crossing of thresholds than with abrupt triggers. Why the Cambro-Ordovician diversification appears to have levelled off remains unclear, however (see Figure 11.1). Possibly, the habitats of the marine realm became "saturated" or possibly there were insufficient nutrients and primary productivity to support further diversification.

11.5.1 Plankton and microfossils

As we saw in Chapter 10, acritarchs dominated the eukaryotic phytoplankton of the Proterozoic. After their near disappearance from the fossil record during the Neoproterozoic, acritarchs rediversified in the fossil record in the Early-to-Middle Paleozoic and were presumably responsible for much of the photosynthesis in the oceans.

Black shales might provide a clue as to the rediversification of acritarchs. Black shales are indicative of low oxygen conditions, which often occurred during the Early-to-Middle Paleozoic, as described previously. Iron is highly soluble under low oxygen conditions, and many types of acritarchs appear related to modern phytoplankton that use iron in their photosynthetic pathway. Thus, iron was likely a biolimiting nutrient during the Early-to-Middle Paleozoic. With the exceptions of the strong deviations toward higher values during the Late Ordovician and Late Devonian, carbon isotope ratios remain relatively low and flat during much of the Early-to-Middle Paleozoic (see Figure 11.1). So, rather than indicating high-primary production during the Early-to-Middle Paleozoic, acritarchs might be indicating much-lowered productivity during this time.

Acritarchs were accompanied by various taxa of zooplankton. These included **radiolarians**, which are single-celled protists that secreted siliceous skeletons and which formed silica-rich cherts (see Chapter 4). Joining radiolarians in the zooplankton were the **graptolites**, which consisted of colonies of tiny polyps with tentacles. Despite their unusual form, graptolites belonged to the Phylum Chordata, which includes the vertebrates (**Figure 11.20**). Initially, graptolites evolved to live attached to the bottom, but many graptolites eventually evolved to join the plankton. Planktonic graptolites might have lived at different depths in the water column, ranging from the surface down to and just below the top of the OMZ. Here, they might have fed on plankton blooms when upwelling brought nutrient-rich waters of the OMZ into the photic zone (recall that nutrients

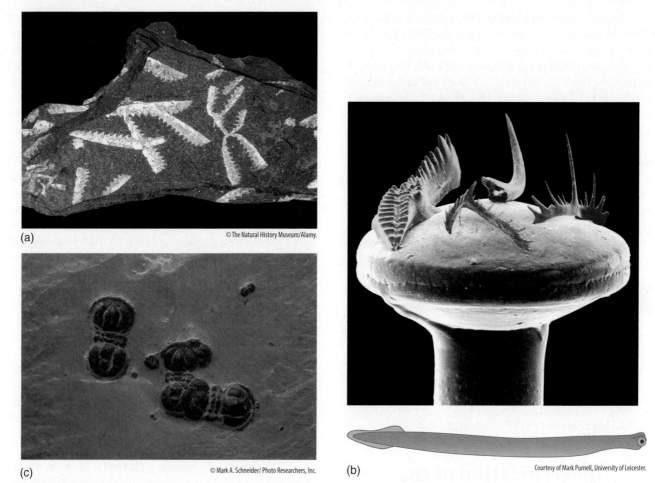

FIGURE 11.20 Planktonic and nektonic taxa of the Early-to-Middle Paleozoic. **(a)** Graptolites: floating colonies near the top of the oxygen minimum zone, stems of actual graptolite colonies, and reconstruction of a small portion of a graptolite colony showing individual polyps, and examples. **(b)** Conodonts: actual specimens and the conodont animal. Although typically found as individual tooth-like structures, in life conodonts occurred as pairs in the mouth of a primitive eel-like chordate. **(c)** Agnostid trilobites; these forms were blind and are thought to have been planktonic because of their widespread distribution in deep-water shales.

are released by the decay of organic matter, like that which occurs in the OMZ). Because they tended to live in the pelagic zone offshore, planktonic graptolites are most often found in sediments like black shales. Graptolites became prominent in the Ordovician and are extremely useful for biostratigraphic zonation and correlation in such sediments during the first half of the Paleozoic. However, they largely disappeared after the Devonian.

Another unusual taxon of microfossils present was the **conodonts**. Conodonts are represented by hard, tooth-like structures composed of calcium phosphate (see Figure 11.20). Until fairly recently, the affinities of conodonts were unknown. However, they were recently found to be associated with the carbon film of an eel-like creature in which the conodonts appear to have lined the mouth. Thus, the actual conodont animal might have been some kind of primitive chordate similar to those preserved in the Burgess Shale (see Figure 10.20). Conodonts first appeared in the Cambrian and lived well past the graptolites hundreds of millions of years later into the Triassic. Based on their distribution in ancient facies, many conodont animals might have been benthic, whereas others appear to have lived in the

plankton or nekton. Despite our lack of knowledge of the actual conodont animal, conodonts have been used extensively for zonation and correlation in the Paleozoic because they produced many species and were widespread in many environments.

Also present in the plankton were trilobites, which belong to the Phylum Arthropoda (see Chapter 10). Some species of trilobites ranged downward in size to small (less than about 1 cm) blind trilobites that lacked eyes (see Figure 11.20). Their pelagic habitat is inferred because they are typically found in deep-water shales. Possibly, these forms lived suspended in the water column below the photic zone. Other species of planktonic trilobites with large, bulbous eyes might have also lived in the plankton below the photic zone. We find similar types of crustaceans (arthropods) with large eyes in the plankton today.

11.5.2 Benthic ecosystems

Trilobites were also among the most prominent benthic taxa of the Cambrian Fauna (**Figure 11.21**). Trilobites tended to be spinose and heavily ornamented during the

Cambrian, but after this they became more "streamlined" in appearance. Some trilobites might have been predatory, but most probably had only very weak powers of swimming and appear to have lived on top of the sediment or just beneath the surface, grubbing in sediments for anything they could ingest into their small mouths. Trilobites exhibited numerous episodes of diversification and extinction, especially during the Cambrian and Ordovician. It is for this reason that they make excellent index fossils and have been used extensively for biostratigraphic zonation and correlation, especially in the Cambrian (**Figure 11.22**). Despite their success during the Cambro-Ordovician, trilobites declined after the Devonian, although they persisted to the end of the Permian Period.

Primitive members of the Phylum Mollusca, which includes clams (Class **Bivalvia**) and snails (Class **Gastropoda**), were also present. All mollusks are characterized by the presence of a muscular "*foot*"; in most bivalves and gastropods the foot is used for crawling (see Chapter 10). Mollusks also included the cephalopods (Class **Cephalopoda**). **Nautiloid** cephalopods appeared by the Late Cambrian and became quite abundant and diverse. These early cephalopods had straight to slightly curved shells; they also possessed well-developed vertebrate-like eyes and so were undoubtedly predatory. Movement was accomplished by rapid muscle contractions of the foot to produce "*jet propulsion*" like that seen in modern squids. Based on the size of their fossils, some of the straight-shelled nautiloids were several meters or more in length, although most were typically much smaller (**Figure 11.23**). The nautiloid shells were subdivided into chambers that were used to regulate gas pressure and buoyancy of the animal in the water column, much like the vest of a scuba diver. Thus, cephalopods could probably range up and down in the water column in pursuit

(a)

© The Natural History Museum/Alamy Stock Photo.

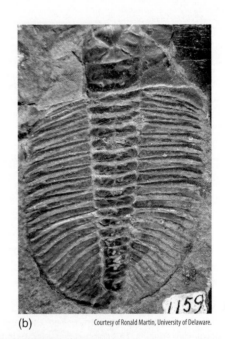

(b)

Courtesy of Ronald Martin, University of Delaware.

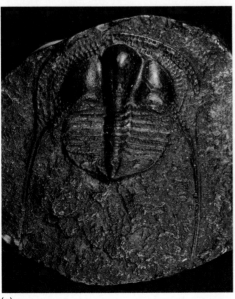

(c)

(d)

© Corbin17/Alamy Stock Photo.

FIGURE 11.21 Examples of trilobite species. **(a)** *Elrathia kingii* from the Middle Cambrian of Utah. **(b)** *Ogyopsis klotzii* from the Middle Cambrian trilobite beds of British Columbia, located near the Burgess Shale quarries. **(c)** *Onnia* sp. from the Ordovician of Morocco. **(d)** *Arctinurus boltoni* from the Silurian.

(continues)

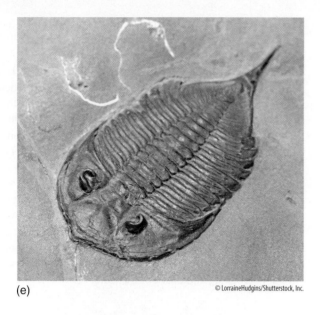

(e)

© LorraineHudgins/Shutterstock, Inc.

(f)

Courtesy of Ronald Martin, University of Delaware.

(g)

Courtesy of Ronald Martin, University of Delaware.

FIGURE 11.21 (Continued) **(e)** *Dalmanites limulurus* from the Silurian of New York state. **(f)** Harpid trilobites were rather small and characterized by elongate spines coming from the cephalon, or head region. **(g)** *Phacops rana*, perhaps one of the most famous and favorite of trilobite species. This species was used to initally demonstrate punctuated equilibria (see Chapter 5).

of prey or to escape predators. Despite their success during the Early Paleozoic, today all that is left of the nautiloids is the pearly *Nautilus* of South Pacific tropical seas.

As the nautiloids began to wane after the Early Paleozoic, they were largely replaced by the **ammonoid** cephalopods in the Devonian. Ammonoids evolved from straight-shelled nautiloids and differed in two important respects: the shell was tightly coiled and the walls that separated the chambers began to become convoluted (**Figure 11.24**). The convoluted chamber walls are thought to have increased the resistance of the shell to implosion by water pressure at great depths. Like nautiloids, the ammonoids were also predatory.

Marine benthic communities of the Paleozoic exhibited another important phenomenon: **tiering** (**Figure 11.25**). Tiering refers to suspension feeding at different levels above

and below the seafloor (that is, within the sediment). Tiering is thought to have been an attempt by organisms to avoid competition for limited resources, especially food. Members of the Cambrian Fauna such as trilobites tended to feed just above bottom or just beneath the sediment surface, where dead organic matter (food) was likely concentrated. As time passed, however, the depth of burrowing by benthic animals increased (see Figure 11.25). Increasing bioturbation by burrowers would have promoted more rapid recycling of biolimiting nutrients from dead organic matter in sediment. Taxa living at the sediment–water interface of the Paleozoic Fauna also began to feed at different levels or tiers above bottom (see Figure 11.25). Prominent among suspension feeders after the Cambrian were the articulate **brachiopods** (**Figure 11.26** through **Figure 11.29**). Like their inarticulate cousins, articulates fed using a tentacle-bearing lophophore (see Chapter 10).

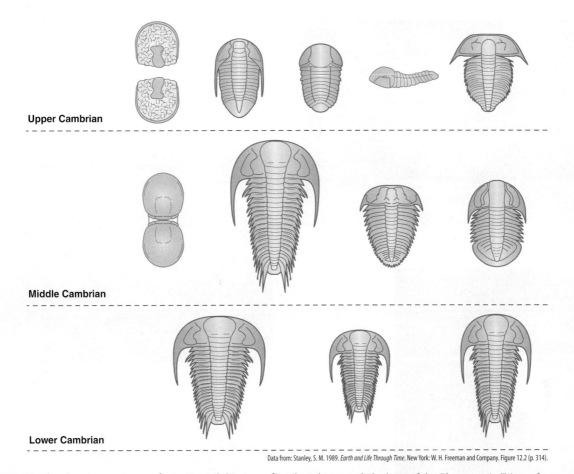

Upper Cambrian

Middle Cambrian

Lower Cambrian

Data from: Stanley, S. M. 1989. *Earth and Life Through Time*. New York: W. H. Freeman and Company. Figure 12.2 (p. 314).

FIGURE 11.22 The abrupt appearance of primitive trilobites was first thought to mark the base of the Phanerozoic ("time of apparent life" or fossil record). However, it is now known that these abrupt appearances were the result of erosion. The subsequent repeated extinction and diversification of trilobites makes them excellent index and zonal fossils for the early Paleozoic.

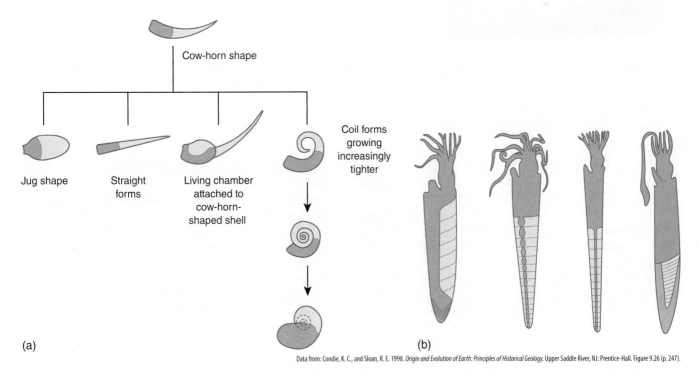

Cow-horn shape

Jug shape

Straight forms

Living chamber attached to cow-horn-shaped shell

Coil forms growing increasingly tighter

(a)

(b)

Data from: Condie, K. C., and Sloan, R. E. 1998. *Origin and Evolution of Earth: Principles of Historical Geology*. Upper Saddle River, NJ: Prentice-Hall. Figure 9.26 (p. 247).

FIGURE 11.23 Coiled and uncoiled nautiloid cephalopods of the Early Paleozoic. Shaded areas indicate the positions of soft parts. The shells were subdivided into chambers that were used to regulate gas pressure and buoyancy of the animal in the water column, much like the vest of a scuba diver.

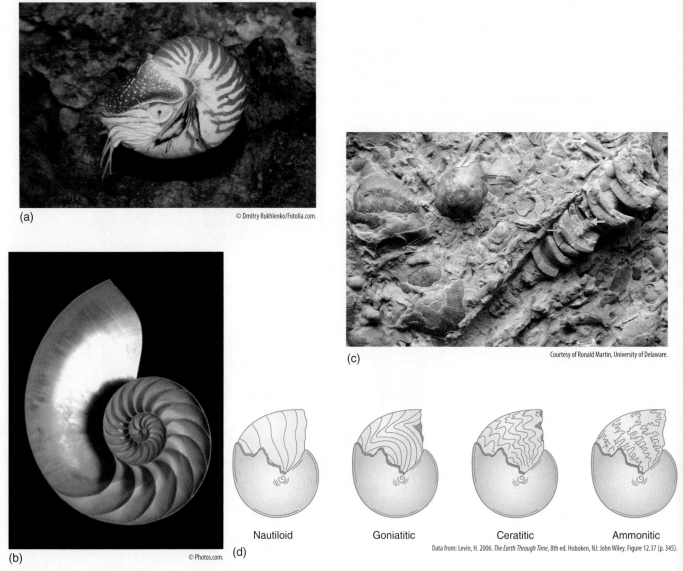

(a)

© Dmitry Rukhlenko/Fotolia.com.

(c)

Courtesy of Ronald Martin, University of Delaware.

| Nautiloid | Goniatitic | Ceratitic | Ammonitic |

Data from: Levin, H. 2006. *The Earth Through Time*, 8th ed. Hoboken, NJ: John Wiley. Figure 12.37 (p. 345).

(b) © Photos.com. (d)

FIGURE 11.24 Cephalopod sutures. **(a & b)** The modern pearly *Nautilus* exhibits the simple suture of nautiloid cephalopds. **(c)** An internal mold of a straight-shelled nautiloid cephalopod, an extinct relative of the modern pearly *Nautilus*, along with strophomenid brachiopods (red arrow) and trepostome bryozoans. Such fossil assemblages are common in rocks of Ordovician age. The nautiloid is only about six inches in length but other shells that have been found are large enough to indicate that some animals reached the approximate size of modern giant squids. **(d)** Ammonoids showed increasing complexity of the sutures.

Articulate brachiopods were often so abundant that they formed the dominant components of limestones during the Early-to-Middle Paleozoic (see Figure 11.26).

A second phylum of lophophore-bearing animals, the **bryozoans**, was represented mainly by colonies of the stick-like **trepostomes**, or "stoney" bryozoans (**Figure 11.30**). The colonies of these animals consisted of countless numbers of microscopic polyps that together secreted calcareous skeletons that held the colonies above bottom or encrusted hard surfaces. Like articulate brachiopods, trepostome bryozoans were often so abundant that their hard parts are also one of the main components of many Early-to-Middle Paleozoic limestones; in fact, articulate brachiopods and trepostomes are often found together.

Two groups of suspension-feeding **echinoderms** were also present: **crinoids** (or "sea lilies") and **blastoids**

(**Figure 11.31** and **Figure 11.32**). These taxa became more prominent after the disappearance of the strange echinoderms of the Cambrian (see Chapter 10). Crinoids fed up to a meter above bottom using a feathery set of arms and were typically attached to the bottom by a stem, resembling an upside-down starfish on a stick. Some later crinoids evolved weak powers of swimming or became planktonic. Blastoids were similar to crinoids but were characterized by a persistent five-fold (pentagonal) symmetry.

The increase in tiering and predation by invertebrates through the Paleozoic suggests that plankton populations and the rain of dead organic matter to the bottom were gradually increasing. In other words, food (energy) availability was slowly increasing above and below bottom, allowing food chains to lengthen and food webs to become more complex. At the same time expansion of

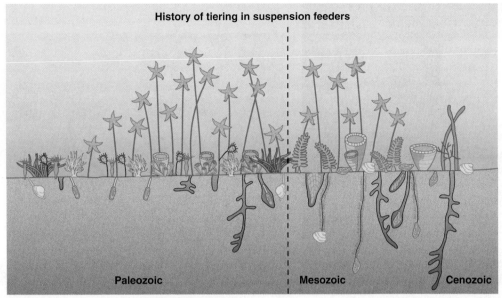

History of tiering in suspension feeders

Paleozoic   Mesozoic   Cenozoic

Data from: Prothero, D. R., and Dott, R. H. 2002. *Evolution of the Earth*, 6th ed. New York: McGraw-Hill. Figure 11.10 (p. 258). From Ausich, W. I., and Bottjer, D. J. 1991. History of tiering among suspension feeders in the benthic marine eocsystem. *Journal of Geological Education, 39*, 315.

FIGURE 11.25 Tiering in Phanerozoic benthic communities. Suspension feeders fed on plankton and suspended organic detritus at different levels in the water column, whereas burrowers moved at different depths in the sediment, either ingesting the sediment or pumping water downward into the burrowers and feeding on plankton and suspended matter.

the biosphere would have increased total living and dead biomass, sequestering nutrients into increasing amounts of organic matter. Rates and depths of bioturbation, the remineralization of dead organic matter, and the recycling of nutrients may have increased accordingly to sustain the biosphere.

11.5.3 Reefs

One of the most remarkable developments of Earth's benthos was the evolution of reefs. We therefore consider these unique ecosystems in a separate section. Reefs are biogenic, wave-resistant structures (see Chapter 4). Reefs became, and continue to be, one of the most diverse ecosystems on Earth. However, the creatures that formed the reefs changed through time, a phenomenon called **ecological replacement**.

Reefs began to proliferate in the warm, clear epicontinental seas that formed as the continents rifted apart in the Cambrian. The first reefs were composed of **stromatolites**, which were commonplace in shallow Cambrian seaways (**Figure 11.33**). Although stromatolites typically formed mound-shaped structures or mats in the fossil record, they sometimes produced much larger structures that were undoubtedly wave resistant (see Figure 11.33). The decline of stromatolites has been attributed to grazing by trilobites and gastropods; however, stromatolites actually began to decline before the appearance of diverse benthos, so the reason for their decline might not be entirely settled.

Another group of creatures, the **archaeocyathids**, appeared in the Early Cambrian and disappeared shortly thereafter (**Figure 11.34**). Archaeocyathids were suspension feeders that resembled sponges (Phylum Porifera), which were also present by this time (see Chapter 10). In

fact, archaeocyathids might have been an early experiment at being a sponge based on the presence of pores in their skeletons; however, the archaeocyathid skeleton lacked the skeletal elements called spicules. Instead, the archaeocyathid skeleton consisted of two perforated cones nested inside each other and anchored to the bottom. Archaeocyathids formed reef-like structures for a time, but the group largely disappeared by the end of the Early Cambrian.

Yet another group of reef-builders appeared in the Middle Cambrian: the **stromatoporoids** (**Figure 11.35**; not to be confused with stromatolites, which were formed by cyanobacteria). It was the stromatoporoids that became the dominant reef builders of the Middle Paleozoic world. Ancient stromatoporoids were colonial creatures that secreted massive skeletons of limestone that formed the main portion of the reef. Stromatoporoids are a somewhat enigmatic group. Because they secreted such large amounts of limestone, much like modern corals, they were once placed in the Phylum Cnidaria, which consists of sea anemones, jellyfish, and corals (see Chapter 10). Despite their prominence during the Early-to-Middle Paleozoic, for many years stromatoporoids were thought to be extinct. Then, they were discovered living at water depths below the main reefs along the north coast of Jamaica (see Figure 11.35). These stromatoporoids were recognized as a distinct class of sponges called the **sclerosponges**.

Other taxa lived among the stromatoporoids and contributed to the formation of reefs. These groups included members of the phylum Cnidaria. **Rugose (or tetra-) corals** are so named because each anemone-like polyp secreted a small cup subdivided by partitions, or septa, arranged in multiples of four (**Figure 11.36**). Some tetracorals, the "horn corals," were solitary and superficially resembled cow

(a)

(b)

(c)

(d)

(e)

(a–e) Courtesy of Ronald Martin, University of Delaware.

FIGURE 11.26 Brachiopods became incredibly diverse during the Ordovician and subsequent periods of the Paleozoic. These photographs represent only a very small sampling of Late Ordovician forms. **(a)** *Rafinesquina* from the Late Ordovician. **(b)** A typical slab of strophomenid brachiopods of Late Ordovician age. **(c)** *Leptaena* was a small genus characterized by its wrinkled shell. **(d)** *Hebertella*. **(e)** Although small, *Resserella meeki* was often very abundant.

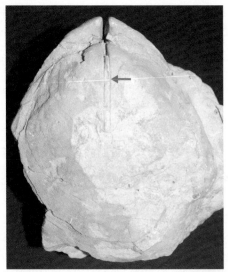

FIGURE 11.27 The brachiopod genus *Pentamerus* was very prominent in the Silurian period. This is an internal mold; the groove (arrow) indicates the position of a structure (now dissolved) that supported the lophophore used in feeding and respiration.

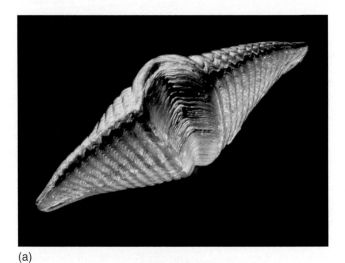

(a)

(b)

FIGURE 11.28 (a) Various species of the Devonian brachiopod *Mucrospirifer*. **(b)** *Paraspirifer* is commonly preserved by pryritization (see Chapter 4).

FIGURE 11.29 *Atrypa* was characterized by a highly convex valve and one highly flattened one on which it is seated in this photograph.

horns, but others were colonial and formed more massive structures. Also present were colonial, chain-like **tabulate corals**, whose corallites are characterized by horizontal partitions (see Figure 11.36).

At times, the stromatoporoid reefs were so common they formed gigantic barrier reefs and atolls in various parts of the world. We can find these reefs today in the interiors of the modern continents because sea level was so high during the Early-to-Middle Paleozoic (**Figure 11.37**). Today, some of these ancient reefs crop out at Earth's surface (**Figure 11.38**), but others still lie buried. By analogy to modern coral reefs, these environments were likely well lit and oligotrophic, or nutrient poor. In modern coral reef environments, nutrient runoff promotes dense populations of plankton. Plankton serve as food for large populations of sponges and clams, which feed on the plankton, overwhelming the reefs in the process.

FIGURE 11.30 Trepostome bryozoans, along with brachiopods, often helped form limestones during the early-to-middle Paleozoic era.

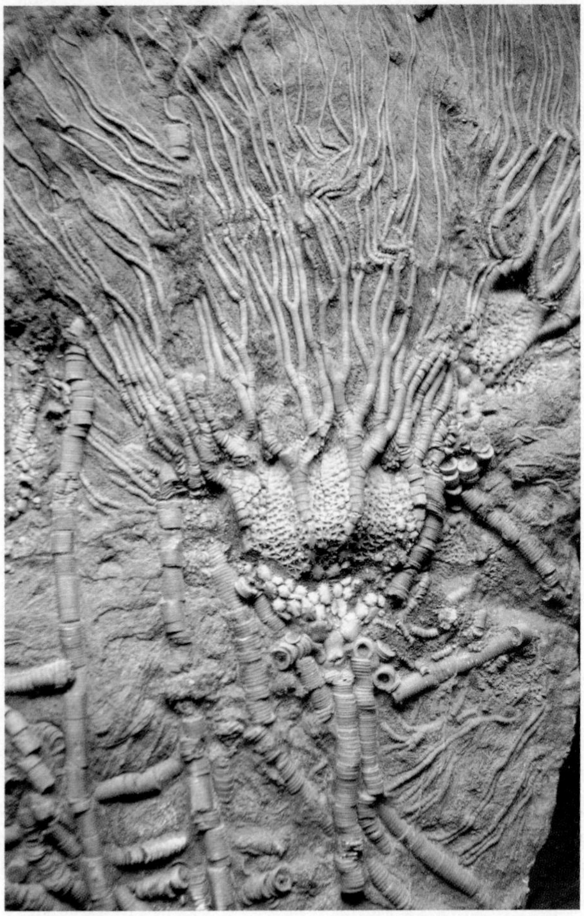

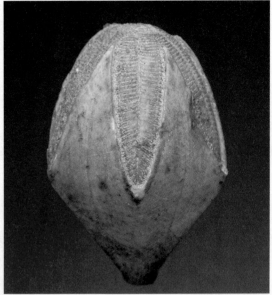

Courtesy of Ronald Martin, University of Delaware.

FIGURE 11.31 *Scyphocrinites elegans* from the Paleozoic of Morocco. Note the feather-like arms used in suspension feeding and which surrounded a mouth. The basic structure of crinoids is like that of an upside-down starfish on a stick.

© Visuals Unlimited, Inc./Mark Schneider.

FIGURE 11.32 Fossil blastoids had a distinct pentagonal symmetry. Resembling crinoids, tentacles extended upward from the feeding grooves and the entire animal was attached to the bottom by a stem.

CONCEPT AND REASONING CHECKS

1. What does each of the following tell you about ancient marine environments during the Early-to-Middle Paleozoic: (a) acritarchs, (b) graptolites, (c) cephalopods, (d) tiering, (e) reefs?

11.6 Marine Realm Invades the Terrestrial Biosphere

11.6.1 Invertebrates

M Animal life, primarily invertebrates, appears to have first diversified in the oceans. Numerous and highly significant changes occurred among the invertebrates somewhat later on land. Like the marine realm, these changes had profound consequences for the structure of ecosystems.

The Phylum Arthropoda, to which trilobites belonged, eventually expanded into the new, wide-open marginal marine, freshwater, and terrestrial habitats of the Old Red Sandstone Continent. Most notable among the arthropods were the **eurypterids** (**Figure 11.39**), which evolved from the same ancestral stock of arthropods as horseshoe crabs (see Chapter 10). Despite their appearance, horseshoe crabs are virtually harmless and have persisted relatively unchanged since the Cambro-Ordovician as living fossils (see Chapter 5). By contrast, eurypterids were predatory. Although most eurypterids grew to only a few centimeters in length, others grew up to 2 meters and possessed large eyes and pincers and, occasionally, stingers. Eurypterids gave rise to aquatic scorpions in the Silurian. Scorpions were, in turn, ancestral to other taxa such as spiders and spider mites, which feed on plants. Mites and small flightless insects were present by at least the Early Devonian, based on the Lägerstatte of the **Rhynie Chert** of Scotland. Burrows, possibly those of a millipede ("thousand-leggers"), have also been reported from soils in the Late Ordovician Juniata Formation of Pennsylvania. The Juniata forms part of the clastic Queenston wedge shed from the Taconic Mountains.

11.6.2 Fish

N Vertebrates belong to the Phylum Chordata. As we have already seen, primitive chordates were present by the Late Proterozoic or Cambrian (see Chapter 10). Not surprisingly, then, the most primitive group of vertebrates, fish (Class Pisces), also appeared in Cambrian marine sediments (**Figure 11.40**). The oldest known fish is *Myllokunmingia,* which was discovered in Early Cambrian sediments of Chengjiang, China, in 1999; these sediments represent a Lagerstätte of soft-bodied forms similar to, but somewhat older than, the Middle Cambrian Burgess Shale. Unfortunately, other remains of early fish from the Cambrian and Ordovician periods are only fragmentary, such as scales and platelets. The earliest fossil fish remains have been found in nearshore marine sediments, whereas fish remains are found

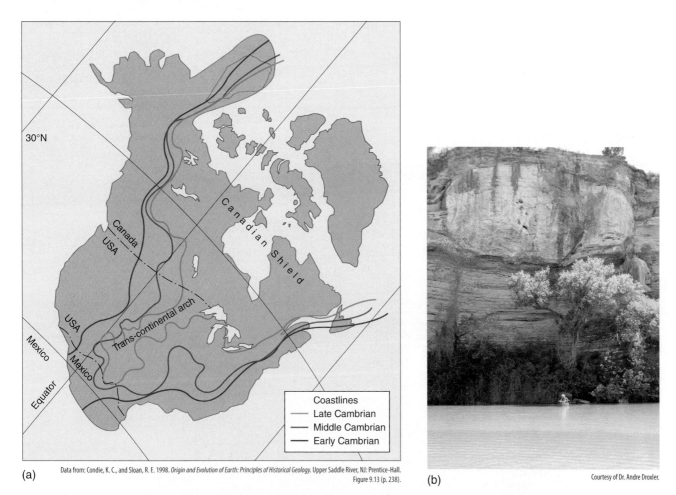

(a) Data from: Condie, K. C., and Sloan, R. E. 1998. *Origin and Evolution of Earth: Principles of Historical Geology.* Upper Saddle River, NJ: Prentice-Hall. Figure 9.13 (p. 238).

(b) Courtesy of Dr. Andre Droxler.

FIGURE 11.33 **(a)** Distribution of stromatolitic limestones in Cambrian seas of North America. The Transcontinental Arch was a gentle uplift that separated eastern and western Laurentia during this time. **(b)** Late Cambrian reef formed by stromatolites, located along the Llano River in the Texas "Hill country."

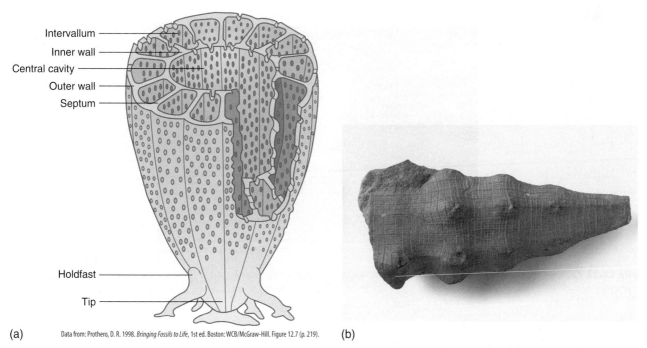

(a) Data from: Prothero, D. R. 1998. *Bringing Fossils to Life*, 1st ed. Boston: WCB/McGraw-Hill. Figure 12.7 (p. 219).   (b)

FIGURE 11.34 Basic structure of **(a)** a sponge-like archaeocyathid. Note the numerous pores. **(b)** The sponge *Hydnoceras* from the Devonian Period.

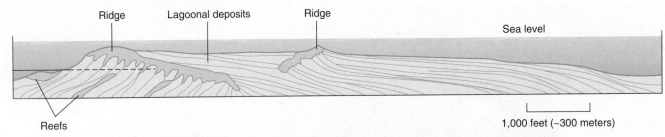

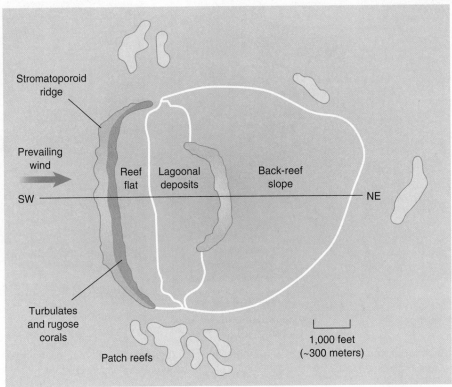

Data from: Stanley, S. M. 1999. *Earth System History*, 3rd ed. San Francisco, CA: W. H. Freman. Figure 14.23 (p. 333). Attributed to Ingels, J. J. C. 1963. American Association of Petroleum Geologists Bulletin, 47, 405–440.

(a)

(b)

Courtesy of W. David Liddell, Utah State University.

FIGURE 11.35 (a) Reconstruction of the Thornton Reef Complex, a stromatoporoid-dominated reef, now located near Chicago, Illinois. **(b)** A sclerosponge from a cave located at 30 meters water depth at Pear Tree Bottom, north coast of Jamaica.

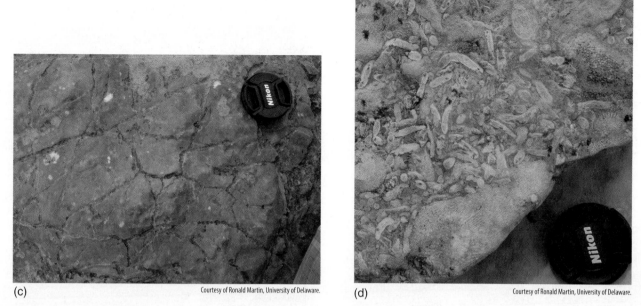

(c)

Courtesy of Ronald Martin, University of Delaware.

(d)

Courtesy of Ronald Martin, University of Delaware.

FIGURE 11.35 **(c)** Close-up of fossil sclerosponge in Devonian reefs of the Louisville Limestone. Compare to part (b). **(d)** Fragmented byrozoans, algae, and rugose corals in the Louisville Limestone (Devonian).

(a)

(b)

(c)

(a–c) Courtesy of Ronald Martin, University of Delaware.

FIGURE 11.36 Corals of the Paleozoic Era. **(a)** Solitary rugose corals. **(b)** Colonial rugose corals. **(c)** Colonial tabulate coral. Each of the small elliptical structures housed a polyp.

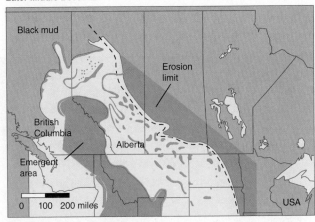

Later Middle Devonian

Black mud

Erosion
limit

British
Columbia

Alberta

Emergent
area

USA

0   100   200 miles

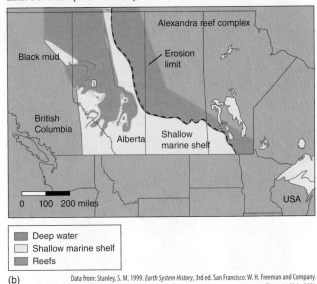

Later Devonian (Late Frasnian)

Alexandra reef complex

Black mud

Erosion
limit

British
Columbia

Alberta   Shallow
marine shelf

0   100   200 miles

USA

☐ Deep water
☐ Shallow marine shelf
☐ Reefs

(b)

Data from: Stanley, S. M. 1999. *Earth System History*, 3rd ed. San Francisco: W. H. Freeman and Company.
Figure 14.29 (p. 339).

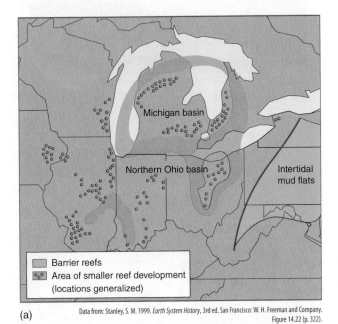

Michigan basin

Northern Ohio basin

Intertidal
mud flats

☐ Barrier reefs
☐ Area of smaller reef development
(locations generalized)

(a)

Data from: Stanley, S. M. 1999. *Earth System History*, 3rd ed. San Francisco: W. H. Freeman and Company.
Figure 14.22 (p. 322).

FIGURE 11.37 The distribution of Middle Paleozoic reefs. **(a)** In the Silurian, the entire state of Michigan was a basin that was more-or-less surrounded by a gigantic atoll-like structure. **(b)** In the Devonian, barrier reefs and atolls occupied a seaway extending through what is now the central United States and western Canada.

Courtesy of Dr. Christopher Kendall.

(a)

Courtesy of Dr. Christopher Kendall.

(b)

FIGURE 11.38 Outcrop of the Devonian reef in the Canning River Basin, Australia. Note steeply dipping forereef sediments adjacent to reef core **(a)** and flat-lying back reef and tidal flat sediments **(b)**.

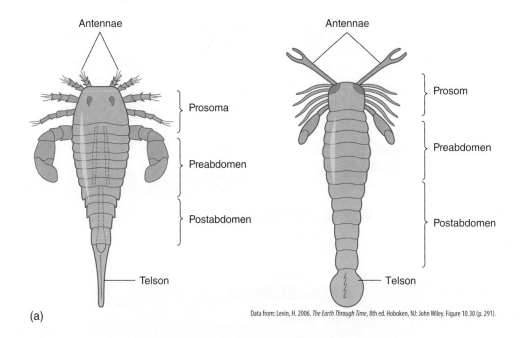

Antennae

Prosoma

Preabdomen

Postabdomen

Telson

(a)

Antennae

Prosom

Preabdomen

Postabdomen

Telson

Data from: Levin, H. 2006. *The Earth Through Time*, 8th ed. Hoboken, NJ: John Wiley. Figure 10.30 (p. 291).

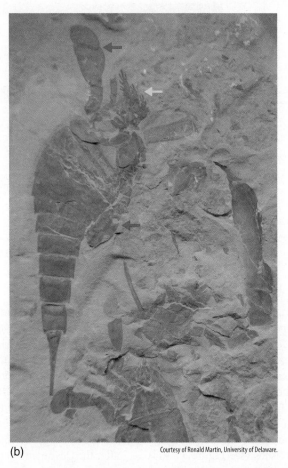

(b)

Courtesy of Ronald Martin, University of Delaware.

FIGURE 11.39 **(a)** A eurypterid or "sea scorpion." **(b)** Eurypterid preserved as carbon film in shale from Herkimer, New York. This specimen is unusual in that the mouth parts are preserved (yellow arrow). Note the large claws called chelipeds (red arrows).

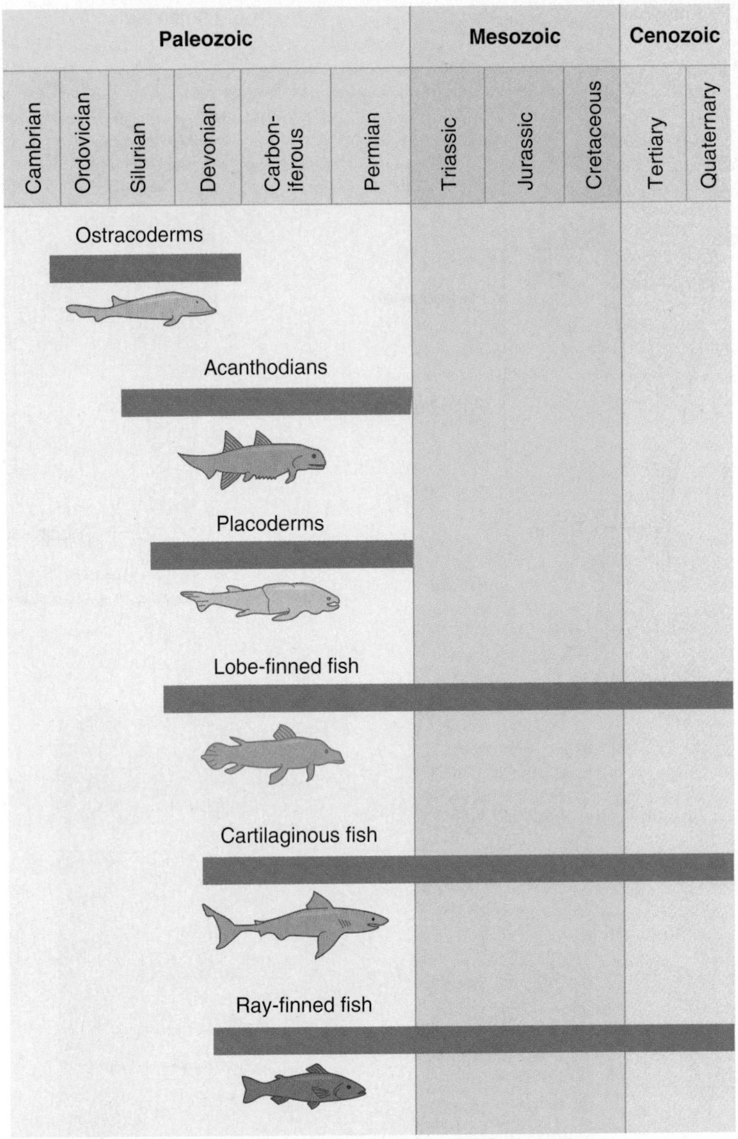

Data from: Monroe, J. S., and Wicander, R. 1997. *The Changing Earth: Exploring Geology and Evolution*, 2nd ed. Belmont, CA: West/Wadsworth. Figure 22.19 (p. 577).

FIGURE 11.40 Evolution of the major groups of fishes.

in freshwater deposits by the Silurian. Other remains are better preserved, like those of *Astraspis* (**Figure 11.41**).

All the earliest fish are placed in the **ostracoderms** ("bony skin"), so named because they possessed a bony skin or covering over their heads (see Figure 11.41). However, the rest of the ostracoderm body lacked a hard internal skeleton and is thought to have been cartilaginous. Ostracoderms were typically small, on the order of 10 to 20 cm in length. Ostracoderms are grouped in the taxon **Agnatha** ("without jaws"), or jawless fishes (see Figure 11.41). Because they lacked jaws, ostracoderms were probably not active predators. Instead, they likely fed on anything they could ingest into their small mouths. Many ostracoderms such as *Hemicyclaspis* probably scurried along the bottom in search of food. These forms had a relatively flat ventral surface, and their eyes were placed on top of their heads, most likely to watch for predators such as cephalopods and, in later times, jawed fish. These particular ostracoderms also

had a tail that was elongated along its upper portion, which caused the fish to move toward the bottom. Another important feature missing from ostracoderms was paired fins. Paired fins are found in later fish taxa and are important to guidance and stability during swimming. However, some ostracoderms such as *Pteraspis* might have been more adept at swimming because they were streamlined and possessed an elongated lower tail that pushed them away from the seafloor. Ostracoderms disappeared from the fossil record after the Devonian. Nevertheless, the agnathans are still represented by the eel-like lamprey and hagfish. Both of these forms eat by attaching themselves to a host fish and feeding on it.

During the Early Silurian, a new taxon of fishes, the **acanthodians**, appeared (see Figure 11.41). Acanthodians are thought to have been predatory for several reasons. Acanthodians were characterized by much-reduced bony armor, scales, a streamlined body shape, and numerous paired fins

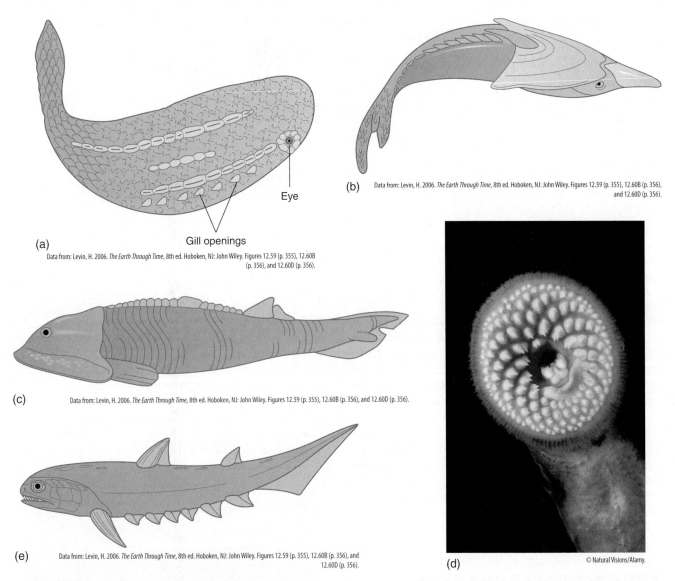

(a)

Eye

Gill openings

Data from: Levin, H. 2006. *The Earth Through Time*, 8th ed. Hoboken, NJ: John Wiley. Figures 12.59 (p. 355), 12.60B (p. 356), and 12.60D (p. 356).

(b) Data from: Levin, H. 2006. *The Earth Through Time*, 8th ed. Hoboken, NJ: John Wiley. Figures 12.59 (p. 355), 12.60B (p. 356), and 12.60D (p. 356).

(c) Data from: Levin, H. 2006. *The Earth Through Time*, 8th ed. Hoboken, NJ: John Wiley. Figures 12.59 (p. 355), 12.60B (p. 356), and 12.60D (p. 356).

(e) Data from: Levin, H. 2006. *The Earth Through Time*, 8th ed. Hoboken, NJ: John Wiley. Figures 12.59 (p. 355), 12.60B (p. 356), and 12.60D (p. 356).

(d) © Natural Visions/Alamy.

FIGURE 11.41 Reconstructions of Early Paleozoic fish **(a)** The ostracoderm *Astraspis*. **(b)** *Pteraspis*, which probably swam above bottom, at least for short distances. **(c)** *Hemicyclaspis*, an ostracoderm that probably lived along the bottom. **(d)** A modern lamprey. Lampreys possess teeth but lack jaws. **(e)** *Climatius*, an Early Devonian acanthodian.

supported by large spines. Thus, acanthodians were probably active swimmers. Acanthodians were most prominent during the Devonian but then declined during the Carboniferous, becoming extinct in the Permian.

Acanthodians also possessed another feature that was significant for the future of all ecosystems: jaws. Jaws are thought to have evolved from gill supports composed of cartilage or bone (**Figure 11.42**). The gill supports were used to support the gill openings, allowing for the exchange of CO_2 and O_2 between the fish's blood and seawater. The bony gill supports were preadapted for relatively easy conversion into jaws by the modification of embryonic developmental pathways involving *Hox* genes (see Chapter 5).

The appearance of jaws opened up an entire new range of ecologic and evolutionary possibilities. Indeed, fish underwent a major diversification during the Devonian. For this reason, the Devonian Period is often referred to as the "Age of Fish." The diversification of jawed fishes included the evolution of shell- and bone-crushing predators that possessed higher metabolic rates. The presence of joints in the gill arches meant the newly evolved jaws could be opened wider for swallowing prey (food). A wider mouth opening also allowed fish to pump more water past the gills for respiration, providing more O_2 for the breakdown of foodstuffs. It might not be coincidental, then, that the diversity of some taxa that were prominent earlier in the Paleozoic, particularly trilobites and ostracoderms, began to decline as new groups of jawed fish appeared.

The evolutionary radiation of jawed fish included a new taxon, the **placoderms**, which appeared during the Late Silurian (see Figure 11.40). Many placoderms were undoubtedly predatory, especially given their relatively large size. One such form was the Late Devonian predator *Dunkleosteus*, which ranged to more than 10 meters in length and possessed a huge, heavily armored head with large, sharp teeth (**Figure 11.43**). However, other placoderms lived on

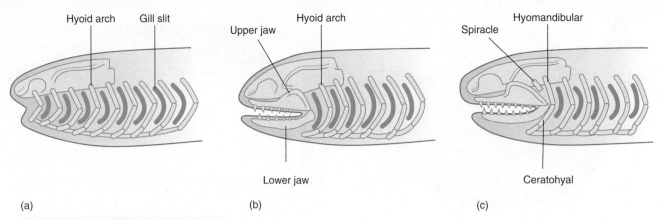

Hyoid arch   Gill slit

(a)

Upper jaw   Hyoid arch

Lower jaw

(b)

Spiracle   Hyomandibular

Ceratohyal

(c)

FIGURE 11.42 The evolution of jaws. **(a)** Jaws are thought to have evolved from bony gill supports by changes in developmental pathways involving *Hox* genes. During the evolution of jaws, the first and second pair of gill supports are thought to have been lost, with the third pair persisting as jaws. **(b)** Gill supports and jaws in the fossil fish, *Acanthodes*. **(c)** The jaw is clearly homologous to the gill supports because both possess the same bones.

the bottom. Most placoderms appear to have lived in freshwater, but some taxa later invaded the shallow seas that flooded the Devonian cratons. Like ostracoderms, they had a bony covering over their heads, but the rest of the body was apparently cartilaginous. Like acanthodians, placoderms peaked in diversity during the Devonian but declined to the end of the Permian, when they became extinct.

Sharks also began to appear by the Middle Devonian and became very prominent in the marine realm by the Late Devonian (**Figure 11.44A**). Sharks belong to the Class **Chondrichthyes**, or **cartilaginous fishes**, which include modern skates and rays. Except for their teeth and vertebrae, sharks and their relatives have cartilaginous skeletons. Despite their soft skeleton, sharks sometimes accompany placoderms as beautifully preserved Lagerstätten in Devonian black shales. The waters above bottom in the Devonian seaways were apparently sufficiently low in oxygen to prevent scavengers and burrowers from destroying the soft remains.

Also appearing in the Devonian were the bony fish, which belong to the Class **Osteichthyes**. One prominent taxon of the Osteichthyes is ray-finned fish (see Figure 11.40). The

fins of these fish are characterized by bony supports that radiate directly from the body (**Figure 11.44B**). Although relatively inconspicuous during the Paleozoic, the ray-finned fish are the ancestors of the bony fish that came to dominate marine and fresh waters during the Mesozoic Era up to the present.

These developments had a profound impact on the evolution and diversification of marine ecosystems (**Figure 11.45**). Predation in the nekton becomes increasingly evident in the

(a)

© Christian Darkin/Photo Researchers, Inc.

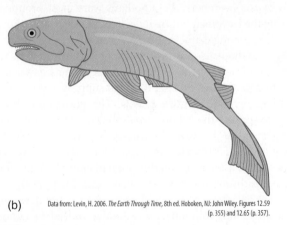

(b)

Data from: Levin, H. 2006. *The Earth Through Time*, 8th ed. Hoboken, NJ: John Wiley. Figures 12.59 (p. 355) and 12.65 (p. 357).

FIGURE 11.44 (a) Reconstruction of the Devonian shark *Cladoselache* based on its Lagerstätten. Note the striking similarity to modern sharks. **(b)** *Cheirolepis*, an ancestral bony fish.

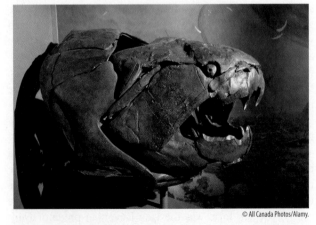

© All Canada Photos/Alamy.

FIGURE 11.43 The skull of *Dunkleosteus*, a Devonian placoderm that grew up to 10 meters (30 feet) in length. Note the well-developed eye, which is typical of active predators.

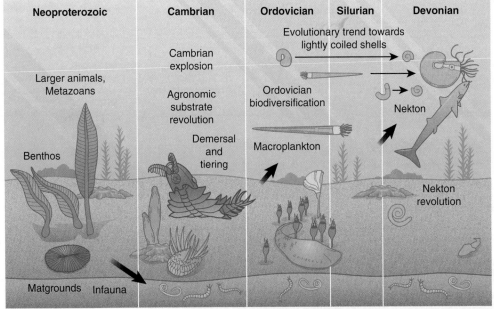

| Neoproterozoic | Cambrian | Ordovician | Silurian | Devonian |

Larger animals, Metazoans

Cambrian explosion

Agronomic substrate revolution

Demersal and tiering

Evolutionary trend towards lightly coiled shells

Ordovician biodiversification

Nekton

Benthos

Macroplankton

Nekton revolution

Matgrounds Infauna

Data from: Klug, C., Kröger, B., Kiessling, W., Mullins, G. J., Servais, T., Frýda, J., Korn, D., and Turner, S. 2010. The Devonian nekton revolution. *Lethaia*, Vol. 43, pp. 465–477.

FIGURE 11.45 The evolution of late Proterozoic through Devonian marine food webs. Note increasing predation in the nekton and increasing depths of bioturbation.

fossil record through the first half of the Paleozoic Era, as does an increasing depth and intensity of bioturbation. Like the increasing height of tiering (see Figure 11.25), these developments again hint at increasing nutrient and food availability at the base of marine food pyramids capable of supporting more diverse food webs. Nutrient availability might have increased as a result of increased weathering and nutrient runoff to the oceans in response to the spread of terrestrial land plants, as we will discuss in a moment. Increasing primary productivity might have also intensified the flux of dead organic matter to the sea bottom, where it could be used as food by burrowing organisms that remineralized the organic matter and recycled nutrients back to the water column.

Another group of **bony fish** also appeared during the Devonian: the **lobe-finned fish** (see Figure 11.40). Unlike the ray-finned fish, the fin supports of lobe-finned fish radiate from a central shaft that is in turn attached to the main body (see Chapter 5). The lobe-finned fish include several taxa. Modern lungfish possess lungs that allow them to gulp air when trapped in stagnant pools of water. Such environmental conditions were probably commonplace in the terrestrial and marginal marine environments of the Old Red Sandstone Continent. The lung structures appear to have evolved from sac-like structures beneath the esophagus. Although they were prominent during the Devonian, today lungfish are only represented by three genera, one each living in freshwater environments of South America, Australia, and Africa. Also belonging to the lobe-finned fishes are the **crossopterygians**. One branch of crossopterygians includes the **coelacanth**. Coelacanths were long thought to be extinct but were discovered living in deep waters off Madagascar (Africa) in the 1930s. Because they have remained relatively unchanged through time, coelacanths are considered living fossils (see Chapter 5).

The other branch of the crossopterygians is the so-called **rhipidistians**. Rhipidistians were the dominant predators in fresh water during the Late Paleozoic, attaining lengths of 2 meters. The rhipidistians had a streamlined body and paired muscular fins they might have used to pull themselves along the bottom or to swim (**Figure 11.46**). The paired fins would have also been useful for crawling between temporary pools of water. Although the taxonomic (evolutionary) affinities of rhipidistians are uncertain, the structural similarities between the crossopterygians and early amphibians are certainly striking (see Figure 5.23); for this reason rhipidistians are thought to have been ancestral to amphibians (see Figure 11.46).

11.6.3 Amphibians and the invasion of land

Unlike the evolutionary stasis of the lobe-finned fishes, their relatives, the rhipidistians, have been implicated as the ancestors of the first terrestrial vertebrates, or **amphibians** (see Figure 11.46). Modern amphibians include the familiar frogs, toads, and salamanders. However, the amphibians of the Paleozoic were much different from their modern counterparts and ranged greatly in size, shape, and niche.

The first amphibians are found in the Upper Devonian Old Red Sandstone of Greenland. These early amphibians already had streamlined bodies and long tails like those of *Ichthyostega* (see Figure 11.46). *Ichthyostega* shows strong similarities to rhipidistians like *Eusthenopteron* (see Figure 11.46). For example, the fins of rhipidistians, which were likely used for walking on land, were likely preadapted for walking for longer intervals by terrestrial vertebrates. The rhipidistian limbs could therefore have easily undergone rapid alteration by mutations in *Hox* genes. Also, the skulls of

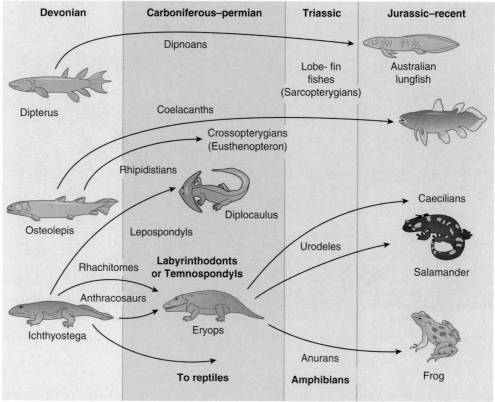

FIGURE 11.46 Evolutionary relationships of amphibians.

Data from: Levin, H. 2006. *The Earth Through Time*, 8th ed. Hoboken, NJ: John Wiley. Figures 12.59 (p. 355) and 12.73 (p. 361).

Ichthyostega and lobe-finned fish have similar shapes and bone structures, whereas the interiors of the teeth are highly crenulated, or "labyrinthine" (hence, the old name of the group: labyrinthodont, or "wrinkled teeth"). The backbone and limbs of rhipidistians evolved to provide support against gravity, whereas primitive lungs could be used for breathing. Thus, *Ichthyostega* and other early amphibians managed to conquer some of the major barriers to the invasion of land: gravity and the breathing of air. Nevertheless, amphibians have always remained tied to water for reproduction throughout their existence.

Ichthyostega so closely resembles rhipidistians that it was long thought to represent a missing link between higher amphibians and lobe-finned fish (see Chapter 5). Recently, however, skeletons of another form have been found that lie closer to the fish origins of amphibians: *Tiktaalik* (**Figure 11.47**). *Tiktaalik* was a crocodile-like creature, with eyes and nostrils on top of its flattened skull; its limbs also strongly resemble those of both lobe-finned fish and early

amphibians. Together these adaptations would have allowed *Tiktaalik* to wade or swim about in shallow water in preparation for ambushing its prey. Not surprisingly, given its habitat preference and mode of life, the remains of *Tiktaalik* were discovered in Late Devonian river deposits.

But the remains of *Tiktaalik* might not represent the earliest evidence for terrestrial tetrapods. Recent impressions found in an abandoned quarry in Poland have been interpreted as tetrapod trackways. If corroborated, these trackways predate *Tiktaalik* by almost 20 million years.

11.6.4 Land plants and the "greening" of the continents

During the Cambrian Period, the continents must have often appeared like moonscapes of bare rock. Only films of cyanobacteria, fungi, and lichens are thought to have existed. Still, these films might have held primitive soils in place.

Based on molecular clocks, all higher land plants are descended from the rather nondescript forms called stoneworts (**Figure 11.48** and **Figure 11.49**). Modern stoneworts thrive in freshwater ponds and streams but were apparently able to invade land because they can survive in relatively dry habitats. Nevertheless, it is thought that the earliest land plants had to remain close to water for reproduction, much like amphibians. Plant fragments resembling cuticles (the outer layer that prevents desiccation) have been found in

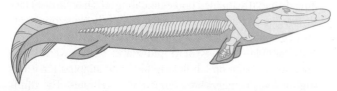

FIGURE 11.47 *Tiktaalik* lies even closer to amphibian origins than do previously discovered skeletons.

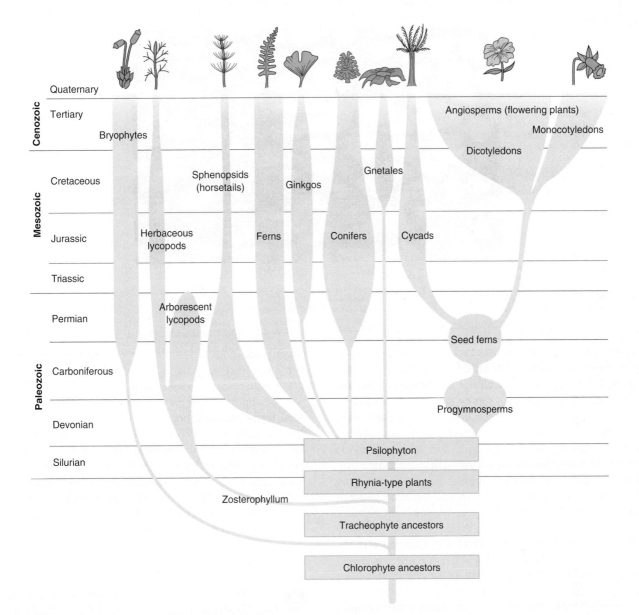

FIGURE 11.48 Evolutionary relationships of land plants.

© Science Photo Library/Alamy.

FIGURE 11.49 A modern stonewort, *Chara fragilis*. Stoneworts ("pondweeds") are usually found in clear, nonflowing fresh water (although some can be found in marine or brackish habitats).

the Middle-to-Late Ordovician (**Figure 11.50**). So, too, have spores, which are reproductive structures used in dispersal. Spores tend to be more abundant and diverse in the early fossil record than other vascular plant structures because they are surrounded by a thick, organic coating.

It was not until the Early Silurian that somewhat more advanced land plants definitely appeared. Their appearance and subsequent radiation involved a major evolutionary innovation: **vascular tissue**. Vascular tissue known as xylem is responsible for conducting water and dissolved mineral nutrients upward from the ground to stems and leaves for photosynthesis, whereas phloem conducts the photosynthetic products (sugars) downward. Vascular tissues have another function: rigidity. The cells of xylem and phloem are each surrounded by cell walls composed of **cellulose** that maintains the integrity of the cells in life. Upon death the cellulose walls remain behind and help stiffen the vascular tissue, giving the plant greater mechanical support.

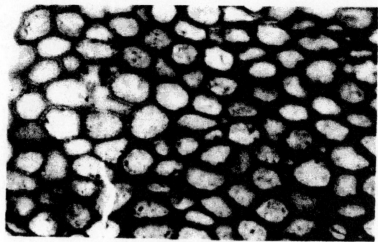

Reproduced from: Gray, J., Massa, D., and Boucot, A. J. 1982. Caradocian land plant microfossils from Libya. *Geology, 10.*

FIGURE 11.50 Cuticle-like structure from the Ordovician, along with spores, represent some of the earliest evidence for the emergence of land plants.

Greater mechanical support was necessary for land plants to grow larger because air is much less dense than water and therefore less supportive. With the appearance of vascular tissue, land plants began to grow upward. Nevertheless, many of the early land plants remained relatively close to the ground because the vascular tissues and root systems remained poorly developed; these taxa also had poorly developed leaves. Instead of true roots, many early taxa produced horizontal shoots that ran along the ground; emerging from the horizontal shoots were vertical stems up to about 30 cm tall that were capped by spore-bearing bodies. One of the best examples is *Rhynia*, from the Early Devonian Rhynie Chert, mentioned earlier (**Figure 11.51**). Coupled with the presence of arthropods, the Rhynie Chert represents one of the earliest terrestrial ecosystems preserved in the fossil record.

By the Middle Devonian, vascular tissues had become a much more prominent component of the stem in taxa such as *Psilophyton* (Figure 11.51). This indicates an increasing ability to conduct water and food. These forms also evolved roots used for anchorage and the uptake of nutrients from soils.

By the Late Devonian vascular plants included true **spore-bearing ferns** and the **sphenopsids**, which are also spore-bearing. Today, sphenopsids are represented by the horsetail, or scouring rush, *Equisetum,* which is commonly found in damp areas and along stream banks (see Figures 11.51 and **11.52**). Sphenopsids are characterized by a joint-stemmed appearance, with the leaves and spore-bearing reproductive structures located around the joints. Ferns and sphenopsids remained relatively small in the Devonian but would grow to tree size in the Carboniferous. Also present by the Late Devonian were the spore-bearing clubmosses, or **lycopods** (see Figure 11.52). Modern clubmosses might only grow to a few centimeters in height, but the ancient lycopods would eventually produce trees up to about 30 meters tall in the Carboniferous (Figure 11.52). Lycopods were able to grow to larger size because they had evolved

larger proportions of vascular tissue, providing greater support and better movement of water and nutrients.

Another major development by the Late Devonian was the appearance of the first seed plants. These were the **seed ferns**, so called because of their fern-like appearance (see Figure 11.51). However, seed ferns were actually primitive gymnosperms, the modern representatives of which include the modern conifers (such as pines and redwoods). Some seed ferns dwarfed spore-bearing ferns, reaching 10 meters or more in height in Late Devonian forests, where they joined lycopod trees (see Figure 11.52). Unlike true ferns, seed ferns did not produce spores and were therefore not dependent on water for reproduction. The evolution of the seed would allow plants to invade drier continental interiors. We will examine this development in greater detail in Chapter 12.

The changes in terrestrial vegetation during the Early-to-Middle Paleozoic changed Earth's landscapes forever. The evolution and diversification of land plants was probably critical for the invasion of land by vertebrates and their subsequent evolution because land plants provided both food and habitat, just as they do in modern ecosystems. Thus, it might not be coincidental that terrestrial vertebrates do not appear to have invaded the interiors of continents, far away from major bodies of water, until the spread of terrestrial forests.

Moreover, the increase in size of land plants meant that root systems for support needed to become stronger. The development of roots had other profound effects on Earth systems. First, deeper root systems promoted deeper weathering below the surfaces of rocks to extract nutrients. Possibly, some of these nutrients "leaked" to the oceans, where they stimulated phytoplankton photosynthesis and increased the rain of dead organic carbon to the ocean bottom. In fact, it has been hypothesized that the black shales of the Devonian described earlier in the chapter might have developed in response to the spread of terrestrial forests and increased nutrient runoff; that is, positive feedback. As sea level continued to rise, the black shales moved up onto the craton, possibly contributing

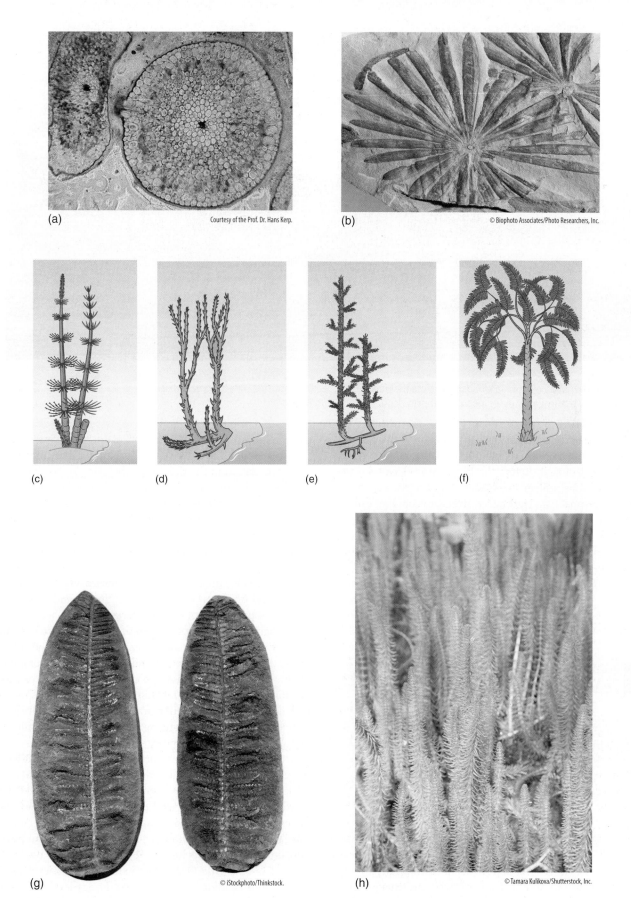

FIGURE 11.51 Early land plants of the Paleozoic. **(a)** Cross-section of a stem of *Rhynia*, from the Early Devonian. This specimen is exceptionally well preserved because it was petrified by waters carrying dissolved silica. **(b, c)** Sphenopsids include *Calamites*, which got up to about 5 meters tall, and the ancient *Annularia*. **(d, e)** Reconstructions of psilopsids. **(f, g)** Reconstruction of a seed fern and fossil specimens. **(h)** Modern club mosses.

© Chase Studio/Photo Researchers, Inc.

FIGURE 11.52 Middle Devonian landscape. Note the presence of trees represented by lycopods and seed ferns. The smaller plants growing close to the water's edge include psilopsids, sphenopsids, and true ferns.

to the Late Devonian extinctions (see section 11.7). Second, better-developed root systems inhibited the rapid physical erosion of sediment. Before the Devonian, braided streams and rivers, consisting mainly of sands and gravels, were much more widespread. Today, braided rivers are common only in relatively barren mountainous regions and indicate rapid erosion and deposition, whereas meandering or winding rivers are much more typical of lower elevations where terrestrial floras are well developed. It is also interesting that the frequency of sandstones declined after the Devonian, whereas well-developed soils, which are composed of weathered rock and dead organic matter from plants (humus), became much more prominent in the geologic record.

11.7 Extinction

Despite its general appearance as a tropical paradise, the Early-to-Middle Paleozoic was punctuated by a number of mass and minor extinctions (see Figure 11.1). The exact cause or causes of these extinctions remain unclear.

During the Cambrian trilobites underwent relatively *minor extinction* a total of five times. In each case extinction was followed by rediversification of the survivors. This is why trilobites have proven so useful for biostratigraphy and correlation in the Cambrian: they are preservable, were widespread, and underwent repeated phases of evolution and extinction to provide time markers.

Several possible causes have been suggested for these repeated extinctions. One was the intrusion of cold waters into warmer, tropical seas, although exactly why this should have happened repeatedly is not clear. Another possible explanation for the trilobite extinctions involves plate movements. As rifting occurred, microcontinents and larger land masses separated from one another and established separate biogeographic provinces characterized by their own endemic faunas. However, as the land masses shifted, collisions might have destroyed the provinces and their habitats. Strong fluctuations in the carbon isotope curve during some of these extinctions (see Figure 11.1) hints that plate movements and changes in ocean circulation and perhaps productivity were somehow involved. Changing continent–ocean configurations might have altered the extent of upwelling regions or the intrusion of cold-water oceanic currents into low latitudes.

The next major extinction of the Early-to-Middle Paleozoic occurred in the Late Ordovician (see Figure 11.1). The Ordovician extinction wiped out approximately 12% of all marine families and 80% of marine genera and is one of the five largest mass extinctions, or **Big Five mass extinctions**, documented in Earth's history. Several causes have been suggested for the Late Ordovician extinction. Sea-level fall is an obvious candidate (see Figure 11.1); sea-level regression could have resulted from both the continental collisions of the Taconic Orogeny and southern hemisphere glaciation. Along with tillites, glaciation and global cooling are indicated by a shift in the oxygen isotope curve to heavier (more positive) values (see Figure 11.1). Sea-level regression would have reduced the habitat area on the continents for marine taxa. In fact, when paleontologists first began examining the occurrences of mass extinctions in the fossil record, sea level was considered a prime cause for all of the Big Five extinctions. However, sea level fell many other times during the Phanerozoic when no extinctions occurred. Thus, although sea-level fall might have been involved in the Late Ordovician extinction, global cooling related to southern hemisphere glaciation is now viewed as a more likely mechanism. Nutrient inputs resulting from Late Ordovician orogeny might have also destabilized ecosystems by causing heightened and prolonged marine photosynthesis above normal levels. Nutrient input from land is suggested by the increase in the ratios of both strontium and carbon stable isotopes (see Figure 11.1).

Finally, there is the Late Devonian marine extinction, which also belongs to the Big Five. The Late Devonian extinction actually occurred slightly before the end of the Devonian (see Figure 11.1). In the seas, the Late Devonian extinction wiped out 33% of marine families, decimating reef faunas and ammonoids. A terrestrial extinction took place somewhat later at the Devonian–Mississippian boundary, killing off the ostracoderms and placoderms.

Like the other extinctions, the Late Devonian extinction also has its share of candidate mechanisms. Sea level was falling, and glaciers might have been present in the southern hemisphere, so global cooling might have been involved. Indeed, following the extinctions, the stromatoporoid reefs were temporarily replaced by sponges that live in cooler water.

This suggests that colder surface waters either moved south or were upwelled in the tropics, possibly in response to glaciation or changes in continent–ocean configurations and ocean circulation patterns. Another possibility is that anoxic waters invaded shallow seaways during sea-level rise and the spread of the OMZ into the continents. In some European localities, the Devonian extinction occurs as two pulses of anoxia, known as the **Kellwasser events**, which are associated with widespread black shale deposition and negative carbon isotope excursions (**Figure 11.53**). The shift to negative values by carbon isotopes reflects the incursions of marine waters enriched in Carbon-12 that was released by organic matter decay.

One intriguing hypothesis for the cause of the anoxia was the spread of land plants. The diversification of land plants into the continents could conceivably have drawn down CO_2 levels in the atmosphere, leading to glaciation. The spread of land plants could have also increased terrestrial weathering and nutrient inputs to the oceans. This is suggested by the slight rise in strontium ($^{87}Sr/^{86}Sr$) isotope ratios in Figure 11.1. Nutrient inputs could have stimulated photosynthesis by phytoplankton, which might have helped to destabilize ecosystems while also increasing the "rain" of dead organic matter through the water column, using up oxygen.

CONCEPT AND REASONING CHECKS

1. Compare and contrast the causes of the Late Ordovician and Late Devonian extinctions.

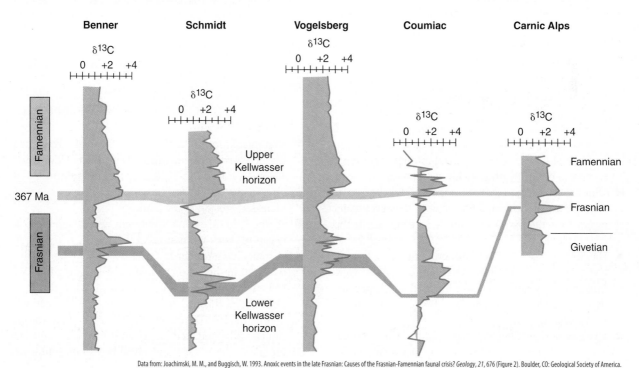

Data from: Joachimski, M. M., and Buggisch, W. 1993. Anoxic events in the late Frasnian: Causes of the Frasnian-Famennian faunal crisis? *Geology, 21*, 676 (Figure 2). Boulder, CO: Geological Society of America.

FIGURE 11.53 Anoxic waters might have decimated marine life in shallow seaways during the Frasnian-Famennian extinctions. In some European sections, the Devonian extinctions occur as two pulses of black shales known as the Kellwasser Events. At the same time, the anoxic waters would have been nutrient-rich from the decay of organic matter. Thus, increased productivity might have also destabilized ecosystems, as suggested by the shift of carbon isotope ratios to more positive values. These shifts were followed by shifts to lesser values, suggesting that marine productivity might have declined dramatically as a result of ecosystem collapse.

SUMMARY

- Earth began its final approach toward a more modern Earth during the Phanerozoic Eon, as Earth's systems continued to evolve. The first major phase of the Phanerozoic spanned the Cambrian through Devonian periods from about 540 to 360 million years ago.
- During the Cambrian and Early-Middle Ordovician, the continents of Laurentia (basically North America), Baltica (Europe), Siberia, and smaller portions of what is now Asia rifted away from Gondwana (Africa, Australia, Antarctica, South America, India, and other parts of Asia) to form the Iapetus and other oceans. As Laurentia and Baltica separated from Gondwana and from each other during the Cambrian and Early Ordovician, several island arcs and microcontinents such as Avalonia developed in the Iapetus and adjacent oceans.
- The rifting of the continents pushed the Paleozoic into a prolonged warm phase for several reasons:
  - Increased mid-ocean ridge volume associated with increased rates of seafloor spreading displaced the oceans into the continental interiors to form epicontinental seas that decreased Earth's albedo.
  - Relatively unimpeded equatorial circulation of ocean circulation currents also allowed tropical waters to warm, whereas increased volcanism associated with rifting pumped CO_2 into the atmosphere, warming Earth.
- The rifting of continents also affected ocean chemistry:
  - Increased rates of seafloor spreading appear to have increased hydrothermal weathering and the production of calcite seas.
  - Widespread deposition of limestones in epicontinental seas might have caused the CCD to shallow.
  - Also, because Earth was so warm, ocean circulation might have been relatively sluggish, so the OMZ was also quite shallow. Because sea level was often high, at times the OMZ spread into epicontinental seas to deposit carbon-rich black shales.
- After Laurentia and other continents had drifted apart during the Cambrian and Ordovician, they eventually began to collide with one another beginning in the Late Ordovician.
- The first phase of mountain building, the Taconic Orogeny, primarily affected what are now New England, Newfoundland, and England and is primarily responsible for the Valley and Ridge and Piedmont provinces.
- The Caledonian Orogeny began in the Late Silurian and extended in time into the Early Devonian, affecting Great Britain, Norway, and eastern Greenland.
- Then, a continuation of the Caledonian Orogeny called the Acadian Orogeny occurred during the Middle-to-Late Devonian:
  - The Acadian Orogeny affected New England and Canada, basically emplacing the central geographic provinces of the Appalachians while redeforming the earlier Taconic terranes represented by the Valley and Ridge and Blue Ridge.
  - During the Acadian Orogeny, Avalonia and other terranes, including what are now the southern British Isles, were trapped between the Laurentia and Baltica plates, closing the Iapetus Ocean to produce the Old Red Sandstone Continent. The Old Red Sandstone Continent was dominated by widespread redbeds deposited in lacustrine and marginal marine environments and continental interiors.
- Terranes were also accreted to the margin of North America by the Antler and Ellesmere orogenies.
- The plate movements and orogenies of the Early-to-Middle Paleozoic are related to changes in ocean circulation, chemistry, and climate:
  - The collisions of Laurentia, Baltica, and other continents account at least in part for uplift and sea-level regression during the Middle and Late Ordovician and the Late Devonian.
  - After the Late Ordovician Taconic Orogeny and glaciation, sea level began to rise and then fell again in the Late Silurian during the Caledonian Orogeny. Sea level then rebounded to somewhat higher levels, only to fall again during the Late Devonian Acadian Orogeny.
- By the Late Ordovician, Gondwana lay over the south pole and became glaciated, which might have enhanced (1) deep-ocean circulation rates, (2) oxygenation of the deep oceans, and (3) the upwelling of nutrients to surface waters, thereby stimulating phytoplankton blooms.
- Phytoplankton productivity was also likely stimulated by increased nutrient inputs from land. During the Taconic Orogeny there was a shift in strontium isotopes ($^{87}Sr/^{86}Sr$) toward heavier ratios, which suggests increased erosion of the land.
- As the Old Redstone Continent was assembled, it began to block the east-to-west flow of surface ocean currents, cooling Earth. Glaciation might again have occurred in the southern hemisphere during the Late Devonian, although in comparison with the Late Ordovician, the evidence for glaciation is less certain. During the Late Devonian $^{87}Sr/^{86}Sr$ and carbon isotope values both increased, suggesting increased nutrient inputs to the oceans and enhanced photosynthesis. At the same time the oceans were probably circulating faster (because of glaciation) and becoming better oxygenated.

- The marine biosphere began to expand and diversify during the Early-to-Middle Paleozoic:
  - In epeiric seas, a Cambrian Fauna dominated by invertebrate trilobites and inarticulate brachiopods gave way to the Paleozoic Fauna dominated by stromatoporoid reefs; articulate brachiopods; clams, snails, and predatory nautiloid; and ammonoid cephalopods.
  - Among the aquatic vertebrates, jawless ostracoderm fish yielded to the jawed acanthodians, placoderms, and sharks.
  - Microfossil groups also diversified but consisted of unusual groups: acritarchs, radiolarians, graptolites, and conodonts.
- On land, plants underwent a dramatic radiation after an evolutionary innovation no less significant than the appearance of the vertebrate jaw: vascular tissue. Like the expanding plankton in the seas, land plants provided habitats and food as vertebrates moved from the seas to land. The limbs and skulls of the earliest known amphibians greatly resembled those of certain primitive fish, suggesting these structures were pre-adapted for movement on land.
- Life suffered a series of setbacks during extinctions in the Cambrian, Late Ordovician, and Late Devonian. The Late Ordovician extinction is thought to have resulted primarily from glaciation and global cooling during an otherwise pronounced greenhouse, whereas the spread of terrestrial forests by the Late Devonian might have caused anoxia in the oceans by increasing continental weathering, nutrient runoff, photosynthesis, and organic decay in the oceans. However, the Late Devonian extinction, especially, might have resulted from more than just anoxia, including orogeny, sea-level change, continental collision, global cooling, nutrient inputs, and impact. Neither of these causes might have been sufficient by themselves, but each could have contributed in one way or another to the extinctions.

KEY TERMS

Acadian Orogeny	calcite seas	graptolites	rhipidistians
acanthodians	Caledonian Orogeny	Great Ordovician Biodiversification Event (GOBE)	Rhynie Chert
Agnatha	Cambrian Fauna		rugose corals
Allegheny Plateau	cartilaginous fishes	Iapetus Ocean	Sauk Sea
ammonoid	Catskill Delta	Kaskaskia Sequence	sclerosponges
amphibians	cellulose	Kazakhstania	seed ferns
Antler Orogeny	Cephalopoda	Kellwasser events	Siberia
Appalachian Mountains	Chondrichthyes	Laurentia	slate belts
archaeocyathids	coelacanth	lobe-finned fish	sphenopsids
Atlantic Coastal Plain	conodonts	lycopods	spore-bearing ferns
Avalonia	crinoids	nautiloid	stromatolites
Baltica	crossopterygians	Old Red Sandstone Continent	stromatoporoids
Big Five mass extinctions	Cumberland Plateau	Osteichthyes	tabulate corals
Bivalvia	echinoderms	ostracoderms	Taconic Orogeny
blastoids	ecological replacement	Paleozoic Fauna	tiering
Blue Ridge Province	Ellesmere Orogeny	Pannotia	Tippecanoe Sea
bony fish	eurypterids	Piedmont	trepostomes
brachiopods	Franklin orogenic belt	placoderms	Valley and Ridge Province
bryozoans	Gastropoda	Queenston Delta	vascular tissue
		radiolarians	

REVIEW QUESTIONS

1. Describe and explain the interactions of Earth's systems in your own words and diagrams.
2. What made Earth warm during most of the Early-to-Middle Paleozoic?
3. What is the basic evidence for high sea level and warm climates during much of the Early-to-Middle Paleozoic?
4. During the Early-to-Middle Paleozoic, what was the effect of continental movements on each of the following: (a) sea level, (b) atmospheric CO_2, (c) ocean currents and circulation? Explain your answers.
5. What factors contributed to widespread limestone deposition in epeiric seas during the Early-to-Middle Paleozoic?
6. Discuss how the climate and oceanographic setting of the Early-to-Middle Paleozoic affected the level of the CCD and oceanic anoxia. Use relevant chemical reactions in your discussion.
7. How have geologic processes affected the present landscape?
8. What is the evidence that sedimentary rocks of the Valley and Ridge Province were deposited in shelf environments?
9. What were the major developments in marine benthos and plankton during the Early-to-Middle Paleozoic, and how were they possibly related to each other?
10. What is the ecologic and evolutionary significance of tiering?
11. Discuss the sequence of ecologic replacement during the evolution of Early-to-Middle Paleozoic reefs.
12. What was the significance of land plants for the evolution of marine and terrestrial faunas?
13. How did evolutionary innovations during the Silurian and Devonian open up new ecologic and evolutionary opportunities?
14. What do you suppose were the effects of these innovations on the biogeochemical cycles of carbon and nutrients?

FOOD FOR THOUGHT:
Further Activities In and Outside of Class

1. What sorts of data can we use to reconstruct the ancient positions of the continents? What are the limitations of the various types of data?
2. Quite possibly one of the reasons atmospheric CO_2 levels were so high during much of the Early-to-Middle Paleozoic was the absence of extensive terrestrial forests. What might have happened to atmospheric CO_2 levels if orogenic activity had been greater during this time? What might have been the effects on oceanic productivity? Justify your answer.
3. In scientific experiments, investigators control all variables except one. If a change in the variable is associated with a corresponding effect, the changes in the variable are commonly regarded as the cause of the effect. Comment on this viewpoint in light of (a) the causes of climate change during the Early-to-Middle Paleozoic and (b) mass extinctions. Can the same effect have more than one cause?
4. What would you have to know to calculate the volume of sediment eroded from a mountain range?
5. Explain the distribution of different genera of lungfish on South America, Australia, and Africa.

Algeo, T., and Scheckler, S. E. 1998. Terrestrial-marine teleconnections in the Devonian: links between the evolution of land plants, weathering processes, and marine anoxic events. *Philosophical Transactions of the Royal Society of London (Series B), 353,* 113–130.

Bambach, R. K. 1999. Energetics in the global marine fauna: a connection between terrestrial diversification and change in the marine biosphere. *Geobios, 32,* 131–144.

Bartels, C., Briggs, D. E. G., and Brassel, G. 1998. *The fossils of the Hunsrück Shale: Marine life in the Devonian.* Cambridge, UK: Cambridge University Press.

Berry, W. B. N., and Wilde, 1978. Progressive ventilation of the oceans: an explanation for the distribution of the lower Paleozoic black shales. *American Journal of Science, 278,* 257–275.

Boss, S. K., and Wilkinson, B. H. 1991. Planktogenic/eustatic control on cratonic oceanic carbonate accumulation. *Journal of Geology, 99,* 497–513.

Clack, J. A. 2004. From fins to fingers. *Science, 304,* 57–58.

Dalziel, I. 1995. Earth before Pangaea. *Scientific American, 272*(1):58–63.

Fortey, R. A. 2004. The lifestyles of the trilobites. *American Scientist, 92,* 446–453.

Karol, K. G., McCourt, R. M., Cimino, M. T., and Delwiche, C. F. 2001. The closest living relatives of land plants. *Science, 294,* 2351–2353.

King, P. B. 1959. *The evolution of North America.* Princeton, NJ: Princeton University Press.

Klug, C., Kröger, B., Kiessling, W., Mullins, G. J., Servais, T., Frýda, J., Korn, D., and Turner, S. 2010. The Devonian nekton revolution. *Lethaia, 43,* 465–477.

Knoll, M. A., and James, W. C. 1987. Effect of the advent and diversification of vascular land plants on mineral weathering through geologic time. *Geology, 15,* 1099–1102.

Lee, J., and Morse, J. W. 2010. Influences of alkalinity and $_pCO_2$ on $CaCO_3$ nucleation from estimated Cretaceous composition seawater representative of "calcite seas." *Geology, 38,* 115–118.

McGhee, G. R. 1996. *The Late Devonian mass extinction: The Frasnian/Famennian crisis.* New York: Columbia University Press.

Niedźwiedzki, G., Szrek, P., Narkiewicz, K., Narkiewicz, M., and Ahlberg. P. E. 2010. Tetrapod trackways from the early Middle Devonian period of Poland. *Nature, 463,* 43–48.

Plank, M. O., and Schenck, W. S. 1998. *Delaware Piedmont geology.* Newark, DE: Delaware Geological Survey Special Publication No. 20.

Raff, R. A. 1996. *The shape of life: Genes, development, and the evolution of animal form.* Chicago: University of Chicago Press.

Rau, D. M. 1991. *Extinction: Bad genes or bad luck?* New York: W. W. Norton.

Rogers, J. J. W., and Santosh, M. 2004. *Continents and supercontinents.* Oxford, UK: Oxford University Press.

Schwartz, J. H. 1999. *Sudden origins: Fossils, genes, and the emergence of species.* New York: John Wiley and Sons.

Servais, T. et al. 2010. The Great Ordovician Biodiversification Event (GOBE): The palaeoecological dimension. *Palaeogeography, Palaeoclimatology, Palaeoecology, 294,* 99–119.

Vermeij, G. J. 1995. *Evolution and escalation: An ecological history of life.* Princeton, NJ: Princeton University Press.

Wilde, P., and Berry, W. B. N. 1984. Destabilization of the oceanic density structure and its significance to marine "extinction" events. *Palaeogeography, Palaeoclimatology, Palaeoecology, 48,* 143–162.

Zimmer, C. 1998. *At the water's edge: Macroevolution and the transformation of life.* New York: The Free Press.

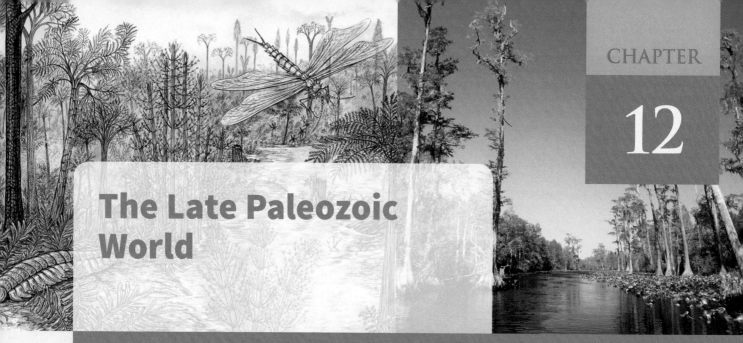

The Late Paleozoic World

MAJOR CONCEPTS AND QUESTIONS ADDRESSED IN THIS CHAPTER

A How did the continent–ocean configurations of this time differ from those of the Early-to-Middle Paleozoic?

B How did the movements of the continents affect orogeny, sea level, and climate?

C What were the effects of continental movements on ocean circulation and chemistry? How did these factors differ from today?

D What were the events leading to the final phases of the formation of Pangaea?

E How did western North America behave during this time?

F How did orogeny affect the other Earth systems?

G How do the rocks deposited during the Late Paleozoic differ from those of the Early-to-Middle Paleozoic?

H How did planktonic and benthic communities of the Late Paleozoic differ from those of the Early-to-Middle Paleozoic?

I What sorts of reefs or reef-like structures replaced those of the Early-to-Middle Paleozoic?

J What new groups of terrestrial plants evolved?

K How did the spread of terrestrial forests affect the biogeochemical cycle of carbon and the concentration of oxygen in Earth's atmosphere?

L When did insects appear, and what were the earliest groups like?

M Which new groups of vertebrates continued to invade the land?

N What evolutionary innovation led to the origin of reptiles and freed them to invade continental interiors?

O Did reptiles begin to evolve mammalian traits during the Late Paleozoic?

P Why did nearly all life go extinct at the end of the Paleozoic Era? Was this the result of multiple causes or one single cause?

CHAPTER OUTLINE

Reconstruction of a coal-forming swamp like those that existed in the eastern U.S. during the Pennsylvanian Period. Modern analogs of the coal-forming swamps include the cypress swamps of the southeastern United States.
Left: © Laurie O'Keefe/Science Source. Right: Courtesy of Ryan Hagerty, U.S. Fish and Wildlife Service.

12.1 Introduction to the Late Paleozoic Era

The Late Paleozoic Era—comprising the Mississippian, Pennsylvanian, and Permian periods—represents the second half of the Paleozoic tectonic cycle (**Figure 12.1**). During this time the collision and suturing of continents that began with earlier Paleozoic orogenies continued; by the end of the Paleozoic further collisions produced a new supercontinent, **Pangaea**. A number of features of Earth's surface, including some of its most famous mountain ranges, were formed by these collisions.

These tectonic events shifted the paleoclimatic setting from the generally warm conditions of the Early-to-Middle Paleozoic to one of more pronounced climatic extremes. Indeed, most of the second half of the Paleozoic is marked by widespread glaciation over the South Pole (see Figure 12.1). Although sea level still remained high during the Late Paleozoic, the continued orogeny and uplift of continents

associated with supercontinent assembly, along with southern hemisphere glaciation, caused sea level to fall more or less imperceptibly toward the end of the Paleozoic Era (see Figure 12.1).

Life also changed dramatically. After the Late Devonian extinctions, the stromatoporoid-dominated reefs of the Silurian and Devonian disappeared forever and were replaced by taxa such as crinoids and algae. Predatory fish also continued to diversify in the marine realm.

On land, plants began to spread into the continental interiors formed by orogeny to produce rather drab coal-forming swamps of grays, greens, and browns. Photosynthesis and carbon storage on land decreased atmospheric carbon dioxide (CO_2) levels by storing carbon in peat (the precursor of coal); this also increased oxygen concentrations in the atmosphere and perhaps also the oceans to much higher levels. Extensive terrestrial forests probably also opened new habitats and niches into which diversified insects, amphibians, and a new group of vertebrates, the **reptiles**. As coal-forming swamps declined during the Permian, they gave way to

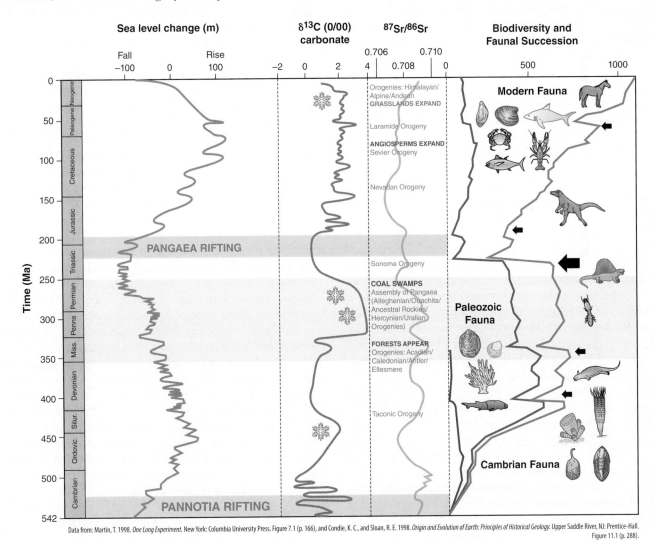

Data from: Martin, T. 1998. *One Long Experiment*. New York: Columbia University Press. Figure 7.1 (p. 166), and Condie, K. C., and Sloan, R. E. 1998. *Origin and Evolution of Earth: Principles of Historical Geology*. Upper Saddle River, NJ: Prentice-Hall. Figure 11.1 (p. 288).

FIGURE 12.1 The physical and biological evolution of the Earth through the Phanerozoic Eon. The lightly shaded area indicates the Mississippian through Permian periods of the Phanerozoic Eon that are discussed in this chapter. Snowflakes indicate times of major glaciation. Carbon isotopes and strontium isotopes are indicated (see Chapter 9 for elementary discussion). Arrows indicate the Big Five mass extinctions.

the **gymnosperms**, which would dominate terrestrial floras through the rest of the Paleozoic and much of the Mesozoic era. The diversification of marine and terrestrial taxa was terminated at the end of the Paleozoic by the largest, and perhaps most complex, of all mass extinctions.

12.2 Tectonic Cycle: Impacts on the Hydrosphere, Atmosphere, and Rock Cycle

12.2.1 Sea level, CO_2, and sedimentary facies

A **B** The continental collisions that had begun during the Early-to-Middle Paleozoic continued to the end of the Paleozoic Era. In the process, the new supercontinent, Pangaea, began to take shape (**Figure 12.2**). As Pangaea was assembled a number of smaller ocean basins that had persisted from the Early-to-Middle Paleozoic were destroyed and were replaced by one large ocean, **Panthalassa**. The large number of continental collisions destroyed both spreading centers and subduction zones while uplifting continental crust. Thus, even though sea level continued to remain relatively high during the Late Paleozoic, the seas drained inexorably from the continents (see Figure 12.1). By the end of the Permian, sea level is thought to have reached its lowest point of the entire Phanerozoic Eon, including the present (see Figure 12.1).

The destruction of spreading centers and volcanic island arcs during the assembly of Pangaea is also thought to have decreased the input of volcanic CO_2 into Earth's atmosphere, cooling the planet. Enormous glaciers also developed on Gondwana as it moved over the South Pole. As the seas drained from the continents, glaciation and the increased exposure of uplifted land at Earth's surface increased the planet's albedo, further cooling Earth through positive feedback.

Superimposed on this prolonged regression of sea level during the Late Paleozoic were two major sea-level rises: the **Kaskaskia** and **Absaroka** (see Chapter 6). The Kaskaskia Sequence had begun earlier in the Paleozoic with clean, well-sorted sandstones that gave way to carbonates, including stromatoporoid-dominated reefs (see Chapter 11). Then, black shales invaded parts of the North American craton during the Late Devonian. Black shale deposition in the Kaskaskia seaway continued into the Early Mississippian, but then began to give way to carbonates. However, limestones were not nearly as extensive as they were during the Early-to-Middle Paleozoic. Moreover, these limestones were not the limestones of stromatoporoid reefs, which had been decimated during the Late Devonian extinctions. The limestones were far different, some consisting of widespread mounds of loose skeletal debris such as crinoidal limestones; other limestones consisted of oölites and some were dolomitic. Cross-bedding and ripple marks indicate that many of these limestones were deposited in relatively wave- and current-swept waters.

The Kaskaskia Sea retreated by the end of the Mississippian, only to be replaced by the Absaroka Sea. The more or less craton-wide unconformity between the Kaskaskia and Absaroka seas basically marks the boundary between the Mississippian and Pennsylvanian periods in North America (called the Early and Late Carboniferous in Europe). The rocks of the Absaroka Sequence reflect the pronounced orogenic activity and erosion of the Late Paleozoic. In the eastern United States sedimentary rocks of this time are generally thickest in the east and southeast and thin westward; these rocks grade from nonmarine terrigenous rocks and coals in the east and southeast to marginal marine rocks such as deltas and more marine-influenced coals in the midwest (see Figure 12.2). Many of the rocks deposited in terrestrial environments were redbeds, indicating that they were exposed to oxygen in the atmosphere.

12.2.2 Ocean circulation and chemistry

C The tectonic cycle that began the Paleozoic Era with rifting continents and rising sea levels began to reverse itself during the last half of the era. Continued continental collision and orogeny during the final assembly of Pangaea likely resulted in slowly falling sea level, lowered CO_2, and higher albedo.

The assembly of Pangaea, which stretched across much of both hemispheres, also continued to block the east-to-west flow of tropical waters around the planet (see Figure 12.2; see Chapter 3). This resulted in an equatorial countercurrent that returned warm waters from the western margin of Panthalassa toward its eastern margin. This countercurrent allowed these waters to be warmed twice, resulting in higher surface water temperatures in the eastern equatorial portion of Panthalassa. Warm waters along the eastern margin of Panthalassa were diverted to the north and south, warming higher latitudes along the western margin of the ocean. As they cooled these western boundary currents turned east. When these waters eventually encountered western Pangaea, portions of the currents turned south toward the equator to complete large gyre systems north and south of the equator, much like those seen in the oceans today. Thus, greater ranges of climatic extremes were established along the margins of Pangaea.

Ocean chemistry also began to change in significant ways in response to changing continent–ocean configurations. Recall that the oceans of the Early-to-Middle Paleozoic Era were thought to have circulated relatively slowly in response to the production of saline (but warm) deep-water masses. By contrast, southern hemisphere glaciers might have increased ocean circulation and oxygenation of deepwater masses rates during the Late Paleozoic (see Chapter 3; compare conditions in Chapter 11).

A strong and persistent rise of strontium isotope ($^{87}Sr/^{86}Sr$) ratios during the Carboniferous also suggests nutrient runoff to the oceans increased. Nutrient runoff most likely increased as a result of orogeny and the spread of land plants (see below); the spread of forests likely increased physical and chemical weathering by plant roots and by increasing transpiration, which would presumably have

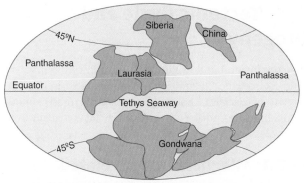

Mississippian (Early Carboniferous, ~350 MYBP)

(a)   Data from: Monroe, J. S., and Wicander, R. 1997. *The Changing Earth: Exploring Geology and Evolution*, 2nd ed. Belmont, CA: West/Wadsworth. Figure 21.3B (p. 530).

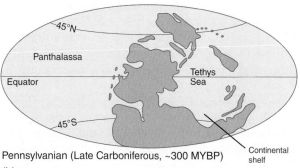

Pennsylvanian (Late Carboniferous, ~300 MYBP)

(b)   Data from: Monroe, J. S., and Wicander, R. 1997. *The Changing Earth: Exploring Geology and Evolution*, 2nd ed. Belmont, CA: West/Wadsworth. Figure 21.4A (p. 531).

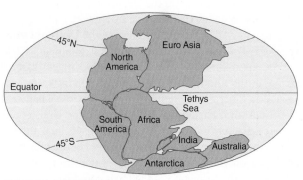

Permian (~225 MYBP)

(c)   Data from: Monroe, J. S., and Wicander, R. 1997. *The Changing Earth: Exploring Geology and Evolution*. 2nd ed. Belmont, CA: West/Wadsworth. Figure 21.4B (p. 531).

(d)   Courtesy of Ronald Martin, University of Delaware.

(e)   Courtesy of Ronald Martin, University of Delaware.

FIGURE 12.2 General paleogeography during the Late Paleozoic. **(a)** Mississippian (Early Carboniferous) period. **(b)** Pennsylvanian (Late Carboniferous) period. **(c)** Late Permian period. **(d)** The so-called "whaleback" anticline (with students seated on its axis) with a syncline positioned immediately above it in the wall opposite. How do you suppose this structural relationship came to be? **(e)** Thrust fault dipping to the right in the center of a small syncline in the Mississippian Mauch Chunk Formation (dark-colored) near Hazleton, Pennsylvania. The upper light-colored unit is the conglomeratic sandstone called the Pottsville Formation of Pennsylvanian age. Relative direction of movement along the fault is indicated by arrows.

accentuated the hydrologic cycle and runoff. The spread of forests would have also undoubtedly sequestered nutrients in living and dead biomass such as peat. However, the *initial* colonization of continental rock surfaces during the Carboniferous by forests and the establishment of more deeply weathered soil profiles might have been accompanied by the greater "leaking" of nutrients to streams and rivers and ultimately the oceans. Together, increased productivity by phytoplankton and, especially, the spread of forests account for the broad increase in carbon isotopes during most of the Late Paleozoic (see Figure 12.1). Despite all the weathering and erosion of the Carboniferous, however, $^{87}Sr/^{86}Sr$ ratios declined during the Permian (see Figure 12.1). Perhaps by this time the continents were moving so close together that interior drainage began to prevent runoff to the oceans.

Orogeny and sea level also increased the probability of the erosion of limestones originally deposited in epicontinental seas during the Early-to-Middle Paleozoic. The erosion of limestones as sea level gradually fell and the delivery of increased amounts of dissolved calcium carbonate to the oceans likely caused the calcite compensation depth (CCD) to deepen at least somewhat. Recall, however, that there is no complete record of deep-ocean sedimentation earlier than the Jurassic because of subduction of Paleozoic seafloor crust, so the hypothesis of the deepening of the CCD during the Late Paleozoic cannot be definitely corroborated.

Another difference between the ocean chemistry of the Early-to-Middle Paleozoic and the Late Paleozoic is that of the mineralogy of calcareous hard parts secreted by benthic invertebrates. Calcareous benthic hard parts secreted during the Early-to-Middle Paleozoic were predominantly of the mineral calcite ("calcite seas"; see Chapter 11). By contrast, those of the Late Paleozoic tended to consist of aragonite. Aragonite crystallizes more easily when the concentration of dissolved magnesium ions (Mg^{2+}) in seawater is relatively high. Because seafloor spreading centers were being destroyed during the assembly of Pangaea, the exchange of Ca^{2+} for Mg^{2+} ions at mid-ocean ridges presumably slowed. Decreased hydrothermal exchange at spreading centers presumably resulted in a greater likelihood of the secretion of aragonitic hard parts or hard parts composed of calcite enriched in magnesium (high-Mg^{2+} calcite). Such seaways are referred to as **aragonite seas**.

CONCEPT AND REASONING CHECKS

1. Make a chart comparing the late Paleozoic with the Early-to-Middle Paleozoic. Label the columns Early-to-Middle Paleozoic and Late Paleozoic. In the rows, indicate (a) greenhouse or icehouse, (b) sea level (high or falling), (c) large continental glaciers (present or absent), (d) calcite compensation depth (shallow or deepening), (e) weathering (fast or slow), (f) calcite or aragonite seas, (g) deep-ocean circulation (Hint: were black shales more or less frequent during this time?), and (h) surface-ocean circulation.
2. Why did these changes occur between the Early-to-Middle and Late Paleozoic? (See also the next sections.)

12.3 Tectonic Cycle and Orogeny

The continuation and eventual completion of the first tectonic cycle of the Phanerozoic Eon culminated in the production of Pangaea. Some of the world's most famous mountain ranges, including the Appalachians, remain from these collisions.

12.3.1 Alleghenian and related orogenies

During the Late Paleozoic, the supercontinent of **Laurasia** began to form as a result of the collisions of North America, Europe, Greenland, and parts of Asia. Asia had begun to form earlier in the Paleozoic (see Figure 12.2). Initially, Siberia collided with Kazakhstan; the resulting continent then moved toward Baltica during the **Uralian Orogeny** of the Permian to produce the Ural Mountains. The Urals lie approximately 1,500 km (~1,000 miles) east of Moscow in the middle of a continent, separating European Russia from Siberia (see Figure 12.2). Recall from Chapter 2 that because of their location, the origin of the Urals could not be satisfactorily explained by the geosynclinal theory, which required mountain building along continental margins; it was only after the acceptance of the theory of plate tectonics that the formation of the Urals could be explained. Much of the remainder of Asia grew as a result of its collision with microcontinents such as Tibet and terranes that had rifted northward from Gondwana.

During this time Gondwana moved north to collide with Laurasia. As it did so, Gondwana rotated in a clockwise direction to produce, more or less in succession, the **Hercynian** (or **Variscan**), **Alleghenian** (or **Appalachian**), and **Ouachita orogenies** (see Figure 12.2).

In North America, the Alleghenian Orogeny marked the third and final episode of uplift in the Appalachian region. During the Alleghenian Orogeny, Gondwana, in particular Africa, collided with the eastern and southeastern margin of Laurentia (see Figure 12.2). This orogenic phase involved repeated thrusting of rocks over several hundred kilometers to produce stacks of thrust sheets, each of Cambrian-to-Devonian age, from New York all the way to Alabama. Unlike the previous Taconic and Acadian orogenies, however, the Alleghenian was not preceded by flysch. Instead, the thrust sheets formed by the Alleghenian Orogeny were eroded and a huge wedge of molasse shed into the cratonic interior as conglomerates and sandstones.

Along the Gulf Coast, South America collided with Laurentia to produce the Ouachita Orogeny (**Figure 12.3**). The collision that formed the Ouachitas is thought to have been an extension of the Alleghenian Orogeny into the southwestern United States because the Ouachita Mountains lie on a line curving east and north into the Appalachians (see Figure 12.2). Before the Ouachita Orogeny the Gulf Coast region was originally a passive margin, with shallow-water sandstones and limestones deposited adjacent to deeper-water

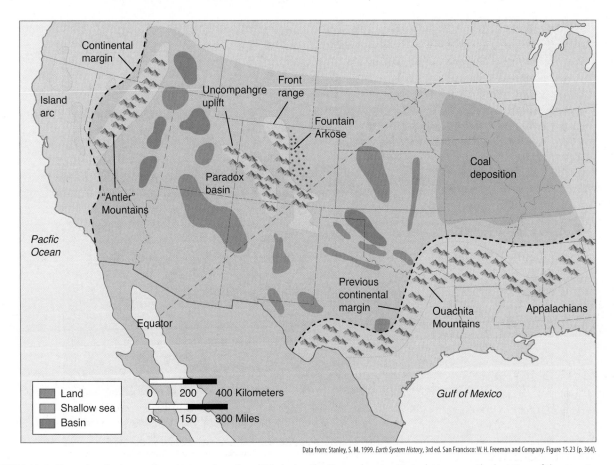

FIGURE 12.3 Tectonism in the southeastern and western U.S. during the Pennsylvanian Period. How was the location of the equator determined?

Data from: Stanley, S. M. 1999. *Earth System History*, 3rd ed. San Francisco: W. H. Freeman and Company. Figure 15.23 (p. 364).

shales and cherts. Beginning in the Mississippian, all this changed. The passive margin was converted to an active one, and a huge clastic wedge of molasse began to form, thickening to the south, during the Pennsylvanian. These sediments were then thrust onto the craton during the Permian. Also caught up in the Ouachita Orogeny were several microplates that would eventually become portions of Central America.

The mountains formed by these orogenies would eventually stretch as a continuous belt across southern Europe to southwestern Laurentia by the Late Paleozoic. In North America, much of the Appalachians uplifted during the Alleghenian Orogeny have since been eroded, with most of the sediments being shed into a foreland basin to the west (see Figure 12.2). Like the Appalachians, the young Ouachitas must have also formed an immense mountain range. Initially, the Ouachitas stretched from what is today Mississippi to west Texas. However, the Ouachitas must have been greatly eroded during the Mesozoic Era because all that is left of this chain today are the Ouachita Mountains of Oklahoma and Arkansas and the Marathon Mountains of west Texas.

In Europe remnants of the Hercynian Orogeny are represented by the **massifs** scattered across western Europe (**Figure 12.4**). The oldest rocks of the massifs themselves are mostly open marine shales and limestones, which give way to coal-bearing and nonmarine deltaic and swamp deposits during the Carboniferous. The rocks of the massifs

were often folded and faulted by the Hercynian Orogeny (see Figure 12.4). Younger Mesozoic and Cenozoic rocks were later deposited around the massifs in places such as the Massif Centrale of southern France.

12.3.2 Ancestral Rockies

Also uplifted during the Pennsylvanian were the **Ancestral Rockies**, which were composed of the **Front Range** (eastern margin of the Rocky Mountains) and Uncompahgre Uplift in Colorado and northern New Mexico (**Figure 12.5**). Unlike the other orogenies of the Paleozoic, these episodes did not occur along continental margins but well within the craton, which is supposed to be stable! Nevertheless, the peaks of the Ancestral Rockies might have been as high as several kilometers.

The Ancestral Rockies are thought to have been formed by high-angle faulting in response to compressive stresses generated during the collision of Gondwana with Laurasia. The uplifts consisted of Precambrian crystalline (mainly metamorphic) basement rocks that were later eroded and rapidly deposited as conglomerates and arkoses along the margin of the modern Front Range (see Figure 12.5). Today, the Precambrian basement is exposed in deep river gorges, whereas the Fountain Arkose is exposed along the modern Front Range (see Chapter 4).

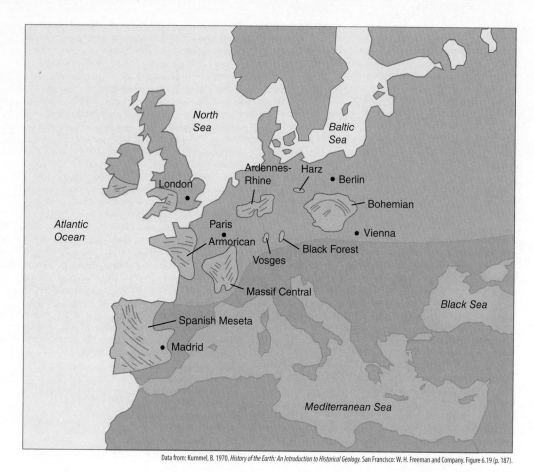

Data from: Kummel, B. 1970. *History of the Earth: An Introduction to Historical Geology.* San Francisco: W. H. Freeman and Company. Figure 6.19 (p. 187).

FIGURE 12.4 Paleozoic massifs of Europe, relics of the Hercynian Orogeny. Trends of the folds produced by the Hercynian Orogeny are shown by the lines (see Chapter 4).

Conglomerates and sands were also shed westward from the Uncompahgre Uplift. These sediments graded west into evaporites of the Paradox Basin (in what is now Utah); the presence of evaporites indicates that the Paradox Basin was a restricted marine basin subject to high rates of evaporation in an arid region. Shales and sandstones, some of which are today exposed in the Grand Canyon, were also deposited in this region in floodplain environments (see Figure 12.3).

As the Pennsylvanian gave way to the Permian, the Uncompahgre Mountains continued to be worn down, shedding arkosic sediments onto the fluvial plains. These sediments eventually gave way to cross-bedded, predominantly dune sands like those of the Coconino Sandstone. Today, the Coconino is visible as a prominent white band near the top of the Grand Canyon (refer to the frontispiece of Chapter 6).

The seas returned to the vicinity of the Grand Canyon again in the Permian, depositing shallow, warm-water limestones of the fossiliferous Kaibab Limestone, which contains the remains of crinoids, brachiopods, clams, and horn corals. The Kaibab Limestone lies at the very top of the Grand Canyon (refer to the frontispiece of Chapter 6) but grades eastward into nonmarine sediments that had continued to be shed from the Uncompahgre Mountains (see Figure 12.5).

12.3.3 Sonoma orogeny

Orogeny also continued in westernmost North America. The western margin of North America had already shifted from a passive to an active continental margin earlier in the Paleozoic (see Chapter 11). The Antler Orogeny, which began by the Late Devonian, continued into the Mississippian, accreting an island arc through what is now Nevada. The western margin of North America extended northward into Canada and Alaska, where the Ellesmere Orogeny, which began by the Late Devonian, accreted one or more other terranes (see Chapter 11).

Western Laurentia was then affected by the **Sonoma Orogeny** during the Late Paleozoic. During the Late Permian and continuing into the Triassic, yet another volcanic island arc called Sonomia accreted to what is now northwest Nevada, southern Oregon, and what are now exposed as the Klamath Mountains in northern California (see Figure 12.2 and **Figure 12.6**). The terrane **Sonomia**, which today represents Nevada, southeast Oregon, and northern California, was added to what is now the western margin of the United States during the Sonoma Orogeny. Although the orogenic activity during the Sonoma Orogeny was similar to that of the Antler Orogeny, the Sonoma Orogeny accreted a much

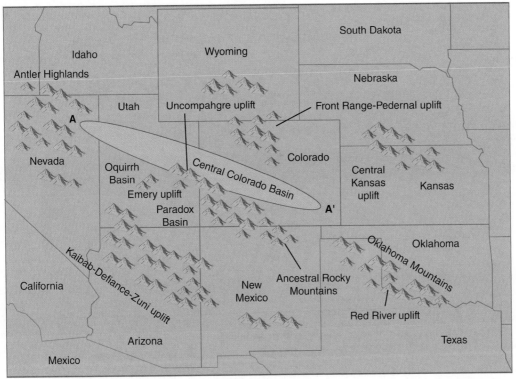

Data from: Wicander, R., and Monroe, J. S. 2000. *Historical Geology: Evolution of Earth and Life Through Time*, 3rd ed. Pacific Grove, CA: Brooks/Cole. Figure 11.11A (p. 284).

(a)

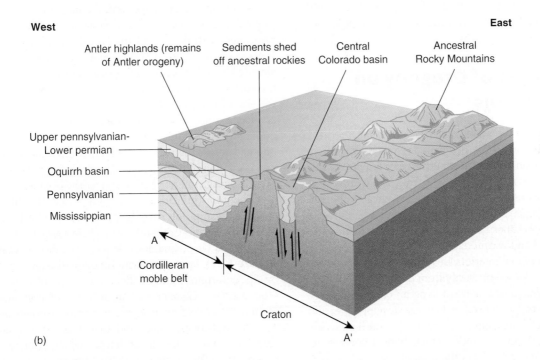

Data from: Wicander, R., and Monroe, J. S. 2000. *Historical Geology: Evolution of Earth and Life Through Time*, 3rd ed. Pacific Grove, CA: Brooks/Cole. Figure 11.11B (p. 284).

(b)

FIGURE 12.5 **(a)** Map of the western United States during the Pennsylvanian from what is now eastern Colorado across the Uncompahgre Uplift to the highlands left from the Antler Orogeny. The area A-A' indicates the area of cross-section shown in part B. **(b)** Cross-section along A-A'.

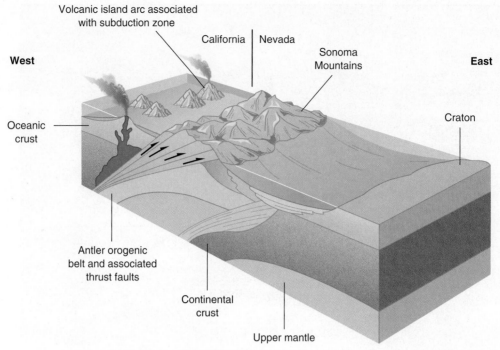

Volcanic island arc associated
with subduction zone

West

California | Nevada

Sonoma
Mountains

East

Oceanic
crust

Craton

Antler orogenic
belt and associated
thrust faults

Continental
crust

Upper mantle

Data from: Monroe, J. S., and Wicander, R. 1997. *The Changing Earth: Exploring Geology and Evolution*, 2nd ed. Belmont, CA: West/Wadsworth. Figure 23.9 (p. 605).

FIGURE 12.6 The accretion of Sonomia.

larger portion of crust to North America. This pattern would continue into the Mesozoic with the Nevadan Orogeny (see Chapter 13).

12.4 Impact of Orogeny on Earth Systems

F G H As we have already seen, the repeated destruction of seafloor spreading centers and the uplift of land by orogeny and supercontinent assembly gradually drove the seas off the continents (see Figure 12.1). As the Late Paleozoic passed, greater climatic extremes also began to develop on Earth. These climates ranged from the southern hemisphere glaciers of Gondwana to warm, humid and warm, dry climates nearer the equator to hot, dry continental interiors (see Figure 12.2). These climatic extremes resulted primarily from the assembly of Pangaea. The movement of low-latitude ocean currents around the globe came to be blocked as Pangaea developed astride the equator, stretching from the north to the south pole (see Figure 12.2). The trade winds, which blow from east to west, pulled waters away from the *western* coast of Pangaea, causing upwelling of cold waters. As the waters moved westward toward the eastern margin of the Pangaea, they would have warmed from the sun. When they reached the *eastern* margin of Pangaea, these warm surface currents were deflected by the eastern margin of Pangaea toward higher, cooler latitudes; here, the water masses released their heat to the surrounding environment. The colder currents then flowed from high latitudes back toward the equator along the western margin of Pangaea (see Figure 12.2).

Rising mountain ranges and the increasing size of continental interiors also produced orographic effects like those associated with the rise of the Himalayas of Asia or Sierra Nevada of the western United States (see Chapter 3). We already know that because of its size, Pangaea nearly spanned the surface of the planet from one pole to the other. Given the size of Pangaea, gigantic monsoons—what some workers have called **megamonsoons**—must have alternately dumped tremendous amounts of rain on mountain chains oriented across the path of the predominant moisture-laden winds coming off Panthalassa. The monsoon carried dry air from Pangaea's interior back out over the oceans.

The moisture-laden winds supported the lush coal-forming swamps of what is now eastern North America during the Permo-Carboniferous. These coals are associated with distinctive sedimentary sequences called **cyclothems** (**Figure 12.7**). Each cyclothem is bounded above and below by disconformities and is therefore a type of depositional sequence (see Chapter 6). At the base of each idealized (complete) cyclothem is a disconformity overlain by nonmarine sandstones, shale, and then freshwater limestone. Above these units lies an underclay, which contains traces of roots from the plants that formed the overlying coal. The underclays are in turn overlain by shallow marine, fossiliferous limestones and shales, suggesting rising sea level. The top of the sequence is marked by another disconformity, after which the sequence of rocks begins again. However, not all units are found in every cyclothem because of erosion. As their name implies, cyclothems resulted from cyclic deposition caused by oscillations in sea level. The most widespread cyclothems are thought to represent the response of sea level and deposition to the waxing and waning of ice sheets on

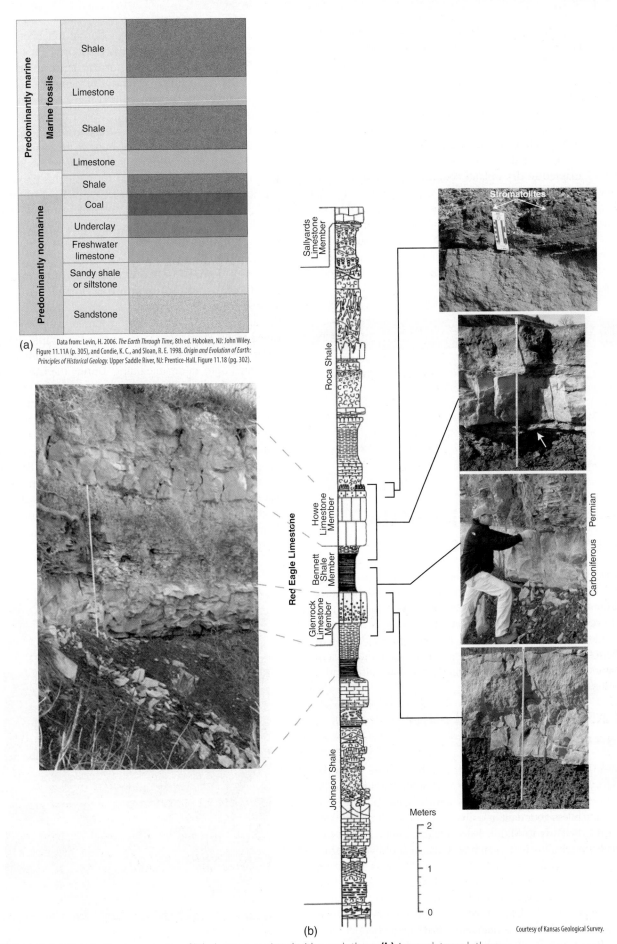

FIGURE 12.7 **(a)** The idealized succession of lithologies associated with a cyclothem. **(b)** A complete cyclothem.

Data from: Levin, H. 2006. *The Earth Through Time*, 8th ed. Hoboken, NJ: John Wiley. Figure 11.11A (p. 305), and Condie, K. C., and Sloan, R. E. 1998. *Origin and Evolution of Earth: Principles of Historical Geology.* Upper Saddle River, NJ: Prentice-Hall. Figure 11.18 (pg. 302).

Courtesy of Kansas Geological Survey.

Gondwana, possibly in response to changes in solar radiation. Because of the low-lying nature of deltas along the ancient coasts, slight changes in sea level due to advances and retreats and glaciers could have easily produced the cyclic patterns of deposition seen in the cyclothems.

As Pangaea continued to be assembled, however, the area of interior drainage increased and coal swamps gave way to drier conditions. Consequently, coal swamps gave way to forests dominated by gymnosperms. Gymnosperms were better adapted to dry upland conditions because they produced seeds, which are resistant to desiccation and also aid in dispersal (see section 12.6). The resulting warm, dry continental interiors of what is now the western United States are reflected by cross-bedded desert sandstones such as the Coconino Sandstone and widespread evaporites that formed in the Paradox Basin located in the lee of the Ancestral Rockies (see Figure 12.5).

12.5 Diversification of the Marine Biosphere

The Paleozoic Fauna, which originated early during the Paleozoic and replaced the Cambrian Fauna, continued to dominate the fossil record until the end of the Permian. However, the dominant members of both the benthos and plankton were much changed by **ecologic replacement** from the first half of the Paleozoic. Members of the Modern Fauna—such as bivalves, gastropods, and fish—also continued to diversify somewhat.

12.5.1 Plankton and other microfossils

The planktonic microfossil taxa of the Late Paleozoic were much changed from those of the Early-to-Middle Paleozoic. Acritarchs continue from the Late Paleozoic but are much less abundant and diverse in the fossil record than during the Early-to-Middle Paleozoic (see Figure 12.1). Normally, the decline in acritarch abundance and diversity during the Late Paleozoic has been attributed to mass extinction (the so-called "phytoplankton blackout") during the Late Devonian. Possibly, the much lower diversity and abundance of acritarchs in the Late Paleozoic reflects increased nutrient runoff to the oceans, perhaps then making the frequent formation of cysts by acritarchs unnecessary. Increased nutrient

runoff to the oceans from land during much of the Late Paleozoic is suggested by increased strontium isotope ratios (see Figure 12.1). Increased nutrient input to the oceans might have resulted from extensive orogeny that occurred as Pangaea was being assembled and the spread of terrestrial forests, which might have also increased rates of continental weathering. Increased rates of oceanic upwelling during the Late Paleozoic in response to southern hemisphere glaciation also would have brought dissolved nutrients already present in deep ocean waters back to the photic zone, thereby stimulating photosynthesis by phytoplankton. Radiolarians and conodonts also continued into the Late Paleozoic but planktonic graptolites died out long before the end of the Devonian.

One of the most remarkable developments in the benthic microfossil record was the diversification of a group of single-celled protists called **foraminifera**. Foraminifera actually appeared late in the Proterozoic or Early Paleozoic during the appearance of metazoans. Foraminifera secrete shells called **tests**. The earliest tests consisted of simple straight or coiled tubes composed of organic layers or sediment grains cemented onto an organic lining. During the Late Paleozoic new taxa of foraminifera appeared that secreted tests of calcium carbonate ($CaCO_3$). Among the most prominent taxa were the **fusulinids** (**Figure 12.8**). Fusulinid tests ranged from round to football-like in shape and grew to at least several millimeters in diameter. The interiors of the tests were subdivided into smaller chambers (see Figure 12.8). In modern reef environments, large foraminiferans having complex test interiors harbor algal symbionts that provide them with food and oxygen and remove waste products, much as they do in modern corals (see Chapter 4). Fusulinids are therefore thought to have also harbored algal symbionts. Fusulinids

(a)

Courtesy of USGS.

FIGURE 12.8 (a) Loose tests of fossil fusulinids.

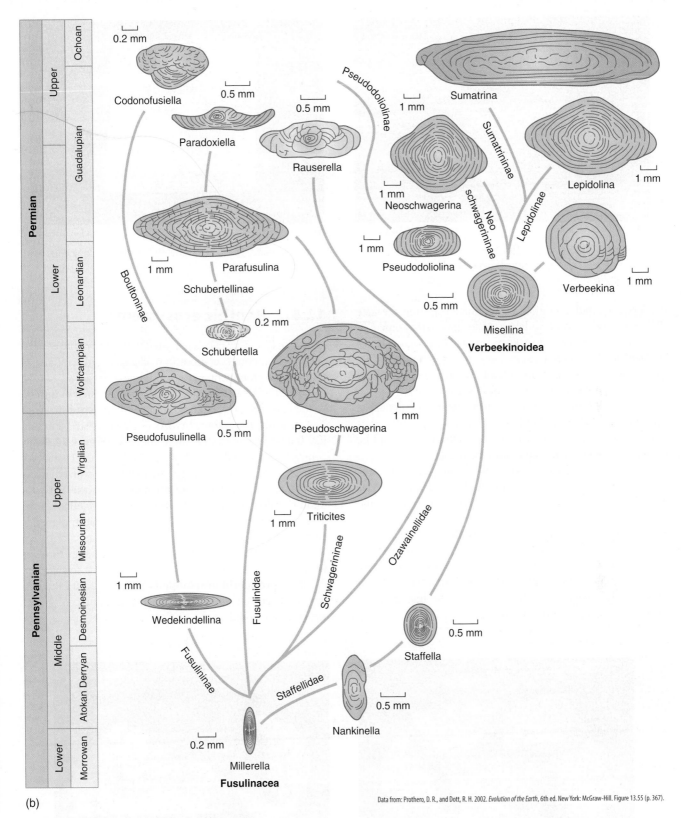

(b)

FIGURE 12.8 (b) Evolution of fusulinid lineages.

Data from: Prothero, D. R., and Dott, R. H. 2002. *Evolution of the Earth*, 6th ed. New York: McGraw-Hill. Figure 13.55 (p. 367).

© Joyce Photographics/Science Source.

(a)

© Science Stock Photography/Science Source.

(b)

FIGURE 12.9 Fenestrate bryozoans produced lacy, fan-like structures, hence their name. The fan-like structure (a) was a colony housing large numbers of polyps; this structure radiated off of a central, corkscrew-like structure (b).

underwent rapid evolution and extinction during the Late Paleozoic and they are widely used for biostratigraphic correlation and zonation of rocks of this age (see Figure 12.8). Indeed, fossil fusulinids were sometimes so abundant they were the dominant components of limestones deposited in sunlit waters of epeiric seaways of this time.

The occurrence of benthic calcareous foraminifera during the Late Paleozoic holds some interesting implications for calcite and aragonite seas. These foraminifera would be expected to have high-magnesium calcite or aragonitic tests. However, fusulinids have low-magnesium and not high-magnesium tests. Given their unicellular grade of organization, foraminifera should have been more susceptible to ionic changes in seawater than multicellular organisms, which possess specialized tissues for controlling ion exchange. Thus, if fusulinids secreted tests of high-magnesium calcite or aragonite as they lived, the tests must have recrystallized to low-magnesium calcite after death. Nevertheless, it is widely accepted that the crystals of the fusulinid test were actually secreted by the animals.

12.5.2 Benthic ecosystems

The Paleozoic Fauna persisted to the end of the era, but its taxonomic composition changed substantially. A number of prominent Early-to-Middle Paleozoic taxa like trilobites and tabulate and rugose corals persisted to the end of the Paleozoic, but often with much diminished diversity. The stick-like trepostome bryozoans also persisted, but fan-like **fenestrate** types were more common (**Figure 12.9**). By contrast, ammonoids, which were hard hit by the Late Devonian extinctions, rediversified quickly and evolved many new taxa, even as the bulkier nautiloids of the Early-to-Middle Paleozoic declined. Along with conodonts and fusulinids, ammonoids are frequently used for biostratigraphic correlation and zonation of Late Paleozoic marine rocks

Articulate brachiopods also rediversified after the Late Devonian extinctions. Prominent among these forms were the spiny **productid brachiopods** (**Figure 12.10**). Productids modified one valve of the shell to produce a cup-like structure anchored to the seafloor by spines. In this, they

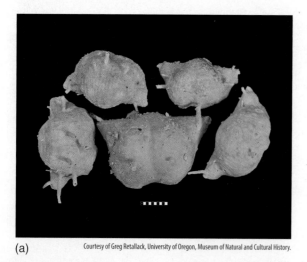

Courtesy of Greg Retallack, University of Oregon, Museum of Natural and Cultural History.

(a)

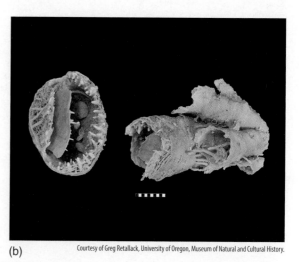

Courtesy of Greg Retallack, University of Oregon, Museum of Natural and Cultural History.

(b)

FIGURE 12.10 Late Paleozoic brachiopods. **(a)** Productid articulate brachiopods, which are characterized by spines that kept them from sinking into sediment. **(b)** Prorichthofenid brachiopods, which grew in clumps.

resembled archaeocyathids or rugose corals (see Chapter 11). However, the cup of the productid shell was covered by a lid formed by the other valve. This basic body structure was modified by some productids into bizarre shapes (see Figure 12.10).

The Late Devonian extinctions also decimated stromatoporoid reefs. Although stromatoporoids have survived to the present (in the form of sclerosponges; see Chapter 11), they never recovered from the Late Devonian extinctions. The stromatoporoid reefs were replaced in the Mississippian by loose mounds of dead skeletal debris. Some mounds were produced by calcareous algae that secrete $CaCO_3$ in life; these calcareous algal mounds were prominent in the shallow, restricted seaways that developed in what is now the western United States in response to orogeny. Today, calcareous algae that live in reef habitats tend to be more prominent when nutrient levels are somewhat elevated, so these ancient environments might have had higher nutrient levels, perhaps as a result of orogeny and weathering.

One of the most prominent producers of reef-like mounds were crinoids. At times during the Mississippian, crinoids literally formed meadows that must have resembled the waves running over the tops of wheat fields. Crinoids were accompanied by the blastoids. Upon death, their calcareous skeletons usually disarticulated quickly, producing large amounts of calcareous crinoidal limestones (**Figure 12.11**). Crinoids and other suspension feeders also continued the trend of increasing tiering above bottom, as the depth of bioturbation continued to increase from earlier in the Paleozoic (see Chapter 11). Increased tiering is consistent with expanding populations of plankton and a rain of dead organic matter that served as food for burrowing organisms.

The best-developed reefs of the Late Paleozoic did not appear until the Permian. The most famous of these is the **Permian Reef Complex**, located in what is now west Texas and New Mexico (**Figure 12.12**). The Permian reefs lay

Courtesy of Ronald Martin, University of Delaware.

FIGURE 12.11 Crinoidal limestone formed by disarticulation of the stems. Cross-sections of stem fragments are also visible.

along the equator, apparently in the lee of the Ouachita Mountains, along the margins of deep-water troughs not unlike the present setting of the Bahamas. The troughs developed in response to the collision of South America with the southern United States during the Alleghenian Orogeny (**Figure 12.13**). Orogenic activity caused block faulting and the formation of deep basins, which filled with shales, cherts (from radiolarians), and sandstones shed from the uplifts further west, while limestones and evaporites were deposited on shallow shelves and in restricted basins behind the reefs.

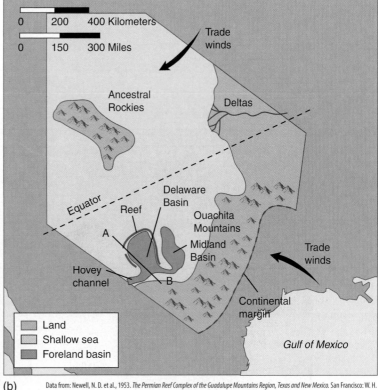

Data from: Newell, N. D. et al., 1953. *The Permian Reef Complex of the Guadalupe Mountains Region, Texas and New Mexico.* San Francisco: W. H. Freeman and Company.

(a) © iStockphoto/Thinkstock. (b)

FIGURE 12.12 The Permian Reef Complex. **(a)** The steep wall is the Capitan Limestone, which basically represents what was the most actively growing portion of the reef. The sloping sediments at the base of the wall are reef talus and fine-grained sediments deposited in deeper waters that once existed adjacent to the front of the reef platform. **(b)** Ancient tectonic setting of the Permian Reef Complex along the margins of deep seaways, not unlike the present setting in the Bahamas. The shallow-water platform and adjacent deep-water basins developed in response to the collision of South America with the southern United States and the resulting Alleghenian Orogeny.

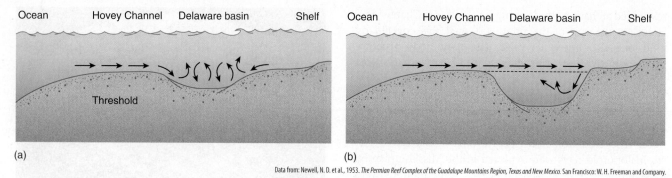

(a)

(b)

Data from: Newell, N. D. et al., 1953. *The Permian Reef Complex of the Guadalupe Mountains Region, Texas and New Mexico.* San Francisco: W. H. Freeman and Company.

FIGURE 12.13 Deepening of the basins adjacent to the reef platform of the Permian Reef Complex restricted the influx of oxygenated waters, while promoting anoxia and the preservation of organic matter that would become petroleum.

The deposition of relatively porous reef limestones adjacent to deep-water sediments rich in dead organic matter was ideal for the formation and entrapment of oil and gas in the reef limestones. The deep-water basins developed into poorly oxygenated silled basins, which restricted circulation and the influx of oxygenated waters. The deeper waters of the basins became relatively anoxic, which preserved organic matter that was eventually transformed into petroleum. As petroleum formed in the deep-water shales and cherts, it migrated into the limestones and sands. Here,

the petroleum was trapped in reservoirs that were sealed by evaporites and dolostones that formed in shallow lagoons and embayments behind the reefs. Consequently, the so-called Permian Basin has been intensively studied by generations of geologists because of its rich petroleum reserves. The rocks have not only been drilled extensively in the subsurface for many decades for oil and gas, but the subsurface formations can also be traced in surface outcrops, making understanding the subsurface formations much easier (**Figure 12.14**).

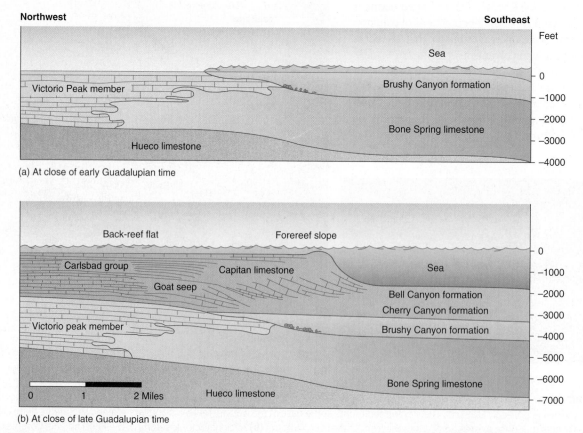

(a) At close of early Guadalupian time

(b) At close of late Guadalupian time

FIGURE 12.14 Cross-section of the Permian Reef Complex showing the different formations and facies changes. The oil and gas derived from the reef limestones was originally sourced from dead pelagic organic matter deposited in the adjacent deep water basin.

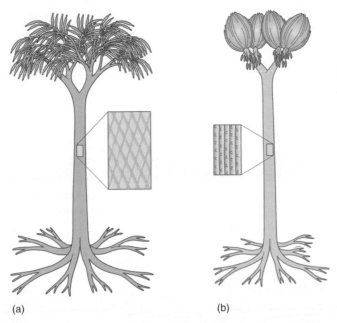

(a)   (b)

FIGURE 12.15 Terrestrial forests became prominent during the last half of the Paleozoic Era (shaded). It was during the Carboniferous that coal-forming swamps dominated by tree-like club mosses, or lycopods, became widespread. Lycopods eventually gave way to gymnosperms during the Permian period.

CONCEPT AND REASONING CHECK

1. Give examples of ecologic replacement in the marine realm from the Early-to-Middle to the Late Paleozoic.

12.6 Diversification of the Terrestrial Biosphere

12.6.1 Terrestrial floras

The spread of terrestrial forests that began by the Late Devonian continued through the Late Paleozoic (**Figure 12.15**; refer to this chapter's frontispiece). However, the relatively high sea level of the Mississippian Period initially slowed the invasion of continental interiors. Then, with further orogenesis and sea-level fall during the Pennsylvanian, foreland basins and other portions of continental interiors began to be populated by terrestrial plants. Among these floras it is the coal-forming swamps of the Late Carboniferous that represent the most spectacular and important development of Late Paleozoic terrestrial floras. Much of the coal originally mined in the United States came from coals in Pennsylvania, West Virginia, and Kentucky (**Figure 12.16**; much of the coal mined in the United States now comes from Paleogene deposits in the western United States because it causes less pollution from mining and burning).

The coal-forming swamps were dominated by several types of spore-bearing plants that appeared during the Devonian. The lycopods that had appeared during the Late

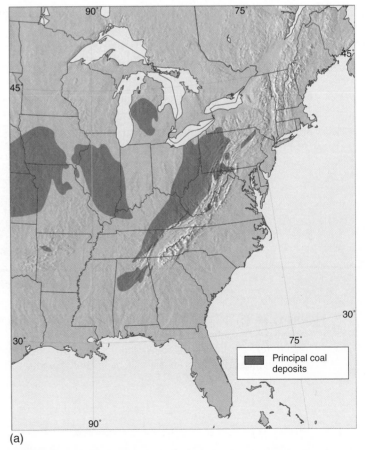

(a)

FIGURE 12.16 (a) Occurrence of major coal-producing regions of the eastern and midwestern portions of the United States.

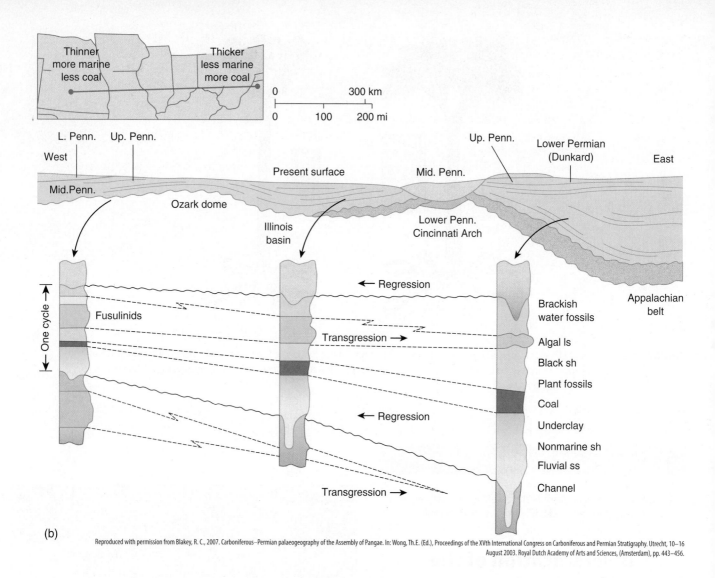

Reproduced with permission from Blakey, R. C., 2007. Carboniferous–Permian palaeogeography of the Assembly of Pangae. In: Wong, Th.E. (Ed.), Proceedings of the XVth International Congress on Carboniferous and Permian Stratigraphy. Utrecht, 10–16 August 2003. Royal Dutch Academy of Arts and Sciences, (Amsterdam), pp. 443–456.

(b)

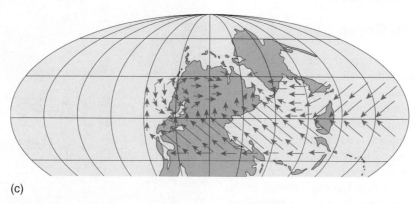

(c)

FIGURE 12.16 (b) Lithocorrelation of an idealized cyclothem across the eastern and middle United States. Note how the strata become more marine in character to the west; this was because the interior of the continent was flooded by the seas. T = Transgression; R = Regression. **(c)** Moisture-laden easterly winds rained down on eastern Pangaea to help produce coal-forming swamps. This resulted in part from the presence of the rising Appalachian mountains to produce an orographic effect. Compare with Figures 12.16a and b.

Devonian diversified to produce trees such as *Sigillaria* and *Lepidodendron* that grew up to 30 meters (~100 feet) tall and were prominent in the swamps (see this chapter's frontispiece). These taxa possessed long, simple leaves to capture sunlight. The leaves were arranged in spiral rows around the trunk, producing distinctive diamond-shaped leaf scars on fossil bark. Lycopods were joined by the spore-bearing sphenopsids, some of which, such as *Calamites,* also grew to the size of trees. Spore-bearing ferns, which required moisture for reproduction (**Figure 12.17**), were also present in the swamps. Sometimes, they also reached the size of trees, but they were also present in the understory beneath the forest's canopy. Still, lycopods and sphenopsids were all tied to water because they produced spores.

What really freed vascular land plants to invade drier continental interiors was the appearance of seeds. In spore-bearing plants, such as true ferns, the sporophyte generation, which is the macroscopic stage we see, matures and

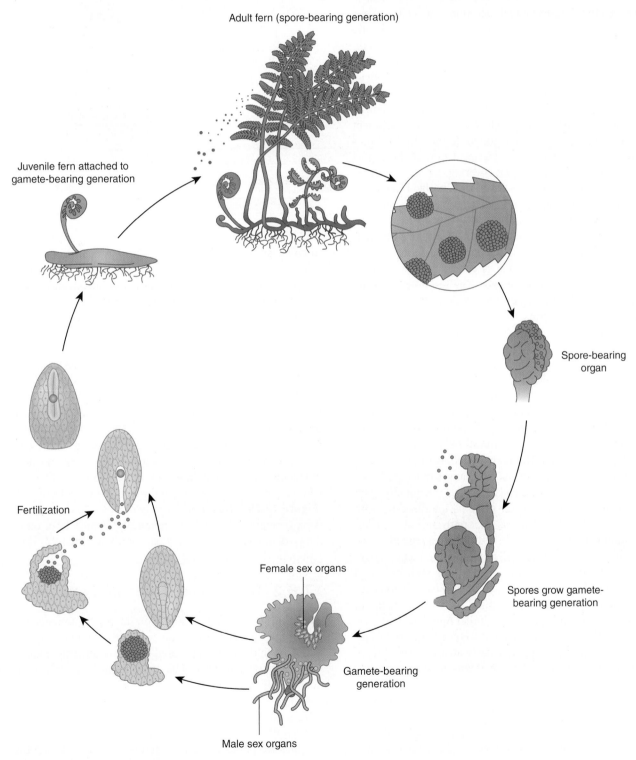

Adult fern (spore-bearing generation)

Juvenile fern attached to gamete-bearing generation

Spore-bearing organ

Spores grow gamete-bearing generation

Fertilization

Female sex organs

Gamete-bearing generation

Male sex organs

FIGURE 12.17 Life cycle of spore-bearing ferns.

produces spores of all one type (homospory); these spores are released to produce the gametophyte, or gamete-bearing generation (see Figure 12.17). The sperm produced by the gametophyte require moisture to travel on the surface of the gametophyte to fertilize the egg, producing the sporophyte generation. By contrast, in seed plants the sperm is encased in a waxy covering to produce pollen (thus omitting the stage that requires water) that is carried by the wind or animals, eventually fertilizing an ovum within another plant. The resulting zygote then produces a hard external coat to form a seed that is resistant to desiccation. The seed can then be dispersed by wind or animals. The fossil record indicates that the appearance of seeds was preceded by an intermediate stage in the Early Devonian characterized by heterospory. In this stage, the spores were retained on the sporophyte, where they matured into two types of spores—megaspores and microspores—that produced, respectively, the female and male portions of the gametophyte. This development mitigated the necessity of moisture for fertilization and provided the opportunity for the encapsulation of the zygote (the fertilized egg) by a resistant covering.

These ancestral lineages gave rise to the progymnosperms during the Middle and Late Devonian. The progymnosperms physically resembled later gymnosperms (which today include the modern conifers) but reproduced by heterospory. In gymnosperms, the sperm is encased in a waxy coating to produce pollen that is carried by the wind to female cones, which bear the eggs. The resulting seeds are borne in exposed positions on the female cones; hence, the name of the gymnosperms: "naked seed."

Seed ferns, which greatly resembled spore-bearing ferns but were actually primitive gymnosperms, appeared in the Late Devonian. Although they were present in the later coal swamps, they were also prominent on higher or drier ground. Seed ferns diversified in the Late Paleozoic to produce the shrub-to-tree like *Glossopteris* on Gondwana. The *Glossopteris* flora also favored cooler, swampy conditions of temperate to higher latitudes rather than the warmer, tropical conditions of lycopod forests. This is evidenced by trees exhibiting rings that indicate distinct seasonal changes in temperature, precipitation, or both. *Glossopteris* was also frequently associated with glaciers, as indicated by its interbedding with glacial tillites. Recall that it was the occurrence of the *Glossopteris* flora, along with fossil reptiles, tillites, the fit of the continents, and the structural trends of mountain ranges on the now-separated continents, that were used by Alfred Wegener to infer the existence of Pangaea (see Chapter 2). Today, the *Glossopteris* flora is found on the continents of Australia, South America, Africa, India, Australia, and Antarctica. But the seeds of *Glossopteris* were too large to be carried by the wind. The only reasonable answer seemed to be that the continents were once joined into the supercontinent of Gondwana.

Seed ferns and other primitive gymnosperms were eventually accompanied by **cordaites** such as the genus *Cordaites* during the Late Carboniferous (**Figure 12.18**). Like the seed ferns cordaites also belonged to the gymnosperms. *Cordaites* reached up to 30 meters in height and was able

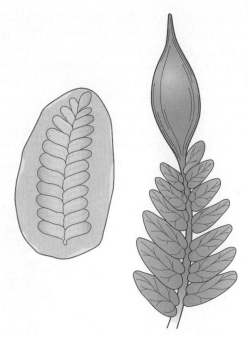

FIGURE 12.18 The fossil cordaite, *Cordaites*, characterized drier upland regions.

to colonize higher, drier ground. Despite their prominence during the Carboniferous, coal-forming swamps began to disappear during the Permian. As we have already discussed, the assembly of Pangaea resulted in vast, dry continental interiors. As a result gymnosperms, including conifers, came to dominate terrestrial floras during the Middle-to-Late Permian. These taxa had needle-like leaves and small seed-bearing cones. The taxa that had previously dominated terrestrial floras—lycopods, sphenopsids, and seed ferns—persisted but became small, creeping forms that lived adjacent to water or in moist habitats. True ferns remained part of the understory of later forests.

Perhaps the most amazing finds in recent years with regard to plant evolution has been the discovery and documentation of a fossil forest in Inner Mongolia of Early Permian age that was buried and preserved by volcanic ash (**Figure 12.19**). This fossil flora is remarkably diverse and is biogeographically and taxonomically distinct from other floras, demonstrating that this part of the world at this time represented a distinct floral realm.

12.6.2 Terrestrial floras and oxygen

The spread of terrestrial forests had a tremendous impact on biogeochemical cycles. Root systems and soil formation undoubtedly increased rates of weathering. Expansion of terrestrial floras also likely led to increased transpiration and the release of water vapor to the atmosphere, which also might have promoted weathering (recall the tremendous increase of strontium isotopes during the Permo-Carboniferous described previously; see Figure 12.1).

Massive photosynthesis by terrestrial forests during the Carboniferous must have had very pronounced effects on the

Wang, J., Pfefferkorn, H. W., Zhang, Y., and Feng, Z. 2012. Permian vegetational Pompeii from Inner Mongolia and its implications for landscape paleoecology and paleobiogeography of Cathaysia. Proceedings of the National Academy of Sciences 109(13):4927–4932, Figure 5.

FIGURE 12.19 Reconstruction of a site of a peat-forming forest of earliest Permian age preserved by a volcanic ash-fall near Wuda, Inner Mongolia, China. The tall brush-like upper-story trees are *Sigillaria* (an extinct lycopsid) that are carrying bundles of cones below their tuft of narrow leaves. The lower-story forest is made up of several species of tree ferns that are characterized by a brown root mantle in the lower part of the stem and dead leaves hanging down. Other smaller trees shown from left to right are *Tingia*, *Pterophyllum*, and *Taeniopteris*. A *Sphenopteris* species appears as a vine on two tree-fern root mantles. An herb layer existed only in some areas and is shown in the right foreground with *Sphenophyllum*, belonging to an extinct group of sphenopsids. The peat was covered most of the time by a few centimeters of standing water protecting it from oxidation. These forests were quite dense and harbored diverse denizens such as dragonflies with one-foot wingspans, 10-foot-long relatives of centipedes, large amphibians, and other ancestors of reptiles and mammals, as described in the text.

biogeochemical cycle of carbon during the Late Paleozoic. In fact, terrestrial photosynthesis probably accounts for much of the shift in the carbon isotope ratios toward higher values during the Permo-Carboniferous (see Figure 12.1). Because of the enormous amounts of carbon sequestered in living forests and dead plant debris on land by photosynthesis, the oxygen content of the atmosphere is calculated to have increased to about 35% during the Pennsylvanian, as opposed to less than 20% before and after this time (**Figure 12.20**).

These percentages were determined by using **mass balance** calculations. In mass balance calculations, changes in the mass of one reservoir (for example, the amount of carbon sequestered in coal) are examined to see how they affect the masses of substances in other reservoirs (for example, oxygen [O_2] in the atmosphere). In the case of coal swamps, the calculations were basically done by (1) assuming the amount of organic matter buried through time depends on changes in the amounts of the main types of sedimentary rocks in which carbon might be found (sandstones, shales, coals, redbeds, etc.), (2) estimating the amount of dead plant debris buried in coals from the distribution and thicknesses of coal-bearing units determined from maps, and (3) using the fact that one molecule of CO_2 combines with one molecule of H_2O during photosynthesis to produce one molecule of O_2:

$$CO_2 + H_2O \rightarrow CH_2O + O_2 \uparrow \qquad (12.1)$$

By estimating the amount of carbon buried in different types of rocks and the rates of deposition or erosion of the rocks themselves, we can estimate the amount of carbon buried through time and therefore the amount of O_2 released to the atmosphere.

Based on such calculations, the massive burial of carbon on land might have pumped sufficient quantities of oxygen into the atmosphere to oxidize iron in terrestrial sediments. This is perhaps another reason (along with widespread continental interiors) why terrestrial redbeds tended to be common later in the Paleozoic. It has been suggested that if it had not been for these sorts of negative feedbacks, especially during the initial spread of terrestrial forests, Earth might have experienced a climatic "overshoot" (see Chapter 10) during the Carboniferous: so much O_2 would have been present in Earth's atmosphere that spontaneous combustion of organic matter would have occurred and Earth would have been incinerated. The tremendous rise in O_2 has been confirmed by studies of charcoal occurrence (an indicator

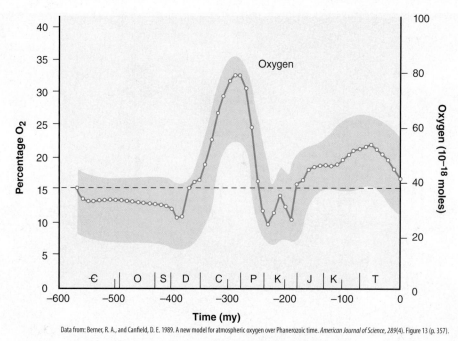

FIGURE 12.20 Oxygenation of the atmosphere during the Pennsylvanian Period by photosynthesis in coal swamps.

Data from: Berner, R. A., and Canfield, D. E. 1989. A new model for atmospheric oxygen over Phanerozoic time. *American Journal of Science, 289*(4). Figure 13 (p. 357).

of wildfires) preserved in bogs; these studies indicate that atmosphere oxygen levels rose to as high as 26%, as compared to the modern concentration of 21%.

12.6.3 Invertebrate life on land

The expansion of terrestrial floras had a profound effect on the evolution of animal life on land. Insects became much more prominent in the fossil record. Insects are basically soft-bodied and so typically decay rapidly. Wingless insects are known before the Carboniferous, but their fossil record is very poor. However, insects have been discovered in exceptionally preserved assemblages, or Lagerstätten, formed in the coal swamps. These Lagerstätten include coal balls, which consist of peat permineralized by calcium and magnesium carbonates, and iron carbonate-rich nodules that formed in the sands and muds of the swamps.

The fossil record of insects from Late Carboniferous (Pennsylvanian) time indicates that a major evolutionary innovation had occurred in insects by this time: wings. Possibly, wings evolved from temperature regulatory organs much earlier (perhaps the Mississippian) by preadaptation (see Chapter 5). However, the wings of early insects did not fold backward over the body. Most of the early winged taxa became extinct although they are still represented today by dragonflies and mayflies. Later insects evolved wings that could be folded back over the body. This was another significant evolutionary innovation because it allowed insects to search for food, or hide from predators, in the cracks and crevices of plant surfaces. These new taxa included cockroaches and other groups of insects known today, such as crickets; some of these taxa evolved specialized mouth parts for sucking plant juices.

Although most Late Paleozoic insects were of normal size (by modern standards), giant dragon-flies with wingspans rivaling that of some birds (up to ~60 cm) existed!

It has been hypothesized that increased oxygen levels during the Late Paleozoic account for the increase in insect size, especially flying forms. An insect's flight muscle is crammed with mitochondria (the sites of aerobic respiration) and is some of the most metabolically active tissue known. Flight velocities in flies, for example, reach 375 cm (about 12 feet) per second, or about 8 mph! The wings of a butterfly beat at a frequency of about 9 per second, whereas those of a dragonfly reach 28, the housefly 180 to 200, and the mosquito 1,024 *per second.* However, insect predators such as birds did not appear until the Mesozoic (see Chapter 13), which also might have also allowed dragonflies to reach large size rather than higher oxygen levels.

Insects eventually evolved the characteristic cycle of **insect metamorphosis**: egg → caterpillar (feeding and growth stage) → pupa (nonfeeding stage during which metamorphosis to the adult takes place) → sexually mature adult with a different diet. Because the stages of insect metamorphosis are characterized by different activities, the evolution of insect metamorphosis undoubtedly contributed to the tremendous evolutionary success of insects by diversifying their food sources and behavior.

12.6.4 Vertebrates

Marine and terrestrial food chains and webs appear to have continued to increase in length and complexity during the Late Paleozoic. Locomotion—both predation and escape from vertebrate predators—was therefore far more important in Late Paleozoic seas than earlier in the Paleozoic. After the Late Devonian extinctions, the heavily armored placoderms were quickly replaced by faster-swimming bony fish and sharks. This implies greater energy (food) availability in the marine realm to support the higher metabolic rates necessary for rapid locomotion and predation.

(a)

© Ken Lucas/Visuals Unlimited, Inc.

(b)

© Kevin Schafer/Alamy.

FIGURE 12.21 Early amphibians. **(a)** *Eryops*. **(b)** *Seymouria*.

On land, amphibians continued to diversify. Coal swamps and other terrestrial forests provided habitats for amphibians as they diversified into herbivores and carnivores, including insect and fish eaters. Many early amphibians were large predators such as *Eryops*, which reached over 2 meters in length (**Figure 12.21** and **Figure 12.22**). These forms, called **temnospondyls**, were previously referred to as **labyrinthodonts** because of the highly folded—or "labyrinthine"—teeth; similar teeth are found in crossopterygian fish thought to have been ancestral to amphibians (see Chapters 5 and 11). Temnospondyls possessed flattened heads, long snouts, limbs positioned at right angles to the body, and eyes on top of the head (see Figure 12.21). Thus, temnospondyls resembled crocodiles and might well have lived like them, waiting to ambush unsuspecting prey in swamps or streams.

Other taxa of amphibians resembled lizards, snakes, or had horns. One particularly unusual but important taxon of amphibians was the **anthracosaurs**. Anthracosaurs, like *Seymouria*, had deep, unflattened heads, and their legs were less sprawling, suggesting they were more agile than temnospondyls, perhaps for predation. These anatomic features are noteworthy because they resemble those of reptiles more than amphibians. Anthracosaurs would therefore appear to lie somewhere among the evolutionary lineage ancestral to reptiles. One of the earliest definite reptiles was *Hylonomus*, the skeleton of which was originally found in fossilized tree stumps of *Sigillaria* from the Early Pennsylvanian of Nova Scotia (see Figure 12.22). Unlike anthracosaurs, though, *Hylonomus* was relatively small and slender with strong jaws. Indeed, one major reason for the success of the reptiles was the development of stronger jaws and teeth that could shred prey. Tearing foodstuffs into pieces increased the surface area of the food for digestion, which meant more energy was released to the consumer.

Another major evolutionary innovation leading to the success of the reptiles was the **amniote egg** (**Figure 12.23**). Like the seed, the amniote egg freed terrestrial vertebrates from water and allowed them to exploit drier habitats and food sources of continental interiors. Amphibians were, and still are, tied to reproduction for laying eggs and the hatching of tadpole larvae, which are subject to enormous losses. By contrast, the amniote egg is a self-contained system that protects and nourishes the embryo until hatching, when it can feed on its own. The amniote egg is also found in birds and primitive mammals. For this reason reptiles, birds, and mammals are sometimes lumped together under the grouping amniotes.

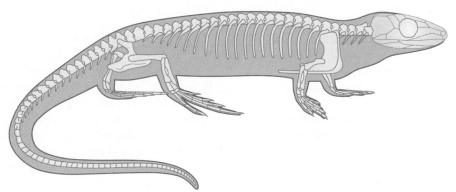

Data from: Prothero, D. R., and Dott, R. H. 2002. Evolution of the Earth. 6th ed. New York: McGraw-Hill. Figure 13.62 upper two figures (p. 374).

FIGURE 12.22 *Hylonomus*. The amniotes had diverged into two main lines by sometime in the Pennsylvanian Period. One lineage—which included reptiles like *Hylonomus*—would lead to the marine reptiles, turtles, lepidosaurs (lizards and snakes), and the archosaurs or "ruling lizards," which include crocodiles and alligators, dinosaurs, and birds.

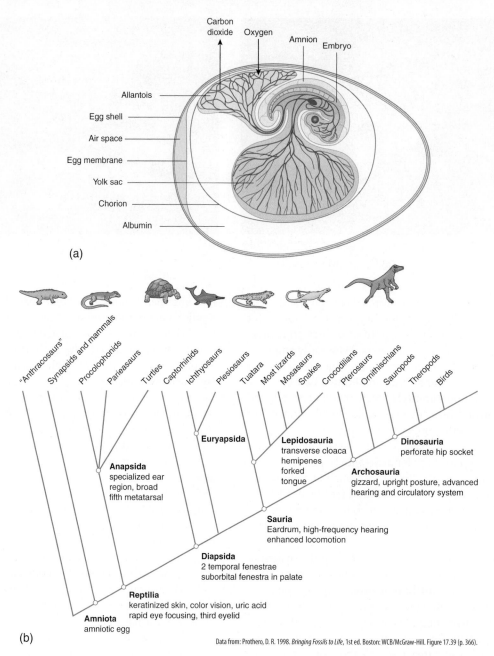

FIGURE 12.23 The amniotes. **(a)** The amniote egg. **(b)** Simplified cladogram of the evolutionary relationships of major taxa of amniotes. Note that anapsids and euryapsids are now considered to be clades within the diapsids based on features other than skull openings (see Figure 12.27).

By the Late Carboniferous (Pennsylvanian Period) the amniotes had diverged into two main lines (see Figure 12.23). (As we will see, we cannot always easily "shoe-horn" the ancient taxa belonging to these lineages into more easily recognized modern taxa, like some of the enigmatic forms that appeared during the Cambrian Explosion described in Chapter 10.) One lineage—which included reptiles like *Hylonomus*—would lead to the marine reptiles, turtles, lepidosaurs (lizards and snakes), and the **archosaurs** ("ruling lizards"). The archosaurs are significant because they include crocodiles and alligators, dinosaurs, and birds.

The other lineage consisted of the **synapsids**, which represent a separate clade that split off very early from the lineage that led to reptiles. Thus, synapsids and reptiles were

contemporaneous in the Paleozoic. The significance of this fact will become evident shortly. The term "synapsid" refers to the number of openings in the skull behind the eye socket; these openings are used for the attachment of jaw muscles (**Figure 12.24**). In the case of the synapsid skull, there is one opening behind the eye socket. (A number of taxa previously established among the synapsids are now considered "paraphyletic" because they do not include the descendants of the taxon's common ancestor [see Chapter 5], but the older names of classification are still useful and are indicated here by quotation marks in the following discussion.)

The fin, or sail-back, synapsids were the prominent terrestrial vertebrate taxon in the Late Paleozoic (these include the paraphyletic "pelycosaurs" of the older literature)

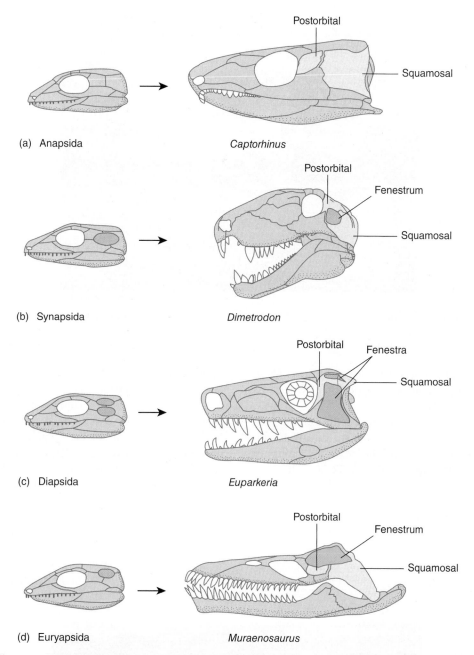

(a) Anapsida   *Captorhinus*

(b) Synapsida   *Dimetrodon*

(c) Diapsida   *Euparkeria*

(d) Euryapsida   *Muraenosaurus*

FIGURE 12.24 Different types of skull openings. **(a)** Anapsid, which was present in some primitive amniotes and the aquatic reptile *Mesosaurus*, is found today only among turtles. **(b)** Synapsid, characteristic of sailback reptiles and also mammals. **(c)** Diapsid, found in lizards and snakes, crocodiles, pterosaurs, dinosaurs, and birds. **(d)** Euryapsid, found in certain marine reptiles.

(**Figure 12.25**). Among the most famous synapsids was *Dimetrodon*, which ranged up to 2 meters in length and became the top carnivores of the Permian. Not all synapsids of the Permian were carnivorous, however. These included the **dicynodonts** ("two dog tooth"), which were large hippopotamus-like herbivores possessing two large canine teeth in the upper jaw. Some, such as the small (less than 1 meter) *Lystrosaurus,* had a beak that was used to crop plants. *Lystrosaurus* is also noteworthy because it is found on different continents that were once joined together as Gondwana. Like the small marine reptile *Mesosaurus*, which lived in fresh to brackish waters, *Lystrosaurus* could not possibly have swam across the Atlantic Ocean.

Like modern reptiles, many synapsids were probably "cold-blooded," or **ectothermic** ("outside temperature"). The body temperature and activity of ectothermic animals strongly depends on the temperature of the external environment. As the external temperature increases, the metabolic reactions involved in energy release, movement, and so on of cold-blooded animals also increase. However, when the temperature of the external environment declines, so too does body temperature. In the case of the sailbacks, the fin might have been used to absorb solar energy or to cool the body by facing it into the wind; it might also have been used in mating displays. By the Late Permian, sailback synapsids, anthracosaurs, and most other amphibians were on the

© Michael Rosskothen/Shutterstock, Inc.

FIGURE 12.25 The carnivorous sailback synapsid, *Dimetrodon*, from the Permian, threatening *Edaphosaurus*, an herbivore. Dimetrodon grew up to ~2 meters (6 feet) in length.

wane. Only the temnospondyls remained prominent. These taxa were replaced by a tremendous radiation of synapsids, most of which were predatory.

In fact, it is believed the main reason the sailback synapsids declined is because one group of sailbacks—called **cynodonts**—evolved into the advanced group of synapsids called the **therapsids** (**Figure 12.26**). The therapsids were of small to medium size and exhibited a number of mammal-like traits (**Figure 12.27**). In the therapsids, these traits included the fusion of the bones in the skull, strengthening it for grasping, tearing, and chewing; larger openings in the skull to accommodate bigger jaw muscles; an enlarged lower jawbone; and the more direct placement of the limbs beneath the body rather than out to the sides for

faster locomotion. Therapsids also exhibited a greater differentiation of the teeth for different functions. Canines were used for stabbing, holding, and tearing prey, whereas multicusped teeth in the cheek region were used for chewing. Some cynodont skulls are even reported to have small holes in the snout that have been inferred as sites of attachment of cat-like whiskers. Because whiskers are really modified hairs, it is possible that cynodonts and thus also therapsids possessed another mammalian trait: a body covering of hair or fur.

The possible possession of hair implies these cynodonts were warm-blooded, or **endothermic** ("inside temperature"), like mammals (see Figure 12.27). Indeed, the fossil record shows the cynodonts ultimately founded the

© Mark Hallett Paleoart/Photo Researchers, Inc.

FIGURE 12.26 Members of the therapsids, the group that gave rise to mammals. A trio of *Cynognathus* have attacked the herbivore *Kannemeyeria*. Note the different appearances of these forms. It is thought that therapsids, like mammals, might have possessed hair and been endothermic.

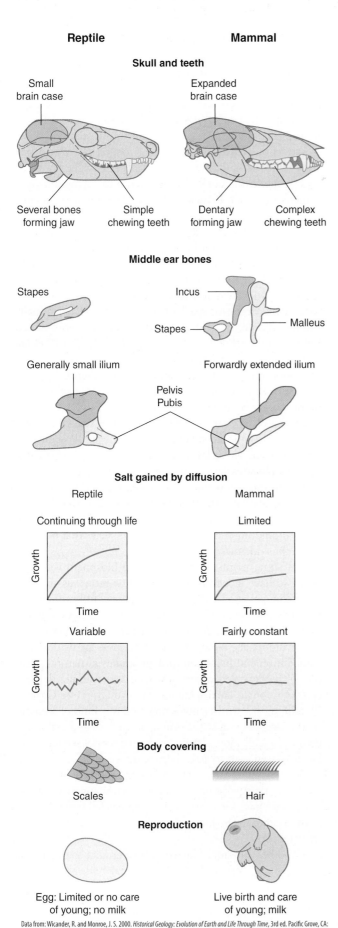

Reptile   Mammal

Skull and teeth

Small brain case — Several bones forming jaw — Simple chewing teeth

Expanded brain case — Dentary forming jaw — Complex chewing teeth

Middle ear bones

Stapes

Incus — Stapes — Malleus

Generally small ilium

Forwardly extended ilium

Pelvis
Pubis

Salt gained by diffusion

Reptile — Continuing through life — Growth / Time

Mammal — Limited — Growth / Time

Variable — Growth / Time

Fairly constant — Growth / Time

Body covering

Scales

Hair

Reproduction

Egg: Limited or no care of young; no milk

Live birth and care of young; milk

Data from: Wicander, R. and Monroe, J. S. 2000. *Historical Geology: Evolution of Earth and Life Through Time*, 3rd ed. Pacific Grove, CA: Brooks/Cole. Figure 15.22 (p. 398).

FIGURE 12.27 Comparison of the traits of reptiles and mammals.

mammals early in the Mesozoic Era (see Chapter 13). Mammals (and also birds) maintain a relatively constant body temperature greater than that of the environment, permitting them to remain active. A very important feature related to endothermy in mammals and birds is the presence of a four-chambered heart. The circulatory arrangement of the heart separates oxygenated blood coming to the heart from the lungs from deoxygenated blood that is returning to the heart from the rest of the body. By contrast, modern reptiles have a three-chambered heart, which allows oxygenated and deoxygenated blood to be mixed before it is pumped to the lungs or the rest of the body. The reptilian heart is therefore less metabolically efficient than the mammalian heart.

More efficient blood flow would have also supplied greater amounts of oxygen to the brain. In fish and amphibians the brain consists of a simple linear arrangement of lobes, or brainstem (**Figure 12.28**). In higher vertebrates the brainstem has become folded and flexed into a cerebrum and cerebellum. The cerebrum is important to mental ability, whereas the cerebellum is involved in coordination and shows particular development in organisms whose movements are quick and require precise coordination. These two regions have become progressively larger in more advanced vertebrates, so that in mammals the cerebrum overlies all other parts of the brain. The outermost portion of the cerebrum, or cortex, has also become thickened and highly convoluted, increasing its surface area. The increase in the size of the cortex parallels the increase in mental abilities of mammals.

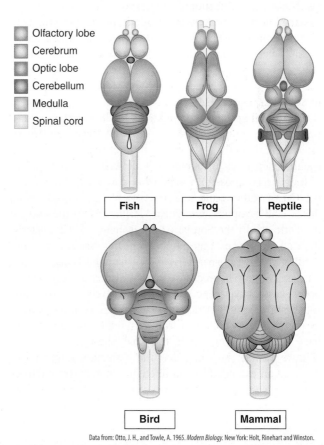

Olfactory lobe
Cerebrum
Optic lobe
Cerebellum
Medulla
Spinal cord

Fish   Frog   Reptile

Bird   Mammal

Data from: Otto, J. H., and Towle, A. 1965. *Modern Biology*. New York: Holt, Rinehart and Winston.

FIGURE 12.28 Comparison of the brain sizes and complexity of representatives of the major vertebrate taxa.

12.7 Multiple Causes of Extinction?

The end of the Permian Period was marked by the greatest of all mass extinctions in Earth's history. Unfortunately, complete sections of outcrops that span the boundary between the Permian and Triassic are not very common, a fact that has long obstructed the study of the Late Permian extinctions. Many marine invertebrate taxa characteristic of the Paleozoic disappeared, although some had long been declining in prominence. Depending on the exact timing and how one tabulates the taxa involved, somewhere between 80% and 95% of marine species disappeared, many of them tropical. Among the taxa that completely disappeared at the end of the Permian were trilobites, rugose and tabulate corals, blastoids, and fusulinids. Other taxa that were severely affected but survived were brachiopods, bryozoans, crinoids, and ammonoids. Bivalves and gastropods were hit less hard, possibly because they burrowed in sediment, which might have provided refuge and food. These taxa rediversified explosively in the Mesozoic, when the Modern Fauna began to dominate the seas (see Chapter 13).

A **fungal spike** is reported to coincide with this extinction on land. Fungal spores are resistant stages, and their rapid increase in the fossil record hints at some catastrophic agent(s) causing deforestation that allowed the rapid spread of fungi (decomposers), which break down dead plant matter. On the other hand, the diversity of sphenopsids, cordaites, and lycopods was on the decline before the end of the Permian. In the Southern Hemisphere, the *Glossopteris* flora on Gondwana was extinguished but was survived by the Permian gymnosperm flora (see Chapter 13).

The vertebrate record on land is also equivocal. These occurrences have been studied extensively in the Karroo Basin of South Africa, where a series of fluvial environments occurs. Many of the vertebrate taxa appear to have been declining before the final phase of extinction (**Figure 12.29**), implying that the extinction might have involved an extended period of environmental stress rather than being sudden. Indeed, a recent study has concluded that the *Dicynodon-Lystrosaurus* zone boundary (see Figure 12.29) is older than the marine extinction. If so, the demise of these vertebrate taxa probably does not coincide with the true end-Permian extinction (see Further Reading). Recently described sections located in the Austrian Alps, Japan, and the Urals (Russia) and remarkably complete sections in South China actually suggest more than one phase of extinction (**Figure 12.30**). The strongest phase occurred right at the end of the Permian, possibly sandwiched between two smaller extinctions that occurred 5 to 10 million years earlier in the Late Permian and 5 to 10 million years later, during the Early Triassic.

The story is also complicated by the possibility that the extinction(s) might have had multiple causes. Sea-level and habitat loss were once considered to be prime candidates for the Permian extinctions, just as they were with the Late Ordovician and Late Devonian extinctions. In the case of the Permian, sea level continued to fall gradually through the period, just as it had been doing more or less through the Carboniferous (see Figure 12.1). Sea-level fall then seemed to accelerate dramatically during the last 2 or so million years of the Permian Period (see Figure 12.1). At this point in time sea level must have been near the edges of the world's continental shelves. The final and main phase of extinction appears to have occurred at the end of the Permian, after the steep fall in sea level began. Possibly, the thinking went, as sea level was falling the biosphere became increasingly stressed because of habitat loss and increased crowding and competition. The assembly of the continents into Pangaea was also thought to have destroyed biogeographic provinces, where many endemic species lived, while promoting the invasion of other species that outcompeted the endemics.

At some point in time, then, the biosphere was supposed to have moved past a point of no return as sea level was falling, beyond which the biosphere could not recover. Initially, this scenario was very appealing. Competition and predation have been documented extensively by ecologists, and evolutionary biologists therefore look for evidence of competition and predation in their studies of natural selection of modern ecosystems. Darwin, for example, thought competition was *the* driving force in evolution (see Chapter 5). Still, most Earth scientists now find this kind of scenario hard to accept because the timing of sea level and extinction do not coincide more closely. Sea level does not seem to have fallen fast enough to have been a direct cause of the Late Permian extinctions; in fact, according to some workers, sea level was rising during the most severe extinction phase. Also, as we have seen, sea level sometimes fell quite dramatically at other times during the Paleozoic, but no extinctions occurred.

Another candidate for the Late Permian extinction is one or more impacts by extraterrestrial bodies. The dust

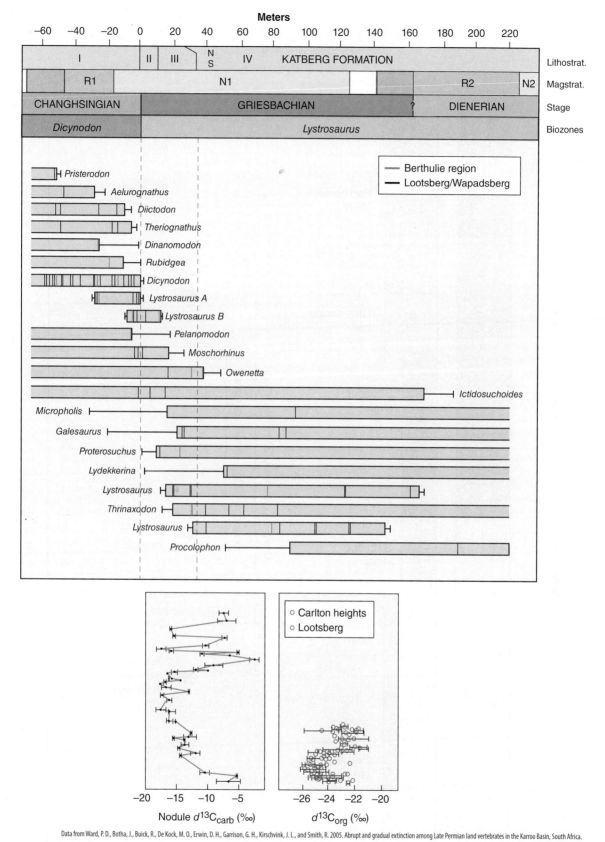

FIGURE 12.29 Paleomagnetic, biostratigraphic, and carbon isotope records across the Permo–Triassic boundary of the Karroo Basin of South Africa. Note in this diagram the gradual disappearance of Permian therapsid taxa below the boundary followed by a more drastic extinction at the boundary (stages are larger biostratigraphic subdivisions based on the zones). Note, too, the appearance of Triassic taxa before the end of the Permian. A recent study has concluded that the *Dicynodon-Lystrosaurus* zone boundary is stratigraphically higher than indicated here but still older than the marine extinction event (see Figure 12.30). If correct, the turnover in vertebrate taxa at this zone boundary probably does not truly represent the biological expression of the terrestrial end-Permian mass extinction, so that the actual Permo-Triassic boundary in the Karroo Basin occurs either higher in the Katberg Formation or is not preserved (see Further Reading).

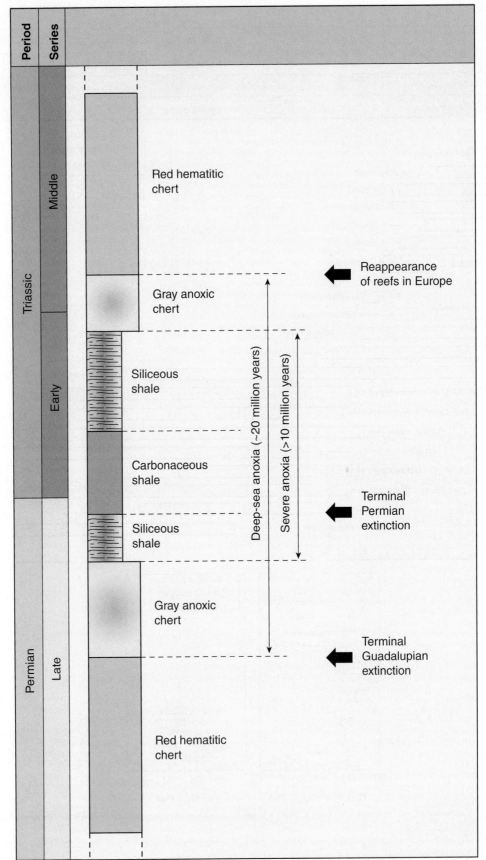

Data from: Holser, W. T., and Magaritz, M. 1992. Cretaceous/Tertiary and Permian/Triassic boundary events compared. *Geochimica et Cosmochimca Acta, 56,* 3301 (Figure 4).

FIGURE 12.30 Events associated with the end-Permian extinctions and recovery in the Mesozoic (stages are larger biostratigraphic subdivisions based on the zones). Iridium is concentrated in the anoxic layers. Although the presence of iridium layers suggests one or more impacts, their occurrence here is more likely the result of concentration by anoxic conditions.

clouds thrown up into the atmosphere would have blocked sunlight from both land and sea, shutting down photosynthesis. In fact, two strong excursions to much lower carbon isotope ratios occur at or near the Permo–Triassic boundary in sections in Austria. Similar excursions occur in terrestrial sections (see Figure 12.29). The rapid shifts of carbon isotope ratios to much lower values signal the rapid shutdown of photosynthesis. Unusually high concentrations of the element **iridium** also have been found in the same layers as the negative carbon isotope shifts in marine sections. Iridium is not normally found in Earth's crust. Other than impact, possible sources of iridium are (1) volcanoes fed from Earth's mantle, which is enriched in iridium, and (2) anoxia, which can concentrate iridium already at the surface. In the case of the Austrian section, it appears that iridium was twice concentrated under anoxic conditions because the iridium is also associated with increased pyrite (iron sulfides), which precipitates in sediments under low oxygen conditions (see Figure 12.30). The other evidence for impact is also questionable. **Microspherules**, which are blebs of molten material that might be indicative of shock (impact) metamorphism (see Chapter 3), have been reported from the South China sections, but the microspherules appear to be of volcanic origin because they are not as severely shocked as would be expected from an impact. Shocked mineral assemblages have also been reported from Australia and Antarctica, but they too do not appear to be as well developed as those thought to result from impact. A crater has also been reported in Late Permian sections in Australia, but it has not been confirmed.

Volcanism could account for the co-occurrence of iridium, shocked mineral assemblages, and anoxia, as well as some other observations. Almost exactly at the end of the Permian, massive outpourings of basalts occurred in what is now Siberia (**Figure 12.31**). These flows were apparently fed by one or more mantle plumes as Siberia passed over them. The flood basalts, or traps, are about 3.9 million cubic kilometers in volume and were erupted in only about 1 million years or less ("trap" refers to the step-like appearance of the lava outcrops). On geologic time scales, 1 million years is fairly rapid, especially given the huge amounts of lava produced. Although dust and aerosols were undoubtedly spewed out during the flows, their cooling effect was apparently only short term at most. However, the enormous volumes of CO_2 pumped into the atmosphere might have warmed Earth quite rapidly and kept it warm.

The sudden warming would also account for other evidence at the end of the Permian. The glaciers that were so prominent in the southern hemisphere disappeared just as the Permian drew to a close. The disappearance of glaciers might in turn have switched the oceans back to an anoxic, stagnant mode after the oceans had become increasingly oxygenated. In fact, red, radiolarian-rich cherts occur just before a first pulse of extinction in sections located in Japan (**Figure 12.32**; see Figure 12.30 for stratigraphic location of the cherts below the Permo–Triassic boundary). The red coloration indicates that sufficient oxygen was present at greater depths in the ocean to oxidize iron in the sediment, whereas the radiolarians suggest nutrients were being

Data from: Renne, P. R., Zichao, Z., Richards, M. A., Black, M. T., and Basu, A. R. 1995. Synchrony and causal relations between Permian-Triassic boundary crises and Siberian flood volcanism. *Science, 269,* 1414 (Figure 1).

FIGURE 12.31 The occurrence of Late Permian flood basalts in modern Siberia and on a reconstruction of Pangaea. These flows were apparently fed by one or more mantle plumes (hot spots) as Siberia passed over them.

Courtesy of Yukio Isozaki.

FIGURE 12.32 Banded, red radiolarian-rich cherts outcropping in Japan. The red coloration indicates that sufficient oxygen was present at greater depths in the ocean to oxidize iron in the sediment, while the radiolarians suggest that nutrients were being upwelled to the photic zone.

upwelled to the photic zone. Then a first, mild pulse of anoxia occurred in conjunction with a first wave of extinction (see Figure 12.30). After this time, the rock record suggests anoxia became more pronounced, especially at the Permo–Triassic boundary, when the second and strongest phase of extinctions occurred. Along with the tremendous amount of volcanism at the Permo–Triassic boundary, the pronounced anoxia accounts for the concentration of iridium and pyrite around the Permo–Triassic boundary in Austria. Anoxia also suggests extensive organic decay or that perhaps nutrients were either not being upwelled into the photic zone or being upwelled only weakly because glaciers had disappeared. Nutrient runoff from land was probably also decreasing, as suggested by the decrease in strontium isotope ratios. Dry continental interiors and interior drainage were probably expanding as Pangaea was undergoing final assembly.

Nevertheless, the sharp shift to lower carbon isotope ratios at the very end of the Permian was so sudden and strong that some workers believe volcanism could not account for the entire shift in isotope values. These workers have instead suggested that Earth experienced a runaway greenhouse but over a much shorter period of time than on Venus (see Chapter 7). In this scenario, volcanism associated with the **Siberian Traps** caused Earth to warm by as much as 6°C. As the warming of Earth's atmosphere extended downward from the ocean surface, the warming melted and released vast reservoirs of frozen methane, or **gas hydrates**, that presumably resulted from the previous burial and decay of dead plant matter on continental shelves. The methane (CH_4) is highly concentrated in the lighter carbon isotope (^{12}C); thus, the release of methane and its oxidation to CO_2 in the atmosphere could have quickly shifted carbon isotope ratios to very negative values, overwhelming Earth's feedback systems (such as weathering or photosynthesis) that normally keep atmospheric CO_2 and climate within a relatively narrow range.

One recent hypothesis has attempted to tie rapid global warming to extinction in another way. The rapid influx of CO_2 to the atmosphere is proposed to have rapidly accelerated weathering rates, increasing soil erosion and the runoff of nutrients to the oceans (**Figure 12.33**). Greater nutrients in the oceans then presumably heightened rates of marine photosynthesis, thereby increasing the pelagic rain of dead organic matter to the oceans and expanding the oxygen minimum zone (OMZ) and black shale deposition, as evidenced by the geologic record. Recent studies have concluded that the warming persisted into the Early Triassic and was so extreme that sea surface temperatures might have reached 40°C; by contrast, modern equatorial sea surface temperatures range from about 25°C to 30°C. Such extreme temperatures appear to have driven terrestrial animals and plants to higher latitudes, where temperatures remained cooler.

At some point, conditions reversed themselves during the Early Triassic, at least in certain parts of the world. In the Japanese sections the conditions above the Permo–Triassic boundary are basically a mirror image of those leading up to the end of the Permian and culminate in the reappearance of red, oxidized cherts. Perhaps the CO_2 pumped into the atmosphere by volcanism caused a massive dose of "acid rain" that increased rates of continental weathering and nutrient input into the oceans, but no one really knows for sure what happened.

So what was *the* cause of the Late Permian extinctions? Obviously, all causal agents described above could have been involved, but some are more important than others. The most obvious candidate now appears to have been massive volcanism that led to anoxia, global warming, and then perhaps methane release, at least right at the end of the Permian. However, the Late Permian was a *very* unusual time in Earth's history. The last time Earth's continents had been clumped together as a supercontinent was about 300 million years earlier, at the end of the Proterozoic. Then, continents began to rift apart and metazoan life seemed to explode onto the evolutionary stage. At the end of the Permian, just the opposite was happening: continents were colliding and metazoan life nearly disappeared. Perhaps the configurations of the continents and oceans played some role in the Late Permian extinctions, but at first glance they would seem to resemble the background scenery on a stage, important but not really noticed, with the other agents (volcanism, anoxia, and so on) acting out their parts.

Nevertheless, history and contingency (see Chapter 1) may well have played a role in the extinctions by establishing the geologic and climatic framework that led to the Permian extinctions. Would the Siberian Traps have erupted onto Pangaea if it hadn't been undergoing assembly? The aggregation of continental crust over a mantle plume might have caused heat to build up to the point that massive eruptions occurred. Would widespread, pronounced anoxia have occurred without the continent–ocean configurations of the Late Permian that, as we have already discussed, affected ocean circulation and chemistry? Once again, although scientists—especially those of the experimental sciences—are taught to search for *the* cause of an effect (in this case, extinction), there could have been multiple causes, although some might have been more important than others.

CONCEPT AND REASONING CHECK

1. How might the hypothesized causes of extinction at the end of the Permian acted synergistically; that is, reinforced one another?

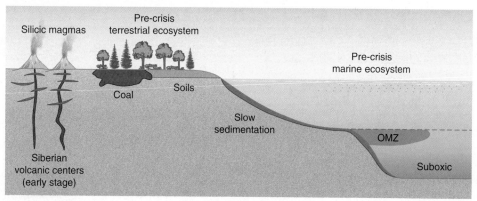

(a) Late permian

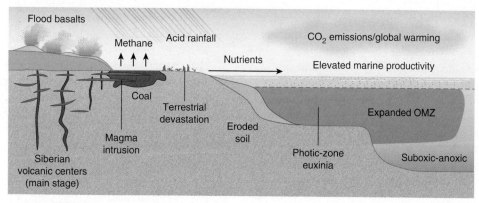

(b) End-permian event

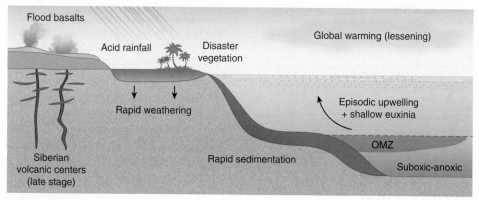

(c) Early triassic

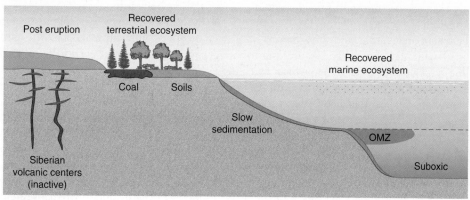

(d) Middle triassic

FIGURE 12.33 Reconstructions of terrestrial-marine interactions during the end-Permian extinction. **(a)** Early stage Siberian Trap volcanism during the Late Permian. **(b)** Main eruptions with attendant environmental effects during the latest Permian. **(c)** Late stage eruptions with lessening environmental effects during the Early Triassic. **(d)** Post-eruption recovery of terrestrial and marine environmental effects during the Early Triassic.

SUMMARY

- The Middle-to-Late Paleozoic Era, spanning the Mississippian through Permian periods from about 350 to 245 million years ago, saw the development of a world substantially different from the Early-to-Middle Paleozoic one.

- During the Late Paleozoic the collision of continents and terranes that had begun during the mid-Paleozoic continued, culminating in the final assembly of the supercontinent, Pangaea, and the corresponding ocean, Panthalassa.

- During the Late Devonian, the western margin of North America shifted from a passive to an active continental margin. The Antler Orogeny, which began by the Late Devonian, continued into the Mississippian, accreting an island arc through what is now Nevada. The western margin of North America extended northward into Canada and Alaska, where the Ellesmere Orogeny, which began by the Late Devonian, accreted one or more terranes. During the Late Permian and continuing into the Triassic, a volcanic island arc accreted in the area of what is now California during the Sonoma Orogeny. The micro-continent Sonomia, which today is represented by Nevada, southeast Oregon, and northern California, added a large portion of landmass to the western margin of the United States during the Sonoma Orogeny.

- A number of other orogenies affected Europe and adjacent regions. During the last half of the Permian, the Uralian Orogeny involved the collision of Baltica and the smaller continent of Kazakhstan. Much of the rest of Asia during this time grew as a result of the collisions of microcontinents, such as Tibet, and terranes that had rifted northward from Gondwana.

- The most extensive phase of mountain building extended across Europe as the Hercynian (Variscan) Orogeny and into North America.

- After the Taconic and Acadian Orogenies, the Alleghenian Orogeny marked the third and final episode of the entire Appalachian Orogeny. During the Hercynian and Alleghenian orogenies, Gondwana, especially Africa, collided with the eastern and southeastern margin of the supercontinent Laurasia, which had formed from the collisions of North America, Europe, Greenland, and parts of Asia that had begun earlier in the Paleozoic. The Alleghenian Orogeny produced thrust sheets over several hundred kilometers that were eroded, with the molasse being shed into the cratonic interior as conglomerates and sandstones.

- Along what is now the Gulf Coast, South America collided with Laurentia during the Ouachita Orogeny, which might have been an extension of the Alleghenian Orogeny into the southwestern United States. Rocks from the Ouachita Orogeny are now exposed in west Texas and Oklahoma. The collision of South America and Laurentia also produced the Ancestral Rockies in Colorado and New Mexico.

- Continental collision affected climate and ocean circulation, so that the Late Paleozoic was a time of increasing climatic extremes. Orogeny uplifted mountains, causing sea level to gradually decline, increasing Earth's albedo and helping the planet to cool. Decreased seafloor spreading pumped less CO_2 into the atmosphere, which also cooled Earth. The assembly of Pangaea diverted warm surface ocean currents to higher latitudes, while uplift and supercontinent assembly produced extensive continental interiors characterized by deserts, restricted seaways, and possibly megamonsoons. Except for the latest Permian, ice caps persisted in the southern hemisphere.

- Supercontinent assembly also affected ocean chemistry. Falling sea level resulted in the decreased deposition and erosion of limestones in epicontinental seaways and perhaps also the deepening of the calcite compensation depth. Decreased seafloor spreading resulted in the production of aragonite seas. Also, because of glaciers and greater temperature gradients between the equator and poles, deep ocean circulation rates and upwelling might have increased.

- These changes dramatically affected the biosphere. The stromatoporoid reefs, which were so widespread during the Silurian and Devonian and that disappeared after the Late Devonian extinctions, were replaced by plant and animal benthos, such as calcareous algae and crinoids, that built mounds of cemented skeletal debris. Nevertheless, some of these reefs, such as the Permian Reef Complex, were quite spectacular. The oceans also teemed with other kinds of life, some of which were able to rebound from the Late Devonian extinctions. These groups included the ammonoids, which were probably preyed upon by increasingly diverse bony fish.

- Even more dramatic developments occurred on land. Lycopods grew upward into the tall trees of coal swamps but would eventually give way to gymnosperms in the Permian. Increased rates of weathering related to the spread of land plants appear to have increased nutrient runoff to the oceans, as evidenced by increased strontium isotope ratios. Increased nutrient levels in turn might have increased plankton populations, including calcareous plankton, helping to deepen the calcite compensation depth.

- Among the vertebrates, amphibians gave way to the reptiles. The evolution of the amniote egg allowed reptiles to diversify into the dry interiors of continents, which expanded in area as Pangaea was being assembled. Initially, the fin-back synapsids were the top carnivores, but even they gave way to the therapsids, which might have been endothermic and ancestral to the mammals.

■ By the end of the Permian the largest of all mass extinctions in Earth's history occurred on land and in the sea. Many groups that had managed to survive the Late Devonian extinctions into the Late Paleozoic perished, including trilobites, rugose and tabulate corals, fusulinids, and other taxa in the seas, whereas on land plant and animal taxa were also hard hit. The exact cause or causes of the Late Permian extinctions is unknown, but present evidence points primarily at anoxia and rapid warming that resulted from massive volcanism and perhaps the oxidation of large amounts of methane release from buried sediments. In the seas, after the extinctions the Modern Fauna, which had lurked in the background during most of the Paleozoic, rediversified in the Mesozoic, never to relinquish its dominance, whereas on land the reptiles would rule in gymnosperm forests.

KEY TERMS

Absaroka	ecologic replacement	iridium	reptiles
Alleghenian (Appalachian) Orogeny	ectothermic	Kaskaskia	Siberian Traps
amniote egg	endothermic	labyrinthodonts	Sonoma Orogeny
Ancestral Rockies	fenestrate	mass balance	Sonomia
anthracosaurs	foraminifera	massifs	synapsids
aragonite seas	Front Range	megamonsoons	temnospondyls
archosaurs	fungal spike	Microspherules	tests
cordaites	fusulinids	Ouachita orogenies	therapsids
cyclothems	gas hydrates	Pangaea	Uralian Orogeny
cynodonts	gymnosperms	Panthalassa	
dicynodonts	Hercynian (Variscan) Orogeny	Permian Reef Complex	
	insect metamorphosis	productid brachiopods	

REVIEW QUESTIONS

1. What are the major differences between the Early-to-Middle Paleozoic and Late Paleozoic in terms of continent–ocean configurations, climate, rocks, and life? Why did these differences occur?

2. Why do earth scientists believe sea level generally fell during the Late Paleozoic?

3. During the Late Paleozoic what was the effect of continental movements on each of the following: (a) sea level, (b) atmospheric CO_2 concentrations, (c) ocean currents and circulation, (d) limestone deposition? Explain your answers.

4. Discuss how the paleogeographic setting of the Late Paleozoic might have affected the activity of chemical reactions involved in determining the level of the calcite compensation depth and oceanic anoxia.

5. Which modern features of Earth were produced during the Late Paleozoic? What were the processes involved?

6. What were the major developments in marine benthos and plankton during the Late Paleozoic?

7. What were the major developments in terrestrial floras, and how did they affect Earth's climate and ocean chemistry?

8. What were the major evolutionary innovations that led to the appearance of reptiles and their radiation into continental interiors?

9. What are the evolutionary relationships between the early reptiles and mammals?

10. What distinguishes reptiles from mammals?

11. Diagram on a time scale the possible causal agents of the Late Permian extinctions, their effects, and interrelationships.

12. How did the causes of the Late Permian mass extinction differ from those of the Late Ordovician and Late Devonian? How were the causes of the extinctions alike? Which taxa were most severely affected by each of these extinctions?

13. Construct a food pyramid for Late Paleozoic marine and terrestrial ecosystems.

FOOD FOR THOUGHT:
Further Activities In and Outside of Class

1. What sorts of rocks became more widespread during the Late Paleozoic, as opposed to the Early-to-Middle Paleozoic? Why?
2. How does the succession of rocks in an idealized cyclothem reflect sea-level change?
3. Given their large size and therefore visibility to any potential predators, how could giant dragonflies have evolved in the Carboniferous?
4. Why is it believed therapsids might have had hair? What other mammalian features might therapsids have possessed?
5. What do you believe were the primary cause(s) of the Late Permian extinctions? Why?
6. Comment on the concept of causality as it is used in science (see Chapter 1) in relation to the Late Permian extinctions.

SOURCES AND FURTHER READING

Algeo, T. J. et al. 2011. Terrestrial-marine teleconnections in the collapse and rebuilding of Early Triassic marine ecosystems. *Palaeogeography, Palaeoclimatology, Palaeoecology, 308,* 1–11.

Baars, D. L. 2000. *The Colorado Plateau: A geologic history.* Albuquerque, NM: University of New Mexico Press.

Baldridge, W. S. 2004. *Geology of the American southwest: A journey through two billion years of plate-tectonic history.* Cambridge, UK: Cambridge University Press.

Benton, M. J. 2003. *When life nearly died: The greatest mass extinction of all time.* London: Thames & Hudson.

Berner, R. A., and Canfield, D. E. 1989. A new model for atmospheric oxygen over Phanerozoic time. *American Journal of Science, 289,* 333–361.

Brand, U., Posenato, R., Came, R., Affek, H., Angiolini, L., Azmy, K., and Farabegoli, E. 2012. The end-Permian mass extinction: A rapid volcanic CO_2 and CH_4-climatic catastrophe. *Chemical Geology, 322–323,* 121–144.

Erwin, D. H. 1993. *The great Paleozoic crisis: Life and death in the Permian.* New York: Columbia University Press.

Farlow, J. O., and Brett-Surman, M. K. 1997. *The complete dinosaur.* Bloomington, IN: Indiana University Press.

Gastaldo, R. A. et al. 2015. Is the vertebrate-defined Permian-Triassic boundary in the Karoo Basin, South Africa, the terrestrial expression of the end-Permian marine event? *Geology,* doi:10.1130/G37040.1.

Glasspool, I. J., and Scott, A. C. 2010. Phanerozoic concentrations of atmospheric oxygen reconstructed from sedimentary charcoal. *Nature Geoscience, 3,* 627–630.

Holser, W. T., Schönlaub, H. P., Boeckelmann, K., and Magaritz, M. 1991. The Permian-Triassic of the Gartnerkofel-1 Core (Carnic Alps, Austria): Synthesis and conclusions: *Abhandlungen der Geologischen Bundesanstalt, 45,* 213–232.

Holser, W. T., Schidlowski, M., Mackenzie, F. T., and Maynard, J. B. 1988. Biogeochemical cycles of carbon and sulfur. In C. B. Gregor, R. M. Garrels, F. T. Mackenzie, and J. B. Maynard (ed.), *Chemical cycles in the evolution of the earth* (pp. 105–173). New York: John Wiley & Sons.

Isozaki, Y. 1997. Permo–Triassic boundary superanoxia and stratified superocean: records from lost deep sea. *Science, 276,* 235–238.

Joachimski, M. M., Lai, X., Shen, S., Jiang, H., Luo, G., Chen, B., Chen, J., and Sun, Y. 2012. Climate warming in the latest Permian and the Permian-Triassic mass extinction. *Geology, 40*(3):195–198.

King, P. B. 1977. *The evolution of North America.* Princeton, NJ: Princeton University Press.

Klug, C., Kröger, B., Kiessling, W., Mullins, G. J., Servais, T., Frýda, J., Korn, D., and Turner, S. 2010. The Devonian nekton revolution. *Lethaia, 43,* 465–477.

Knoll, M. A., and James, W. C. 1987. Effect of the advent and diversification of vascular land plants on mineral weathering through geologic time. *Geology, 15,* 1099–1102.

Pittermann, J. 2010. The evolution of water transport in plants: an integrated approach. *Geobiology, 8,* 112–139.

Reichow, M. K. et al. 2002. $^{40}Ar/^{39}Ar$ dates from the west Siberian basin: Siberian flood basalt province doubled. *Science, 296,* 1846–1849.

Renne, P. R. 2002. Flood basalts—bigger and badder. *Science*, *296*, 1812–1813.

Renne, P. R., Zichao, Z., Richards, M. A., Black, M. T., and Basu, A. R. 1995. Synchrony and causal relations between Permian-Triassic boundary crises and Siberian flood volcanism. *Science*, *269*, 1413–1416.

Servais, T., Martin, R. E., and Nützel. 2016. The impact of the 'terrestrialisation process' in the late Palaeozoic: pCO_2, pO_2, and the 'phytoplankton blackout.' *Review of Palaeobotany and Palynology*, *224*, 26–37.

Shen, J., Algeo, T. J., Hu, Q., Zhang, N., Zhou, L., Xia, W., Xie, S., and Feng, Q. 2012. Negative C-isotope excursions at the Permian-Triassic boundary linked to volcanism. *Geology*, *40*(11):963–966.

Sobolev, S. V., Sobolev, A. V., Kuzmin, D. V., Krivolutskaya, N. A., Petrunin, A. G., Arndt, N.T., Radko, V. A., and Vasiliev, Y. R. 2011. Linking mantle plumes, large igneous provinces and environmental catastrophes. *Nature*, *477*, 312–316.

Sobolev, S. V. et al. 2011. Linking mantle plumes, large igneous provinces and environmental catastrophes. *Nature*, *477*, 312–316.

Stokstad, E. 2003. Ancient weapons of mass destruction: methane gas? *Science*, *301*, 1168.

Sun, Y., Joachimski, M. M., Wignall, P. B., Yan, C., Chen, Y., Jiang, H., Wang, L., and Lai, X. 2012. Lethally hot temperatures during the Early Triassic greenhouse. *Science*, *338*, 366–369.

Wang, J., Pfefferkorn, H. W., Zhang, Y., and Feng, Z. 2012. Permian vegetational Pompeii from Inner Mongolia and its implications for landscape paleoecology and paleobiogeography of Cathaysia. *Proceedings of the National Academy of Sciences*, *109*(13):4927–4932.

Ward, P. D. 2005. Abrupt and gradual extinction among Late Permian land vertebrates in the Karroo Basin, South Africa. *Science*, *307*, 709–714.

Wilson, J. L. 1975. *Carbonate facies in geologic history*. New York: Springer Verlag.

Yang, Y. G. et al. 2000. Pattern of marine mass extinction near the Permian-Triassic boundary in south China. *Science*, *289*, 432–436.

Mesozoic Era

(A) How were the Mesozoic and Early-to-Middle Paleozoic alike? How were they different?

(B) What were the positions of the continents during the Mesozoic Era?

(C) How did the movements of the continents affect orogeny, sea level, and climate?

(D) What were the effects of continental movements on ocean circulation and chemistry? How did these conditions differ from today?

(E) Why was the deposition of black shales less widespread during the Mesozoic than during the Early-to-Middle Paleozoic?

(F) What happened in eastern North America tectonically after the formation of the Appalchians?

(G) Where did orogeny occur in the west, and why did the sites of mountain building shift there?

(H) How did orogeny in the west affect sea level and sedimentation?

(I) What new and old groups of plankton diversified, and how did this affect the geologic cycle of carbon?

(J) What new groups evolved within the marine benthos of the Modern Fauna?

(K) Why did sessile benthos retreat back toward the sediment water interface and burrow deeper into sediment?

(L) Why are calcareous plankton largely absent from the fossil record of the Paleozoic?

(M) Which group replaced the reef-building corals for a time in the Cretaceous?

(N) What new groups of fish and marine reptiles appeared in the oceans?

(O) What were forests like in the Mesozoic, and how did they influence the evolution of insects?

(P) Were dinosaurs as diverse as mammals?

(Q) Were dinosaurs warm-blooded?

(R) Did "flying" reptiles really fly?

(S) From which dinosaur taxa did birds evolve, and why did feathers appear?

(T) Was there more than one cause for each of the Late Triassic and Late Cretaceous extinctions?

CHAPTER OUTLINE

Artist's reconstruction of *Microraptor gui*, showing four wings and feathers. Possibly this species used its wings to glide between trees, like flying squirrels.

© Journal Nature/Portia Sloan/Getty Images.

13.1 Introduction to the Mesozoic Era

A The Mesozoic Era marks the beginning of the second tectonic cycle of the Phanerozoic. The Mesozoic also represents the beginning of the transition of Earth from the ancient world of the Paleozoic to the more modern world of the Cenozoic Era.

Like the Paleozoic, this transition is marked by major tectonic events (**Figure 13.1**). The continents that had collided to produce Pangaea by the end of the Paleozoic Era began to move toward their present-day positions as Pangaea began to rift apart in the Triassic. As the Atlantic Ocean began to open, the seas once again flooded the continents, and a great circumequatorial seaway called the **Tethys** stretched more-or-less unobstructed around Earth. By the end of the Mesozoic new mountain ranges were also being uplifted and eroded, especially in western North America, as the Appalachians began to be worn down.

As its name implies, the Mesozoic ("middle life") also represents a distinct shift of the biosphere away from that of the Paleozoic ("ancient life") toward that of the Cenozoic ("recent life"). After the Late Permian extinctions there was a tremendous rediversification of the biosphere. The Modern Fauna began to diversify in the Mesozoic after the demise of most of the Paleozoic Fauna during the end-Permian extinctions. New taxa appeared in the geologic record among the marine plankton, benthos, and nekton, including many new predators.

Major changes to the biosphere occurred on land, as well. Gymnosperms, which had appeared during the Late Paleozoic, dominated; however, many of these taxa were unlike the conifers we see today. Roaming the forests and swamps were the dinosaurs, whereas their reptilian cousins dominated the seas and began to radiate into the air (refer to this chapter's frontispiece). The first birds evolved no later than the Jurassic from the same stock as the dinosaurs. Mammals mostly lurked in the background; their chance would finally come with mass extinction and the demise of the dinosaurs at the close of the Cretaceous Period.

13.2 Tectonic Cycle: Impacts on the Hydrosphere, Atmosphere, and Rock Cycle

13.2.1 Rifting of Pangaea

B C The positions of the continents during the Mesozoic differed markedly from those at the beginning of the first tectonic cycle of the Phanerozoic (**Figure 13.2**). Most of Gondwana (Africa, Antarctica, Australia, India, and South America) remained in the southern hemisphere, whereas Laurasia (North America, Europe, and Siberia) lay in the northern hemisphere. As a result of the rifting, the Tethys Seaway formed.

The rifting of Pangaea occurred in several phases (see Figure 13.2). During the Late Triassic, the Tethys Seaway began to open as a kind of wedge, or reentrant, between Laurasia and Gondwana, along what is now southern Europe and North Africa. The Tethys gradually widened and deepened from east to west along weaknesses in Earth's crust left by the Hercynian Orogeny of the Carboniferous (see Chapter 12). The incipient Tethys continued to deepen and spread westward, eventually separating North and South America.

Somewhat later, North America began to rift from South America as the incipient central Atlantic Ocean began to open. As North America continued to rift from South America, the surface waters of the Tethys were pushed along by the trade winds into the central Atlantic. At the same time, Pacific Ocean water spilled into what would become the Gulf of Mexico, which at this time was a restricted basin lying in low latitudes (see Figure 13.2).

Eventually, Gondwana began to break apart. During the Late Triassic and Jurassic, Australia and Antarctica drifted away from the rest of Gondwana, and the two continents continued to remain sutured until the Late Cretaceous. India also rifted from Gondwana and began to move north toward its present position. The drowned remains of an ancient microcontinent (dubbed "Mauritia") that rifted from the island of Madagascar as India moved away might occur beneath the

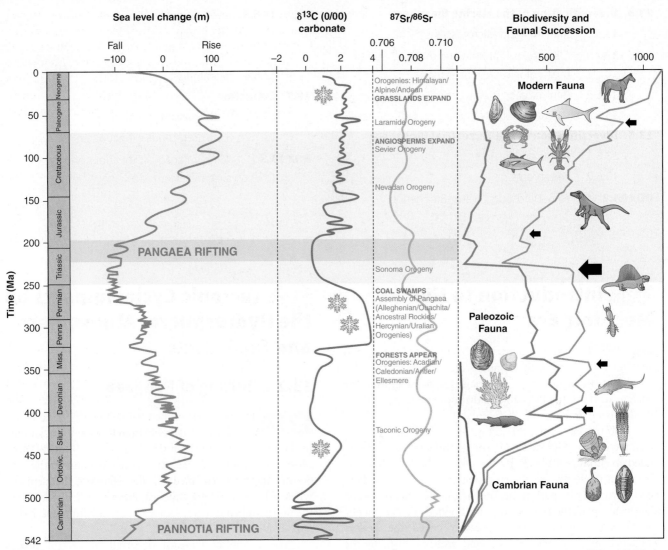

Data from: Martin, T. 1998. *One Long Experiment*. New York: Columbia University Press. Figure 7.1 (p. 166), and Condie, K. C., and Sloan, R. E. 1998. *Origin and Evolution of Earth: Principles of Historical Geology*. Upper Saddle River, NJ: Prentice-Hall. Figure 12.1 (p. 323) and Figure 13.1 (p. 352).

FIGURE 13.1 The physical and biologic evolution of Earth through the Phanerozoic Eon. This chapter discusses the Triassic, Jurassic, and Cretaceous periods of the Mesozoic Era (shaded). Snowflakes indicate times of major glaciation. Carbon isotopes and strontium isotopes are indicated (see Chapter 9 for an elementary discussion). Arrows indicate the Big Five mass extinctions.

Indian Ocean, based on radiometric dates of ancient zircons brought to the surface by volcanoes and found in beach sand on the modern island of Mauritius. The stretching and thinning of the crust associated with rifting are thought to have sunk fragments of crust that comprised an island or archipelago approximately three times the size of the modern island of Crete in the eastern Mediterranean Sea.

The South Atlantic Ocean finally began to open during the Late Jurassic, as South America rifted from Africa. At the same time, the eastern portion of the Tethys began to close. This portion of the seaway was caught in a pincer-like movement as Laurasia rotated clockwise and Africa moved northward, leaving behind the ancestral Mediterranean to the west.

Pangaea continued to rift apart during the Cretaceous. In the northern hemisphere, North America and Europe, which had begun to separate during the Jurassic, continued

to separate. Eastern Greenland began to rift away from northern Europe and then North America, whereas the British Isles moved away from continental Europe and the North Sea began to open. The South Atlantic Ocean continued to widen as South America and Africa continued to move apart. By the Late Cretaceous, Australia and Antarctica had separated as they continued to move toward their present positions, whereas India continued to move north toward lower latitudes to eventually collide with southern Asia.

13.2.2 Sea level, CO$_2$, and sedimentary facies

D **E** Despite the differences in continent–ocean configurations, the results of the rifting that began with the initiation of the second tectonic cycle were largely the same as those of the first cycle: a Late Paleozoic-to-Early

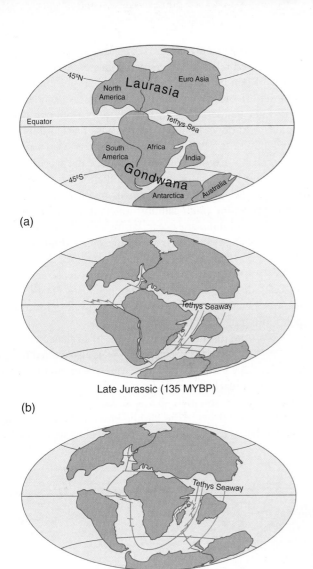

(a)

Late Jurassic (135 MYBP)

(b)

Late Cretaceous (65 MYBP)

(c)

Data from: Monroe, J. S., and Wicander, R. 1997. *The Changing Earth: Exploring Geology and Evolution*, 2nd ed. Belmont, CA: West/Wadsworth. Figure 23.2 (p. 599).

FIGURE 13.2 The changing continent–ocean configurations of the Mesozoic world. **(a)** Triassic. **(b)** Late Jurassic. **(c)** Cretaceous.

Mesozoic world characterized by climatic extremes (see Chapter 12) gave way to one of widespread seas and overall warmth due to the production of mid-ocean ridges, subduction, and volcanism (see Figure 13.1). Sea level continued to remain quite high through much of the Mesozoic, especially after the Triassic Period. Seafloor spreading centers appear to have again displaced water from the ocean basins up onto the continents to produce epeiric seas.

Smaller-scale fluctuations in sea level were again superimposed on the broad transgression that resulted from rifting and seafloor spreading. The Absaroka Sequence continued from the Late Paleozoic (see Chapter 12) to the Early Jurassic. After a major sea-level fall, the Absaroka Sequence was succeeded by the **Zuni Sequence**, which lasted into the Early Paleocene. However, sea level did not reach the levels of the Early-to-Middle Paleozoic until the Cretaceous. The delay of sea-level rise might have resulted from widespread orogeny

in western North America and the erosion of the Appalachians and Ouachitas that shed sediment into the seaways that would have otherwise spread across the continents.

Widespread rifting and volcanism are again indicated by an increase in the volume of igneous rocks in the geologic record (see Figure 13.1). Increased rifting, volcanism, and seafloor spreading associated with the breakup of Pangaea imply that large amounts of carbon dioxide (CO_2) were once again pumped into the atmosphere (see Figure 13.1). Relatively high atmospheric CO_2 levels and epeiric seas, which undoubtedly reduced Earth's albedo, no doubt promoted warm conditions during most of the Mesozoic. However, atmospheric CO_2 levels were probably lower during the Mesozoic than during the Early-to-Middle Paleozoic because of photosynthesis by terrestrial forests, which had begun to spread across the continents during the Carboniferous (see Figure 13.1). Atmospheric CO_2 levels eventually declined during the latter part of the Cretaceous after peaking earlier in the Mesozoic.

Like the Early Paleozoic, epeiric seas of this time can be inferred from the dominant sedimentary facies. Limestones—especially in the form of calcareous oozes—were widespread in the epeiric seas, especially during the Cretaceous. Evaporites were also prevalent at times because of the overall climatic warmth. In fact, evaporite basins formed along the early Tethys and included what is now the entire Gulf of Mexico! Evaporite deposition in the Gulf was so prolonged, the entire basin is underlain by the Louann Salt, which is thousands of meters thick (**Figure 13.3**). Some rift basins that formed along the margins of the Tethys and the Atlantic at relatively high latitudes (up to 30 degrees or more) also filled with evaporites. Evaporite deposition

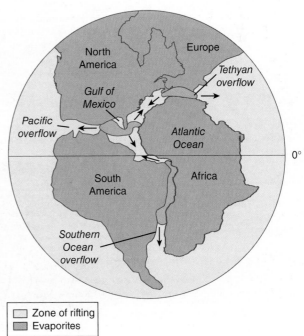

Zone of rifting
Evaporites

Data from: Monroe, J. S., and Wicander, R. 1997. *The Changing Earth: Exploring Geology and Evolution*, 2nd ed. Belmont, CA: West/Wadsworth. Figure 23.3 (p. 600).

FIGURE 13.3 Restricted basins formed as Pangaea rifted apart, contributing to evaporite (salt, anhydrite) formation.

continued into the Early Cretaceous in other areas, suggesting that widespread, isolated basins persisted until this time.

Tropical marine faunas and terrestrial floras (coals) also extended to relatively high latitudes in both hemispheres (see Figure 13.2). Not surprisingly, there is little if any evidence for tillites. Both lines of evidence, along with widespread evaporites, suggest that Earth was too warm to sustain permanent ice caps; however, it is possible that sea ice waxed and waned seasonally.

13.2.3 Ocean circulation and chemistry

Ⓕ Changing continent–ocean configurations during the Mesozoic also promoted warm conditions by affecting ocean circulation. The widening of the Tethys as Pangaea rifted apart led to the more or less circumglobal Tethys at low latitudes (**Figure 13.4**). Recall that as trade winds drive surface ocean currents from east to west, the surface currents tend to warm as they are exposed to the sun. Stable isotope values from fossils of organisms that once lived in the pelagic zone indicate that during the Jurassic and Cretaceous, surface water temperatures were as warm as 25°C to 30°C near the equator and 12°C to 15°C at 60 to 70 degrees North and South latitudes.

Deep-ocean circulation during the Mesozoic probably also involved relatively warm water masses. Oxygen isotope values of benthic fossils indicate that bottom temperatures were on the order of about 15°C. By contrast, modern bottom water temperatures such as those of Antarctic Bottom Water range down to about 4°C because these waters originate adjacent to polar ice caps (see Chapter 3). Thus, there was probably insufficient ice to produce deep ocean water masses like those of today. Instead, like the Early-to-Middle Paleozoic, deep-ocean water masses might have originated from hypersaline (high-salinity) waters in low latitudes that "leaked" from incipient ocean basins or hallow continental shelves where evaporation was widespread (**Figure 13.5**). Because of their higher salt concentrations, hypersaline surface waters would have been relatively dense and prone to sink, despite their warmer temperatures. These deep, warm-water masses might have later upwelled toward the poles, bringing warmth to higher latitudes.

Ⓖ And just like the Early-to-Middle Paleozoic oceans, Jurassic and Cretaceous oceans also tended toward well-developed anoxia and the widespread deposition of carbon-rich black shales. Large amounts of petroleum, generated from organic-rich shales of this time, are produced today from oil fields in the Middle East and South America. The tendency toward **anoxia** was probably augmented by

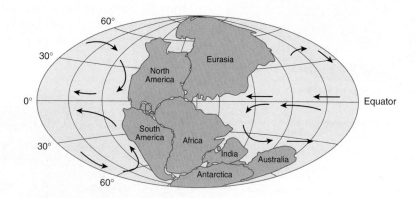

(a) Triassic period

Data from: Monroe, J. S., and Wicander, R. 1997. *The Changing Earth: Exploring Geology and Evolution*, 2nd ed. Belmont, CA: West/Wadsworth. Figure 23.4A (p. 601).

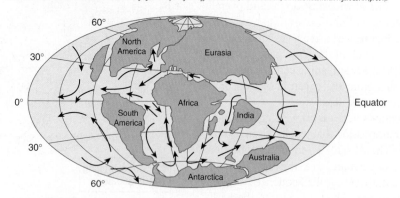

(b) Cretaceous period

Data from: Monroe, J. S., and Wicander, R. 1997. *The Changing Earth: Exploring Geology and Evolution*, 2nd ed. Belmont, CA: West/Wadsworth. Figure 23.4B (p. 601).

FIGURE 13.4 Changing paleoceanographic conditions from the Triassic into the Cretaceous. **(a)** Triassic. **(b)** Cretaceous.

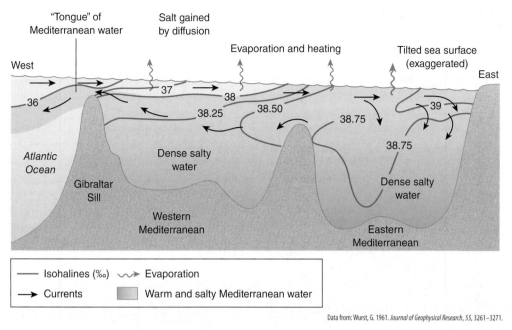

"Tongue" of
Mediterranean water

Salt gained
by diffusion

Evaporation and heating

Tilted sea surface
(exaggerated)

West

East

36

37

38

38.25

38.50

38.75

39

38.75

Atlantic
Ocean

Gibraltar
Sill

Dense salty
water

Dense salty
water

Western
Mediterranean

Eastern
Mediterranean

—— Isohalines (‰)   ⌇⤳ Evaporation

⟶ Currents   ▨ Warm and salty Mediterranean water

Data from: Wurst, G. 1961. *Journal of Geophysical Research, 55,* 3261–3271.

FIGURE 13.5 The formation of dense, saline "Meddies," which flow from the Mediterranean Sea into the Atlantic Ocean through the Straits of Gibraltar. During the Mesozoic, dense, saline water masses may have originated in a similar manner to drive deep ocean circulation.

warm deep waters in the oceans (which hold less oxygen) and sluggish ocean circulation (which decreases the chance for surface waters to pick up oxygen before redescending; Chapter 3).

The tendency toward black shale deposition during the Mesozoic might also be related to tectonics. Black shale deposition was especially widespread during the middle part of the Cretaceous, from about 125 to 90 million years ago. This time corresponds to a prolonged interval of normal orientation of Earth's magnetic field (**Figure 13.6**). Some workers believe that perhaps a gigantic "superplume" originated near Earth's core–mantle boundary, affecting Earth's magnetic field and preventing it from flip-flopping between normal and reversed states. The superplume is also thought to have generated massive amounts of crust in the ocean basins, such as the modern Ontong-Java Plateau of the western Pacific Ocean (**Figure 13.7**). The production of such massive amounts of ocean crust might have promoted both degassing of CO_2 from Earth's mantle and higher sea level. Both conditions might have made Earth more prone to a well-developed oxygen minimum zone (OMZ).

The anoxic episodes of the Mesozoic were, however, short lived compared with those of the Early-to-Middle Paleozoic. In fact, the Mesozoic anoxic events were sufficiently short that they are referred to as **oceanic anoxic events (OAEs)**. These events generally lasted less than a million years. Some oceanic anoxic events appear to have been global events, whereas others appear to have been more localized.

Why were these events relatively short-lived? One possibility is that the atmosphere and oceans had become

increasingly oxygenated after the spread of terrestrial forests. An increase in dissolved oxygen in the oceans would have helped to counteract the production of a well-developed OMZ like that which existed during the Early-to-Middle Paleozoic. The Mesozoic anoxic events might also represent times of relatively high productivity resulting from upwelling. The black shales associated with the oceanic anoxic events are often laminated, consisting of relatively thin organic-rich black layers sandwiched between thin lighter layers. The laminations suggest that conditions during each anoxic episode actually alternated between anoxia and the presence of at least some oxygen. Such conditions are found today in upwelling regions off continental margins, especially in silled basins where bioturbation is limited because of low oxygen levels. Interestingly, all oceanic anoxic events are associated with shifts in carbon isotope ratios, suggesting changing rates of marine photosynthesis. Strontium isotope ratios, which reflect the relative input of [87]Sr from land and [86]Sr from hydrothermal exchange, were low during the early Mesozoic (see Figure 13.1). Perhaps greater seawater exchange with warm mafic ocean crust during these particular intervals dissolved more iron—a limiting nutrient of photosynthesis through hydrothermal weathering (see Chapter 2)—into the oceans, stimulating photosynthesis. In fact, green algal plankton like those prominent during the Early-to-Middle Paleozoic seem to become more prominent during OAEs. Massive volcanism has also been reported to coincide with some of these OAEs; the carbon dioxide pumped into the atmosphere may have accelerated the hydrologic cycle and weathering, delivering nutrients to the oceans.

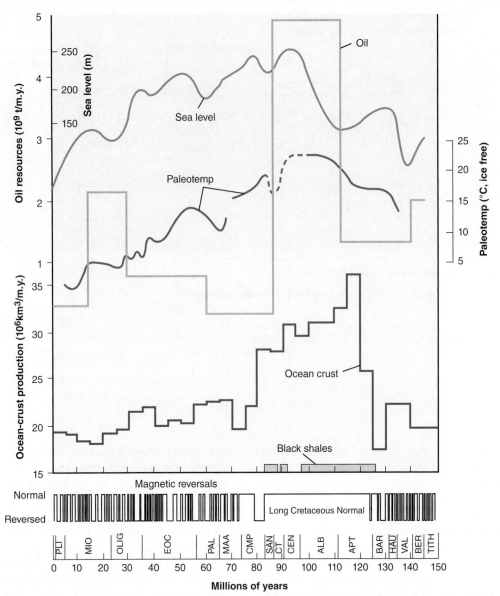

FIGURE 13.6 The mid-Cretaceous superplume episode and the occurrence of black shales. Black shale deposition was especially prominent about 110 and 90 million years ago. Note the occurrence of oil during the Cretaceous, especially during the superplume episode.

As we have already seen, much of the Mesozoic, especially the Cretaceous Period, was a time of widespread limestone deposition. Based on studies of the calcium carbonate ($CaCO_3$) content of deep-sea cores and calculations of the depths at which the sediments in the cores were originally deposited, the calcite compensation depth (CCD) lay at approximately 3 to 3.5 km deep. The CCD might actually have begun to deepen earlier, during the Late Paleozoic, as sea level fell and, presumably, shallow-water limestones were eroded (see Chapter 12). The expansion of calcareous plankton in the Mesozoic might have accelerated the deepening of the CCD by sending increasing amounts of $CaCO_3$ to the deep ocean bottom, not just to the bottom of the shallow epeiric seas that had flooded the continents.

Much of the Mesozoic was also a time of calcite seas. This is in basic agreement with higher rates of seafloor spreading as Pangaea rifted apart. The continued rifting

of Pangaea would have decreased magnesium ion (Mg^{2+}) concentrations in seawater, promoting calcite seas (see Figure 13.1). However, as the Mesozoic continued, strontium ratios began to increase, perhaps because of widespread orogeny and continental weathering.

CONCEPT AND REASONING CHECKS

1. How is the Mesozoic similar to the Early-to-Middle Paleozoic in terms of (a) greenhouse conditions, (b) sea level (high or falling), (c) large continental glaciers (present or absent), (d) CCD (shallow or deep), (e) weathering (fast or slow), (f) calcite or aragonite seas, (g) deep-ocean circulation (Hint: were black shales more or less frequent during this time?), and (h) surface ocean circulation?
2. Why are the two intervals similar in these respects?

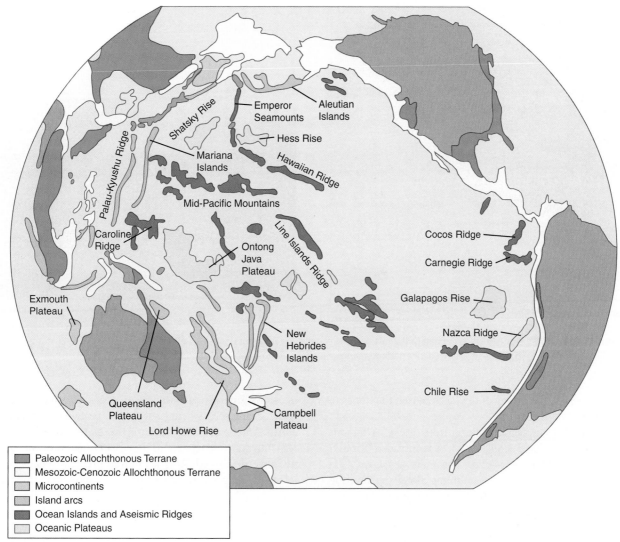

Data from: Condie, K. C., and Sloan, R. E. 1998. *Origin and Evolution of Earth: Principles of Historical Geology.* Upper Saddle River, NJ: Prentice-Hall. Figure 12.7 (p. 328).

FIGURE 13.7 The modern Pacific Ocean basin showing the Ontong-Java Plateau and the terranes of Panthalassa.

Legend:
- Paleozoic Allochthonous Terrane
- Mesozoic-Cenozoic Allochthonous Terrane
- Microcontinents
- Island arcs
- Ocean Islands and Aseismic Ridges
- Oceanic Plateaus

13.3 Tectonic Cycle and Orogeny

13.3.1 Eastern North America

The eastern portion of North America remained relatively quiet after the repeated Paleozoic upheavals that produced the Appalachians. Instead of orogenesis along an active margin, the Appalachians now largely underwent erosion along a passive one. Much of the sediment was shed to the east, producing thick sedimentary sequences that built out the continental shelf along the passive margin as the Atlantic Ocean opened.

The initial rifting of Pangaea along the Atlantic margin of North America is indicated by a variety of geologic features. An entire series of rift basins left from this tectonic phase extend up and down much of the eastern seaboard. The rift basins are found from North Carolina into eastern Canada and range as far west as Virginia and Pennsylvania (**Figure 13.8**).

These rift basins outcrop on land, but they are also buried beneath the continental shelf far offshore. Like their modern analog, the East African Rift Valley (see Chapter 2), the rift basins formed lakes that filled with sediments shed from the margins of the basins. The sediments of the rift basins belong to the **Newark Group** and are predominantly Triassic in age and up to several kilometers in thickness. The sediments of the Newark Group consist of brown to reddish conglomerates, arkoses, and shales that were oxidized to their reddish color by exposure to the atmosphere. (During the 19th century, these rocks were quarried near Newark, New Jersey, for many residences—called "brownstones"—in New York City.) Sometimes, the shales of the Newark Group exhibit cyclic deposition, indicating expansion and contraction of the lakes. Impressions of raindrops and mud cracks have also been found, indicating ephemeral (temporary) wet and also warm, dry conditions; such a climate might be expected given the restricted nature of the basins and the location of North America close to the equator. Dinosaur tracks are sometimes exposed, but hard parts are typically not found,

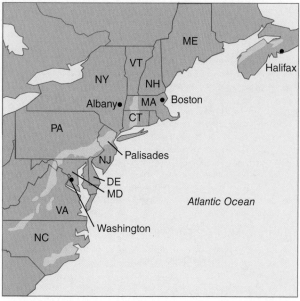

(a) Data from: Monroe, J. S., and Wicander, R. 1997. *The Changing Earth: Exploring Geology and Evolution*, 2nd ed. Belmont, CA: West/Wadsworth. Figure 23.8A (p. 605).

(b) Courtesy of Ronald Martin, University of Delaware.

FIGURE 13.8 Features of the North American east coast associated with the rifting of Pangaea. **(a)** Map of Triassic rift basins along the east coast of North America. **(b)** Reddish, oxidized lake sediments near Reading, Pennsylvania, that filled rift basins.

possibly because the drying of the lakes caused the bones to disintegrate.

Rifting also emplaced mafic igneous rocks. A famous example is the Palisades Sill, which was intruded approximately 190 million years ago. The Palisades now form prominent cliffs of fence-like columns along the Hudson River above New York City (palisade for "fence-like"); the columnar appearance resulted from the joints formed in the magma as it cooled. Here, the rifting and intrusions are superimposed on rocks of Precambrian and Early Paleozoic age such as the Fordham Gneiss in the Bronx and the Manhattan Schist. Like a number of major rivers in the world, the Hudson River flows down an aulacogen floored by Triassic sediments that lie between the Palisades Sill on the west and metamorphic rocks on the east that are remnants of the Late Ordovician Taconic Orogeny (see Chapter 11). Other mafic intrusions associated with Triassic rifting are found much further to the west at Gettysburg, Pennsylvania, and were defended by Union forces during the battle there.

13.3.2 Cordilleran Orogenic Belt

The western margins of the Americas belong to what is called the **Cordilleran Orogenic Belt**, or just the **Cordillera**. The Cordillera is one of the longest mountain chains in the world, extending from Alaska down the western margin of North America as the Rockies, on through Central America, and then into South America as the Andes. The Cordillera is about 1,000 km (about 600 miles) wide and is still active today, as evidenced by volcanism and earthquakes associated with the Pacific Ring of Fire (see Chapter 2).

Unlike eastern North America, the western margin of the Americas was quite active orogenically during the Mesozoic. After the Early-to-Middle Paleozoic, the western margin

of North America had become an ocean–continent margin, with oceanic crust being subducted eastward beneath the North American plate. Because the North American plate was moving west relative to the ocean crust, orogenic activity in western North America moved progressively eastward from the deep-sea trench across the continental slope and shelf onto the craton.

This style of tectonism was basically a continuation of the style of tectonics that began in the Devonian with the Antler Orogeny (see Chapter 11). To the west, Panthalassa, the huge ocean formed by the assembly of Pangaea (see Chapter 12), contained a series of subduction zones, island arcs, oceanic plateaus, seamounts, and microcontinents, similar to the Pacific today. A number of these island arcs and microcontinents accreted as terranes to the western margin of North America during the Mesozoic (and Cenozoic) eras as North America moved over the trenches and their volcanic island arcs, colliding with the terranes like a continental bulldozer (**Figure 13.9**). As the terranes accreted to North America, some appear to have moved northward along strike-slip faults; thus, rather than subducting directly beneath each other at right angles, the plate margins moved past each other at oblique angles. Based on polar wandering curves and fossil assemblages, these terranes might have traveled up to thousands of kilometers.

13.3.3 Orogenic episodes

The first major orogenic episode of the Mesozoic in western North America was really a continuation of the Sonoma Orogeny from the Permian. During this orogeny the **Golconda terrane**, representing an accretionary wedge, was caught between the collision of the **Roberts Mountain terrane** (which had accreted during the Antler Orogeny;

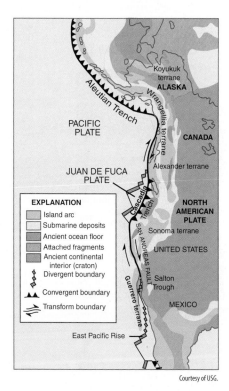

FIGURE 13.9 The accretion of terranes to western North America during the Mesozoic.

see Chapter 11) and the microcontinent of Sonomia. Today, Sonomia is represented by western Nevada, southeastern Oregon, and northern California.

The next phase of mountain building, sometimes referred to as the **Cordilleran Orogeny**, ranged from the Jurassic Period into the Eocene Epoch of the Cenozoic Era. In a sense, the Cordilleran Orogeny was somewhat like the uplift of the Appalachians in that several phases of uplift in different regions were involved: the **Nevadan**, **Sevier**, and **Laramide**.

The Nevadan Orogeny began in the Jurassic in western North America (**Figure 13.10**). During the Nevadan Orogeny enormous batholiths were emplaced as ocean crust was subducted to sufficient depth to produce magma. The Nevadan Orogeny can be viewed as a continuation of the tectonic style that accreted the Antler and Klamath Mountains in the Paleozoic (see Chapters 11 and 12). In fact, the ages and lithologies of the northwest portion of the Sierra Nevada resemble those of the Klamaths. Many of the plutons emplaced during the Nevadan Orogeny were later exposed by faulting and erosion to form the Sierra Nevada (Spanish for "snowy mountains") of eastern California, which includes Yosemite, Kings Canyon, Sequoia, and Mineral King national parks (**Figure 13.11**). These magma bodies fed a volcanic arc from which sediments were shed to the west into a deep forearc basin. These sediments comprise the Great Valley Group, which consists of conglomerates, sands, shales, and turbidites (see Figure 13.10). Today, highly folded rocks of the Great Valley Group occupy the Great (Central) Valley of California.

Further west and spanning about 150 million years, the **Franciscan mélange** was forming from about the middle Mesozoic (late Jurassic) to early Tertiary (mainly Eocene; see Figure 13.10). This unit consists of a jumbled mass of volcanic breccias, pillow lavas, ophiolites, turbiditic graywackes and mudstones, deep-sea cherts derived from deep-sea oozes comprised of microfossil radiolarian and diatoms, and blueschist (see Figure 13.10). The presence of blueschist indicates the Franciscan mélange was originally forming as a massive submarine fan complex associated with the subduction zone that was producing the volcanic arc to the east. As the North American plate continued to move west during the Late Jurassic and Cretaceous, ocean crust of the Farallon Plate continued to be subducted beneath North America. Eventually, the Franciscan mélange and the Great Valley Group were both folded and thrusted onto the continent. Portions of the Franciscan mélange were also uplifted much later in the Cenozoic Era to form a long range of very steep hills that today represent the **Coast Ranges** that run along either side of the San Andreas fault system located to the west of the Great Valley (**Figure 13.12**). It is also now recognized that at least nine different terranes occur within the Franciscan mélange based on radiolaria, paleomagnetic analyses, and studies of the mineral content of the graywackes. These terranes represent remnants of several oceanic plates that eventually collided with North America. A similar style of orogeny occurred south all the way into Chile. During the same time a large exotic terrane called **Wrangellia** extending from Washington into Alaska was accreted. This terrane actually consisted of a series of smaller terranes that had accreted to form Wrangellia in the Triassic before its collision with North America in the Jurassic.

The Nevadan Orogeny was followed by the Sevier Orogeny in the Cretaceous. During the Cretaceous, the orogenic belt in western North America became quite broad. Subduction of the Franciscan mélange and volcanism continued nearer the coast, but igneous activity gradually shifted eastward into Nevada and Idaho (**Figure 13.13**). One hypothesis proposed to explain the eastward shift in tectonic activity is a decrease in the angle of plate subduction. A decreased angle would have allowed the leading edge of the plate to be located much further inland before melting than if the plate had descended at a steeper angle (see Figure 13.13). Possibly, the North American plate was moving west too fast for the subduction of ocean crust underneath it to keep pace. The Sevier Orogeny culminated in extensive folding and thrusting as far east as the Idaho–Wyoming border, to the south into Utah, and to the north through Montana and well into Canada (see Figure 13.2). Today, this deformation is exposed in spectacular outcrops in western North America. Further to the west and north in Canada, terranes continued to be accreted (see Figure 13.13).

The Laramide Orogeny began to the east of the Sevier Orogeny in the vicinity of New Mexico, Colorado, and Wyoming during the Late Cretaceous. The Laramide Orogeny continued well into the Cenozoic Era and was responsible for many of the features of the Rocky Mountains seen today. We will examine the Laramide Orogeny in greater detail in Chapter 14.

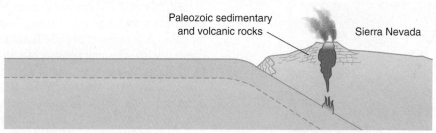

Paleozoic sedimentary
and volcanic rocks

Sierra Nevada

1. Late Triassic-Early Jurassic

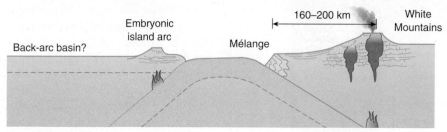

Back-arc basin?

Embryonic
island arc

Mélange

160–200 km

White
Mountains

2. Middle Jurassic

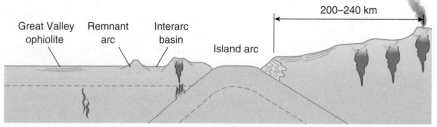

Great Valley
ophiolite

Remnant
arc

Interarc
basin

Island arc

200–240 km

3. Early Late Jurassic

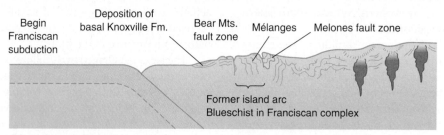

Begin
Franciscan
subduction

Deposition of
basal Knoxville Fm.

Bear Mts.
fault zone

Mélanges

Melones fault zone

Former island arc
Blueschist in Franciscan complex

4. Late Jurassic-Nevadan orogeny

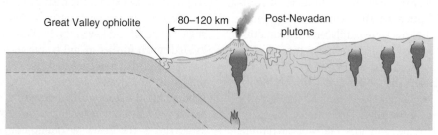

Great Valley ophiolite

80–120 km

Post-Nevadan
plutons

5. Latest Jurassic

(a)

Data from: Monroe, J. S., and Wicander, R. 1997. *The Changing Earth: Exploring Geology and Evolution*, 2nd ed. West/Wadsworth. Belmont, CA: West/Wadsworth. Figure 23.10 (p. 606).

FIGURE 13.10 Tectonics and sedimentation along the Pacific margin of California during the Nevadan Orogeny. **(a)** Cross-sections of events.

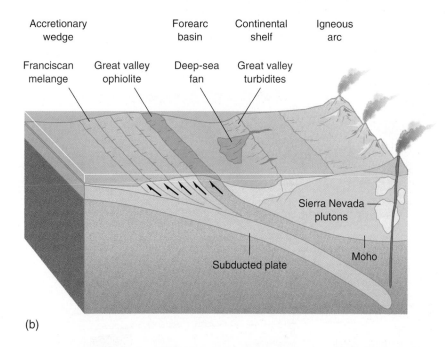

Accretionary wedge   Forearc basin   Continental shelf   Igneous arc

Franciscan melange   Great valley ophiolite   Deep-sea fan   Great valley turbidites

Sierra Nevada plutons

Moho

Subducted plate

(b)

(c)

Courtesy of Ronald Martin, University of Delaware.

FIGURE 13.10 (b) Three-dimensional reconstruction. **(c)** Franciscan mélange exposed along the northern California coast.

Courtesy of Ronald Martin, University of Delaware.

FIGURE 13.11 Granitic plutons emplaced during the Nevadan Orogeny are now exposed in the Sierra Nevada of eastern California.

Courtesy of Ronald Martin, University of Delaware.

FIGURE 13.12 California coast ranges looking south toward San Francisco.

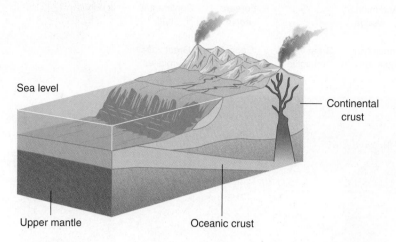

Sea level

Continental crust

Upper mantle

Oceanic crust

(a) Data from: Monroe, J. S., and Wicander, R. 1997. *The Changing Earth: Exploring Geology and Evolution*, 2nd ed. Belmont, CA: West/Wadsworth. Figure 23.14A (p. 608).

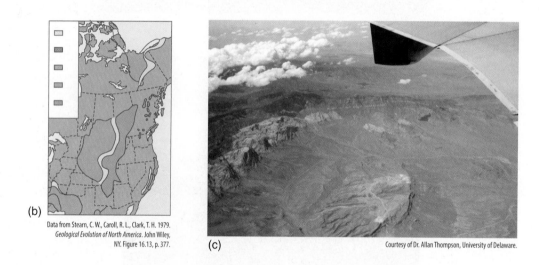

(b) Data from Stearn, C. W., Caroll, R. L., Clark, T. H. 1979. *Geological Evolution of North America*. John Wiley, NY. Figure 16.13, p. 377.

(c) Courtesy of Dr. Allan Thompson, University of Delaware.

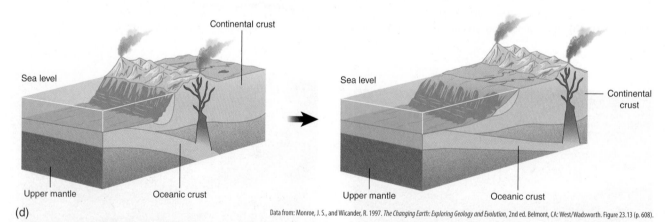

(d) Data from: Monroe, J. S., and Wicander, R. 1997. *The Changing Earth: Exploring Geology and Evolution*, 2nd ed. Belmont, CA: West/Wadsworth. Figure 23.13 (p. 608).

FIGURE 13.13 The Sevier Orogeny. **(a)** Reconstruction of the Sevier Orogeny. **(b)** Location of the Sevier orogenic belt in relation to modern physiographic provinces. The origins of these provinces are discussed in Chapters 14 and 15. **(c)** The Keystone Thrust Fault exposed west of Las Vegas, Nevada. The dark-colored Paleozoic rocks have been thrust over lighter-colored Mesozoic rocks. **(d)** A change in the angle of plate subduction beneath North America is hypothesized as the reason for the shift of orogenic activity from the western margin to the interior of North America.

13.4 Orogeny, Sea Level, and Sedimentation

During the Early Triassic, many continental interiors remained isolated from sources of moisture because of the continued existence of Pangaea. As a result, the relatively dry, arid conditions of the Permian persisted through the Triassic and into the Jurassic in western North America. Many of the rock units left from this time are various types of redbeds that have been eroded into the spectacular scenery of the "Red Rock Country" of the American southwest (**Figure 13.14**, **Figure 13.15**, and **Figure 13.16**). The red-to-brownish mudstones of the Early Triassic Moenkopi Formation, for example, were deposited in stream channels, floodplains, and temporary pools of water; these rocks can contain desiccation cracks—attesting to the dry environment—along with dinosaur tracks. Spectacular exposures of this and other formations occur in Monument Valley, which occurs along the crest of a gentle, eroded uplift near the Arizona–Utah border within the Navajo Indian nation. The upper portion of the **Chinle**, called the **Shinarump Conglomerate**, often forms a resistant cap on top of the Chinle. Volcanic ashes are also frequently interspersed in the Chinle. The Chinle is

© dibrova/Shutterstock, Inc.

FIGURE 13.14 Formations of Monument Valley in southeast Utah. The steep cliffs consist of the DeChelly (pronounced "DeShay") Sandstone of Middle Permian age, whereas the rocks at the base of the cliffs represent the Organ Rock Shale (also Permian). The Moenkopi Formation occurs as shales above the DeChelly Sandstone and are sometimes capped by gravels of the Shinarump Member of the Triassic Chinle Formation.

especially famous in the Petrified National Forest of Arizona. Here, petrified wood was formed by the weathering and dissolution of silica from the volcanic ashes and its reprecipitation in the pore spaces of the plants; at times, the colors and patterns are almost psychedelic. The trees were mostly gymnosperms and included conifers and short, palm-like cycads. Some trees were preserved in place, whereas others appear to have accumulated as debris during floods. Among the vertebrates were amphibians, crocodile-like forms, and primitive reptiles that include crocodiles (and similar forms) and the ancestors of birds and dinosaurs.

(a) Courtesy of Ronald Martin, University of Delaware.

(b) © Paul B. Moore/Shutterstock, Inc.

FIGURE 13.15 (a) Exposures of the Triassic Chinle Formation on the O'Keeffe Ranch in northern New Mexico. The American artist Georgia O'Keeffe lived nearby and painted spectacular exposures like these. The coloration of the sediments results from volcanic ash falls.
(b) Permineralized tree stumps found in the nearby Petrified Forest (see Chapter 4).

(a)

Courtesy of Ronald Martin, University of Delaware.

(b)

© Kenneth Keifer/Shutterstock, Inc.

FIGURE 13.16 The Navajo Sandstone (Jurassic). **(a)** Cross-bedding in ancient dune sands of the Navajo Sandstone exposed at Checkerboard Mesa near the entrance to Zion National Park in southern Utah. **(b)** Valley of Zion National Park as seen from Angel's Landing. Approximately the upper half of the valley walls where the cliffs are steepest consist of the Navajo Sandstone.

These deposits gave way in the Early Jurassic to desert sandstones. Among the most famous of these is the **Navajo Sandstone** (see Figure 13.16). The Navajo Sandstone consists of seemingly endless numbers of massive cross-beds composed of almost pure quartz sand deposited in environments like those of the modern Sahara Desert. Today, the Navajo Sandstone outcrops over a wide area of the American southwest, including northern Arizona and Utah. Nowhere is the Navajo better exposed than in the spectacular cliffs of Zion National Park of southern Utah (see Figure 13.16).

During the Middle Jurassic the **Sundance Sea** twice invaded the foreland basin east of the main orogenic belt from the present Arctic and northern Pacific (**Figure 13.17**). The sediments of the corresponding Sundance Formation consisted of sands and shales shed from highlands to the west, along with interbedded marine limestones and evaporites. The Sundance Sea eventually retreated northward in response to uplift and deformation associated with the Nevadan Orogeny and was replaced by large volumes of nonmarine molasse, mainly sandstones and mudstones, shed from the uplift to the west (see Figure 13.17). One of the most famous formations laid down during this time is the dinosaur-rich Morrison Formation, which formed a huge clastic wedge of uplands and plains coming off the uplifts (see

Figure 13.17). Although the dinosaur skeletons are often disarticulated, they are sometimes concentrated together, suggesting that perhaps the animals died in a flood and their carcasses were carried downstream, where they became lodged against one another.

The eastward shift in subduction associated with the Sevier Orogeny eventually led to the formation of another foreland basin by the end of the Early Cretaceous. This basin was intermittently flooded from the north by the **Mowry Sea** during the mid-Cretaceous, when a global transgression flooded the continents. Black shale deposition was widespread in the Mowry Sea, suggesting circulation was relatively restricted. By the Late Cretaceous the Mowry Sea had established contact with waters encroaching from the Gulf of Mexico, which was itself part of the warm Tethys Seaway. This **Cretaceous Interior Seaway** stretched for almost a thousand miles from what is now the Rocky Mountains to the eastern margin of the Great Plains, splitting North America in two (**Figure 13.18**).

Like the Sundance Sea, the Cretaceous Interior Seaway gradually filled with sediment. The foreland basin deepened to the west, where sediment accumulation was greatest, and shallowed gradually to the east (see Figure 13.18). Cyclic deposition in response to changing sea level, sediment

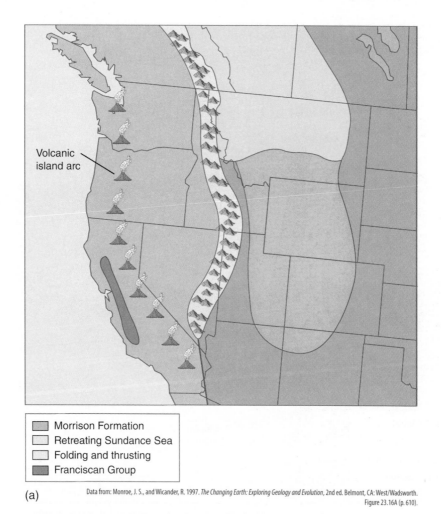

Volcanic island arc

Morrison Formation
Retreating Sundance Sea
Folding and thrusting
Franciscan Group

(a)

(b)

FIGURE 13.17 Seaways, tectonics, and sedimentation in the interior of western North America during the Jurassic. **(a)** The Morrison Formation, which is molasse shed from the west that filled the Sundance Sea. **(b)** Exposure of the Morrison Formation.

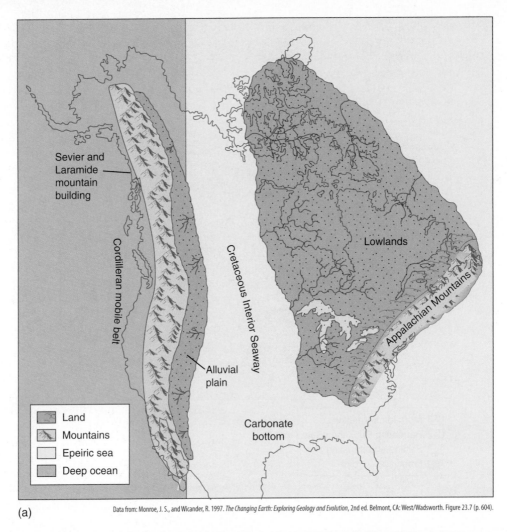

Data from: Monroe, J. S., and Wicander, R. 1997. *The Changing Earth: Exploring Geology and Evolution*, 2nd ed. Belmont, CA: West/Wadsworth. Figure 23.7 (p. 604).

(a)

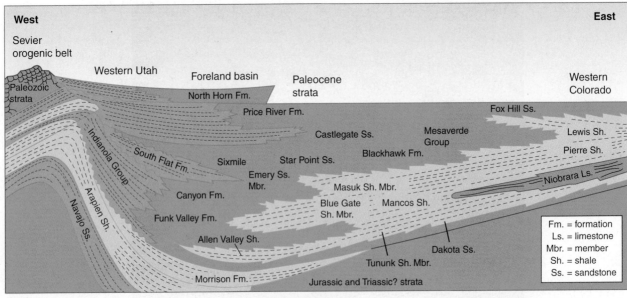

Data from: Monroe, J. S., and Wicander, R. 1997. *The Changing Earth: Exploring Geology and Evolution*, 2nd ed. Belmont, CA: West/Wadsworth. Figure 23.17 (p. 610).

(b)

FIGURE 13.18 Seaways, tectonics, and sedimentation in the interior of western North America during the Cretaceous. **(a)** The Cretaceous Interior Seaway, which stretched from the Arctic to the Gulf of Mexico, cut North America in two. **(b)** Cross-section of sediment shed off the Sevier uplift into the Cretaceous seaway. Notice how the total thickness of the strata thins toward the eastern, shallower margin of the seaway, where the Niobrara Limestone occurs, far away from the Sevier Uplift and the source of terrigenous sediment in the basin. Terrigenous sediments would have otherwise "diluted" the limestone (see Chapter 4).

supply, and subsidence of the foreland basin is often evident. Transgressive sands and conglomerates of the Dakota Sandstone were initially widespread early in the basin's history but eventually gave way to chalks and carbon-rich black shales indicative of more open marine conditions. Later, alluvial fans, conglomerates, and sandstones were shed into the western part of the basin, near where the Sevier Orogeny was most active. These sediments grade eastward into those of the Mesaverde Group, which consist of nearshore shallow marine sandstones; these rocks are today exposed in the Anasazi cliff dwellings of Mesa Verde National Park, in southwestern Colorado (refer to the frontispiece of Chapter 1). Even further east, far from shore and the Sevier uplift, these deposits grade into shales and chalks indicative of less terrigenous input and more open marine conditions. Among the most famous and widespread of these formations are the **Niobrara Chalk** and the **Mancos** and **Pierre** shales, which range from the Great Plains to the west and south (see Figure 13.18). The Cretaceous Interior Seaway was eventually driven off the North American craton by the Laramide Orogeny, which began in the Late Cretaceous and continued into the Cenozoic (see Chapter 14).

CONCEPT AND REASONING CHECKS

1. The Cretaceous Interior Seaway represents a foreland basin. Where was the foreland basin in North America during the formation of the Appalachian Mountains?
2. How do foreland basins form?

13.5 Diversification of the Marine Biosphere

After the Late Permian extinctions, the **Modern Fauna** composed of **bivalves**, **gastropods**, sea urchins, crabs and lobsters, and bony fish diversified tremendously, especially during the Jurassic and Cretaceous. Nektonic Paleozoic taxa such as ammonoids and sharks also rediversified during the Mesozoic and were joined by reptilian predators. These developments suggest increasing predation and perhaps a further increase in the complexity of marine ecosystems. In fact, the diversification of the Modern Fauna was so pronounced that it has been called the **Mesozoic marine revolution**.

13.5.1 Plankton and microfossils

Two major planktonic taxa persisted from the Paleozoic into the Mesozoic: **conodonts** and **radiolarians**. However, conodonts disappeared by the end of the Triassic period, whereas radiolarians flourished. The radiolarians were joined in the plankton by a number of newly evolved planktonic taxa; these taxa became critical links in the biogeochemical cycles of carbon and silica and have persisted to the present (**Figure 13.19**).

Among these taxa were the calcareous plankton. By the Late Triassic the **coccolithophorids** had appeared in the phytoplankton and **planktonic foraminifera** among the zooplankton (see Figure 13.19). Planktonic foraminifera evolved from benthic species. By the Cretaceous these calcareous taxa, especially the coccolithophorids, formed extensive oozes (chalks) in shallow seaways all over the world; today, these oozes are exposed in localities such as the White Cliffs of Dover along the English Channel (see Chapter 1). As calcareous plankton became increasingly diverse and abundant, the dominant site of limestone deposition shifted from the shallow-water cratons to the continental slope and eventually the deep sea (**Figure 13.20**). Thus, it appears the deep-sea link of biogenic calcareous oozes in the long-term carbon cycle, and its feedback on atmospheric CO_2 concentrations, began to be established no later than the Cretaceous.

As noted in previous chapters, dramatic changes in the concentration of atmospheric CO_2 are inferred from Earth's geologic history. Enormous volumes of volcanic rock erupted onto Earth's surface at times in the ancient past because of the processes of plate tectonics. Because modern volcanoes pump CO_2 into the atmosphere, we can deduce that the ancient eruptions must have spewed enormous volumes of the same gas into the atmosphere. The CO_2 then remained in the atmosphere for long periods of time, spanning tens of millions of years. Based on the greenhouse effect, the CO_2 warmed the Earth, speeding up—through positive feedback—the chemical reactions involved in the weathering of rocks on land (see Chapter 4). As rock surfaces are exposed to air and water, they undergo chemical weathering. Chemical weathering results from weak acids produced by the decay of dead organic matter (for example, humic acids resulting from the decay of dead plant matter) and from the dissolution of atmospheric CO_2 into rainwater to produce carbonic acid. Ions such as calcium and phosphate are dissolved from the minerals forming the rock, causing the rocks to break down. The ions are, in turn, carried in dissolved form by rivers and groundwater to the oceans. As these reactions were speeded up, they then began to act as negative feedback: much of the CO_2 pumped into the atmosphere by volcanism was slowly used up by weathering, eventually lowering atmospheric concentrations of CO_2 and lowering Earth's surface temperature. The CO_2 in the atmosphere is transferred to and sequestered (or stored) in pelagic limestone (calcium carbonate, or $CaCO_3$) produced by coccolithophorids and planktonic foraminifera that are deposited in the deep sea as $CaCO_3$. Here, they might eventually be carried into deep-sea trenches, where they are subducted and heated. The CO_2 stored in the limestone (as carbonate) is released and vented back to the atmosphere through volcanoes. These processes therefore act as positive feedback by increasing levels of CO_2 in the atmosphere. The shift of limestone deposition from the epeiric seas on the cratons to open oceans also deepened the CCD (**Box 13.1**).

Two other distinctive phytoplankton taxa appeared in the geologic record during the Mesozoic. **Diatoms** produce siliceous shells, or frustules. Diatoms first appeared during the Jurassic but did not become prominent until the

Mid-to-Late Cretaceous. Perhaps not coincidentally, it was about this time that **angiosperms**, which include flowering plants and grasses, began to diversify (see Figure 13.1; see section 13.6). The increase in diatom diversity and the spread of angiosperms in the Cretaceous might not be coincidental. Modern diatoms tend to thrive in nutrient-rich conditions, and plant litter of modern angiosperms (such as dead leaves) decays relatively rapidly, releasing nutrients, some of which might have found their way into runoff to the oceans. Like the acritarchs of the Paleozoic, **dinoflagellates** produce cysts that are preserved in the fossil record (see Figure 13.19). There are scattered reports of dinoflagellates, or dinoflagellate-like fossils, from the Paleozoic and the Late Proterozoic based on geochemical studies of Paleozoic sediments. It has therefore been suggested that acritarchs were ancestral to dinoflagellates. Whether or not they are related to acritarchs, dinoflagellates did not become prominent in the fossil record until the Mesozoic.

The diversification of new phytoplankton taxa during the Meso-Cenozoic parallels the diversification of the Modern Fauna and a general increase of strontium isotope ratios (suggesting enhanced nutrient runoff in response to widespread orogeny). Possibly, these new phytoplankton lineages were more nutrient-rich and could support greater primary productivity at the base of food pyramids. These nutrients would have been made available to consumers higher in food pyramids during the remineralization of organic matter by consumers and decomposers. Greater productivity would have supported larger populations that could spread over larger areas and become genetically isolated, promoting the diversification of new taxa. Greater primary productivity would have also supported increasing metabolic rates, especially of predators, which became more prominent in the Modern Fauna. A similar phenomenon might have occurred during the expansion of metazoans during the late Proterozoic Eon, the Cambrian explosion, and the Great Ordovician Biodiversity Event (see Chapters 10 and 11).

13.5.2 Benthic ecosystems

The lack of a well-developed benthic marine fauna in the Early Triassic is indicated by the brief appearance of

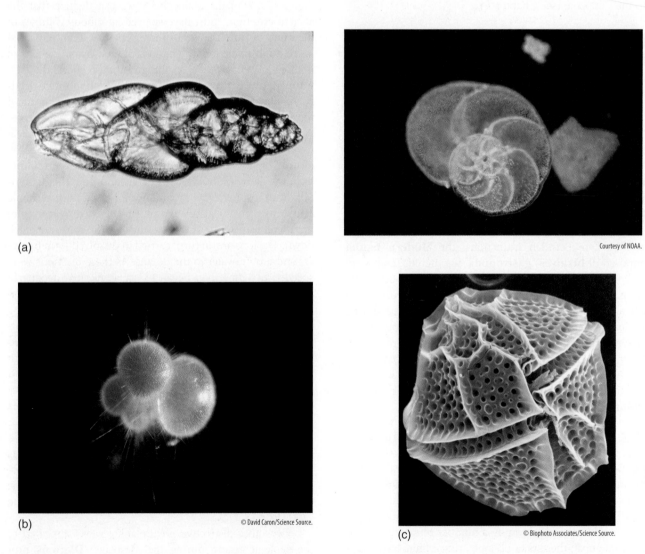

(a)

Courtesy of NOAA.

(b) © David Caron/Science Source.

(c) © Biophoto Associates/Science Source.

FIGURE 13.19 Microfossils of the Mesozoic Era. **(a)** Benthic foraminifera. **(b)** Planktonic foraminifera. **(c)** Dinoflagellates, which can produce benthic stages.

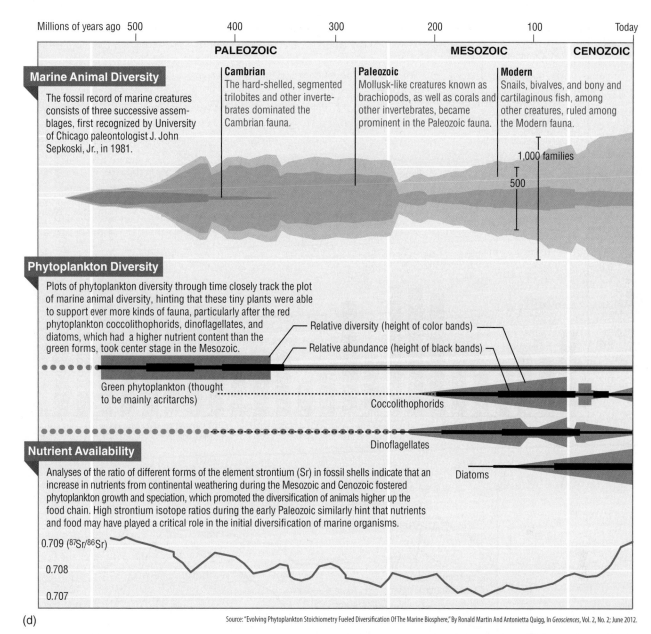

Millions of years ago 500   400   300   200   100   Today

PALEOZOIC   MESOZOIC   CENOZOIC

Marine Animal Diversity

The fossil record of marine creatures consists of three successive assemblages, first recognized by University of Chicago paleontologist J. John Sepkoski, Jr., in 1981.

Cambrian
The hard-shelled, segmented trilobites and other invertebrates dominated the Cambrian fauna.

Paleozoic
Mollusk-like creatures known as brachiopods, as well as corals and other invertebrates, became prominent in the Paleozoic fauna.

Modern
Snails, bivalves, and bony and cartilaginous fish, among other creatures, ruled among the Modern fauna.

1,000 families

500

Phytoplankton Diversity

Plots of phytoplankton diversity through time closely track the plot of marine animal diversity, hinting that these tiny plants were able to support ever more kinds of fauna, particularly after the red phytoplankton coccolithophorids, dinoflagellates, and diatoms, which had a higher nutrient content than the green forms, took center stage in the Mesozoic.

Relative diversity (height of color bands)

Relative abundance (height of black bands)

Green phytoplankton (thought to be mainly acritarchs)

Coccolithophorids

Dinoflagellates

Nutrient Availability

Analyses of the ratio of different forms of the element strontium (Sr) in fossil shells indicate that an increase in nutrients from continental weathering during the Mesozoic and Cenozoic fostered phytoplankton growth and speciation, which promoted the diversification of animals higher up the food chain. High strontium isotope ratios during the early Paleozoic similarly hint that nutrients and food may have played a critical role in the initial diversification of marine organisms.

Diatoms

0.709 (^{87}Sr/^{86}Sr)

0.708

0.707

(d)

Source: "Evolving Phytoplankton Stoichiometry Fueled Diversification Of The Marine Biosphere," By Ronald Martin And Antonietta Quigg, In *Geosciences*, Vol. 2, No. 2; June 2012.

FIGURE 13.19 (d) The occurrence of phytoplankton during the Phanerozoic Eon. Note how the diversification of phytoplankton taxa during the Meso-Cenozoic parallels the diversification of the Modern Fauna. Possibly, these new phytoplankton lineages were more nutrient-rich and could support greater secondary productivity at the base of food pyramids. Greater productivity would have supported larger populations that could spread over larger areas and become genetically isolated, promoting the diversification of new taxa. Greater primary productivity also would have supported increasing metabolic rates, especially of predators, which became more prominent in the Modern Fauna.

stromatolites. Stromatolites were prominent during the Proterozoic and Early Paleozoic, after which they largely disappeared from the fossil record, possibly because of increasing grazing (see Chapter 11). Similarly, stromatolites largely disappeared as the Triassic wore on, again suggesting the reexpansion of marine faunas.

As stromatolites declined, tiering again became prominent above and below bottom among benthos such as crinoids and bivalves. Increased tiering hints at rebounding food availability above and below bottom in the form of plankton and organic detritus. Bivalves, for example,

reradiated in the Mesozoic and were much more successful than in the Paleozoic. Some bivalves were **epifaunal** (living on surfaces), such as oysters, which appeared in the Jurassic. But other bivalves were **infaunal** (living beneath the surface). Infaunal bivalves evolved well-developed structures called **siphons** used in feeding and respiration by the animal even as it lives well beneath the sediment surface (**Figure 13.21**). Siphons also permitted infaunal bivalves to avoid predation.

By contrast, tiering among crinoids and other epifaunal suspension feeders began to decline later in the

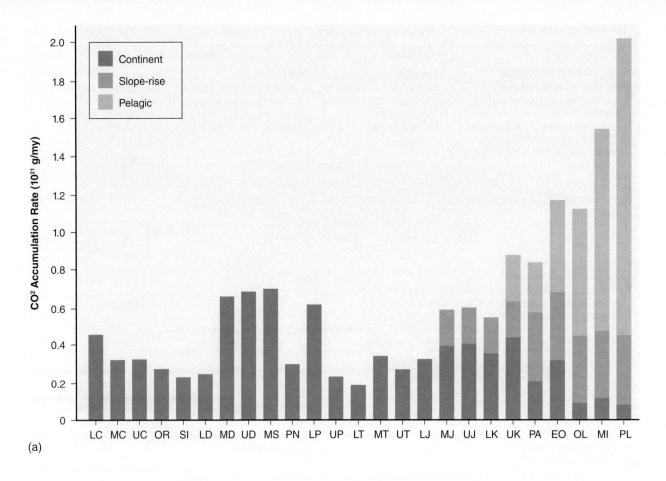

(a)

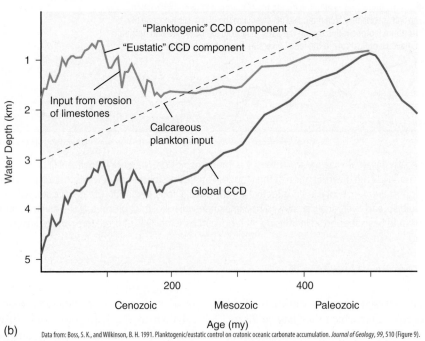

(b)

Data from: Boss, S. K., and Wilkinson, B. H. 1991. Planktogenic/eustatic control on cratonic oceanic carbonate accumulation. *Journal of Geology, 99,* 510 (Figure 9).

FIGURE 13.20 (a) Shift in the dominant sites of limestone deposition during the Phanerozoic. Limestone deposition is shown in terms of the amount of atmospheric CO_2 sequestered in limestones as $CaCO_3$. The Paleozoic was dominated by shallow-water limestones, but the increasing abundance of calcareous plankton beginning in the Mesozoic began to shift limestone deposition into the deep sea. **(b)** The shift of limestone deposition on the cratons to open-ocean oozes deepened the CCD. Note how deep-sea calcareous oozes (a) and deepening of the CCD (b) parallel one another.

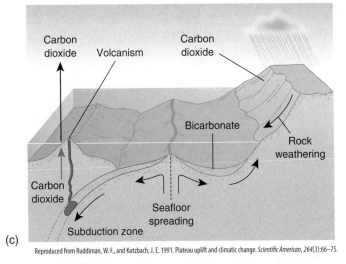

(c)

Reproduced from Ruddiman, W. F., and Kutzbach, J. E. 1991. Plateau uplift and climatic change. *Scientific American, 264*(3):66–75.

FIGURE 13.20 **(c)** The geologic cycle of carbon, in which the regulation of atmospheric CO_2 by positive and negative feedback occurs on long scales of geologic time and involves the sequestration of carbon in limestones by marine plankton.

BOX 13.1 Calcareous Plankton and the Evolution of the Geologic Cycle of Carbon

The appearance in the fossil record of calcareous plankton—coccolithophorids and planktonic foraminifera—is typically stated matter-of-factly as having occurred in the Late Triassic. Given the importance of calcareous plankton in the geologic cycle of carbon and that other planktonic taxa existed in the Paleozoic, we must ask: Why did it take so long for calcareous plankton to appear in the fossil record? The easiest and most widely accepted explanation is that calcareous plankton did not *evolve* until the Mesozoic, when they appeared in the geologic record. Let's examine this explanation in light of the method of multiple working hypotheses and multiple causation (see Chapter 1).

There are actually a few scattered reports of coccolithophorids, or coccolithophorid-like objects, from the Early-to-Middle Paleozoic. Most of these reports are considered unreliable because the fossils were not properly described (no figures of them were ever published). Other reports of these taxa from the Pennsylvanian Period are typically discounted as contamination from much younger sediments; however, some workers regard these reports as being reliable. And there are more recent reports of calcareous microfossils from the Silurian.

If coccolithophorids or some form of calcareous plankton were present during the Paleozoic, why are they virtually absent from the fossil record during this era? First, with the exception of some ophiolites, the oldest ocean floor known is Jurassic in age; thus, any calcareous oozes that might have been present in the Paleozoic were subducted and destroyed.

Another possible clue to the existence of calcareous plankton during the Paleozoic comes from the nature of the other plankton that was present. During the Early-to-Middle Paleozoic, planktonic taxa of the fossil record—acritarchs, radiolarians, graptolites, and perhaps also some conodonts—were noncalcareous and therefore resistant to dissolution. Possibly, then, the CCD was so shallow during the Early-to-Middle Paleozoic that any calcareous plankton living in the open ocean dissolved. Recall from Chapter 4 that the CCD represents the equilibrium level in the oceans between limestone deposition and dissolution. A shallow CCD might have resulted from the widespread limestone deposition during this time. Widespread limestone deposition might have resulted from negative feedback on atmospheric CO_2 by continental weathering. Still, atmospheric CO_2 levels might have remained so high,

(continues)

especially during the Early-to-Middle Paleozoic when forests had not yet become established (see Chapter 12), that they could have promoted dissolution of almost all calcareous microfossils in the epeiric seas that flooded the continents.

Yet another hypothesis is based on the nutrient preferences of modern coccolithophorids. Modern species of this group tend to live in ocean waters characterized by intermediate levels of nutrients. Perhaps, then, nutrient levels were too low to sustain large populations of coccolithophorids or other calcareous plankton until the Mesozoic. Increased weathering rates might have resulted from the warm, humid conditions of the later Mesozoic and extensive terrestrial floras. The resulting nutrients would have run off into the shallow epicontinental seaways where plankton were able to live in a widespread photic zone.

Another explanation is that oxygen levels in the oceans increased during the Mesozoic, judging from the more restricted nature of black shale during the Mesozoic as compared with the Paleozoic. Higher oxygen levels are thought to have promoted the presence of dissolved molybdenum, a biolimiting

trace element used in the photosynthetic pathways of not only coccolithophorids, but also diatoms and dinoflagellates. Calcite seas might have further promoted calcification of the marine plankton lineages that led to coccolithophorids.

All these conditions (and potential causes) might have begun to change *before* the Mesozoic, during the Late Paleozoic. The widespread orogeny, sea-level fall, and the spread of terrestrial forests during the Late Paleozoic would have increased rates of erosion and weathering (including limestones) and nutrient runoff in epeiric seas that still covered much of the continents. This is corroborated by the broad rise in strontium isotopes and the increase in tiering of the marine benthos and presumably the availability of food above (plankton) and below (dead organic matter) the seafloor during the Late Paleozoic (see Chapter 12). The possible expansion of coccolithophorids in the Late Paleozoic, along with the erosion of shallow-shelf limestones, might have provided positive feedback that caused the CCD to begin to deepen, eventually leading to the more widespread deposition and preservation of calcareous deep-sea oozes in the Mesozoic.

FIGURE 13.21 Siphons allowed bivalves to burrow deeply beneath the sediment–water interface while maintaining contact with ocean water above bottom. Oxygenated water and food are pumped downward through an incurrent siphon and waste products are eliminated through an excurrent siphon leading back to the surface.

Courtesy of Claire Fackler, NOAA National Marine Sanctuaries.

FIGURE 13.22 Cheilostome bryozoans tend to have a bushy appearance and became prominent beginning in the Mesozoic.

Mesozoic. The decline of epifaunal tiering as the Mesozoic progressed is thought to have resulted from the diversification of fish and other predators, which would have cropped **sessile** (attached) taxa such as crinoids because they were so vulnerable to predation. Today, modern sessile crinoids are commonly found in deeper waters where predation by fish is less severe; other crinoids crawl along the bottom or swim weakly. Some taxa of brachiopods also rediversified in the Mesozoic, but they never again became as prominent as they had been in the Paleozoic. Like crinoids, modern brachiopods are epifaunal and have largely retreated into deep water or caves, most likely to escape predation.

However, new taxa of the brachiopods' lophophorate cousins, the bryozoans, appeared in the Jurassic and began to diversify by the Late Cretaceous. These bryozoans belonged to the group called the **cheilostomes** (**Figure 13.22**). Instead of the upright branches produced by their Paleozoic ancestors and begging to be cropped by predators, cheilostomes lived close to the surface and encrusted hard substrates. Today, cheilostomes commonly encrust shells, wharf pilings, and the hulls of boats.

Increasing predation in the marine realm is indicated by other trends in the fossil record. Echinoderm lineages such as **echinoids** (sea urchins and sand dollars) radiated into benthic environments. Initially, echinoids lived at the sediment surface (**Figure 13.23a**), but through time they burrowed deeper into the sediment, possibly seeking refuge from predators such as crabs and lobsters (**Figure 13.23b**). Crabs and lobsters have large claws that are used for cracking calcareous shells, such as those of echinoids and gastropods. By the Cretaceous, gastropods themselves became predators, when the **neogastropods** appeared (**Figure 13.24**). The feeding structure known as the **radula**, which originally evolved in gastropods to scrape algae off rocks, was modified in neogastropods to bore through calcareous shells and in some cases to inject poison into prey. Compared with the Mesozoic, the holes produced by neogastropods are virtually absent from shells of Paleozoic bivalves and brachiopods.

Among other notable developments in the benthos during the Mesozoic was the rediversification of reefs. A new order of reef-forming corals, the **Scleractinia**, appeared along the margins of the Tethys (**Figure 13.25**). The scleractinians were both solitary and colonial and might have evolved from naked, anemone-like ancestors. Scleractinians remained the dominant reef builders of the Meso-Cenozoic, with one exception: the **rudists** (**Figure 13.26**). The rudists were a bizarre group of bivalves that grew up to approximately 1 meter in length. Rudists first appeared in the Jurassic and temporarily displaced the scleractinians from reef-building habitats during the mid-Cretaceous. Unlike the symmetric shells of most bivalves, the rudists possessed a conical or coiled shell, consisting of a conical valve capped by another valve that acted as a "lid." During the mid-Cretaceous, rudists formed reefs in warm waters all over the world that have yielded large amounts of petroleum.

The reason(s) rudists displaced scleractinians from their habitat is unknown. Because they were bivalves, rudists might have been suspension feeders. However, because they lived in clear, tropical waters of reef habitats, some workers have suggested that they might have also possessed symbiotic algae. If so, a symbiotic relationship with algae would have allowed the rudists to secrete $CaCO_3$ more rapidly that scleractinians, but no one really knows for sure. The demise of the rudists is also enigmatic. No one knows why, but rudists became extinct millions of years before the mass extinction that decimated marine and terrestrial ecosystems at the end of the Cretaceous (see below).

Among the nekton, the ammonoids survived from the Paleozoic. Species of ammonoids underwent repeated cycles of evolutionary diversification and extinction during the Mesozoic, making them excellent markers for biostratigraphic zonation. Some of them also became quite large, with the shells alone reaching about 2 meters in diameter (**Figure 13.27**)! Many of these species had highly complex suture patterns, from which it has been inferred that these species could dive to great water depths.

Ammonoids were also joined by another group of cephalopods, the **belemnoids** (see Figure 13.27). Like ammonoids, belemnoids moved by means of jet propulsion, but belemnoids were more closely related to the modern squid. The belemnoids had a reduced shell that was completely internal to the soft parts. Shell reduction might have been an adaptation for faster locomotion, either to catch prey or to prevent predation by fish and the newly evolved nektonic reptiles.

13.5.3 Marine vertebrates

Vertebrate biotas also underwent dramatic changes during the Mesozoic. Among fish, sharks survived into the Mesozoic, rediversified, and became more modern in appearance. Skates and sting rays, both relatives of sharks that feed on shellfish with flattened, pavement-like teeth, also appeared in the Mesozoic. Bony fish survived from the Paleozoic. These forms initially diversified into primitive forms with peg-like teeth that were probably used for

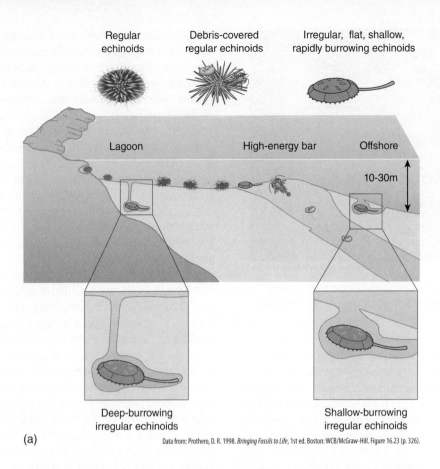

Regular echinoids

Debris-covered regular echinoids

Irregular, flat, shallow, rapidly burrowing echinoids

Lagoon

High-energy bar

Offshore

10-30m

Deep-burrowing irregular echinoids

Shallow-burrowing irregular echinoids

(a)

Data from: Prothero, D. R. 1998. *Bringing Fossils to Life*, 1st ed. Boston: WCB/McGraw-Hill. Figure 16.23 (p. 326).

(b)

Courtesy of Ronald Martin, University of Delaware.

FIGURE 13.23 Echinoid evolution. **(a)** Sea urchins and sand dollars live at the sediment surface or in cavities in rocks. However, echinoids have evolved for burrowing into shallow sediment, possibly as a refuge from predation. **(b)** Sand dollars live just beneath the sediment-water interface while sea urchins may bore into rocks using their spines.

Courtesy of G. P. Schmahl, Flower Garden Banks National Marine Sanctuary/NOAA.

FIGURE 13.24 Neogastropods, which are predatory, can inject venom into their prey.

crushing shells. By the Cretaceous, however, bony fish belonging to the **teleosts** became prominent. Today, teleosts are the dominant fish taxon in marine and freshwater.

Some of the most remarkable developments in marine vertebrates involved the reptiles. **Marine reptiles** of the Mesozoic belong to the group known as the **euryapsids**,

which are characterized by an opening in the skull high above the cheek region (see Chapter 12). The euryapsids appear to have been derived from the **diapsids** (which includes the dinosaurs and pterosaurs; see below); in the case of the euryapsids the lower opening of the diapsid skull became open at its base and was effectively lost. Among the earliest euryapsids were the **placodonts** and **nothosaurs** (**Figure 13.28** and **Figure 13.29**). The placodonts resembled turtles but lived more like walruses by digging shellfish such as clams out of the seafloor and crushing them. Nothosaurs were crocodile-like in appearance but had paddle-like limbs. Neither placodonts nor nothosaurs survived the Triassic.

Before their demise, the nothosaurs gave rise to another group of euryapsids: **plesiosaurs** (**Figure 13.30**). Based on their long jaws and sharp teeth, plesiosaurs were undoubtedly carnivorous, attaining lengths up to around 10 meters (30–40 feet). By contrast to modern marine reptiles such as turtles, which come ashore to lay eggs (**oviparous**), recent studies indicate that plesiosaurs were **viviparous**, giving birth to live young.

Joining the plesiosaurs were the **ichthyosaurs** ("fish lizards"), which appeared during the Early Triassic (**Figure 13.31**). Ichthyosaurs greatly resembled dolphins,

(a)

Courtesy of Brent Deuel/NOAA.

(b)

© John A. Anderson/Shutterstock, Inc.

(c)

Courtesy of Ronald Martin, University of Delaware.

(d)

Courtesy of Ronald Martin, University of Delaware.

FIGURE 13.25 Representatives of the Scleractinia or hexacorals, the dominant reef-frame builders of most of the Meso-Cenozoic. **(a)** Individual polyps of a colonial coral. **(b)** A type of "brain coral" in which the polyps are interrconnected by their gut cavities to share food. **(c)** Cross-section of a Pleistocene patch reef, exposed at Windley Key State Park, Florida Keys. The park was originally a quarry. The vertical holes were used for blasting, whereas the smaller holes were used for sampling. **(d)** Close-up of a coral exposed at the quarry.

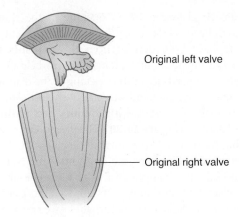

Original left valve

Original right valve

Radiolites mammilaris

(a) Data from: Clarkson, E. N. K. *Invertebrate Palaeontology and Evolution.* London: George Allen & Unwin. Figure 8.13G (p. 155).

(b) Courtesy of T. Steuber, Institut für Geologie und Mineralogie, Universität Erlangen, Nürnberg.

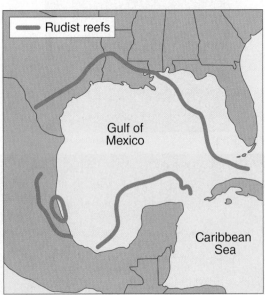

— Rudist reefs

Gulf of Mexico

Caribbean Sea

(c) Data from: Bebout, D. G., and Loucks, R. G. Lower Cretaceous reefs, south Texas. In: Scholle, P. A., Bebout, D. G., and Moore, C. H. (eds.) *Carbonate Depositional Environments. AAPG Memoir 33.* Figure 1 (p. 441). Tulsa, OK: American Association of Petroleum Geologists (AAPG).

FIGURE 13.26 Rudists. **(a)** Basic structure showing cone-like base and "lid." **(b)** Rudists in outcrop. **(c)** Distribution of rudist reefs along eastern North America and the Gulf of Mexico.

porpoises, and tunas in their stream-lined shape for rapid swimming because of convergent evolution (see Chapter 4). Fossils with hind limbs indicate that ichthyosaurs originated from terrestrial reptiles; these forms branched off from the base of the lineage that gave rise to lizards and snakes. The vestigial hind limbs of early ichthyosaurs indicate their early niche. Early ichthyosaurs had relatively narrow backbones, suggesting they moved their entire bodies in a wave-like manner, like that of lizards and snakes. Such a means of locomotion would have been most suitable along the bottom in shallow water. Later ichthyosaurs had thickened backbones, which promoted faster, more efficient back-and-forth propulsion by the tail alone. Advanced ichthyosaurs were probably full-time open-ocean dwellers. In fact, recent studies of fossil Lagerstätten indicate that, like plesiosaurs, ichthyosaurs were viviparous. Ichthyosaurs could probably also dive to 1,000 meters or more, feeding on fish and perhaps cephalopods. It has been calculated that ichthyosaurs had raised metabolic rates and could swim at speeds about equivalent to modern tuna (~1.5–2 meters per second) but less than some modern whales (3 or more meters per second). Despite their later evolutionary success, most likely as top predators in marine food chains, ichthyosaurs became extinct during the mid-Cretaceous.

Ichthyosaurs were replaced by the **mosasaurs**, sea-going lizards that reached up to about 15 meters in length (**Figure 13.32**). Mosasaurs appear to have evolved from a group of lepidosaurs that today are represented by the monitor lizard and Komodo dragon of southeast Asia and Australia. The modern Komodo dragon can reach approximately 3 meters in length, but mosasaurs were up to 10 meters long, possessing paddle-like limbs, numerous sharp conical teeth, and a long, lizard-like tail fin. Like plesiosaurs and ichthyosaurs, mosasaurs were undoubtedly carnivorous and viviparous.

CONCEPT AND REASONING CHECKS

1. Give examples of ecologic replacement in the marine realm from the Paleozoic to the Mesozoic.
2. What major evolutionary innovations occurred during the Mesozoic?
3. Describe the geologic cycle of carbon in terms of positive and negative feedback.
4. Give examples of directionality in marine ecosystems during the Mesozoic.

13.6 Diversification of the Terrestrial Biosphere

13.6.1 Plants and insects

Early Triassic land floras continued to be dominated by gymnosperms such as seed ferns, which had spread during the Late Paleozoic. Land plant communities of the Early Mesozoic also contained seedless vascular plants such

Data from: Clarkson, E. N. K. *Invertebrate Palaeontology and Evolution*. London: George Allen & Unwin. Figure 8.27 (p. 182) and Batt, R. J. 1989. Ammonite morphotype shell distribution in the western interior Greenhorn Sea and some paleoecological implications. *Palaios, 4,* 41 (Figure 12). Tulsa OK: Society for Sedimentary Geology (SEPM).

(a)

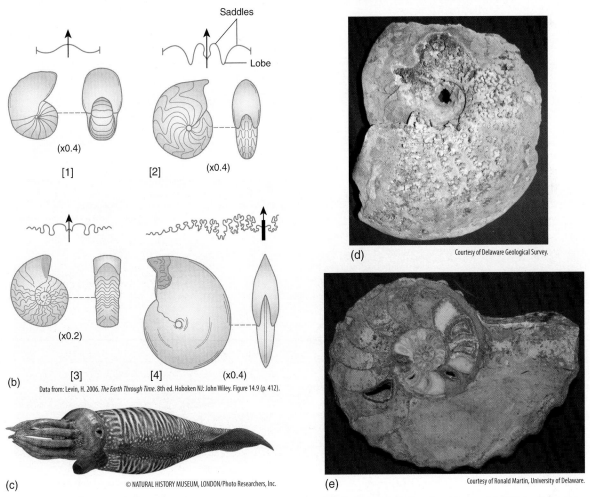

(b) Data from: Levin, H. 2006. *The Earth Through Time*. 8th ed. Hoboken NJ: John Wiley. Figure 14.9 (p. 412).

(c) © NATURAL HISTORY MUSEUM, LONDON/Photo Researchers, Inc.

(d) Courtesy of Delaware Geological Survey.

(e) Courtesy of Ronald Martin, University of Delaware.

FIGURE 13.27 Cephalopods of the Mesozoic. **(a)** Some types of ammonites and habitats. Note the distribution of the different species (indicated by shell types) according to habitat depth. **(b)** Increasing complexity of sutures in cephalopods from nautiloid [*1*] through goniatite [*2*] and ceratite [*3*] to ammonite [*4*]. **(c)** Reconstruction of swimming belemnoids, which possessed a highly reduced, internal cigar-like shell. **(d)** Internal mold of an ammonite from the Cretaceous. Note the very complex, moss-like patterns left behind by the sutures. **(e)** Cross-section of the Cretaceous ammonoid *Acanthoceras wintoni* showing spiral addition of chambers, most of which have been filled with sediment.

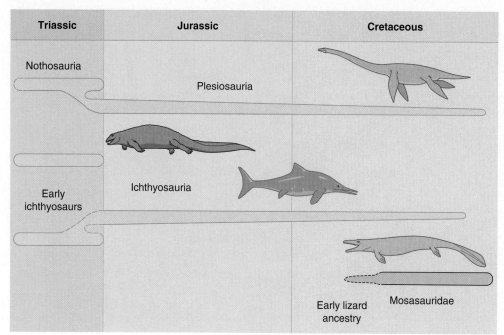

Data from: Levin, H. 2006. *The Earth Through Time*, 8th ed. Hoboken, NJ: John Wiley. Figure 14.44 (p. 435).

FIGURE 13.28 Stratigraphic distribution of marine reptiles.

Data from: Levin, H. 2006. *The Earth Through Time*, 8th ed. Hoboken, NJ: John Wiley. Figure 14.45 (p. 435).

FIGURE 13.29 Placodonts and nothosaurs.

(a) Data from: Levin, H. 2006. *The Earth Through Time*, 8th ed. Hoboken, NJ: John Wiley. Figure 14.46 (p. 436).

(b)   Courtesy of Ronald Martin, University of Delaware.

FIGURE 13.30 **(a)** Not all plesiosaurs had long necks, although this is how they are commonly portrayed. **(b)** This keichosaur from the Gyanling Formation, Xingyi, China is about 215 million years old. It lived during the Triassic Period and was a member of marine reptiles called nothosaurs that had originally evolved from land-dwelling reptiles (much like whales; see Chapter 15). Keichosaurs ranged in length from a few centimeters in length to about 6 meters. This specimen is about a dozen centimeters long.

as spore-bearing ferns (which are still with us today), horsetails, and club mosses. Unlike their ancestors of the Permo-Carboniferous, these lycopods and horsetails were mostly much smaller, creeping forms during the Mesozoic.

By the Middle Triassic, seed ferns began to be replaced by new taxa of gymnosperms: the **cycads**, **cycadeoids**, **ginkgos**, and **conifers** (**Figure 13.33**). Cycads were most prevalent during the Jurassic and have survived to the present. Today, cycads are found in the subtropics and tropics and resemble palm trees (which are angiosperms, or flowering plants); however, modern cycads are greatly reduced in diversity from their peak in the Jurassic. Cycadeoids were closely related to cycads but became extinct. Although

modern forms greatly resemble those of the Mesozoic and would therefore qualify as "living fossils," modern cycads actually evolved only about 12 million years ago. Ginkgos were prevalent during the Mesozoic Era but nearly disappeared during the Cenozoic. Modern ginkgos are yet another example of a living fossil. Long thought to be extinct, they were discovered living in China and have since been imported to other countries. Although they are gymnosperms, they resemble hardwood trees like maples. In the fall, their distinctive leaves change to a brilliant yellow color before they are shed. Although most families of modern conifers were present in the Triassic, they did not become prominent until the Cretaceous. Conifers and other gymnosperms began to slowly yield to angiosperms by the mid-to-late Cretaceous. Recently discovered Chinese Lagerstätten about 125 million years old suggest the ancestor might have been an aquatic, "weedy" herb capable of rapid recovery after grazing.

Angiosperms have several advantages over gymnosperms that probably account for their tremendous evolutionary success (**Figure 13.34**). Angiosperms rapidly develop a food supply for the seed after fertilization; the seed can therefore reach reproductive maturity quickly, sometimes within weeks. By contrast, gymnosperms can take many months to develop the food supply for their seeds. Second, unlike gymnosperms, which depend on wind-borne pollen, angiosperms also spread pollen by attracting insects with colorful flowers, nectars, and fruits. Insects serve as the primary means of cross-pollination as a result of the process of **coevolution**. Coevolution is really a kind of symbiosis. It is also a kind of positive feedback: if a new species of insect evolved, it could feed on certain plants with distinctive flowers. Similarly, new types of flowers or nectars might attract different insects. A number of insect taxa, including bees and butterflies, evolved during the Cretaceous and became specialized for feeding on certain plants, thereby promoting genetic isolation and speciation. By the Late Cretaceous, primitive representatives of modern tree taxa were present, including the sycamore, magnolia, palm, oak, walnut, and birch. Although angiosperms did not come to dominate terrestrial floras until during the Cenozoic Era, they account for more than 90% of all modern terrestrial plant species. With the spread of trees, it became increasingly likely for insects to be trapped and preserved in amber, or fossilized tree sap. Although the earliest known amber comes from the Late Carboniferous, amber did not become prevalent until the Cretaceous.

13.6.2 Vertebrates

A few **temnospondyl** ("labyrinthodont") amphibians persisted from the Paleozoic Era into the Triassic Period, after which they declined through the rest of the Mesozoic. But most notable from a modern viewpoint is the appearance of frog-like skeletons in the Triassic and definite frog skeletons in the Jurassic.

Among the reptiles, turtles appeared in the Triassic, some of which later evolved into giants several meters in length during the Cretaceous. Also present were various

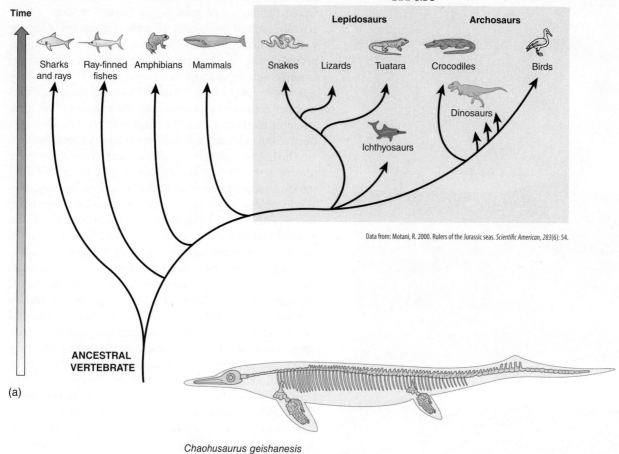

DIAPSIDS

Lepidosaurs

Archosaurs

Time

Sharks and rays

Ray-finned fishes

Amphibians

Mammals

Snakes

Lizards

Tuatara

Crocodiles

Birds

Dinosaurs

Ichthyosaurs

ANCESTRAL VERTEBRATE

(a)

Data from: Motani, R. 2000. Rulers of the Jurassic seas. *Scientific American, 283*(6): 54.

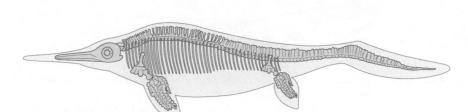

Chaohusaurus geishanesis
0.5 to 0.7 meter
Lived 245 million years ago (Early Triassic)

Mixosaurus cornalianus
0.5 to 1 meter
Lived 235 million years ago (Middle Triassic)

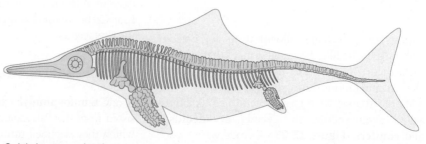

Ophthalmosaurus icenicus
3 to 4 meters
Lived 165 million to 150 million years ago (Middle to late Jurassic)

(b)

Data from: Motani, R. 2000. Rulers of the Jurassic seas. *Scientific American, 283*(6): 54.

FIGURE 13.31 Ichthyosaurs. **(a)** The evolutionary relation of ichthyosaurs to other vertebrates. **(b)** Evolution of ichthyosaur body shape and fins.

(c)

FIGURE 13.31 (c) The ichthyosaur *Grendelius* snatching a cephalopod.

types of diapsids (see Chapter 12). The diapsids include two main lineages (**Figure 13.35**). One lineage includes the snakes and lizards (lepidosaurs), which arose in the Mesozoic and appear to have been ancestral to the ichthyosaurs and plesiosaurs, as previously described. The other line includes the **archosaurs** ("ruling lizards"), which include

© Stephen J. Krasemann/Science Source.

FIGURE 13.32 The skeleton of a mosasaur, which was basically a large sea-going lizard.

crocodiles and related taxa (**Figure 13.36**). As the archosaurs diversified in the Triassic, terrestrial reptiles came to be dominated by primitive pig-like, herbivorous archosaurs called *rhynchosaurs* that might have dug up roots for food. Also present were larger carnivores and the crocodile-like *phytosaurs,* which differed from true crocodiles by having their nostrils on top of the head and not at the end of a snout. True crocodiles appeared in the Late Triassic and some became seagoing, with paddle-like appendages.

One branch of the archosaurs, the *Ornithodira*, contains the **pterosaurs**, birds, and **dinosaurs** (**Figure 13.37** and **Figure 13.38**). The term "dinosaur" was originally coined in the 19th century to include exceptionally large "pachydermous" (elephant-like) reptiles of the Mesozoic Era, but it now also includes large predators. Dinosaurs arose from small (up to ~1 meter tall), **bipedal** ancestors, meaning they walked or ran on their hind legs, much like turkeys or other flightless birds. The small (1 meter) bipedal *Eoraptor* most closely approximates the common ancestor of dinosaurs that led to their tremendous diversification (see Figure 13.38). Well-preserved dinosaur skeletons from about 230 million years ago (Late Triassic) indicate the early diversification of dinosaurs occurred rapidly (see Figure 13.37). However, the initial radiation of dinosaurs might have occurred as much as 15 million years earlier in the Triassic because even the most fundamental adaptations for herbivory and carnivory appeared fairly early. *Eoraptor,* for example, had jaws adapted for grasping prey and an elongate hand with three functional digits for grasping and raking, whereas the small (1 meter) *Pisanosaurus* possessed teeth adapted for slicing plant matter.

The initial diversification of dinosaurs into their major taxa (described shortly) is now thought to have occurred in South America. After their rapid initial diversification, dinosaurs became widespread on some continents, including North America, China, and Australia, even ranging into what are now polar regions. Despite their wide distribution, dinosaur faunas on the different continents during the Late Triassic and Jurassic remained relatively uniform. Most likely, the continents were still too close to result in substantial

(a)

Data from: Stanley, S. M. 2009. *Earth System History*, San Francisco: W. H. Freeman and Company. Figure 16.11 (p. 382).

(c)

(b)

(d)

FIGURE 13.33 Mesozoic plants. **(a)** Reconstruction of a Mesozoic forest with cycads. **(b)** Modern cycad. **(c)** Leaves of *Ginkgo*, a living fossil. **(d)** A modern *Ginkgo* tree.

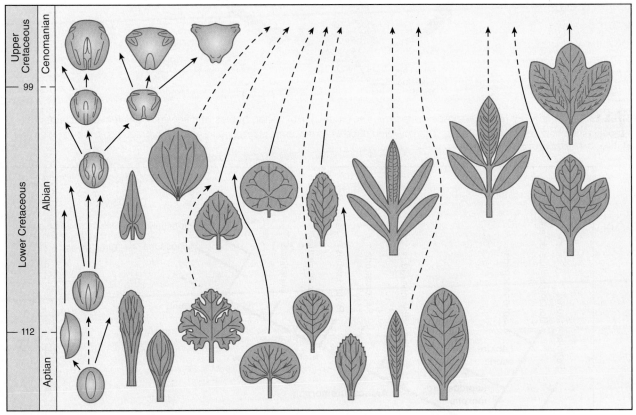

(b)

FIGURE 13.34 (a) Reconstruction of *Archaefructus sinensis*, which might be ancestral to all other angiosperms. The delicate stems, float-like structures, and the presence of fossil fish in the same deposits suggest that the plant lived in aquatic habitats, where it might have been less subject to competition from gymnosperms. **(b)** The diversification of angiosperms began during the Early Cretaceous.

(continues)

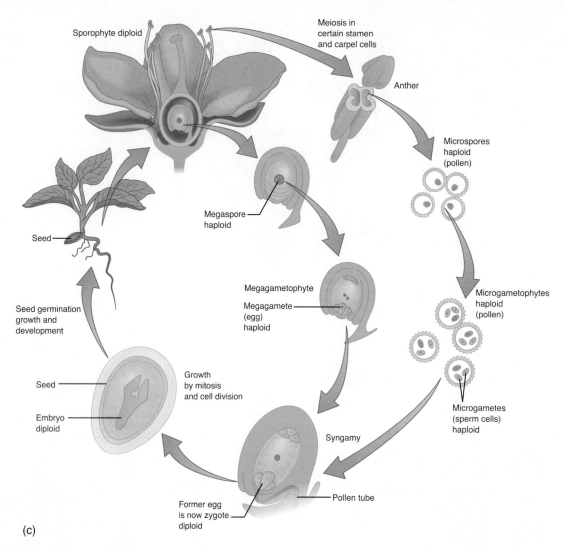

(c)

FIGURE 13.34 (Continued) **(c)** The angiosperm life cycle. The production of nectar, aromas, and flowers of different colors promotes cross-pollination between plants by insects. This coevolution between flowering plants and insects has promoted genetic variation and evolution.

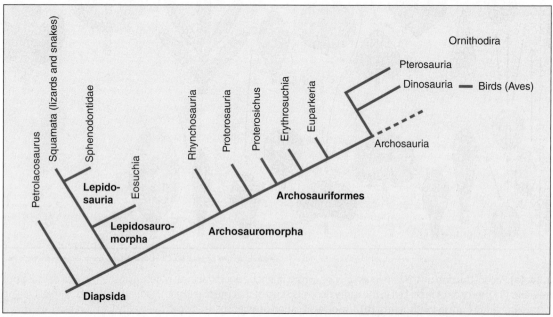

Data from: Parrish, J. M. 1997. Evolution of the archosaurs. In: Farlow, J. O., and Brett-Surman, M. K. (eds.), *The Complete Dinosaur*. Bloomington, IN: Indiana University Press. Figure 15.3 (p. 193).

FIGURE 13.35 Evolutionary relationships of diapsid reptiles.

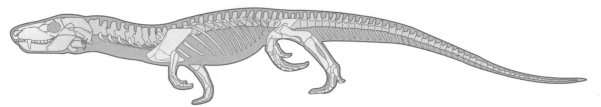

Data from: Parrish, J. M. 1997. Evolution of the archosaurs. In: Farlow, J. O., and Brett-Surman, M. K. (eds.), *The Complete Dinosaur*. Bloomington, IN: Indiana University Press. Figure 15.4 (p. 195).

FIGURE 13.36 An early archosaur, *Proterosuchus*.

geographic isolation. Dinosaurs became highly differentiated during the Cretaceous, though, suggesting geographic isolation and speciation as a result of biogeographic provinciality and regional extinction and replacement on the different continents related to the later, greater separation of the continents by oceans. Nevertheless, studies of dinosaur phylogeny indicate that polar land connections persisted between Asia and western North America into the Cretaceous. Similar studies indicate the continents in the northern and southern hemispheres also remained connected across the Tethys into the Cretaceous.

Dinosaurs by no means represent just a handful of big, dumb "lizards." They were seemingly as diverse as mammals were to become and, in a number of ways, like them in behavior (**Box 13.2**). Current estimates of dinosaur diversity during the Mesozoic range from 900 to 1,200 genera and 1,100 to 1,500 species, but with the caveat that many more species no doubt remain to be discovered. Figure 13.38 gives just an inkling of their diversity. Some of the richest dinosaur assemblages in North America date from the Jurassic Morrison Formation and the Cretaceous of Alberta, Canada, South America, and China.

By the Late Triassic, dinosaurs had diverged into two groups based on the structure of the pelvis: the **Ornithischia** and the **Saurischia** (**Figure 13.39**). The pelvis of reptiles (and also mammals) consists of three bones: ilium, ischium, and pubis. In the Ornithischia, or "bird-hipped" dinosaurs, the ischium and pubis are fused together and point backward. This pelvic arrangement allows a stronger pull of leg muscles toward the rear of the animal. (The resemblance of the ornithischian pelvis to that of birds is superficial; as we will see in a moment, other evidence indicates birds were derived from the saurischians.) All ornithischians were herbivorous, with most possessing a horny bill for cropping plants and a cheek, or holding space adjacent to the rows of teeth. In the Saurischia, or "lizard-hipped" dinosaurs, the pubis points forward and the ischium toward the rear. The saurischians included both herbivores and carnivores.

Ornithischians are quite rare as fossils during the Late Triassic, when they are represented primarily by isolated teeth. However, they became quite diverse during the Jurassic and Cretaceous. The first well-preserved skeletons occur in the Early Jurassic, by which time the major clades were already established. During the Middle Jurassic to Late Cretaceous these clades gave rise to a number of taxa. One major lineage included the familiar stegosaurs and ankylosaurs. The **stegosaurs** appeared by the Middle Jurassic and were the earliest armored dinosaurs (see Figure 13.37 and

Figure 13.40). Unlike early bipedal forms such as *Pisanosaurus* and *Lesothosaurus*, stegosaurs were quadrupedal (i.e., they walked on all four feet). Most stegosaurs had hind limbs longer than the forelimbs and long narrow skulls, whereas along their backs ran rows of bony plates that graded into pointed spines over the tail. The bony plates might have been used for body temperature control or for protection, whereas the spiked tails were probably whipped about for defense. The **ankylosaurs** appeared somewhat later than stegosaurs and were derived from them (see Figure 13.37 and Figure 13.40). Ankylosaurs took the armor of stegosaurs one step further by evolving smaller plates to create a solid shield over the neck and trunk. By the end of the Jurassic some had also evolved a club at the end of the tail.

A second major lineage of ornithischians was the **ornithopods** (see Figure 13.40). Ornithopods ranged from bipedal to quadrupedal. Although they were herbivores, some developed long forelimbs tipped with claws. Ornithopods reached their greatest diversity during the Late Cretaceous with the appearance of the **hadrosaurs**, or "duck-billed" dinosaurs, so-named because of the appearance of their snout. Hadrosaurs possessed tooth plates used for grinding the vegetation of drier upland regions. Some duck-bills also possessed large hollow crests on their skulls that suggest sexual dimorphism: males had the largest crests, females smaller, and juveniles none. Because the crests were larger in males, it is thought they served for signaling; reconstructions of the tubes yield different musical notes depending on tube size. Thus, young, females, and males might have been able to recognize one another by sound. On the other hand, if the tubes had evolved for respiration or thermoregulation, as has also been suggested, they likely would have been the same size in both sexes.

Branching off from the ornithopods was another large group, the **marginocephalians**. The marginocephalians were characterized by a bony fringe on the rear margin of the skull. Marginocephalians are represented by two main taxa known only from northern hemisphere continents and mainly from the Late Cretaceous of western North America. The **pachycephalosaurs**, or "bone-headed" dinosaurs, received their nickname from the thick bony covering over the skull that was sometimes heavily ornamented by numerous protuberances. It is thought that the covering protected the brain as animals rammed against each other head-to-head during mating rituals or territorial defense, much like modern antelope or deer. The other major taxon of marginocephalians was the **ceratopsians**. The stereotypical ceratopsians were quadrupedal and resembled the modern

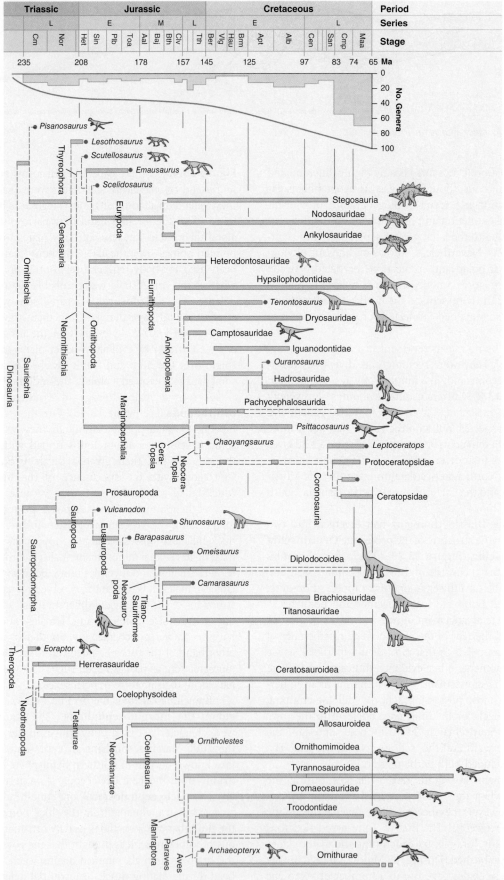

FIGURE 13.37 Evolution of dinosaurs during the Mesozoic. Blue portions of bars represent known ranges; red portions, missing ranges; and dashed bars, ranges extended by fragmentary or undescribed specimens.

Data from: Sereno, P. 1999. The evolution of dinosaurs. *Science*, *284* (June 25), Figure 1 (p. 2138).

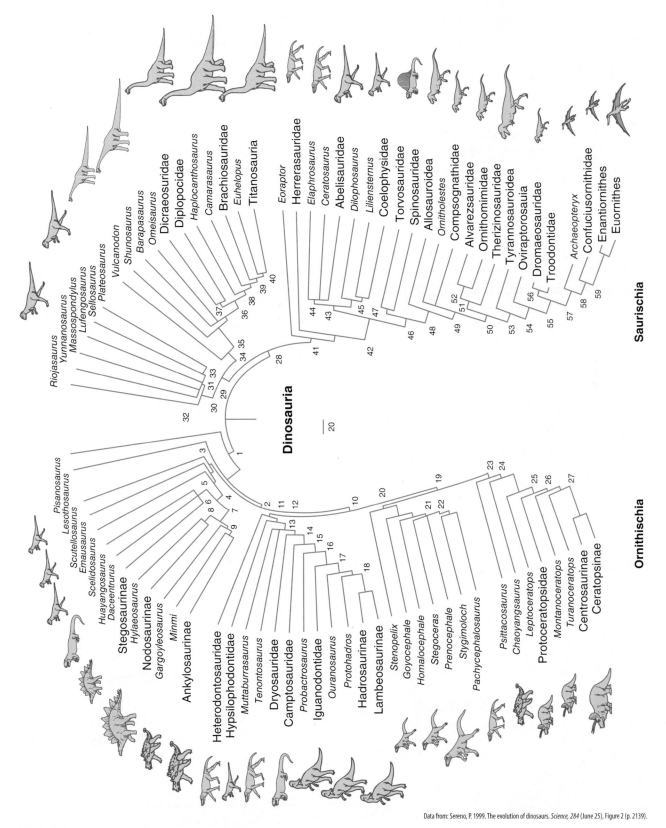

FIGURE 13.38 The tremendous diversity of ornithischians and saurischians. This figure only hints at the enormous diversity of dinosaurs, as many more taxa will undoubtedly be discovered.

Data from: Sereno, P. 1999. The evolution of dinosaurs. *Science, 284* (June 25), Figure 2 (p. 2139).

BOX 13.2 Were Dinosaurs "Warm-Blooded"?

Until a few decades ago, dinosaurs were widely regarded as big, dumb lizards. However, some workers are no longer sure dinosaurs should even be placed in the reptiles because dinosaurs appear to have exhibited a number of characteristic mammalian traits (see Chapter 12). One feature of dinosaurs that might have given them a competitive advantage over many reptiles and amphibians was that their legs were placed directly beneath their bodies rather than at right angles. This limb arrangement is like that of mammals and undoubtedly facilitated locomotion, including running (see Chapter 12). Other species seemed to have moved in herds or packs. Fossil skeletons of the carnivorous *Allosaurus* killed by a flood range in size from 2 to 15 meters. Impressions of footprints in the *Allosaurus* trackways indicate that small dinosaurs traveled in the center of herds surrounded by adults. Other species, both saurischian and ornithischian, laid and brooded eggs in nests in large colonies like those discovered in Late Cretaceous rocks of Montana. Based on the fossils recovered from nests, hatchlings were nurtured by the parents until about 1 to 2 meters in size. Hatchlings grew rapidly, which implies that they had relatively rapid metabolic rates.

By far the most contentious issue about dinosaur metabolism is whether dinosaurs were "cold-blooded" (ectothermic) or "warm-blooded" (endothermic). One argument for endothermy stems from the great size of some dinosaurs. For blood to reach the brain the heart would need to be four-chambered, as it is in warm-blooded birds and mammals. One way to test such hypotheses about dinosaur metabolism is to examine bone structure. Bones are highly vascularized, and the bone structure of modern reptiles, birds, and mammals are distinctive. Comparisons of the bone structure of modern reptiles (all ectotherms) with birds and mammals (both endotherms) and dinosaurs indicate that dinosaurs were not necessarily strictly ectothermic or endothermic; instead, this particular study found that dinosaur metabolism and growth rates ranged between that of modern reptiles and birds and mammals. Possibly, then, some dinosaurs were able to exert greater physiologic control over their body temperature, but their ability to do so also might have depended on evolutionary constraints such as body size and ecologic niche.

More recent studies have offered conflicting results. However, one of the most recent based on a large database compiled from various studies, concluded that, like the study described here, dinosaurs' body temperatures were intermediate between those of ectotherms and endotherms, or what the authors termed "mesothermy," meaning that, in general, dinosaurs were intermediate between ectotherms and endotherms with respect to body temperature regulation.

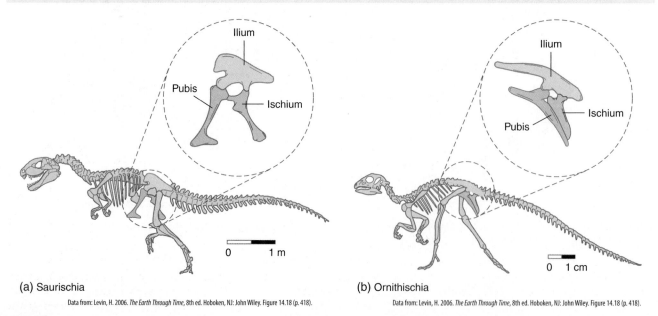

(a) Saurischia

Data from: Levin, H. 2006. *The Earth Through Time*, 8th ed. Hoboken, NJ: John Wiley. Figure 14.18 (p. 418).

(b) Ornithischia

Data from: Levin, H. 2006. *The Earth Through Time*, 8th ed. Hoboken, NJ: John Wiley. Figure 14.18 (p. 418).

FIGURE 13.39 Structure of the pelvis in saurischian and ornithischian dinosaurs.

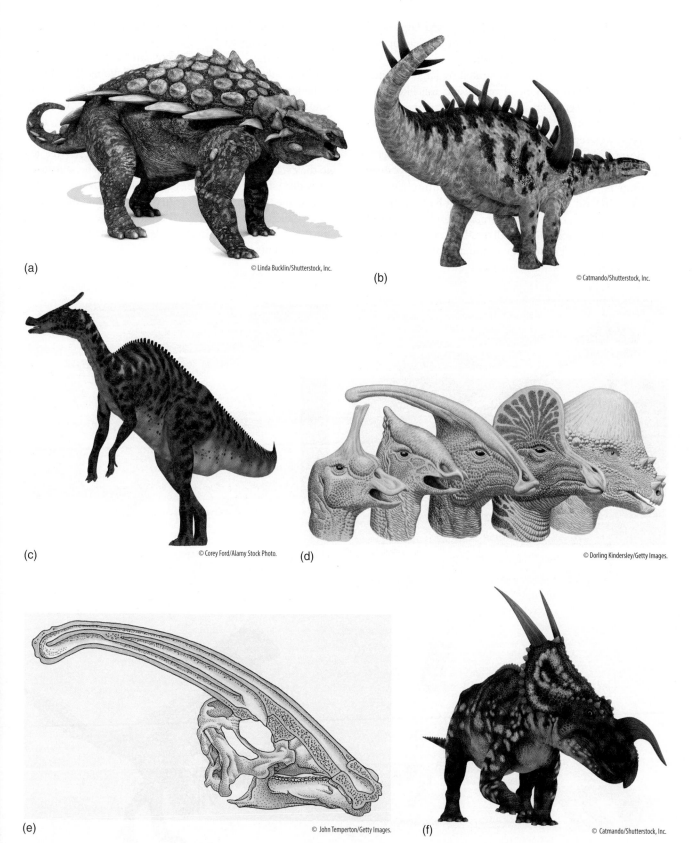

(a)

© Linda Bucklin/Shutterstock, Inc.

(b)

© Catmando/Shutterstock, Inc.

(c)

© Corey Ford/Alamy Stock Photo.

(d)

© Dorling Kindersley/Getty Images.

(e)

© John Temperton/Getty Images.

(f)

© Catmando/Shutterstock, Inc.

FIGURE 13.40 Representative ornithischian dinosaurs, all of which were herbivorous. **(a)** Ankyosaurs were characterized by numerous bony protuberances over their entire body. **(b)** Stegosaurs were characterized by one or more rows of bony plates along the back. **(c)** Hadrosaurs are also known as "duck-billed dinsoaurs" for obvious reasons. These might have been able to stand on their hind legs using their tail as a counterweight while they fed on vegetation higher in treetops. **(d)** Pachycephalosaurs exhibited a variety of skull crests, including the so-called "bone-headed" dinosaurs (far right). **(e)** Cross-section of a pachycephalosaur bony rest that might have been used in breathing or mating activities. **(f)** Ceratopsians resemble the modern rhinoceros but had a bony head shield, sometimes quite elaborate, extending back over the neck.

rhinoceros (see Figure 13.40). They possessed large skulls and one or more horns on the head or snout. Ceratopsians also had a bony fringe projecting from behind the skull over the neck that was sometimes penetrated by large openings. The fringe was filled with blood vessels and was likely used for temperature regulation.

The other main branch of the dinosaurs, the saurischians, was represented by two major lineages. Among the saurischians, the *Sauropodomorpha* represent a second group of herbivorous dinosaurs, and their early diversification is as ancient as that of the ornithischians. By the Late Triassic the sauropodomorphs had diversified into two main groups: *prosauropods* and **sauropods** (**Figure 13.41**). Prosauropods diversified rapidly with little modification to become the dominant large herbivores

on land from the Late Triassic through the Early Jurassic throughout Pangaea.

By contrast, sauropods were rare in the Early Jurassic, when the ornithischians were undergoing their major radiation. The earliest known sauropod, *Vulcanodon,* from southern Africa indicates that moderate-sized sauropods had already adopted a quadrupedal posture by the time they appeared. Despite their eventual disappearance, sauropodomorphs and prosauropods evolved features that led to the evolutionary success of the sauropods. The sauropods then diversified rapidly after the prosauropods disappeared to become the dominant large herbivores of the Middle and Late Jurassic, especially in the northern hemisphere. Sauropods later became the dominant large herbivores on southern continents in the Cretaceous.

(a) © LindaMarieB/Getty Images.

(b) © danku/Getty Images.

(c) © Dorling Kindersley/Getty Images.

(d) © Warpaint/Shutterstock, Inc.

FIGURE 13.41 Representative saurischian dinosaurs. **(a)** Sauropods were all herbivorous, in a sense representing the Mesozoic equivalent of enormous cows or giraffes. **(b)** Theropods were all carnivorous or carrion eaters. **(c)** Tyrannosaurus attacking ankylosaur. **(d)** Deinonychus had grasping forelimbs and large, retractable claws on its hind feet that were presumably used to slash open the bellies of prey as it hung onto the prey's side.

Some sauropods had short necks, but a number also evolved long necks (probably by duplication of *Hox* genes) and gigantic size (see Figure 13.41). One form, *Diplodocus*, which was common in North America, had 15 elongate neck vertebrae and a whip-like tail composed of 80 vertebrae. The vertebrae in the neck were perforated, decreasing their weight, which would have otherwise made the neck impossible to hold up. *Apatosaurus* was up to about 20 to 25 meters long and is estimated to have weighed upward of 27,000 kilograms (about 27 tons)! (*Apatosaurus* at one time included the genus *Brontosaurus*, but *Brontosaurus* has again recently been separated out as a distinct genus. *Apatosaurus* is considered to have been stouter and more robust than *Brontosaurus*.) One of the largest sauropods was up to 35 meters long and weighed in at 90,000 kilograms (about 90 tons). This form was initially named, appropriately enough, *Seismosaurus* (since reclassified into the genus *Diplodocus*). Sauropods were the dinosaur equivalents of mammals such as cows, elephants, and giraffes. Long necks presumably allowed sauropods to graze high up in the canopies of forests, especially if they used their long tails as counterweights to stand on their hind legs. Their long necks would have allowed them to sweepback and forth across treetops without having to move their enormous bodies and consume energy. Early sauropodomorphs had two sacral vertebrae linking the backbone to the rear legs in the pelvic region, whereas the pelvis of prosauropods was strengthened by three vertebrae. This allowed a jump in size to much larger forms weighing thousands of kilograms. Indeed, given the immense size of some species, many sauropods must have consumed vegetation constantly, perhaps up to a ton a day in the largest forms, as evidenced by the presence of teeth that were continuously replaced because they were worn down from chewing. Digestion was aided by "gizzard stones" (found associated with skeletons) like those in birds, which help grind up food. Despite being large, ungainly herbivores, adult sauropods were probably simply too big to be attacked by predators such as theropods (described momentarily), but the young would have been susceptible. The bone structure of sauropods, however, suggests that they grew much faster than modern reptiles (although perhaps not as fast as mammals); fast growth rates implies some level of endothermy (warm-bloodedness; see Box 13.2).

The other major group of saurischians was quite different from the sauropods. These were the **theropods** (see Figure 13.41). The earliest theropods appeared in the Triassic and were rather tall, bipedal, lightweight runners such as *Eoraptor*. Like the ornithischians, early theropods underwent rapid diversification during the Late Triassic and Early Jurassic. During this time, most theropods belonged to the *ceratosaurs*. *Coelophysis*, for example, was a bipedal, lightweight runner ranging up to about 2.5 meters long but weighing only 20 kilograms (45 pounds). Ceratosaurs persisted into the Cretaceous in Europe and in the southern hemisphere in South America, India, and the island of Madagascar off the southeast coast of Africa. By contrast, the other major groups of theropods that arose during the Late Triassic had diversified on all continents by the Middle Jurassic.

Most later theropods belong to the *allosaurids* (such as *Allosaurus*; see Figure 13.38) and the *coelurosaurids*. Of the two, the coelurosaurs were the most diverse. The coelurosaurs included the ostrich-like *ornithomimids*, the deep-snouted *oviraptors*, and the **tyrannosaurids** (see Figure 13.41). The tyrannosaurids included the well-known *Tyrannosaurus rex* ("tyrannical lizard king"), which existed during the Late Cretaceous. *T. rex* had an enormous head up to about 1.75 meters (5–6 feet) long, reached up to 6 meters in height (about 18–20 feet), and weighed about 6,000 kilograms (6 tons)! Among the theropods, only the allosaurids rivaled the tyrannosaurids in size. *Tyrannosaurus's* relatively small forelimbs remain an enigma, however, because its forelimbs could not even reach its mouth! Possibly, the forelimbs were reduced to help balance the front half of the body and its enormous head against its large tail.

Tyrannosaurus was long regarded to have been an active predator and therefore a swift runner. This implies it must have had an elevated metabolic rate (see Box 13.2). However, other workers have suggested that *Tyrannosaurus* might have also been a scavenger. Recent studies of the relationship of muscle mass to body weight and speed suggest a 6,000-kilogram *Tyrannosaurus* could not have carried sufficient muscle to run faster than about 40 km/h (~25 mph), which is substantially less than earlier estimates of up to 72 km/h (~40 mph). To sprint at 72 km/h, *T. rex* would have needed muscles in each leg equal in volume to 43% of its entire body weight! Still, *T. rex* must have been able to move fast enough to chase down certain prey.

Certainly one of the most fearsome coelurosaurs was *Deinonychus* ("terrible claw"), which belonged to the *dromaeosaurs* (see Figure 13.41). *Deinonychus* ranged up to about 3 meters long and weighed 75 kilograms. It had a long, stiff tail that served as a counterweight, allowing it to remain balanced as it ran swiftly. *Deinonychus* could probably also jump. It had grasping forelimbs and large, retractable claws on its hind feet that were presumably used to slash open at the bellies of prey as it hung onto the prey's side. In the movie, *Jurassic Park*, it was even suggested that such animals hunted in packs, like wolves! There is, however, no evidence for this.

CONCEPT AND REASONING CHECKS

1. Is coevolution a form of positive feedback?
2. Describe the basic ecologic niches of the *major* taxa of ornithischian and saurischian dinosaurs (stegosaurs, sauropods, tyrannosaurids, etc.) in terms of the modern mammals of the African Serengeti.
3. Were dinosaurs all definitely warm-blooded? What does the study described in Box 13.2 tell you about the corroboration or rejection of a hypothesis (see Chapter 1)? Is corroboration or rejection of a hypothesis always clear-cut?
4. Dinosaurs are considered by some workers to be a taxon wholly separate from reptiles and to have a taxonomic status equal to reptiles or mammals. Why do you suppose this is?

13.6.3 Evolution of flight

Pterosaurs

S Reptiles not only invaded the land and sea but also the air in the form of pterosaurs. Pterosaur wing spans ranged from about 20 cm (10 inches) to well more than 10 meters (30 feet) by the end of the Cretaceous!

Despite their bird-like appearance, a number of traits betray pterosaurs' reptilian origins and their possible niches. Like reptiles, many pterosaurs had tails and teeth, but like birds their bones were hollow for increased vascularization and decreased weight. Also, unlike birds, pterosaurs did not possess feathers. The wings consisted of skin stretched between the body and the highly elongated fourth digit of the original reptilian claw (**Figure 13.42**). The remaining claws were present at the front of the wing. Also, rather than being active fliers like birds, many pterosaurs, especially the large forms, might have soared on wind currents. A gliding mode of flight is suggested by the pterosaurs lack of a large breastbone, which is used for the attachment of flight muscles in actively flying birds. Instead, pterosaurs, especially the largest forms, might have glided along on "thermals," similar to the behavior of modern condors and hang-glider enthusiasts who jump off cliffs.

Based on their large eyes and stomach contents (which often consisted of fish spines and scales), pterosaurs were probably predatory. Some might have either skimmed the sea surface for fish, like modern pelicans or sea gulls, diving into the water after prey. However, the apparent lack of well-developed flight muscles would seemingly have made it difficult for them to exit the water. Larger pterosaurs might have been scavengers, feeding on carrion.

Birds

T Despite their extinction at the end of the Cretaceous, the dinosaurs appear to still be with us in the form of birds. The first definite birds appear in the fossil record by the Late Jurassic and are represented by **Archaeopteryx** ("ancient wing"), which was preserved in the Lagerstätte, the Solnhofen Limestone, of Germany (**Figure 13.43**). The Solnhofen is a fine-grained limestone that appears to have been deposited in a quiet, back-reef setting, enhancing fossil preservation potential.

Archaeopteryx has long been considered a classic missing link between dinosaurs and birds (see Chapter 5). It exhibits both traits of birds (impressions of feathers in the limestone) and reptiles (claws on the wings, teeth, long tail; see Figure 13.43). Moreover, if it were not for the impressions of feathers found with the skeletons, *Archaeopteryx* would likely have been classified as a dinosaur! The hypothesis that birds are

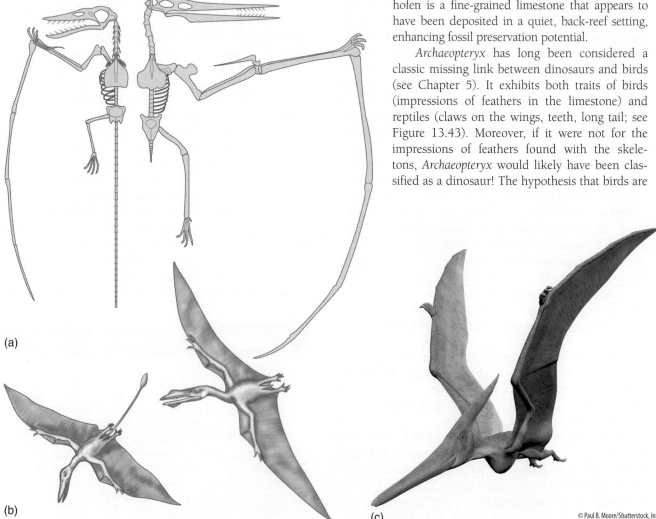

(a)

(b)

(c)

© Paul B. Moore/Shutterstock, Inc.

FIGURE 13.42 Comparison between rhamphorhynchoids, a lineage of pterosaurs (left in **a** and **b**) and pterodactyloids, a lineage of pterodactyls (right in a and b). Rhamphorhynchoids were tailed with wing spans of 0.3 to 2.1 meters. Pterodactyloids had very short tails, and wing spans from 15 centimeters to as much as 12 meters. **(c)** Reconstruction of *Pteranodon*, perhaps the largest Late Cretaceous pterodactyl.

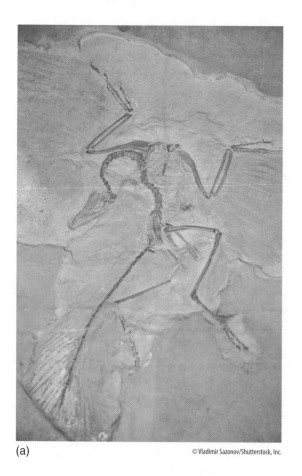

(a)

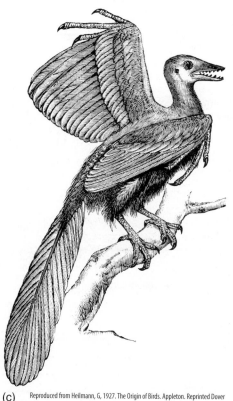

(c)

(b)

FIGURE 13.43 *Archaeopteryx*. (**a** & **b**) Fossil skeleton and reconstruction. Note the overall reptilian appearance of the skeleton. If it were not for the impressions of feathers, *Archaeopteryx* would be classified as a bipedal reptile like that shown in part (**c**), although *Archaeopteryx* would have been small, about the size of a chicken.

(continues)

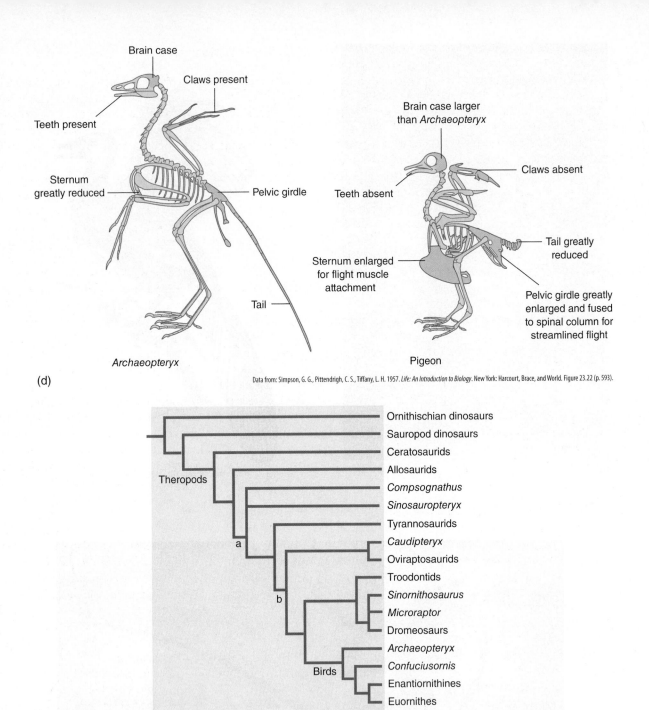

Brain case

Claws present

Teeth present

Sternum
greatly reduced

Pelvic girdle

Tail

Archaeopteryx

Brain case larger
than *Archaeopteryx*

Claws absent

Teeth absent

Tail greatly
reduced

Sternum enlarged
for flight muscle
attachment

Pelvic girdle greatly
enlarged and fused
to spinal column for
streamlined flight

Pigeon

(d)

Data from: Simpson, G. G., Pittendrigh, C. S., Tiffany, L. H. 1957. *Life: An Introduction to Biology*. New York: Harcourt, Brace, and World. Figure 23.22 (p. 593).

Ornithischian dinosaurs

Sauropod dinosaurs

Ceratosaurids

Allosaurids

Compsognathus

Sinosauropteryx

Theropods

Tyrannosaurids

a

Caudipteryx

Oviraptosaurids

Troodontids

Sinornithosaurus

b

Microraptor

Dromeosaurs

Archaeopteryx

Confuciusornis

Birds

Enantiornithines

Euornithes

(e)

FIGURE 13.43 (Continued) **(d)** Reconstruction of *Archaeopteryx* compared to that of a pigeon. Note the small size of the breastbone in *Archaeopteryx*, which suggests that it had weak powers of flight. **(e)** A cladogram which traces the relationship of birds through other taxa back to the Ornithischia and Saurischia. Branch (a) indicates the presence of simple feathers; branch (b) indicates the presence of more complex (basically modern) feathers. As the name implies, Euornithes ("true birds") refers to the most recent common ancestor of a clade of flying dinosaurs which are represented only by true birds.

derived from dinosaurs is not new and was first suggested in the 1860s by the American vertebrate paleontologist, O. C. Marsh, and the British paleontologist and close colleague of Charles Darwin, Thomas Huxley. Huxley was the first to note the similarity of *Archaeopteryx* and the small bipedal dinosaur *Compsognathus*, which grew to about the size of a chicken. In fact, two of the original seven skeletons of *Archaeopteryx* were initially identified as the bipedal theropod, *Compsognathus*, which grew to about the size of a chicken (see Figure 13.43). However, in the fossil record *Compsognathus* is contemporaneous with, not older than, *Archaeopteryx*, so *Compsognathus* cannot be ancestral to birds. The current consensus among paleontologists, corroborated by recent studies of small theropod dinosaurs that share avian features with *Archaeopteryx*, is that *Archaeopteryx* did indeed evolve from small bipedal theropods, most likely coelurosaurs (see Figure 13.35) and that it was in fact not a true bird but a flying dinosaur!

Bipedalism was nevertheless important to the origin of birds because it freed the forelimbs for the development of wings. Upright posture in bipedal ancestors necessitated rearrangements of the foot, ankle, knee, and hip for stability and the bearing of weight by the skeleton when not flying. Recent studies of pigment-bearing structures (called melanosomes) of a form transitional between dinosaurs and birds *Anchiornis huxleyi* (named after Thomas Huxley), indicate that they could be quite colorful. Birds evolved an ornithischian pelvis with the loss of the tail; this achieved greater stability during flight by allowing the legs to be carried horizontally nearer the body's center of gravity. By the Cretaceous most birds had lost their teeth. *Confuciusornis* from the lower Cretaceous of China, for example, retained the large claws of *Archaeopteryx* but had already lost its teeth, indicating that this characteristic appeared early during avian evolution (see Figure 13.35). Enantiornithines represent a lineage exhibiting many features of modern birds—claws allowing them to perch, and well-developed flight—but they did not survive the end-Cretaceous mass extinction. Other lineages became active fliers, in which feathers on the hind legs also served for flight (see this chapter's frontispiece), whereas others evolved into flightless diving birds or wading shorebirds like egrets. In fact, the oldest bird fossil with the greatest resemblance to modern birds is a loon-like shorebird from the Cretaceous of China.

But how and why did feathers, and then wings, evolve? After all, half a wing, for example, is just as useless as, say, half an eye (see Chapter 5). There are several hypotheses about the origin of flight in birds. Although structures interpreted as feathers have been found before the Jurassic, the interpretation of these structures as feathers has been disputed, and no other earlier bird skeletons have been discovered. However, feathers have been found to be associated with other nonavian dinosaurs, suggesting feathers originally evolved for body temperature regulation, the brooding of eggs, or perhaps mating displays rather than flight. *Archaeopteryx* also had a very small breastbone, suggesting that it was a poor flier, at best. Some paleontologists also believe the structure of the wing bones in *Archaeopteryx* indicates the wings could not be folded or rotated; rotation of the wing is necessary for the recovery stroke during flapping. Initially, then, wings might have been used to glide between trees (refer to this chapter's frontispiece). Another possibility is that flapping of wings helped smaller bipedal dinosaurs maintain stability or gain speed while running over the ground. Another intriguing hypothesis is that wings were used to run up steeply inclined surfaces, as in modern partridges.

CONCEPT AND REASONING CHECKS

1. How were pterosaurs like birds? How were they different?
2. How did *Archaeopteryx* resemble modern reptiles? How did it differ?
3. What does *Archaeopteryx* represent in terms of evolution?
4. The Solnhofen Limestone is an example of what kind of preservation?

13.6.4 Mammals

True mammals appeared by the end of the Triassic. Mammals evolved from the therapsids, which appeared during the Late Paleozoic and barely survived the Late Permian extinctions. Therapsids might have possessed several mammalian traits: endothermy and hair, legs positioned underneath the body, and possibly a four-chambered heart.

After their appearance mammals evolved a number of distinctive features during the Mesozoic that would appear to have supported higher metabolic rates. During the Mesozoic the distinctive mammalian jaw, which is composed of a single bone called the **dentary**, evolved, as did the mammalian inner ear (**Figure 13.44**). The single dentary bone is stronger than a jaw bone composed of several components (as in the reptiles), giving it greater strength for chewing. As the teeth became more specialized for stabbing, tearing, slicing, and chewing in mammals, a greater range of jaw motions developed. By contrast, in cynodonts, which were the last major therapsid group to appear in the geologic record (Late Permian), the jaw still consisted of several bones. In addition, mammals possess a secondary palate, which separates the nasal passages from the mouth, allowing mammals to eat and breathe at the same time (see Figure 13.44). The brain also enlarged through time, perhaps for greater nervous control and coordination of the jaw muscles. However, the lack of teeth in specimens of young mammal fossils suggests teeth were unnecessary in young mammals. Perhaps, then, young fed on milk supplied by the mother after birth, another mammalian trait.

The earliest mammals were carnivorous, about the size of mice or rats, which they generally resembled. However, recent studies indicate that by the Early Cretaceous, some early mammals were up to a meter long (about one-and-a-half times the size of an opossum) and might have resembled a Tasmanian devil. These forms preyed on baby and perhaps even larger dinosaurs based on the recent discovery of ceratopsian remains within the stomach region of fossil mammals. Nevertheless, the mammals do not appear to have been a serious threat to the dominance of the dinosaurs, most likely because of competition and predation by dinosaurs.

By the Late Cretaceous mammals were dominated by insect-eating **insectivores**, which might not be entirely coincidental given the coevolution of angiosperms and insects. Modern insectivores include shrews, moles, and hedgehogs. It is from these taxa that mammals appear to have diversified after the final demise of the dinosaurs in the Late Cretaceous. We examine mammalian evolution in greater detail in Chapters 14 and 15.

CONCEPT AND REASONING CHECKS

1. How do mammals and modern reptiles differ?
2. What is the advantage of (a) a four-chambered heart and (b) legs placed beneath the body?
3. The mammals are placed in a larger group with reptiles and birds called the amniotes. Why? (See Chapter 12.)

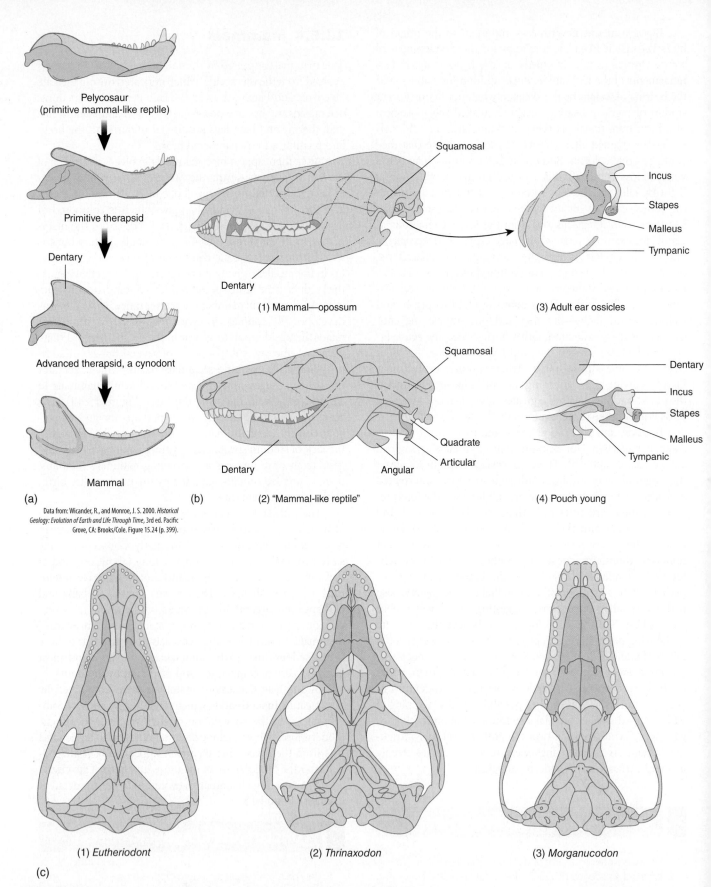

(a)

Pelycosaur
(primitive mammal-like reptile)

Primitive therapsid

Dentary

Advanced therapsid, a cynodont

Mammal

Data from: Wicander, R., and Monroe, J. S. 2000. *Historical Geology: Evolution of Earth and Life Through Time*, 3rd ed. Pacific Grove, CA: Brooks/Cole. Figure 15.24 (p. 399).

Squamosal

Dentary

(1) Mammal—opossum

Squamosal

Quadrate

Articular

Dentary

Angular

(2) "Mammal-like reptile"

(b)

Incus

Stapes

Malleus

Tympanic

(3) Adult ear ossicles

Dentary

Incus

Stapes

Malleus

Tympanic

(4) Pouch young

(1) *Eutheriodont*

(2) *Thrinaxodon*

(3) *Morganucodon*

(c)

FIGURE 13.44 Mammalian evolution. **(a)** Evolution of the mammalian jaw. **(b)** Evolution of the inner ear. Mammalian evolution. **(c)** Enlargement of the secondary palate allowed mammals to eat and breathe at the same time.

13.7 Extinction

U A number of minor extinctions took place during the Mesozoic. These minor extinctions, especially those of the Cretaceous, might be related to changes in ocean circulation, climate, and oceanic anoxia such as the oceanic anoxic events described earlier in the chapter. Significantly, two of the Big Five mass extinctions took place during the Mesozoic Era.

13.7.1 Late Triassic extinctions

Two of the largest mass extinctions of the Phanerozoic occurred in the Mesozoic during the Late Triassic and Cretaceous periods. The Late Triassic extinction affected many diverse taxa: the therapsids nearly went extinct, conodonts and placodonts died out, and bivalves, gastropods, ammonoid and nautiloid cephalopods, plesiosaurs, and ichthyosaurs all nearly disappeared.

The Late Triassic extinction also appears to have had a profound effect on dinosaur evolution. The rapid rise (geologically speaking) of dinosaurs cannot be explained simply by the increasing specialization for different niches or outcompeting contemporaneous tetrapods such as synapsids and other archosaurs. The most plausible hypothesis for the dinosaurs' rapid rise to dominance might lie in mass extinction in the Late Triassic. Extinction of other taxa would have opened up habitats and niches, promoting the rapid diversification of dinosaurs.

Despite its importance, however, the mechanism(s) of extinction at the end of the Triassic is poorly understood. Iridium layers have been reported from several localities at this time, and therefore the obvious candidate is an impact. Recent radiometric dates point to the **Manicouagan crater** in Quebec (Canada), once thought to be too old to have caused the Late Triassic extinction, as being suspect (**Figure 13.45**). Also, carbon isotopes first increased and

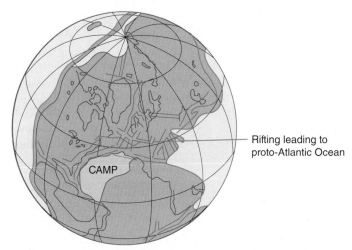

Data from: Hesselbo, S. P., Robinson, S. A., Surlyk, F., and Piasecki, S. 2002. Terrestrial and marine extinction at the Triassic-Jurassic boundary synchronized with major carbon-cycle perturbation: A link to initiation of massive volcanism? *Geology, 30,* 251 (Figure 1). Boulder, CO: Geological Society of America.

FIGURE 13.46 The Central Atlantic Magmatic Province formed as the Atlantic Ocean opened during the rifting of Pangaea.

then dropped off near the end of the Triassic, suggesting a change in ocean circulation or productivity.

Recently, earth scientists have become interested in volcanism as a causal agent of extinction at the end of the Triassic. During the rifting of Pangaea there were massive outpourings of flood basalts in the incipient Atlantic basin referred to as the **Central Atlantic Magmatic Province** (**Figure 13.46**). The massive outpourings of flood basalts resemble those of the Siberian Traps associated with the Late Permian extinctions (see Chapter 12); a recent study dating the eruptions indicates that they occurred in four pulses spanning approximately 600,000 years. Because volcanoes are fed from the mantle, volcanism accounts for the iridium layers at the Triassic–Jurassic boundary.

Increased volcanism would have been expected to increase CO_2 levels in the atmosphere. Decreased carbon isotope values (as detected) would seem to confirm this supposition (but these could have resulted from impact, as well). Comparison of the density of stomata on leaves of fossil cycads and ginkgos with their modern representatives also appears to confirm increased CO_2. Stomata density of modern representatives of these taxa decreases with increasing CO_2; stomatal density also decreases across the Triassic–Jurassic boundary, suggesting Earth warmed by several degrees Celsius. In contrast, recent calculations based on stable isotopes in ancient soils indicate that atmospheric CO_2 levels did not increase substantially during the Late Triassic, thereby seemingly excluding rapid global warming like that which apparently occurred at the end of the Permian. However, even more recent studies suggest massive injection of CO_2 into the atmosphere by the oxidation of methane.

13.7.2 Late Cretaceous extinctions

In contrast to the Late Triassic extinction, the Late Cretaceous extinction finally snuffed out the dinosaurs, which, according to a recent study, may have been in decline for

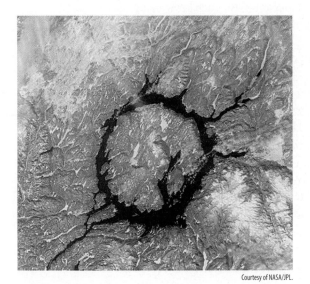

Courtesy of NASA/JPL.

FIGURE 13.45 The Manicouagan crater of Quebec, Canada, which is regarded as a candidate for the Late Triassic extinctions. The crater is now filled by a lake.

tens-of-millions of years before the end of the Cretaceous period (see Further Reading). In any case, the demise of the dinosaurs gave the mammals the opportunity to diversify in the Cenozoic (see Chapter 14). There is now a general consensus that one or more meteors hit Earth at the end of the Cretaceous, and that impact was the primary cause of the final demise of the dinosaurs and other taxa on land and in the sea. The sudden injection of dust high into Earth's atmosphere is thought to have caused rapid global cooling, which many taxa on land and in the sea could not tolerate. Indeed, the Late Cretaceous extinctions are associated with one or more iridium layers that presumably settled out with the dust after the impact(s). More important, well-developed shocked mineral assemblages and microtektites occur at the end of the Cretaceous Period. The intensity of metamorphism of the shocked minerals could only have occurred by a sudden increase in intense pressures and temperatures like those generated during an impact (see Chapter 3 and **Box 13.3**).

BOX 13.3 Late Cretaceous Extinctions and the Scientific Method

Most mass extinctions appear to be somehow related to the tectonic cycle. However, the Late Cretaceous extinctions involved—and may well have resulted primarily from—an impact, as indicated by the occurrence of shocked mineral assemblages (**Box Figure 13.3A**). Whereas the impact hypothesis certainly arouses our imaginations, how the hypothesis came to be widely accepted by the scientific community is also a prime example of how scientific investigation works (see Chapter 1). Moreover, the corroboration of the hypothesis paved the way for the acceptance of extraterrestrial impacts as important—even extraordinary—agents of geologic, climatic, and biospheric change. It also radically altered—once and for all—earth scientists unquestioned acceptance of Lyell's dogma of slow, gradual change to a broader doctrine that recognized that Earth systems processes vary through time and in rate (see Chapter 1).

Initially, a dark sedimentary layer containing a high concentration of the element iridium was found near Gubbio, Italy, almost by accident (see Chapter 1). The iridium layer also occurred at the time of the mass extinction at the end of the Cretaceous Period about 65 million years ago, during which dinosaurs and many other organisms became extinct. Iridium is not normally found in rocks of Earth's crust and could have come from only two sources: volcanoes fed by the mantle, which is enriched in iridium, or from an extraterrestrial body. The hypothesis was that the iridium layer was generated by a meteor enriched in iridium. The impact presumably threw a gigantic dust cloud into Earth's stratosphere that suddenly cooled the planet, causing extinction; the blockage of sunlight also shut down marine photosynthesis causing a **Strangelove Ocean** (named after the character of the same name in a famous movie) in which there was a sudden, strong shift in carbon isotope ratios to much lower values (see Chapter 9; **Box Figure 13.3B**).

A prediction made from the hypothesis was that if an impact were responsible for the Late Cretaceous extinctions, an iridium layer should be found all over the world in rocks of exactly the same age. Scientists tested the hypothesis by exploring for the iridium layer all over the world, on land and in deep-sea cores, where the rocks were of the right age. The hypothesis was corroborated: the Late Cretaceous iridium layer is now known not only from Gubbio, Italy, but also from Stevns Klint (Steven's Cliff) near Copenhagen, Denmark; El Kef, Tunisia, in north Africa; and El Mimbral, Mexico (to name only a few of the more famous and intensively studied localities), as well as in many deep-sea cores (see Box Figure 13.3B).

© Gl0ck/ShutterStock, Inc.

BOX FIGURE 13.3A An artist's visualization of the impact of an asteroid with Earth.

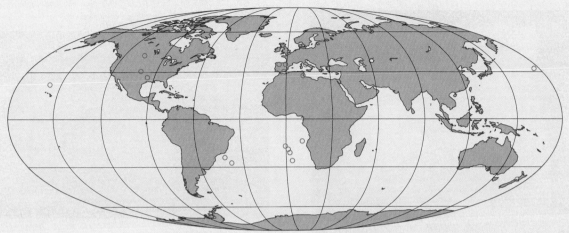

Data from: Levin, H. 2006. *The Earth Through Time*, 8th ed. Hoboken NJ: John Wiley. Figure 14.64 (pg. 445).

BOX FIGURE 13.3B Sites at which the Late Cretaceous iridium layer has been found, including Gubbio (Italy), Stevns Klint (Denmark), and El Kef (Tunisia). The iridium layer has also been identified in a number of deep-sea cores.

But if a meteor hit Earth, the question remained: where was the crater? Several candidate craters were already known, such as the Manson crater buried beneath the surface of Iowa, but they were not quite the right age. Then, like the discovery of the iridium layer at Gubbio, serendipity occurred. Many years before the discovery of the iridium layer in Italy, petroleum companies had explored in the Yucatan (Mexico) peninsula for oil because there were geologic structures beneath the surface that might hold petroleum. The oil companies even drilled a few exploratory wells, but they did not find promising economic "shows" and plugged and abandoned the wells, which were then forgotten. Years later, it was realized that the structure in the Yucatan might be an impact crater, and the original well samples were located in storage.

The *Chicxulub Crater*, as it is now known, is about 150 to 200 kilometers in diameter and is buried in the subsurface of the Yucatan Peninsula, Mexico (**Box Figure 13.3C** and **Box Figure 13.3D**). The size of the crater has been used to infer the events immediately before and after the impact.[1] The meteor is estimated to have been about 10 kilometers across, or about a kilometer higher than Mt. Everest, and approached Earth at a speed of about 30 kilometers/sec, or about 100 times the speed of sound. This is equivalent to 1,000 times the velocity of a speeding car or about 150 times that of a jet airliner! Depending on the exact velocity and angle of approach of the

meteor, it would have taken only 1 or 2 seconds for it to penetrate Earth's atmosphere.

Because of its speed, the air in front of the impactor would have been tremendously compressed, producing an enormous sonic boom, as the air was heated to a temperature of four to five times that of the sun. This produced a blinding flash of light during the 1 or 2 seconds the object penetrated the atmosphere. Animals living just over the horizon from the impact saw a flash of light, quickly followed by violent seismic (earthquake) waves. Then, the sky would have changed from a faint glow to more

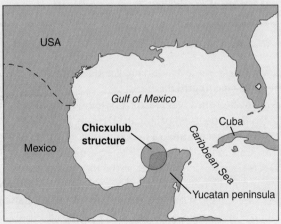

Data from: Levin, H. 2006. *The Earth Through Time*, 8th ed. Hoboken, NJ: John Wiley. Figure 14.66 (p. 445).

BOX FIGURE 13.3C Location of the Chicxulub crater on the Yucatan peninsula, Mexico. Based on the rock record, the meteor hit in a shallow seaway that existed here at the end of the Cretaceous.

(continues)

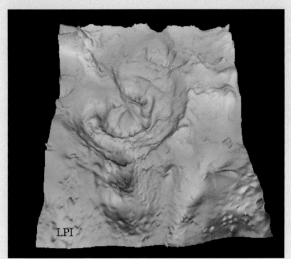

Courtesy of Virgil L. Sharpton, Lunar Planetary Institute.

BOX FIGURE 13.3D Magnetic anomaly map of crater.

Courtesy of Ron Martin.

BOX FIGURE 13.3E A possible tsunami deposit generated by the Late Cretaceous impact in the Yucatan (Mexico). This site is located along the north coast of Cuba. An alternative explanation is that this section represents a submarine slump.

intensely red, eventually becoming incandescent like a light bulb, as forests were immolated and impact ejecta rained down.

The impact is estimated to have released an amount of energy equivalent to 100 million megatons of TNT (1 megaton equals 1 million tons), or one billion times the energy of the first atomic bombs and about 10,000 times that of the entire world's nuclear arsenal at the peak of the Cold War. The fireball of the impact was so great it blew a hole through the atmosphere into outer space!

At the time of the impact the Yucatan region was a shallow seaway. Thus, the impact is thought to have generated one or more tsunamis, perhaps up to 1 kilometer high, that roared across Mexico, the Caribbean, and into the interior of North America, ripping up the sea bottom and redepositing layers of poorly sorted sedimentary debris ranging from mud to large blocks. Jumbled, poorly sorted deposits of rocks and sediments—like those that would have been deposited by a tsunami—have in fact been found from this time in various places (**Box Figure 13.3E**).

[1]Much of the description of the events immediately before and after the impact is derived from Alvarez, W. 1997. *T. rex and the crater of doom*. Princeton, NJ: Princeton University Press.

Nevertheless, sea level was falling and possibly Earth was cooling somewhat toward the end of the Cretaceous, and both factors might have already stressed ecosystems before their final demise. Huge outpourings of lava, the **Deccan Traps** of India, also occurred about the time of the extinctions (**Figure 13.47**; the term "trap" refers to the step-like appearance of the volcanics). The Deccan Traps formed as the Indian subcontinent rifted away from Gondwanaland and moved over the **Réunion hotspot** (see Chapter 2). Massive volcanism is thought to have pumped enormous amounts of CO_2, dust, and iridium into the atmosphere, causing climate change and perhaps making the biosphere more susceptible to impact. Thus, an impact might have been the last straw that broke the back of the biosphere and caused the final collapse of ecosystems at the end of the Cretaceous. Other workers, however, point out the Deccan Traps represent relatively gentle fissure eruptions,

which tend to flow out over the Earth's surface rather than exploding violently into the atmosphere. Fissure eruptions are therefore presumably incapable of pumping enormous amounts of iridium and dust into the atmosphere; explosive volcanism is required to pump material high enough into the atmosphere for it to spread over the planet.

CONCEPT AND REASONING CHECKS

1. Volcanism has been implicated in several mass extinctions. Which ones?
2. Diagram the test of a meteor impact as the causal agent of the Late Cretaceous mass extinction in terms of the scientific method diagrammed in Chapter 1 (see Box 13.3).

(a)

© Hemera/Thinkstock.

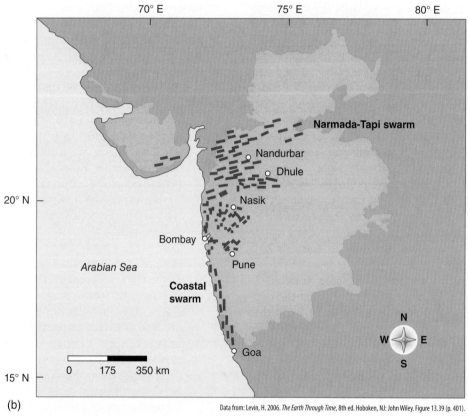

(b)

Data from: Levin, H. 2006. *The Earth Through Time*, 8th ed. Hoboken, NJ: John Wiley. Figure 13.39 (p. 401).

FIGURE 13.47 (a) The Deccan Traps of India. **(b)** Location of the Deccan Traps.

SUMMARY

- The Mesozoic Era represents "middle Earth": the mid-point of Earth's transition from the ancient world of the Paleozoic to the more modern world of the Cenozoic.
- As Pangaea rifted apart, continents began to slowly move toward their modern positions and the modern configuration of the oceans was initiated. During the Late Triassic, the Tethys Seaway began to open between Laurasia and Gondwana, along with what is

now southern Europe and North Africa, and gradually widened and deepened from east to west. North America also began to rift from South America during the Late Triassic and Early Jurassic as the central Atlantic Ocean began to open, whereas the Tethys continued to deepen and spread westward, separating North and South America. The Tethys was eventually transformed into a circumequatorial seaway.

- In the southern hemisphere South America and Africa remained sutured. So, too, did Australia and Antarctica as they drifted away from the rest of Gondwana, whereas India began to move northward. South America began to rift from Africa during the Late Jurassic, as the south Atlantic Ocean opened. At the same time the eastern portion of the Tethys began to close, leaving behind the ancestral Mediterranean to the west. By the Late Cretaceous Australia and Antarctica separated as they began to move toward their present positions, whereas India continued to move north. The Atlantic Ocean also continued to widen. South America and Africa continued to move apart, as eastern Greenland began to rift away from northern Europe.

- The rifting of Pangaea resulted in more-or-less long-term high sea-level and warm conditions. Like the Early-to-Middle Paleozoic, seafloor spreading centers appear to have displaced water from the ocean basins up onto the continents to produce epicontinental seas. In North America, these seaways at times extended from what is now the Gulf of Mexico through the western United States to the Arctic Circle. Increased rifting, volcanism, and seafloor spreading associated with the breakup of Pangaea indicate that large amounts of CO_2 were pumped into the atmosphere. However, atmospheric CO_2 levels were probably lower during the Mesozoic than those of the Early-to-Middle Paleozoic because of photosynthesis by widespread terrestrial forests. Besides evaporites, both tropical marine faunas and floras (coals) extended to relatively high latitudes in both hemispheres; there is little evidence for permanent polar ice caps during the Mesozoic.

- Thus, the oceans were probably warm and their circulation relatively sluggish, as suggested by oxygen isotope ratios. Ocean circulation is thought to have depended primarily on the production of hypersaline waters in restricted basins and shallow shelves because permanent polar ice caps were absent. Because the oceans were relatively warm and circulated relatively slowly, black shales were sometimes prominent, especially during the mid-Cretaceous superplume episode. The rifting of Pangaea also would have decreased magnesium concentrations in seawater, generally promoting calcite seas.

- During the Mesozoic the eastern portion of North America became a passive margin and remained relatively quiet, tectonically speaking. The initial rifting along the Atlantic margin of North America is indicated by rift basins filled with poorly sorted lake sediments. Other basins filled with evaporites along the margins of the Tethys and Atlantic Ocean, including the entire Gulf of Mexico. The occurrence of evaporites ranging from low-to-high latitudes suggests relatively warm climates, which would have promoted high rates of evaporation.

- By contrast, the western margin of the Americas was quite active tectonically. A number of island arcs and microcontinents accreted as terranes to the western margin of North America during the Mesozoic. As the terranes accreted to North America, they appear to have moved northward along faults for thousands of kilometers.

- The first major orogenic episode of the Mesozoic in western North America was a continuation of the Sonoma Orogeny into the Triassic from the Permian. During this orogeny, an island arc (Golconda terrane) and microcontinent (Sonomia) accreted to form what are now western Nevada, southeastern Oregon, and northern California, whereas other terranes accreted in areas ranging from western Canada to Alaska by the Late Triassic.

- Rapid accumulation of terranes occurred through the Jurassic in western North America, stretching from Alaska to Mexico.
  - Thickening of the crust during the Jurassic might have initiated the Nevadan Orogeny, during which enormous volumes of granitic plutons were intruded from Baja, California, to Alaska, whereas the turbidite-rich Great Valley sequence was shed westward into a deep-water basin adjacent to the uplift. As this was occurring, island arcs accreted to western North America, the subduction zone moved westward, and the Franciscan mélange was added to the western margin of North America along with the Great Valley. Parts of the Franciscan mélange were later uplifted to form the Coast Ranges.
  - Microcontinents continued to accrete along the western margins of North America during the Cretaceous. However, the site of orogenic activity in the western United States shifted to the east with the Sevier Orogeny. The Sevier Orogeny might have been initiated by the collision of the microcontinent Wrangellia in the vicinity of the southwestern United States and the Baja peninsula before Wrangellia moved northward toward its final location. Possibly, the eastward shift in orogenic activity during the Sevier Orogeny occurred because of a shallowing of the angle of plate subduction.
  - By the Late Cretaceous, the Laramide Orogeny had also begun in the vicinity of the Colorado Plateau and would continue into the Cenozoic Era.

- Life also diversified tremendously during the Mesozoic. Modern groups of plankton—the coccolithophorids, planktonic foraminifera, dinoflagellates, and diatoms—became established in the fossil record. Coccolithophorids and planktonic foraminifera, both members of the calcareous plankton, began to shift the site of limestone deposition from the continents to the deep sea, thereby establishing the link of calcareous oozes in the long-term carbon cycle. Calcareous oozes were also widespread in epicontinental seas, especially by the Cretaceous. Although the CCD remained relatively shallow, it was still several kilometers deeper than during the Paleozoic. As the Mesozoic continued, strontium ratios began to increase, suggesting increased nutrient runoff because of orogeny and weathering, despite extensive seafloor spreading. The appearance of diatoms in the fossil might be also related to the evolution of angiosperms on land, continental weathering, and the runoff of increasing amounts of nutrients such as silica to the oceans.

■ The evolution of marine life indicates increasing predation in the seas. Marine invertebrates were dominated by members of the Modern Fauna that radiated into the ecospace left vacant by the Late Permian extinctions. These forms included bivalves with longer siphons, carnivorous gastropods, sea urchins, crustaceans, and fish. Accompanying the Modern Fauna were members of the Paleozoic Fauna that rediversified in the Mesozoic, especially the ammonoids, which were accompanied by a group of cephalopods with reduced shells, the belemnoids. Also present were the marine reptiles: initially placodonts and nothosaurs, which gave way to ichthyosaurs, plesiosaurs, and mosasaurs.

■ Significant changes in the biosphere also took place on land.

○ The gymnosperms, which appeared during the Late Paleozoic, dominated terrestrial floras but gave way to angiosperms in the Cretaceous, perhaps as the result of coevolution with insects.

○ Among the vertebrates, the dinosaurs dominated. Dinosaurs were composed of two lineages: the Saurischia, or lizard-hipped dinosaurs, and the Ornithischia, or bird-hipped dinosaurs. The Saurischia included both the large herbivorous sauropods and theropods such as the fearsome *Tyrannosaurus rex*. The dominant ornithischians of the Jurassic were the stegosaurs, which were eventually replaced by ornithischians belonging to the ankylosaurs, ornithopods, and ceratopsians in the Cretaceous.

■ Flight among the vertebrates also evolved. Pterosaurs radiated into the air but were probably gliders. The first bird, *Archaeopteryx,* appeared in the Late Jurassic, although feathers might have initially evolved for body temperature regulation.

■ A number of minor and two mass extinctions took place during the Mesozoic. The minor extinctions during the Cretaceous might be related to changes in ocean circulation, climate, and oceanic anoxia. Although impacts have been implicated in the Late Triassic and Late Cretaceous extinctions, volcanism might also have been involved, just as it apparently was during the Late Permian extinctions. Only the Late Cretaceous mass extinction has been strongly associated with impact, based on shocked mineral assemblages and microtektites.

KEY TERMS

angiosperms	Cordilleran Orogenic Belt	hadrosaurs	nothosaurs
ankylosaurs	Cordilleran Orogeny	ichthyosaurs	oceanic anoxic events (OAEs)
anoxia	Cretaceous Interior Seaway	infaunal	Ornithischia
Archaeopteryx	cycadeoids	insectivores	ornithopods
archosaurs	cycads	Laramide	oviparous
belemnoids	Deccan Traps	Mancos shale	pachycephalosaurs
bipedal	dentary	Manicouagan crater	Pierre shale
bivalves	diapsids	marginocephalians	placodonts
Central Atlantic Magmatic Province	diatoms	marine reptiles	planktonic foraminifera
ceratopsians	dinoflagellates	Mesozoic marine revolution	plesiosaurs
cheilostomes	dinosaurs	Modern Fauna	pterosaurs
Chinle	echinoids	mosasaurs	radiolarians
Coast Ranges	epifaunal	Mowry Sea	radula
coccolithophorids	euryapsids	Navajo Sandstone	Réunion hotspot
coevolution	Franciscan mélange	neogastropods	Roberts Mountain terrane
conifers	gastropods	Nevadan	rudists
conodonts	ginkgos	Newark Group	Saurischia
Cordillera	Golconda terrane	Niobrara Chalk	sauropods

Scleractinia	siphons	teleosts	tyrannosaurids
sessile	stegosaurs	temnospondyl	viviparous
Sevier	Strangelove Ocean	Tethys	Wrangellia
Shinarump Conglomerate	Sundance Sea	theropods	Zuni Sequence

REVIEW QUESTIONS

1. During the Mesozoic what was the effect of continental movements on each of the following: (a) sea level, (b) atmospheric CO_2 concentrations, (c) ocean currents and circulation, (d) limestone deposition? Explain your answers.

2. What are the different lines of evidence indicating the Mesozoic was generally warm?

3. Discuss how the paleogeographic setting of the Mesozoic affected the activity of chemical reactions involved in determining the level of the CCD and oceanic anoxia.

4. How did the climatic conditions of the Jurassic and Cretaceous periods (Mesozoic Era) and Ordovician through Devonian periods (Paleozoic Era) resemble each other and how did they differ in terms of climate and life? Make a chart comparing the two time intervals. Put the time intervals at the top in separate columns and then down the left side of the chart, list the following: (a) continental movements, (b) sea level, (c) atmospheric CO_2, (d) ocean circulation, (e) black shales, (f) limestones, (g) plankton, (h) CCD, (i) marine benthos, (j) terrestrial plants, (k) terrestrial animals.

5. How might the generation of anoxia during the Cretaceous have differed from that of the Early-to-Middle Paleozoic?

6. How did the styles of orogeny in North America during the Paleozoic and Mesozoic resemble one another? How did they differ?

7. Which modern features of Earth were produced during the Mesozoic? What were the processes involved?

8. What is the evidence for increasing predation in the marine realm during the Mesozoic?

9. What were the major developments in marine benthos and plankton during the Mesozoic? Why?

10. Give examples of positive and negative feedback in the physical and biologic realms during the Mesozoic.

11. What were the major developments in terrestrial floras, and how did they affect Earth's climate and ocean chemistry during the Mesozoic?

12. Diagram the evolutionary relationships of the major groups of dinosaurs through the Mesozoic.

13. What are the relationships of the following modern groups to Mesozoic reptiles: snakes, lizards, turtles, crocodiles?

14. What traits exhibited by dinosaurs are normally associated with birds and mammals? What is the evidence for these traits in dinosaurs?

15. Make a chart of Mesozoic extinctions and possible causal agents. Rank each causal agent as to its relative role in each of the extinctions.

FOOD FOR THOUGHT:
Further Activities In and Outside of Class

1. Identify some potential terranes in the modern Pacific Ocean. How long do you believe it would take for them to accrete somewhere in the Americas given present rates of seafloor spreading? (Hint: examine the timing of terrane accretion in the Mesozoic Era.)

2. Why do you suppose the terranes of western North America appear to have been "stretched"?

3. What genetic mechanism accounts for the large number of vertebrae in the neck of sauropods, the changes in the limbs of ichthyosaurs, and of the evolution of the mammalian jaw and inner ear?

4. It has been stated that dinosaurs and birds were not reptiles and that birds are dinosaurs. Do you agree or disagree? Why or why not?

5. Despite the tremendous amount of vacant ecospace left by the Late Permian extinctions, evolutionary experimentation comparable with the Cambrian explosion did not occur during the Mesozoic. Why?

6. The communities formed by dinosaurs vaguely resembled those found today in the African savannah, with herds of herbivores like antelope, zebra, and elephants, as well as top carnivores like lions. Which evolutionary mechanisms explain the resemblance between dinosaur and mammal communities separated in time by more than 65 million years?

SOURCES AND FURTHER READING

Allmon, W. D., and Martin, R. E. Seafood through time revisited: the Phanerozoic increase in marine trophic resources and its macroevolutionary consequences. *Paleobiology*, 40, 256–287.

Alvarez, W. 1997. *T. rex and the crater of doom*. Princeton, NJ: Princeton University Press.

Barthel, K. W., Swinburne, N. H. M., and Morris, S. C. 1994. *Solnhofen: A study in mesozoic palaeontology*. Cambridge, UK: Cambridge University Press.

Berner, R. A., and Canfield, D. E. 1989. A new model for atmospheric oxygen over Phanerozoic time. *American Journal of Science*, 289, 333–361.

Blackburn, T. J., Olsen, P. E., Bowring, S. A., McLean, N. M., Kent, D. V., Puffer, J., McHone, G., Rasbury, E. T., and Er-Touhami, M. 2013. Zircon U-Pb geochronology links the end-Triassic extinction with the Central Atlantic Magmatic Province. *Science*, 340, 941–945.

Cheng, Y. N., Wu, X. C., and Ji, Q. 2004. Triassic marine reptiles gave birth to live young. *Nature*, 432, 383–386.

Chinsamy, A., and Dodson, P. 1995. Inside a dinosaur bone. *American Scientist*, 83(2):174–180.

Dial, K. P. 2003. Wing-assisted incline running and the evolution of flight. *Science*, 299, 402–404.

Eagle, R. A., et al. 2011. Dinosaur body temperatures determined from isotopic (^{13}C–^{18}O) ordering in fossil biominerals. *Science*, 333, 443–445.

Erickson, G. M., Rogers, K. C., and Yerby, S. A. 2001. Dinosaurian growth patterns and rapid avian growth rates. *Nature*, 412, 429–433.

Farlow, J. O., and Brett-Surman, M. K. (ed.). 1997. *The complete dinosaur*. Bloomington, IN: Indiana University Press.

Fastovsky, D. E., and Weishampel, D. B. 1996. *The evolution and extinction of the dinosaurs*. Cambridge, UK, Cambridge University Press.

Grady, J. M., et al., 2013. Evidence for mesothermy in dinosaurs. *Science*, 344, 1268–1272.

Godefroit, P., Cau, A., Dong-Yu, H., Escuillie, F., Wenhao, W., and Dyke, G. 2013. A Jurassic avialan dinosaur from China resolves the early phylogenetic history of birds. *Nature*, doi:10.1038/nature12168.

Harden, D. R. 1998. *California Geology*. Upper Saddle River, NJ: Prentice-Hall. 479.

Heeren, F. 2011. Rise of the titans. *Nature*, 475, 159–161.

Hesselbo, S. P., Robinson, S. A., Surlyk, F., and Piasecki, S. 2002. Terrestrial and marine extinction at the Triassic-Jurassic boundary synchronized with major carbon-cycle perturbation: a link to initiation of massive volcanism? *Geology*, 30, 251–254.

Hu, Y., et al. 2005. Large Mesozoic mammals fed on young dinosaurs. *Nature*, 433, 149–152.

Hutchinson, J. R., and Garcia, M. 2002. *Tyrannosaurus rex* was not a fast runner. *Nature*, 415, 1018–1021.

Johnson, C. C. 2002. The rise and fall of rudist reefs. *American Scientist*, 90(2):148–153.

Jones, T. D., et al. 2000. Nonavian feathers in a Late Triassic archosaur. *Science*, 288, 2202–2205.

Kerr, R. A. 2000. Did ancient volcanoes drive ancient extinctions? *Science*, 289, 1130–1131.

Klug, C., Kröger, B., Kiessling, W., Mullins, G. J., Servais, T., Frýda, J., Korn, D., and Turner, S. 2010. The Devonian nekton revolution. *Lethaia*, 43, 465–477.

Larson, R. L. 1991. Geological consequences of superplumes. *Geology*, 19, 963–966.

Leckie, R. M., Bralower, T. J., and Cashman, R. 2002. Oceanic anoxic events and plankton evolution: biotic response to tectonic forcing during the mid-Cretaceous. *Paleoceanography*, 17, 13-1–13-29.

Li, Q., et al. 2010. Plumage color patterns of an extinct dinosaur. *Science*, 327, 1369–1372.

Martin, R. E., Quigg, A., and Podkovyrov, V. 2008. Marine biodiversification in response to evolving phytoplankton stoichiometry. *Palaeogeography, Palaeoclimatology, Palaeoecology*, 258, 277–291.

Martin, R. E., and Quigg, A. 2013. Tiny plants that once ruled the seas. *Scientific American*, 308, 40–45.

Motani, R. 2000. Rulers of the Jurassic seas. *Scientific American*, 283(2):52–59.

Motani, R. 2002. Scaling effects in caudal fin propulsion and the speed of ichthyosaurs. *Nature*, 415, 309–312.

Nagalingum, N. S., Marshall, C. R., Quental, T. B., Rai, H. S., Little, D. P., and Mathews, S. 2011. Recent synchronous radiation of a living fossil. *Science*, 334, 766–767.

Nesbitt, S. J., Smith, N. D., Irmis, R. B., Turner, A. H., Downs, A., and Norell, M. A. 2009. A complete skeleton of a Late Triassic saurischian and the early evolution of dinosaurs. *Science*, 326(5959):1530–1533.

Norell, M., Ji, Q., Gao, K., Yuan, C., Zhao, Y., and Wang, L. 2002. "Modern" feathers on a non-avian dinosaur. *Nature*, 416, 36–37.

Padian, K., and Chiappe, L. M. 1998. The origin of birds and their flight. *Scientific American*, 281(2):38–47.

Pennisi, E. 2003. Uphill dash may have led to flight. *Science*, *299*, 329.

Perkins, S. 2002. No Olympian: analysis hints *T. rex* ran slowly, if at all. *Science News*, *161*(9, March 2):131.

Prum, R. O., and Brush, A. H. 2003. Which came first, the feather or the bird? *Scientific American*, *288*(3): 84–93.

Reid, R. E. H. 1997. Dinosaurian physiology: the case for "intermediate" dinosaurs. In J. O. Farlow and M. K. Brett-Surman (ed.), *The complete dinosaur* (pp. 449–473). Bloomington, IN: Indiana University Press.

Rowe, T. 1999. At the roots of the mammalian family tree. *Nature*, *398*, 283–284.

Ruben, J. A., et al. 1999. Pulmonary function and metabolic physiology of theropod dinosaurs. *Science*, *283*, 514–516.

Ruhl, M. et al. 2011. Atmospheric carbon injection linked to end-Triassic mass extinction. *Science*, *333*, 430–434.

Sakamoto, M., Benton, M.J., and Venditti, C. Dinosaurs in decline tens of millions of years before their final extinction. *Proceedings of the National Academy of Sciences U.S.A.*, *113*, 5036–5040.

Schaller, M. F., Wright, J. D., and Kent, D. V. 2011. Atmospheric pCO$_2$ perturbations associated with the Central Atlantic magmatic Province. *Science*, *331*, 1404–1409.

Sephton, M. A., et al. 2002. Carbon and nitrogen isotope disturbances and an end-Norian (Late Triassic) extinction event. *Geology*, *30*, 1119–1122.

Sereno, P. 1999. The evolution of dinosaurs. *Science*, *284*, 2137–2147.

Shipman, P. 1998. *Taking wing: Archaeopteryx and the evolution of bird flight*. New York: Touchstone Books.

Sigloch, K., and Mihalynuk, M. G. 2013. Intra-oceanic subduction shaped the assembly of Cordilleran North America. *Nature*, *496*, 50–56.

Steinthorsdottier, M., Jeram, A. J., and McElwain, J. C. 2011. Extremely elevated CO$_2$ concentrations at the Triassic/Jurassic boundary. *Palaeogeography, Palaeoclimatology, Palaeoecology*, *308*, 418–432.

Stokstad, E. 2000. Feathers, or flight of fancy? *Science*, *288*, 2124–2125.

Stokstad, E. 2001. Doubts raised about dinosaur heart. *Science*, *291*, 811.

Stokstad, E. 2002. *T. rex* was no runner, muscle study shows. *Science*, *295*, 1620–1621.

Stokstad, E. 2002. Fossil plant hints how first flowers bloomed. *Science*, *296*, 821.

Sun, G. et al. 2002. Archaefructacea, a new basal angiosperm family. *Science*, *296*, 899–904.

Tanner, L. H., Hubert, J. F., Coffey, B. P., and McInerney, D. P. 2001. Stability of atmospheric CO$_2$ levels across the Triassic/Jurassic boundary. *Nature*, *411*, 675–677.

Torsvik, T. H., Amundsen, H., Hartz, E. H., Corfus, F., Kusznir, N., Gaina, C., Doubrovine, P. V., Steinberger, B., Ashwal, L. D., and Jamtveit, B. 2013. A Precambrian microcontinent in the Indian ocean. *Nature Geoscience*, *6*, 223–227.

Vajda, V., Raine, J. I., and Hollis, C. J. 2001. Indication of global deforestation at the Cretaceous-Tertiary boundary by New Zealand fern spike. *Science*, *294*, 1700–1702.

Wilkinson, B. H., and Walker, J. C. G. 1989. Phanerozoic cycling of sedimentary carbonate. *American Journal of Science*, *289*, 525–548.

Witmer, L. H. 2001. Nostril position in dinosaurs and other vertebrates and its significance for nasal function. *Science*, *293*, 850–853.

Zheng, X., Zhou, Z., Wang, X., Zhang, F., Zhang, X., Wang, Y., Wei, G., Wang, S., Xu, X. Hind wings in basal birds and the evolution of leg feathers. *Science*, *339*, 1309–1312.

The Cenozoic Era: The Paleogene Period

MAJOR CONCEPTS AND QUESTIONS ADDRESSED IN THIS CHAPTER

A What orogenic and tectonic events brought Earth's surface closer to its present state?

B Why did the site of mountain building shift to the interior of western North America?

C Do the mountains of the western interior represent the actual mountains that formed during the Laramide Orogeny?

D What was the climate like during the Paleogene?

E How did movements of the continents affect sea level, ocean chemistry, climate, and the evolution of new taxa?

F Was the Earth first fried and then frozen during the Paleogene, and what is the evidence?

G What new groups of plankton and marine mammals replaced the plankton and marine reptiles of the Mesozoic?

H How did climate change affect terrestrial plant evolution, and how, in turn, did this affect the evolution of terrestrial mammals?

I Why is the Cenozoic called the "Age of Mammals"?

J How did Paleogene mammals differ from later mammals of the Neogene?

K How did climate change affect the evolution of mammals?

L What sorts of strange birds evolved into the niches previously occupied by theropod dinosaurs?

M Were there multiple causes of extinction again?

CHAPTER OUTLINE

The Badlands of South Dakota have yielded rich assemblages of fossil mammals and are among the most famous fossil-collecting localities for Cenozoic mammals in the world.

© Welcomia/Shutterstock, Inc.

14.1 Introduction to the Cenozoic Era

The Cenozoic Era is the third and final era of the Phanerozoic Eon. The Cenozoic spans the transition from the first half of the tectonic cycle that began in the Mesozoic to the second half in the Neogene. It is during the Cenozoic that Earth began to evolve toward its truly modern state, both physically and biologically (**Figure 14.1**). This means that, increasingly, we will return to the concepts of plate tectonics, orogeny, climate and ocean circulation, and continent–ocean configurations as they were portrayed in Part I for Earth's systems.

During the Cenozoic the Atlantic Ocean continued to open and the tropical Tethys Seaway closed, continental uplift occurred in many parts of the world, and many of the highest mountain chains we see today began to form. The continent of Antarctica began to move over the South Pole by the Late Eocene and Oligocene epochs, and a continental ice cap began to grow. As a result, sea level began to fall, and the modern circulation of the oceans began to be established.

The biosphere also became more modern during the Cenozoic ("recent life"). The Modern Fauna continued to diversify in the marine realm, and angiosperms spread over the land. Mammals underwent an explosive diversification within the forests, replacing the niches left vacant by the extinction of the dinosaurs, whereas in the seas, marine reptiles were replaced by early whales. For this reason the Cenozoic is sometimes referred to as the "Age of Mammals." A major turnover of mammals also occurred in response to the Late Eocene–Oligocene climate change involving Antarctica. The forests that spread over Earth during the Paleocene and Eocene were populated by "archaic" mammalian taxa. As Antarctica moved over the South Pole and ice sheets began to develop, Earth began to dry out during the Oligocene. Forests then began to give way to grasslands populated by more modern taxa of mammals as the archaic taxa died out.

This climatic transition is recognized by the subdivision of the Cenozoic into two major periods: the Paleogene (comprising the earlier-recognized Paleocene, Eocene, and Oligocene epochs) and the Neogene (comprising the Miocene, Pliocene, and Pleistocene epochs). We primarily turn our attention to the Paleogene in this chapter.

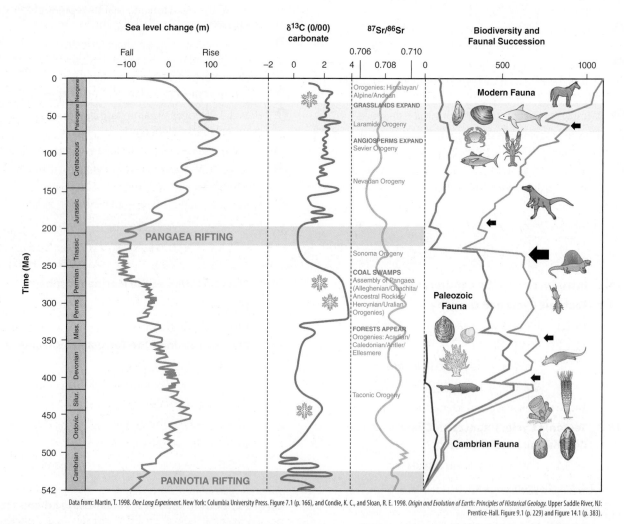

Data from: Martin, T. 1998. *One Long Experiment*. New York: Columbia University Press. Figure 7.1 (p. 166), and Condie, K. C., and Sloan, R. E. 1998. *Origin and Evolution of Earth: Principles of Historical Geology*. Upper Saddle River, NJ: Prentice-Hall. Figure 9.1 (p. 229) and Figure 14.1 (p. 383).

FIGURE 14.1 The physical and biologic evolution of the Earth through the Phanerozoic Eon. This chapter discusses the Paleogene period of the Cenozoic Era (shaded). Snowflakes indicate times of major glaciation. Carbon isotopes and strontium isotopes are indicated (see Chapter 9 for an elementary discussion). Arrows indicate the Big Five mass extinctions.

14.2 Tectonic Cycle and Orogeny

14.2.1 Europe and Asia

A The rifting of Pangaea and the movement of the continents that began in the Triassic Period continued during the Paleogene (**Figure 14.2**). As the Atlantic Ocean continued to open, Panthalassa, the ocean that originated during the formation of Pangaea and whose remnant is today recognized as the Pacific Ocean, began to close.

So, too, did the circumglobal Tethys Seaway. The destruction of the Tethys began with extensive orogeny across southern Europe and Asia. India, which had rifted away from Gondwanaland, plowed into Asia to begin the **Himalayan Orogeny**, the uplift of Tibet, and the closure of the Tethys (**Figure 14.3**). This collision began in the Paleocene, about 60 million years ago, but uplift continues to the present. Today, the Himalayas are the highest continental mountain chain in the world, with Mt. Everest reaching nearly 9 kilometers (about 5.5 miles) in elevation. Initially, during the Eocene, shallow-water limestones were deposited on the Indian shelf. As India moved toward Asia, however, the northern plate margin of India was subducted beneath Asia, replacing the limestone with a volcanic island arc system, graywackes, turbidites, and other deep-water sediments. When the continental crust of India and Asia actually began to collide, subduction and volcanism ceased. India began to wedge downward beneath Asia but could not be subducted deeply because India consists of relatively buoyant continental crust. As collision and compression continued, the relatively resistant continental crust fractured into huge thrust sheets, thickening the continental crust near the surface and pushing it upward through isostatic readjustment (see Chapter 2).

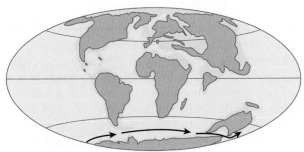

Middle Eocene (50 million years ago)

Data from: Stanley, S. M. 1999. *Earth System History*, 3rd ed. San Francisco: W. H. Freeman and Company. Figure 18.16 (p. 506).

FIGURE 14.2 Continent positions during the Middle Eocene as the continents continued to move toward their present positions and the oceans began to more closely assume their modern configurations.

The Himalayan Orogeny had a profound effect on the climate of Asia by initiating the monsoon that moved over the Indian subcontinent (see Chapter 3). Uplift and rainfall in turn affected orogeny and erosion. Rocks and sediments such as shallow-water limestones, turbidites, and ophiolites were all uplifted and eroded. By the Miocene, these sediments began to be shed into the northern Indian Ocean on either side of India to form huge deltas and submarine fans.

The **Alpine Orogeny** contributed to the shrinking of the Tethys to the west. This orogenic belt is more or less continuous with the Himalayan belt of deformation that extended through the rest of Asia (see Chapter 2). The Alpine Orogeny extended along the southern margin of Europe all the way into Asia Minor (the region between the Black Sea and the Mediterranean, including most of Turkey). The Alpine Orogeny is responsible for a number of noteworthy modern mountain ranges: the Pyrenees, which separate Spain from France; the Alps of southern Europe (e.g., France, Switzerland, Austria, Italy); the Apennines, which form the "backbone" of Italy, the Carpathians and Balkans of eastern Europe, which were formed along a marginal basin of the Tethys called the **para-Tethys** during the Alpine Orogeny; and the Caucasus Mountains, which lie between the Black and Caspian seas. Portions of northwest Africa, such as the Atlas Mountains, were also formed by this orogeny.

The Alpine Orogeny resulted from the eventual movement of the African plate north toward Europe. Caught between Europe and Africa were a number of smaller plates that were also deformed and uplifted. Deformation was most pronounced during the Eocene and Miocene, producing huge thrust sheets and overturned folds called nappes (see Chapter 2). Tectonic activity has continued to the present day in portions of southern Europe and the Middle East, explaining the widespread occurrence of earthquakes and volcanism in this region.

Rotation of the plates during their movement appears to have also induced rifting across northern and southern Europe. Some of the resulting rift basins lie buried beneath the North Sea, where they are exploited for oil and gas. Others serve as major conduits for rivers such as the Rhine of western Germany and the Rhône of France. The sides of these river valleys can be quite steep and serve as excellent natural fortifications; the eastern margin of the Rhine Graben, for example, was heavily defended by the Germans against the Allies as World War II drew to a close.

14.2.2 The Pacific Rim

Orogeny outside North America

As the Atlantic continued to open, the Pacific Ocean shrank. As a result, many of the volcanic island arcs and back-arc basins (such as the Aleutian Islands, Sea of Japan) that we see today in the Pacific began to form by plate subduction (see Chapter 2). Australia and Antarctica also continued to move toward their present positions. During the Late Triassic and Jurassic, Australia and Antarctica had drifted away from Gondwana but continued to remain sutured during the Mesozoic. During the Late Paleocene–Early Eocene these

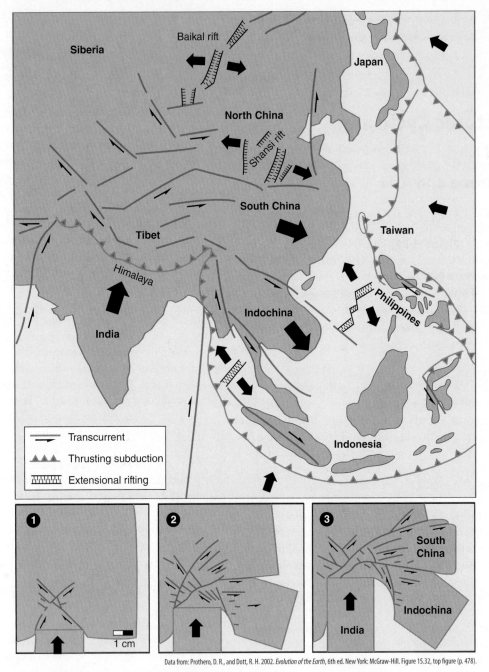

Legend:
- →→ Transcurrent
- ▲▲▲ Thrusting subduction
- WWWW Extensional rifting

1 cm

Data from: Prothero, D. R., and Dott, R. H. 2002. *Evolution of the Earth*, 6th ed. New York: McGraw-Hill. Figure 15.32, top figure (p. 478).

FIGURE 14.3 Major phases of the Himalayan Orogeny. Note the major fault zones established during mountain building.

two continents finally rifted apart, with Antarctica moving toward the South Pole. As we will see later, these movements had profound consequences for ocean circulation, climate, and the evolution of mammals.

The processes and events occurring along the margins of the Pacific Ocean also affected the American Cordillera. Along the west coast of South America, the **Andean Orogeny** continues to the present day. This orogeny actually began during the Cretaceous, possibly as the result of the opening of the South Atlantic Ocean. Today, the Andes are the highest mountains in the western hemisphere, counting nearly 50 peaks with elevations greater than 6,000 meters. As the Andes formed, igneous activity shifted toward the interior of South America, resulting in the intrusion of igneous rocks

called granitoids. Granitoids have unusual chemical compositions and frequently contain fragments of mafic and ultramafic rocks that can represent older oceanic crust. Today, granitoids are found only in regions where relatively young, hot oceanic crust is rapidly subducted, such as along the margin of southern Chile in South America. (This type of ocean crust is thought to have been prominent much earlier during Earth's history because of greater heat flux from within Earth's interior [see Chapter 7].) A foreland basin lay to the east as the Andes rose. This foreland basin was connected to the Atlantic by an arm that ran through a failed rift valley, or aulacogen, that filled with molasse shed from the Andes. Today, this rift valley is occupied by the Amazon River (see Chapter 2).

Western North America

Tectonic activity coupled with erosion and deposition during the Cenozoic Era also led to many of the broad features we see today in North America. Much of the orogenic activity in western North America stemmed from the Laramide Orogeny, the last major orogenic episode involving the North American Cordillera. The Laramide Orogeny began in the Late Cretaceous but continued well into the Eocene Epoch. During the Late Cretaceous a subduction zone existed along the western margin of North America. Here, the North American plate overrode the Farallon plate in the Jurassic, resulting in the emplacement of the batholiths of the Sierra Nevada (see Chapter 13). As the North American plate continued to move west over the Farallon plate in the Paleocene and Eocene, igneous activity ceased in the Sierra Nevada region (**Figure 14.4**). The Sierra Nevada remained as a highland, though, shedding deltaic sediments and turbidites westward into what is today the Central Valley of California, which was then a forearc basin (see Chapter 13).

The Laramide orogenic belt formed far inland to the east of the earlier limits of the Nevadan and Sevier orogenies (see Figure 14.4). The Laramide orogenic belt exhibited several other curious features. First, a volcanic arc ran through western Montana, Idaho, and eastern Washington and north into Canada, whereas another arc extended south from the New Mexico region into Mexico. Deformation occurred just to the east of each of these volcanic arcs, producing thrust sheets and folds that rival the nappes of the Alps. Second, separating the two belts of volcanism and deformation was an area distinguished by relatively little igneous activity and consisting

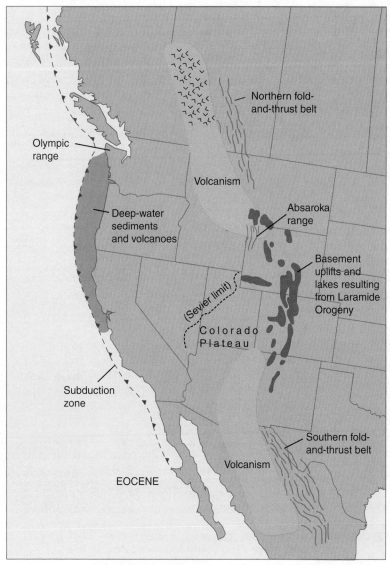

Data from: Stanley, S. M. 1999. *Earth System History*, 3rd ed. San Francisco: W. H. Freeman and Company. Figure 18.21 (p. 512).

FIGURE 14.4 Distribution of volcanism and tectonism in the western United States during the Paleogene, including the ancestral Olympic Mountains, the Cascades, and the Laramide Orogeny. Notice how the Laramide Orogeny is located even farther east than the limit of the previous Sevier Orogeny of the Cretaceous period (Chapter 13). This is another reason why it is thought that the angle of subduction of the Farallon plate shallowed (see the text for further discussion). Tectonic activity and volcanism shifted back to the west later during the Paleogene and Neogene (see Chapter 15).

of mostly vertical uplifts of Precambrian and Paleozoic crystalline rocks. Interestingly, this area was located about where the Ancestral Rockies had risen 200 million years earlier in the Late Paleozoic (see Chapter 12). This area was also located to the east of the older Sevier Orogeny, which had been centered on Wyoming and Colorado, and was separated from the Sierra Nevada highlands far to the west by a broad, tectonically quiescent region.

Why did orogeny shift so far eastward after the Sevier Orogeny? Remember that the Sevier orogenic belt has been hypothesized to have shifted eastward in response to a shallowing of the angle of subduction of the Farallon plate (see Chapter 13). It has therefore been hypothesized that this tectonic behavior continued into the Paleogene; thus, volcanic activity

and compressive forces are thought to have shifted further eastward as the North American plate continued to move over the Farallon plate after the Sevier Orogeny. Previously, some workers suggested the shallow angle of subduction of the Farallon plate might have been due to a mantle plume pushing upward on it, preventing it from being subducted as quickly. However, evidence for such a plume has not been found. Instead, vertical uplift due to compressive forces associated with subduction might have occurred at relatively shallow depth.

The vertical uplift associated with the central portion of the Laramide belt created a series of isolated, fault-bounded **intermontane** ("between mountains") **basins** stretching from southern Montana through Wyoming, Colorado, Utah, and New Mexico (**Figure 14.5**). Today, the remnants

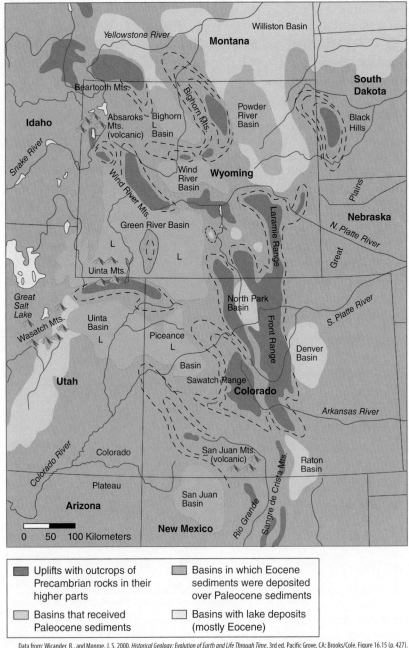

Uplifts with outcrops of Precambrian rocks in their higher parts

Basins that received Paleocene sediments

Basins in which Eocene sediments were deposited over Paleocene sediments

Basins with lake deposits (mostly Eocene)

Data from: Wicander, R., and Monroe, J. S. 2000. *Historical Geology: Evolution of Earth and Life Through Time*, 3rd ed. Pacific Grove, CA: Brooks/Cole. Figure 16.15 (p. 427).

FIGURE 14.5 Uplifts and basins associated with the Laramide Orogeny. Many basins filled by large lakes began to form adjacent to uplifts in response to the Laramide Orogeny.

(a)

Courtesy of Ronald Martin, University of Delaware.

(b)

Courtesy of National Park Service.

FIGURE 14.6 **(a)** Varves represent seasonal deposition, much like tree rings. In this specimen, each varve consists of a thin, dark winter layer and a thicker, lighter summer layer. The dark layer consists of fine terrigenous sediment, whereas the lighter layer represents spring and summer productivity by phytoplankton. **(b)** The Green River formation of the western United States is famous for its varved sediments, fossil fish, and plants.

of these uplifts, which consist of Precambrian igneous and metamorphic rocks at their centers, are exposed in such places as the Black Hills of South Dakota (see this chapter's frontispiece) and the Front Range of the Rockies, west of Denver, near where the Ancestral Rockies of the Late Paleozoic had once existed (see Chapter 12). Erosion kept pace with these uplifts, filling the adjacent basins with thousands of feet of sediment. Consequently, rather than rugged, jagged peaks, the Rockies in this region at this time consisted of much smaller, eroded peaks and knobs protruding from the surrounding sediment that otherwise buried them.

These intermontane basins that formed adjacent to the uplifts behaved independently of one another. Some of the basins developed swamps, where extensive coal sometimes accumulated, such as in the Powder River Basin of northeast Wyoming, where the coals are strip-mined. Other basins developed into gigantic lakes, the sediments of which often exhibit annual varves related to seasonal changes in plankton productivity and sedimentation. Among the most famous of the lacustrine deposits is the Eocene Green River Formation,

which is famous for its varved sediments, which are thought to have formed by seasonal deposition, and exceptionally preserved fish and plant fossils (**Figure 14.6**). Some lake deposits are so rich in organic carbon that they have been considered as a potential source of petroleum. Some of the most spectacular exposures of the former lake basins consist of conglomerates, sandstones, and freshwater limestones of the Eocene Wasatch Formation found in Bryce Canyon National Park in southeast Utah (**Figure 14.7**). The rocks at Bryce Canyon weather in an unusual manner to produce what are called "hoo doos," which make the Canyon look like, in one author's description, "a giant, haunted chess board."

Many of these uplifts were later beveled beginning in the Oligocene, so that today the ranges have a relatively subdued or even flat topography (**Figure 14.8**). (This is why many pioneers moved westward through southern Wyoming to avoid the more rugged terrane further south.) As the ranges were beveled, sediment and volcanic ash accumulated to the east in remnant seaways to form fluvial deposits and floodplains that spread into the Great Plains of what are

(a)

Courtesy of Ronald Martin, University of Delaware.

(b)

Courtesy of Ronald Martin, University of Delaware.

FIGURE 14.7 Lacustrine (lake) deposition was widespread as a result of the Laramide Orogeny. **(a)** Bryce Canyon National Park consists of exposures of calcareous sediments of the Eocene Wasatch Formation deposited in a lacustrine basin associated with the Laramide Orogeny. **(b)** The unusual formations are referred to as "hoo doos."

© John McLaird/ShutterStock, Inc.

FIGURE 14.8 Erosional surface in the Rocky Mountains formed as a result of continued uplift and erosion during the Eocene. This erosion beveled many areas that had been previously uplifted by the Laramide Orogeny.

now Wyoming, central and eastern Colorado, Nebraska, and the western Dakotas. Today, these deposits are exposed in localities like the highly eroded Badlands of South Dakota, where a treasure trove of highly fossiliferous mammal assemblages occurs (refer to this chapter's frontispiece).

Volcanism was renewed along much of the Cordillera by the Late Eocene and Oligocene to the west of the Laramide belt (**Figure 14.9**). A volcanic belt extended from western Canada through central Washington and Oregon into Nevada, Utah, and southwest Colorado, and then south into New Mexico, Arizona, and Mexico. The renewal of volcanism to the west and northwest of the Laramide Orogeny suggests that the angle of subduction of the Farallon plate increased. This volcanism intruded the deep marine basins of western Oregon and spread north through Washington to produce a volcanic arc that foreshadowed the modern Cascades. Here, a fore-arc basin existed in Washington and Oregon, where graywackes, turbidites, and other deep-water sediments accumulated; these rocks were eventually deformed by subduction to produce the Olympic and Cascade Mountains.

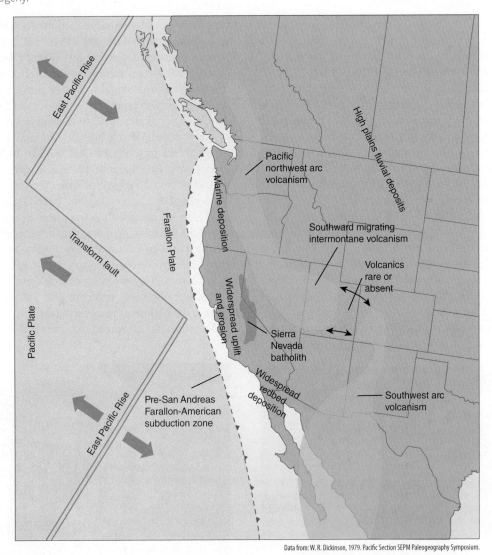

Data from: W. R. Dickinson, 1979. Pacific Section SEPM Paleogeography Symposium.

FIGURE 14.9 Renewal of volcanism to the west of the Laramide Orogeny. Notice how western North America is approaching the subduction zone associated with the Farallon plate. The North American plate will eventually collide with the East Pacific Rise to produce the San Andreas fault system (see Chapter 15).

Igneous activity also became widespread elsewhere. Volcanism extended into the tectonically quiescent region separating the Sierra Nevada from the Laramide orogenic belt. In southwest Colorado, volcanism produced the San Juan Mountains. Remnants of this volcanic activity can be seen at Shiprock in northwest New Mexico. Shiprock consists of an ancient volcanic vent and its feeder dikes that have been exposed by erosion of the less resistant surrounding rocks. Volcanism also occurred to the north and east. Among these localities is Yellowstone National Park in northwest Wyoming (recall that Yellowstone overlies a hotspot; see Chapter 2). In the park itself, 27 separate volcanic layers, sometimes with buried trees of the original forests standing upright, are preserved. Further to the east in Wyoming is Devils Tower, which was intruded about 45 to 50 million years ago (**Figure 14.10**) and featured as the landing site for the gigantic spaceship in the movie *Close Encounters of the Third Kind*. The distinctive columns of Devils Tower formed as the rock cooled beneath Earth's surface, forming joints between the columns.

14.2.3 Gulf and Atlantic coastal plains

The late Cretaceous regression led to the retreat of the Cretaceous Interior Seaway—now called the **Tejas Seaway**—from most of the craton. As previously described, sediments from the Laramide uplifts were shed eastward into the seaway and deposited in marginal-marine to fully marine environments.

© iStockphoto/Thinkstock.

FIGURE 14.10 Devils Tower in Devils Tower National Park was intruded about 45 to 50 million years ago in what is now northeast Wyoming during the Laramide Orogeny. The rocks formed polygonal-shaped columns as they cooled beneath the Earth's surface.

In the Paleocene, a remnant of this seaway still existed in the Dakotas region. With time, the Tejas Seaway retreated south into the **Mississippi Embayment**, which extends along the course of the Mississippi River from southern Illinois to the Gulf of Mexico (**Figure 14.11**). Like the Amazon River, the Mississippi Embayment is thought to correspond to an ancient aulacogen buried deep beneath the surface.

As the Tejas Seaway retreated, sediments derived from the Laramide uplifts, the continental interior, and the Appalachians deposited thick sections of sediment during the Eocene. Although the Appalachians had already been greatly eroded, rejuvenation through isostatic adjustment resulted in further uplift and erosion during the Cenozoic. As a result thick wedges of sediment several kilometers or more built out along the continental margins and shelves (**Figure 14.12**). These sediments—mostly sand, silt, and clay—were deposited in a broad range of environments, from marginal marine to open marine, which today comprise the Gulf and Atlantic Coastal plains.

The gradual withdrawal of the Tejas from the North American craton is indicated by the increasingly younger ages of sediments exposed at the surface toward the Gulf of Mexico (see Figure 14.11). In the western Gulf of Mexico, clastic sediments eventually came to lie on top of the evaporites of the **Louann Salt** that formed in the Gulf of Mexico when a gigantic evaporite basin formed during the rifting of Pangaea (see Chapter 13). The weight of the overlying sediment caused the evaporites to deform and move upward, creating traps for oil and gas (see Chapter 17). By contrast, broad carbonate shelves developed along much of the western Florida shelf and the coast of Mexico because of the lack of terrigenous input from the surrounding low-lying areas of the coastal plain.

CONCEPT AND REASONING CHECKS

1. Why are so many mountain chains so much higher than the Appalachians?
2. How does orogeny in western North America resemble that of eastern North America during the Paleozoic? How does it differ?
3. Where are the foreland basins for the Andes? The Himalayas?
4. Where was much of the sediment shed from the Laramide Orogeny deposited?
5. Give examples of how ancient orogenies influence modern landscapes.

14.3 Tectonic Cycle: Impacts on Climate, Ocean Circulation, and Chemistry

14.3.1 Climate and ocean circulation

D E F The apparent rapid cooling that took place during the Late Cretaceous was supplanted

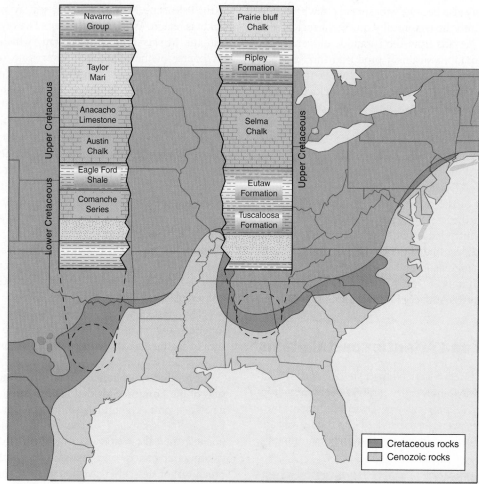

Data from: Levin, H. 2006. *The Earth Through Time*, 8th ed. Hoboken, NJ: John Wiley. Figure 13.29 (pg. 393).

FIGURE 14.11 Distribution of sediments exposed in the Mississippi Embayment. Note the extent of Cretaceous rocks found upriver. Cenozoic rocks become progressively younger toward the Gulf of Mexico, suggesting that sea level was in general retreating as the embayment filled.

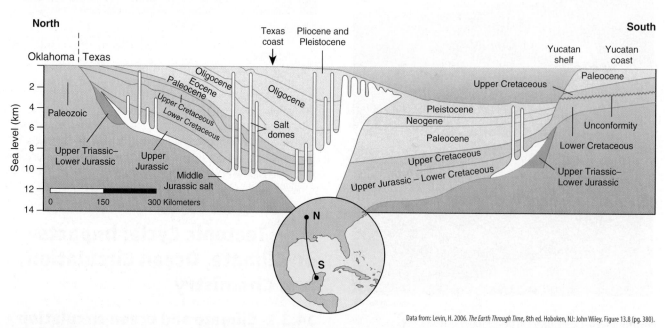

Data from: Levin, H. 2006. *The Earth Through Time*, 8th ed. Hoboken, NJ: John Wiley. Figure 13.8 (pg. 380).

FIGURE 14.12 Cross-section of the Gulf Coast margin in the vicinity of Texas and Louisiana. Note the tremendous thicknesses of sediments, which were eroded from the western Appalachians and western uplifts. The white finger-like upward projections are intrusions of evaporite sediments caused by the pressure of the overlying sediment and are associated with petroleum reservoirs (see Chapter 17).

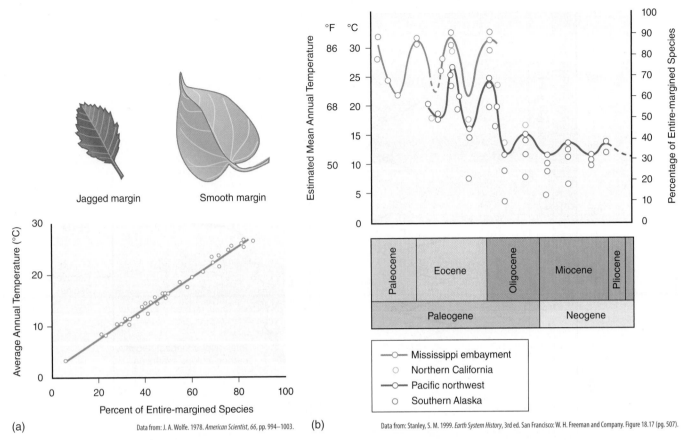

(a)

Data from: J. A. Wolfe. 1978. *American Scientist, 66*, pp. 994–1003.

(b)

Data from: Stanley, S. M. 1999. *Earth System History*, 3rd ed. San Francisco: W. H. Freeman and Company. Figure 18.17 (pg. 507).

FIGURE 14.13 Relationship between the margins of plant leaves and paleotemperature. Note the strong cooling during the latest Eocene and Oligocene when ice sheets were forming on Antarctica as the continent continued to move over the South Pole (see discussion in text). **(a)** Relation between average annual air temperature to leaf margins in modern angiosperms. **(b)** Air temperature during the Cenozoic based on the relationship in part A.

in the Early Paleogene by widespread, warm climates. Palms and cycads like those of the Mesozoic existed up to 60 degrees north latitude, whereas temperate plants spread north of the Arctic Circle. The leaf margins of angiosperms also indicate that the Late Paleocene to Early Eocene was marked by a major warming (**Figure 14.13**). The margins of modern angiosperm leaves exhibit a significant relationship to temperature that has been used to infer past temperatures (see Figure 14.13). Warm, humid climates are associated with large, thick leaves with smooth margins. Such leaves also have a pointed "drip tip" that allows them to shed the abundant rainfall that occurs under such conditions. By contrast, plants from cooler, drier climates do not have a sufficiently long growing season to form large, thick leaves. Instead, they produce leaves that are smaller and thinner, have jagged edges, and are shed every fall.

The rapid warming at the Paleocene–Eocene boundary, which occurred over a span as short as a few tens of thousands of years, is now referred to as the **Paleocene–Eocene Thermal Maximum (PETM)** (**Figure 14.14**). Exactly why the Early Eocene became so warm is unclear. Several mechanisms have been proposed, none of which is necessarily mutually exclusive. An increase in atmospheric carbon dioxide (CO_2) levels associated with rifting is an obvious suspect. Recall that there was widespread plate reorganization in the Pacific and Indian oceans as Pangaea continued to rift apart,

whereas the collision of India with Asia caused some spreading centers to cease and others to originate. Recall, too, that Australia and Antarctica began to rift apart about this time (see Figure 14.2). Another possible source of CO_2 was the continued rifting that eventually connected the Arctic Ocean to the North Atlantic. As the Atlantic basin continued to open, the mid-Atlantic Ridge moved northward like a zipper (see Figure 14.2). Eventually, Canada and Greenland rifted apart to form the Labrador Sea. Greenland and Scandinavia also began to move apart, further opening the North Atlantic. Rifting was marked by the massive eruption of flood basalts in this region; indeed, their volume of about 2 million cubic kilometers was only slightly less than those of the Deccan lavas at the end of the Cretaceous! Much of the CO_2 might have eventually been locked up in coals (i.e., negative feedback).

A third possibility is that ocean circulation was driven by the production of warm, hypersaline water masses in shallow shelf seas and isolated basins like that proposed for the Early-to-Middle Paleozoic and the Mesozoic (see Chapters 11 and 13). Remember the Tethys was closing and the Mediterranean (a remnant of the Tethys) was becoming an isolated basin, where such waters could have been produced. Upwelling of these water masses at higher latitudes would have warmed adjacent land and air masses while also bringing massive amounts of dissolved carbon to the ocean surface, where it could be oxidized to CO_2 and warm the

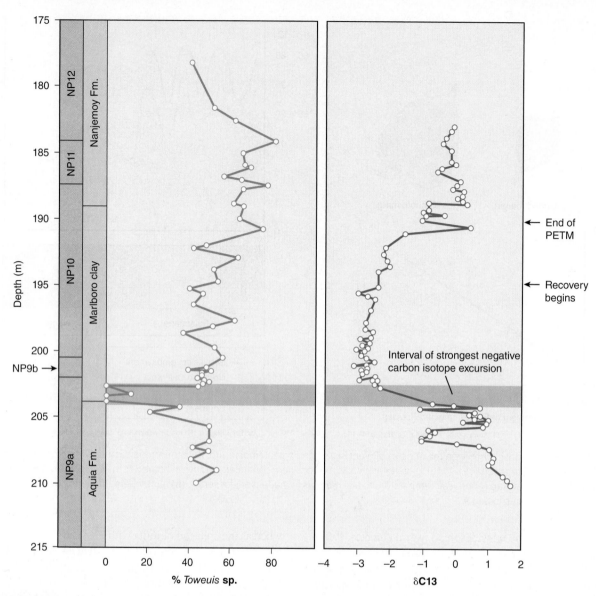

FIGURE 14.14 The Paleocene–Eocene Thermal Maximum (PETM). Such a strong negative excursion of carbon isotopes is interpreted as the release of massive amounts of CO_2 to the atmosphere by the oxidation of methane hydrates exposed to the atmosphere. Note the increase in abundance of the coccolithophore as the PETM comes to an end. This might indicate that the surface oceans were becoming less acidic from the CO_2.

Earth. Such a change in the deep oceans is suggested by the extinction of benthic foraminifera living in deep, cold waters at the end of the Paleocene and the diversification of warm-water plankton.

Another, more recent hypothesis that has attracted a great deal of attention postulates that the PETM resulted from a massive release of methane (CH_4) into the atmosphere. This hypothesis is based on a sharp decrease of carbon isotopes to lower values (the "carbon isotope excursion" or CIE of Figure 14.14). It is thought that such a sharp downward spike could have resulted only from the rapid input of CO_2 into the atmosphere. The carbon isotope spike could conceivably have resulted from the release of CH_4 from continental shelves during global (eustatic) sea-level fall (which would have removed the overlying weight of the water); however, only relatively minor sea-level changes occurred around the Paleocene–Eocene boundary based on global sea-level

curves. It has instead been suggested that widespread volcanism like that described earlier might have intruded volcanic vents into carbon-rich, methane-bearing sediments on the shelves. The CH_4 had presumably been generated long before by the burial and decay of organic matter.

The PETM itself lasted only about 120,000 years but was followed by a series of shorter negative carbon isotope excursions—called **hyperthermals**—during the next several million years. Most if not all of the hyperthermals appear to have lasted less than 200,000 years and might have also resulted from CH_4 release. A more recent hypothesis suggests instead that dissolved organic carbon stored in deep ocean water masses was repeatedly oxidized to CO_2 by enhanced ocean circulation events. The hyperthermals, in turn, led up to the Early Eocene Climate Optimum (EECO) at about 52 to 50 million years when Earth's average temperature rose as much as 12°C (based on negative shifts of oxygen isotope

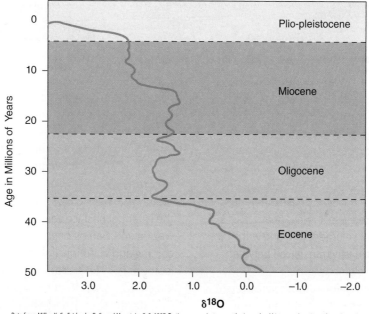

FIGURE 14.15 Cenozoic oxygen isotope curve (see Box 14.1). Compare to Figure 14.13.

ratios). The EECO represents the culmination of warming that started in the Paleocene, so Earth might have been sufficiently warm to have been ice free. Earth's temperature then declined and rose again to the Middle Eocene Climate Optimum (MECO) at 41.5 Ma (again based on oxygen isotopes). However, this time there was no corresponding shift in $\delta^{13}C$. This suggests the trigger for the MECO differed from that of previous intervals, perhaps representing a short-term rise in CO_2 levels that did not result from another release of methane.

Fossil floras and a shift of oxygen isotopes to higher values both indicate that later, during the Middle-to-Late Eocene (about 37 to 38 million years ago), Earth's temperature

plunged by as much as 10°C in as little as 15 million years (**Figure 14.15**; **Box 14.1**). This drastic cooling is thought to have resulted from the initial movement of Antarctica over the South Pole after it had rifted from Australia. As Antarctica moved into higher latitudes, a polar ice cap began to grow. The growth of ice sheets is confirmed by the occurrence of ice-rafted debris in deep-sea cores taken off Antarctica.

By about this time shallow-water circulation between Antarctica and the South Tasman Rise (located between Antarctica and Australia) began, marking the initiation of the **Antarctic Circumpolar Current** (**Figure 14.16**). The Antarctic Circumpolar Current was critical to glaciation on Antarctica for several reasons. First, the current isolated

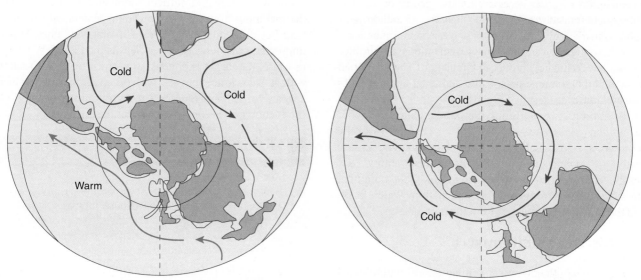

FIGURE 14.16 Movement of Antarctica over the South Pole initiated the growth of a polar ice cap. As Antarctica moved nearer the South Pole, submarine ridges around Antarctica sank, allowing the continent to be isolated from warm waters by the Antarctic Circumpolar Current.

BOX 14.1 Oxygen Isotopes, Ice Volume, and Temperature

We introduced the use of stable isotopes in deciphering nutrient and carbon cycling in Chapter 10. Recall that stable isotopes do not undergo radioactive decay. However, the isotopes of the same element behave somewhat differently in chemical reactions; isotopes of the same element can be separated by their atomic mass because they have different numbers of neutrons. Because it is less heavy, the "lighter" isotope of an element (the one with fewer neutrons) is more easily transferred between reservoirs involving the element.

Here, we introduce oxygen isotopes, which are used to study the behavior of the hydrologic cycle, especially changes in ice volume, through time. Oxygen consists of two isotopes of interest to earth scientists: ^{16}O and ^{18}O. Like carbon and strontium (Chapter 10), the different mass numbers of oxygen are indicated next to the symbol for the element, for example ^{16}O and ^{18}O. In this case, oxygen-16 (^{16}O) and oxygen-18 (^{18}O) both have the same number of protons (atomic number 16), but ^{18}O has two more

neutrons. Both isotopes are present in seawater. Because ^{16}O is lighter than ^{18}O, however, water containing ^{16}O is more easily incorporated into ice, leaving behind seawater enriched in ^{18}O.

By calculating the ratios of the two oxygen isotopes ($^{18}O/^{16}O$)—usually denoted as $\delta^{18}O$ (read as "delta O 18")—in fossil shells scientists can determine the extent of past ice volumes and therefore the waxing and waning of glaciers in response to global cooling and warming. If the ratio of the two stable isotopes goes up, the inference is that ice volume increased, taking ^{16}O out of seawater and increasing the ratio of ^{18}O-enriched water left behind in the oceans. Conversely, if the ratio decreases, it is said to become "lighter," or more "negative"; in this case, it is generally inferred that conditions were warmer and that some ice has melted, putting ^{16}O back into ocean water and decreasing the ratio. Oxygen isotope ratios can also be calibrated to determine temperatures, as also discussed in the text.

Antarctica from the influence of warm-water currents, allowing the continent to remain cold. Second, the essentially circular, continuous flow of the current around Antarctica kept the current's waters from being deflected toward lower latitudes, maintaining a relatively steep latitudinal temperature gradient that also kept Antarctica cold. Third, the opening of the South Tasman Rise allowed the current to bring moisture to Antarctica that was necessary for the growth of ice. The subsequent transition to the Early Oligocene 33 million years ago was marked by the expansion of glaciers on Antarctica.

These events were in turn of tremendous significance for sea level, climate, and ocean circulation. The continued growth of the Antarctic polar cap sequestered huge volumes of water in ice and resulted in the largest sea-level regression of the entire Cenozoic during the Middle Oligocene about 30 million years ago (see Figure 14.1). The growth of the ice sheet also increased the density of surrounding surface waters by cooling them and increasing their salinity. As a result, cold oxygenated water began to sink to the bottom of the ocean, the forerunner of Antarctic Bottom Water, flushing the much warmer waters from the ocean basins.

14.3.2 Ocean chemistry

Changes in continent–ocean configurations undoubtedly affected the oceans' chemistry. Although shallow-water limestone deposition was still widespread during the Paleocene and Eocene, sea-level fall associated with orogeny and

glaciation caused the erosion of continental shelves and shallow-water limestones, which might have contributed to the deepening of the calcite compensation depth (CCD). Nutrient inputs associated with uplift and erosion also would have stimulated the growth of calcareous (and other) plankton farther from shore, deepening the CCD. Rising strontium isotope ratios during this time also hint at increased continental weathering and nutrient runoff and decreased hydrothermal input from seafloor spreading centers, which might have begun to shift calcite oceans to aragonite ones. Superimposed on this longer-term cycle was the PETM. Whatever its cause, the PETM is of concern not just for its association with global warming but also with ocean acidification, which appears to be happening as a result of anthropogenic CO_2 production (see Section 14.6 later in the chapter).

CONCEPT AND REASONING CHECKS

1. How are the Paleocene and Eocene like the Mesozoic Era in terms of (a) greenhouse or icehouse, (b) sea level (high or falling), (c) large continental glaciers (present or absent), (d) CCD (shallow or deep), (e) calcite or aragonite seas, (f) deep-ocean circulation (Hint: were black shales more or less frequent during this time?), and (g) surface-ocean circulation?
2. How did the glaciation of what is now North Africa during the Late Ordovician resemble that of Antarctica?

14.4 Diversification of the Marine Biosphere

14.4.1 Microfossils and other invertebrates

G After the Late Cretaceous extinctions, the Modern Fauna composed of bivalves, gastropods, sea urchins, crabs and lobsters, and bony fish rediversified and rose to its current prominence. However, the well-developed tiering of marine communities above bottom that was so prominent in the Paleozoic and Mesozoic was now a thing of the past, most likely because of increased cropping and predation. Gone, too, were the ammonites (which disappeared at the end of the Cretaceous); however, their cousins, the belemnoids, are thought to have persisted into the Eocene, perhaps because they had a reduced shell that increased their ability to escape predators. Interestingly, although the Scleractinia were now free to reoccupy the niches left vacant by the demise of rudist bivalves, reefs were not prominent during the Paleogene.

Joining the marine benthos was a new taxon of benthic foraminifera called **nummulitids** (**Figure 14.17**). Nummulitids were larger, pancake-shaped foraminifera that grew to the size of pennies or even quarters. They had complex interiors, with smaller compartments joined by tunnel-like connections, and lived in shallow seaways stretching from Europe to Indonesia. Like fusulinids, nummulitids are therefore thought to have harbored algal symbionts. During the Eocene and Oligocene, nummulitids were so abundant that they often formed the dominant components of limestones; some of these limestones were quarried to build the Egyptian pyramids. Larger foraminifera were also prominent in the extreme western Tethys, now known as the Caribbean, but these forms belonged to somewhat different taxa.

Various taxa of plankton that were so prominent during the Cretaceous and that had been hard hit during the Late Cretaceous extinctions—coccolithophorids, foraminifera, and radiolaria—rediversified in the Paleogene. Interestingly, planktonic foraminifera underwent iterative evolution (see Chapter 5) and rediversified to produce test shapes similar to those of their ancestors of the Mesozoic; this implies the same natural selective pressures were at work. By contrast, dinoflagellates and diatoms were not as severely affected by the Late Cretaceous extinctions. Both taxa produce resistant cysts that might have allowed them to better survive extinction.

Despite their rediversification, the diversity of planktonic taxa remained relatively flat or declined from the Mesozoic. One hypothesis for the relatively low diversity of coccolithophorids during much of the Paleogene is that magnesium/calcium (Mg/Ca) ratios in seawater were still too high; coccolithophorids produce low Mg-calcite shells. However, this hypothesis does not account for the relatively low diversity of other noncalcareous planktonic taxa during much of the Paleogene. Possibly, rates of nutrient runoff (as suggested by relatively low and constant $^{87}Sr/^{86}Sr$ ratios) or upwelling (because of sluggish ocean circulation) were

(a) © Sinclair Stammers/Science Source.

(b) Courtesy of Scott Fay.

FIGURE 14.17 (a) Fossil nummulitid foraminifera. **(b)** Today, nummulitids have dwindled to one genus, *Heterostegina*, shown here. Unlike its ancestors, the test (shell) of *Heterostegina* is quite thin and the genus is found living in deeper, poorly lit waters in front of reefs and related environments of the Gulf of Aqaba (Middle East), the Caribbean Sea, and the Pacific Ocean.

initially too low to sustain plankton populations sufficiently abundant and widespread to produce a diverse fossil record until later in the Paleogene and Neogene.

CONCEPT AND REASONING CHECKS

1. Why is iterative evolution among the plankton an example of natural selection?
2. How do nummulitids resemble the fusulinids of the late Paleozoic?

14.4.2 Vertebrates

H **I** The habitats of the marine reptiles were left wide open with their demise at the end of the Cretaceous. These habitats were filled by the **mammals** (**Figure 14.18**).

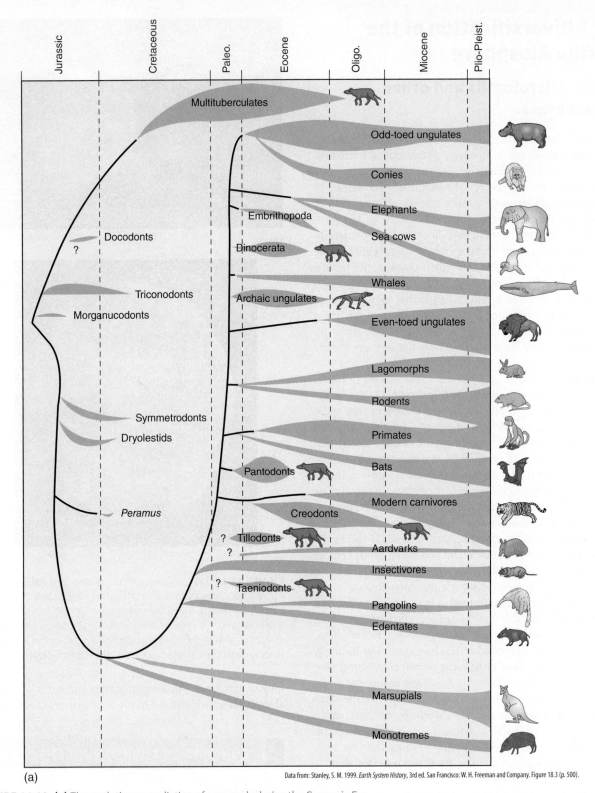

Jurassic | Cretaceous | Paleo. | Eocene | Oligo. | Miocene | Plio-Pleist.

Multituberculates

Odd-toed ungulates

Conies

Docodonts
?

Embrithopoda

Elephants

Dinocerata

Sea cows

Triconodonts

Archaic ungulates

Whales

Morganucodonts

Even-toed ungulates

Lagomorphs

Rodents

Symmetrodonts

Primates

Dryolestids

Pantodonts

Bats

Peramus

Modern carnivores

Creodonts

? Tillodonts

Aardvarks

?

Insectivores

? Taeniodonts

Pangolins

Edentates

Marsupials

Monotremes

(a)

Data from: Stanley, S. M. 1999. *Earth System History*, 3rd ed. San Francisco: W. H. Freeman and Company. Figure 18.3 (p. 500).

FIGURE 14.18 (a) The evolutionary radiation of mammals during the Cenozoic Era.

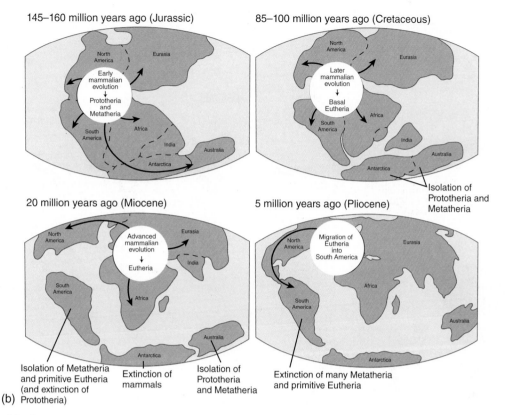

145–160 million years ago (Jurassic)

85–100 million years ago (Cretaceous)

20 million years ago (Miocene)

5 million years ago (Pliocene)

Isolation of Prototheria and Metatheria

Isolation of Metatheria and primitive Eutheria (and extinction of Prototheria)

Extinction of mammals

Isolation of Prototheria and Metatheria

Extinction of many Metatheria and primitive Eutheria

(b)

FIGURE 14.18 (b) Effects of the continental drift on the dispersion and isolation of major lineages of mammals over the 140 My spanning the Jurassic to the Pliocene.

One of the most prominent developments in the marine realm in this regard was the appearance of whales and their relatives. Whales belong to the mammalian order **Cetacea**, which also includes dolphins and porpoises. Whales evolved from a group of bear- or wolf-like hoofed carnivores called **mesonychids (Figure 14.19)**. The evolution of whales from terrestrial vertebrates is indicated by the recent discovery of whale-like ancestors with vestigial hind legs. The presence of legs in early cetaceans suggests they moved back and forth between land and sea much like seals (although seals belong to a different order of mammals, the Carnivora; see the discussion that follows). The presence of vestigial hind legs also indicates early cetaceans were probably paddlers. With time, the hind limbs became increasingly vestigial as cetaceans evolved into fully marine creatures; these forms resembled marine reptiles such as plesiosaurs and mosasaurs because of convergent evolution.

By the Oligocene whales had split into two main taxa that have persisted up to the present. The toothed whales reflect the evolutionary heritage of cetaceans: carnivory. Toothed whales prey on fish and squid and include the giant sperm whales, dolphins, killer whales, and porpoises. Dolphins and porpoises also greatly resemble the extinct marine ichthyosaurs because of convergent evolution. Occasionally, whales are found with the sucker marks of giant squid left from battle. The other line of whales includes the gigantic baleen whales. Baleen whales strain plankton (such as copepods and other crustaceans) from ocean water using a gigantic comb- or curtain-like structure in their mouths consisting of the substance baleen.

Appearing by the Late Oligocene to Early Miocene were the seals, sea lions, and walruses. These taxa evolved from a bear-like ancestor belonging to the mammalian order Carnivora (which includes cats and dogs). As we will see, the Order Carnivora and many other mammalian orders underwent tremendous diversification and evolutionary change during the Paleogene.

CONCEPT AND REASONING CHECKS

1. Why are vestigial structures evidence of natural selection?
2. What do the fossils with vestigial pelvises represent in terms of evolution?

14.5 Diversification of the Terrestrial Biosphere

14.5.1 Plants

After the drastic cooling associated with the Late Cretaceous extinctions, both gymnosperms and angiosperms (flowering plants) rediversified and became widespread. The Paleocene and especially the Eocene were relatively warm, humid times that were a continuation of those of the Late Mesozoic. Such conditions favor the growth of subtropical to tropical forests. Places like Wyoming, Montana, and the Dakotas that we today associate with the

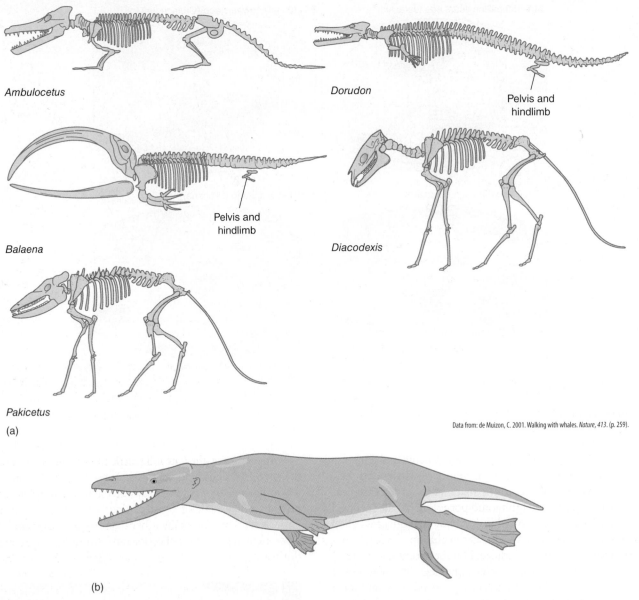

FIGURE 14.19 **(a)** Evolution of whales from mesonychids. **(b)** Reconstruction of the Eocene whale, *Rodhocetus*, about 47 million years ago.

Ambulocetus

Dorudon

Pelvis and
hindlimb

Balaena

Pelvis and
hindlimb

Diacodexis

Pakicetus

(a)

Data from: de Muizon, C. 2001. Walking with whales. *Nature, 413*. (p. 259).

(b)

climatic extremes of dry, arid conditions in summer and blizzards in winter were much like the modern tropical forests of Panama. Tropical and subtropical floras even expanded to high latitudes inside the Arctic Circle, with temperate floras occurring further north. Similar conditions occurred in Antarctica.

During this time modern families of angiosperms (flowering plants) evolved. Consequently, by the beginning of the Oligocene forests had taken on a more modern appearance because about half of all modern plant genera were present. Among the new angiosperm genera were the grasses. Early grasses could not tolerate continued grazing by herbivores and appear to have lived in moist or swampy conditions in forests. However, during the Oligocene a major adaptive breakthrough occurred: **continuous growth**. The evolution of continuous growth allowed grasses to grow back rapidly after being grazed (it is also why lawns must be mowed frequently).

The evolution of continuous growth was critical to the spread of grasslands because grasslands began to replace forests in response to drier climatic conditions during the Oligocene. Initially, savannahs (mostly grassland with some trees, like that of East Africa) appeared, followed by steppes and prairies. Continuous growth also had a profound effect on the evolution of mammals, as described in the next section.

14.5.2 Early evolution and diversification of mammals

Paleocene vertebrate faunas were dominated by mammals. To be sure, reptiles like turtles and crocodiles were present, but it was the mammals that dominated vertebrate faunas after the demise of the dinosaurs. As we saw in Chapter 13, mammals of the Late Triassic and Early Jurassic were descended from the therapsids. Therapsids might have

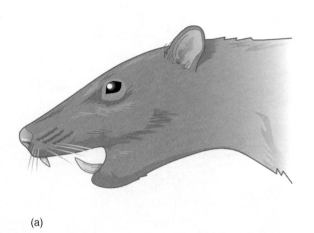
(a)

(b)

FIGURE 14.20 Reconstructions of multituberculates of the Cenozoic Era.

possessed a number of mammalian traits such as endothermy, hair, and the positioning of the legs underneath the body; it is for this reason the therapsids and related lineages were sometimes referred to in older literature as "mammal-like reptiles."

Several evolutionary lineages—some might say "experiments"—of mammals evolved from the therapsids during the Mesozoic (**Figure 14.20**). One line, the *triconodonts*, was represented mostly by shrew- to rat-like creatures with differentiated teeth. That line is now extinct; however, this group also included larger mammals that preyed upon some dinosaurs or their young (see Chapter 13). A second lineage was the squirrel or rodent-like *multituberculates*, which possessed long prehensile tails for their arboreal (tree-dwelling) life (see Figure 14.20). Multituberculates get their name from their molar teeth, which consisted of long rows of teeth with multiple bumps for grinding. They also possessed blade-shaped premolars, which might have been used to crack or slice through tough nuts and seeds, and chisel-shaped incisors, like those of rodents, which might have helped them gnaw seeds and fruits (like modern squirrels). A third lineage, the *monotremes* or egg-laying mammals, is represented today only by the duckbilled platypus and the spiny anteater found in Australia and New Guinea.

By the mid-Cretaceous mammals had diverged into the two main taxa still with us today: marsupials and placentals. Modern **marsupials** include the opossum, kangaroo, and koala, among others. Marsupials produce eggs that hatch inside the mother's womb; here the embryo attaches to the uterus, grows further, and is then born in a highly immature state. The immature embryo then moves to the mother's external pouch, where it grows further. Modern marsupials are characteristic of southern hemisphere continents, which were originally united as Gondwana in the southern hemisphere. In fact, the biogeographic distribution of marsupials can be traced to the rifting of this supercontinent. Marsupials originated in either North or South America during the Cretaceous and then migrated to Australia. As Gondwana continued to rift apart, the continents eventually separated and Australia and South America each developed its own marsupial fauna through convergent evolution. Other unique marsupial forms then evolved such as the kangaroo, which seems to fill the same ecologic niche as hoofed animals like deer.

Many of the marsupials were the ecologic counterparts of the **placental** mammals that evolved in the northern hemisphere through convergent evolution. Placentals are so named for the presence of the *placenta* in the mother, which accumulates waste products and is commonly known as the "afterbirth" in humans. In the placentals, the embryo is retained in the uterus for a much longer period of time, allowing it to grow to a more advanced state before birth. The earliest definite placentals were the insectivores, which include the modern shrew. It is from the insectivores or an insectivore-like group that other placental mammals are descended. The recently discovered skeleton of *Eomaia* ("dawn mother") in China from Early Cretaceous lake deposits (which also contain remains of feathered dinosaurs) is shrew-like in appearance (**Figure 14.21**). The skeleton of *Eomaia* extends the record of early placental mammals backward in time 40 to 50 million years, which is about 40 million years earlier than estimates from molecular clocks.

After the Late Cretaceous extinctions, terrestrial mammals began to diversify into the habitats left vacant by the demise of the dinosaurs, just as cetaceans did in the seas after the extinction of marine reptiles. Like their Mesozoic ancestors, Paleocene mammals remained relatively small and rather unspecialized, walking on all four legs.

By the Late Paleocene, however, a tremendous adaptive radiation of mammals and other vertebrates had begun that

Courtesy of Nobumichi Tamura.

FIGURE 14.21 Reconstruction of the recently discovered earliest-known placental mammal, *Eomaia*.

eventually led to about two dozen orders within the Class Mammalia (see Figure 14.18). Among the terrestrial taxa that diversified about this time were the orders **Artiodactyla** ("even-toed" hoofed mammals, or ungulates, which today include deer and many domesticated mammals such as cows), **Perissodactyla** ("odd-toed" ungulates that include the horse and rhinoceros and, as we shall see, many extinct taxa), **Proboscidea** (elephants and their relatives), **Sirenia** (manatees, or sea cows), **Carnivora** (dog- and cat-like creatures like the mesonychids, ancestral dogs and cats themselves, and bears, raccoons, and seals), **Rodentia** (rodents such as rats and mice) and **Lagomorpha** (rabbits), **Primates** (lemurs, monkeys, apes, and humans), and **Chiroptera** (bats). And these are just the orders with living representatives!

Based on the timing of first appearances in the fossil record on different continents, the Late Paleocene radiation of mammals appears to have begun in Asia. The taxa quickly migrated to North America and then Europe, indicating that they must have used the Bering land bridge between North America and Asia. Mammals would also migrate across other land bridges linking Canada, Greenland, and Europe prior to the opening of the Labrador Sea and northernmost Atlantic Ocean. The primary migration occurred within 10,000 years after the Paleocene–Eocene Thermal Maximum.

14.5.3 Archaic mammals

Ⓜ Mammals of the Cenozoic consist of two basic types: **archaic** and **modern**. Archaic taxa dominated through the Paleocene and Eocene. These taxa were either primitive representatives of orders that have persisted to the present or belonged to taxa that later became extinct. Although recognizable as mammals, archaic mammals looked quite different from more modern taxa, which came to dominate in the Neogene.

Archaic taxa were adapted for living in the lush, tropical forests and jungles of the Paleocene and Eocene and were quite diverse. Arboreal mammals included the multituberculates (noted earlier) and small lemur-like primates. The primitive Eocene primates had the well-developed eyes, opposable thumbs for grasping, and the long tails typical of modern lemurs (**Figure 14.22**) and probably fed on leaves, fruits, and seeds. Today, lemurs are restricted chiefly to the island of Madagascar, located off the east coast of Africa. The migration of mammals from Asia to North America introduced rodents and rabbits to forests in North America, where they replaced the multituberculates and lemur-like taxa by the end of the Eocene.

Many of the taxa that lived on the ground belonged to the **ungulates**, or hoofed mammals. The archaic ungulates of the Paleogene were dominated by the **perissodactyls**. All perissodactyls appear to have descended from a common ancestor that initially inhabited Asia during the Late Paleocene. Many early perissodactyls tended to be small. Early horses like *Protorohippus*, for example, were present in the forests but were no larger than a small dog. Horses migrated from China to North America in the Early Eocene, where they continued to flourish through the rest of the Tertiary Period. The perissodactyl ancestors of the tapir and rhinoceros also

© iStockphoto/Thinkstock.

FIGURE 14.22 Modern lemurs.

appeared in the Early Eocene, looking very much like early horses. Tapirs became the dominant herbivores by the Late Eocene (**Figure 14.23**). They evolved to small-to-medium sized, heavily built herbivores characterized by a long snout or proboscis. Tapirs were found in North America and Europe

© lightpoet/Shutterstock, Inc.

FIGURE 14.23 A modern tapir.

during the Paleogene but then died out there. However, they are still found in Central and South America, Malaysia, and Sumatra (Indonesia) and still occupy the same basic niche they have for tens of millions of years. Rhinoceroses, on the other hand, diversified. Early rhinoceroses were small and slim runners (much like their horse relatives), hornless (unlike modern forms), and had four toes on their front feet. Others became short-legged aquatic forms. Today, the rhinoceroses of the Old World tropics are all that is left of a once important group. Modern rhinos are large, ungainly, three-toed animals with one or two horns on top of their skulls that consist not of bone or horn but of hairs glued together.

Despite the relatively small size of many perissodactyls, some grew to large size. The **brontotheres** (or titanotheres) of the Eocene grew to the size of cows. Some developed blunt, paired horns on their snouts, but fossils from Mongolia show that some evolved structures resembling battering rams (**Figure 14.24a**). The rhinoceros-like **uintatheres** were the largest mammals of the middle Eocene (**Figure 14.24b**). They possessed tusks and had three pairs of knob-like horns running down the sides of the skull.

Some archaic (non-perissodactyl) taxa of the Paleogene were even more bizarre. The largest ground-dwelling herbivores of the Early Eocene were the **pantodonts**. Pantodonts are a zoological puzzle whose evolutionary relationships are obscure. For example, the pantodont genus *Coryphodon* had hooves but was not related to the ungulates (**Figure 14.25**). It had a large head and its molar (back) teeth resembled those of insectivores. Nevertheless, it might have eaten leaves rather than insects. Other bizarre groups were the **taeniodonts** and **tillodonts**, which grew to the size of bears. Even though some of these forms had claws (see Figure 14.25), they were likely herbivorous, as some had chisel-like incisor (front) teeth like those of rodents.

Although perissodactyls dominated the archaic ungulates, artiodactyls were present. The first artiodactyl was *Diacodexis*, which appeared in the Early Eocene of what is now Pakistan (**Figure 14.26**). This genus was about the size of a jackrabbit. Prominent among the artiodactyls of the Middle Eocene and Oligocene were the **oreodonts** (**Figure 14.27**), which persisted into the Late Miocene. Oreodonts might be distantly related to camels (which are also artiodactyls), but they resembled small sheep. They remained confined to North America throughout their existence.

Given the prominence of herbivores in the Paleogene forests, not surprisingly archaic carnivores lurked in ambush. The early members of the Carnivora appeared in the Late Paleocene, but most were weasel-like and did not become particularly prominent until much later in the Oligocene. By contrast, one extinct group, the **creodonts**, was as big as modern wolves and possessed sharp, shearing, and stabbing teeth (**Figure 14.28**). Also present were the hoofed mesonychids (which gave rise to whales).

14.5.4 Climate change and mammals

The Late Eocene and Oligocene epochs mark the transition of mammals from archaic to modern. As described

(a)

(b) © National Geographic Image Collection/Alamy.

FIGURE 14.24 Some of the large archaic mammals that developed during the Paleogene. **(a)** Brontotheres with a "battering ram" type of structure on their snouts. **(b)** The rhinoceros-like uintatheres, here shown with some small horses, were the largest mammals of the middle Eocene. They possessed tusks and had three pairs of knob-like horns running down the sides of the skull.

earlier in the chapter, this was the time when the growth of southern hemisphere glaciers caused Earth to dry out, affecting its vegetation. With the expansion of southern hemisphere ice caps and the drying out of Earth, the lush, tropical forests of the Paleocene and Eocene became increasingly restricted to low latitudes. Initially, forests became more open and then gradually became more and more mixed with grasslands, until completely giving way in many places to temperate grasslands or savannahs like those in Africa.

(a)

(b)

(c) © Tom McHugh/Science Source.

FIGURE 14.25 Some more bizarre archaic mammals.
(a) Pantodonts. **(b)** Taeniodonts. **(c)** Tillodonts.

Data from: Stanley, S. M. 1999. *Earth System History*, 3rd ed. San Francisco: W. H. Freeman and Company. Figure 18.8 (p. 502). Attributed to Rose, K. D. 1982. Skeleton of *Diacodexis*, oldest known artiodactyl. *Science, 216*, 621–623.

FIGURE 14.26 *Diacodexis*, the first known artiodactyl from the Early Eocene.

FIGURE 14.27 Oreodonts might be distantly related to camels and grew to the size of small sheep.

Data from: Kummel, B. 1970. *History of the Earth: An Introduction to Historical Geology.* San Francisco: W. H. Freeman and Company. Figure 14-26 (p. 526).

FIGURE 14.28 Creodonts were among the dominant terrestrial carnivores during the Paleogene.

With the loss of forests came the loss of habitat and food sources for many of the archaic mammals that had evolved in the Paleogene: arboreal taxa like the multituberculates, archaic ungulates, and the mesonychid and creodont carnivores all disappeared. Archaic predators did not rely so much on running as ambushing their prey. However, with the spread of the open spaces of grasslands, the techniques of archaic predators gave way to those of more advanced predators (such as dogs and cats) adapted for hunting in open spaces. These techniques included fast running and better eyesight for stalking prey.

Changes also occurred among the ungulates. With the drying out of Earth during the Oligocene, perissodactyl-dominated faunas began to give way to artiodactyls by the process of **ecologic replacement**. In fact, perissodactyls are fairly rare in modern natural environments. The evolutionary changes in the artiodactyls during their diversification are exemplified by the evolution of the horse that, of course, belongs to the perissodactyls. But the exception proves the rule. Horses not only managed to survive the drastic climate changes that occurred during the Oligocene, but their evolution demonstrates just how widespread and intense the selective pressures were on all ungulates, not just perissodactyls, and other mammals.

Horses and artiodactyls countered the newly evolving predators with the same adaptations as their predators: fast running and good eyesight (**Figure 14.29**). The loss of forests left fewer places to hide, and the only choice for survival was to run. In early perissodactyls the "hooves" consisted of several toes that provided stability on soft, moist forest floors (see Figure 14.29). These sorts of hooves were poorly adapted for running over open, dry, hard ground like that of grasslands. A solid hoof is much better adapted for swift running. Through the Oligocene (and Miocene), then, the side toes of horses and other perissodactyls were used less and less for running over the hard, open ground that was becoming increasingly common. Accordingly, the toes gradually became vestigial (see Figure 14.29). Oligocene horses like *Mesohippus* and *Miohippus* had only three toes on the front and hind feet. Three-toed horses continued to dominate in the Miocene, although the toes continued to be reduced. The genus *Equus* evolved in the Pliocene and has a single functional toe and hoof on each foot. Species of this genus include the modern horse, zebras, and donkey, which are capable of swift running.

In effect, perissodactyls came to walk (or run) on their toes. You can demonstrate this by a simple experiment. Place your hand flat on a table; this represents the primitive

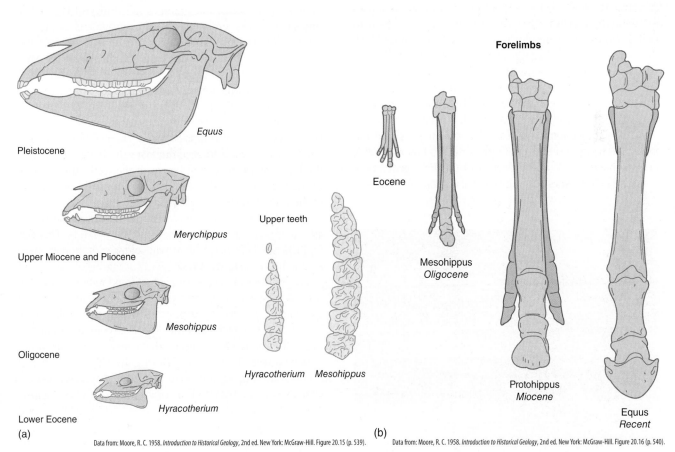

Data from: Moore, R. C. 1958. *Introduction to Historical Geology*, 2nd ed. New York: McGraw-Hill. Figure 20.15 (p. 539).

Data from: Moore, R. C. 1958. *Introduction to Historical Geology*, 2nd ed. New York: McGraw-Hill. Figure 20.16 (p. 540).

FIGURE 14.29 Changes in the limbs, skull, and teeth during horse evolution in response to the spread of grasses. Other perissodactyls and artiodactyls underwent similar skeletal modifications in response to climate change and the spread of grasslands. **(a)** Note the increased complexity and height of the crowns of teeth for grinding abrasive grasses, and backward shift in eye socket to accommodate the increased size of teeth. **(b)** Evolution of the horse limb.

state of perissodactyls in which all toes touch the ground. Now, slowly raise your wrist off the table. Notice how contact between the thumb and table is lost; this represents the earliest stage of horse evolution in which the "thumb" had already been lost. Raise your hand a bit more and contact between the little finger and table is lost; this represents the "three-toed" stage of *Miohippus* and *Mesohippus*. Raising your hand further leaves contact between only the middle finger and table, essentially representing a "one-toed" condition like that of *Equus*.

Another adaptation for swift running was an increase in body size and an increase in the length of the legs, increasing the stride. *Miohippus* and *Mesohippus* were larger than their Eocene ancestors, reaching about the size of a collie. Especially noteworthy in this regard was the largest land mammal that ever lived, the hornless rhinoceros *Paraceratherium* (**Figure 14.30**), which belonged to the perissodactyls and lived during the Oligocene in Asia. *Paraceratherium* reached 6 meters tall at the shoulder, allowing it to browse on the tops of trees, and weighed up to 20,000 kilograms (roughly about 20 tons)! These particular rhinos were the dominant herbivores of the Oligocene and Miocene. Also reaching large size among the perissodactyls were the **chalicotheres** (**Figure 14.31**). Chalicotheres appeared in the Eocene and were about the same size as the earliest horses but grew to large size during the Oligocene and later. In fact, chalicotheres strongly resembled horses but instead of bearing hooves the digits bore claws! The claws might have been used to pull down tree branches for browsing, but it is more likely they were used for digging up roots and tubers. Chalicotheres were present in North America and Eurasia during the Eocene but were rare in North America thereafter. None survived in North America beyond the Miocene, but they persisted in Asia and Africa into the Pleistocene.

Other adaptations occurred as grass was used as food and involved changes to the skull and teeth (see Figure 14.29). Like so many other perissodactyls of the Paleogene, early horses were **browsers**, feeding on forest vegetation. However,

FIGURE 14.31 Chalicotheres grew to a large size. Although they were perissodactyls, they possessed claws, not hooves.

horses and artiodactyls tended more and more toward the herbivorous type known as **grazers** in response to the spread of grasslands. Grasses are rich in a hydrated form of silica (SiO_2) known as opaline silica, or just opal; opal occurs in grasses as microscopic granules called **phytoliths**. These tiny silica bodies are highly abrasive (imagine chewing fine sand or silt) and are responsible for changes in the skull and tooth structure of horses and other mammals. By the Miocene horses and other ungulates were characterized by teeth with higher crowns, longer roots, and more complex ridges for chewing the highly abrasive grasses. Teeth also evolved continuous growth; as the upper surfaces wore down, the teeth grew upward to replace them. The jaw bone and face lengthened forward to accommodate the increasing size of the teeth. With increasing height of the teeth, the position of the eye socket in the skull also had move backward out of the way of the back rows of teeth.

In the past, these evolutionary changes were portrayed as one of gradual, progressive evolutionary trends. This was typified nowhere better than with older evolutionary trees for the horses (**Figure 14.32**). Early scenarios of horse evolution proposed a long, slow gradual trend in their evolution toward the present in the style of phyletic gradualism (see Chapter 5). However, we now know that the evolution of horses was, like so many other species, typified by branching evolution involving species that coexisted in time, if not in habitat (see Figure 14.32). During the Miocene, in North America alone there were up to 12 species. A number of them actually evolved back to browsing in forests, retaining low-crowned teeth and toes, and also spread to the Old World from North America. Most of these species died out by the end of the Miocene, leaving only *Equus*. Horses eventually died out in North America, too, presumably in response to climate change, but were later reintroduced by the Spanish conquistadors. The horses that we see today, both domesticated and wild, are descended from these introductions.

Data from: Levin, H. 2006. *The Earth Through Time*, 8th ed. Hoboken, NJ: John Wiley. Figure 16.43 (p. 513).

FIGURE 14.30 Reconstruction of the hornless rhinoceros *Paraceratherium*, formerly known as *Baluchitherium*, which lived in the Oligocene.

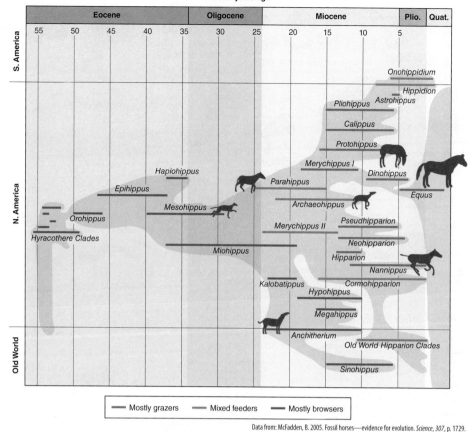

Data from: McFadden, B. 2005. Fossil horses—evidence for evolution. *Science, 307*, p. 1729.

FIGURE 14.32 Divergent—or branching—evolution of the horse in response to global cooling and the spread of grasslands. Note the shift from browsers to predominantly grazers.

14.5.5 Diversification of modern mammals

Artiodactyls such as deer, antelope, bison, camels, and a number of domesticated mammals such as cows, sheep, goats, and pigs all diversified tremendously during the Oligocene and Miocene. Many of these taxa underwent a reduction in toes similar to those in perissodactyls. Returning to our earlier hand experiment, begin with the stage at which the thumb is already lifted off the table. This stage was characteristic of many early artiodactyls and is still found in pigs and some other modern artiodactyls. Now, with your four fingers touching the table, shift them about until they rest comfortably on the table. Then, continue to lift your hand. Eventually, the second (index) and fifth (little) fingers will be raised above the table, with only the third (middle) and fourth digits still touching. The hooves of many common artiodactyls such as sheep and cows in reality consist of the third and fourth digits closely applied together, whereas the second and fifth digits were lost.

Among the artiodactyls that diversified so tremendously were the **ruminants**. Ruminants chew a cud composed of grass and other vegetation and possess a four-chambered stomach that harbors symbiotic ciliated protists that break down the cellulose of plant cell walls. Vegetation is chewed,

sent to the stomachs, partially digested, and then regurgitated and chewed a second time before it is swallowed again and finally digested. Modern ruminants include bison, camels, cattle, sheep, goats, deer, antelope, and giraffes. Camels and deer came to dominate artiodactyls in North America, whereas antelope, bison, cattle, and sheep dominated in the Old World. Camels had migrated to North America in the Eocene, where they eventually formed large herds but eventually died out there during the Pleistocene. Before camels died out in North America, they migrated to Asia and Africa, where they still exist, and to South America, where they evolved into llamas and alpacas. The U.S. cavalry briefly experimented with using camels instead of horses in the American west during the 19th century. Bison and sheep migrated to North America during the Pleistocene Ice Ages.

Also noteworthy are the elephants, which belong to the Order Proboscidea and which originated in Africa in the Eocene. The same ancestral stock gave rise to the manatees (sea cows) and dugongs of shallow subtropical and tropical waters (**Figure 14.33**). Early elephants looked much like tapirs. One line stemming from these tapir-like ancestors was the **mastodons**, which possessed four tusks and a short trunk and were browsers (**Figure 14.34**). By the Miocene numerous extinct lineages of the mastodons had spread to Europe and Asia (including Russia and Siberia) and North America.

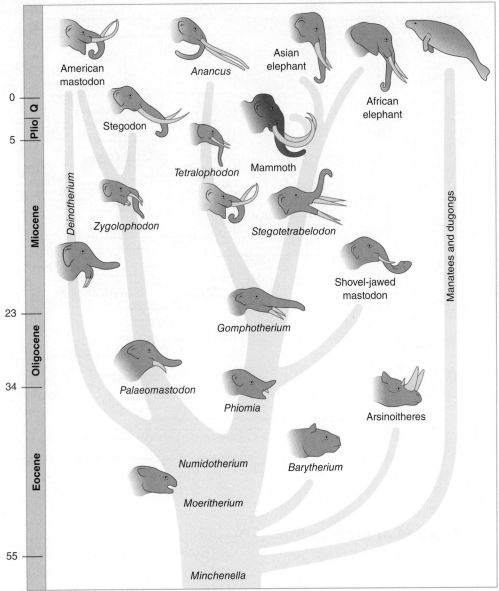

Data from: Prothero, D. R., and Dott, R. H. 2002. *Evolution of the Earth*, 6th ed. New York: McGraw-Hill. Figure 15.48 (p. 498).

FIGURE 14.33 The evolution of the Order Proboscidea, which includes elephants, mastodons, and sirenians (sea cows and dugongs).

By contrast, some artiodactyls underwent relatively little evolution. Pigs and their relatives such as peccaries are the most primitive of artiodactyls in many ways. They have retained four toes (although the side toes are much reduced), and the limbs are still short, reducing speed. The diet has also remained a relatively primitive, omnivorous type; swine will eat anything from potatoes to rattlesnakes. Pigs and their closest fossil relatives are confined to the Old World. In addition to wild boars, from which domesticated pigs are derived, a number of distinct living forms include the wart hogs of Africa. A distant relative of pigs is the hippopotamus. Although in earlier times the hippos ranged into localities as far afield as England and China, they are now found living only in tropical Africa.

14.5.6 Birds

What would Earth be like if birds, and not mammals, had replaced the dinosaurs after the Cretaceous? Despite the tremendous ecologic and evolutionary opportunities

FIGURE 14.34 A mastodon.

available to birds after the Late Cretaceous mass extinction, birds were slow to take off in the Paleocene. Flying birds, especially **passerine** (or song) **birds**, were not particularly prominent. Most birds were wading birds that lived along the shore.

Still, birds are notable during the Paleogene because of the evolution of large, flightless forms. Like marsupials, these forms developed independently on southern hemisphere continents as Pangaea rifted apart. Each Gondwana continent is characterized by its own form such as the ostrich in Africa, the emu of Australia, and the rhea of South America. Joining these taxa during the Eocene were tall (up to about 2.5 meters), flightless predatory birds like those belonging to the genus *Diatryma*. *Diatryma* undoubtedly snatched up its prey with its large beak, bearing a striking resemblance to the bipedal predatory theropods of the Mesozoic (**Figure 14.35**). Fortunately, these avian predators eventually died out in North America and most other continents. However, they persisted as late as the Pliocene in South America, which remained isolated from North America until a land bridge with Central America was established (see Chapter 15).

Diatryma gigantea *Aepyornis maximus* (elephant bird) *Dinornis maximus* (giant moa)

Data from Feduccia, A., 1980. *The Age of Birds*. Harvard University Press, Cambridge, MA.

FIGURE 14.35 Reconstructions of some extinct large, flightless birds showing their relative sizes. *Diatryma* was an early Cenozoic bird; the others date from the much later Pleistocene.

14.6 Extinction: Glaciers, Volcanoes, and Impacts

Q Like the Cretaceous Period, paleoclimatic change during the Paleogene has been heavily studied because of the implications for anthropogenic warming. As we have already seen, there are a number of important paleoenvironmental changes that occurred during the Paleogene (see Figure 14.1), but these environmental changes are not considered to have resulted in any mass extinctions of biotas on land or in the oceans (as they did during the Big Five extinctions). In fact, the extinctions that occurred during the transition from the Eocene to the Oligocene are generally regarded as *minor* extinctions (see Chapter 5).

We have already noted the significant warming that occurred during the PETM. This warming seems to coincide with the initial spread of mammals into North America and Europe. Warming might have occurred in response to an increase in atmospheric CO_2 due to seafloor spreading or, as seems to be currently favored, the massive release of methane from buried organic matter on continental shelves and its subsequent oxidation to CO_2.

As mentioned previously, the PETM is also of concern because it appears to provide an ancient analog for the acceleration of global warming by anthropogenic fossil fuel combustion and possible ocean acidification. There is a distinct PETM assemblage of calcareous nannofossils that appear during the peak warmth and then disappear. This assemblage consists of nannofossils that have normal morphologies (for example, *Coccolithus* and *Toweius* of Figure 14.14) and that seem to be strongly temperature controlled, and also of nannofossils that have abnormal morphologies that appear during maximum warmth and then disappear abruptly as temperatures begin to cool. These nannofossils might have actually been responding to ocean acidification caused by higher CO_2 levels rather than temperature.

Significant cooling occurred about 10 to 12 million years later across the Middle-to-Late Eocene boundary (about 37 to 38 million years ago). This cooling appears to coincide with the initial formation of glaciers on Antarctica and is marked by the largest extinctions during the Cenozoic Era. In the marine realm, warm-water plankton were affected by extinction. On land, cooling in the southern hemisphere marked the initiation of shrinking forests and spreading grasslands. This cooling is associated with significant turnover of terrestrial mammals, as described earlier, but the changes were drawn out over millions of years.

Only 1 to 2 million years later after this initial cooling, as many as four extraterrestrial impacts occurred (**Figure 14.36**). **Microtektites**, which are associated with shock metamorphism (see Chapter 3), are found today at the same stratigraphic level in deep-sea cores from a large area of the Pacific Ocean. These microtektite fields are known as **strewn fields** (**Figure 14.37**); the shape and direction of the strewn field and the chemistry of the microtektites can be used to infer the potential sites of impact based on the chemical composition of the underlying rock. Possible tsunami-related deposits have been traced as far away as the mouth of the Chesapeake Bay, where boulders larger than 2 meters in diameter and shocked quartz grains have been found, and a crater has also been detected and well defined in the subsurface of the Chesapeake Bay (**Figure 14.38**). In fact, this impact structure is now considered to be the fourth largest meteor crater in North America. But despite the strong evidence for impacts, they seem to have had little effect on the biosphere as compared with those at the end of the Cretaceous.

Similarly, the so-called **Terminal Eocene Event**, which occurred only 1 to 2 million years later, was marked by only minor extinctions of planktonic foraminifera and some other taxa such as radiolarians. The surviving species of foraminifera underwent iterative evolution (see Chapter 5) during the Neogene (see Figure 14.36). The extinctions associated with the Terminal Eocene Event most likely resulted from further climate change rather than impact. By about this time the Antarctic Circumpolar Current, which contributed to the formation of ice sheets on Antarctica, was initiated. The subsequent transition to the Early Oligocene 33 million years ago was marked by massive cooling, expanding glaciers on Antarctica, and increasing production of Antarctic Bottom Water (see Chapter 2). The continued growth of glaciers on Antarctica resulted in the largest sea-level regression of the entire Cenozoic during the Middle Oligocene about 30 million years ago (see Figure 14.1). Possibly, extinction at this time was subdued (as compared with the Big Five) because the biosphere had not had sufficient time to "rebound" from earlier extinctions. Many taxa adapted to warmer climates, and forests had already been decimated and replaced by taxa adapted to colder climates.

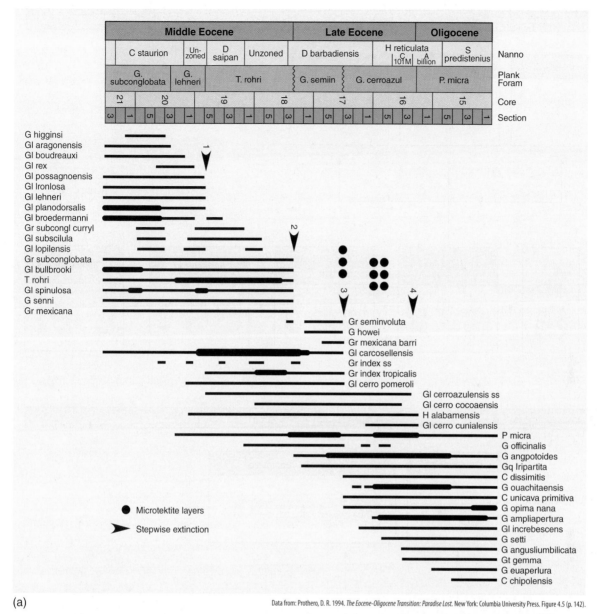

(a)

Data from: Prothero, D. R. 1994. *The Eocene-Oligocene Transition: Paradise Lost.* New York: Columbia University Press. Figure 4.5 (p. 142).

FIGURE 14.36 (a) Multiple extinctions of planktonic foraminifera in deep-sea cores during the Middle-to-Late Eocene. Note the association of extinctions with microtektite layers.

(continues)

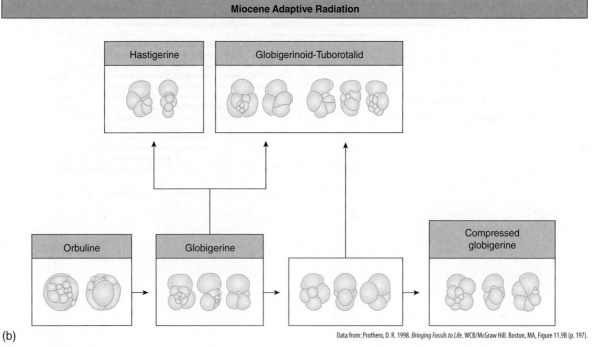

Data from: Prothero, D. R. 1998. *Bringing Fossils to Life*. WCB/McGraw Hill. Boston, MA, Figure 11.9B (p. 197).

(b)

FIGURE 14.36 (Continued) **(b)** Iterative evolution is like convergent evolution (Figure 4.14) but occurs among closely related taxa, such as the species of planktonic foraminifera shown here.

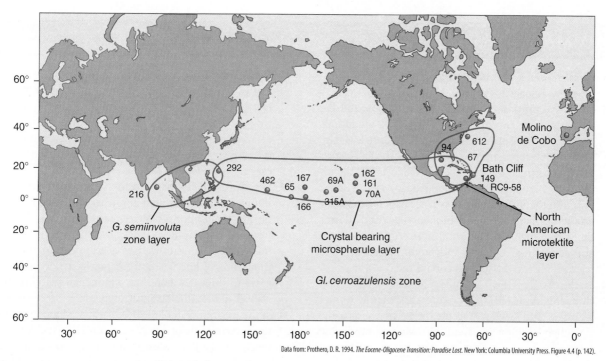

FIGURE 14.37 Late Eocene microtektite strewn fields resulted from one or more impacts during the Late Eocene. Numbers indicate deep-sea cores in which the microtektites were found.

Data from: Prothero, D. R. 1994. *The Eocene-Oligocene Transition: Paradise Lost.* New York: Columbia University Press. Figure 4.4 (p. 142).

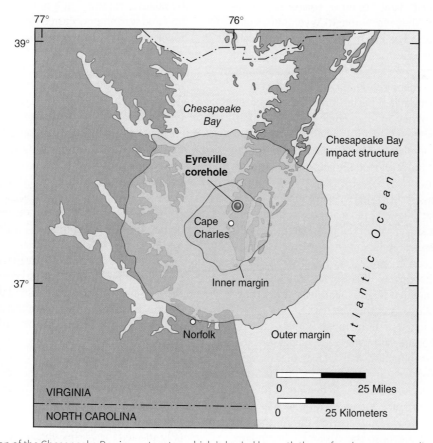

FIGURE 14.38 Location of the Chesapeake Bay impact crater, which is buried beneath the surface by younger sediments.

- The Paleocene, Eocene, and Oligocene epochs comprise the Paleogene Period.
- The initial portion of the Paleogene composed of the Paleocene and Eocene epochs is basically a continuation of the Mesozoic greenhouse. Nevertheless, significant developments occurred during the Paleogene. Cooling began in the Late Eocene–Oligocene, coinciding with the movement of Antarctica over the South Pole and the initiation of a polar ice cap in the southern hemisphere. Mountain building occurred all over the world, including the Himalayan, Alpine, Andean, and Laramide Orogenies, and produced many of the features observed at Earth's surface today.
- In the case of the Laramide Orogeny, folding and thrusting occurred well to the east of the former Nevadan and Sevier orogenic belts, north into the Canadian Rockies, and south into Mexico, producing huge thrust sheets. However, the central portion of the Laramide orogenic belt centered in the Rockies of Wyoming and Colorado was characterized by relatively little igneous activity and mostly vertical uplifts.
- Whatever its cause, the vertical uplift associated with the central portion of the Laramide belt created a series of isolated, fault-bounded basins stretching from southern Montana through Wyoming, Colorado, Utah, and New Mexico. Erosion kept pace with the uplifts, so the Rockies consisted of small, eroded peaks and knobs protruding from the surrounding sediment that otherwise buried them. Adjacent to the uplifts, down-faulted basins formed that behaved independently of one another. Some of the intermontane basins developed swamps, where extensive coal sometimes accumulated, whereas others, such as the Green River Basin, developed into gigantic lakes, with varved sediments.
- Many of the ranges and basins were eventually beveled beginning in the Oligocene, so many of the mountains in these areas today have subdued topography. As the ranges were beveled, sediment and volcanic ash accumulated to the east in remnant seaways to form fluvial deposits and floodplains that spread into the Great Plains of what are now Wyoming, central and eastern Colorado, Nebraska, and the western Dakotas.
- Volcanism was renewed along much of the Cordillera by the Late Eocene and Oligocene, suggesting the angle of plate subduction had increased. The deep marine basins of western Oregon were intruded, and volcanism spread from central Oregon north through Washington to produce the ancestral Cascades. Farther to the east, volcanism occurred during the Late Eocene in what are now the Absaroka Mountains, where Yellowstone National Park is found. Volcanism also resumed to the south in southern Colorado, northern Arizona, and New Mexico.
- Sediment from the Laramide uplifts was also shed eastward into the southern and eastern margins of North America, which was flooded by the Tejas Seaway. As the Tejas Seaway began to retreat, sediments derived from the Laramide uplifts and the continental interiors and the western sides of Appalachians were carried to the Gulf of Mexico.
- The Appalachians had already been greatly eroded, but rejuvenation resulted in further uplift and erosion during the Cenozoic. Rivers also carried sediment from the eastern side of the Appalachians to the Atlantic Ocean. As a result wedges of sediment up to many thousands of kilometers thick that generally thicken seaward built out along the continental margins and shelves of the Atlantic and Gulf coasts.
- In the western Gulf of Mexico, clastic sediments eventually came to lie on top of evaporites that had formed when the Gulf of Mexico was a gigantic evaporite basin. Terrigenous sediments were largely confined to the northern and western margins of the Gulf of Mexico, whereas broad carbonate shelves developed along Florida and the coast of Mexico.
- Much of the Paleocene and Eocene was characterized by relative warmth, but a number of significant coolings and warmings occurred during this time based on fossil plants and oxygen isotope ratios. An especially warm episode across the Paleocene–Eocene boundary, called the Paleocene–Eocene Thermal Maximum, might have resulted from the rapid release of methane from buried organic matter on continental shelves and its subsequent oxidation to CO_2.
- Fossil floras and oxygen isotopes both indicate that Earth's temperature plunged in the Late Eocene and Oligocene. By the Late Eocene–Oligocene, Antarctica had moved over the South Pole and a permanent polar ice cap began to grow because Antarctica was progressively isolated from warmer ocean currents by the Antarctic Circumpolar Current. The growth of the Antarctic polar cap sequestered huge volumes of water in ice, drawing down sea level during the Oligocene and producing cold, oxygenated water masses that began to sink to the bottom of the ocean, flushing the much warmer waters from the ocean basins.
- Changes in continent–ocean configurations also occurred in the Arctic during the Paleogene. As the Atlantic basin continued to open, the mid-Atlantic Ridge moved northward like a zipper so that cold waters from the Arctic began to spill southward into the Atlantic basin.
- Changes in continent–ocean configurations undoubtedly affected the oceans' chemistry. Although shallow-water limestone deposition was still widespread during the Paleocene and Eocene, sea-level fall associated with orogeny and glaciation caused the erosion of

continental shelves and shallow-water limestones, which might have contributed to the deepening of the CCD. Nutrient inputs associated with uplift and erosion would have also stimulated the growth of calcareous (and other) plankton farther from shore, deepening the CCD. Rising strontium isotope ratios during this time also suggest decreased hydrothermal input from seafloor spreading centers, which might have begun to shift calcite oceans to aragonite ones as Earth began the transition to cooler, drier conditions during the late Paleogene and Neogene.

■ After the Late Cretaceous extinctions, the Modern Fauna composed of bivalves, gastropods, sea urchins, crabs and lobsters, and bony fish rediversified and rose to its current prominence. Various taxa of plankton that were so prominent during the Cretaceous—coccolithophorids, foraminifera, and radiolaria—that had been hard hit by the Late Cretaceous extinctions rediversified in the Paleogene, although their diversity remained relatively flat or declined from the Mesozoic.

■ The Cenozoic Era is sometimes referred to as the "Age of Mammals." Terrestrial mammals also underwent a tremendous evolutionary diversification into the habitats and niches left vacant by the extinction of the dinosaurs at the end of the Cretaceous. Primitive whales diversified in the seas, whereas perissodactyls (odd-toed ungulates, or hoofed mammals) expanded during the Paleocene and Eocene. The movement of Antarctica over the South Pole began to dry out Earth. As a result, the forests began to shrink and grasslands expand and the mammals began to shift from forest-dwelling browsers dominated by perissodactyls to artiodactyls (even-toed ungulates) that graze on grasslands.

■ Although one or more impacts occurred during the Late Eocene, extinctions were relatively minor. Instead, it appears that it was tectonism that had the most long-lasting effect on global climate and biotas. Sea level fell, and there was a dramatic turnover in deep-sea benthic oraminifera as deep-ocean circulation increased and deep-water masses cooled and became more oxygenated.

KEY TERMS

Alpine Orogeny

Andean Orogeny

Antarctic Circumpolar Current

archaic mammals

Artiodactyla

brontotheres

browsers

Carnivora

Cetacea

chalicotheres

Chiroptera

continuous growth

creodonts

ecologic replacement

grazers

Himalayan Orogeny

hyperthermals

intermontane basins

Lagomorpha

Louann Salt

mammals

marsupials

mastodons

mesonychids

microtektites

Mississippi Embayment

nummulitids

oreodonts

Paleocene–Eocene Thermal Maximum (PETM)

pantodonts

para-Tethys

passerine birds

Perissodactyla

perissodactyls

phytoliths

placental

Primates

Proboscidea

Rodentia

ruminants

Sirenia

strewn fields

taeniodonts

Tejas Seaway

Terminal Eocene Event

tillodonts

uintatheres

ungulates

REVIEW QUESTIONS

1. What was the effect of the uplift of the Himalayas on global climate? (See also Chapter 3.)
2. During the Paleogene what was the effect of continental movements on each of the following: (a) sea level, (b) atmospheric CO_2 concentrations, (c) ocean currents and circulation, (d) oxygenation of the oceans? Explain your answers.
3. Why did such tremendous diversification of mammals occur during the Paleogene? Why did the ungulates change from being dominated by perissodactyls to domination by artiodactyls?
4. What are the different lines of evidence indicating that overall the Paleogene was relatively warm?

What is the evidence for cool episodes during the Paleogene?
5. Which modern features of Earth were produced during the Mesozoic? What were the processes involved?
6. What were the major developments in marine benthos and plankton during the Mesozoic?
7. What were the major developments in terrestrial floras during the Paleogene, and why did they happen? How did these events affect the evolution of mammals?
8. What is a vestigial structure? What is the significance of vestigial structures in the evolution of marine mammals?
9. What was the cause(s) of the Late Eocene extinctions?

FOOD FOR THOUGHT:
Further Activities In and Outside of Class

1. Compare the Late Eocene and Late Cretaceous extinctions in terms of mechanisms. Do you believe the Late Eocene and Late Cretaceous extinctions resemble each other or are very different? Why?
2. How do ancient geologic structures affect modern topography and climate? Give examples.
3. What do you suppose accounts for the resemblance of large, flightless birds to bipedal theropods?
4. Why are archaic mammals recognizable as mammals?

5. Why did ruminants undergo such tremendous evolution as grasslands spread?
6. Does the evolution of archaic mammals represent parallel or convergent evolution? What about the evolution of modern mammals?
7. Give as many examples of ecologic replacement as you can think of that occurred during the Cenozoic Era.

SOURCES AND FURTHER READING

Beard, C. 2002. East of Eden at the Paleocene/Eocene boundary. *Science*, *295*, 2028–2029.

Bowen, G. J., et al. 2002. Mammalian dispersal at the Paleocene/Eocene boundary. *Science*, *295*, 2062–2065.

de Muizon, C. 2002. Walking with whales. *Nature*, *413*, 259–260.

Foreman, B. Z., Heller, P. L., and Clementz. 2012. Fluvial response to abrupt global warming at the Palaeocene/Eocene boundary. *Nature*, *491*, 92–95.

Gingerich, P. D., Haq, M. ul, Zalmout, I. S., Khan, I. H., and Malkani, M. S. 2002. Origin of whales from early artiodactyls: hands and feet of Eocene Protocetidae from Pakistan. *Science*, *293*, 2239–2242.

Ji, Q. et al. 2002. The earliest known eutherian mammal. *Nature*, *816*, 816–822.

Norman, D. 1994. *Prehistoric life: Rise of the vertebrates*. New York: Macmillan.

Murphy, J. B., Oppliger, G. L., Brimhall, G. H., and Hynes, A. 1999. Mantle plumes and mountains. *American Scientist*, *87*(2):146–153.

Prothero, D. R. 1994. *The Eocene-Oligocene transition: Paradise lost*. New York: Columbia University Press.

Romer, A. S. 1953. *Man and the vertebrates.* Chicago: University of Chicago Press.

Sexton, P. F. et al. 2011. Eocene global warming events driven by ventilation of oceanic dissolved organic carbon. *Nature, 471,* 349–353.

Sexton, P. F., Norris, R. D., Wilson, P. A., Pälike, H., Westerhold, T., Röhl, U., Bolton, C. T., and Gibbs, S. 2011. Eocene global warming events driven by ventilation of oceanic dissolved organic carbon. *Nature, 471,* 349–352.

Stanard, A. 2004. Desert idylls: Wanderings among the saguaros and hoodoos of the Southwest. *Philadelphia Inquirer,* June 13, N6.

Stokstad, E. 2002. "Fantastic" fossil helps narrow data gap. *Science, 296,* 637.

Svensen, H. et al. 2004. Release of methane from a volcanic basin as a mechanism for initial Eocene global warming. *Nature, 429,* 542–545.

Thewissen, J. G. M., Williams, E. M., Roe, L. J., and Hussain, S. T. 2002. Skeletons of terrestrial cetaceans and the relationship of whales to artiodactyls. *Nature, 413,* 277–281.

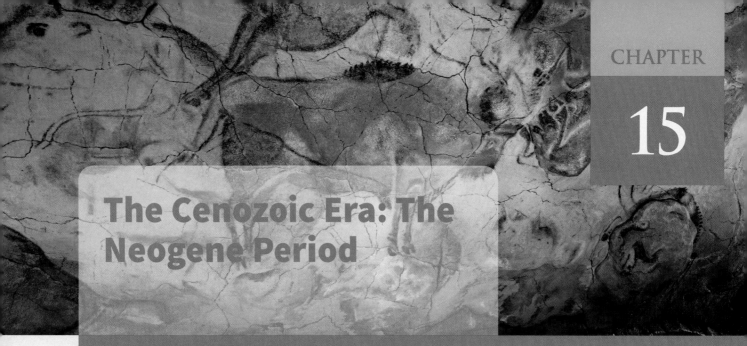

The Cenozoic Era: The Neogene Period

A How did Neogene Earth evolve to become modern Earth as humans know it?

B How did the western United States evolve geologically?

C How did the San Andreas Fault system originate?

D Did the west widen in response to the formation of the San Andreas Fault system?

E How did the movements of the continents affect orogeny, sea level, ocean chemistry, sedimentation, and climate elsewhere during the Neogene?

F What caused northern hemisphere glaciers to advance and retreat during the Pleistocene Epoch?

G How and why did ecosystems change on land and in the oceans as Earth moved from the Paleogene into the Neogene?

H What have grasses got to do with it?

I How are humans like and different from other mammalian groups?

J Was human evolution punctuated or gradual?

K Why did ancestral humans leave Africa?

CHAPTER OUTLINE

Art from the cave walls of Altamira, located on the north coast of Spain. One of many images, this art dates from the Upper Paleolithic (or Late Stone Age), dating between 40,000 and 10,000 years ago, approximately when human behavior was becoming more modern.

© age fotostock/Alamy Stock Photo.

15.1 Introduction to the Neogene

The Neogene Period spans about the past 25 million years and represents the last major phase of Earth's history up to the present. During the Neogene the configuration of the continents and oceans and many of the mountain ranges and other geologic features reached their present, familiar form. Many of the phenomena represent a continuation of systems processes that began in the Paleogene, but other processes seem to have changed dramatically.

In western North America, the tectonic style of collision and subduction to produce mountain ranges shifted to one involving lateral movements to produce great fault systems. The history of these processes is quite complex and by no means resolved, and we again use the method of multiple working hypotheses (see Chapter 1) to examine the various mechanisms proposed for the evolution of western North America.

The Neogene also marks the continued cooling of Earth that began in the Paleogene during the second tectonic cycle of the Phanerozoic Eon (**Figure 15.1**; see Chapter 14). Not only did glaciers persist on Antarctica, but ice eventually formed in the northern hemisphere several million years ago in response to the closing of the Isthmus of Panama. As we will see, northern hemisphere glaciers then advanced and retreated in response to changes in the solar flux to Earth's surface. Changes in the solar flux were not due to the sun itself but to natural changes in the behavior of Earth in its orbit around the sun. We examine the evolution of hypotheses proposed for the advance and retreat of these glaciers, as well.

Finally, the waxing and waning of glaciers might have influenced one of the most significant events in Earth's biotic history: the evolution and spread of humans (refer to this chapter's frontispiece). Here, too, a number of hypotheses have been proposed. As human populations expanded and spread over the planet, they had an increasing impact on the environment, especially in the 19th and 20th centuries. This

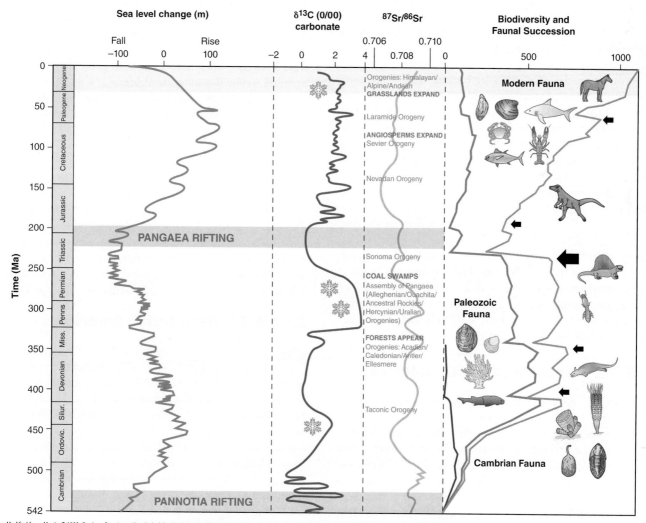

Modified from: Martin, T. 1998. *One Long Experiment.* New York: Columbia University Press. Figure 7.1 (p. 166) and Condie, K. C., and Sloan, R. E. 1998. *Origin and Evolution of Earth: Principles of Historical Geology.* Upper Saddle River, NJ: Prentice-Hall. Figure 14.1 (p. 383).

FIGURE 15.1 The physical and biologic evolution of the Earth through the Phanerozoic Eon. This chapter discusses the Neogene period of the Cenozoic Era (shaded). Snowflakes indicate times of major glaciation. Carbon isotopes and strontium isotopes are indicated (see Chapter 9 for elementary discussion). Arrows indicate the Big Five mass extinctions.

15.2 Tectonics and Sedimentation

15.2.1 Europe, Asia, and Africa

(A) In Europe and Asia, the tectonic phenomena that began in the Paleogene (see Chapter 14) continued through the Neogene. The Tethys Seaway continued to close with the ongoing collision of India and Asia to form the Himalayas (see Chapter 14) and the crushing of smaller plates between the African plate and southern Europe to raise the Alps. Together, these collisions produced a vast belt of mountains spanning southern Europe, the Middle East, and Asia that is Eurasia's equivalent of the American Cordillera.

Tectonic closure of the eastern portion of the Tethys running through Asia and the closure of the Straits of Gibraltar opening to the Atlantic Ocean continued to turn the Tethys into a series of inland lakes by the Late Miocene, about 6 million years ago. Only the Black, Caspian, and Aral seas, along with the Mediterranean, are left from the once-great Tethys (**Figure 15.2**). At the same time, sea level was falling, possibly because of the growth of ice on Antarctica. As a result, circulation in the Mediterranean became highly restricted. Just like the Gulf of Mexico during the rifting of Pangaea (see Chapter 13), enormous volumes of salt were deposited during the so-called "Messinian salinity crisis." Then, sea level began to rise in the Early Pliocene, reflooding the Mediterranean and probably transforming the Straits of Gibraltar into a gigantic waterfall that far surpassed Niagara Falls today.

Elsewhere, rift valleys and volcanoes (such as Mt. Kilimanjaro) originated during the Pliocene in east Africa in response to the appearance of a triple junction. As the Arabian Peninsula rifted toward what is now Iran, the narrow, deep-water troughs of the Red Sea and Gulf of Aden began to open (see Chapter 2), but as Arabia continued to move northeastward, it eventually collided with Iran to produce the Zagros Mountains (see Chapter 2). To the west, in what is now Iraq, the Fertile Crescent, or "Cradle of Civilization," would appear millions of years later.

15.2.2 Central and South America

One of the most significant events that affected climate beginning in the Neogene—and continuing up to the present—occurred in Central America. About 3 to 3.5 million years ago, the remnant of the Tethys that ran between North and South America and extended into the Pacific Ocean was closed by a land bridge now represented by Central America (**Figure 15.3**).

This land bridge developed from a volcanic arc that originated when a newly developing subduction zone connected the separate subduction zones located along the western margins of North and South America. As this new subduction zone formed, the smaller Caribbean plate, which was originally a portion of Pacific Ocean crust, was cut off. In the process, what are now the islands of the Greater Antilles—Cuba, Puerto Rico, Hispaniola (the Dominican Republic and Haiti), and Jamaica—moved from the southern end of the North American Cordillera into the Caribbean Sea. The Caribbean Sea is therefore also a remnant of the ancient Tethys like the Mediterranean and the great inland seas of Eurasia. As these islands moved toward their modern positions, newly developed transform faults along the margin of the Caribbean plate transferred Cuba to the North American plate (see Figure 15.3).

In South America, the Andes continued to rise, and the sediment shed to the east continued to fill the foreland basin of Amazonia with molasse. Today, the Amazon River, which drains the great Amazon rain forest, flows down an aulacogen left over from the rifting of Pangaea that began approximately 200 million years earlier.

15.2.3 Atlantic and Gulf Coasts of North America

The erosion and sedimentation that began in the Paleogene along the Atlantic and Gulf Coasts continued into the Neogene. These ancient geologic settings prefigured the modern coasts of these regions. Although much of the Appalachians were strongly eroded during the Paleogene, they were once again rejuvenated during the Miocene as a result of isostatic uplift (**Figure 15.4**). Much of the sediment was shed onto the passive margin of eastern North America into a series of shallow embayments (see Figure 15.4) separated by gentle uplifts. At the same time, the Gulf of Mexico continued to serve as a gigantic trap for sediment shed during the Laramide Orogeny and later tectonism in the western United States. Initially, sediments of Paleogene age were deposited in deep-water environments that are now buried beneath the surface on land. Later, tremendous volumes of sediment were again flushed into the Gulf of Mexico by numerous rivers, building the continental margin out and over the Jurassic Louann Salt, which continued to subside and deform into salt diapirs to form traps for oil and gas (see Chapter 17).

15.2.4 Western North America

(B) The unusual tectonism of the Paleogene continued in the western United States (**Figure 15.5**). The Rockies were again uplifted by several kilometers during the Miocene while being simultaneously beveled. In the case of the **Front Range** just west of Denver, total uplift has been estimated at about 7 kilometers! As uplift occurred, the softer sedimentary cover that had been eroded and redeposited from Paleogene uplifts was also stripped away. Many of the river sands and gravels were shed south to the Gulf of Mexico and east to form the Great Plains. Among these deposits the Ogallala Formation is noteworthy. Today, the Ogallala is the primary **aquifer** (water-conducting sedimentary layer) for much of the Great Plains from Texas to the Dakotas. During the Pliocene, block faulting in the Teton Mountain Range of western Wyoming exposed Archean rocks from which the

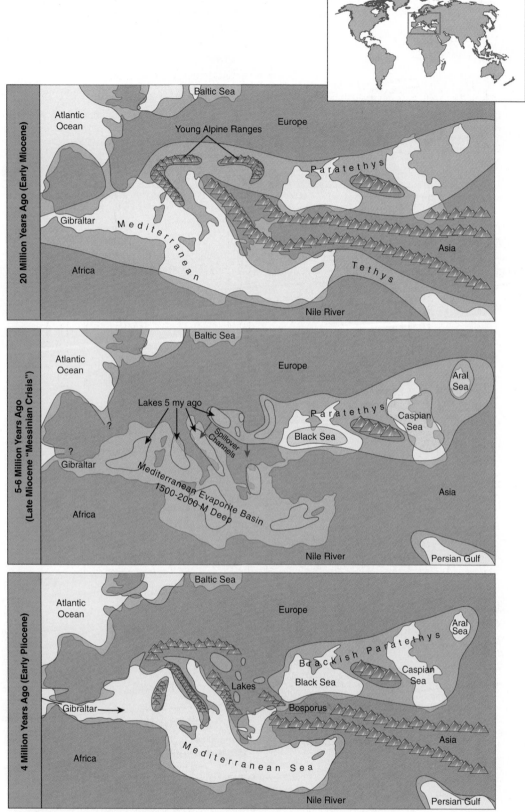

Data from: Brinkman, R. 1960. *Geologic Evolution of Europe*. Stuttgart: Verlag, New York: Hafner Publishing; Wills, L. J. 1951. *A Palaeogeographical Atlas of the British Isles*. London: Blackie; and Hsü, K. J. 1978. When the Black Sea was drained. *Scientific American, 238*, 52–63.

FIGURE 15.2 The Black, Caspian, and Aral Seas, along with the Mediterranean, are remnants of the ancient Tethys and ParaTethys seaways that were destroyed by the collision of India with Asia and the northward movement of the African plate against southern Europe.

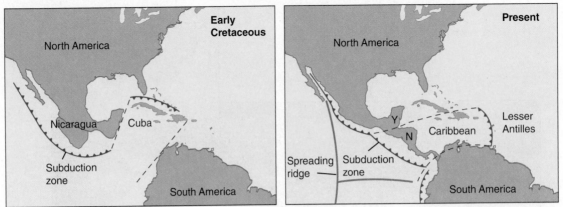

FIGURE 15.3 Origin of the Caribbean Sea and the Greater Antilles. The formation of a subduction zone between North and South America resulted in the isolation of a portion of Pacific Ocean crust to produce the Caribbean plate. A volcanic island arc that produced a land bridge between North and South America formed along the site of the new subduction zone. The rise of the Isthmus of Panama closed off the Tethys and is thought to have diverted warm, tropical surface currents to the north, where their moisture then fed the growth of northern hemisphere ice caps.

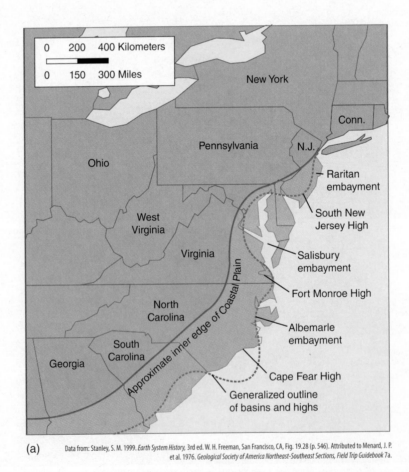

FIGURE 15.4 Rejuvenation and erosion during the Cenozoic produced the modern topography of the Appalachians. **(a)** Much of the sediment eroded from the Appalachians during the Cenozoic was shed into a series of shallow embayments along the Atlantic Coast. This sediment comprises the youngest portion of the Atlantic Coastal Plain, which extends offshore as the continental shelf.

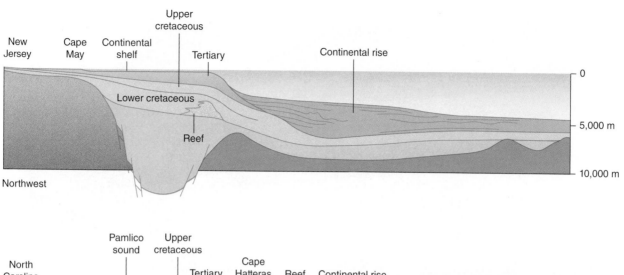

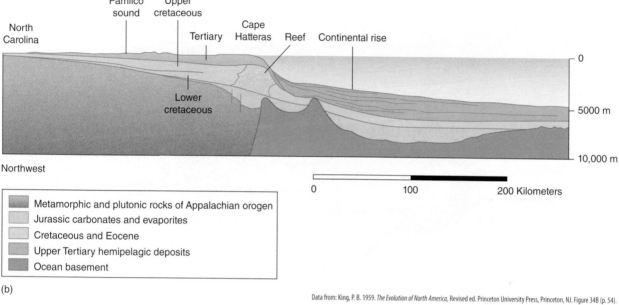

Metamorphic and plutonic rocks of Appalachian orogen
Jurassic carbonates and evaporites
Cretaceous and Eocene
Upper Tertiary hemipelagic deposits
Ocean basement

(b)

Data from: King, P. B. 1959. *The Evolution of North America,* Revised ed. Princeton University Press, Princeton, NJ. Figure 34B (p. 54).

(c)

FIGURE 15.4 (b) Cross-section of the Atlantic margin of the United States off Cape Hatteras, North Carolina. Note that the margin is a passive one. Also note the presence of Jurassic carbonates and evaporites left from the rifting of Pangaea and the presence of a reef of Cretaceous age that grew along the margin of (then) North America as the Atlantic Ocean continued to open. **(c)** Today, Miocene deposits—some of which are highly fossiliferous—are exposed in places such as Calvert Cliffs along Chesapeake Bay. These sediments were deposited in the shallow embayments and on the shelves that formed along the Atlantic Coast.

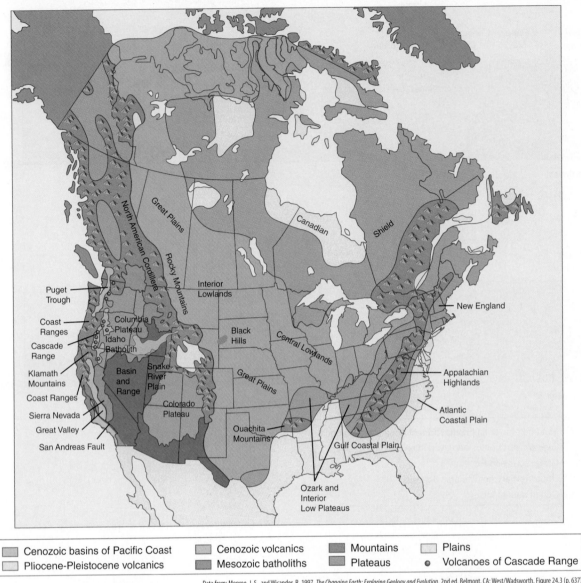

| Cenozoic basins of Pacific Coast | Cenozoic volcanics | Mountains | Plains |
| Pliocene-Pleistocene volcanics | Mesozoic batholiths | Plateaus | Volcanoes of Cascade Range |

Data from: Monroe, J. S., and Wicander, R. 1997. *The Changing Earth: Exploring Geology and Evolution*, 2nd ed. Belmont, CA: West/Wadsworth. Figure 24.3 (p. 637).

FIGURE 15.5 Major physiographic provinces of the lower 48 states of the United States reflect both relatively recent and ancient geologic events.

much younger overlying sediments were also eventually stripped away (**Figure 15.6**).

Gentler uplift also occurred further south in the region of the Colorado Plateau (see Figure 15.5). Here, the tectonic mechanism also remains puzzling. The Colorado Plateau embraces the "Four Corners" region of southeast Utah, southwest Colorado, northwest New Mexico, and northeast Arizona (see Figure 15.5). During the Late Miocene and Pliocene, the Colorado Plateau rose as much as 1.5 kilometers as a single gigantic block. Despite the uplift, the strata remained relatively horizontal or were only gently folded. Nevertheless, the uplift caused rivers such as the Colorado to erode downward beginning in the Pliocene (about 5 million years ago) to produce spectacular features like the Grand Canyon. The same broad uplift coupled with downcutting by the Virgin River also produced the spectacular cliffs of Zion National Park, exposing the ancient sand dunes of the Jurassic Navajo Sandstone (see Chapter 13). Late Paleozoic, Mesozoic, and Paleogene rocks in Arches, Bryce, Canyonlands, and Mesa Verde parks, as well as in Monument Valley in southeast Utah, were also exposed by this uplift (see Chapter 13). Many of the rocks now exposed in the Colorado Plateau region were originally deposited in shallow to marginal marine, freshwater, and terrestrial settings. Their exposure has caused the iron to oxidize to beautiful red, orange, and brown hues; hence, the nickname of the Four Corners region: "Red Rock country."

15.2.5 West Coast of North America

As uplift occurred over large areas of western North America's interior, subduction continued along what was then the West Coast as the North American plate moved over the Farallon plate (**Figure 15.7**). Some idea of the amount of movement

(a)

Courtesy of NPS.

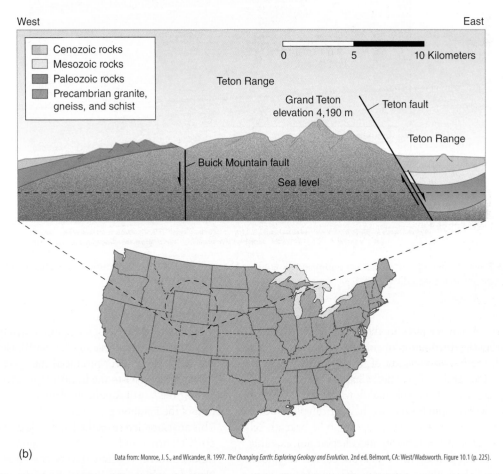

(b)

Data from: Monroe, J. S., and Wicander, R. 1997. *The Changing Earth: Exploring Geology and Evolution.* 2nd ed. Belmont, CA: West/Wadsworth. Figure 10.1 (p. 225).

FIGURE 15.6 The Teton Mountain Range. **(a)** View of the Grand Tetons, looking west. The steep east face is an eroded fault scarp, exposing Precambrian rocks up to ~2.8 billion years old (late Archean). The rugged peaks, not unlike those seen in the Alps of southern Europe, result from erosion, especially by glaciers. The Snake River lies in the foreground and flows through the Snake River volcanic plain described in the text. **(b)** Generalized cross-section of the Teton Range. Note the normal fault dipping steeply to the east; the fault forms the eastern face of the range shown in part A of the figure. Paleozoic rocks dip more gently to the west, but these and older formations have been stripped away by erosion in the Park.

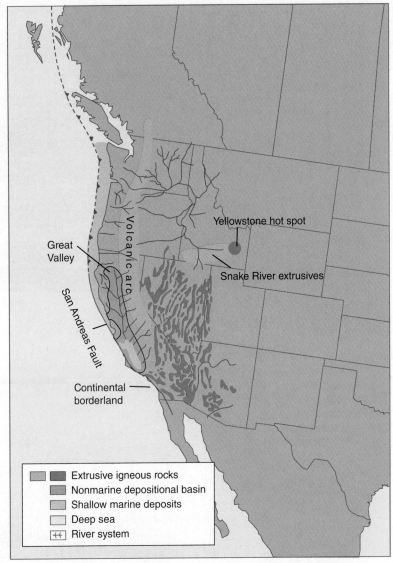

Great
Valley

Volcanic arc

San Andreas Fault

Yellowstone hot spot

Snake River extrusives

Continental
borderland

Extrusive igneous rocks
Nonmarine depositional basin
Shallow marine deposits
Deep sea
River system

Data from: Stanley, S. M. 1999. *Earth System History*. 3rd ed. San Francisco: W. H. Freeman and Company. Figure 19.26 (p. 542). Stated to be modified from Armentrout, J. M.,
Cole, M. R., and TerBest, H., Jr. (Eds.) 1979. Cenozoic paleogeography of the western United States. SEPM Pacific Coast Paleography Symposium 3, 297–323.

FIGURE 15.7 The changing tectonic setting and paleogeography of the western United States during the Neogene. Note the widening of
Nevada, as volcanism increased.

of the North American plate to the west and northwest is
apparent from the distribution of volcanic eruptions emanating from the Yellowstone region of northwestern Wyoming
(see Figures 15.5 and 15.7; see also Chapter 2). The progressively younger ages of the volcanic flows from west to east
suggest they were erupted as the North American plate moved
west, perhaps over a hotspot that appears to lie beneath Yellowstone. In eastern Washington and Oregon, for example,
are found the Columbia River plateau basalts of Middle
Miocene Age. These flows represent massive fissure eruptions stacked on top of one another that cover hundreds of
thousands of square kilometers (see Figures 15.5 and 15.7).
By the Late Miocene and Pliocene, the site of the eruptions
had shifted eastward (relative to plate motion) into southern
Idaho to form the Snake River plain (see Figures 15.5 and
15.7). Today, volcanic activity is located in the vicinity of Yellowstone National Park, where the spectacular hot springs

indicate the presence of a magma source beneath the surface
(refer to the frontispiece of Chapter 8). Here, too, erosion by
the Yellowstone River produced the spectacular Yellowstone
River Gorge (refer to the frontispiece of Chapter 8).

As the North American plate moved westward, subduction of the Farallon plate beneath it produced a volcanic arc.
This arc extended from the Pacific Northwest into southeastern California and Nevada (see Figure 15.7). At the same
time sediments continued to be shed from the rising **Sierra
Nevada**, which had formed from the uplift and intrusion
of granitic plutons originally emplaced during the Jurassic
Nevadan Orogeny (see Chapter 13). The sediments from the
rising Sierra continued to be shed westward into what is now
the Central Valley of California, which at this time consisted
of a series of deep-water basins like those now located off
the coast of southern California (see Chapter 4). Today, these
ancient basins are rich in petroleum formed from the dead

FIGURE 15.8 The eastern face of the Sierra Nevada is essentially an enormous fault scarp that marks the western edge of the Basin and Range.

remains of plankton. During the Pliocene the granitic plutons of the Sierra Nevada were finally exposed by faulting that tilted them to the west. Despite its spectacular elevation, the east-facing fault scarp of this gigantic block is only about 5 million years old (**Figure 15.8**)!

Further west, subduction continued to form the **Coast Ranges** of California. These mountains, which are more like large hills, are largely composed of the deep-water

Franciscan mélange that originally formed in the Mesozoic (see Chapter 13). Outcrops of the Franciscan, consisting of well-stratified but often deformed diatomaceous sediments, sometimes lying on top of serpentinized seafloor crust, are commonplace along the coast north of San Francisco and elsewhere along the Pacific Coast.

15.2.6 Evolution of the San Andreas Fault system

In the Late Oligocene (roughly 30 million years ago) the portion of the North American plate represented today by central and southern California began to encounter a transform—or offset—in the seafloor spreading center known as the **East Pacific Rise** (**Figure 15.9**). The East Pacific Rise had originated earlier and was the site of formation of the **Farallon plate**, which had been spreading and subducting to the southeast beneath the North American plate (see Chapter 14). At the same time, the **Pacific plate** spread from the East Pacific Rise to the west-northwest. As the North American plate encountered the transform, subduction of the Farallon plate beneath the North American plate ceased. So too did volcanism in the volcanic arc located to the east. In the process, the old Farallon plate eventually split into two plates: the **Juan de Fuca plate** to the north, which is still being subducted beneath the North American plate along the Pacific Northwest, and the **Cocos plate** to the

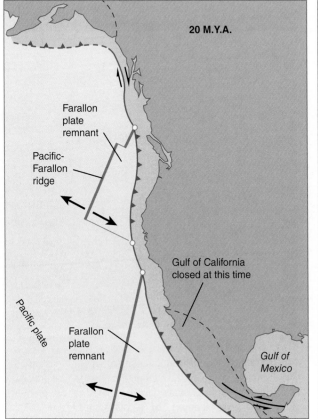

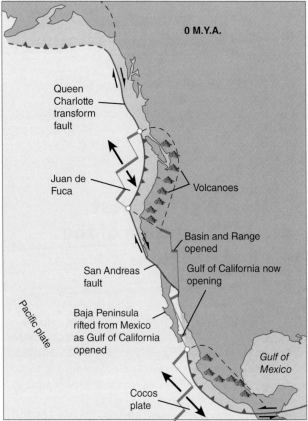

Data from: Monroe, J. S., and Wicander, R. 1997. *The Changing Earth: Exploring Geology and Evolution*, 2nd ed. Belmont, CA: West/Wadsworth. Figure 24.11 (p. 642).

FIGURE 15.9 Evolution of the San Andreas Fault. It is hypothesized that the overriding of the East Pacific Rise by the North American plate resulted in rifting of the Baja Peninsula from mainland Mexico and the formation of the San Andreas Fault.

south along Mexico and Central and northern South America. Eventually, the North American plate began to encounter the Pacific plate. But, unlike the Farallon plate, the Pacific plate was not subducted. Instead, the northwest lateral movement of the Pacific plate along the old transform became dominant. As a result, the western and southwestern portion of California began to move in a northwesterly direction relative to the North American plate (see Figure 15.9).

This change in plate motions is thought to have resulted in the transformation of the transform fault into the **San Andreas Fault** system (see Figure 15.9). The San Andreas Fault is a *right* lateral strike-slip fault. This means if an observer were to stand on the sliver of crust on the west side of the San Andreas Fault and look east, the North American plate would appear to be moving horizontally to the right (or southeast in this case) relative to the crust west of the fault on which the observer was standing (see Figure 15.9). Tectonism associated with the San Andreas fault system is responsible for the series of deep-water basins and islands located off the coast of California (see Chapter 4). Further south, as the North American plate continued to override the Pacific plate and the East Pacific Rise, seafloor spreading associated with the rise began to pull the sliver of Mexico known as the Baja Peninsula away from the mainland to form the Gulf of California (see Figure 15.9). Today, the East Pacific Rise disappears beneath the North American continent, possibly against a transform fault associated with the San Andreas Fault system.

CONCEPT AND REASONING CHECKS

1. How did the tectonic and climatic evolution of western North America initially resemble that of the Himalayas (see Chapter 2)? How did it differ?
2. Diagram how a transform fault could turn into a strike-slip fault.

15.3 How Was the West Widened? Evolution of the Basin and Range

D We are now in a position to examine how much of the American West and Southwest, lying between the Rockies to the east and the Sierra Nevada to the west, came into existence. Much of this region is encompassed by the **Basin and Range Province** (see Figure 15.5; **Figure 15.10**) This province extends from eastern California, Oregon, and Idaho across all of Nevada to western Utah and to the south through Arizona (all the way to Mexico City) and east into New Mexico to include the Rio Grande Rift (see Figure 15.10). Although the **Rio Grande Rift** has normally been considered to be a separate extensional province, some workers have recently included the Rio Grande Rift in the Basin and

(a) © Nagel Photography/Shutterstock, Inc.

(b) Courtesy of Ronald Martin, University of Delaware

FIGURE 15.10 The Basin and Range. **(a)** This photo is very representative of the Basin and Range in Nevada. The province consists of a seemingly endless series of flat-bottomed valleys (grabens) bordered by uplifts (horsts), all desert. **(b)** The beginnings of the Rio Grande river and rift near Taos, New Mexico.

Range because the origins of both features seem to be related. Today, the Rio Grande River flows from the southern Rockies down the rift before it breaks out of the desolate, beautiful Big Bend country of west Texas, turning southeast along the Texas–Mexico border to flow to the Gulf of Mexico (see Figure 15.10).

15.3.1 Background

The Basin and Range is an extensional tectonic regime. This is indicated by a series of north–south-trending **horst-and-graben** structures (see Figure 15.10; see Chapter 2). The grabens represent valleys formed by the downward movement of crust along normal faults relative to the intervening horsts that separate the valleys. Extension is also suggested by two other basic geologic observations: the crust underlying the Basin and Range is relatively thin (about 20 to 30 km) and volcanism has been widespread.

The Basin and Range records a complex history of tectonism and sedimentation. Some of the most critical evidence for identifying the mechanisms responsible for the formation of the Basin and Range are the timing and direction of extension. Extension appears to have begun about 55 to 45 million years ago (early Paleogene) in what is now southern British Columbia (Canada) and northern Washington state. Large-scale extension then shifted south through most of the eastern two-thirds of the northern Basin and Range, as evidenced by (1) faulting, erosion, and sedimentation; (2) the general north–south orientation of the directions of igneous dikes that filled the cracks and fissure systems during extension; and (3) radiometric dates indicating that extension and volcanism moved south. The radiometric dates suggest that extension began in southern Idaho about 45 million years ago, 39 million years ago in eastern Nevada, and no earlier than about 35 million years in southern Nevada. By contrast, extension in the southern Basin and Range began no earlier than 25 million years ago.

15.3.2 Hypotheses for the formation of the Basin and Range

Although the data point to extension as the means by which the Basin and Range formed, the exact mechanisms causing extension are still poorly understood. Numerous hypotheses have therefore been proposed for the extension of the Basin and Range. We discuss each of the hypotheses and the geologic evidence for and against them in the context of the method of multiple working hypotheses (see Chapter 1). Consideration of the different hypotheses offers a prime opportunity for students to understand the behavior and complexity of Earth systems—in this case, just a single system, the solid Earth system—and reconstructing its history. In formulating and testing the hypotheses, we must consider the possible types of forces involved. These forces potentially deformed the rocks and produced the igneous activity in the Basin and Range. We must also compare—or test— these possible forces against what is observed in the geologic record to see if the forces can account for the geologic

observations. As we will see, none of the hypotheses satisfactorily explains all the observations, and it is quite possible that more than one mechanism or some combination explains the evolution of the Basin and Range.

Early hypotheses

Several early hypotheses for the origin of the Basin and Range proved popular but have since been rejected. One involved crustal extension caused by the overriding of the East Pacific Rise by the North American plate, as just described. This hypothesis was widely embraced as the theory of plate tectonics was accepted. It was believed the disappearance of a portion of the East Pacific Rise beneath the North American continent meant that extension associated with seafloor spreading continued to occur beneath North America. However, based on the geologic evidence previously cited, the site and timing of collision of the North American plate with the East Pacific Rise (about 30 million years ago) is inconsistent with the initiation of extension over much of the rest of the Basin and Range, which began much earlier (about 45 million years ago) and much farther to the north and northeast.

Another hypothesis involved the formation of an incipient back-arc basin by subduction of the Farallon plate. One version of this hypothesis suggested fragments of the Farallon plate remained sufficiently warm and buoyant (because they had only been fairly recently subducted) to cause uplift and extension of the overlying crust (**Figure 15.11**). However, in these cases (1) the Basin and Range should be associated with a subduction zone, which it is not, and (2) a relatively distinct volcanic arc should occur rather than the more widespread volcanism that has been documented.

Involvement of the San Andreas Fault system

Because early hypotheses for the origin of the Basin and Range proved unsatisfactory, other, more complex hypotheses were formulated. A third hypothesis proposed that extension was caused by the right-lateral movement of the San Andreas Fault system described earlier. Studies of the modern North American–Pacific plate motion suggested that some of the force generated by the right-lateral motion of the San Andreas Fault was transferred to the Basin and Range, causing it to stretch apart.

However, there are problems with this hypothesis, as well. As we have already seen, the timing and site of initiation of the San Andreas system is inconsistent with the initiation of extension over much of the rest of the Basin and Range. Also, the forces generated by right-lateral movement have been calculated to only be strong enough to account for extension in the western 100 to 200 kilometers of the northern Basin and Range. Thus, at best, the force produced by movement along the San Andreas appears to augment other forces affecting the Basin and Range but is not the prime extensional force itself.

Volcanism and uplift

A fourth hypothesis is based on the fact that basaltic volcanism has been widespread through much of the Basin and

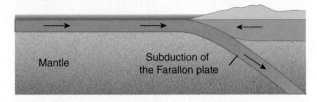

65–70 million years ago

Mantle

Subduction of
the Farallon plate

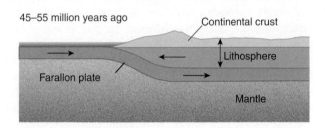

45–55 million years ago

Continental crust

Lithosphere

Farallon plate

Mantle

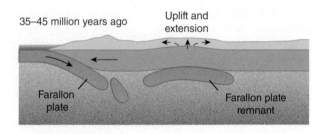

35–45 million years ago

Uplift and
extension

Farallon
plate

Farallon plate
remnant

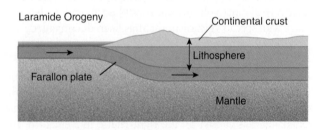

Laramide Orogeny

Continental crust

Lithosphere

Farallon plate

Mantle

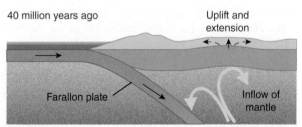

40 million years ago

Uplift and
extension

Farallon plate

Inflow of
mantle

Data from: Wicander, R., and Monroe, J. S. 2000. *Historical Geology: Evolution of Earth and Life Through Time*, 3rd ed. Pacific Grove, CA: Brooks/Cole. Figures 16.13 and 16.14.

FIGURE 15.11 Hypotheses for the widening of the western United States. These particular hypotheses all involve an increase in the buoyancy beneath the crust of this region. A remnant of the Farallon plate was presumably buoyant because it was still warm, thereby decreasing its density. A steepening of the angle of plate subduction following the Laramide Orogeny may have allowed warmer, more buoyant mantle to flow in above the plate.

Range (see Figure 15.7). This suggests the crust is mechanically weak and the mantle or the mantle portion of the lithosphere supplies the basalts and is involved in extension. Perhaps, according to this hypothesis, the presumed shallow angle of the Farallon plate associated with the Laramide Orogeny steepened beginning about 40 million years ago. Then, as the angle of the slab increased, the warmer, buoyant mantle filled in above the cooler crustal slab, causing uplift and extension as the lithosphere was stretched apart (see Figure 15.11). However, an increase in the Farallon slab's angle without changing the direction in which the slab was being subducted would produce an east-to-west migration of extension and volcanism across the Basin and Range. This is inconsistent with the general migration of volcanism and extension from north to south described earlier.

Mantle uplift

A fifth hypothesis is based on the buoyancy of a warm mantle plume that might have caused uplift and extension of the crust. In fact, it has been suggested that the presumed Yellowstone mantle plume is responsible for extension, especially in the northern Basin and Range, as the North American plate overrode the presumed hotspot (recall from Chapter 14 that a mantle plume has also been suggested to have been involved in the Laramide Orogeny). However, the pattern of extension in the northern Basin and Range is from north to south, whereas, as we have already seen, the Yellowstone plume (if it in fact exists) has moved eastward relative to the North American plate.

Slab-gap hypothesis

A sixth hypothesis—the **slab gap hypothesis**—represents a variation on several of the preceding hypotheses (**Figure 15.12**). According to this hypothesis, as the crustal block to the west of the ancestral San Andreas Fault moved northwest, it caused the Basin and Range province to expand like a spreading hand fan (**Figure 15.13**). This caused the Sierra Nevada, the Central Valley of California, and the mountain ranges extending into the Pacific northwest to swing to the west around the "hinge" of the fan, which was located in the Pacific northwest. In the process, what is now the state of Nevada and adjacent areas "stretched" into the wider region now recognized as the Basin and Range (see Figure 15.12).

The name of the hypothesis refers to the mechanism proposed for the spreading and the corresponding extension of the Basin and Range. Normally, a plate descending into a subduction zone continues to descend beneath the overriding plate. This should produce compression on the overriding plate rather than extension. Instead, the slab-gap hypothesis proposes that the *descending* remnants of the Farallon plate left behind an exposed region of mantle formerly associated with the spreading center and that the exposed mantle was in direct contact with the *overriding* North American plate. This mantle material then upwelled and spread beneath the North American plate because the mantle material is warm

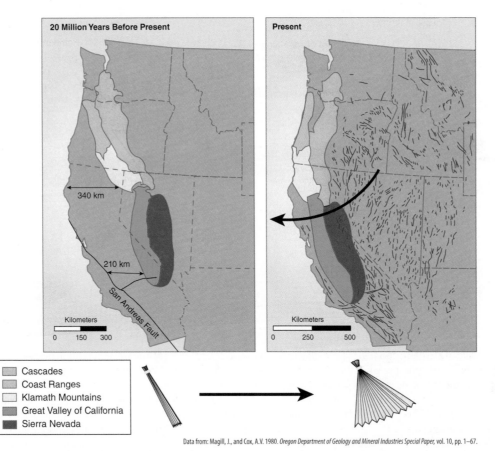

20 Million Years Before Present

340 km

210 km

San Andreas Fault

Kilometers
0   150   300

Present

Kilometers
0   250   500

Cascades
Coast Ranges
Klamath Mountains
Great Valley of California
Sierra Nevada

Data from: Magill, J., and Cox, A.V. 1980. *Oregon Department of Geology and Mineral Industries Special Paper*, vol. 10, pp. 1–67.

FIGURE 15.12 The slab-gap hypothesis explains a number of features associated with the geologic evolution of the American West. Because a wedge of ocean crust was not being subducted beneath the North American plate east of the San Andreas Fault, no volcanism could occur south of the point of contact between the North American and Pacific plates, which today occurs in the vicinity of Mt. Lassen and Mt. Shasta in northern California. To the north, however, subduction of the remnant of the Farallon plate—the Juan de Fuca plate— produces the Cascades. Because of the relatively thin crust that overlies the mantle, according to the slab-gap hypothesis, the Basin and Range is associated with extensive volcanism.

and buoyant, causing extension (the spreading of the fan in Figure 15.12) to produce the Basin and Range.

This hypothesis accounts for a number of geologic observations. First, movement of the southern portion of the "spreading fan" accounts for the formation of deep-water basins in southern California during the Neogene, as previously discussed. These basins are basically rift basins that formed as blocks of crust moved downward along normal faults.

Second, the volcanic activity of the Cascade Mountains decreases to the south. The Cascades of Washington and Oregon seem to stop north of the much older granitic intrusions of the Sierra Nevada, where volcanism long ago ceased (see Figure 15.7). Why should this be? According to the slab-gap hypothesis, the reason is that the North American plate came to lie directly on top of the mantle because the Farallon plate was moving sideways instead of being subducted beneath North America. Thus, ocean crust is not being subducted beneath the North American plate east of the San Andreas Fault; therefore, no volcanism can occur south of the point of contact between the North American and Pacific plates. Today, this point lies in the vicinity of the southernmost volcanoes of the Cascades: Mt. Lassen and Mt. Shasta in northern California. To the north the remnant

of the Farallon plate—the Juan de Fuca plate—is still being subducted (hence the volcanism in the Cascade range), and to the south the other remnant of the Farallon plate—the Cocos plate—also continues to be subducted.

According to some workers the expansion of the "slab gap" by movement along the San Andreas Fault system might also be responsible for the later uplift of the Colorado Plateau and the Rockies described earlier. At the point of contact between the North American and Pacific plates in the slab gap, the North American plate could have been buoyed upward because it presumably came to lay directly on top of warm, buoyant mantle. As the North American plate continued to move over the slab gap, the slab gap and associated mantle would have penetrated farther eastward beneath continental crust causing upward movement.

Despite the appeal of the slab-gap hypothesis (because it seems to explain so many geologic features), it too has its drawbacks. First, the actual dimensions of the slab gap, especially the location of its eastern margin, are poorly understood. Second, the overall west-to-east direction of extension proposed by the hypothesis is not consistent with the overall direction of migration of extension and volcanism within the Basin and Range previously described.

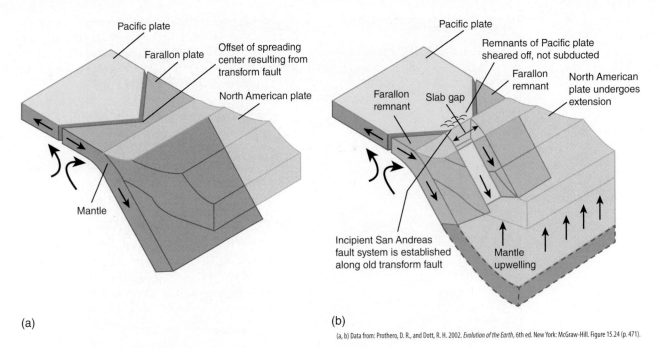

(a)

(b)

(a, b) Data from: Prothero, D. R., and Dott, R. H. 2002. *Evolution of the Earth*, 6th ed. New York: McGraw-Hill. Figure 15.24 (p. 471).

FIGURE 15.13 Origin of the Basin and Range according to the slab-gap hypothesis. **(a)** The western portion of the North American plate is approaching the subduction zone into which the Farallon plate is descending but the North American plate is about to encounter an offset (transform margin) in the spreading center that is producing the Farallon plate and the Pacific plate. The Pacific plate is moving to the west-northwest from the spreading center. **(b)** The North American plate begins to encounter the corner of the offset in the spreading center. As it does so, the Farallon plate splits in two, initiating the "gap" in the Farallon plate, and the North American plate begins to encounter the Pacific plate. The remnants of the Farallon plate continue to subduct so that as the North American plate continues to move westward, the "gap" between the two remnants of the Farallon plate widens. However, the Pacific plate is not subducted because it is moving laterally to the west-northwest from the spreading center along the incipient San Andreas fault system (the old transform fault). The North American plate eventually begins to encounter deeper mantle material that has moved upward beneath the spreading center to form ocean crust. This mantle material is then subducted and is sufficiently buoyant (because it is hot and less dense) to rise up beneath the North American plate, causing fracturing and extension of the crust of the Basin and Range, which is colder and more brittle.

Gravitational collapse

One of the most recent hypotheses for the origin of the Basin and Range involves the gravitational collapse of the lithosphere. Based on recent reinterpretations of fossil plant evidence, it has been suggested that during the Early-to-Middle Cenozoic the western United States was a much wider and higher mountain belt, with elevations up to 2 to 3 kilometers. These elevations would have been up to about 1.5 kilometers higher than modern elevations and would have rivaled—or even surpassed—those of the modern Andes.

According to this hypothesis, the presumed great height and width of the western United States resulted from the Sevier and Laramide orogenies (see Chapters 13 and 14). These mountain-building episodes would have resulted in compression, thrusting, and crustal thickening, producing a thick lithosphere that could have buoyed up high mountain ranges. Ironically, then, the great compression and uplift of previous orogenies would have caused the ancient mountain ranges of the Basin and Range to collapse on themselves and spread out! However, calculations indicate that the buoyancy forces alone would have been insufficient to raise up the mountains to such a height that they could "fall back down" again. So, perhaps the mantle was also involved. According to this modification of the hypothesis, the gradual

conduction of heat from the mantle into the crust beneath the mountains warmed and weakened the crust.

CONCEPT AND REASONING CHECKS

1. How does the timing of volcanism constrain the various hypotheses for the widening of the western United States?
2. What are the advantages of the slab-gap hypothesis? What are the disadvantages?
3. The Cascade Mountains represent a continental volcanic arc. How is the evolution of the San Andreas Fault system thought to have inhibited subduction and volcanism south of the Cascades?

15.4 Climate, Ocean Circulation, and Chemistry

E As the modern continent–ocean configurations finally appeared, the modern climate regime also began to take hold. Fossil floras and oxygen isotopes both indicate Earth's temperature decreased in the Late Eocene and Oligocene in response to the movement of Antarctica over the

South Pole and the development of thick continental glaciers there (see Chapter 14).

The range of climate in different regions tended to become even more extreme during the Neogene in response to glaciation and tectonism. Deserts began to appear in Tibet, as the Tibetan Plateau became increasingly isolated from the moisture of the Indian Ocean (see Chapter 2). At the same time the uplift of the Himalayas was strengthening the Asian monsoon, as evidenced by the appearance of modern rain forests in southeast Asia (see Chapter 3).

The grasslands of the Great Plains also expanded (see Chapter 14). The interior of western North America might have dried out for several reasons. Like the Tibetan Plateau, the Colorado Plateau underwent uplift and is also isolated by mountain ranges on its sides, contributing to a monsoon-style effect. The Basin and Range might have also dried out in part because of a rain shadow (orographic effect) produced by the rise of the Sierra Nevada. Forests were present in the Basin and Range during the Miocene as the Sierras were uplifted, and the spectacular east face of the Sierra Nevada that forms the western boundary of the Basin and Range formed in the Pliocene, as discussed earlier, placing the Basin and Range in its shadow. The advance of northern hemisphere glaciers (discussed in a moment) also might have helped to dry out the atmosphere.

On the other hand, the positioning of the interior of South America and equatorial Africa beneath major rainfall belts led to the development of tropical rain forests in the Amazon and tropical Africa (see Chapter 3). Given the amount of rainfall and the proximity of the Himalayas to southeast Asia and the Andes to the Amazon, respectively, both regions are characterized by some of the highest sedimentation rates in the world (**Figure 15.14**). Not surprisingly, then, during the Neogene strontium isotope ratios ($^{87}Sr/^{86}Sr$) underwent a dramatic rise (see Figure 15.1), reflecting in part increased uplift and erosion all over the world.

Ice caps eventually began to grow over the North Pole late in the Neogene, based on the appearance of ice-rafted debris about 2.8 million years ago in deep-sea cores taken off the British Isles. This follows at least several hundred thousand years after the rise of the Isthmus of Panama, which is thought to have blocked the flow of warm-water currents from the Atlantic into the Pacific (see Figure 15.3). The diversion of warm waters northward into the Gulf of Mexico and eventually into the western Atlantic Ocean strengthened the **Gulf Stream**. The Gulf Stream is a major ocean current that transports warm, moist air to higher latitudes, where it precipitates as snow and ice. The presence of ice caps over both poles led to the establishment of more vigorous ocean circulation patterns and water masses such as North Atlantic Deep Water (see Chapter 3).

CONCEPT AND REASONING CHECK

1. How and why does the Neogene differ from the earlier Paleogene in terms of climate, sea level, and ocean circulation?

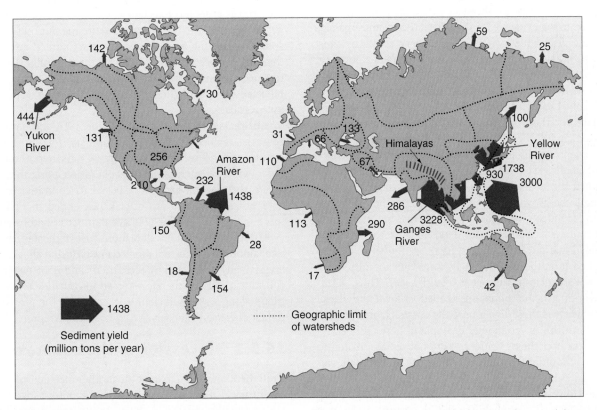

FIGURE 15.14 Modern sedimentation rates. Note the large sediment influxes associated with the uplift of the Himalayas and the Amazon River, which drains the Andes.

15.5 The "Ice Ages": Evolution of a Theory

15.5.1 Background

F The growth of northern hemisphere ice sheets heralded the beginning of the Pleistocene Epoch, or what most people know as the "Ice Ages." The northern hemisphere glaciers did not simply remain stationary after they formed. Rather, they advanced and retreated—waxed and waned—repeatedly during the Pleistocene.

Although the advance and retreat of glaciers in the northern hemisphere are now widely accepted, it has not always been so. Two basic observations served as clues to the previous advance of glaciers. **Glacial erratics** represent rocks of lithologies greatly different from the underlying bedrock and thus must have been derived from elsewhere, in this case the Canadian Shield. Also, deposits of poorly sorted sediment called till form features termed **moraines** which were interpreted to have been left behind as the ice sheets melted. The presence of glacial erratics and associated scratches and grooves in the adjacent bedrock (**Figure 15.15**) suggested to some early 19th-century natural historians working in the Alps and the French Jura that mountain glaciers had once existed at much lower elevations. One Swiss scientist—Louis Agassiz, who was already quite famous for his work on fossil fishes—originally scoffed at the idea but eventually accepted the evidence and became one of the strongest advocates of the "glacial theory." Agassiz even went so far as to suggest Europe had at one time been covered by vast ice sheets.

Despite the evidence for the glacial theory and Agassiz's fame, the glacial theory was received very coldly by most scientists because the theory was simply too radical for them to accept. Ironically, catastrophism was used by some scientists to explain the deposits in relation to Noah's flood, the so-called **diluvial theory**. (The term "drift" used by the diluvians for these supposed flood deposits has stuck in scientific circles down to the present, but the term is now used to denote sediment of glacial origin found on the continents.) Charles Lyell, for example, in his *Principles of Geology*, explained the deposits as ice-rafted debris deposited from floating icebergs when sea level was much higher; this conclusion "fit" nicely with the diluvial theory, but was hardly "uniformitarian"! Some glacial deposits close to the coasts contained marine shells, suggesting the sediments had been deposited from the sea (the shells were later found to have been reworked from the sea bottom by ice). Thus, scientists working in the mountains tended to favor the glacial theory, whereas those living near the coasts favored marine deposition (this is yet another example of how the method of multiple hypotheses works!). It did not seem to matter to some that in order for icebergs to float to the mountain valleys, sea level would have needed to be about 1,500 meters higher! However, the evidence for ancient ice sheets eventually became so overwhelming that the glacial theory was widely accepted by the 1860s, including by Charles Lyell.

15.5.2 The eccentricity of James Croll

What, then, caused the glaciers to advance and retreat? Some scientists turned to astronomic causes for an explanation. Scientists knew that Earth's orbit varied from nearly circular to slightly elliptical—or eccentric—because each of the planets in the solar system exerts a gravitational attraction on Earth (**Figure 15.16**). Earth's orbit changes cyclically from circular to elliptical and back again with a duration or period of about 100,000 years. This means that during a period of 100,000 years the orbit cycles from circular to slightly eccentric and back again. Today, Earth's orbit has an **eccentricity** of about 1% (from a circular orbit), but it has varied from near 0% to as much as 6% over periods of 100,000 years.

James Croll, a shy Scotsman interested in philosophy, became interested in this problem. During his life, Croll was a millwright, carpenter, proprietor of a tea shop, innkeeper, and insurance salesman, professions at which he either failed or from which he had to resign because of his wife's health. About 1860, Croll secured a position as a janitor at the Andersonian College and Museum in Glasgow, Scotland, where he had access to an outstanding library and became interested in the glacial theory. Croll reasoned that Earth warmed and cooled depending on its distance from the sun, which would account for the advance and retreat of glaciers thought to occur about every 100,000 years. Croll thought that when Earth's orbit is circular or nearly so, winters are relatively mild (such as today) because Earth maintained a relatively constant distance from the sun and remained relatively warm (see Figure 15.16). On the other hand, when the orbit is at its most eccentric, exceptionally cold winters occur because the winter solstice in the northern hemisphere (when the northern hemisphere is tilted away from the sun) also occurred far away from the sun. After they were initiated, glaciers would continue to advance because they reflected more sunlight, further cooling Earth through positive feedback.

In the succeeding decades of the 19th and early 20th centuries, evidence began to accumulate for multiple glaciations, which seemed to confirm Croll's hypothesis. But other evidence, such as the rates of erosion along rivers near where the glaciers had been, began to contradict it. This data indicated that the last glacial cycle ended about 10,000 to 15,000 years ago, far younger than Croll had calculated (~80,000 years). Later calculations also indicated that the total amount of solar radiation received during an *entire year* is essentially unaffected by eccentricity and remains virtually constant. Croll was therefore forced to conclude that eccentricity alone was insufficient to account for the growth and retreat of glaciers, and his theory fell into disrepute.

15.5.3 Precession of the equinoxes

Croll therefore began to look for another mechanism besides eccentricity to cause the advance and retreat of glaciers. He eventually concluded the phenomenon known as the **precession of the equinoxes** affects the amount of solar

(a)

(b)

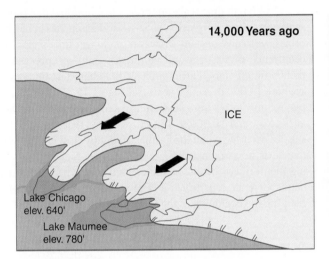

14,000 Years ago

ICE

Lake Chicago
elev. 640'

Lake Maumee
elev. 780'

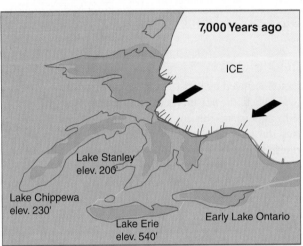

7,000 Years ago

ICE

Lake Stanley
elev. 200'

Lake Chippewa
elev. 230'

Lake Erie
elev. 540'

Early Lake Ontario

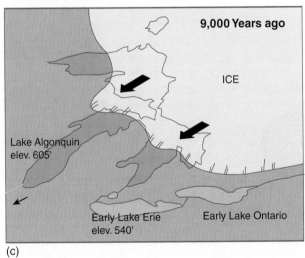

9,000 Years ago

ICE

Lake Algonquin
elev. 605'

Early Lake Erie
elev. 540'

Early Lake Ontario

(c)

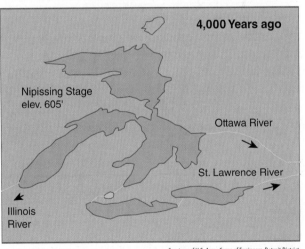

4,000 Years ago

Nipissing Stage
elev. 605'

Ottawa River

St. Lawrence River

Illinois
River

FIGURE 15.15 Evidence for the advance of northern hemisphere glaciers. **(a)** Poorly-sorted sediment deposited by a glacier at its terminus in what is called a "terminal moraine." Till exposed in a terminal moraine. **(b)** A glacial erratic. **(c)** The Great Lakes were created by glaciers scouring out the underlying bedrock. This figure shows the retreat of northern hemisphere glaciers after their last major advance.

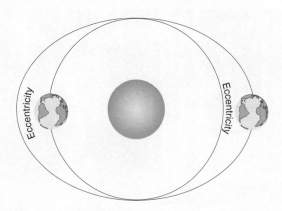

FIGURE 15.16 Slight variations in the eccentricity of the Earth's orbit cause the amount of solar radiation reaching the Earth to fluctuate. The eccentricity (or "ellipticity") is actually highly exaggerated in this and other figures. Eccentricity has a cycle of about 100,000 years, meaning that it takes about 100,000 years for the Earth's orbit to go from nearly circular to its most elliptical state and back again.

radiation received by Earth. Precession—or **wobble**—has to do with where the seasons occur in Earth's orbit.

To understand this concept we must first examine what makes the seasons change (**Figure 15.17**). Each year the northern hemisphere passes through a **summer solstice** on about June 21, when the maximum intensity of solar radiation occurs highest in the northern hemisphere and the northern hemisphere experiences its longest day of the year;

a corresponding winter solstice occurs in the southern hemisphere, when the southern hemisphere experiences its shortest day of the year. After June 21 the northern hemisphere begins to lean away from the sun because of Earth's continued movement in its orbit around the sun. By about September 22 the sun's rays have, in effect, migrated southward over Earth's surface and become concentrated at the equator; hence, the origin of the term "autumnal equinox," when the concentration of solar radiation is equal in both hemispheres. As Earth moves further along in its orbit, the net effect is to make the northern hemisphere lean even farther from the sun. Eventually, we experience **winter solstice** in the northern hemisphere (summer solstice in the southern hemisphere) about December 21. Then, Earth moves gradually toward its position at about March 20 (spring equinox), by which time the northern hemisphere is warming up and the southern hemisphere is cooling down.

On human time scales, or even a few hundred to a thousand years, the occurrence of solstices and equinoxes in Earth's orbit is virtually constant. However, on longer time scales the occurrence of solstices and equinoxes shift—or **precess**—their position along Earth's orbit with a characteristic period of about 19,000 to 23,000 years (see Figure 15.17). For example, we now know from *hindcasting* (as opposed to *forecasting*) back to around 10,000 to 15,000 years ago that the summer solstice occurred when the Earth was slightly closer to the Sun than it is now so that the amount of solar radiation reaching the northern hemisphere increased somewhat. Northern hemisphere ice caps began a general retreat about this time, marking the beginning of the **Holocene**, or "wholly recent."

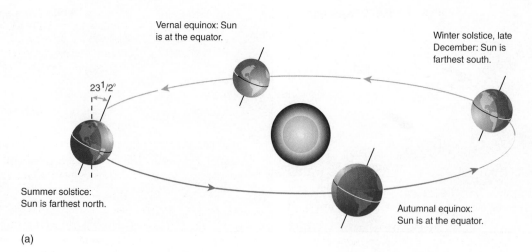

(a)

FIGURE 15.17 Precession of the equinoxes, or "wobble." **(a)** The occurrence of the seasons as the Earth currently orbits the sun. As the Earth orbits the sun, the northern hemisphere varies between the extremes of summer (northern hemisphere "leaning" toward the sun) and winter (northern hemisphere "leaning" away from the sun). The greatest amount of solar radiation reaches the northern hemisphere in summer, with June 21 being the longest day of the year (summer solstice). On this day, the sun's rays are concentrated at the highest latitude in the northern hemisphere. The Earth's continued movement along its orbit then causes the sun's rays to move toward the equator; the amount of solar radiation reaching both hemispheres is equal by about September 22 (autumnal equinox). As the Earth continues to orbit, the sun's rays become most concentrated at their highest latitude in the southern hemisphere on December 21 (northern hemisphere winter solstice), which is also the shortest day of the year in the northern hemisphere. As the Earth continues to orbit, the sun's rays begin to move back toward the equator, reaching the spring equinox about March 20.

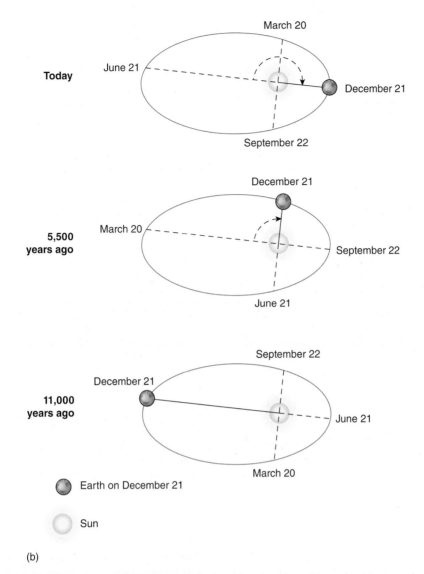

(b)

FIGURE 15.17 (b) The angle of the Earth's axis of rotation is relatively constant on short time scales. However, the positions in the Earth's orbit at which the equinoxes and solstices occur shifts over thousands of years, which is called the precession of the equinoxes. Today, northern hemisphere summer (when the northern hemisphere is tilted toward the sun) occurs when the Earth is actually slightly farther from the sun, but ~11,000 years ago, northern hemisphere summer occurred when the Earth was closest to the sun. The stage intermediate between 11,000 years ago and today is also shown.

15.5.4 Milutin Milankovitch and obliquity

The study of the astronomic forcing of the growth and retreat of ice sheets was further stimulated early in the 20th century by the Serbian astronomer, Milutin Milankovitch. Milankovitch decided to make a mathematical theory of the climate of Earth, Mars, and Venus his life's work. He began by developing a mathematical theory that allowed calculation of the solar flux for any given latitude and season.

When he had accomplished this, he decided to develop a mathematical description of Earth's ancient climates, but he immediately ran into a problem: each latitude and each season at each latitude has its own solar radiation history. Which solar radiation history should he use? Croll had earlier decided that diminished radiation during the winter at high latitudes was

the critical factor promoting the growth of ice (see the earlier discussion). However, Milankovitch concluded that changes in winter radiation had little effect on snow accumulation. He instead decided it was a decrease in solar radiation at high latitudes during the summer that was critical for ice growth; any decrease in summer radiation would prevent melting, permitting glaciers to grow in the long term.

Milankovitch also found that the **obliquity** or **tilt** of Earth's axis of rotation strongly affected his calculations (**Figure 15.18**). Obliquity should not be confused with precession, or wobble. Obliquity refers to changes in the angle of Earth's axis of rotation in the range of about 22 to 25 degrees from the vertical and has a period of approximately 40,000 years for the completion of one full cycle through the total range of about 3 degrees. The northern hemisphere of Earth receives greater amounts of sunlight

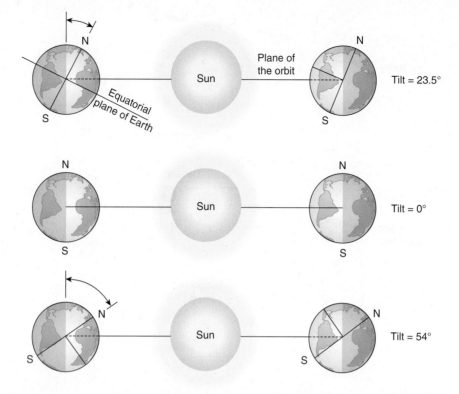

FIGURE 15.18 Effect of obliquity or tilt of the Earth's axis of rotation on the distribution of solar radiation and heat on the Earth's surface. On relatively short time scales, the angle of the Earth's axis of rotation is constant; currently the angle is 23.5°. However, the angle of tilt varies between about 22° and 24.5° from the vertical with a cycle of about 40,000 years. Thus, it takes about 40,000 years for the angle to vary from 22° to 24.5° and back to 22°. When the northern hemisphere is tilted most strongly toward the sun (24.5°), it receives more solar radiation on average than when it is tilted less strongly. The angle of tilt affects the climatic extremes of the seasons; the greater the angle, the more pronounced the seasons.

when Earth is tilted toward the sun and lesser amounts when it is tilted away (see Figure 15.18).

Milankovitch's theory, which was finally published in 1924, made testable predictions. Geologists and glaciologists such as the noted American geologist T. C. Chamberlin (who originally advocated the method of multiple working hypotheses described in Chapter 1) immediately began to search the geologic record. The Great Lakes were ultimately recognized to have been scoured out by glacial activity (see Figure 15.15), as were many spectacular mountain features, such as Yosemite Valley in the Sierra Nevada of California (**Figure 15.19**). Terraces in certain areas of the American West were also recognized as corresponding to the levels of ancient lakes formed from glacial meltwater, such as those of ancient **Lake Bonneville**, the precursor of the modern Great Salt Lake (**Figure 15.20**). Terraces formed by a combination of tectonic uplift and sea level were also later documented in New Guinea (see Figure 15.20). Thus, at first, Milankovitch's calculations seemed to agree with the actual record and received enthusiastic support from meteorologists such as Alfred Wegener of continental drift fame (see Chapter 2).

However, like Croll's theory, evidence began to accumulate that seemed to contradict it. For example, in the United States four major tillites were recognized. Similarly, four major glacial intervals, named after the river valleys where they were first described, were recognized in Europe. Some geologists immediately correlated the American and European deposits, although absolute dating techniques were still far too imprecise to come to such a conclusion. Other geologists suggested the number of tillites did not match the number predicted by Milankovitch's theory. By 1955 radiocarbon dating indicated the last glacial interval occurred

© Carl Dawson/Shutterstock, Inc.

FIGURE 15.19 During the Pleistocene, the Sierra Nevada plutons were carved by mountain glaciers to form the valley of Yosemite National Park and spectacular features such as Half Dome, seen on the right at the edge of the valley.

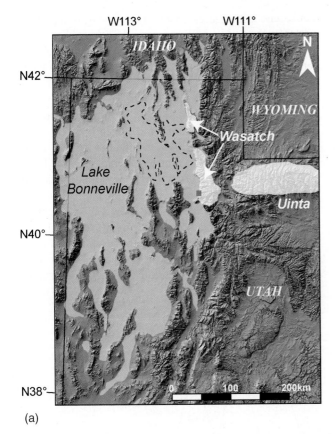

FIGURE 15.20 **(a)** Extent of the ancient Lake Bonneville in relation to the modern Great Salt Lake (dashed lines). Positions of the Wasatch and Uinta mountains and Salt Lake City (red square) are also shown. **(b)** Uplifted Pleistocene reef terraces in New Guinea. Note that the older terraces occur higher than younger terraces because of tectonic uplift of land while sea level rose and fell to form the terraces.

about 18,000 years before present, which was about 7,000 years too young based on Milankovitch's calculations. Astronomic forcing of glacial cycles once again fell by the wayside.

CONCEPT AND REASONING CHECKS

1. Do the cycles of solar insolation described above alone explain the *initial* formation of northern hemisphere ice sheets? What had to have happened to cause the ice sheets to form in the first place?
2. Diagram and explain the precession of the equinoxes in your own words.
3. What is the difference between the obliquity of Earth's axis of rotation and the precession of the equinoxes?

15.5.5 Planktonic foraminifera and the oxygen isotope curve

As support for Milankovitch's theory waned, other workers began to examine deep-sea cores from the North Atlantic that contained fossil assemblages of planktonic foraminifera. Different species of planktonic foraminifera (which float in the pelagic zone close to the surface) are characteristic of different surface water masses that have slightly different temperatures (see Chapter 4). These workers hypothesized that by determining changes in the abundances of foraminifera in cores, one could infer the temperature of the overlying water masses through time while the sediment (and foraminifera) of the core had settled to the bottom of the ocean. Based on these data, these workers determined the surface water masses of the North Atlantic had shifted back and forth during the Pleistocene in response to the repeated advances and retreats of sea ice over the North Atlantic.

About the time climatic evidence from planktonic foraminifera began to appear, the technique of using **oxygen isotopes** was being developed (see Box 14.1 for discussion of oxygen isotopes). The resulting oxygen isotope curve was initially interpreted as reflecting ancient surface water temperatures, with warm intervals denoted by odd numbers and cool intervals by even numbers (**Figure 15.21**). However, the paleotemperatures based on the oxygen isotope curve and planktonic foraminifera disagreed. This led to yet another controversy about which curve was "correct": the one for planktonic foraminifera or the one for oxygen isotopes. It was not realized until the late 1960s that the oxygen isotope curve for the Neogene, and for the Pleistocene in particular, indicates primarily ice volume, not temperature. New dating techniques that permitted the determination of ages older than those available from carbon-14 became available and were used to date terraces in ancient coral reefs of the Caribbean (Barbados) and New Guinea (Pacific Ocean); these terraces correspond to ancient sea levels and their dates agreed with calculations based on precession.

Then, in 1973, oxygen isotope curves from different deep-sea cores were calibrated to ages derived from reversals of Earth's magnetic field (see Figure 15.21; see Chapter 6). These studies demonstrated that the activity of the glaciers was affecting oxygen isotopes worldwide. Thus, not only was the astronomic theory for glacial advances and retreats again firmly established, it was then realized the oxygen isotope stages could be used for precise studies of stratigraphic correlation and climate change all over the world (see Chapter 6).

Moreover, when the curves for eccentricity, obliquity, and precession are added together (commonly referred to as **Milankovitch cycles**), they produce a curve that bears a remarkable similarity to the oxygen isotope curve during the Pleistocene (see Figures 15.21). Eccentricity, obliquity, and precession can add or subtract from the effects of each other to produce all sorts of gradations of ice advance and retreat between extreme cold and extreme warmth (**Figure 15.22**). Also, sudden **terminations**—rapid shifts to lower values that indicated rapid warming and melting of ice sheets—are followed by gradual shifts to higher values, indicating the gradual growth of glaciers until the next termination occurs. However, the numbering scheme for the oxygen isotope curve works only back in time to the middle of the Pleistocene Epoch, or about 0.7 to 0.8 million years ago. Before this time, the isotope stages are much less distinctive. Later work using statistical techniques showed the shift in the character of the isotope curve occurs because Milankovitch cycles during the first half of the Pleistocene are dominated by precession, whereas those during the last half of the epoch are dominated by obliquity, which causes greater fluctuations in solar radiation and ice volume.

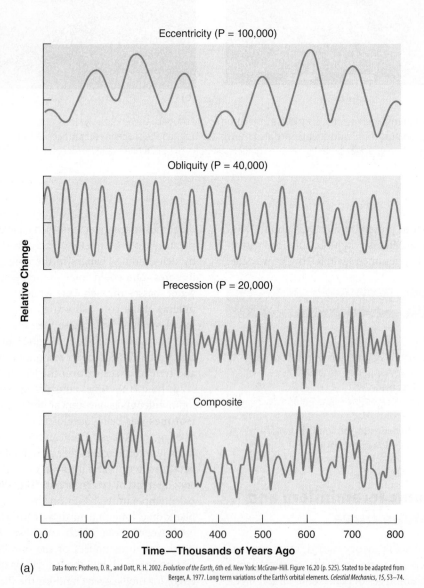

(a)   Data from: Prothero, D. R., and Dott, R. H. 2002. *Evolution of the Earth*, 6th ed. New York: McGraw-Hill. Figure 16.20 (p. 525). Stated to be adapted from Berger, A. 1977. Long term variations of the Earth's orbital elements. *Celestial Mechanics*, 15, 53–74.

FIGURE 15.21 (a) The combination of the Milankovitch cyclicities accounts for the behavior of the oxygen isotope curve during the Pleistocene.

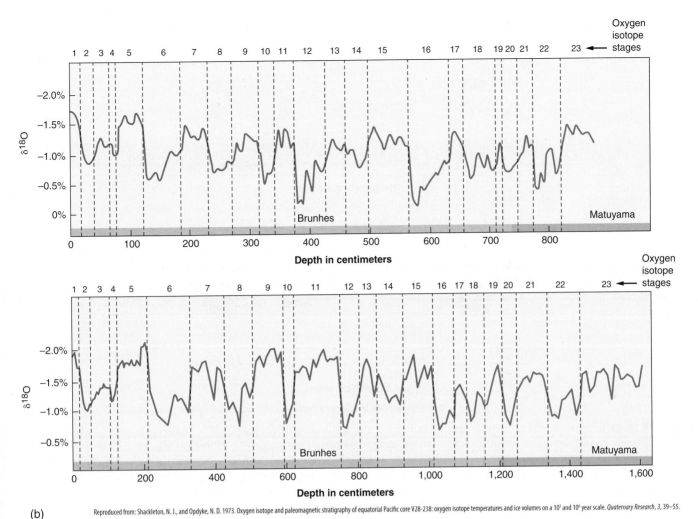

(b)

Reproduced from: Shackleton, N. J., and Opdyke, N. D. 1973. Oxygen isotope and paleomagnetic stratigraphy of equatorial Pacific core V28-238: oxygen isotope temperatures and ice volumes on a 10⁵ and 10⁶ year scale. *Quaternary Research, 3,* 39–55.

FIGURE 15.21 (b) The correlation of oxygen isotope stages using paleomagnetic changes of reversals between the Brunhes and Matuyama intervals indicated that changes in oxygen isotopes occurred rapidly throughout the oceans. Such rapid changes could occur only via ocean circulation, not temperature. Compare to part a.

Nevertheless, Milankovitch cycles are not the only factor involved in the advance and retreat of glaciers. Ice cores indicate the concentrations of carbon dioxide (CO_2) and methane changed naturally through time as glaciers waxed and waned (see Chapter 1). These gases are greenhouse gases and likely contribute to both natural and anthropogenic global warming. We examine the mechanisms of anthropogenic climate change, how they are intertwined with natural climate change, and their potential consequences in greater detail in Chapter 16.

CONCEPT AND REASONING CHECKS

1. How was the hypothesis that ice volume determines the behavior of oxygen isotopes tested?
2. How was the hypothesis of the synchronicity of oxygen isotopes tested?

15.6 Neogene Life

15.6.1 Marine life

The diversification of the Modern Fauna of marine invertebrates and vertebrates, which began in the Mesozoic and continued through the Paleogene, appears to have continued through the Neogene to the present. However, it is unclear whether the apparent diversification in the fossil record is real or an artifact of better preservation toward the present because there has been less time for erosion to occur (see Chapter 5).

Among the plankton, distinct populations began to be strongly associated with certain water masses because of the water masses' temperature and salinity characteristics. The broad modern patterns of calcareous (coccolithophorids, foraminifera) and siliceous oozes (diatoms, radiolarians) and red clay on the deep ocean floor also began to be established (see Chapter 4); these patterns likely resulted

Data from: van Andel, T. J. 1994. *New Views on an Old Planet*. Cambridge, UK: Cambridge University Press. Figure 5.3 (p. 93).

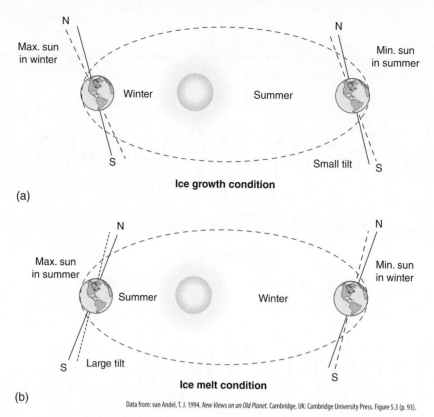

FIGURE 15.22 Eccentricity, obliquity, and precession combine to change the amount of solar radiation reaching the Earth's surface. Two extreme conditions are shown: ice growth and ice melt. The solid and dashed lines indicating the Earth's axis of rotation represent the extremes of the range of the angle of obliquity. Note how in the ice growth condition, the Earth is closest to the Sun in winter but is tilted away from the Sun during winter, whereas in summer the Earth is farthest from the Sun even though it is tilted toward the Sun. Together, these conditions minimize incoming solar radiation and heating in the northern hemisphere and thus melting; glaciers will therefore tend to grow and advance. Just the opposite conditions prevail during ice melt conditions because during northern hemisphere summer, Earth is closest to the sun, thereby maximizing solar radiation and heating of the northern hemisphere; glaciers will tend to melt and retreat.

from a combination of more vigorous deep-ocean circulation, upwelling of nutrient-rich waters like those off Antarctica, and deepening of the calcite compensation depth in response to sea-level fall and continued expansion of calcareous plankton. Diatoms, especially, underwent a tremendous diversification during the Neogene. Marine vertebrates such as whales took on a clearly modern appearance and distribution as the Neogene progressed. However, some taxa, such as corals, became restricted toward lower latitudes because of glaciation and the cooling of water masses closer to the poles.

15.6.2 Land plants

H Grasses, particularly herbs, diversified into thousands of species as Earth dried out as a result of glaciation. Herbs are small, nonwoody angiosperms that include lettuce and sunflowers. Many of the taxa are weeds (such as the infamous crabgrass of lawns) adapted for the rapid invasion and colonization of barren areas after disturbances such as fires.

A significant change in the grasses also occurred with regard to their ability to extract CO_2 from the atmosphere. During the Late Miocene about 6 million years ago, there was a distinct shift in carbon isotopes to higher values (see Figure 15.1). This shift is thought to reflect the expansion of what are called **C4 grasses** in place of **C3** types. C4 grasses extract more of the heavier isotope of carbon (^{13}C) during photosynthesis, which is thought to account for the shift in carbon isotope values. Genetic mutation in the DNA responsible for the enzymes of photosynthetic pathways, climate change, or both might have triggered the shift from C4 to C3 grasses. Possibly, C4 grasses were responding to lower CO_2 levels in the atmosphere that resulted from increased carbon burial in expanding grasslands; C4 grasses can tolerate lower atmospheric CO_2 levels than can C3 forms. C4 grasses are also more tolerant of dry conditions like those associated with the cooler, drier climate of the Neogene.

The spread of grasses is also thought by some workers to have contributed to the tremendous diversification of diatoms noted above. Recall that grasses contain microscopic phytoliths of opaline silica (see Chapter 14). During decay, this form of silica is released to water and carried to the oceans. Thus, the spread of grasses might have stimulated diatom diversification by supplying silica, a biolimiting nutrient (see Chapter 2) of shell construction by diatoms. Indeed, it has been calculated that modern diatoms use silica so ravenously that without rapid recycling in the upper water column, all dissolved silica in the world's oceans would be used up in a few hundred years!

15.6.3 Terrestrial vertebrates

The expansion of grasses had a tremendous effect on terrestrial communities and the diversification of different types of vertebrates. Many species of vertebrates we see today originated during the Neogene. Rats and mice, which eat seeds, became more prominent; so too did frogs, which eat insects. Many of our common songbirds, which eat both seeds and insects, also appeared. Snakes, which prey on small mammals, amphibians, and birds, also expanded. The changes in **ungulates** that began during the Eocene–Oligocene transition continued. The diversity of horses and other perissodactyls continued to decline during the Neogene. Simultaneously, artiodactyls diversified in response to the cooler, drier climates and the spread of grasslands. New species of antelope, deer, and cattle, for example, all appeared.

Some of the most iconic mammals of the Neogene and especially the Pleistocene are the proboscideans **mammoths** and mastodons. Mastodons inhabited North and Central America beginning about the late Miocene whereas mammoths appeared somewhat later in the Pliocene (living into the Holocene) and first appeared in Africa and then spread northward into Europe, Asia, and North America. Although somewhat similar in appearance, the two groups are only distantly related. Mastodons resembled modern Asian elephants, possessing a long, low body, and had long tusks and cusped teeth. Mammoths on the other hand developed high skulls and had much longer curved tusks and ridged teeth; species living at higher, colder latitudes also developed long hair (hence the term "woolly mammoth"). Mammoths, so named because they belong to the particular genus *Mammuthus,* evolved into grazers in the Pleistocene. Mammoths migrated back and forth repeatedly between northern continents and Africa. Consequently, the remains of this group are sometimes found entombed in ice, at the bottoms of sinkholes into which they fell, or mummified within caves. In some cases, mammoths were so abundant their tusks were used to build huts by early humans. Based on cave paintings, they were both admired and feared by prehistoric humans.

Carnivores—particularly wild dogs, saber-toothed cats, hyenas, and bears—preyed upon the herbivorous artiodactyls. Many of these predators—and their prey—are preserved in the famous Lagerstätte of the La Brea tar pits of southern California. These localities are thought to have served as watering holes during the Pleistocene. However, oil seeps made them treacherous for both prey and predators, which became mired in the deposits and died.

15.6.4 Evolution of humans[1]

By far the most significant evolutionary development during the Neogene (from our standpoint anyway!) was the evolution and eventual appearance of our **species:** *Homo sapiens* ("wise," "prescient," or "intelligent man," depending on the translation). Humans belong to the Order **Primates** of the Class Mammalia (**Figure 15.23**). Besides humans, modern primates are represented by lemurs, tarsiers, monkeys, chimpanzees, and other apes. Modern primates are mostly tropical to subtropical in distribution and are mostly, if not entirely, arboreal (tree-dwelling). Each limb typically possesses five digits, of which the thumb (or big toe) is opposed to the other digits; the presence of an opposable thumb (or big toe) means it can be touched to the other digits, enabling the hand (or foot) to grasp (**Figure 15.24**). Also, the eyes are typically directed forward for **stereoscopic** (three-dimensional) **vision**; the head can easily turn on the neck; the brains are relatively large; and most are highly social.

It has long been realized that humans are related by evolution to the other primates, despite the anatomic differences between the groups (**Figure 15.25**). Indeed, about 98.5% of the DNA in humans and chimpanzees is identical. The theory that humans evolved slowly and gradually from an ape-like ancestor is descended from Charles Darwin, who was compelled to propose it based on his theory of evolution. Before publication of *On the Origin of Species* in 1859, scientists, even those who favored some form of evolution, widely regarded the origin of humans as distinct from "lower" animals. Darwin therefore applied his theory reluctantly to the evolution of humans, knowing full well the reaction it would provoke from the lay public and many scientists alike. He also feared that if he discussed the subject too much in the *Origin,* it would provoke a strong backlash against his entire theory. Nevertheless, Darwin realized that if all other creatures had evolved, humans must have evolved, too, and by the same processes as all other organisms. To propose otherwise would make no sense in terms of a scientific theory and would subvert the entire theory of evolution. But it was not until 12 years after the publication of the *Origin,* in 1871, that Darwin published *On the Descent of Man.*

Evolutionary relationships of humans

The first actual primate fossils now known are the teeth of the earliest Paleocene genus *Purgatorius.* However, molecular clocks suggest primates originated earlier in the Cretaceous, at about 80 to 90 million years before present. If the clocks are accurate, there was a much longer interval of evolution of early primates than previously thought, and this interval occurred before the demise of the dinosaurs. Fortunately, the fossil record of primate evolution begins to improve in the Paleocene. These forms belong to the rodent-like genus *Plesiadapis,* which has been found in both the United States and

[1] The author extends his sincere gratitude to Drs. Karen Rosenberg and Thomas Rocek of the Department of Anthropology, University of Delaware, for their review of Section 15.6.4. Any errors are, however, attributable to the author.

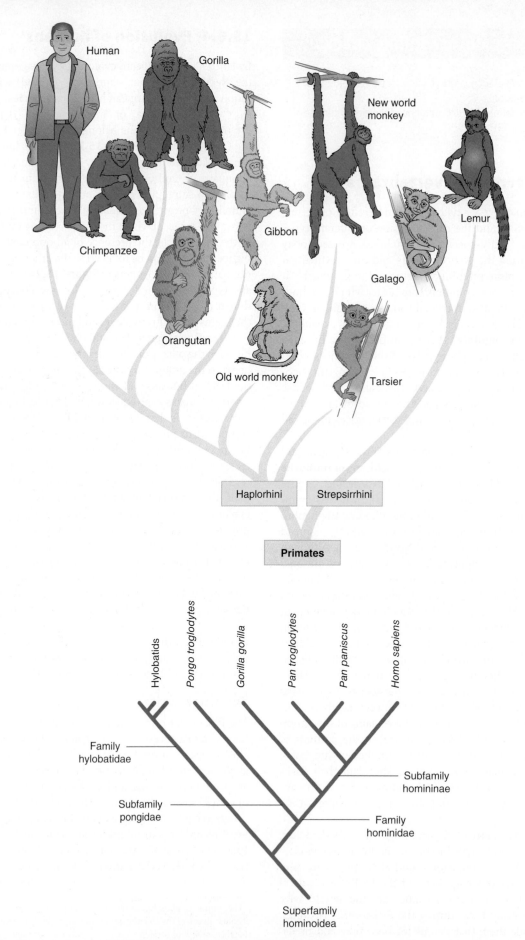

FIGURE 15.23 The evolutionary relationships of humans to other primate lineages.

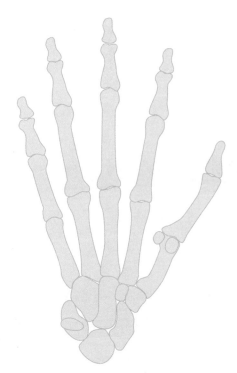

FIGURE 15.24 An opposable thumb and stereoscopic (three-dimensional) vision are two features of early primate lineages that played a major role in the later evolution of humans. This is an image of the right hand of the Eocene prosimian *Eurolopemus*. Note the strong resemblance to a human hand.

Europe. Later, in the Eocene, lemur-like forms belonging to the genus *Notharctus* evolved in the Old and New World.

Modern primates are now classified into two broad taxa that diverged about 77 million years ago based on molecular evidence, although evidence from the fossil record indicates that this divergence happened in the Eocene:

■ The **strepsirrhines**, or wet-nosed primates (based on the occurrence of a sensitive wet area—the rhinarium—located at the end of the snout of dogs and many other mammals), are represented by small, nocturnal lemurs found in Madagascar and the lorises of Africa and southeast Asia (see Figure 15.23).
■ The **haplorhines**, or dry-nosed primates, are represented today by three evolutionary lineages: New World monkeys (found in Central and South America and parts of Mexico), Old World monkeys (found in Africa and Asia), and apes, including the lesser apes in Asia (gibbons and siamangs) and the great apes in Africa and Asia (humans, chimpanzees, gorillas, and orangutans; see Figure 15.23). The recently discovered *Archicebus achilles* (~55 million years ago, Paleocene) might be the oldest known primate belonging to the haplorhines, although other workers think it might have diverged soon after the appearance of the haplorhines and been ancestral only to tarsiers (**Figure 15.26**). It is nevertheless thought to have resembled the last common ancestor of all haplorrhines and of all primates, as well. Haplorhines diversified in the Oligocene, and by the end of the Oligocene the earliest apes

had split off from the Old World monkeys. New World monkeys are descended from African ancestors that colonized South America about 40 million years ago.

Traditionally the great apes (chimpanzees, gorillas, and orangutans) were classified together and separately from humans, but as the evolutionary relationships between humans and the great apes (especially chimpanzees) became more closely established, it became increasingly apparent that previous classifications needed to be revised. Great apes and humans are now referred to as *hominids* (the Family *Hominidae* of older classifications) and humans and their closest extinct relatives are known as hominins. Humans and their closest extinct relatives are now placed in the "tribe" *Hominini*, which occurs between the subfamily and genus categories. The other two tribes of African apes, *Panini* (chimpanzees) and *Gorillini* (gorillas; see Figure 15.23), are most closely related to humans, whereas the Asian great ape, the orangutan (*Pongo*), is more distantly related.

During the Miocene, as the Earth continued to cool, Africa and Arabia continued to move north, eventually colliding with southern Asia. These plate movements contributed to the closure of the Tethys Seaway and caused the regional climate to shift from moist, tropical forests to more arid grasslands studded with lakes enclosed by rift valleys. During this time, new genera appeared that would give rise to the ape group including chimpanzee and gorilla ancestors and early humans. The collision of Africa, Arabia, and Eurasia also established connections between land masses in East Africa that would allow these new taxa to migrate out of Africa. Because East Africa was (and still is) volcanically active, we can accurately date the timing of these originations by using ash layers, as we will see.

CONCEPT AND REASONING CHECKS

1. What features unite humans with other haplorhines?
2. How did plate tectonics influence the evolution of primates?
3. How do humans differ from the other Great Apes anatomically?

Evidence for human origins

Ever since Darwin, various hypotheses have been proposed for the evolution of humans. As you might suspect based on the story so far, the history of human origins is a complex one that is still not fully understood, full of evolutionary branches and cul-de-sacs. The "lumping" and "splitting" of fragmentary remains into different species by different workers (see Chapter 5) have also augmented controversies about the fossil lineages leading to modern humans, as we will see momentarily. The resulting reconstructions of human origins are therefore seemingly constantly being revised and debated, especially after new specimens are discovered.

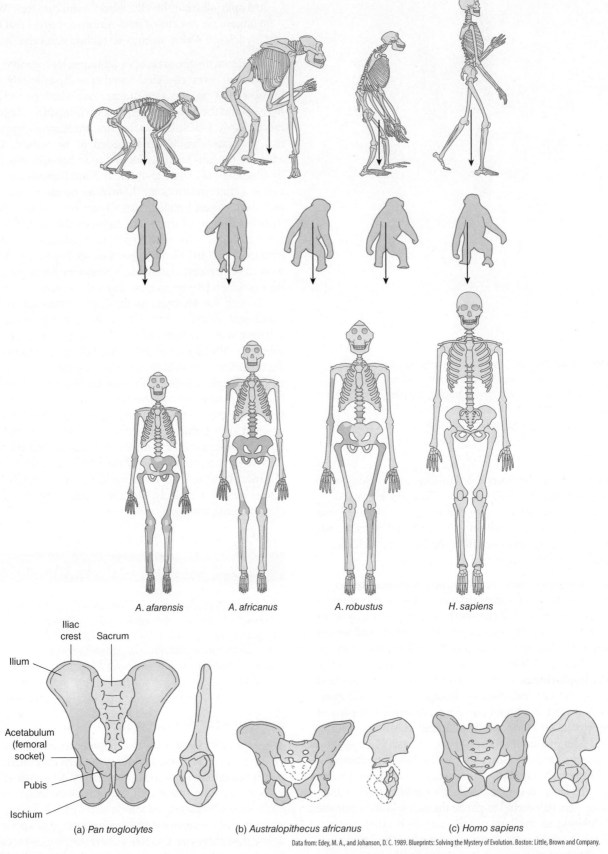

A. afarensis   A. africanus   A. robustus   H. sapiens

Iliac crest   Sacrum
Ilium
Acetabulum (femoral socket)
Pubis
Ischium

(a) *Pan troglodytes*   (b) *Australopithecus africanus*   (c) *Homo sapiens*

Data from: Edey, M. A., and Johanson, D. C. 1989. Blueprints: Solving the Mystery of Evolution. Boston: Little, Brown and Company.

FIGURE 15.25 Humans and apes compared. The curved back of apes makes them lean forward to support the body when walking on all fours. The pelvis also leans forward and is long to support the body when walking on all fours. By contrast, except as infants, humans walk upright. As a result, the backbone is S-shaped so that the weight of the upper body is supported by the hips, and the pelvis is shorter to support the gut. The feet, toes, and legs are longer for walking and balance. Note also the nearly vertical forehead in humans and the lack of a prominent bony ridge over the eyes.

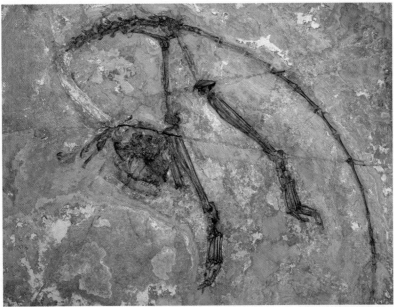

FIGURE 15.26 Reconstruction of the oldest known complete primate *Archicebus achilles* from approximately 55 million years ago. *Archicebus achilles* lived in trees along a lake in what is now China, when the climate was much warmer. The species was less than 3 inches long and belongs to the same taxon as tarsiers, but also lies close to the base of the branch that eventually gave rise to humans. Although humans evolved in Africa almost 50 million years later, the discovery of *Archicebus* supports the theory that primates originally evolved in Asia.

The **single species hypothesis** dominated debates about human origins from the 19th century into the 1950s and 1960s. This hypothesis proposed that there was only sufficient carrying capacity on Earth to accommodate one species of human. In this view, ancient hominins were thought to have evolved slowly and gradually through time by phyletic gradualism (see Chapter 5), giving rise to a series of gradational forms that culminated in modern humans like the series of forms depicted in the upper portion of Figure 15.25. Because *H. sapiens* has largely had Earth to itself for about the last 35,000 years, it is not hard to understand that our "aloneness" contributed to this anthropocentric ("human-centered") view. Support for the single species hypothesis began to weaken in the late 1970s because new fossils were unearthed that indicated more than one species of ancestral humans had existed and, as we will see, *perhaps coexisted,* on Earth.

Even more recent discoveries continue to indicate that human evolution was much more complex than the scenario of the single species hypothesis. A nearly complete skull, two lower jaw fragments, and three teeth assigned to a new species, *Sahelanthropus tchadensis*, were discovered in Chad (**Figure 15.27**). The fossils of *Sahelanthropus* are associated with others such as fish, crocodiles, turtles, hippopotamuses, monkeys, and antelopes, suggesting that the *Sahelanthropus* lived near a lake surrounded by grasslands and forests. The fossils of *Sahelanthropus* are highly significant for several reasons. First, the fossils are quite old: 6 to 7 million years (Late Miocene), predating other long-known fossils related to human origins by 2 to 3 million years. Second, the skull has a small braincase (340–360 cc^3 as opposed to a rough average of 1,400 cc^3 in modern

humans) with a prominent brow ridge like that of chimpanzees, but the face and teeth are like those of bigger-brained hominins occurring at about 1.75 million years; this combination of traits was completely unexpected. Third, as the scientific name indicates, the fossils were discovered in the Sahel region, not east Africa, where so much of early human evolution is thought to have occurred. (The term "sahel" refers to the region of Africa transitional between the Sahara Desert of northern Africa and the humid tropical regions nearer the equator.) Finally, *Sahelanthropus* indicates that the divergence of humans and chimpanzees would seem to agree with ages predicted by molecular clocks. It is unclear, however, if this species was bipedal or not because post-cranial skeletal material has not yet been discovered; however, the foramen magnum (the large opening in the skull where the spine enters) lies at the base of the skull, indicating that the head was held upright and suggesting that *Sahelanthropus* might have been bipedal, although the cranial base is very crushed. Although suggestive, then, neither of these findings is definitive. If *Sahelanthropus* was in fact bipedal, it represents the earliest known putative hominin and suggests that hominins evolved much earlier and over a much wider geographic area than had previously been thought.

Two species found in younger deposits provide stronger evidence for bipedalism: the chimpanzee-sized *Orrorin tugenensis* is dated at 6.1 to 5.8 million years (see Figure 15.27). Its hind limbs indicate that it was bipedal on the ground, whereas its forelimbs indicate that it also lived arboreally. *Ardipithecus kadabba* is somewhat younger (5.8–5.2 million years) and shares more features with later australopithecines (the "southern apes" of Africa; see the discussion that follows),

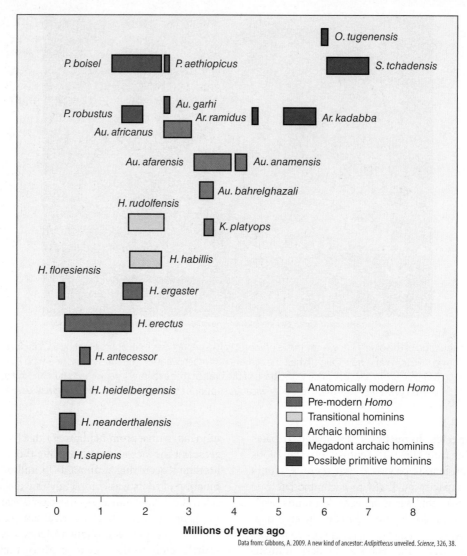

P. boisei
O. tugenensis
P. aethiopicus
S. tchadensis
P. robustus
Au. garhi
Ar. ramidus
Au. africanus
Ar. kadabba
Au. afarensis
Au. anamensis
Au. bahrelghazali
H. rudolfensis
K. platyops
H. habillis
H. floresiensis
H. ergaster
H. erectus
H. antecessor
H. heidelbergensis
H. neanderthalensis
H. sapiens

Anatomically modern *Homo*
Pre-modern *Homo*
Transitional hominins
Archaic hominins
Megadont archaic hominins
Possible primitive hominins

0   1   2   3   4   5   6   7   8

Millions of years ago

Data from: Gibbons, A. 2009. A new kind of ancestor: *Ardipithecus* unveiled. *Science*, 326, 38.

FIGURE 15.27 Species from the fossil record identified as hominins plotted in relation to their estimated or known ages. Abbreviations for genera are *Ar, Ardipithecus; Au, Australopithecus; H, Homo; K, Kenyanthropus; O, Orrorin; P, Paranthropus; S, Sahelanthropus. P. boisei* and *P. robustus* are discussed as species of *Australopithecus* in the text.

although its larger canine teeth make it more difficult to determine its place in the lineages leading to humans. *Ardipithecus ramidus* at 4.4 million years ago is better known than *A. kadabba* and has equivocal but intriguing evidence for a transitional (early) form of bipedalism. If *Sahelanthropus, Orrorin,* and *Ardipithecus* were bipedal (and therefore hominins), the split between the ancestors of humans and chimpanzees had already occurred by the time they lived. If the split occurred 4.4 million years ago, they are not considered to be hominins.

Although *Ardipithecus ramidus* (or "Ardi") is not an australopithecine (it has small canines), it might have been on the line to australopithecines given the evidence for its early form of bipedalism (see Figure 15.27). Ardi roamed Ethiopia's Afar Rift Valley when it was covered by rain forests and not desert about 4.4 million years ago. Ardi weighed about 50 kilograms (100 pounds) and was about 4 feet tall. Although her skeleton indicates she could climb among trees on all fours, she probably spent more time walking upright on the ground (**Figure 15.28** and **Figure 15.29**). In other

words, Ardi had evolved away from the last common ancestor of humans and chimpanzees but still had ape-like features. Ardi had ape-like long arms, short legs, and a grasping big toe for climbing among tree branches. At the same time, Ardi possessed relatively short palms and flexible fingers, indicating that although she climbed trees, she was probably less comfortable doing so because she did not swing through them like living apes. Her pelvis indicates she could also walk upright more efficiently than living apes. Although her face sloped forward like that of apes, it did not slope forward as much; her upper canine teeth are also more like those of modern humans than the long, pointed ones of apes. Significantly, Ardi indicates that after gorillas had split off from the line leading to humans and chimps, humans and chimps later diverged (as noted earlier) from a common ancestor by 6 to 7 million years ago rather than evolving from a chimp-like creature; thereafter, the lineages leading to African apes and humans evolved independently (see Figure 15.23). Ardi therefore appears to lie somewhere between *Sahelanthropus* and *Australopithecus* (see Figure 15.27).

Data from: Gibbons, A. 2009. A new kind of ancestor: *Ardipithecus* unveiled. *Science*, 326, 36.

FIGURE 15.28 Reconstruction of *Ardipithecus ramidus*. Although her skeleton indicates she could climb among trees on all fours, Ardi probably spent more time walking upright.

In any case, australopithecines are the first unequivocal bipeds. They appeared by 4.2 million years ago and then began to diversify between this time and 2.5 million years ago (**Figure 15.30** compare Figure 15.27). The australopithecines were up to about 4.5 feet tall and weighed about 30 to 50 kilograms (65 to 100 pounds); their skulls were characterized by a prominent **supraorbital ridge** over the

eyes and a sloping face (**Figure 15.31**). Despite their small size, they were strongly built. At least two forms of *Australopithecus*, *A. robustus* and *A. boisei* (sometimes put into the genus *Paranthropus*), were relatively robust forms that would eventually prove to be evolutionary dead-ends after *Australopithecus* began to diversify (see Figure 15.30). More "gracile" (slender) forms of *Australopithecus* such as *A. anamensis* (found in Kenya and dating to 4.2 to 3.9 million years ago) appear to have been more closely related to the main line of human evolution (see Figure 15.30).

Appearing somewhat later was *Australopithecus afarensis*. *A. afarensis* is best known from the famous "Lucy" skeleton (named after the Beatles' song *Lucy in the Sky with Diamonds*) discovered in the Afar Depression in Ethiopia in 1974 (**Figure 15.32**). *A. afarensis* appears to have retained stronger adaptations to climbing than modern humans, resembling a "bipedal chimpanzee," meaning that it probably moved back and forth between trees and the ground. *A. afarensis* did not have opposable big toes (Ardi did), whereas the pelvis was wide (like that in humans), indicating that *A. afarensis* was an obligate biped or was fully adapted to walking upright. In fact, fossil footprints of two individuals spectacularly preserved under a layer of 3.7-million–year-old volcanic ash (see Figure 15.32) and the anatomy of the feet indicate that this species had a striding bipedal gait like that of modern humans.

One of the most famous of the gracile species of *Australopithecus* is *A. africanus* (**Figure 15.33**). This species was named by the paleoanthropologist Raymond Dart in 1925 after the front part of a skull and most of the lower jaw of a 6-year-old child were discovered in a quarry in the Cape Province of South Africa and which came to be known as the "*Taung child.*" It was this discovery that set off more exploration in South Africa because Taung child's brain and teeth resembled those of humans more than those of apes.

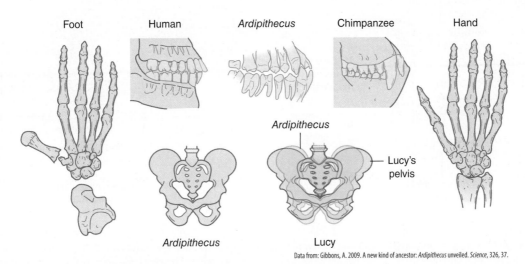

Data from: Gibbons, A. 2009. A new kind of ancestor: *Ardipithecus* unveiled. *Science*, 326, 37.

FIGURE 15.29 *Ardipithecus* had an opposable toe (left) and flexible hand (right); her canines (top center) are sized between those of a human (top left) and chimp (top right); and the blades of her pelvis (lower left) were broad like Lucy's.

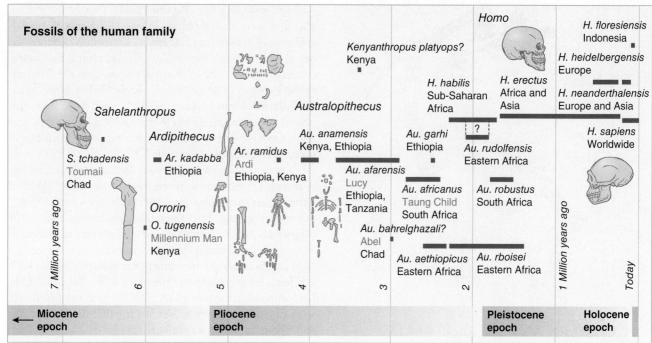

FIGURE 15.30 Geologic ranges, fossil remains, and location of discovery of species leading to modern humans.

The evolution of bipedalism might have been selected for by a number of factors. These include: carrying young, foraging for food, or body temperature regulation (by exposing less of the back to the sun). Bipedalism might have also allowed early humans to expend less energy on foraging for food or evading predators, for example. At walking rates, bipedalism is less expensive energetically than quadrupedalism, or "walking on all fours," like that of chimpanzees, gorillas, and orangutans. These taxa all evolved in and still occupy dense forests, in which movement over as little as a mile a day is all that is necessary to find enough to eat. By contrast, much of early human evolution took place in more open woodlands and grasslands, where it is harder to find food and one is more exposed to predators.

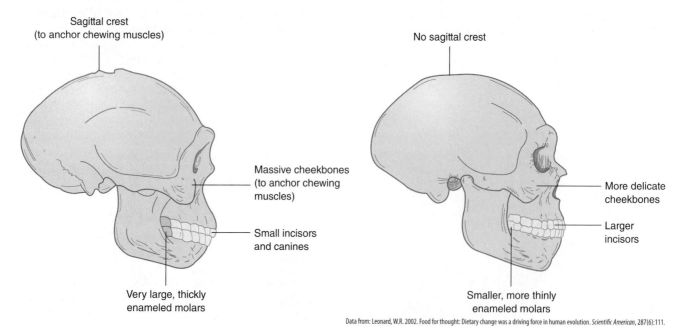

Data from: Leonard, W.R. 2002. Food for thought: Dietary change was a driving force in human evolution. *Scientific American*, 287(6):111.

FIGURE 15.31 Comparison of the skulls of *Australopithecus boisei*, and "Peking man," which belongs to the species *Homo erectus*. The skull of *A. boisei*, which was an evolutionary "dead end" in human evolution, exhibits adaptations for eating tough, fibrous plant foods: a sagittal crest for the attachment of jaw muscles (which must have been quite large for chewing large amounts of tough food), massive cheekbones (also for anchoring jaw muscles), large, thick molars, and small incisors and canines. By contrast, the more delicate features of *H. erectus* suggest that it consumed larger quantities of meat.

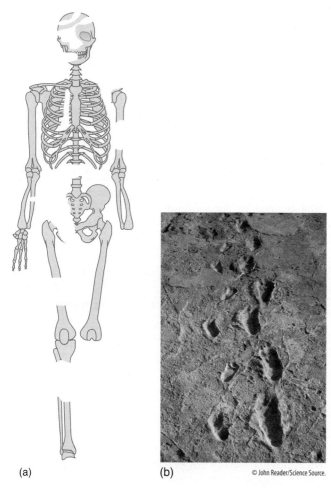

(a) (b) © John Reader/Science Source.

FIGURE 15.32 **(a)** The "Lucy" skeleton, which dates from ~3.2 million years in Ethiopia, represents *Australopithecus afarensis* and might have resembled a "bipedal chimpanzee." **(b)** Footprints of individuals thought to belong to *A. afarensis* made in volcanic ash dated to 3.2 million years ago.

© Pascal Goetgheluck/Science Source.

FIGURE 15.33 The skull of the "Taung Child," *Australopithecus africanus*.

Fossils assigned to a new species, *Australopithecus sediba*, appear transitional to the genus *Homo*, which first appears in the fossil record of East Africa at about 2.5 million years ago. Fossils of *A. sediba* were found in South Africa, have been dated to almost 2 million years old, and are remarkably complete because the individuals fell into a deep cave and were rapidly buried. The hands for example have, like apes, long fingers for climbing trees but they also have an opposable thumb that could have been used for manipulating tools. The brain's frontal cortex, used in thinking, had also become more human-like. However, like other hominin fossils, the question of the relation of *A. sediba* to later forms remains open but some workers insist that South Africa, and not East Africa, was the cradle of human evolution because the caves in which *A. sediba* was discovered are located in South Africa.

Perhaps even more significant than the origin of bipedalism is that the increase of brain size in *Homo* followed the appearance of bipedalism. It was long thought that an increase in brain size led to *Homo*. So now the question has become: did bipedalism lead to an increase in brain size and to language and tool use? Early hominins lived among patchy woodlands and dry grasslands where food supplies would have been seasonal. Such an environment would have selected for hominins who could utilize a wide range of food resources and search for them over long distances. Bipedalism and an upright stance while retaining the ability to climb might have supported such a lifestyle while also increasing predator avoidance.

Homo was initially similar in size to the australopithecines but had a larger brain (600–700 cc³) and smaller molars. At least six separate species of *Homo* evolved in parallel between 1.6 and 2.5 million years ago. (Note to reader: Because of the "lumping" and "splitting" of fossil species noted earlier, there is a great deal of controversy about some of the species of *Homo* that have been recognized and there are strong arguments that this proliferation of taxa is not real. It is clear today that there were certainly times when there were multiple species of hominin and perhaps even of *Homo* living side by side, but there remains a great deal of controversy about how many and what they should be called.) *Homo rudolfensis* came on the scene by 2.4 and 1.8 million years ago and *H. habilis* between 2.0 to 1.6 million years ago (see Figure 15.30). The similarity of the anatomy of the hands and feet of *H. habilis* to that of australopithecines indicates that *H. habilis* could climb trees.

It is thought that *H. rudolfensis* and *H. habilis* were members of the lineage that gave rise to two new species of *Homo* that appeared perhaps no later than 500,000 and 100,000 years, respectively, after *H. rudolfensis* and *H. habilis*. *H. ergaster* was taller than its ancestors, had a larger brain than them, and used fire and produced large hand axes. *H. erectus* lived contemporaneously with *H. ergaster*. Its brain volume overlapped that of *H. ergaster*, and it also made tools, but it had heavier brow ridges and a thicker skull than its contemporary species. *H. erectus* was originally described about the beginning of the 20th century from a skull cap and thighbone found in Java (Indonesia) dated at about 1 to

0.7 million years ago. The species was later found in China dating to about 0.5 million years ago, where it was dubbed "Peking Man" after the former name for the capital of China (see Figure 15.31). These fossils were believed to be sufficiently different from each other and other species known at the time that they were originally assigned to separate genera and species: *Sinanthropus pekinensis* and *Pithecanthropus erectus*, respectively. A large braincase assigned to *H. erectus* was also found at Olduvai Gorge in Africa. *H. erectus* in Africa is exemplified by "Turkana Boy," now called Nariokotome Boy, the nearly complete skeleton of an adolescent dated at 1.5 to 1.6 million years old discovered near Lake Turkana in Kenya in 1984 (**Figure 15.34**). Nariokotome Boy had long, slender limbs like modern Africans adapted to living in a hot, dry savannah. Remains found in 2004 on the island of Flores in Indonesia have been interpreted as either a relict population of dwarf *H. erectus* or a new species, *Homo floresiensis*. In either case, this form lived until as recently as several tens-of-thousands of years ago.

The variability of *H. erectus* fossils has resulted in the recognition of a separate species, *Homo heidelbergensis* from

Swanscombe, England, and Steinheim, Germany, at about 200,000 years ago; this species exhibits features intermediate between *H. erectus* and *H. sapiens*. However, it is possible that *H. heidelbergensis* arose from *H. ergaster* rather than *H. erectus*, because *H. erectus* persisted well after *H. heidelbergensis* arose. In Europe, *H. heidelbergensis* might have given rise to *Homo neanderthalensis* (**Figure 15.35** and **Figure 15.36**). The remains of Neanderthal man were originally discovered in the Neander River Valley of Germany but have since been found all over Europe and western Asia, dating back between 200,000 and 30,000 years ago. Despite their relatively primitive-looking skull, the Neanderthals were relatively large-brained. In fact, the brain case of Neanderthals equaled or surpassed that of *H. sapiens*. Neanderthals persisted in Spain and Gibraltar as recently as 35,000 years ago, and it has been suggested from DNA evidence that the two species (Neanderthals and modern humans) interbred, though the extent of gene flow between them remains unclear. Neanderthals used sophisticated stone tools and appear to have practiced cannibalism, based on cut marks on dismembered Neanderthal bones found next to hearths. Neanderthals practiced at least some rituals (they buried their dead) and participated in some symbolic activities such as use of beads and collecting of feathers.

Our species, *H. sapiens*, has been traced back to a lineage that contained *H. erectus*. It is difficult, however, to determine the exact lineage to which *H. erectus* belonged because *H. erectus* exhibits variability and subtle differences from *H. sapiens* that possibly resulted from potentially

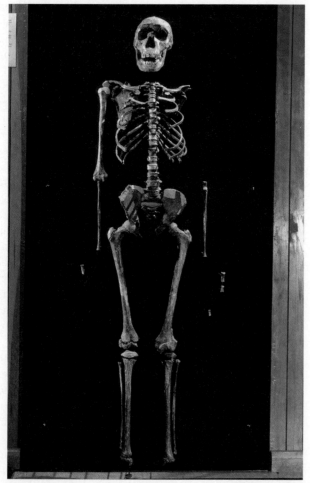

© Danita Delimont/Alamy.

FIGURE 15.34 Fossilized skeleton of "Turkana Boy," which is about 1.6 million years old. Turkana Boy is representative of the first hominids possessing modern skeletons.

© Cro Magnon/Alamy Stock Photo.

FIGURE 15.35 Reconstruction of Neanderthal man.

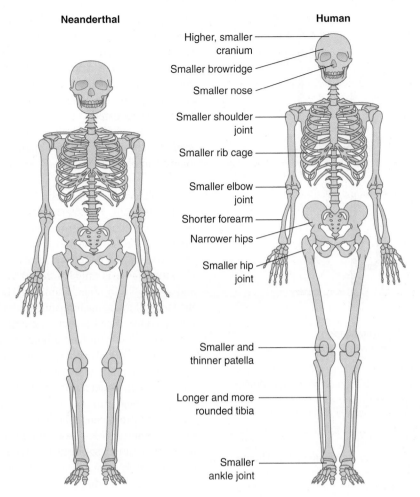

Neanderthal

Human

Higher, smaller cranium

Smaller browridge

Smaller nose

Smaller shoulder joint

Smaller rib cage

Smaller elbow joint

Shorter forearm

Narrower hips

Smaller hip joint

Smaller and thinner patella

Longer and more rounded tibia

Smaller ankle joint

FIGURE 15.36 Comparison of modern and Neanderthal skeletons.

non-interbreeding populations inhabiting different geographic areas. Two hypotheses have been proposed for the origin and dispersal of *Homo sapiens*. One, called the **multiregional evolution hypothesis** (**Figure 15.37**) proposes that modern humans evolved from *H. erectus* after *H. erectus* originated in, and then migrated out of, East Africa to Europe and Asia about 1 million years ago. According to this hypothesis, *H. sapiens* originated from the various regional populations of *H. erectus* that continued to exchange genes through time. Thus, *H. sapiens* would have arisen across the Old World from *H. erectus* as populations interbred and continued to undergo evolutionary change. The second hypothesis, called the **single origin** or **Out of Africa hypothesis** (see Figure 15.37), is based on the first appearance of *H. sapiens* in east Africa, after which our species dispersed to other continents and replaced *H. erectus*.

Despite all the uncertainties of the fossil record of human evolution, one thing stands out: human evolution resembled a branching tree or better, a bush; in other words, human evolution was punctuated (see Chapter 5). Think of it: we are accustomed to the existence of distinct human races. Subspecies (sometimes called races) are slightly different groups that belong to the same **species**, but all races belonging to the same species can still interbreed (see Chapter 5).

Genetic evidence such as the human genome project has shown that the human species is not particularly diverse genetically and that human "races" previously recognized in different parts of the world overlap so much in their genetic makeup that the biological subspecies (sometimes called "race") concept cannot be usefully applied to humans. But in the case of ancestral humans, *wholly different species (which could not interbreed) lived contemporaneously in the same place.*

CONCEPT AND REASONING CHECKS

1. Which hypothesis makes more sense in terms of evolutionary theory: the multiregional or the single-origin hypothesis?
2. What features set *Homo* apart from *Australopithecus*?

Why did hominins leave Africa?

 Perhaps rapid climate change also influenced human evolution. Forests were becoming increasingly restricted as grasslands spread in response to global cooling that resulted from the movement of Antarctica over the south pole during the Eocene-Oligocene. So by the time

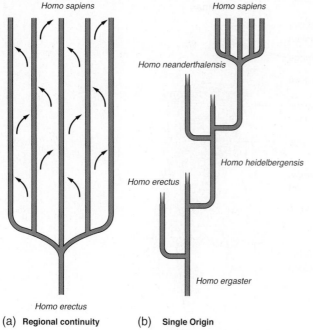

Homo sapiens

Homo sapiens

Homo neanderthalensis

Homo heidelbergensis

Homo erectus

Homo erectus

Homo ergaster

(a) **Regional continuity**   (b) **Single Origin**

Data from Tattersall, I. 1997. Out of Africa again . . . and again? *Scientific American*, 279(4): 67.

FIGURE 15.37 Comparison of two hypotheses of human evolution. **(a)** The regional continuity, or multiregional, hypothesis states that all modern humans are descended from *H. erectus*, but separate regional populations were established that continued to interbreed sufficiently so that all populations evolved into *H. sapiens* more-or-less simultaneously. **(b)** The single-origin, or out-of-Africa, hypothesis states that *H. sapiens* descended from a single ancestral population that originally evolved in one place, most likely Africa. In this view, *H. erectus* is an offshoot of the main evolutionary line that migrated to Asia.

early hominins were evolving, glaciation in the northern hemisphere had begun to affect Africa. The Sahara desert, for example, began to expand by this time, as evidenced by sand grains found in deep-sea cores in the Atlantic Ocean off West Africa that were blown in by the trade winds. According to the **turnover pulse** hypothesis, there was a major pulse of mammal evolution in Africa beginning about 2.7 million years ago, perhaps only a few hundred thousand years after the closure of the Isthmus of Panama. According to this hypothesis climate change yet again triggered both extinction and speciation, which, taken together, caused evolutionary turnover. African antelope and other grazing mammals, for example, are thought to have evolved from a browsing to a grazing habit because of climate change. It was also about this time that *Australopithecus afarensis* was splitting into gracile and robust lines. The robust australopithecines possessed large molars, premolars, and jaw muscles, most likely for chewing grasses, not moist forest vegetation. However, other workers believe the evolutionary turnover is not confined to a relatively narrow 200,000- to 300,000-year interval of rapid change; instead, it might have been spread through a million years of the Pliocene, from about 3 to 2 million years ago. A recent study based on carbon isotopes of tooth enamel of distantly related fossil Old World monkeys indicates that one genus gradually shifted its diet

from C3 to C4 grasses from about 4 to 1 million years ago, which spans the onset and expansion of northern hemisphere glaciers, but similar studies of fossil human ancestors have not yielded a clear trend. But, if climate change influenced human evolution in such a significant way, why did early hominins presumably walk around in forests before grasslands appeared?

Perhaps another possible cause of the human diaspora is related to the evolution of the brain. As the australopithecines evolved into *H. erectus* and *H. erectus* perhaps into *H. sapiens*, brain volume increased three to four times, from about 385 cubic centimeters to about 1,400 cubic centimeters in modern humans. The total energy requirements of the human body at rest is no greater than that of other mammals of the same size, but even at rest, brain metabolism accounts for 20% to 25% of an adult human's energy needs; by contrast, the same figure is only 8% to 10% in nonhuman primates and 3% to 5% in other mammals. Thus, as humans evolved they might have come more often to eat meat, which is a more highly concentrated source of energy than plant matter. In other words, humans had to hunt other animals for food.

Still, as noted above, about 98.5% of the DNA in humans and chimpanzees is identical; thus, physical or biochemical traits (such as metabolism), although necessary, are alone insufficient to account for the initiation of the rise of humans to prominence. Perhaps, then, the evolution of the human brain also had something to do with making and using increasingly sophisticated tools. However, in the case of *Homo*, tools were still more elaborate than any other primates' tools when the first *Homo* left Africa.

Possibly, the evolution of human behavior is also involved. Although there are reports of ritual burial by Neanderthals, burial of objects in preparation for an afterlife do not definitely occur until after *H. sapiens* appeared. *H. sapiens* brought with it more sophisticated tools and art, including cave paintings, musical instruments, and jewelry. Put another way, modern humans had begun to evolve the symbolic thought, and perhaps also language, that characterizes our species. Language, thought, and creativity go hand-in-hand because they give us the prescience, or forethought (*H. sapiens*), that allows us to ask questions like "what if?"

15.7 Extinction

Although the Neogene is not widely known for extinction, extermination did occur nonetheless. A significant turnover in large ungulate grazers occurred during the Late Miocene in both North America and Asia. Significantly, many survivors of the extinctions were all forms with very high-crowned teeth, suggesting extinction was related to changes in vegetation. It was about the Late Miocene that C4 grasses began to replace C3 taxa. C4 grasses have much larger amounts of silica embedded in their tissues than do C3 forms. Silica is of course highly abrasive, and this property is in part responsible for the evolutionary changes seen in perissodactyls and artiodactyls during the Eocene–Oligocene transition (see Chapter 14). In the case of the Late Miocene extinctions, it is believed that species with lower-crowned teeth were unable

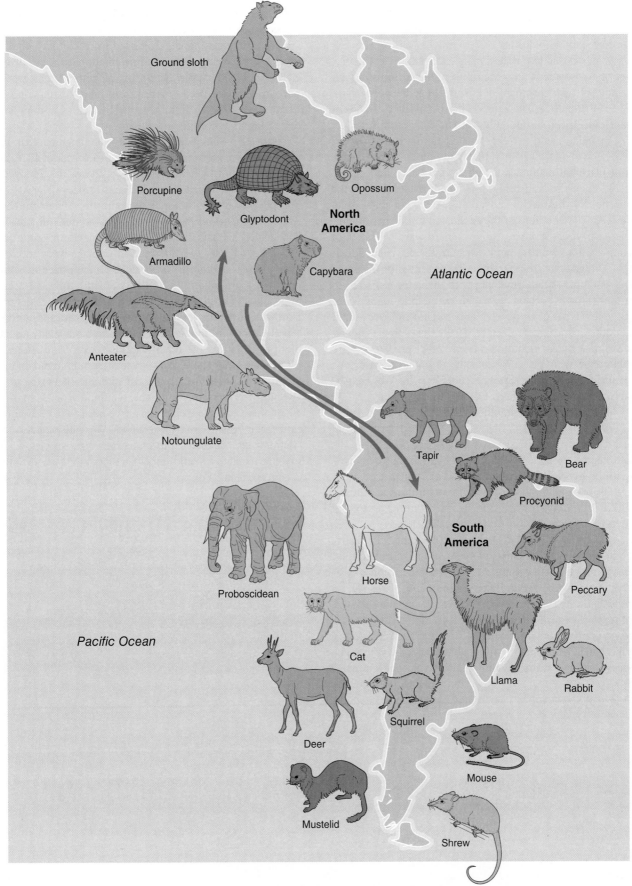

FIGURE 15.38 The Great American Interchange. The exchange was primarily one way, with North American placental mammals migrating into South America. Relatively few South American marsupials migrated into North America.

Data from Pough, F. H., Janis, C. M., and Heiser, J. B. 1998. *Vertebrate Life*. New York: MacMillan Publishers, and Marshall, L. G., *American Scientist*, 76(1988): 380–388.

to adapt to the increase in silica in their food. In effect, the teeth of these species might have worn down to the point that they starved to death.

A later extinction, that of marsupials in South America, might have occurred after the uplift of the Isthmus of Panama and the establishment of a land bridge between North and South America. This episode has been called the **Great American Interchange** (**Figure 15.38**). South America, like Australia, came to be populated by marsupials in the early Cenozoic (see Chapter 14). Also present were a handful of placental mammals including armadillos and porcupines. When the land bridge was finally established in the Pliocene, North America had far fewer tropical habitats than South America because of the expansion of northern hemisphere ice sheets. Consequently, the northward migration of mammals from South America was greatly restricted. Today, in the United States these forms are represented by the opossum, porcupine, and armadillo, with the latter restricted to warmer climates such as those of Texas and Florida. By contrast, many types of artiodactyls and predators moved south. In South America, marsupials that were the ecologic equivalents of placentals, including cats and other predators, were replaced by their placental cousins. Although the obvious conclusion is that the placentals "outcompeted" the marsupials, no one knows for sure if this is what happened.

Mastodons and mammoths also died out in North America, Europe, and Asia as recently as 25,000 years ago, during the last glacial interval. Extinction might have been due to climate change, predation by ancient humans, or both. According to the **overkill hypothesis**, these and other taxa might have disappeared in response to increased hunting as humans arrived in various parts of the world about 10,000 years ago (see Chapter 17).

However, recent studies that have examined the distribution of the fossil remains of large mammals through time across Europe and Asia indicate that the extinction of different species occurred at different times and places. These studies indicate that the woolly mammoth actually survived in the northeastern Siberian Arctic until about 4,000 years ago, making it contemporaneous with the Bronze Age Xia Dynasty in China. Similarly, the Irish elk, which was thought to have died out at the beginning of the Holocene, survived in western Siberia to about 7,500 years before present. Thus, it seems more likely that the geographic ranges of different species of large mammals decreased in response to climate cooling or warming, which in turn might have affected vegetation and food supply. Decreasing geographic range would have made these species more susceptible to extinction (see Chapter 5). Possibly, human activities such as hunting and deforestation delivered the final blows.

Indeed, it has been repeatedly documented that the extermination of numerous species took place after the arrival of colonists in various parts of the world. Today, for example, only two types of elephants remain of what was once a diverse and widespread group: the African and the Asian. Elephants are being increasingly confined to smaller and smaller habitats due to human population pressure and poaching for tusks, which are used to make ivory trinkets. Chimpanzees and gorillas, whose heads are prized by some as trophies, are suffering a similar fate. Thus, there are strong indications that this so-called **Sixth Extinction** results primarily from habitat loss due to human activities. The term "Sixth Extinction" implies that the loss in biodiversity resembles in magnitude the other "Big Five" mass extinctions of the Phanerozoic. There is also the matter of hominid extinctions. All hominid species described earlier persisted for a few hundred thousand to roughly a million years or so. And all of these species except us—*H. sapiens*—have become extinct for reasons not understood. We will examine the relationships between rapid climate change, species diversity and habitat loss, and human activity in greater detail in Chapters 16 and 17.

CONCEPT AND REASONING CHECK

1. Could the spread of humans over the globe have been in response to climate change, thereby explaining the overkill hypothesis as primarily anthropogenic in origin? How could you test this hypothesis? Do you believe the hypothesis is testable given the nature of the human fossil record? If you could test the hypothesis, do you believe you'd get a definitive answer?

- The same basic tectonic phenomena that began in the Paleogene continued through the Neogene, along with several significant developments. During the Neogene the Tethys Seaway continued to close with the continued collision of India and Asia to form the Himalayas and the crushing of smaller plates between the African plate and southern Europe to form the Alps. These collisions produced the Alps and Himalayas. In the Pliocene, the Arabian Peninsula rifted toward what is now Iran to open the narrow, deep-water troughs of the Red Sea and Gulf of Aden. As Arabia moved northeast, it collided with Iran to produce the Zagros Mountains.

- In Central America, the remnant of the Tethys left between North and South America was closed by the rise of the Isthmus of Panama about 3 to 3.5 million years ago. During this time the Greater Antilles moved toward their modern positions, and newly developed transform faults along the margin of the Caribbean plate transferred Cuba to the North American plate. In South America, the Andes continued to rise, obliterating a foreland basin with molasse and leading ultimately to the Amazon rain forest.

- In North America, the major physiographic provinces continued to evolve. Although the Rockies, which formed during the Laramide Orogeny, continued to be beveled, they appear to have again been uplifted beginning in the Miocene. Uplift extended into the Colorado Plateau of the "Four Corners" region, which rose as a single block and caused rivers such as the Colorado to erode downward to produce features such as the Grand Canyon.

- The subduction zone along the western margin of North America persisted as the North American plate moved over the Farallon plate. The Columbia River plateau basalts erupted in the Middle Miocene in eastern Oregon and Washington as the North American plate moved over what might be a hotspot. By the Late Miocene and Pliocene, the eruptions shifted eastward into southern Idaho to form the Snake River plain and then Yellowstone.

- As the North American plate moved to the northnorthwest, subduction of the Farallon plate beneath it produced a volcanic arc to the east that extended from the Pacific Northwest into southeastern California and Nevada. To the south, subduction also continued to form the Coast Ranges of California.

- Sediments also continued to be shed from the rising Sierra Nevadas into the central valley of California, which consisted of a series of deep-water basins that eventually filled with molasse. The granitic plutons emplaced during the Nevadan Orogeny were eventually exhumed as a huge fault block that tilted to the west.

- In the Late Oligocene, the portion of the North American plate represented today by central and southern California began to encounter a transform offsetting the East Pacific Rise. As the North American plate encountered the transform, subduction of the Farallon plate ceased, as did volcanism further east. The Farallon plate eventually split into two plates: the Juan de Fuca plate to the north, which is still being subducted beneath the North American plate along the Pacific Northwest, and the Cocos plate to the south along Mexico and Central America. Eventually, the Pacific plate began to encounter the North American plate, but the Pacific plate was not subducted. Instead, northwest movement of the Pacific plate along the transform became dominant, resulting in the origin of the San Andreas Fault. Further south, as the North American plate continued to override the Pacific plate and the East Pacific Rise, seafloor spreading pulled the Baja Peninsula away from mainland Mexico to form the Gulf of California.

- These changes in plate boundaries are thought to be related to the formation of the Basin and Range province. Possibly, as the crust west of the San Andreas moved northwest, it caused the ancestral Basin and Range province to spread like a fan, causing the Sierra Nevadas, the Great Valley, and the mountain ranges extending into the Pacific northwest to swing to the west. The lack of subduction might explain the cessation of volcanism at the southern terminus of the Cascades in northern California. This hypothesis might also explain the uplift of the Colorado Plateau and the Rockies; at the point of contact between the North American and Pacific plates, the North American plate might lay on top of warm mantle material, which could have buoyed up the overlying continental crust.

- Erosion and sedimentation continued along the Atlantic and Gulf Coasts. Although much of the Appalachians were strongly eroded during the Paleogene, they were rejuvenated during the Miocene as a result of isostatic uplift. Much of the sediment was shed into a series of shallow embayments along the passive margin of eastern North America. The Gulf of Mexico also continued to serve as a trap for sediment shed during the Laramide Orogeny and the rejuvenation of the Rockies. Enormous volumes of sediment were flushed into the Gulf of Mexico, causing the continental margin to prograde over the Louann Salt, which continued to deform into salt diapirs and form traps for oil and gas.

- Modern ocean circulation, chemistry, and climate regime also began to be established during the Neogene. During the Neogene strontium isotope ratios underwent a dramatic rise, reflecting increased uplift and erosion all over the world. Also, the calcite compensation depth tended to deepen as sea level fell.

- A permanent polar ice cap began to grow on Antarctica as it continued to move over the South Pole, which led to a lowering of sea level.

- Ice caps also appeared in the northern hemisphere, based on ice-rafted debris about 2.8 million years old in deep-sea cores taken off the British Isles. This followed about 700,000 years after the rise of the Isthmus of Panama, which is thought to have blocked the flow of warm-water currents from the Atlantic into the Pacific. The diversion of warm waters northward into the Gulf of Mexico and the western Atlantic Ocean transported warm, moist air to higher latitudes, where it began to precipitate as snow and ice. With the appearance of ice caps in the northern hemisphere, production of North Atlantic Deep Water began, which, along with the production of Antarctic Bottom Water, resulted in increasing oxygenation of the deep oceans.

- After the northern hemisphere ice sheets were established, they waxed and waned in response to the amount of solar radiation reaching Earth's surface. Cycles of solar radiation are thought to represent Milankovitch cycles of solar radiation that occur at three dominant periods: (1) eccentricity (or "ellipticity") of Earth's orbit that completes one cycle about every 100,000 years; (2) obliquity ("tilt") of Earth's axis toward or away from the sun, which affects the amount of solar radiation impinging on the northern hemisphere and which completes one cycle every 40,000 years; and (3) precession of the equinoxes about every 19,000 to 23,000 years, which affects the occurrence of the seasons within Earth's orbit around the sun.

- The growth of ice sheets in the northern hemisphere might have caused forests to further shrink and grasslands to expand. With the continued expansion of grasses came further diversification of artiodactyls. Similarly, according to the turnover pulse hypothesis, shrinking forests caused ancient human ancestors, or hominids, which were arboreal, to leave forests for grasslands. On the other hand, human evolution might have resulted from changes in behavior.

- In either case, our ancestors eventually began to walk upright. As hominids evolved, different populations might have coexisted in the same habitats and might have migrated a number of times to Asia and eventually to Europe. As humans spread across the globe during the Holocene, they might have contributed to the demise of different species of large mammals and flightless birds.

KEY TERMS

aquifer	Gulf Stream	oxygen isotopes	species
Basin and Range Province	haplorhines	Pacific plate	stereoscopic vision
C3 grasses	Holocene	precess	strepsirrhines
C4 grasses	horst-and-graben	precession of the equinoxes	summer solstice
Coast Ranges	Juan de Fuca plate	primates	supraorbital ridge
Cocos plate	Lake Bonneville	Rio Grande Rift	terminations
diluvial theory	mammoths	San Andreas Fault	tilt
East Pacific Rise	Milankovitch cycles	Sierra Nevada	turnover pulse
eccentricity	moraines	single origin ("Out of Africa") hypothesis	ungulates
Farallon plate	multiregional evolution hypothesis	single species hypothesis	winter solstice
Front Range	obliquity	Sixth Extinction	wobble
glacial erratics	overkill hypothesis	slab gap hypothesis	
Great American Interchange			

1. How do Paleogene and Neogene events differ from each other? Make a chart labeled Paleogene and Neogene across, and down the left side of the chart list the following: (a) continental movements, (b) sea level, (c) atmospheric CO_2, (d) ocean circulation, (e) oxygen, (f) plankton, (g) calcite compensation depth, (h) terrestrial plants, and (i) terrestrial animals.

2. On a map of North America or another continent(s) of the world, find the following features that formed during the Neogene and discuss how they formed: Amazon rain forest, Amazon River, Arabian Peninsula, Aral Sea, Basin and Range, Cascade Mountain Range, Coast Ranges, East African Rift Valley, Front Range, Greater Antilles, Gulf of California, Himalayas, Isthmus of Panama, Mississippi River, Rio Grande Rift, San Andreas Fault, Sierra Nevadas, and Yellowstone hotspot.

3. What was the effect of the uplift of the Himalayas on global climate? (See also Chapter 2.)

4. What was the effect of the rise of the Isthmus of Panama on global climate and life on land and in the oceans?

5. How might have tectonism contributed to the growth of glaciers over both poles of Earth?

6. What is the evidence for the advance and retreat of northern hemisphere glaciers from land? From the deep sea?

7. How do the three major Milankovitch frequencies interact to produce climate change? Do all three frequencies always accentuate glaciation or warming? Why or why not?

8. How does the evolution of humans resemble that of other taxa, such as the horse? What factors contributed to the evolution of humans?

9. Evaluate the different hypotheses for human evolution for their strengths and weaknesses: multiregional, single species, and turnover pulse.

10. What is the difference between a species and a race?

11. What was happening on Earth about 10,000 to 11,000 years ago?

FOOD FOR THOUGHT:
Further Activities In and Outside of Class

1. Construct a table of the hypotheses described in the text for the origin of the Basin and Range. List the hypotheses down the left side and place a heading at the top titled "Evidence." Include in that column both the geologic evidence and the forces inferred from the evidence. Then, to the right place a column titled "Success of the Hypothesis" with two columns underneath for each of the main regions of the Basin and Range emphasized in the text: Northern (N) and Southern (S). For each hypothesis, indicate whether it satisfactorily explains the evidence within the region (+), does not explain it one way or the other (0), or contradicts it (–). Discuss your results in terms of the method of multiple working hypotheses (see Chapter 1).

2. Which normally causes sea level to change faster: the advance and retreat of glaciers or the movements of continents?

3. Describe the tectonics and sedimentation of the western United States in terms of the method of multiple working hypotheses (see Chapter 1).

4. How did preadaptation in early primates later affect the evolution of humans?

5. Why is the fossil record of humans so hotly debated?

6. Of the different hypotheses for human evolution, which one do you favor and why?

7. What is the significance of the finding of *Plesiadapis* in both North America and Europe?

8. Quite frequently, new species of humans are based on a single fossil fragment such as a jaw fragment. How can an entire new species be inferred from a single fragmentary fossil? (Hint: See Chapter 4 for Cuvier's correlation of parts.)

Alroy, J. 2001. A multispecies overkill simulation of the end-Pleistocene megafaunal mass extinction. *Science, 292*, 1893–1896.

Asfaw, B. et al. 2002. Remains of *Homo erectus* from Bouri, Middle Awash, Ethiopia. *Nature, 416*, 317–320.

Baldridge, W. S. 2004. *Geology of the American Southwest: A journey through two billion years of plate-tectonic history*. Cambridge, UK: Cambridge University Press.

Bower, B. 1999. DNA's evolutionary dilemma. *Science News, 155*(6, February 6):88–90.

Bower, B. 2001. Fossil skull diversifies family tree. *Science News, 159*(12, March 24), 180.

Bower, B. 2002. Evolution's surprise: Fossil find uproots our early ancestors. *Science News, 162*(2):19.

Brunet, M. et al. 2002. A new hominid from the Upper Miocene of Chad, central Africa. *Nature, 418*, 145–151.

Cerling, T. E., Chritz, K. L., Jablonski, N. G., Leakey, M. G., and Manthi, F. K. 2013. Diet of *Theropithecus* from 4 to 1 Ma in Kenya. *Proceedings of the National Academy of Sciences*, www.pnas.org/cgi/doi/10.1073/pnas.1222571110

Chester, Stephen G. B., et al. Oldest known euarchontan tarsals and affinities of Paleocene *Purgatorius* to Primates. *Proceedings of the National Academy of Sciences, 112*.5(2015):1487–1492.

Dayton, L. 2001. Mass extinctions pinned on ice age hunters. *Science, 292*, 1819.

Demenocal, P. B. 2011. Climate and human evolution. *Science, 331*, 540–542

Gibbons, A. 2011. Skeletons present an exquisite paleo-puzzle. *Science, 333*, 1370–1372.

Guo, Z. T. et al. 2002. Onset of Asian desertification by 22 Myr ago inferred from loess deposits in China. *Nature, 416*, 159–163.

Imbrie, J., and Imbrie, K. P. 1979. *Ice ages: Solving the mystery*. Short Hills, NJ: Enslow Publishers.

Leakey, R. and Lewin, R. 1995. *The Sixth Extinction: Patterns of life and the future of humankind*. New York: Doubleday Dell.

Leakey, M. G. et al. 2001. New hominid genus from eastern Africa shows diverse middle Pliocene lineages. *Nature, 410*, 433–440.

Leakey, M. G., Spoor, F., Dean, M. C., Feibel, C. S., Antón, S. C., Kiarie, C., and Leakey, L. N. 2012. New fossils from Koobi Fora in northern Kenya confirm taxonomic diversity in early *Homo*. *Nature, 488*, 201–204.

Leonard, W. R. 2002. Food for thought: Dietary change was a driving force in human evolution. *Scientific American, 287*(6):106–115.

Monastersky, R. 1996. Out of arid Africa. *Science News, 150*(5, August 3), 74–75.

Monastersky, R. 1999. The killing fields: What robbed the Americas of their most charismatic mammals? *Science News, 156*(23):360–361.

Murphy, J. B., Oppliger, G. L., Brimhall, G. H., and Hynes, A. 1999. Mantle plumes and mountains. *American Scientist, 87*(2):146–153.

Ni, Xijun, et al. "The oldest known primate skeleton and early haplorhine evolution." *Nature, 498*.7452 (2013):60–64.

Pastor, J., and Moen, R. A. 2004. Ecology of ice-age extinctions. *Nature, 431*, 639–640.

Roberts, R. G. et al. 2001. New ages for the last Australian megafauna: Continent-wide extinction about 46,000 years ago. *Science, 292*, 1888–1892.

Ruddiman, W. F., and McIntyre, A. 1976. Northeast Atlantic paleoclimate changes over the past 600,000 years. In R. M. Cline and J. D. Hays (eds.), *Investigation of Late Quaternary paleoceanography and paleoclimatology* (pp. 111–146). Special Publication 145. Boulder, CO: Geological Society of America.

Sonder, L. J., and Jones, C. H. 1999. Western United States: How the west was widened. *Annual Review of Earth and Planetary Sciences, 27*, 417–462.

Stringer, C. 2003. Out of Ethiopia. *Nature, 423*, 692–695.

Stuart, A. J., Kosintsev, P. A., Higham, T. F. G., and Lister, A. M. 2004. Pleistocene to Holocene extinction dynamics in giant deer and woolly mammoth. *Nature, 431*, 684–689.

Tattersall, I. 1997. Out of Africa again … and again? *Scientific American, 279*(4):60–67.

Tattersall, I. 2000. Once we were not alone. *Scientific American, 282*(1):56–62.

Tavaré, S., Marshall, C. R., Will, O., Soligo, C., and Martin, R. D. 2002. Using the fossil record to estimate the age of the last common ancestor of extant primates. *Nature, 416*, 726–729.

Templeton, A. R. 2002. Out of Africa again and again. *Nature, 416*, 45–51.

Vekua, A. et al. 2002. A new skull of early *Homo* from Dmanisi, Georgia. *Science, 297*, 85–89.

Wilson, E. O. 1992. *The diversity of life*. Cambridge, MA: Belknap Press.

PART IV

HUMANS AND THE ENVIRONMENT

Part IV assesses natural environmental variation during the time called the Holocene, or "wholly recent." The Holocene is really a continuation of the processes of the Pleistocene. Humans just happen to have spread over the planet during the Holocene, so we have adopted the anthropocentric notion that Earth during this time is the way things have always been. But it was not, because rapid climate change occurred even during the Holocene. Chapter 16 examines the processes of rapid climate change on time scales shorter than those of Milankovitch frequencies during this time. It is these natural processes that have most significantly affected humans during the Holocene.

These are also the processes that humans have begun to affect. In other words, humans are living the "theory" of Earth; humans are a part of Earth systems, not separate from them. We are part of what has been dubbed the "Anthropocene." Chapter 17 examines human impacts on these processes and the sociopolitical consequences of natural and anthropogenic climate change. Chapter 18 then examines what science can and cannot tell us about natural and human impacts on the environment in the context of some of the broad themes of Earth history and scientific method discussed in this book.

Rapid Climate Change During the Holocene

MAJOR CONCEPTS AND QUESTIONS

A What is the significance of the Holocene?

B How did the Holocene begin, and what was the Younger Dryas?

C What is the legend of Noah's Flood, and what is the most recent hypothesis for its origin? What are the drawbacks of this hypothesis?

D How is ocean circulation involved in rapid climate change?

E What are some examples of catastrophic climatic and geologic change during the Holocene?

F How is glacial melting involved in rapid climate change?

G How is the sun potentially involved in rapid climate change?

H What are teleconnections, and how are they involved in rapid climate change?

I Is climate change irreversible?

CHAPTER OUTLINE

Studies of rapid climate change have consistently implicated ocean circulation, the formation and melting of ice, shifts in major air currents, and, increasingly, the behavior of the sun. Major questions persist as to how these components of Earth systems interact to produce climate, and how human activities such as fossil fuel combustion will affect natural climate processes now and in the future.

© David Greitzer/Shutterstock, Inc.

16.1 Introduction to the Holocene

A The last 10,000 years of the Pleistocene are commonly referred to as the Holocene ("wholly recent") Epoch. Quite often, the Holocene is simply referred to as the "Recent." The Holocene is when human civilization began to spread over the globe.

The Holocene was long thought to be a time of relatively mild climate compared with previous glacial intervals of the Pleistocene. Why? Most likely because many experts thought Earth was basically able to regulate itself to overcome environmental disturbances and to maintain relatively constant environments for many thousands of years. The deep-seated notion of Lyell's brand of uniformitarianism, which maintained that the forces of nature do not vary, was still very much in vogue (see Chapter 1).

However, during the second half of the 20th century, scientists began to change their minds when they realized from the accumulating evidence that natural, *rapid* climate change had occurred repeatedly all over the planet during the last major interglacial interval: oxygen isotope stage 5 (**Figure 16.1**; see Chapter 15). Stage 5 had long been regarded as another mild and climatically stable time, representing natural "baseline" environmental conditions (those not affected by humans) against which potential anthropogenic effects of fossil fuel combustion could be assessed. The rapid climate fluctuations during stage 5 were indicated by changes in the concentrations of the greenhouse gases, carbon dioxide (CO_2), and methane (CH_4) trapped in bubbles in ice cores taken through glaciers in Greenland and Antarctica (see Figure 16.1; see Chapter 1). The gas composition of the air bubbles represents the atmosphere's composition when the bubbles were trapped in the ice nearer the surface. The findings in the ice cores forced scientists to confront an issue they had not previously considered to affect humans, namely that natural climate change during the Holocene has often been quite abrupt and rapid. Such rapid climate change is now understood to occur at **sub-Milankovitch time scales**; this means that climate change occurred over time spans less than a few thousand years all the way down to durations as short as decades or even less; that is, time spans comparable with life spans of human and human cultures.

These findings further confounded the study of human impacts on climate. CO_2 and CH_4 have of course also been produced in tremendous quantities by anthropogenic activities through fossil fuel combustion, agriculture, and deforestation. It is the potential effects of the human contribution to natural levels of these gases that remain uncertain and that are the subject of much scientific study.

16.2 Beginning of the Holocene

16.2.1 Sea-level rise

B Like the retreats, or terminations, of northern hemisphere glaciers earlier during the Pleistocene, the Holocene is also marked by sea-level rise (**Figure 16.2**). The release of water from northern hemisphere glaciers is augmented in some localities by isostatic rebound (**Figure 16.3**). As the overlying weight of ice is removed the land rises, which should, seemingly, cause a regression. However, as rebound occurs, the land in front of the glacier can be pulled downward, causing sea level to rise (see Figure 16.3).

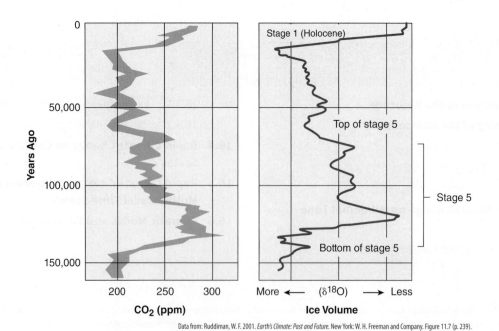

Data from: Ruddiman, W. F. 2001. *Earth's Climate: Past and Future.* New York: W. H. Freeman and Company. Figure 11.7 (p. 239).

FIGURE 16.1 Climatic events from the late Pleistocene through the Holocene. The record begins with the last major interglacial (oxygen isotope stage 5, about 125,000 years before present). The oxygen isotope record of a deep-sea core (right) is compared with CO_2 concentrations in an ice core from Vostok, Antarctica. Note the fluctuations in CO_2 levels during, for example, oxygen isotope stage 5, which was long thought to be climatically stable.

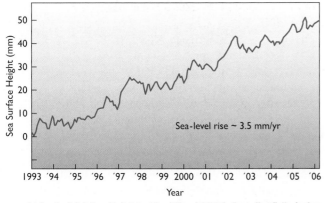

FIGURE 16.2 Sea-level rise since 1993.

Data from: King, M. D., Parkinson, C. L., Partington, K. C., and Williams, R. G. 2007. *Our Changing Planet: The View from Space.* Cambridge University Press.

Isostatic rebound is very pronounced in certain parts of the world, such as Scandinavia and along the middle Atlantic and northeast Atlantic coasts of North America, where it continues to affect the rates of sea-level rise. Rapid sea-level rise is also thought by some workers to have occurred catastrophically in other localities for various reasons (**Box 16.1**).

Catastrophic sea-level rise might also occur in the future because of the collapse of Antarctic ice sheets. Antarctica currently stores about 90% of the world's ice, and if it all were to melt, global sea level would rise by nearly 200 feet! The ice sheet is up to about 4 kilometers (2.5 miles) thick and consists of two main parts: the east and west Antarctic ice sheets. Each ice sheet is dome-shaped, so ice flows outward as streams from the centers at rates up to a few kilometers per year. When the ice reaches the coasts, it forms ice shelves around much of Antarctica's margin (**Figure 16.4**). The ice shelves themselves contain relatively little water, so if they melted there would be little impact on sea level. However, the ice shelves dam the main sheets; thus, if the shelves melt, the remaining ice sheets might also break up rapidly, flooding coasts all over world.

16.2.2 Younger Dryas

The Holocene actually began twice. As the world began to warm during the Holocene, the world suddenly turned cold again during the interval called the **Younger Dryas**. Evidence for the Younger Dryas originally came from studies of fossil pollen in lakes and bogs. *Dryas* is a small flowering plant that lived above the Arctic Circle. About 13,000 years ago, pollen from *Dryas* and other plants

adapted to the cold tundra became abundant as forests retreated southward from advancing glaciers. Planktonic foraminifera adapted to cold water also appeared in deep-sea cores at lower latitudes, indicating that cold polar waters had moved south (see Chapter 15). Overall, the Younger Dryas lasted roughly 1,000 years, but records of ice accumulation from cores taken in Greenland indicate that its initiation occurred within a century, with much of the initial cooling occurring in perhaps as little as a decade (**Figure 16.5**)! The climatic effect of the Younger Dryas was far reaching, as evidenced by the advance of mountain glaciers as far away as New Zealand that are now represented only by isolated terminal moraines in mountain valleys.

To understand the climatic events associated with the Younger Dryas, we must first examine ocean circulation on longer Milankovitch time scales and its effect on greenhouse gases in the atmosphere. On average, the water masses of the modern oceanic conveyor completely circulate once about every 1,000 years (see Chapter 3). Many scientists have concentrated on the oceans as a source of rapid climate change because of their relatively rapid (geologically speaking) circulation rate. Two deep-water masses appear to be primarily involved in rapid climate change: Antarctic Bottom Water (AABW) and North Atlantic Deep Water (NADW; see Figure 16.6a). AABW sinks off the continent of Antarctica and then flows over the floors of ocean basins all over the world (see Chapter 3). As AABW circulates throughout the ocean basins, it picks up dissolved nutrients released by the remineralization of dead organic matter settling through the water column or lying on the bottom and from animal respiration (**Figure 16.6**). When AABW upwells into the photic zone off Antarctica, its nutrient-rich waters promote photosynthesis by plankton, especially diatoms. The photosynthesis in the surface waters around Antarctica takes CO_2 out of the atmosphere and stores it in organic matter, some of which settles to the ocean bottom. By contrast, NADW is produced primarily off the coasts of Greenland and Iceland in the North Atlantic (see Chapter 3).

During past interglacial (warm) intervals like oxygen isotope stage 5 (see Chapter 15), NADW flowed to the south through the Atlantic basin, just as it does today. NADW flows only along the bottom of the Atlantic Ocean basin, so it has less time to pick up nutrients from decaying organic matter and respiration than does AABW. NADW eventually upwells off Antarctica to replace AABW as it sinks. As NADW upwells it mixes with AABW and "dilutes" the nutrients in AABW because NADW is nutrient-poor. The dilution of nutrients in AABW decreases photosynthesis around

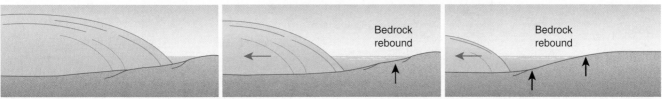

Data from: Ruddiman, W. F. 2001. *Earth's Climate: Past and Future.* New York: W. H. Freeman and Company. Figure 14.9 (p. 310).

FIGURE 16.3 Collapse of a glacial forebulge. The underlying rock rebounds as the weight of the overlying ice is removed.

BOX 16.1 Noah's Flood?

Ⓓ Noah's flood is one of the most widely reported sea-level changes and has long intrigued archaeologists and Earth scientists. The account of the flood in the Old Testament of the Bible is well known, but similar stories were recounted by the Sumerians about 4000 BC as the tales of Gilgamesh, and by the Babylonians about 2000 BC.

The intense level of interest in identifying plausible mechanisms for the biblical account of Noah's flood has, not surprisingly, generated a number of hypotheses for its cause. It has long been thought that

Noah's flood occurred in lower Mesopotamia in the vicinity of the confluence of the Euphrates, Tigris, and Kurun rivers (where the Sumerians lived). This region has existed for many thousands of years as a marshy area just a few feet above sea level at the northern margin of the Arabian Gulf (**Box Figure 16.1A**). Lower Mesopotamia lies in a tectonically active region, so flooding could have occurred as the result of a tsunami ("tidal wave") generated by an earthquake. However, because the region lies close to sea level, flooding could also have occurred because of a storm surge,

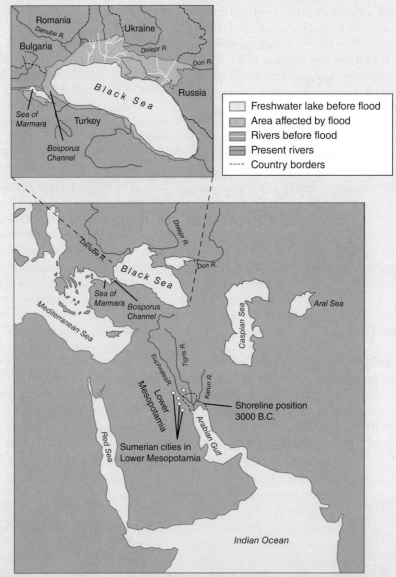

	Freshwater lake before flood
	Area affected by flood
	Rivers before flood
	Present rivers
----	Country borders

Data from: Coe, A. L. (ed). 2003. *The Sedimentary Record of Sea-Level Change.* Cambridge, UK: Cambridge University Press. Figure 3.1 (p. 35).

BOX FIGURE 16.1A General geography of the Black Sea and surrounding Middle Eastern region today.

BOX 16.1 Noah's Flood? (Continued)

heavy rainfall, or a rise in sea level due to the melting of glaciers.

Recently, the so-called flood hypothesis has suggested that Noah's flood did not occur in Mesopotamia but in the Black Sea (Box Figure 16.1A). According to the flood hypothesis, at the beginning of the Holocene the Black Sea was a gigantic freshwater lake into which drained waters from the ice sheets of Scandinavia and Russia far to the north. Nevertheless, according to the flood hypothesis the water level in the Black Sea was about 140 meters below its present level. It is thought that as the glaciers melted and sea level rose during the Holocene, the transgression increased the level of the Mediterranean Sea to the Straits of Bosporus, which connect the Mediterranean with the Black Sea. These straits acted as a dam to the rising waters of the Mediterranean until they finally broke through, catastrophically, about 7,200 years ago (based on ^{14}C dates). It has been estimated that when the Straits of Bosporus were breached, the flow equaled that of 200 Niagara Falls, and in just a few months the waters entering the Black Sea rapidly flooded more than 100,000 square kilometers of the shelves along the Black Sea that had previously been subaerially exposed.

The proponents of the flood hypothesis claim the flooding accelerated the migration of Neolithic peoples and the spread of farming into the interior of Europe, Asia, Egypt, and Mesopotamia from the Black Sea region. The recent finding of the remains of human settlements about 100 meters below the present level of the Black Sea would seem to support the hypothesis. However, the ruins do not definitely "prove" that *rapid* flooding occurred because the

settlements could have been flooded if the level of the Black Sea rose rapidly or more slowly. Also, some recent studies suggest that the Straits of Bosporus, through which the flood waters presumably entered from the Aegean Sea, might only be 7,000 to 5,000 years old. Thus, the straits might be too young to have connected the Black Sea to the Aegean.

The geologic record in cores for rapid flooding 7,200 years ago also provides conflicting evidence. On the one hand, freshwater gastropods and sedimentary structures such as mud cracks and cross-bedded sand dunes indicate subaerial to freshwater environments at the base of the cores. Higher in the cores, layers of abraded shells of bivalves that live in fresh waters suggest high-energy conditions associated with strong currents that might be associated with rapid flooding. Above this level, fossils of organisms living in brackish to fully marine waters suggest the establishment of a connection between the Black Sea and the Mediterranean Sea. On the other hand, dinoflagellate assemblages suggest that the Black Sea was flowing into the Sea of Marmara by at least 10,500 to 9,500 years ago, several thousand years before the flood was to have taken place. Also, Mediterranean species of euryhaline (salinity-tolerant) clams and foraminifera are found at the base of other cores taken within the Black Sea at about 27,000 years ago. The presence of these taxa in the cores at this time suggests that there might have already been some connection between the Black Sea and the Mediterranean. Moreover, the cores indicate nine transgressions during the past 800,000 years, suggesting repeated connections between the Black Sea and other water bodies long before the presumed Noah's flood.

Antarctica, allowing more CO_2 to remain in the atmosphere. Increased levels of CO_2 in the atmosphere in turn promote warming, like that of the Holocene and previous interglacials (see Figure 16.6b). Conversely, if the production of NADW were to slow, less NADW would reach Antarctica and dilute the nutrients in AABW. This would effectively increase nutrient concentrations in AABW and promote greater photosynthesis, taking more CO_2 out of the atmosphere and leading to colder glacial intervals (see Figure 16.6).

How, then, could the production of NADW have slowed to produce glacial intervals? Quite likely, sea ice grew out over the sites of NADW production off Greenland

and Iceland. We know that sea ice repeatedly formed and retreated in this region because of the layers of **ice-rafted debris (IRD)** found in deep-sea cores. The presence of sea ice over the North Atlantic's surface would have slowed the production of NADW.

So what happened during the Younger Dryas? As the northern hemisphere ice sheets began to retreat from the Great Lakes region in response to Milankovitch cycles, large amounts of meltwater are thought to have been trapped behind ice and rock dams in a huge lake called Lake Agassiz. Lake Agassiz was at least twice the volume of the Caspian Sea, which is the largest lake of the modern world

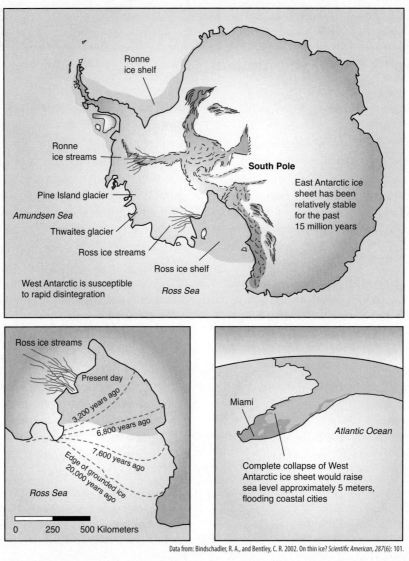

Data from: Bindschadler, R. A., and Bentley, C. R. 2002. On thin ice? *Scientific American, 287*(6): 101.

FIGURE 16.4 Sea level and Antarctic ice sheets and ice shelves. Complete collapse of the west Antarctic ice sheet would raise sea level about 5 meters, inundating many low-lying areas such as the southern third of Florida and the east and Gulf coasts of the United States.

(**Figure 16.7**). It has therefore been hypothesized that the "trigger" for the Younger Dryas was the influx of an enormous volume of fresh water from Lake Agassiz after the collapse of ice and rock dams that helped form the lake. In one scenario, an enormous pulse of meltwater was sent down the St. Lawrence Seaway to marine waters adjacent to Greenland. No evidence of such a massive outflow through the St. Lawrence Seaway has been found, however. The recent discovery of gravels associated with a regional (i.e., widespread) erosion surface in the Mackenzie River basin of northern Canada suggests the outflow was instead into the Arctic Ocean and then flowed into the North Atlantic (see also **Box 16.2**). Meltwater pulses—like those that presumably came from Lake Agassiz—into marine waters are also evidenced in deep-sea cores by a very strong, rapid shift to low oxygen isotope values (**Figure 16.8**). We know this based on oxygen isotope records. Ice has a very low oxygen isotope value of about 30% because it primarily stores oxygen enriched in the light isotope of oxygen (^{16}O). After the

Holocene warming began, there was indeed such a shift to low values in the North Atlantic, where NADW forms. According to this scenario the fresh water would have lain on top of the seawater because fresh water is less dense than seawater. This would have prevented the formation and sinking of NADW, much like sea ice during glacial intervals (**Figure 16.9**). It has been suggested that these sorts of events might have occurred more than once, prior to the Holocene.

CONCEPT AND REASONING CHECKS

1. How might rapid climate change account from some ancient myths or teachings?
2. How was a broader view of uniformitarianism than Lyell's uniformitarianism involved in Bretz's reasoning? (See BOX 16.2.)
3. What is the evidence for the presumed cause of the Younger Dryas?

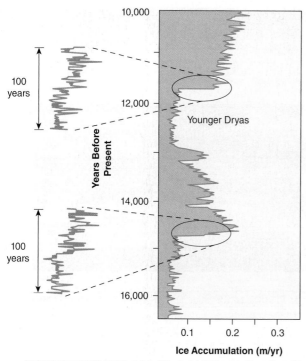

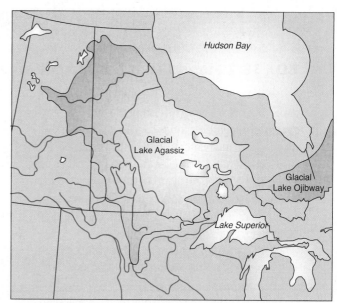

Data from: Ruddiman, W. F. 2001. *Earth's Climate: Past and Future.* New York: W. H. Freeman and Company. Figure 14.10 (p. 311).

FIGURE 16.7 Huge lakes of meltwater formed in the vicinity of the modern Great Lakes as ice sheets in the northern hemisphere began to retreat at the beginning of the Holocene.

Data from: Ruddiman, W. F. 2001. *Earth's Climate: Past and Future.* New York: W. H. Freeman and Company. Figures 14.7 (p. 308, cut Ca in dust) and 14-6 (p. 307).

FIGURE 16.5 The occurrence of the Younger Dryas and its relation to ice accumulation.

Rapid Climate Change on Millennial Time Scales

As we have now seen, rapid climate change occurred during the Holocene, despite its overall warmth and mild conditions. Many of these processes occur on sub-Milankovitch scales, meaning that they occur on time scales of about a thousand years or less. Sometimes, the processes occur on time spans as short as decades. We discuss a variety of processes that act over different durations of time and that involve ocean–atmosphere interactions resembling those described for the Younger Dryas. These are the processes that will affect future generations of humans most directly in the coming centuries and, conversely, these are the processes that might be most affected by anthropogenic change.

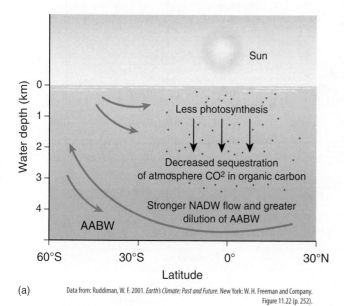

(a)

Data from: Ruddiman, W. F. 2001. *Earth's Climate: Past and Future.* New York: W. H. Freeman and Company. Figure 11.22 (p. 252).

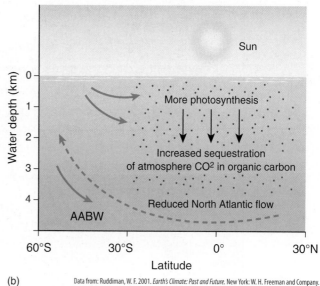

(b)

Data from: Ruddiman, W. F. 2001. *Earth's Climate: Past and Future.* New York: W. H. Freeman and Company. Figure 11.22 (p. 252).

FIGURE 16.6 Changes in ocean circulation and photosynthesis between interglacial (e.g., modern) and glacial conditions. **(a)** Modern (and interglacial) conditions, in which nutrient-poor NADW upwells off Antarctica and dilutes nutrient-rich AABW. As a result, photosynthesis by phytoplankton decreases and atmospheric CO_2 increases. **(b)** Glacial conditions, during which NADW production slows. As a result, AABW is diluted less by NADW, more photosynthesis occurs, and atmospheric CO_2 decreases.

BOX 16.2 Floods That Carved the West[1]

Many, but not all, workers accept the hypothesis that the Younger Dryas was caused by the massive discharge of meltwater from a huge glacial lake into the North Atlantic. However, the flood hypothesis (see Box 16.1) has been met with considerable skepticism. A similar hypothesis was proposed years ago for the American West and also met with considerable skepticism but is now widely accepted.

This particular hypothesis was proposed by the American geologist J. Harlan Bretz. Bretz noticed unusual features on geologic maps of the region he believed could only be explained by a massive discharge of water. The features noted by Bretz included gigantic potholes (**Box Figure 16.2A**), dry waterfalls, and an unusual area of about 2,000 square miles called the Scablands that stretch from eastern Washington state all the way to the Cascades (**Box Figure 16.2B**). Here, the overlying soil and other sediment had been stripped away, exposing basaltic

Courtesy of Tom Foster.

BOX FIGURE 16.2A Huge potholes gave Bretz his first clues as to the flooding.

bedrock (**Box Figure 16.2C**). Gigantic granite erratics were also located as far away as the Willamette Valley near Portland, Oregon. These erratics could only have been carried far from their source area (perhaps northern Idaho) because they did not match the basaltic bedrock where they were found. Because of

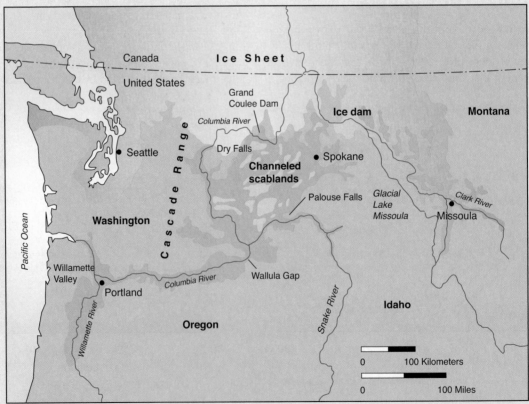

Data from: Parfit, M. 1995. The floods that carved the west. *Smithsonian Magazine, 26*(1): 50.

BOX FIGURE 16.2B Reconstruction of the geography of the Pacific Northwest during the time of Glacial Lake Missoula.

BOX 16.2 Floods That Carved the West[1] (Continued)

© Jpiks/Shutterstock, Inc.

BOX FIGURE 16.2C The ledges on the mountains above Missoula, Montana, mark the positions of ancient shorelines of Glacial Lake Missoula.

their size, perhaps icebergs transported them to their final resting place. Also present were isolated mounds of gravel and soil several miles long and up to 500 feet high. These mounds resemble the sand and gravel bars found along modern streams, except they far surpassed the size of battleships!

Bretz presented his work as early as 1923 and was severely criticized. This was due, fundamentally, to the indoctrination of generations of geologists who had been taught Charles Lyell's view of slow, gradual change (see Chapter 1). Lyell's views were deeply entrenched in the psyche of these geologists. Thus, most workers, many of whom had never even visited Bretz's study areas, insisted that the features Bretz observed could be explained by stream processes we normally observe today. Even one of Bretz's own students, who later became a famous glacial geologist, attacked his hypothesis, stating the features were produced by "leisurely streams with normal discharge." The primary criticism of Bretz's argument centered on the mechanism. From where had the water come? In other words, like Darwin for evolution (see Chapter 5) and Wegener for continental drift (see Chapter 2), Bretz was unable to propose a plausible mechanism for the massive discharge of water he proposed because of Lyell's strict brand of uniformitarianism (see also Chapter 1).

Then, in 1940 another geologist, Joseph Pardee of the U.S. Geological Survey, gave a talk on an ancient glacial lake called Glacial Lake Missoula. Most workers knew of Glacial Lake Missoula, but no evidence for massive discharge had ever been described. Pardee believed this lake had been formed by an ice dam over what is now the Clark Fork River in the vicinity of Missoula, Montana. Based on the ancient shorelines preserved on the sides of the mountains above Missoula, he concluded the volume of Lake Missoula was about that of Lake Erie and Lake Ontario combined, or about 500 *cubic miles*! Pardee described enormous ripple marks (actually water-formed dunes) up to 50 feet high and having wavelengths (distance from crest to crest of two adjacent ripple marks) ranging from 200 to 500 feet! Ripple marks of this size could only have been made by a massive discharge of water, such as that hypothesized for the collapse of the ice dam at Lake Missoula. Because this lake began just a few miles upstream of the channeled scablands, these waters would have discharged into this region. Here was the source of the water—and the mechanism—for Bretz's flood.

As it burst from the lake, a gigantic wall of water up to 1,500 feet deep flowed outward at speeds up to 60 mph. The discharge was 10 times greater than all the rivers in the world today, or greater than 600 million cubic feet *per second*! Half a day later, the flood rammed into the narrow Wallula Gap further west. Here, the water backed up for a few days to produce another lake about 3,500 square miles in size. At the same time water continued down the Columbia River Gorge, leaving behind icebergs and rocks. The entire flood was over in less than a week.

But the story does not end here. The lake that formed at Wallula Gap records multiple layers of sediment, suggesting that multiple floods—perhaps as many as 100—occurred. This might explain why there are so many parallel shorelines recorded on the hills above Missoula. When an ice dam burst at Lake Missoula, another one formed until the glaciers melted sufficiently to finally recede from the area.

[1] Condensed from Parfit, M. 1995. The floods that carved the west. *Smithsonian Magazine, 26*(1), 48–59.

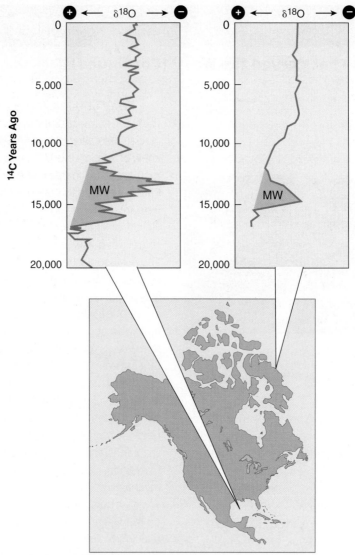

Data from: Ruddiman, W. F. 2001. *Earth's Climate: Past and Future*. New York: W. H. Freeman and Company. Figure 14.5 (p. 306).

FIGURE 16.8 Simultaneous meltwater pulses in the Gulf of Mexico and North Atlantic at the beginning of the Younger Dryas as recorded by negative excursions of oxygen isotope curves.

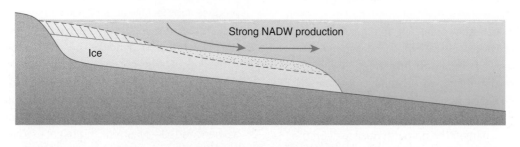

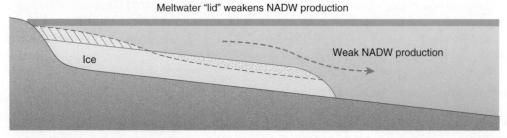

FIGURE 16.9 A meltwater layer like that which formed during the Younger Dryas is not dense enough to sink and prevents the formation of NADW until the freshwater is sufficiently mixed with normal marine water.

16.3.1 Rapid climate change involving the oceans

Evidence indicates that meltwater pulses to the North Atlantic, like those described for the Younger Dryas, occurred repeatedly during the Holocene. During the first half of the Holocene the edge of the northern hemisphere ice sheet fluctuated back and forth in the vicinity of the Great Lakes region, between about 43 and 49 degrees North latitude. Fluctuations of the edge of the ice sheet within this range of latitude are thought to have released meltwater into the North Atlantic by different routes, slowing NADW production (**Figure 16.10**).

In effect, the behavior of the ice sheet acted like a switch, sending meltwater down different river systems at different times (**Figure 16.11**). In the scenario shown in Figure 16.11, meltwater first flowed down the Hudson River, causing a shutdown in NADW production. This in turn cooled temperatures, as described previously, leading to the growth of ice sheets. As the ice sheets advanced over the drainage basin of the Hudson River, the freshwater was routed (switched) to the Mississippi River. This gave the meltwater layer left in the North Atlantic time to mix and dissipate, leading to an increase in NADW production, warmer temperatures, and the retreat of the ice margin to the north. Eventually, as the ice sheets retreated further to the north during the Holocene, the main outlet for meltwater shifted back to the Hudson Strait and the Arctic Ocean (see Figure 16.11).

There is a great deal of other evidence for rapid climate change during the Holocene after the Younger Dryas. Layers of IRD occur at fairly discrete intervals in North Atlantic deep-sea cores. These IRD layers are called **Heinrich events** after their discoverer, Hartmut Heinrich, a graduate student (**Figure 16.12**). Like the Younger Dryas, the Heinrich events are associated with the appearance of cold-water planktonic foraminiferal assemblages in deep-sea cores. Heinrich events are also associated with finer-grained sediments, suggesting less bottom scour and sediment winnowing by NADW, as would be expected if the production of NADW slowed during glacial advances. The Heinrich events occur about every 1,500 years. However, Heinrich events not only occur during the Holocene but also during glacial intervals before the Holocene. Thus, Heinrich events appear to *act independently of ice volume*, meaning that the ice itself is not causing the

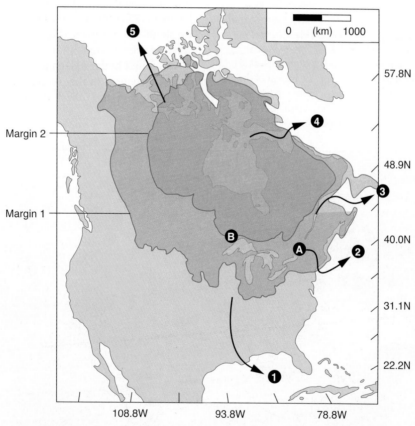

Reproduced from: Goldenberg, S. B., Landsea, C. W., Mestas-Nuñez, A. M., and Gray, W. M. 2001. The recent increase in Atlantic hurricane activity: Causes and implications. Stanley B. Goldenberg, et al., *Science, 293*, 474–479. Reprinted with permission from American Association for the Advancement of Science (AAAS).

FIGURE 16.10 The paths of meltwater input to the North Atlantic varied during the Holocene according to the position of the ice sheet. Like a switch, if ice blocked one river system, meltwater was diverted down another. (1) Mississippi River, (2) Hudson River, (3) St. Lawrence River, (4) Hudson Strait (adjacent to Hudson Bay), (5) Arctic Ocean. Margins 1 and 2 indicate maximum extent of ice sheets at approximately 21,000 and 13,000 years BP, as the ice sheet retreated. A = eastern outlet from the southern Great Lakes region to the Hudson River. B = eastern outlet from Lake Agassiz basin to the St. Lawrence River. Fluctuations in the margin of the ice sheet caused meltwater to be routed between the two northeastern outlets and the southern outlet (Mississippi River).

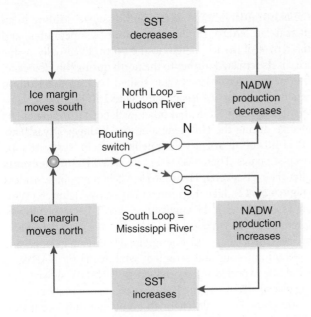

Data from: Clark, P. U., Marshall, S. J., Clarke, G. K. C., Hostetler, S. W., Licciardi, J. M., and Teller, J. T. 2001. Freshwater forcing of abrupt climate change during the last glaciation. *Science*, 293 (13 July): 286 (Figure 3) DOI: 10.1126/science.1062517.

FIGURE 16.11 Switching of meltwater routes during the Holocene (refer also to Figure 16.10). In this example, meltwater is first sent down the Hudson River (routing switch goes to N), causing a shutdown in NADW production, which in turn cools sea surface temperatures (SSTs), leading to the advance of ice. As the ice sheet advances over the drainage basin of the Hudson River, freshwater routing then switches to the Mississippi River(S); the meltwater layer in the North Atlantic has time to mix and dissipate, leading to an increase in NADW production and SSTs, and the retreat of the ice margin northward. As the ice sheets retreat further to the north, the main outlet for meltwater shifts back to the Hudson Strait and the Arctic Ocean (routing switch moves back to N).

climate change; rather, the growth and retreat of ice sheets are the responses of the ice sheets to climate change caused by other factors that operate during *both* glacial and interglacial intervals.

The intensity of Heinrich events also seems to vary. This is indicated by the kinds of IRD found in cores (see Figure 16.12). In fact, the debris is sufficiently distinctive that it acts as a tracer that can be matched to the source rock from which it was eroded, or provenance. During well-developed Heinrich events limestones from the Hudson Bay region in northern Canada were eroded and redeposited in the deep North Atlantic. Less pronounced events were represented by grains of volcanic rocks eroded further south from Iceland and by grains of the Old Red Sandstone from Greenland and Europe.

16.3.2 Rapid climate change on land

Rapid climate change during the Holocene was by no means confined to the oceans. **Speleothems** (cave-deposited limestone called travertine) provide detailed records of changes in temperature and precipitation that reflect shifts in air masses over land. For example, stalagmites from a cave in northeast Iowa dated radiometrically indicate that temperature in that area was relatively cool from about 7,800 years to 5,900 years before present (BP) and then warmed rapidly by about 3°C until about 3,600 years BP. After this, cooling by about 4°C resumed (**Figure 16.13**). These temperature changes are inferred from changes in the oxygen isotope record found in the stalagmites.

The changes in the oxygen isotope record of the speleothems correspond to well-documented changes from forest

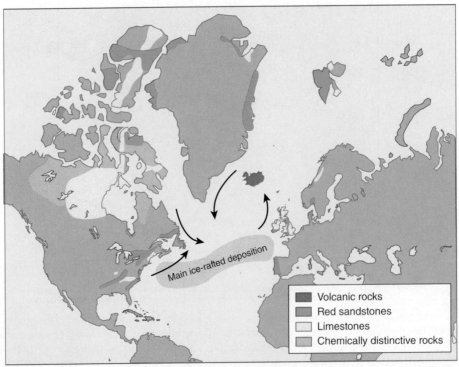

Data from: Bond, G., Kromer, B., et al. Persistent solar influence on North Atlantic climate during the Holocene, *Science* 7, 294, 2130–2136; published online 15 November.

FIGURE 16.12 Heinrich (ice-rafted debris) events. Heinrich events as indicated by percent iron-stained grains and ^{14}C production.

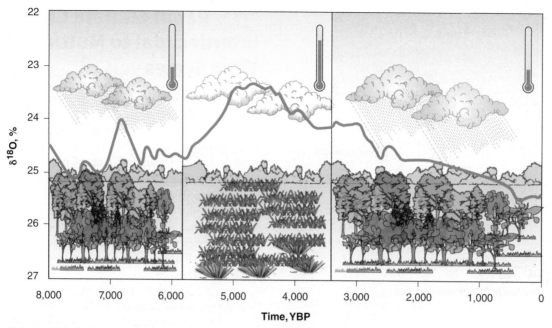

Data from: Dorale, J. A., Gonzalez, L. A., Reagan, M. K., Pickett, D. A., Murrell, M. T., and Baker, R. G. 1992. A high-resolution record of Holocene climate change in speleothem calcite from Cold Water Cave, northeast Iowa. *Science, 258* (4 December): 1626–1630.

FIGURE 16.13 Rapid climate change on land during the Holocene is evidenced by the oxygen isotope record of a speleothem from northern Iowa. This area was relatively cool from about 7,800 years to 5,900 years BP, then warmed rapidly by about 3°C, until about 3,600 years BP, after which cooling by about 4°C occurred. The oxygen isotope record of the speleothems corresponds to changes from forest to prairie grasses and back to forest. The shifts in vegetation, in turn, reflect changes in precipitation, as shown.

to prairie grasses and back to forest (see Figure 16.13). The shifts in vegetation in turn reflect changes in precipitation: forests tend to occur when cooler, wetter conditions prevail, whereas prairie grasses occur under warmer, drier conditions. The shifts in the speleothem and vegetation records have been used to infer that humid air from the Gulf of Mexico dominated in this region from 7,800 to 5,900 years BP, allowing temperate forests to prevail. This air mass was replaced by drier air from the Pacific that allowed prairies to spread. By 3,000 years BP, cooler air moving down from the Arctic pushed out the drier Pacific air, allowing coniferous forests to dominate.

CONCEPT AND REASONING CHECKS

1. Why are the oceans thought to be involved in rapid climate change?
2. What is the significance of the provenance of IRD for studies of rapid climate change?

16.4 Rapid Climate Change on Centennial Time Scales

F Newly emerging evidence indicates that climate fluctuates significantly on centennial (century) scales. Some possible mechanisms acting on these time scales might be related to climate change on millennial scales. One of the most exciting developments regarding rapid climate change on these shorter time scales has been the growing realization that the sun might be involved.

Changes in solar radiation are associated with changes in the numbers of *sunspots*. Exactly how sunspots form is not understood, but they appear to be related to gigantic magnetic storms on the sun's surface. Individual sunspots can range in diameter from 10,000 to 50,000 kilometers (6,200–31,000 miles); some reach 160,000 kilometers (100,000 miles), which is more than 12 times Earth's diameter! Sunspots are relatively cool regions on the surface of the sun (although they are still intensely hot), making them appear darker than the rest of the sun's surface. It has long been known that an 11-year cycle in sunspot activity occurs, with the number—or frequency—of sunspots varying from as few as 5 to as many as 100.

Longer cycles of reduced sunspot frequency, or **sunspot minima**, have also been documented. These longer periods of sunspot minima might occur during prolonged intervals of harsher climate such as the **Little Ice Age**. This climatic episode ranged from about 1300 AD to about 1850 to 1880 AD, when winters in northern Europe were more severe than they are now. The longer periods of sunspot minima have been well documented by historical records of literature and art from this time.

These longer periods of sunspot *minima* are evidenced by *increasing* amounts of carbon-14 (**Figure 16.14**). Carbon-14 (^{14}C) forms in the upper atmosphere as a result of the bombardment of nitrogen gas by cosmic rays (see Chapter 6).

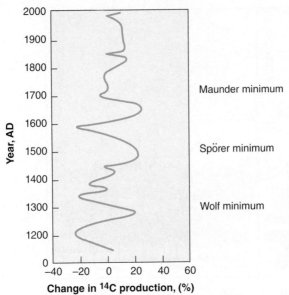

Data from: Black, D. E., Peterson, L. C., Overpeck, J. T., Kaplan, A., Evans, M. N., and Kashgarian, M. 1999. Eight centuries of North Atlantic ocean atmosphere variability. *Science*, 286(26 November): 1709–1713.

FIGURE 16.14 The formation of ^{14}C in relation to sunspot minima and climate change. Wolf (1290–1350 AD), Spörer (1420–1540 AD), and Maunder (1645–1715 AD) noted the increased production of ^{14}C during long-term sunspot minima.

Increased sunspot frequency is associated with increased solar wind, which is the flux of subatomic and ionized particles streaming off the surface of the sun into space (see Chapter 7). However, cosmic ray bombardment of Earth and ^{14}C production are *inversely* related to the intensity of sunspot activity because increased sunspot activity *reduces* the amount of cosmic rays bombarding Earth's atmosphere. This is because Earth is shielded by its magnetic field; during increased sunspot activity, the magnetic field interacts with the increased solar wind to partially prevent cosmic rays from reaching Earth. *Reduced* sunspot frequency and *reduced* solar wind are therefore associated with the greater bombardment of Earth by cosmic rays and the *increased* production of ^{14}C.

Despite these documented fluctuations in solar radiation, until fairly recently most earth scientists were skeptical that changes in solar radiation on such short time scales could significantly affect Earth's climate. The basic criticism was that there was no obvious way in which small changes in solar radiation could affect climate. Some sort of "amplification mechanism" or positive feedback is required, but no such natural amplifier has definitively been identified. Nevertheless, recent computer simulations of sunspot cycles indicate that a decrease of only about 0.1% in solar radiation over an 11-year interval can affect the atmosphere's temperature and, potentially, climate.

CONCEPT AND REASONING CHECKS

1. Explain how sunspots affect the production of ^{14}C.
2. What obstacle must be overcome to integrate the sun's behavior in rapid climate change on Earth?

16.5 Rapid Climate Change on Interdecadal to Multidecadal Time Scales

Rapid climate change on interdecadal to multidecadal time scales is primarily an attempt by Earth's systems to redistribute heat, sometimes many thousands of miles away, by atmospheric and oceanic currents. In the process, major wind systems, the redistribution of moisture, and the frequency of storms are affected. Because the behavior of the convection cells associated with these rapid climate changes fluctuate back and forth over different parts of Earth's surface, they are frequently referred to as **oscillations**. These climatic oscillations have received increasing attention in just the past decades because of the obvious implications for human welfare.

The study of these oscillations is complicated because they do not act in isolation. Rather, they and other ocean–atmosphere phenomena over the planet are all linked by long-distance **teleconnections** ("tele" refers to "transmission over distance," as in a telephone). Still, exactly how and to what extent these oscillations feed back on each other remains unclear. Other questions that remain unresolved are how these oscillations and teleconnections might be influenced by changes in solar radiation and anthropogenic activities. A number of these oscillations are now recognized, such as the North Atlantic Oscillation and the Pacific Decadal Oscillation, but their behavior and teleconnections are complex and poorly understood. Let's take a moment to examine the best understood of such oscillations: **El Niño**.

El Niño occurs in the southern Pacific Ocean about every 3 to 10 years, but some El Niños (such as the one in 1982–1983) are much stronger than others (**Figure 16.15**). El Niño conditions become most pronounced during late southern hemisphere winter (August). However, the first hint of the onset of El Niño conditions often occurs during the preceding southern hemisphere summer (December), hence, the name El Niño, "the boy child," for the birth of Jesus Christ.

El Niño is frequently referred to as the **El Niño-Southern Oscillation**, or **ENSO** for short, because it involves yet another large atmospheric convection cell called the Southern Oscillation (**Figure 16.16**). The behavior of this convection cell involves the behavior of the trade winds. Under normal conditions, trade winds blow from east to west across the Pacific (see Figure 16.16). This pulls surface waters away from the coast of South America, causing cold, nutrient-rich waters to upwell at the surface. These nutrient-rich waters stimulate plankton blooms that support large fishing grounds. The trade winds then push the upwelled water toward the western Pacific, where it "piles up." In fact, sea level in the western Pacific is normally slightly higher than in the eastern Pacific.

El Niño results in several related effects. Normally, as upwelled waters move from east to west in the tropical Pacific, the currents warm because they are exposed to the sun as they travel from east to west. Under El Niño

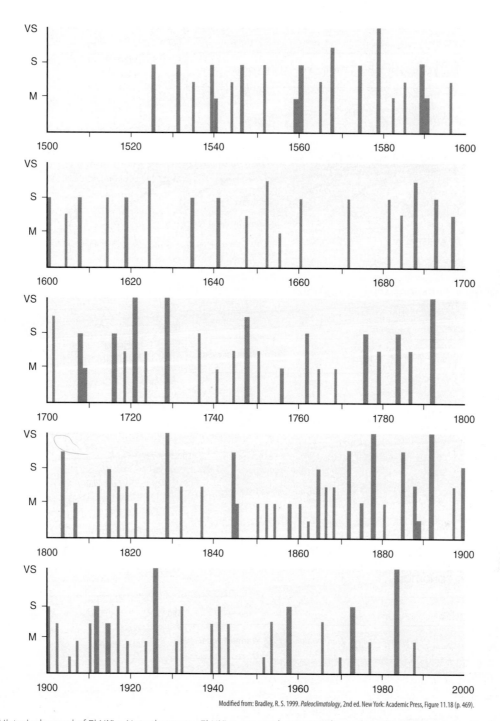

FIGURE 16.15 Historical record of El Niño. Note that some El Niños are much stronger than others. M = Medium intensity. S = Significant. VS = Very Signficant.

Modified from: Bradley, R. S. 1999. *Paleoclimatology*, 2nd ed. New York: Academic Press, Figure 11.18 (p. 469).

conditions the trade winds weaken, allowing warm water piled up in the western Pacific to flow east, where it disrupts upwelling, fishing grounds, and weather systems as far away as North America (see Figure 16.16). Also, under normal conditions, as the trades move from east to west, they carry moisture away from South America, producing abnormally dry conditions along the coasts of Peru and Chile. During El Niño years, however, warm, moisture-laden air moves eastward, where it eventually encounters the Andes; here, the orographic effect releases much of the moisture as torrential rains and floods.

The effects of El Niño can be very pronounced far from South America. El Niño affects the behavior of the **jet stream**, the "river" of air flowing from west to east in the upper atmosphere over North America. During an El Niño, the jet stream intensifies, bringing strong storms, coastal erosion, flooding, and mudslides to southern California and rainfall and flooding to the American southwest (**Figure 16.17** and **Figure 16.18**). As the storms track eastward, they can replenish their moisture with the warm, humid air of the Gulf of Mexico. This moisture is then dumped in the southeast United States during storms that

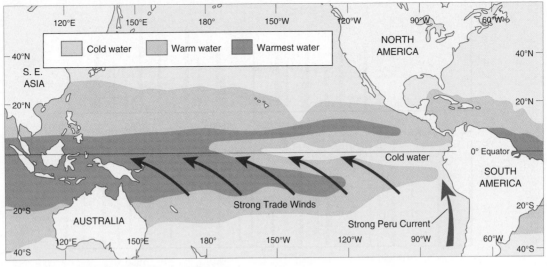

(a) NORMAL OCEANOGRAPHIC CONDITIONS

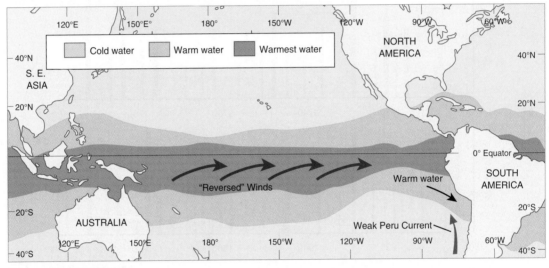

(b) EL NIÑO CONDITIONS

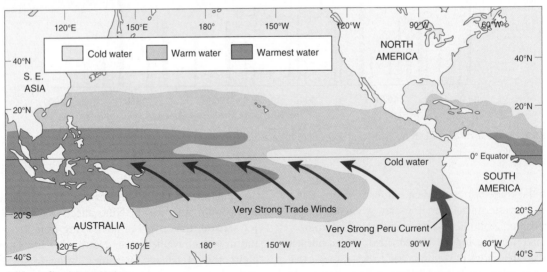

(c) LA NIÑA CONDITIONS

FIGURE 16.16 The El Niño-Southern Oscillation (ENSO). **(a)** Under normal conditions, trade winds blowing from east to west upwell deep waters off the coast of South America. Because these waters are nutrient-rich, they sustain fishing grounds. Upwelled waters are then pushed across the Pacific, where the water "piles up." As surface waters move from east to west in the tropical Pacific, they also warm. **(b)** El Niño conditions. Trade winds weaken, allowing warm water piled up in the western Pacific to flow east, where it disrupts upwelling, fishing grounds, and weather systems as far away as North America. **(c)** La Niña.

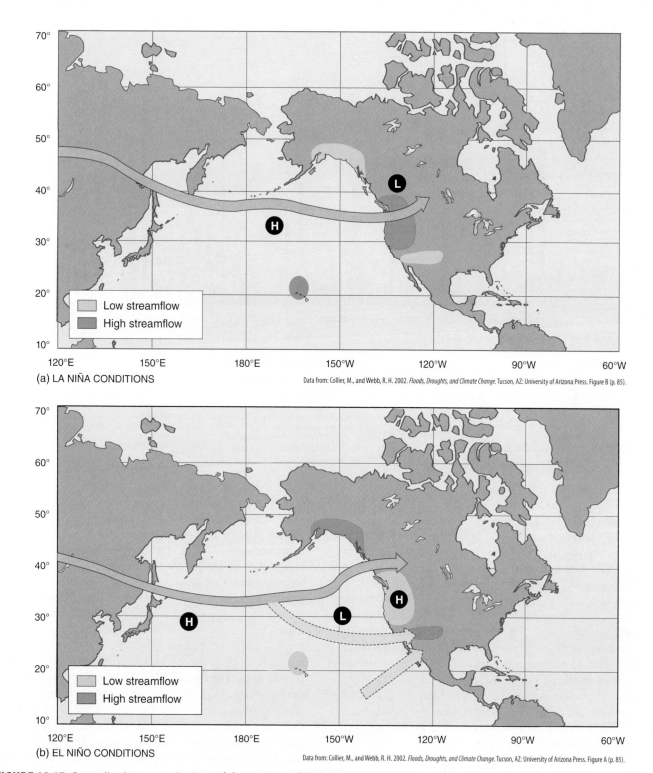

(a) LA NIÑA CONDITIONS

Low streamflow
High streamflow

Data from: Collier, M., and Webb, R. H. 2002. *Floods, Droughts, and Climate Change.* Tucson, AZ: University of Arizona Press. Figure B (p. 85).

(b) EL NIÑO CONDITIONS

Low streamflow
High streamflow

Data from: Collier, M., and Webb, R. H. 2002. *Floods, Droughts, and Climate Change.* Tucson, AZ: University of Arizona Press. Figure A (p. 85).

FIGURE 16.17 Generalized storm tracks during **(a)** La Niña and **(b)** El Niño events. H and L indicate high and low pressure air systems, respectively. Fluctuations in rainfall indicated for the middle of the Pacific are measured on Hawaiian Islands.

can also spawn tornadoes. At the same time, the development of hurricanes is suppressed in the Atlantic because the southern branch of the jet stream tends to shear off the tops of developing storms. By contrast, the northern branch of the jet stream can bring drought to the Pacific Northwest, western Canada, and the midwestern United States.

La Niña events sometimes follow El Niño events. During La Niñas, conditions reverse themselves, but to the extreme

(see Figure 16.17a). A strong cell of descending air develops in the South Pacific that promotes the development of strong trade winds from the southeast. Consequently, upwelling along the coast of South America is greatly strengthened and a tongue of cold water extends far westward from the eastern Pacific. Like El Niño, the climate effects of La Niñas are widespread. The dry, descending air associated with La Niñas means that there is little evaporation and therefore

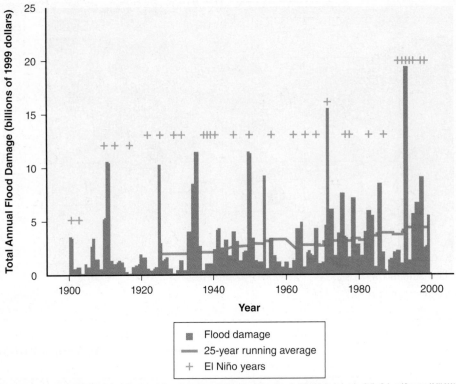

Data from: Collier, M., and Webb, R. H. 2002. *Floods, Droughts, and Climate Change*. Tucson, AZ: University of Arizona Press. (p. 108). Stated to be based on Pielke, R. A., and Downton, M. W. 2000. Precipitation and damaging floods: Trends in the United States 1932–1997. *Journal of Climate, 13,* 3625–3637.

FIGURE 16.18 Flood damage in the U.S. 1903–1999 (in billions of 1999 dollars). Running average (green line) has doubled in the past 75 years. Spikes tend to correlate with El Niño years.

little precipitation; thus, areas that have been deluged during El Niños might suffer drought.

The cooling in the Pacific during La Niñas also means heat is transferred elsewhere, usually the Indian Ocean. Here, the warmer surface water temperatures promote changes in evaporation and precipitation patterns as far away as Australia and South Africa. The cooler temperatures in the Pacific are associated with colder winters and heavier snowfalls over North America because shifts in wind systems allow Arctic air to move farther south. Also, hurricanes develop more easily in the warm tropical Atlantic because of a shift in the jet stream associated with La Niña. These storms then move west and northwest (because of the trade winds) toward the Gulf and east coast of the United States, sometimes with disastrous consequences for life, property, and local economies.

CONCEPT AND REASONING CHECK

1. Diagram the processes involved in an El Niño event.

16.6 Climatic Modes and Climatic Irreversibility

As we have seen, changes in ocean circulation and ocean–atmosphere interactions are intimately involved in the rapid shift of Earth's climate from one mode to another. Even within relatively warm, seemingly benign intervals such as oxygen isotope stage 5, which represents the last major interglacial of the Pleistocene, abrupt climate shifts occurred. Nevertheless, after the climate shifts have occurred, they can last for up to a *millennium or more*. Given the complexity of the systems involved in regulating Earth's climate, it is difficult to predict the exact mode Earth's climate will exhibit in the immediate future.

We can get a general idea of how climate behaves in response to disturbances by using what is called a **hysteresis loop** (**Figure 16.19**). A hysteresis loop is a kind of feedback loop. In some cases, climate might be moving in a particular direction, but as long as it does not move too far in a particular direction, it can return to its starting point if the forces (changes in CO_2, solar radiation, meltwater inputs, and so on) causing the climate shift are relaxed. This behavior is like a car (representing climate) moving on a two-way street (see Figure 16.19a). In other words, the car—and climate—can reverse itself easily.

However, if climate is pushed too far in a particular direction, climate cannot simply turn around and go back the way it came. Extending the analogy of a car as climate, if the car makes a right turn onto a one-way street, it simply cannot turn around (see Figure 16.19b). Instead, the car (climate) can only return to its original state by another route. For example, in the case of meltwater inputs to sites of NADW formation, relatively small freshwater inputs will probably have little effect on NADW formation and climate.

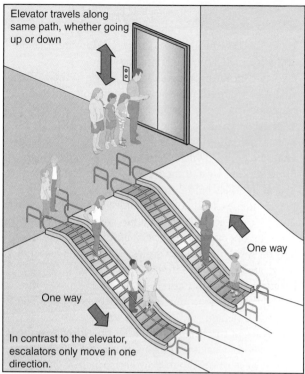

Elevator travels along same path, whether going up or down

One way

One way

In contrast to the elevator, escalators only move in one direction.

(a) Data from: Alley, R. B. 2000. *The Two-Mile Time Machine: Ice Cores, Abrupt Climate Change, and Our Future*. Princeton, NJ: Princeton University Press. Figure 14.1 (p. 151).

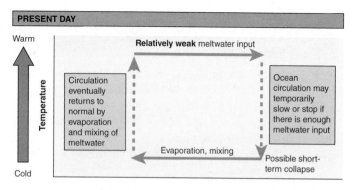

PRESENT DAY

Warm

Temperature

Cold

Relatively weak meltwater input

Circulation eventually returns to normal by evaporation and mixing of meltwater

Ocean circulation may temporarily slow or stop if there is enough meltwater input

Evaporation, mixing

Possible short-term collapse

Ocean Circulation unable to re-establish for much longer time

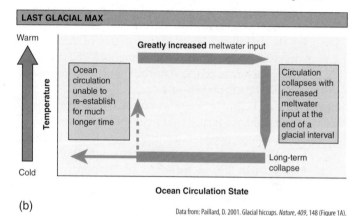

LAST GLACIAL MAX

Warm

Temperature

Cold

Greatly increased meltwater input

Ocean circulation unable to re-establish for much longer time

Circulation collapses with increased meltwater input at the end of a glacial interval

Long-term collapse

Ocean Circulation State

(b) Data from: Paillard, D. 2001. Glacial hiccups. *Nature, 409,* 148 (Figure 1A).

FIGURE 16.19 Climate modes and hysteresis loops. **(a)** How a hysteresis loop works. Any time you are on the elevator, you can reverse direction pretty much at any point and go back the way you came. However, if you take the escalator, you can only return to your initial position by a new path, as shown. **(b)** Hysteresis loop for climate (temperature) stability as a function of NADW production in the North Atlantic Ocean. The modern North Atlantic has two basic climate modes. Top (present day): If meltwater input exceeds a threshold value, ocean circulation jumps (dashed line pointing down) from the upper warm mode (unperturbed present-day state) to the lower, colder mode (solid blue line), where NADW slows. Climate can only return (dashed line pointing up) to the warm mode (upper line) when meltwater has mixed or evaporated sufficiently. Note that both of these more moderate warm and cold modes could occur under modern conditions. Bottom (last glacial maximum): Like the escalators in part A, it takes a much larger disturbance like a large influx of meltwater to shut down modern NADW formation for longer periods of time (solid line pointing to the left). This is presumably what happened during glacial intervals when much greater volumes of meltwater would have been released. Such a large influx of meltwater would require a much longer time to mix and dissipate, allowing NADW production to begin to resume (dashed line pointing upward).

But, if a large freshwater input occurs, such as that which apparently happened during the Younger Dryas, climate might shift to a new mode for the foreseeable future. In the case of large freshwater inputs to the North Atlantic, summers in Ireland would resemble those of Spitsbergen (northern Greenland) and winters in London those of Siberia!

In the future such meltwater inputs could occur in response to the combustion of fossil fuels. Indeed, humans might be pushing Earth's climate closer and closer toward a threshold beyond which climate will rapidly shift to a new mode that might or might not bode well for us. Based on the hysteresis loop, it is conceivable that, given the input of CO_2 into Earth's atmosphere from fossil fuel combustion, CO_2 might accumulate in Earth's atmosphere with little or no effect until a threshold is crossed and climate suddenly shifts into an entirely new mode. This is certainly what ice core records of CO_2 suggest has happened in the past (see

Figure 16.1). So, too, do the Younger Dryas and Heinrich events.

We might be confronted with such a climatic threshold in the near future. Recent data suggest that freshening of the North Atlantic has occurred over about the last 40 years of the 20th century in the vicinity of Greenland and Iceland. If the glaciers melt sufficiently, NADW production might suddenly shut down, which could send the northern hemisphere into a deep freeze after a prolonged interval of warming due to fossil fuel combustion.

CONCEPT AND REASONING CHECK

1. What are the climatic implications of a hysteresis loop?

SUMMARY

- The last 10,000 to 15,000 years of the Pleistocene represent the transition to the Holocene Epoch.
- Climate change during the Holocene occurred at sub-Milankovitch time scales, meaning that climate change occurred over time spans less than a few thousand years all the way down to durations as short as decades or less.
- The beginning of the Holocene was marked by rapid climate change, as ice sheets began to retreat from the northern United States and northern Europe in response to Milankovitch forcing. As the ice covering the continents retreated, the release of the overlying weight allowed the underlying rock and soil to undergo isostatic rebound, resulting in sea-level rise in various parts of the world.
- As Earth began to warm during the Holocene, the planet suddenly turned cold again during the Younger Dryas. As the northern hemisphere ice sheets began to retreat from the Great Lakes region in response to Milankovitch forcing, meltwater was trapped behind ice and rock dams in a huge lake called Lake Agassiz. Based on very strongly negative oxygen isotope ratios, it has been hypothesized that the Younger Dryas was caused by the collapse of the ice and rock dams of Lake Agassiz, which sent a large pulse of meltwater down the St. Lawrence to high-latitude sites adjacent to Greenland, where NADW forms. The freshwater would have lain on top of the seawater, preventing the formation and sinking of NADW.
- Decreased production of NADW would have effectively increased nutrient levels around Antarctica, promoting greater photosynthesis, the drawdown of atmospheric CO_2, and cooling. After the meltwater layer had mixed sufficiently with normal marine water, NADW production resumed, diluting AABW and allowing CO_2 levels to increase.
- Evidence indicates that meltwater pulses to the North Atlantic occurred repeatedly during the Holocene, implying the same basic process that caused the Younger Dryas to recur during the Holocene. These events are also indicated by layers of IRD called Heinrich events, which occur about every 1,500 years. However, the ice-rafting events also occurred during glacial intervals before the Holocene, suggesting they occur independently of ice volume. Thus, the growth and retreat of ice sheets might be caused by other factors that operate during glacial and interglacial intervals on sub-Milankovitch scales. Possibly, these events are related to changes in solar radiation.
- Rapid climate change also results from so-called atmospheric oscillations and their teleconnections. The best-documented oscillation, El Niño, occurs every few years in the Pacific Ocean. During El Niños, the trade winds weaken, allowing warm water piled up in the western Pacific to flow east, where it disrupts upwelling, fishing grounds, and weather systems as far away as North America. El Niños can be followed by a severe reversal of conditions called La Niñas.
- Most climate scientists agree the anthropogenic production of CO_2 has caused some of the warming during the 20th century. Anthropogenic warming might dominate climate during the next few centuries or longer. Moreover, based on Earth's past climatic behavior it is conceivable that, given the input of CO_2 into Earth's atmosphere from fossil fuel combustion, CO_2 might accumulate in Earth's atmosphere with little relative effect until a threshold is crossed and the climate suddenly shifts into an entirely new mode.

KEY TERMS

El Niño-Southern Oscillation (ENSO)

Heinrich events

hysteresis loop

ice-rafted debris (IRD)

jet stream

La Niña

Little Ice Age

oscillations

speleothems

sub-Milankovitch time scales

sunspot minima

teleconnections

Younger Dryas

REVIEW QUESTIONS

1. What processes cause sea level to change on relatively short geologic time scales?
2. Give some examples of rapid climate change that occur on different time scales.
3. Discuss the evidence for what caused the Younger Dryas.
4. How does photosynthesis off Antarctica presumably take up atmospheric CO_2?

5. Draw cross-sections and corresponding map views of the Atlantic Ocean, indicating the involvement of deep- and surface-ocean currents during warm (interglacial) and cool (glacial) conditions.
6. Why are atmospheric phenomena important in short-term climate change?
7. How does ENSO work?

FOOD FOR THOUGHT:
Further Activities In and Outside of Class

1. Why don't scientists believe Earth will simply warm gradually in response to increasing CO_2 in the atmosphere?
2. Cite evidence discussed in this chapter that you could use to argue *against* anthropogenic warming.
3. What do you consider to be the best evidence *for* anthropogenic warming?
4. What is the significance of the climatic fluctuations found during oxygen isotope stage 5?

5. If anthropogenic warming is actually occurring, do you believe all localities on the planet will warm uniformly? Why or why not?
6. What do you suppose are some of the sociopolitical consequences of rapid climate change, whether natural or anthropogenic (see Chapter 17)?
7. What is the significance of the Little Ice Age relative to anthropogenic climate change?

SOURCES AND FURTHER READING

Alley, R. B. 2000. *The two-mile time machine: Ice cores, abrupt climate change, and our future*. Princeton, NJ: Princeton University Press.

Alley, R. B. et al. 2003. Abrupt climate change. *Science, 299*, 2005–2010.

Barnett, T. P., Pierce, D. W., and Schnur, R. 2001. Detection of anthropogenic climate change in the world's oceans. *Science, 292*, 270–274.

Bazzaz, F. A., and Fajer, E. D. 1992. Plant life in a CO_2-rich world. *Scientific American, 1*, 68–74.

Bindschadler, R. A., and Bentley, C. R. 2002. On thin ice? *Scientific American, 287*(6):98–105.

Bond, G. et al. 2001. Persistent solar influence on North Atlantic climate during the Holocene. *Science, 294*, 2136.

Caviedes, C. N. 2001. *El Niño in history: Storming through the ages*. Gainesville, FL: University Press of Florida.

Chisholm, S. W., Falkowski, P. G., and Cullen, J. J. 2001. Discrediting ocean fertilization. *Science, 294*, 309–310.

Clark, P. U. et al. 2001. Freshwater forcing of abrupt climate change during the last glaciation. *Science, 293*, 283–287.

Clarke, G., Leverington, D., Teller, J., and Dyke, A. 2003. Superlakes, megafloods, and abrupt climate change. *Science, 301*, 922–923.

Cobb, K. M., Charles, C. D., Cheng, H., and Edwards, R. L. 2003. El Niño/Southern Oscillation and tropical Pacific climate during the last millennium. *Nature, 424*, 271–276.

Collier, M., and Webb, R. H. 2002. *Floods, droughts, and climate change.* Tucson, AZ: University of Arizona Press.

Crowley, T. J. 2000. Causes of climate change over the past 1,000 years. *Science, 289,* 270–277.

Davidson, E. A., Trumbore, S. E., and Amundson, R. 2000. Soil warming and organic carbon content. *Nature, 408,* 789–790.

Dickson, R. et al. 1996. Long-term coordinated changes in the convective activity of the North Atlantic. *Progress in Oceanography, 38,* 241–295.

Dickson, B. et al. 2002. Rapid freshening of the deep North Atlantic Ocean over the past four decades. *Nature, 416,* 832–837.

Dorale, J. A. et al. 1992. A high-resolution record of Holocene climate change in speleothem calcite from Cold Water Cave, northeast Iowa. *Science, 258,* 1626–1630.

Enfield, D. B., Mestas-Nuñez, A. M., and Trimble, P. J. 2001. The Atlantic multidecadal oscillation and its relation to rainfall and river flows in the continental U.S. *Geophysical Research Letters, 28*(10):2077–2080.

Fagan, B. 2000. *The Little Ice Age: How climate made history, 1300–1850.* New York: Basic Books.

Goldenberg, S. B. et al. The recent increase in Atlantic hurricane activity: Causes and implications. *Science, 293,* 474–479.

Gupta, A. K., Anderson, D. M., and Overpeck, J. T. 2003. Abrupt changes in the Asian southwest monsoon during the Holocene and their links to the North Atlantic Ocean. *Nature, 421,* 354–357.

Harvell, C. D. et al., 2002. Climate warming and disease risks for terrestrial and marine biota. *Science, 296,* 2158–2162.

Houghton, J. T. et al. (ed.). 2001. *Climate change 2001: The scientific basis. Intergovernmental Panel on Climate Change (IPCC).* Cambridge, UK: Cambridge University Press.

Hurrell, J. W. 1995. Decadal trends in the North Atlantic Oscillation: Regional temperatures and precipitation. *Science, 269,* 676–679.

Hurrell, J. W., Kushnir, Y., and Visbeck, M. 2001. The North Atlantic Oscillation. *Science, 291,* 603.

Keigwin, L. D. 1996. The Little Ice Age and medieval warm period in the Sargasso Sea. *Science, 274,* 1504–1508.

Lajeunesse, P. 2012. A history of outbursts. *Nature Geoscience, 5,* 846–847.

Lamb, H. H. 1979. Climatic variation and changes in the wind and ocean circulation: The Little Ice Age in the northeast Atlantic. *Quaternary Research, 11,* 1–20.

Lean, J., Beer, J., and Bradley, R. 1995. Reconstruction of solar irradiance since 1610: Implications for climate change. *Geophysical Research Letters, 22,* 3195–3198.

Lean, J., and Rind, D. 2001. Earth's response to a variable sun. *Science, 292,* 234–236.

Levitus, S. et al. 2001. Anthropogenic warming of Earth's climate system. *Science, 292,* 267–270.

Murton, J. B., Bateman, M. D., Dallimore, S. R., Teller, J. T., and Yang, Z. 2010. Identification of Younger Dryas outburst flood path from Lake Agassiz to the Arctic Ocean. *Nature, 464,* 740–743.

Neelin, J. D. and Latif, M. 1998. El Niño dynamics. *Physics Today, 51*(12):32–36.

Nigam, S., Narlow, M., and Berbery, E. H. 1999. Analysis links Pacific decadal variability to drought and streamflow in United States. *Eos, 80*(51):621.

Oren, R. 2001. Soil fertility limits carbon sequestration by forest ecosystems in a CO_2-enriched atmosphere. *Nature, 411,* 469–472.

Paillard, D. 2001. Glacial hiccups. *Nature, 409,* 147–148.

Philander, G. 1989. El Niño and La Niña. *American Scientist, 77*(5):451–459.

Ryan, W. B. F. et al. 2003. Catastrophic flooding of the Black Sea. *Annual Review of Earth and Planetary Sciences, 31*(1):525–554.

Sandweiss, D. H. et al. 2001. Variation in Holocene El Niño frequencies: Climate records and cultural consequences in ancient Peru. *Geology*, 29(7):603–606.

Sarmiento, J. L., and Gruber, N. 2002. Sinks for anthropogenic carbon. *Physics Today*, 55(8):30–36.

Schlesinger, W. H., and Lichter, J. 2001. Limited carbon storage in soil and litter of experimental forest plots under increased atmospheric CO_2. *Nature, 411*, 466–469.

Smil, V. 1999. *Energies: An illustrated guide to the biosphere and civilization*. Cambridge, MA: MIT Press.

Stott, L. et al. 2002. Super ENSO and global climate oscillations at millennial time scales. *Science, 297*, 222–226.

Wallace, J. M., and Thompson, D. W. J. 2002. Annular modes and climate prediction. *Physics Today*, 55(2):28–33.

Weart, S. 2004. The discovery of rapid climate change. *Physics Today, 56*(8):30–36.

Wofsy, S. C. 2001. Where has all the carbon gone? *Science, 292*, 2261–2263.

Woodhouse, C. A., and Overpeck, J. T. 1998. 2,000 years of drought variability in the central United States. *Bulletin of the American Meteorological Society, 79*, 2693–2714.

Zahn, R. 2003. Monsoon linkages. *Nature, 421*, 324–325.

The Anthropocene: Humans as an Environmental Force

MAJOR CONCEPTS AND QUESTIONS ADDRESSED IN THIS CHAPTER

A How has climate affected humans and civilization in the past?

B How did society come to depend on fossil fuels, especially petroleum?

C How do we find petroleum?

D What are the alternatives to fossil fuels, and what are their advantages and drawbacks?

E What are the political and socioeconomic consequences of burning fossil fuels?

F How are humans affecting biodiversity, and why should we care?

G How does this affect me personally?

H What can you do?

CHAPTER OUTLINE

Deforestation of tropical rainforest.

© guentermanaus/Shutterstock, Inc.

17.1 Introduction: From Geohistory to Geopolitics

How many of our present problems arise from not under-standing our environment and making unrealistic demands upon it?
 —Hubert Lamb, *Climate, History, and the Modern World*

One way to view the discipline of history is the way most people view it: as the "past," over and done with. It might then seem odd at first glance—even to your instructor—to insert a discussion of human impacts in a book on "Historical Geology." But there is another way to view history: as the waves of history breaking on the seemingly solid shores of our "present," our "normal." Those sands shift, sometimes unpredictably, as we attempt to control them, just as they have in the past. The history of civilization reveals the tremendous impact of contingency—accidents, coincidences, and other unforeseen events—on human enterprises and institutions. The history of Earth, as emphasized in this book, has been no different. As noted at the outset of this book (see the frontispiece in Chapter 1), humans are not immune from natural climate change. Conversely, as noted in the preface to Section IV, *humans are now the major geo-logic and climatic force on this planet* because of the rapidity of anthropogenic change during the so-called "Anthropocene." In fact, humans have been affecting the natural environment for thousands of years. Witness for example, the infilling of the ancient natural harbors of Troy, Miletus, and Ephe-sus (in what is now western Turkey) by sediment derived from the introduction of goats and the resulting deforesta-tion of the surrounding land (**Figure 17.1**). But human populations have expanded so rapidly in the past centuries that the repercussions of anthropogenic impacts are now being felt on a global scale. If we normalize the age of Earth (5 billion years in round numbers) to one year, the last cen-tury of human population expansion and impact represents only a little more than one-half second! Indeed, the rapidity and complexity of anthropogenic interactions with natural processes are what founded the science of earth systems in the first place (see Chapter 1). In this view, all history—as an old adage says—is contemporary history; shouldn't we try to learn from history? Understanding where we have come from can tell us what we *can* and *cannot* do, not necessarily to foresee and forestall the hurdles that will face us, but to meet and then cope with them as we adapt, and about what we ought to do or not do, morally and ethically.

As we will see in this chapter, *geohistory has become geopolitics*. The Earth's history has set the constraints on the modern conditions of countries throughout the world: their dependence on various sources of energy, their biodiversity, water availability, and patterns of poverty.

17.1.1 Climate, history, and the modern world

Most scientists agree that Earth warmed naturally dur-ing the 20th century possibly as a result of solar activity

(a) Courtesy of Ronald Martin, University of Delaware.

(b) Courtesy of Ronald Martin, University of Delaware.

(c) Courtesy of Ronald Martin, University of Delaware.

FIGURE 17.1 The effects of deforestation on the infilling of ancient harbors. **(a)** Troy. The broad flat plain in the distance represents the ancient harbor, as does that of **(b)** Miletus. **(c)** Ephesus. The edge of the ancient harbor is indicated by the buildings in the distance (yellow arrow), where ships docked and unloaded. Goods were then brought into the city along the broad street leading into the distance.

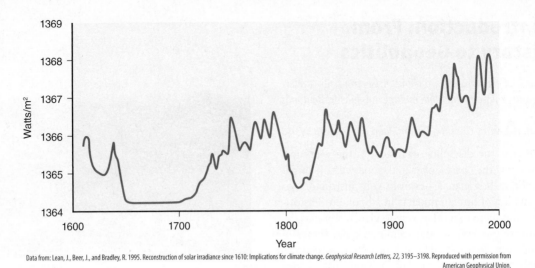

Data from: Lean, J., Beer, J., and Bradley, R. 1995. Reconstruction of solar irradiance since 1610: Implications for climate change. *Geophysical Research Letters, 22,* 3195–3198. Reproduced with permission from American Geophysical Union.

FIGURE 17.2 Total solar radiation from 1610 to present. Notice the increased solar irradiance and warming beginning in the late 1800s and continuing through the 20th century.

(**Figure 17.2**). But natural warming is likely compounded with the burning of fossil fuels and the production of carbon dioxide (CO_2) so that natural warming is being greatly exacerbated beyond normal levels by the burning of fossil fuels. Atmospheric CO_2 levels rose steadily through the last half of the 20th century and are projected to continue to rise into the foreseeable future (see Chapter 1). Calculations indicate that about 55% of anthropogenic CO_2 has so far accumulated in the atmosphere. Today, CO_2 levels stand at about 400 ppm. This concentration is over 100 ppm greater than the 280 ppm recorded by ice cores (see Chapter 16) during the last major interglacial (oxygen isotope stage 5) and over 180 ppm greater than during the last major glacial (200 ppm) just before the beginning of the Holocene.

The burning of fossil fuels supplies energy for society. Indeed, civilization depends on the flux of energy and matter, just like natural systems (see Chapter 1). In this chapter, we examine the consequences of, and some possible remedies for, the use of fossil fuels. Alternative energy sources must be developed, if for no other reason than fossil fuels will eventually run out. At the present time, however, alternative energy sources cannot supply the energy necessary for society. Thus, we will continue to be dependent on fossil fuels for the near-term future. Increasing energy use revolves around one primary driving force: human population growth (see Chapter 1). Thus, human population growth must slow; otherwise, continued demands on other natural resources such as land, air, and water will continue. Water resources is an especially critical global issue.

Human population growth also decreases biodiversity by destroying habitat, especially in tropical regions, where biodiversity is greatest. At the same time, resource consumption in developed countries, which mostly lie in temperate regions where biodiversity is normally lower, must also be curbed or natural resources will run out.

17.2 Examples of Human–Climate Interactions During the Last Millennium

A We have already examined one possible effect of climate on humans: the evolution of humans themselves (see Chapter 15). Nevertheless, scientists have only recently begun to realize just how important, complex, and sometimes subtle the interactions between humans and climate are.

For example, beginning about 1300 AD, Earth entered the climatic phase known as the Little Ice Age. Based on historical records and records of ancient climate, the Little Ice Age represents only the most recent of a series of cold periods that have occurred about every 1,000 to 1,500 years before and during the Holocene (see Chapter 16). Although temperatures fell at most only a few degrees during that period, and even fluctuated at times, this temperature behavior was enough to have enormous consequences on everyday life. Viking settlements were established in Greenland between about 800 and 1200 AD during a warmer phase known as the **Little Climate Optimum** (or **Medieval Warm Period**). But by about 1200 AD, near the beginning of the Little Ice Age, the settlements were abandoned because the oceans became covered with ice year round, preventing ships from bringing iron and wood from Iceland and Norway.

These climate changes also affected continental Europe. Winters in Europe during the Little Ice Age were especially harsh (**Figure 17.3**). Up to 80% of the human population in Europe was engaged in subsistence agriculture, and this standard of living was easily subject to the collapse and famine brought on by sudden, markedly colder or wetter conditions associated with storms and killing frosts. By about the middle of the 1800s, though, mountain glaciers that had closely encroached on villages during the Little Ice Age

© The Art Gallery Collection/Alamy.

FIGURE 17.3 Peter Brueghel's "Hunters in the Snow," painted in 1565, depicts the harsh winters during the Little Ice Age.

began to retreat (**Figure 17.4**). Although no one realized it at the time, the end of the Little Ice Age had begun.

17.3 Brief History of the Growing Dependence on Fossil Fuels

The beginning of the warming that marked the end of the Little Ice Age is thought to have resulted from natural causes. However, by the middle-to-late 19th century

(a)   Courtesy of Mauri S. Pelto.

(b)   Courtesy of Mauri S. Pelto.

FIGURE 17.4 The Whitechuck Glacier in Glacier Peak Wilderness in 1973 (top) and in 2006 (bottom).

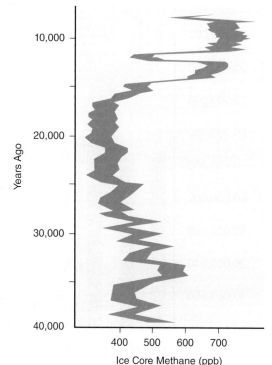

Data from: Chappellaz, J., Barnola, J. M., Raynaud, D., Korotkevich, Y. S., and Lorius, C. 1990. Ice-core record of atmospheric methane over the past 160,000 years. *Nature, 345,* 127–131.

FIGURE 17.5 Natural (pre-Industrial Revolution) and anthropogenic production of methane (CH_4). Based on ice core records, the natural level of methane is ~700 ppm or less.

human activities were quickly gaining momentum. The ongoing Industrial Revolution was heavily reliant on fossil fuels, especially coal, the burning of which pumped CO_2 into the atmosphere.

Besides CO_2, another primary greenhouse gas is methane (CH_4), the concentration of which skyrocketed beginning in the 19th century (**Figure 17.5**). When methane is exposed to oxygen in the atmosphere, much of it is oxidized to produce CO_2. The steep rise in anthropogenic methane resulted primarily from the spread of agriculture as human populations expanded. Cattle and other livestock are ruminants (see Chapter 15) that produce tremendous amounts of methane as they digest plant matter in a series of stomachs. Rice paddies also produce large amounts of methane. Other sources of methane include natural gas that is burned as it is released from oil wells and refineries, biomass burning, and sewage treatment.

The production of CO_2 accelerated through the 20th century, as civilization shifted toward its current reliance on petroleum. It was the development of the internal combustion engine in the late 1800s that paved the way (no pun intended) for the primary consumer of oil—the automobile—because gasoline was found to be the best fuel for powering internal combustion engines. Cars were such a curiosity in 1896 that they were used in circus acts. Nevertheless, by 1912 electrified assembly lines began to mass produce cars, and by 1995 the world had more than *500,000,000* of them (**Figure 17.6**)!

Road building zoomed upward in the United States with the adoption of the automobile. The initiation of the

Global Automobile Production

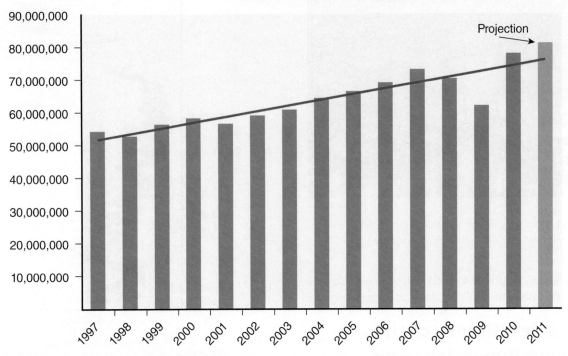

FIGURE 17.6 Global automobile production since 1997, according to the International Organization of Motor Vehicle Manufacturers (OICA), a federation of automobile manufacturers based in Paris.

interstate system by the Eisenhower Administration in the 1950s—primarily for national defense during the Cold War—promoted the use of the automobile, human settlement and business, and the reorganization of the American lifestyle. By 1990, the United States had 5.5 million kilometers of surfaced roads, much of which were built from 1920 to 1980. Automobiles were no longer a luxury but a necessity.

Today, oil production in the United States occurs primarily in the vicinity of the Gulf Coast (mainly Texas and Louisiana), Alaska, portions of the Rocky Mountains, and southern California. As the size of oil reservoirs became smaller and smaller in the United States, they and other western countries became increasingly dependent on imports of foreign oil, especially from the Middle East, where huge reserves exist (**Figure 17.7**). Many of these countries belong to the cartel called the **Organization of Petroleum Exporting Countries**, or **OPEC**, which is a group (consisting in this case of countries in the Middle East, along with Venezuela) that was formed to regulate world petroleum prices and output. Although OPEC has generally been favorably inclined toward the West in the past, it has at times used oil as a political weapon and caused oil shortages, such as the so-called "energy crises" of the 1970s, by drastically cutting production. The energy crises, although severely disrupting economies and lifestyles, also had a positive effect, for a short time, by encouraging the development of lighter and more fuel-efficient automobiles during the subsequent decades (**Figure 17.8**).

Nevertheless, countries such as the United States remain heavily dependent on petroleum. Moreover, as developing countries such as China and India and their enormous human populations merge into the global economy, they have also come to demand more and more petroleum.

Thus, society continues to seek more oil. Enormous, untapped reserves are estimated to exist in Siberia, for example. In the United States, debate about domestic petroleum exploration centers on opening the **Arctic National Wildlife Refuge**. But, even if opened to exploration immediately, the Arctic National Wildlife Refuge would not produce oil for at least five years, with peak flows of no more than one million barrels per day by about two decades after that. This would be equivalent to only about 4% of U.S. *daily* oil consumption today. At the same time, American oil consumption is projected to rise from 19.5 million barrels per day to well more than 20 million barrels per day during the present decade. Moreover, world oil production has been predicted to peak during the first half of the 21st century. Even though world production was initially predicted to peak as early as the 1970s, the peak has continually shifted into the future. (**Figure 17.9**).

This is because petroleum companies have been very successful at developing innovative, cost-effective technologies such as **directional drilling** that effectively increases supplies of petroleum by extracting it more cheaply from reservoirs (**Box 17.1**). New, unforeseen fields also continue to be discovered, even in the United States. Most recently, the process of **fracking**, which is based on petroleum

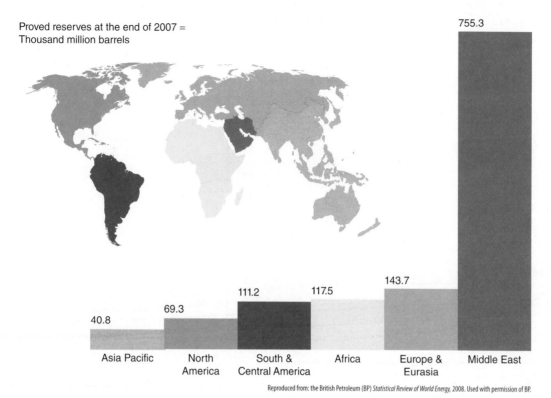

Proved reserves at the end of 2007 =
Thousand million barrels

755.3

143.7

111.2   117.5

69.3

40.8

Asia Pacific   North   South &   Africa   Europe &   Middle East
               America   Central America   Eurasia

Reproduced from: the British Petroleum (BP) *Statistical Review of World Energy*, 2008. Used with permission of BP.

FIGURE 17.7 Oil reserves of the world. Relative proportions of crude oil reserves held by countries or regions as of 2007. Volumes in thousands of millions (or billions) of barrels.

exploration and development technologies, is now being used in certain states (such as Texas, Pennsylvania, and the Dakotas) to increase natural gas supplies. In this process, pressurized water is pumped down a well into subsurface gas-rich shales, causing the formations to fracture and release gas, which is then recovered at the surface. Because of these technologic innovations (see Box 17.1), it is now projected that petroleum production in the Americas will surpass the Middle East sometime in the 2020s. The United States alone has been projected to harbor up to two trillion barrels of oil in "unconventional" reservoirs (those difficult to tap), Canada another 2.4 trillion, and South America more than 2 trillion barrels. However, the profitability of such enterprises is heavily dependent on the price of petroleum, which can fluctuate dramatically. Also, the fracking process has been claimed to contaminate groundwater supplies with noxious gases, toxic chemicals from drilling fluids, and sediment, and to produce earthquakes in otherwise tectonically quiescent areas.

CONCEPT AND REASONING CHECKS

1. How is civilization like an open system?
2. What are the sociopolitical consequences of continued reliance on foreign sources of petroleum?
3. Further development of petroleum sources in the United States has been proposed to lessen our reliance on foreign sources of petroleum. Will this necessarily work in the long term? What is "long term," in your view?

17.4 Alternative Energy Sources and Technologies

Coal is a fossil fuel proposed as a "clean" energy source to supply power plants. When burned, however, coal produces CO_2 and sometimes other gases such as sulfur dioxide, which dissolves in rainwater to produce acid rain that pollutes lakes and streams. Power plants in the United States that burn sulfur-rich coal are required by the U.S. Environmental Protection Agency to have special "scrubbers" to prevent the production of such gases. Coal is also normally strip-mined, which can pollute rivers and streams with sulfur precipitates sometimes referred to as "yellow boy."

The apparent solution to the current environmental conundrum is, of course, to stop burning fossil fuels altogether. However, even if all fossil fuel combustion were stopped immediately, it would take centuries and perhaps a millennium or more for *natural* systems to sequester the CO_2 out of the atmosphere and store it elsewhere such as in peats, forests, or the deep ocean because the processes involved act relatively slowly on human time scales.

Whether anthropogenic warming is actually occurring, alternative energy sources and technologies must be developed to replace fossil fuel–based technologies if society is to continue as we know it and if the United States wishes to maintain its national security (**Figure 17.10**). In the sections that follow, we consider the advantages and disadvantages of some alternative energy strategies, although the discussion is by no means exhaustive.

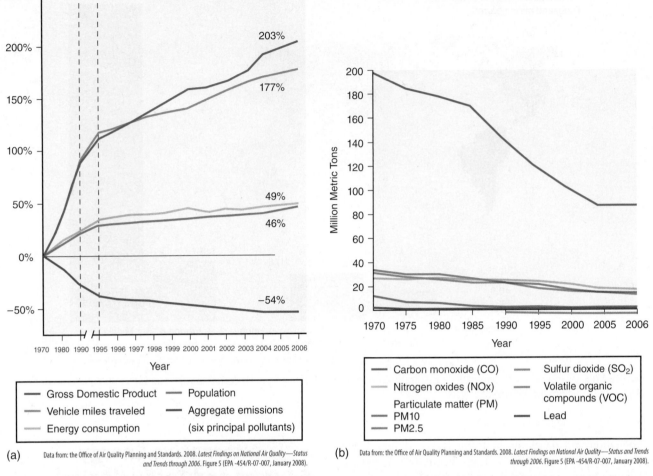

(a) Data from: the Office of Air Quality Planning and Standards. 2008. *Latest Findings on National Air Quality—Status and Trends through 2006.* Figure 5 (EPA -454/R-07-007, January 2008).

(b) Data from: the Office of Air Quality Planning and Standards. 2008. *Latest Findings on National Air Quality—Status and Trends through 2006.* Figure 5 (EPA -454/R-07-007, January 2008).

FIGURE 17.8 **(a)** Although the Gross Domestic Product (GDP), vehicle miles traveled, energy consumption, and human population in the United States have all increased over the past several decades, emissions of major pollutants (except CO_2) declined as a result of increasing fuel-efficiency of automobiles and pollution controls on factories, cars, and power plants. **(b)** Total pollution emissions of all major pollutants except CO_2 in the United States since 1970.

17.4.1 Wind, geothermal, and solar energy

Alternative energy sources such as wind, solar, and geothermal (which uses steam from natural sources) all emerged during the late 20th century. However, the use of wind and geothermal power is largely restricted to certain areas with appropriate environmental conditions: offshore "wind farms" located in coastal regions or areas of high geothermal gradients such as northern California, where natural steam sources occur (the steam is used to power turbines). The technology to develop solar power as a viable alternative to fossil fuels is still emerging, but the generation of solar power is becoming increasingly widespread by power companies and individual home and business owners.

17.4.2 Hydroelectric power

Hydroelectric generation of power has long been used in the Western World and is considered a "mature" power-generating technology. Hydroelectric plants rely on the flow of water to power turbines and are of two basic types: high-head plants, which run water stored at great heights behind large dams, and low-head (or "run-of-the-river") types, which run large volumes of water through turbines submerged in a river. Hydroelectric generation now accounts for about 18% of all energy generated worldwide and about two-thirds of Canada's electric production. Some of the largest dams in the United States were built during the 1930s as part of Franklin Roosevelt's Works Progress Administration to keep as many Americans employed as possible during the Great Depression. Some of these dams include the high-head Boulder Dam near Las Vegas, Nevada.

Hydroelectric power generation has become much more widespread in recent years in developing nations. These projects have not been without problems. Aswan Dam, located in Egypt on the Nile River and finished in 1970, is of the high-head type and, as a result, has caused a number of environmental and cultural problems. The ancient Nile normally flooded every year, bringing nutrient-rich silts to the adjacent floodplain, where agriculture thrived. Originally built in part to prevent massive flooding of villages that later

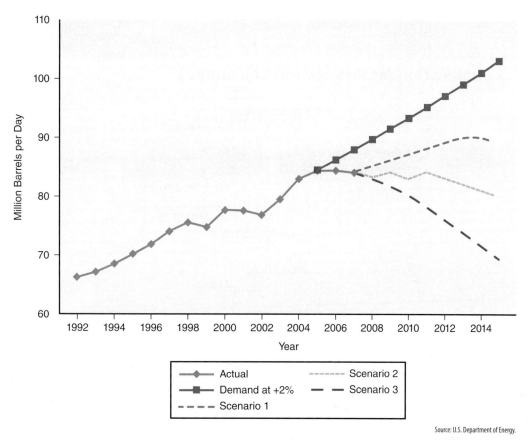

Source: U.S. Department of Energy.

FIGURE 17.9 Projected oil production for various scenarios. A barrel equals 42 U.S. gallons.

BOX 17.1 Exploration for Petroleum

Petroleum, or oil and gas, is a biogenic product of dead marine plankton and terrestrial plant material that has been buried and decayed (**Box Figure 17.1A** and **B**). Much of the *exploration* for, and *production* of, oil and gas in the United States now occurs in waters of the Gulf of Mexico, a hundred miles or more off the coasts of Texas and Louisiana.

The oil in the Gulf of Mexico formed many millions of years ago. The organic matter that forms petroleum is relatively fine-grained, so it tends to settle out with mud. During burial of the muds the dead organic matter is heated to about 100°C, whereupon it begins to liquefy and *migrate*. As it migrates from the shales, the petroleum eventually encounters sands in which the oil continues to migrate until it is *trapped* (**Box Figure 17.1C**). The *reservoir* sands in the traps being explored and produced today in the Gulf of Mexico are typically no more than 30 to 50 meters thick and lay thousands of meters below the sea bottom. Many of

these sands were deposited in the channels and lobes of submarine fans formed by turbidity currents. During low sea-level stands, deltas prograded across the shelves and shed sediment—in the form of turbidity currents—onto the continental slope (see Chapter 4).

In the Gulf Coast, petroleum traps typically form along faults. In the process shales, which are impervious to fluid flow, can come to lie against the sands, preventing further migration (see Box Figure 17.1C). Sediment tends to collect on the side of the fault that is moving downward relative to the other side. As this happens the sedimentary section on the downthrown block tends to thicken or "grow"; hence, the name for these faults: **growth faults**.

The faults formed because of the phenomenon of **salt tectonics**. Much of the Gulf of Mexico is underlain by salt and anhydrite because at one time (when Pangea was rifting apart) the Gulf of Mexico was a gigantic evaporite basin (see Chapter 13). When

(continues)

BOX 17.1 Exploration for Petroleum (Continued)

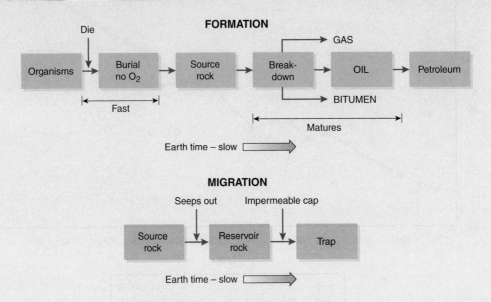

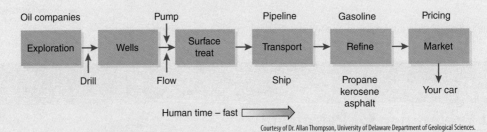

Courtesy of Dr. Allan Thompson, University of Delaware Department of Geological Sciences.

BOX FIGURE 17.1A A flow chart for petroleum formation, migration, exploration, and production. Note the time scales involved.

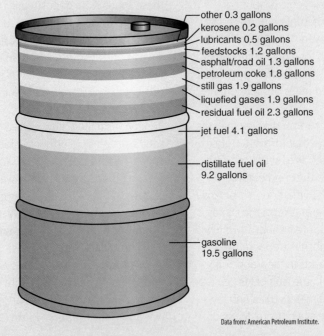

Data from: American Petroleum Institute.

BOX FIGURE 17.1B Products refined from a 42-gallon barrel of oil.

BOX 17.1 Exploration for Petroleum (Continued)

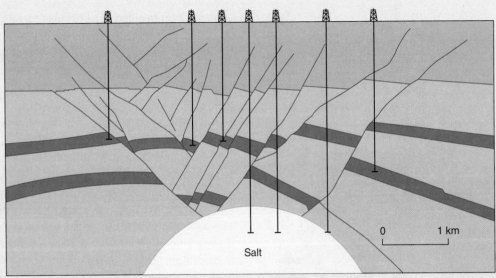

Data from: Wicander, R., and Monroe, J. S. 2000. *Historical Geology: Evolution of Earth and Life Through Time*, 3rd ed. Pacific Grove, CA: Brooks/Cole. Figure 16.30B (p. 442).

BOX FIGURE 17.1C Petroleum traps in the Gulf Coast. Such traps usually form at a fault plane where impervious shales butt up against sands along a growth fault, preventing further petroleum migration. The salt diapirs form as salt deforms from the overlying weight of sediment. As they push upward, the diapirs cause faults to occur, which is where petroleum traps tend to occur. Note how the petroleum-bearing layers (darker layers), called reservoirs, butt up against shale (lighter color) along the faults, trapping the oil there because shale is impermeable.

large amounts of sediment are deposited on top of evaporites they deform like modeling clay, only much more slowly on geologic scales of time. As the evaporites deform they begin to push upward ever so slowly to produce irregular, pillar-like structures called **diapirs** (see Box Figure 17.1C; see Chapter 13). The upward movement of the diapirs causes the growth faulting because blocks of sediment above and around the salt are forced to move past one another.

Exploring for new petroleum reservoirs in the Gulf of Mexico starts when federal and state governments announce that certain areas will be opened for exploration. These areas are subdivided into a series of lease blocks of about 10 square miles. Geologists for various companies then typically construct cross-sections of the area of interest. Using these data, petroleum geologists for each company determine each lease block's potential for producing oil and gas. Areas interpreted to be geologically favorable for producing petroleum are called "prospects." Each company then submits a secret bid for each prospect, based on its assessment, to the government agencies involved. The bids for each lease block can range

into the many tens of millions of dollars and can vary radically depending on each company's evaluation of the prospect. The company with the highest bid gets the opportunity to drill for oil and gas in the particular lease block; in other words, the company gets to "rent" the lease block and does not own it.

In terms of the scientific method (see Chapter 1), then, the geologic cross-sections are used to formulate a hypothesis (where petroleum is likely to be found) and the hypothesis is then tested by drilling. Drill rigs are sometimes leased from service companies. Increasingly, however, because of the tremendous water depths involved, the rigs must be specially constructed; these rigs can cost hundreds of millions of dollars and must be towed in large pieces to the well site, where they are assembled. At the well site, the rigs may be anchored to the bottom on "legs" or pillar-like structures to keep them stationary (**Box Figure 17.1D**). If the water is sufficiently deep, however, the rig floats at the surface, with its position over the well site maintained by the sensors of special dynamic positioning systems.

Because of their expense, offshore rigs are designed so that multiple wells can be drilled from the

(continues)

BOX 17.1 Exploration for Petroleum (Continued)

same platform to different reservoirs ("targets") using directional drilling. The technology of directional drilling allows wells to be drilled at very high angles from the same platform (**Box Figure 17.1E**) and has been a tremendous boon to the exploration for small reservoirs because it cuts drilling expenses.

Given the large and numerous expenses involved, it is not uncommon for a petroleum oil company to invest many tens or even hundreds of millions of dollars or more in exploring an oil and gas field before even the first drop of petroleum is found. In fact, it is now quite common for two or more companies to form partnerships in the exploration and production of a particular lease block. That being said, there are no guarantees of success. Indeed, the success rate for finding economic petroleum reservoirs is about 30% to 40% today. Those containing "subeconomic" quantities of oil and gas or fresh water or saltwater are called "dry holes" and are plugged with cement and abandoned. If, however, the reservoir appears promising, further drilling determines the reservoir's real extent and thickness and therefore the volume of reserves potentially present. Eventually, petroleum will be pumped from the field onto tankers or through a pipeline to refineries where it will be processed to produce gasoline, heating oil, jet fuel, and so on. Petroleum is also used as a "feedstock" to make plastics, upholstery, appliance parts, and aspirin, just to name a few products.

The price of oil is normally quoted in dollars per barrel (one barrel is equal to 42 U.S. gallons), which is about half the volume of an average bathtub. Why 42 gallons? Why not an even 40 or 50? Possibly because oil used to be transported in horse-drawn wagons, so there was probably a lot of spillage and seepage with open barrels that contained about 40 gallons. To compensate for this loss, the oil producer might have had to credit the transporter with an extra 2 gallons for every 40 gallons of sales. The price of oil varies daily, especially with political events, when it might "spike" to much higher prices, significantly affecting the global economy.

© AP Photos.

BOX FIGURE 17.1D An oil rig in the Gulf of Mexico.

Courtesy of The National Energy Technology Laboratory/U.S. Department of Energy.

BOX FIGURE 17.1E Directional drilling from an oil well.

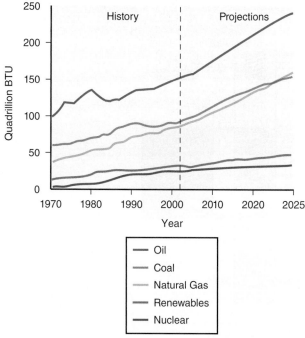

Data from: U.S. Department of Energy, Energy Information Administration, *International Energy Annual*, 2002, 2003 (May–July 2005), 2005, and *System for the Analysis of Global Energy Markets*, 2005 and 2006.

FIGURE 17.10 Historical and projected global energy sources.

developed along the river, flooding caused by the Aswan Dam displaced approximately 60,000 people (although new, planned settlements were established in their place). The lake behind the dam also submerged ancient archaeological sites, whereas the silt originally deposited on the floodplain was trapped behind the dam, lowering the water storage capacity. The recent construction of the Three Gorges site on the Yangtze River is also of the high-head type and is the largest hydroelectric project in history. The dam is so large that the hydroelectric plant would be able to light up Boston, New York, and Washington, DC, combined! However, the project has displaced about 1.2 million people, many of whom have had families living in the river valley for generations. The dam has also flooded archaeological and cultural sites and, like other dams of its type, caused drastic changes to the surrounding ecosystems.

17.4.3 Nuclear energy

For a time, nuclear power also appeared promising, but **nuclear accidents** in the United States (for example, **Three Mile Island** near Harrisburg, Pennsylvania), the United Kingdom, the **Chernobyl** disaster near Kiev (Ukraine) in 1986, and, most recently, in 2011, at the **Fukushima** power plant in Japan (as a result of flooding by a tsunami) have hindered its development. At Chernobyl only 31 people died quickly, but countless others among the approximately *800,000* workers and soldiers forced to clean up after the disaster have died or will die from various types of cancer; hundreds of thousands of others in the surrounding population were also affected. The total release of radiation from Chernobyl was hundreds of times greater than that given off

by the nuclear bombs dropped on Hiroshima and Nagasaki to end World War II and everyone in the northern hemisphere received at least a tiny dose of radiation from the Chernobyl debacle. The full extent and cost of the Japanese disaster are yet to be determined.

There is also the question of what to do with spent radioactive material from a nuclear reactor core (**Figure 17.11**). Fuel rods enriched in the radioactive element uranium comprise the reactor core. Water flows through the core and is heated to produce steam that powers turbines, which, in turn, power generators that produce electricity. New designs using radioactive pellets and different fluids for cooling appear to substantially improve safety. However, when the fuel is spent the radioactive waste that is left must either be reprocessed or disposed of; the problem is that although the fuel is eventually no longer useful for energy production, it is still radioactive. The only solution is to store the radioactive waste for thousands of years in designated storage sites until the waste has completely decayed. Despite this drawback, the alternative is on-site storage of nuclear waste near current facilities all across the country.

Yucca Mountain, located about 150 km northwest of Las Vegas, Nevada, had been designated as a national nuclear waste repository to counter storage at multiple sites (**Figure 17.12**). However, highly vocal opposition from state representatives and environmentalists led to the abandonment of the project despite its completion. Originally, about 77,000 tons of nuclear waste, mainly in the form of reduced uranium, or UO_2, had been destined for storage at Yucca Mountain over a period of 24 years (UO_2 is the mineral uraninite, which has been used as an indicator of oxygen levels in the early Earth's atmosphere; see Chapter 9). The repository was then supposed to be monitored for 300 years before being sealed off. Federal law required that the site remain safe for 10,000 years, which is about the duration of the entire Holocene epoch and longer than the entire span of recorded human civilization!

Approximately 20 years of study by government agencies had affirmed that Yucca Mountain was a safe repository. Nevertheless, a number of serious questions were raised about the safety of the proposed site. First, the waste was to be transported by rail to Yucca Mountain from all over the country; 90% of the waste was to come from commercial nuclear reactors and the rest from federal weapons research facilities. Transportation by such means was of course subject to accidents and spillage. Second, the waste was to be buried in tunnels about 300 meters beneath the ground's surface in special containers made of metal alloy. This level is about 300 meters above the **water table**, which is the level below which the ground is permanently saturated with water. The rationale was that burying the waste above the water table would prevent leakage by any groundwater contamination. However, even in the presence of minor amounts of moisture and oxygen, UO_2 is unstable and is oxidized to a more soluble form (see Chapter 9). Thus, some critics of the Yucca Repository suggested it would actually be safer to store the waste below the water table. Third, two fault lines run nearby and a small earthquake was reported 11 miles away in June 2002 and another

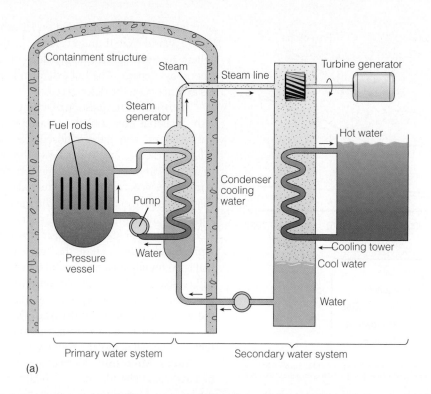

(a)

(b)

Courtesy of Mark Marten/U.S. Department of Energy.

FIGURE 17.11 **(a)** Basic structure of a nuclear reactor. Fuel rods enriched in the radioactive element uranium comprise a reactor core through or around which water flows and heats to produce steam that powers turbines, which in turn power generators that produce electricity. The fuel rods, along with control rods, can be inserted or withdrawn from the reactor core as desired, depending upon the demand for electricity. Cooling towers release the steam that is generated. **(b)** View into a containment structure. The reactor core is located under approximately 11 meters of water.

FIGURE 17.12 Yucca Mountain Nuclear Waste Repository had been proposed as the national storage site for nuclear waste, but the project has been abandoned for environmental and political reasons.

affected the area in 1992. Fourth, the area was close to one of the youngest volcanoes (about 80,000 years old)—Little Skull Mountain—in the United States. Some scientists concluded that Yucca Mountain overlies a region of hot mantle. (Recall from Chapter 15 that the Basin and Range region and nearby areas are tectonically active.) Thus, the likelihood of future volcanic activity might be much higher than review panels had stated, perhaps as high as 11 to 15 events per million years. Although this might not sound particularly threatening, *based on the study of the history of the Earth*, volcanism and earthquakes could happen there quite unexpectedly during such a long interval of time. Volcanism and earthquake activity also opened up the possibility of fracturing the near-surface crust, allowing the downward leakage of contaminated water.

17.4.4 Methane gas hydrates

Another potential energy source is **gas-hydrate deposits** or **clathrates** that are buried beneath continental margins (like those thought to have caused the Paleocene–Eocene Thermal Maximum; see Chapter 14). Similar deposits are also buried on land beneath **permafrost** (perennially frozen subsoil) at high latitudes. Clathrates consist of frozen water in which reside gas molecules such as methane, which is derived from the decay of dead organic matter.

However, two main obstacles stand in the way of clathrates as energy sources: technologic and environmental. High subsurface pressures and the detection of economic deposits worth drilling make finding and developing clathrates unlikely at the present time, although technologic barriers (such as directional drilling) have been overcome in the past by petroleum companies (see Box 17.1). Also, the burning of methane and other gases trapped in clathrates will pump more CO_2 into the atmosphere. As we saw in Chapter 14 the massive release of such deposits has been implicated in rapid climate change in the geologic past.

17.4.5 Biofuels

Biofuels were one of the earliest forms of energy production used by early humans. Wood-burning stoves are still widely used in rural portions of North America, and wood-burning (and deforestation) is commonplace in underdeveloped countries where other energy sources are scarce. Today, biofuels encompass a much broader range of sources. "Landfill gas" (mainly methane) is now used to generate electricity by some municipalities, and wood waste is sometimes used by paper and pulp mills to generate electricity.

One of the most widely available biofuels today is ethanol, or grain alcohol. Ethanol is produced from corn and other grains by the same processes used to produce alcoholic beverages: fermentation and distillation. Today, ethanol commonly makes up to 10% of gasolines marketed in the United States. However, large amounts of energy are necessary for corn production (in the form of tractor fuel and fertilizer production) and to separate alcohol from water during distillation. One report indicates that between 29% and 57% more fossil energy is required than is produced by fermentation in the form of ethanol, depending on the type of biomass chosen for fermentation. Large areas of land formerly used for food production are also now devoted to ethanol production, which some claim has driven up food prices. Thus, ethanol production from grain sources might not be sustainable in the long run. Other feedstocks, such as vegetable oil (from sunflowers and soybeans), for the production of biodiesel fuels might hold more promise.

17.4.6 Automobiles

Automobiles are one of the prime users of fossil fuels as well as one of the major contributors to air pollution. One of the most promising recent technologic innovations in this regard is **hybrid cars**. On highways hybrid cars are powered by a gasoline engine, but in the stop-and-go traffic of cities they are powered by electricity generated during braking, thereby decreasing air pollution. In "grid-independent" or "stand-alone" hybrid cars, the energy produced during breaking and normally lost as heat is used to power a generator that produces electricity that recharges a battery. However, hybrid cars are really a stopgap measure until other technologies are developed because they still use gasoline. In "grid-dependent" or "plug-in" vehicles, the electricity that recharges the battery is, today, likely generated at a power plant that runs on fossil or nuclear fuel. For this technology to make environmental change, then, the electricity will eventually need to come from sources not powered by fossil fuels. Still, with the current technology these vehicles have a range of at least several tens of miles when running on electricity alone and are useful for short commutes or as fleet vehicles. When necessary the distances could be increased beyond this range using gasoline.

Another promising technology for future automobile transportation is **hydrogen fuel cells**. Hydrogen fuel cells convert hydrogen gas into electricity cleanly by splitting hydrogen into protons (H^+) and electrons (**Figure 17.13**). The electrons are then used to power an electric motor, which drives the automobile. Then, the electrons return to the fuel cell where they are combined with the protons and oxygen and emitted to the environment as water vapor. Hydrogen fuel cells are up to 60% efficient, as compared with gasoline-powered and diesel-powered internal combustion engines, which top out at 22% and 45% efficiency, respectively. A basic logistical problem remains with hydrogen cells, however: fuel availability. To become commercially feasible, hydrogen-powered ground transportation will need to carry sufficient hydrogen to cover the long distances of gasoline-powered engines (with the exception of fleet vehicles used by, say, government agencies or public transportation). Although gasoline is three to four times as heavy as

liquid hydrogen, gasoline still provides over three times the energy per unit volume that liquid hydrogen does.

This presents a classic "chicken-or-egg" problem. Given current technology, the widespread adoption of hydrogen-powered automobiles will require refueling stations, just as gasoline-powered ones do now. But, adequate numbers of refueling stations will require sufficient numbers of hydrogen fuel cell–powered automobiles to justify their existence.

17.5 Consequences of Fossil Fuel Combustion

Between 1890 (at the end of the Little Ice Age) and 1990, average surface temperatures on Earth increased by 0.3°C to 0.6°C. Instead of steadily rising during the 20th century, however, temperatures actually fluctuated. The temperature increases occurred in two surges: between 1910 and 1940 and after 1975. But from 1940 to 1975 average temperatures declined slightly.

Given that climate varies on millennial to decadal scales (see Chapter 16), how much of the warming during the last part of the 20th century is due to natural climate change and how much is due to fossil fuel combustion and other anthropogenic activities such as deforestation, and what are the potential effects on climate? The answer is that no one knows exactly. Despite the uncertainty of the effect of anthropogenic CO_2 on global warming, global temperatures during the last half of the 20th century exceeded those of the previous millennium, including the Little Climate Optimum. Moreover, the years from 2001 to 2010 are all the warmest on record. The effects of this warming—anthropogenic or not—may well dominate climate during the next few centuries or longer. Not only will developed nations be affected, but so too will underdeveloped and developing ones. What will be the possible consequences of climate change for natural ecosystems, economic policy, nationalism, poverty, famine, terrorism, and war?

17.5.1 Carbon dioxide: Temperature rise and ocean acidification

Data analyses and climate models indicate that as Earth warms, climate is more likely to fluctuate dramatically on

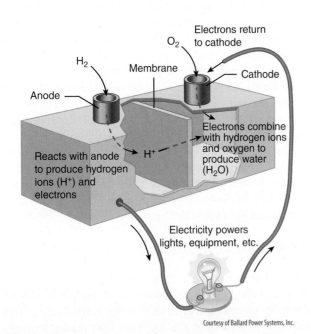

O₂

Electrons return to cathode

H₂

Membrane

Cathode

Anode

Reacts with anode to produce hydrogen ions (H^+) and electrons

H^+

Electrons combine with hydrogen ions and oxygen to produce water (H_2O)

Electricity powers lights, equipment, etc.

Courtesy of Ballard Power Systems, Inc.

FIGURE 17.13 How a hydrogen fuel cell works.

time scales less than a decade long. Thus, one possible effect of global warming will be more intense, more frequent, and longer lasting heat waves in the United States and Europe beginning in the second half of the 21st century. Climate models indicate that these heat waves are associated with a particular atmospheric circulation pattern that is intensified by global warming.

Other climate models have developed scenarios for future $2XCO_2$ and $4XCO_2$ worlds in which atmospheric CO_2 levels are doubled or quadrupled, respectively, from pre-industrial values of about 280 ppm (**Figure 17.14**). Both scenarios predict peak CO_2 levels and temperatures 200 to 300 years in the future. The $2XCO_2$ scenario indicates a 2.5°C rise by 2200 AD, whereas the $4XCO_2$ scenario indicates a 5°C rise. Despite their differences, both scenarios predict temperatures well above the approximately 0.5°C rise of the late 20th century.

Assuming that the projections are correct, the $4XCO_2$ world will resemble the warm conditions of the Paleocene–Eocene world, whereas the $2XCO_2$ world will be intermediate between the warmer conditions of the Paleogene and the glacial conditions of the Neogene (see Chapters 14 and 15). However, given the uncertainties of projected temperatures, which are based on CO_2 levels inferred from the geologic record and climate models, future temperature rises could range from a much smaller 1.5°C rise all the way up to 8 to 10°C! In any case, most temperature change due to anthropogenic warming will occur at higher latitudes because the tropics are already quite warm and cannot absorb much more heat (**Figure 17.15**).

Are there natural sinks that can absorb the CO_2 and reduce its accumulation in the atmosphere? Yes, but not any time soon, at least on human time scales. As we have already seen from examples of Earth's ancient history, continental weathering acts far too slowly, over millions to tens of millions of years, to have a significant effect on CO_2 emissions.

What about the oceans? Measurements of surface waters indicate that about 25% to 30% of anthropogenic CO_2 has been absorbed by the surface ocean. The surface waters eventually sink to greater depths. There, the CO_2 dissolves in water to produce carbonic acid (H_2CO_3), which, in turn, dissolves calcium carbonate ($CaCO_3$), causing the calcite compensation depth (CCD) to rise. However, given that the oceans have only taken up about 25% to 30% of anthropogenic CO_2 and the circulation rate of the oceans is on the order of 1,000 years, it will probably take several millennia for the oceans to take up all the anthropogenic CO_2 that will accumulate in the atmosphere.

In the meantime, CO_2 will continue to dissolve into the surface waters of the ocean, causing them to acidify. This appears to already be affecting coral reefs, which are being affected by other anthropogenic factors, as well (see below), and calcareous plankton, which are very sensitive to CO_2 levels. Recall the Paleocene-Eocene Thermal Maximum (PETM) discussed in Chapter 14. If calcareous plankton are dissolving before they reach the ocean bottom, the deep oceans might be unable to take up excess CO_2 as just described. The loss of both of these taxa could exacerbate

warming. Blooms of the modern coccolithophore *Emilania huxleyi* can cover areas greater than 100,000 square kilometers and produce significant amounts of the compound dimethyl sulfide, which, in turn, seeds cloud formation. Clouds of course reflect sunlight back into space because they increase the Earth's albedo, cooling the planet. Without coccolithophores, then, the Earth would absorb more solar energy than it already does.

Another possible sink is terrestrial biomass. Indeed, some workers have suggested that as CO_2 levels continue to rise, the process of CO_2 fertilization will begin to counteract rising CO_2 emissions. According to **CO_2 fertilization**, rising CO_2 levels will actually stimulate photosynthesis because land plants are presently not photosynthesizing at "full capacity." There is some evidence that CO_2 fertilization is happening, but it is not conclusive. Field experiments involving the exposure of small patches of forest and other terrestrial ecosystems exposed to elevated CO_2 for extended periods show a rapid decrease in CO_2 uptake after the initial fertilization effect. This suggests terrestrial plants could initially serve as a temporary sink for increasing CO_2, but then atmospheric CO_2 levels would begin to increase again. The extent of CO_2 fertilization also appears to vary from one region to another. For example, variation in the nutrient availability and moisture content of soils and atmospheric CO_2 affect CO_2 uptake by terrestrial vegetation. And although CO_2 is stored long term in woody tissues, it also accumulates in leaves, which decay rapidly once shed; such rapid decay can limit CO_2 uptake by terrestrial vegetation. Another mitigation technique currently being proposed is **CO_2 sequestration** in underground storage reservoirs (**Figure 17.16**).

On the other hand, the **iron hypothesis** suggests that biologic uptake of CO_2 could be accelerated by adding iron to the photic zone of the oceans (**Figure 17.17**). Iron is a significant biolimiting nutrient of photosynthesis by marine phytoplankton (see Chapter 3). Field experiments on iron fertilization in the Pacific Ocean have demonstrated that iron is indeed a limiting nutrient, even in upwelling regions; when iron was added to relatively small areas of surface waters (~100 square kilometers), phytoplankton blooms occurred. However, there can be significant drawbacks to iron fertilization. For one thing, the addition of iron to the ocean might be impractical on large scales. Also, artificial ocean fertilization might affect biologic communities in unintended ways because of the larger rain of organic matter to the ocean bottom. Moreover, the organic matter would decay as it settled, releasing CO_2 to deep water masses. These water masses would eventually upwell to the surface, releasing the CO_2 back to the atmosphere. More recent studies suggest another problem with this approach: that it might stimulate blooms of particular phytoplankton such as dinoflagellates that produce known neurotoxins. Such blooms could affect coastal fisheries and associated populations of vertebrates such as sea lions and birds. These blooms are thought by some to have triggered the abnormal behavior of birds along the California coast in 1961 that served as the basis for Alfred Hitchcock's film, *The Birds*, released in 1963.

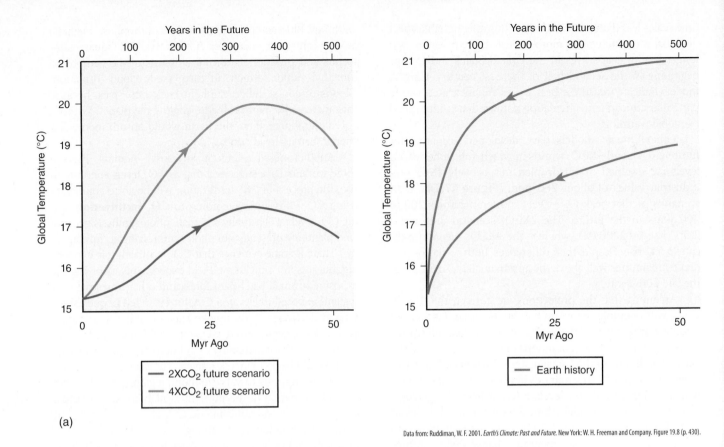

(a)

2XCO₂ future scenario
4XCO₂ future scenario

Earth history

Data from: Ruddiman, W. F. 2001. *Earth's Climate: Past and Future.* New York: W. H. Freeman and Company. Figure 19.8 (p. 430).

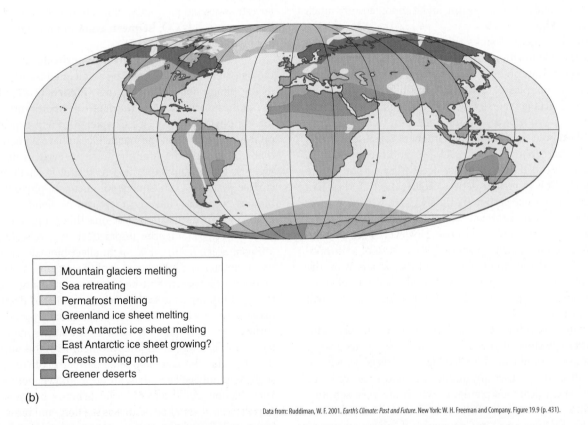

Mountain glaciers melting
Sea retreating
Permafrost melting
Greenland ice sheet melting
West Antarctic ice sheet melting
East Antarctic ice sheet growing?
Forests moving north
Greener deserts

(b)

Data from: Ruddiman, W. F. 2001. *Earth's Climate: Past and Future.* New York: W. H. Freeman and Company. Figure 19.9 (p. 431).

FIGURE 17.14 **(a)** Predicted CO_2 concentrations and global temperatures in $2XCO_2$ and $4XCO_2$ worlds as compared to those levels reached earlier in Earth's history. **(b)** Effect of $2XCO_2$ levels on the world's biomes. This world will resemble that of the Pliocene or Late Miocene of 5–10 million years ago. There will be less sea ice and permafrost and fewer and smaller alpine glaciers, whereas deserts might become greener because of shifts in precipitation patterns.

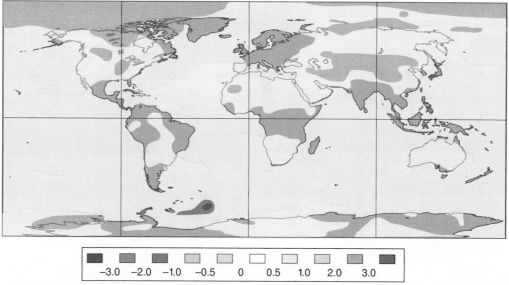

Data from: Houghton, J., Ding, Y., Griggs, D. J., Noguer, M., van der Linden, P. J., Dai, X., Maskell, K., and Johnson, C. A. 2001. *Climate Change 2001: The Scientific Basis*. Cambridge, UK: Cambridge University Press. p. 535.

FIGURE 17.15 Global warming will cause temperatures at higher latitudes to increase. The figure is the mean of three different model runs using slightly different starting conditions.

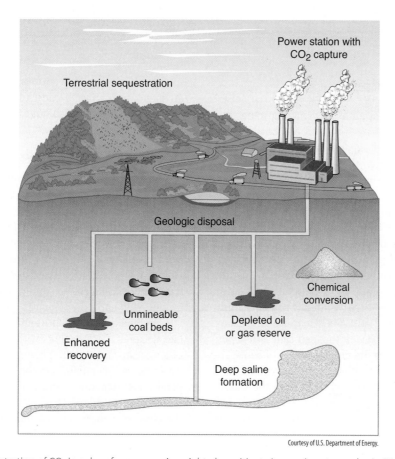

Courtesy of U.S. Department of Energy.

FIGURE 17.16 The sequestration of CO_2 in subsurface reservoirs might also mitigate increasing atmospheric CO_2 levels.

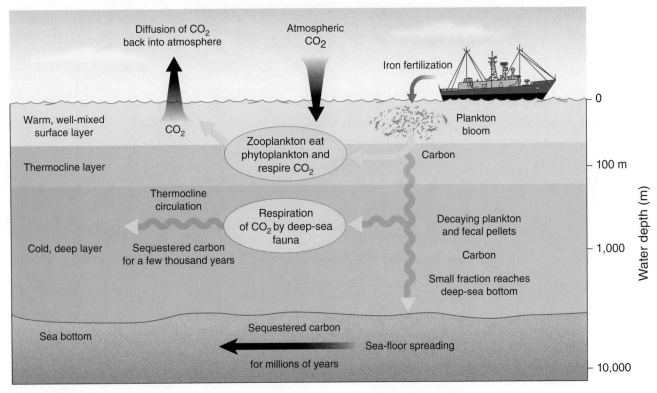

FIGURE 17.17 Iron fertilization might eventually help mitigate increasing atmospheric CO$_2$ levels.

17.5.2 Sea-level rise

One obvious consequence of anthropogenic warming is of course accelerated sea-level rise. Sea-level rise can result in part from the warming and expansion of the oceans, but most would likely result from the melting of the ice caps. Antarctic ice sheets have been collapsing for several decades, and it has recently been documented that Arctic ice sheets are also melting. Mountain glaciers, which are retreating all over the world, might also be harbingers of the future.

Sea level is currently rising at a rate of about 3 mm per year, depending on location. Although 3 mm per year does not sound like much, when multiplied by a decade, the total rise is 30 mm (3 cm), or a little more than an inch. Given that coastal regions such as the river deltas of Bangladesh, not to mention the Atlantic and Gulf Coastal plains, have very gentle slopes, sea-level rise in these so-called "hotspot" coastal regions will resemble the rapid flooding of continental margins by an epeiric sea, although on a smaller spatial scale. In fact, the Atlantic coast stretching from the Carolinas to Massachusetts has been deemed a "hotspot" of sea-level rise because, based on tide-gauge records from 1950 to 2009, sea-level is rising three to four times faster there than it is globally.

Coastal regions are often heavily populated, and in some countries extensively developed, so that sea-level rise will potentially cause severe economic and personal hardship. Solutions include engineering features such as sea walls and the pumping of sand from offshore onto beaches to replenish them after erosion. However, these solutions are largely temporary because of persistent sea-level rise and erosion, and they are also quite expensive.

17.5.3 Storms

Although most temperature change due to anthropogenic warming will occur at higher latitudes, even a slight warming of the tropics can yield sufficient heat to drive the production of a larger number of more intense storms or an increase in the frequency or intensity of other storm-related phenomena such as El Niño (**Figure 17.18**; see Chapter 16). Coupled with sea-level rise, storms will only increase the amount of coastal erosion and economic destruction.

El Niño and other climatic oscillations alter weather on large spatial scales in the Americas, Europe, and elsewhere (see Chapter 16). Exactly how these climatic oscillations and other ocean–atmosphere phenomena will respond to global warming is uncertain. Geologic records from ice cores, coral reefs, tree rings, and anthropogenic shell accumulations called **middens** (basically, refuse heaps) found along the coast of South America indicate significant changes in El Niño on longer time scales. These records suggest that the frequency of El Niño events might have increased about 7,000 years ago, and especially by 3,200 to 2,800 years ago, when temples along the Peruvian coast were abandoned. More recent studies that extend back through the last glacial maximum of the Pleistocene to about 70,000 years ago indicate that during cool intervals El Niño–like conditions

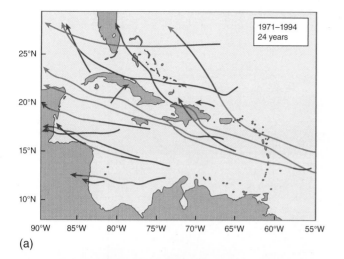

(a)

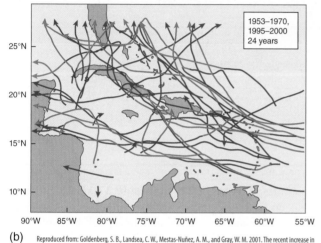

(b)

Reproduced from: Goldenberg, S. B., Landsea, C. W., Mestas-Nuñez, A. M., and Gray, W. M. 2001. The recent increase in Atlantic hurricane activity: Causes and implications. *Science, 293*, 474–479. Reprinted with permission from American Association for the Advancement of Science (AAAS).

FIGURE 17.18 The frequency of major hurricane landfalls along the U.S. east coast during **(a)** cooler and **(b)** warmer years.

dominated for millennia (not just a few years), whereas more "normal" El Niños or cooler La Niñas dominated during brief warm intervals.

17.5.4 Fisheries

El Niño–related phenomena—especially water temperatures—off the coast of South America cause the massive die-offs of plankton, fishes, and their predators and the collapse of economies related to fishing. Increasing evidence also points to the effects of other atmospheric oscillations like El Niño on marine ecosystems in the North Atlantic. Compounding any natural effect of the North Atlantic oscillations or other climate oscillations on marine ecosystems and fisheries is the finding that over the past half-century populations of most of the large predatory fish such as tuna, swordfish, and marlin have declined by as much as 90% as the result of commercial overfishing (**Figure 17.19**). Consequently, beginning in 2007 the U.S. Congress imposed catch limits on overfished species and populations of a handful of species, but so far only a small number of them have begun to rebound. Bluefin tuna, haddock, albacore, and gag have begun to rebound in certain regions.

17.5.5 Precipitation patterns

Shifts in precipitation patterns will also likely occur in response to warming (**Figure 17.20**), with obvious consequences for the global hydrologic cycle. Shifts in precipitation patterns might adversely affect agriculture in this country. Thus, the midwestern United States might no longer be the "breadbasket of the world"; this development would, like petroleum supplies, have serious consequences for sociopolitical stability in different parts of the world.

Portions of this region, including much of Nebraska and western Kansas, Oklahoma, and the panhandle of Texas, are underlain by the **Ogallala Formation**. The Ogallala consists of sands and gravels shed eastward from the Rockies during the Cenozoic Era (see Chapter 14). These sediments comprise the Ogallala Aquifer, which is the largest aquifer in the

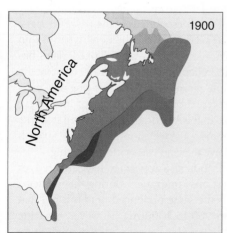

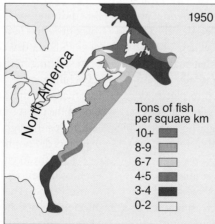

Tons of fish
per square km

10+	
8-9	
6-7	
4-5	
3-4	
0-2	

Data from: Malakoff, D., and Pauly, D. 2002, *Science, 296*, 458–461.

FIGURE 17.19 The collapse of fisheries off the Atlantic coast of the United States is attributed to overfishing.

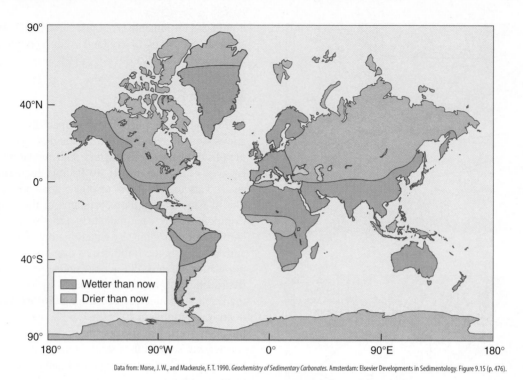

Data from: Morse, J. W., and Mackenzie, F. T. 1990. *Geochemistry of Sedimentary Carbonates.* Amsterdam: Elsevier Developments in Sedimentology. Figure 9.15 (p. 476).

FIGURE 17.20 Potential changes in soil moisture patterns on a warmer Earth.

United States, underlying about 450,000 square kilometers (~175,000 square miles) of the High Plains (**Figure 17.21**). The water in aquifers is "fossil water." The Ogallala's water accumulated during the past 10,000 to 25,000 years by the process of infiltration. With the advent of the gasoline-powered engine, farmers began pumping water out of the Ogallala at faster and faster rates. By the late 1970s, Ogallala water accounted for about 20% of all the irrigated land in the United States, and a large percentage of the nation's wheat, corn, alfalfa, and cotton was grown with its water. At present rates of extraction, there is perhaps only a few decades' worth of water left in the Ogallala and certainly less than a century's worth. A shift in rainfall patterns in response to global warming will only exacerbate the situation.

This is precisely what happened during the **Dust Bowl** of the 1930s, when topsoil was stripped away by gigantic dust storms (see Figure 17.21). This phenomenon was caused by severe drought exacerbated by decades of poor farming practices, such as deep plowing (which allowed soils to dry out) and lack of crop rotation, that maximized soil erosion. Farmland was abandoned as hundreds of thousands of people (known as "Okies") migrated to places like California, where the ongoing Great Depression presented them with few opportunities to make a living. The misery of these people was chronicled in painful detail in John Steinbeck's novel *The Grapes of Wrath.*

Long-term shifts in rainfall could also result in widespread drought. In fact, most people do not realize it, but potable (drinkable) water is becoming an increasingly scarce resource (**Figure 17.22**). Although world population "only" tripled from 1900 to 1995, global water demand increased six times during the same period! Severe impacts on water

availability and potability will increase as population growth continues (**Figure 17.23** and **Table 17.1**). Much of the demand for water occurred in response to increasing industrial activity and irrigation-intensive agriculture, which now supplies about 40% of the world's food crops. These activities taint drinking water supplies with industrial pollutants, pesticides, and fertilizers, along with sewage. Even in the developing world up to 90% of wastewater, including sewage, is discharged directly into rivers and streams without treatment. If current trends of population growth and water consumption continue, by 2025 at least 3.5 billion people (about half of the projected world population) will live in river basins where water scarcity affects household and economic activity.

Prolonged drought in the United States is by no means uncommon. Tree-ring records indicate the Lost Colony of Roanoke (1587–1589) and Jamestown (1606–1612) were established during some of the driest periods of the past 800 years; thick tree rings indicate relatively wet conditions, whereas thin rings indicate dry conditions. The western United States has been suffering from a drought that began in 1999 (despite the drenching rains in more recent years). But tree-ring data indicate that even more severe droughts have occurred in the past. One of the driest times in the American west occurred from the late 1500s through the 1600s, which might be why Spanish explorers did not believe the land was worth colonizing.

The Mayan civilization also appears to have suffered, or even collapsed, in response to drought during the Medieval Warm Period. This is the same prolonged mild interval that affected Europe from 900 to 1300 AD. However, tree-ring records indicate that a period of elevated aridity and drought occurred during this time throughout the western United States. Drought from about 750 to 900 AD in particular is

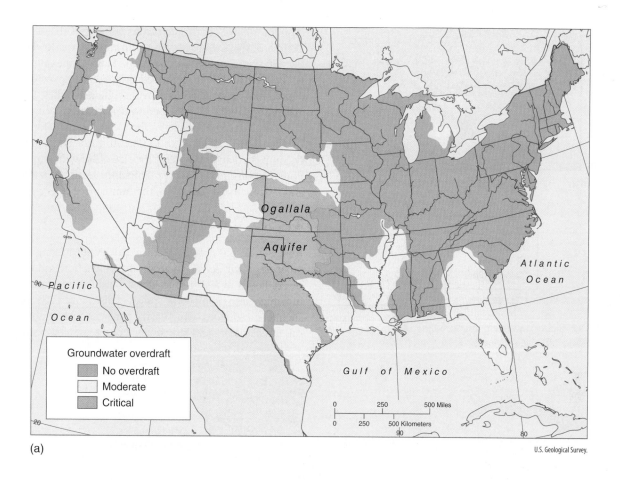

(a)

U.S. Geological Survey.

(b)

Courtesy of the Library of Congress, Prints & Photographs Division [reproduction number fsa-8b29516].

FIGURE 17.21 **(a)** Distribution of the Ogallala Formation in the subsurface of the High Plains of the United States. **(b)** The Dust Bowl. A photo montage of an approaching dust storm outside Stratford, Texas, in 1935.

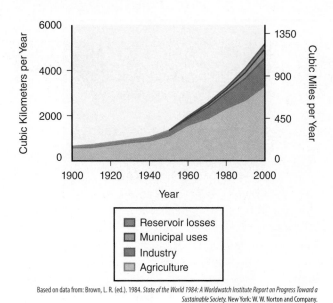

Based on data from: Brown, L. R. (ed.). 1984. *State of the World 1984: A Worldwatch Institute Report on Progress Toward a Sustainable Society.* New York: W. W. Norton and Company.

FIGURE 17.22 Global water demand is rising.

thought to have contributed to social stresses that led to the decline of the Classic Mayan civilization in the lowlands of the Yucatan Peninsula of Mexico; Mayan rulership might have finally been undermined when existing ceremonies and technologies failed to provide sufficient water. Drought is also thought to have helped to destabilize the civilization of the Anasazi (the "Ancient Ones") by about 1200 AD (see the frontispiece of Chapter 1). The Anasazi civilization once encompassed an area at least the size of New England in the Four Corners region where Colorado, Utah, New Mexico, and Arizona meet today (see the frontispiece of Chapter 1).

17.5.6 Disease

Ironically, despite drought predicted to develop in some regions, other areas will see an increase in precipitation and possibly severe floods in response to global warming. Increased moisture might also promote the spread of disease-carrying **vectors** such as mosquitoes, which carry parasites such as those that cause malaria. Increased frequency of disease, for example, is

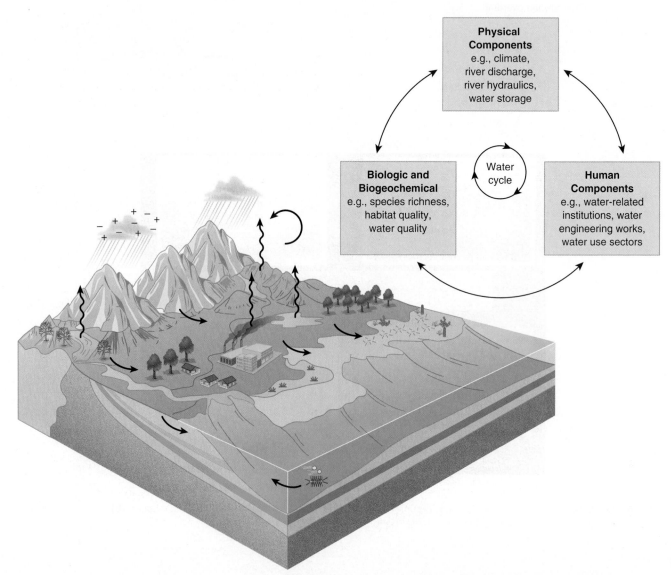

Data from: Vörösmarty, C. et al. 2004. Humans transforming the global water system, *Eos, 85*(48): 509 (Figure 1). Washington, DC: American Geophysical Union.

FIGURE 17.23 Human impacts on the global hydrologic cycle. See Table 17.1 for description of items in the figure.

TABLE 17.1

Examples of Major Anthropogenic Impacts on the Global Hydrologic Cycle* (Compare with Figure 17.23)		
Issue	Agents of Change	Impacts and Potential Global Significance
Climate change	Emissions of greenhouse gases lead to sustained rise in surface air temperature and other atmospheric changes through the instrumental record.	Possible accelerated hydrologic cycle; increased frequency of extreme events; loss of snow cover/mountain ice; more frequent, intense ENSO-like events; unknown threshold linking increased Eurasian runoff to potential shutdown of ocean circulating with rapid northern hemisphere cooling; sea-level rise affects sensitive coastal areas, such as deltas, that depend on a balance of river basin and ocean processes.
Basin-scale water balance changes	Climate change and variability; deforestation; wetland drainage; irrigation; evaporation in reservoirs; cooling water loss; interbasin transfers.	In heavily populated and irrigated basins, withdrawal can exceed river flow and water is reused many times, yielding public health and pollution problems; groundwater mining in arid areas and coastal salinity intrusion; water use/diversions create artificially dry river systems with severed connections to ocean, with global contributing basin area under the threat of ×10M km².
River flow regulation	Dams; interbasin transfers; locks; stream channelization, leveeing, human settlement of floodplains.	Pandemic waterworks reconfigure natural hydrographs; impoundment stabilizes flows for human use, but changes habitat and migration paths of aquatic organisms; globally 45,000 large and 800,000 small dams; area of basins heavily regulated rivals free-flowing systems; global 2- to 3-fold increase in residence time of continental runoff in river corridors.
Sediment fluxes	Elevated erosion from deforestation, grazing, agriculture, mining, construction; reduced flux via reservoir siltation and reduced river discharge.	Local-scale order-of-magnitude increases in erosion, depending on management and globally a factor 3 to 4 increase over natural state; loss of agricultural fertility upstream; habitat destruction, loss of reservoir investments through siltation; erosion of deltas and shorelines; artificial sediment retention 30% of natural global flux.
Chemical pollution	Industrialization, mining, urbanization without adequate treatment; industrial agriculture; atmospheric transport and deposition (acid rain).	Impacts across full land-to-ocean continuum; nutrient loads accelerate eutrophication; stoichiometric shifts yield coastal zone anoxia, harmful algal blooms, fisheries loss; xenobiotic distribution now global; worldwide river nitrogen transport to ocean increased 2- to 3-fold over pristine condition, with 10x increases in some industrialized regions.
Microbial pollution	Increases in fecal contamination from uncontrolled urbanization and animal husbandry.	Waterway pollution followed by downstream use yields major global public health threat; under control in most advanced countries; increasing faster than population in most regions, particularly around megacities.
Biodiversity changes	Pollution from agriculture, cities, mining; fragmentation of waterways, thermal pollution; introduction of exotics from globalization.	Threats to habitat and pollution have caused widespread species loss (e.g., >120 North American freshwater fish species since 1900); introduction of exotics changes local plant and animal communities; invasions by exotic species now widespread.

Data from: Vörösmarty, C. et al. 2004. Humans transforming the global water system, *Eos*, *85*(48): 520 (Table 1). Washington, DC: American Geophysical Union.

associated with modern El Niño events along the west coast of South America. More widespread pathogens might also affect other terrestrial species, not just humans.

17.6 Biodiversity and Extinction

Global warming will also alter habitat on regional to global scales. As predicted, tropical species are beginning to expand their geographic ranges into temperate latitudes; no one knows what the consequences of these species invasions will be for ecosystem stability. Recent evidence indicates significant range shifts of nearly 300 species toward the poles, averaging 6.1 kilometers (about 4 miles) per decade, and the advancement of spring seasonal events by an average of 2.3 days. Although these rates do not sound particularly rapid, over relatively long spans of time (even by human standards) some Arctic seabirds now breed nearly a month earlier than a few decades ago and some species of butterflies have shifted their biogeographic ranges northward by almost 60 miles in the last century.

Moreover, a large number of ecologists and paleontologists believe we are now in the midst of what they have termed the **Sixth Extinction**, an anthropogenic mass extinction that rivals the Big Five mass extinctions already discussed. The noted conservation biologist, Edward O. Wilson, once estimated the number of species that disappears *each year at 27,000*; *each day*, 74; and *each hour*, 3. However, his estimate was optimistic because it was based only on the species that are known to have become extinct. Another calculation indicates that at current rates of extinction, *all* terrestrial species will disappear within 200 years.

One of the biggest problems facing conservation biologists in documenting and preventing extinction is that species are disappearing faster than they can find and describe them, especially in the tropics. Indeed, tropical rain forests are among the most diverse of all ecosystems on the planet. There are roughly 1.4 million recorded species, most of which live in tropical rain forests, and recent estimates place the total number of species at 8.7 million. When species disappear as rapidly as they are now, the chances of collapse of ecosystems is greatly increased. Moreover, based on the fossil record it might take many thousands of years or longer for the biosphere to recover from a mass extinction. This amount of time totals more than all the generations of humans that have occurred since the beginnings of ancient civilization.

One thing that all extinctions share in common, no matter what the mechanism or how unusual it is, including the present one, is the loss of habitat. The number of species (S) present in an area (A) is directly proportional (α) to the area. This relationship is called the **species–area relationship**:

$$S \alpha A \qquad (17.1)$$

The species–area relationship basically states that the larger the area, the more species that live in it. This relationship was used to make the estimates of aforementioned extinction rates. Despite the mathematical simplicity of the species–area relationship, it is extremely complex because habitat area represents a large number of interacting factors: diversity of habitat types, number of niches, and nutrient and food availability, among others, all of which interact with environmental tolerance, biogeographic distribution, dispersal ability, and other traits (see Chapter 3).

According to ecologic studies, localized populations called **demes**, or **peripheral isolates**, exist on the edge of the main parental population (see Chapter 5). These smaller populations come and go like blinking Christmas tree lights, even without any disturbance. If the peripheral populations of a species are lost, this leaves only the main parent population. If the habitat of the main population shrinks sufficiently, the main population can become so small that random fluctuations in its size might cause it to "blink off" permanently. Small populations can also suffer from decreased genetic variation because of inbreeding and the spread of deleterious recessive mutations that further decrease the chances of survival.

The primary effect of humans on biodiversity is habitat loss resulting from the growth and spread of human populations (**Figure 17.24**). Human population growth has entered its most rapid, or exponential, phase (see Chapter 1). It took tens of thousands of years to reach the first 1.6 billion people by the year 1900. During the 20th century, more people were added to the human population than in all of humanity's previous history. By 1950 the population had grown to 2.5 billion. Persons born in 1950 then saw the world's population double from 2.5 billion to 5 billion by 1990.

It has been predicted that by the year 2050 the world's human population will have doubled from about 5.3 billion in 1990 to 10 billion. Although human population growth has slowed somewhat, 200,000 people, or the equivalent of a small city, are still added each day. Unlike populations of plants and animals in the wild, then, human population growth is beginning to resemble that of bacteria. In fact, if 20th-century rates of population growth had occurred since the beginnings of agriculture thousands of years ago, Earth would now be covered in a squiggling mass of human flesh, thousands of light years in diameter and expanding outward with a velocity many times greater than the speed of light.[1]

Most of the population growth is occurring, and will continue to occur, in underdeveloped countries. Many of these countries are located in the tropics, where biodiversity is greatest (**Figure 17.25**; see Chapter 3); thus, biodiversity is most likely to be degraded in tropical terrestrial and marine environments.

---

[1]Paraphrased from McNeill, J. R. 2000. *Something new under the sun: An environmental history of the twentieth-century world.* New York: W.W. Norton, p. 9.

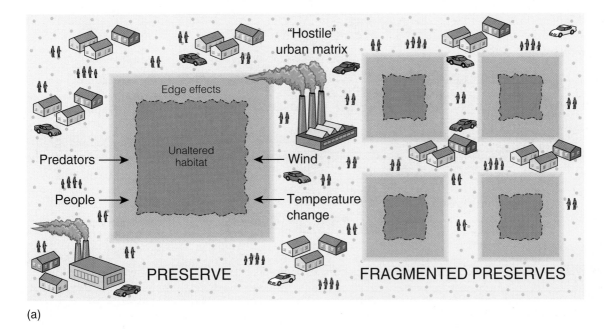

(a)

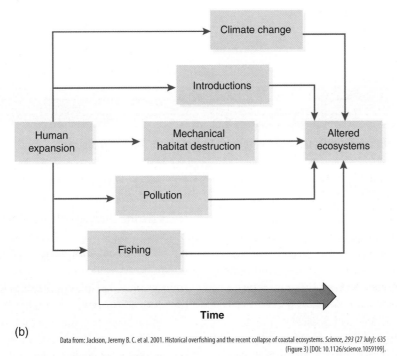

(b)

Data from: Jackson, Jeremy B. C. et al. 2001. Historical overfishing and the recent collapse of coastal ecosystems. *Science*, *293* (27 July): 635 (Figure 3) [DOI: 10.1126/science.1059199].

FIGURE 17.24 (a) Anthropogenic processes affecting habitat area and fragmentation. Although the total amount of area remains the same in both the left and right halves of the figure, the habitat area on the right is much more greatly affected by what are called "edge effects." Compare the chapter frontispiece. **(b)** Anthropogenic effects on habitat area.

However, there is little incentive for people in many underdeveloped countries to care about the state of the environment when they can barely eke out a living for themselves and their families (**Box 17.2**). In some cases, governments have encouraged families to leave poverty-stricken cities and move to the country to farm. Farming in these regions often involves **slash-and-burn** techniques to clear tropical forests (refer to this chapter's frontispiece). As we have already seen, these soils are nutrient-depleted (see Chapter 4). The soils are quickly exhausted and the farmers move on to new areas,

where the entire process is repeated. To aggravate matters further, when large areas of rain forest are cut down, drought ensues; transpiration normally perpetuates precipitation and the hydrologic cycle through positive feedback. Deforestation and colonization can also destroy the cultures of indigenous tribes.

Among marine habitats, coral reefs by far and away harbor the greatest biodiversity. Reef ecosystems all over the world are being destroyed. In fact, because the number of species living on reefs are highly diverse and tend to have

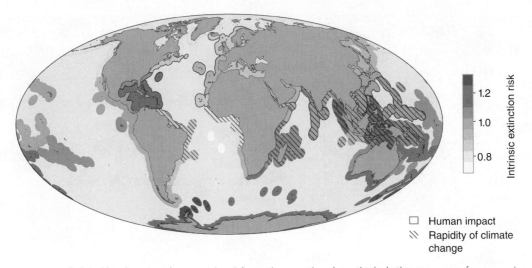

FIGURE 17.25 Hotspots of global biodiversity where species richness is exceptional, particularly the presence of rare species.

BOX 17.2 Tropical Biodiversity and Tropical Economies

It is now estimated that the world harbors somewhere between 8 and 10 million species. Tropical regions harbor most of the world's biologic diversity in rain forests and coral reefs. These same regions are also at risk of environmental degradation because of high human population densities. Unfortunately, the same factors that promote high biodiversity in the tropics also promote high population densities and adversely affect the national economies of tropical countries.

The overriding factor is climate, which is a function of atmospheric circulation and the hydrologic cycle. In tropical regions, climate has adversely affected soil fertility, birth and death rates, and access to cheap transportation. For example, tropical soils are easily depleted of nutrients because of warm temperatures and rainfall, both of which accelerate chemical weathering and the leaching of nutrients. Also, in tropical regions with extreme wet and dry seasons, farmers must also contend with both floods and drought. Both phenomena curtail agricultural production. Lowered agricultural production tends to lower the size of urban populations, which in turn hinders technologic advances in manufacturing and services.

Large portions of many tropical regions are also landlocked. Unfortunately, seaways and navigable rivers are the cheapest methods of transportation of goods and services. Thus, many tropical nations have a low **Gross Domestic Product**, which is the monetary value of all the goods produced and services provided in a country for a specified period of time (**Box Figures 17.2A** and **B**).

Tropical regions also tend to be characterized by a high incidence of disease. Malaria, for example, is transmitted by mosquitoes, which lay their eggs in quiet bodies of water. Widespread disease and early mortality hinder a nation's economic performance by reducing worker productivity. Societies with high incidences of child mortality also tend to have high birth rates because high birth rates tend to guarantee that at least some children survive to adulthood. However, with so many children families cannot afford to invest in education, and the role of women in society is hindered because they are busy with family responsibilities. Such countries often do not have access to family counseling services or contraceptives, which are often doled out at the political whims of the more developed nations, especially those of the United States.

BOX 17.2 Tropical Biodiversity and Tropical Economies (Continued)

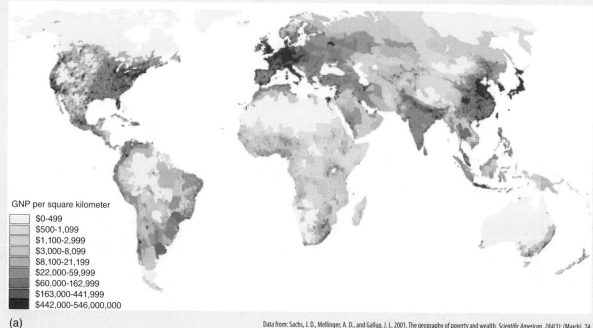

GNP per square kilometer

- $0-499
- $500-1,099
- $1,100-2,999
- $3,000-8,099
- $8,100-21,199
- $22,000-59,999
- $60,000-162,999
- $163,000-441,999
- $442,000-546,000,000

(a)

Data from: Sachs, J. D., Mellinger, A. D., and Gallup, J. L. 2001. The geography of poverty and wealth. *Scientific American, 284*(3): (March), 74.

BOX FIGURE 17.2A Comparison of **(a)** Gross National Product (GNP) density versus **(b)** major climate zones. GDP density is a measure of economic activity by area and shows the impacts of geography on the economy of an area. It is calculated by multiplying GDP per capita (per person) of an area by the population density of that area and is expressed as GDP per square kilometer. Note the concentration of people and GDP density along coasts with harbors and navigable waterways leading to the oceans. Note also the high GDP density of the eastern half of the United States and the west coast compared to the rest of the country, much of which occurs at relatively high elevations. The GDP density of the eastern half of the United States resembles that of Europe and eastern Asia.

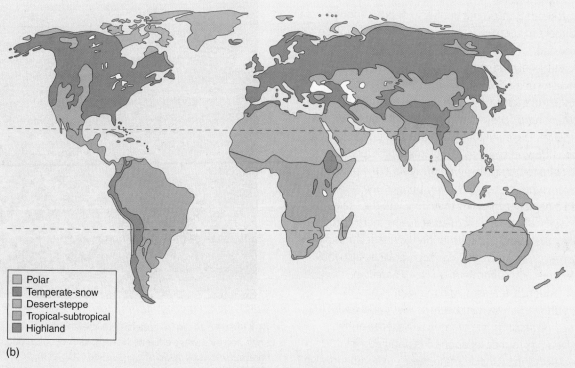

- Polar
- Temperate-snow
- Desert-steppe
- Tropical-subtropical
- Highland

(b)

BOX FIGURE 17.2B Major climate zones. Note the general correspondence to GDP density.

fairly narrow environmental tolerances for factors such as water temperature, light, and nutrient levels, reefs are sensitive indicators of environmental change. Reefs may well represent a kind of "*miner's canary,*" an analogy of the use of canaries to warn of poisonous or explosive gas leaks in mines; the birds quickly died from inhalation of the gas, warning the miners to get out.

The degradation of reefs has typically been attributed to excessive sedimentation and nutrients from sewage, fertilizers, deforestation, and runoff all associated with human encroachment. Similar impacts are occurring in estuaries and coastal waters (**Box 17.3**). These processes all contribute to one result: **eutrophication**, or nutrient enrichment. Under eutrophic conditions, plankton populations increase dramatically; the waters become turbid, which decreases light penetration; and opportunistic taxa that feed on the expanding plankton populations invade the reef and outcompete the slower-growing corals. Also contributing to the degradation of reefs is overfishing and the collection of shells and coral for sale in tourist shops.

17.7 Closer to Home

F "Well," you say, "what does this have to do with me? Most biodiversity loss is occurring in the tropics, far away from where I live."

Many of the species that have or will go extinct in the future do indeed reside in the tropics. Thus, destruction of tropical habitat is likely to cause far more extinction than in temperate regions, such as the United States, simply because tropical regions harbor greater biodiversity. For example, there are only a few stands of virgin woodlands left in the United States, among them the Adirondacks (upper New York State), the Great Smoky Mountains (eastern Tennessee), and parts of the western United States. The rest is mostly secondary and tertiary growth. However, only four species of

BOX 17.3 Marginal Marine Environments and Human Disturbance

Marginal marine environments are quite fragile. Because they lie at the transition between land and sea, marginal marine environments are subject to even small changes in sea level. Thus, marginal marine environments are at risk to sea-level transgression caused by anthropogenic warming and the melting of ice caps. If sea level rises too fast because of an anthropogenic sea-level rise, marginal marine environments might not unable to deposit enough sediment to keep up with the sea surface and will be "drowned."

Marginal marine environments such as coral reefs are being degraded by other human activities (**Box Figure 17.3A**). In coral reef environments nutrient runoff promotes dense populations of plankton. The plankton serve as food for large populations of sponges and clams, which feed on the plankton, overwhelming the reefs in the process. Certain types of these invasive sponges and clams are also *bioeroders*, meaning they bore into the skeletons of corals, weakening them and causing them to topple, killing the corals and contributing to the decimation of the reef. Suspended sediment in runoff can also foul the corals and suffocate them (see Chapter 4).

Estuaries, marshes, and mangroves are exceptionally important because they serve as breeding grounds for commercial fisheries and as feeding grounds for enormous populations of migratory birds. Sediments in these environments also

© iStockphoto/Thinkstock.

BOX FIGURE 17.3A A modern coral reef that has been impacted by sewage effluent. Note the lack of coloration of the coral skeletons by symbiotic algae (see Chapter 4).

BOX 17.3 Marginal Marine Environments and Human Disturbance (Continued)

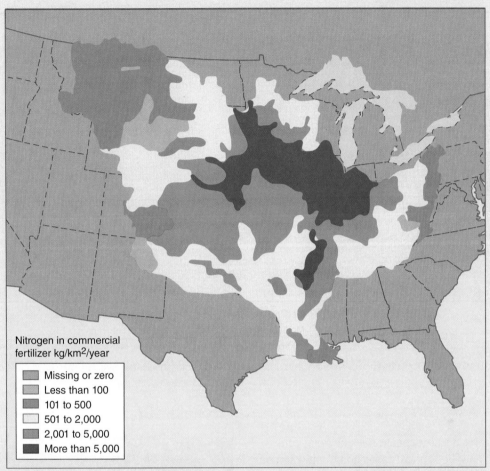

Data from: Raloff, J. 2004. Limiting dead zones: How to curb river pollution and save the Gulf of Mexico. *Science News, 165*(24)(June 12): 378.

BOX FIGURE 17.3B Nitrogen in commercial fertilizer (in kilograms per square kilometer per year) applied by farmers to croplands lying within the Mississippi River watershed. Approximately 15% of the applied nitrogen drains into rivers that feed the Mississippi, which in turn drains into the Gulf of Mexico.

filter out pollutants such as pesticides from contaminated waters before they reach the sea. Nevertheless, environments of this sort are often regarded as a nuisance because they harbor large populations of insects (such as mosquitoes). As a result, these environments are often drained, destroying them.

Several major marginal marine environments have been heavily affected by human activities (**Box Figures 17.3E**). The Mississippi Delta has been subject to extensive canal development and dredging to tow oil rigs to and from the Gulf of Mexico. The canals and dredging have permitted saltwater to intrude farther and farther inland as the delta drowns. On the east coast of the

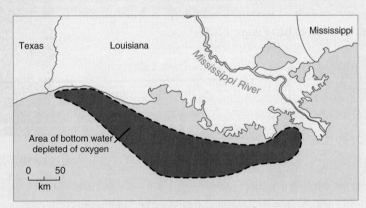

HYPOXIC WATER OFF LOUISIANA

Data from: U.S. Congress, Office of Technology Assessment. 1987. *Wastes in Marine Environments*, OTA-0-334. U.S. Government Printing Office.

BOX FIGURE 17.3C The "dead zone" in the Gulf of Mexico caused by nutrient-rich runoff from the Mississippi River. In recent years, this dead zone has reached about the size of New Jersey.

(continues)

BOX 17.3 Marginal Marine Environments and Human Disturbance (Continued)

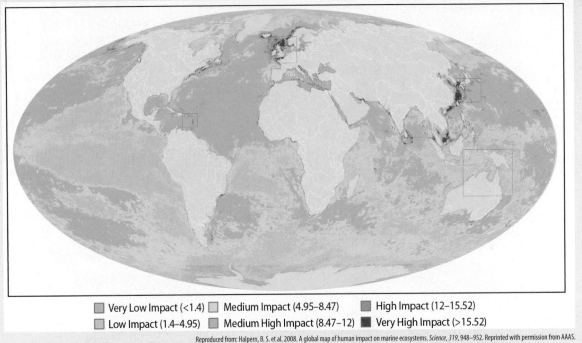

☐ Very Low Impact (<1.4) ☐ Medium Impact (4.95–8.47) ■ High Impact (12–15.52)
☐ Low Impact (1.4–4.95) ☐ Medium High Impact (8.47–12) ■ Very High Impact (>15.52)

Reproduced from: Halpern, B. S. et al. 2008. A global map of human impact on marine ecosystems. *Science, 319,* 948–952. Reprinted with permission from AAAS.

BOX FIGURE 17.3D The frequency of red tides and other harmful algal blooms (HABs) is reported to have increased in recent years, possibly as the result of nutrient runoff from land and coastal pollution.

United States the Chesapeake Bay and many other estuaries have been heavily affected by deforestation, pollution, and fishing ever since colonists came to North America. Deforestation has resulted in erosion, sedimentation, and increased inputs of nutrients to the bay (**Box Figure 17.3F**). On the West Coast, estuaries like San Francisco Bay have also been used as landfills (**Box Figure 17.3G**).

Blooms of planktonic algae can also occur in marginal marine environments. These blooms are thought to be caused by the runoff of nutrients from agricultural and industrial operations, sewage effluent, and deforestation. Such blooms can include the *red tides* caused by dinoflagellates. The plankton blooms use up available oxygen, killing off many different kinds of organisms. In fact, nutrient runoff has resulted in more than 30 so-called *dead zones* in the oceans all over the world, including off the coasts of Texas and Louisiana, in the Gulf of Mexico, and Oregon.

Courtesy of P. Alejandro Diaz.

BOX FIGURE 17.3E A red tide.

BOX 17.3 Marginal Marine Environments and Human Disturbance (Continued)

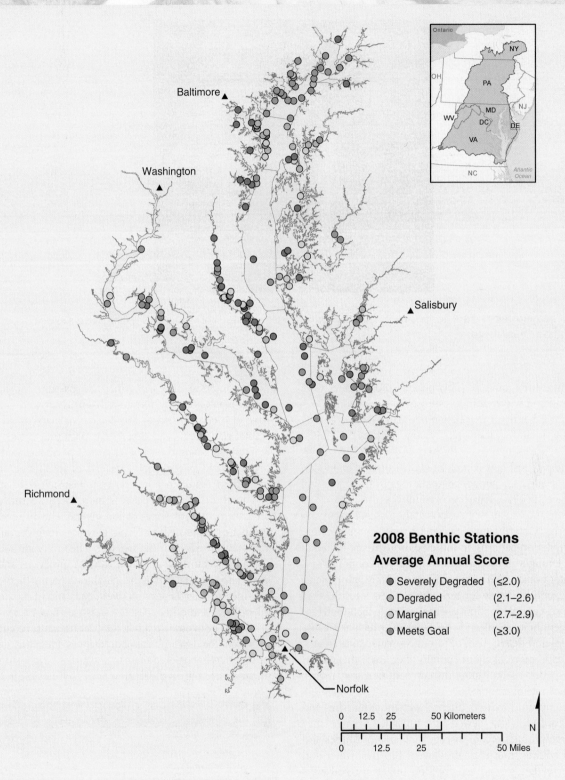

Baltimore ▲

Washington ▲

Richmond ▲

Salisbury ▲

Norfolk ▲

2008 Benthic Stations

Average Annual Score

- ● Severely Degraded (≤2.0)
- ○ Degraded (2.1–2.6)
- ○ Marginal (2.7–2.9)
- ● Meets Goal (≥3.0)

0   12.5   25   50 Kilometers

0   12.5   25   50 Miles

N

BOX FIGURE 17.3F Chesapeake Bay has been heavily affected by human activity, including activity occurring hundreds of miles away because the bay receives waters from the entire Susquehanna River drainage basin.

(continues)

BOX 17.3 Marginal Marine Environments and Human Disturbance (Continued)

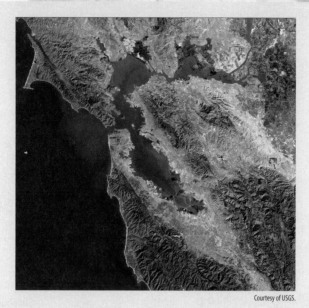

Courtesy of USGS.

BOX FIGURE 17.3G Much of San Francisco Bay has been filled in as result of human activity (gray and gray-green areas along the coasts).

birds went extinct, quite possibly because most species were widespread and survived in forested areas elsewhere.

On the other hand, amphibian populations are declining worldwide, including temperate regions such as the United States. The factors involved in the decline of amphibian populations indicate just how complicated the causes of extinction can be. Elevated sea surface temperatures in the equatorial Pacific since the 1970s have been identified as the prime cause of declining amphibian populations in the Cascade mountain range of the western United States. Changes in temperature and precipitation patterns, in this case drought, are linked to El Niño. Drought in turn decreases the depths of ponds and other water bodies used as breeding grounds by amphibians, which depend on water for reproduction. As a result, eggs are more susceptible to ultraviolet radiation, which increases their susceptibility to infection by parasites, which also cause deformities and death. Also, runoff into smaller lakes, rivers, and ponds is more likely to contain toxic trace metals and pesticides from air and groundwater pollution that reduce the survival rate of embryos.

To compound matters further, wetlands (marshes and swamps) are typically viewed as a nuisance because they breed large populations of insects such as mosquitoes and flies and so are drained or destroyed. But freshwater and saltwater wetlands serve as stops on migratory pathways used by birds. Coastal wetlands also serve as natural filters by helping to purify waters before they enter estuaries, as breeding grounds for fish and invertebrates (many of which have commercial value), and to protect coasts from erosion.

Agriculture and the American passion for beautiful lawns maintained by nitrate-rich fertilizers and pesticides have also contributed to the decline of marshes and other coastal ecosystems (see Box 17.3). With the spread of well-maintained lawns also came increased pavement in the form of roads and parking lots. Instead of infiltration into the ground to maintain aquifers, runoff enriched in nitrogen and toxic organic chemicals is more likely to run directly off pavement into streams and rivers. The nitrogen then stimulates the growth of plants growing along the landward and seaward edges of the marshes, effectively shrinking the marsh. The nitrogen-enriched runoff from lawns and farmlands also stimulates massive plankton blooms that result in red tides and dead zones off river deltas and in estuaries and inland bays and other coastal waters (see Box 17.3).

It has been estimated that fertilizer use must be cut in half to prevent the development of dead zones. Programs are under study by conservation groups to encourage farmers to use less nitrate-rich fertilizers and to prevent less soil erosion and runoff.

CONCEPT AND REASONING CHECKS

1. Besides species area, how are humans degrading environments?
2. What is meant by the term "miner's canary"? Give possible examples.
3. Could the loss of seemingly innocuous species result in ecosystem collapse? If so, how (see Chapter 2)?
4. What is the root source of environmental degradation?

17.8 Why Should We Care, and What Can We Do?

G Why *should* we care about issues like environmental quality and biodiversity? Perhaps for any number of reasons, some of which are purely selfish. An old argument is that some species produce chemical compounds that are of great medical significance. However, After the medicinal properties of a natural compound are discovered, they can be copied and the species from which the compound was originally recovered is no longer needed. In the future, computer models and biotechnology will probably be able to synthesize the compounds we will need to fight disease.

But there are other important reasons:

1. *Ecosystem productivity rises in conjunction with diversity.* For example, the value of sustained (year-to-year) yields from tropical rain forests in terms of food, fiber, oil, and other marketable compounds far exceeds that from the production of timber by clear-cutting of forests.
2. *Related to productivity and biodiversity are* **ecosystem services**, *in which ecosystems are viewed in a utilitarian manner.* Such effects include, for example, the filtration effects of wetlands (see Box 17.3).
3. *Better stewardship of Earth's resources.* Most of the world's economic growth has occurred because of population growth. Moreover, much of the economic growth has been concentrated in developed nations, which have pursued policies of profligate energy use and pollution. For every person in the world to reach present levels of consumption in the United States, with current technology, would require four more planet Earths. The increase in the number of households due to urban sprawl in both developed and underdeveloped nations has resulted in higher per capita (per person) resource consumption all over the world. Thus, stewardship must occur in both developed and underdeveloped nations.
4. *Ethics.* For this basic reason, many scientists and an increasing number of religious leaders believe we really have no right, other than that conferred by us upon ourselves, to do what we are doing to the planet. Indeed, humans evolved by the same basic processes as all other species and are therefore a part of Earth's ecosystems, not above them.

The global community must consider the following:

1. *Third-world countries need the support of the developed world in overcoming the economic barriers imposed by climate.* Movement away from agricultural economies that are largely constrained by climate and diversification into manufacturing and service sectors will help.
2. *Changing consumer habits and lowering fossil-fuel consumption of developed nations (especially the United States) and developing ones (such as China) is a monumental task given that these countries are already comfortably ensconced in their lifestyles.*

3. *Human population growth must slow.* The good news is that population growth is slowing, at least in developed nations. And even in many developing nations, increased access to education has meant that women are increasingly opting to have fewer children and families are concentrating on the quality of life. Still, the human population will continue to swell in coming decades.

Several ongoing developments might help in this regard:

1. *Instead of using Gross Domestic Product as an indicator of economic growth, countries could use a* **genuine progress indicator**, *which includes estimates of the environmental costs of economic activity.* Already, a growing number of economists, politicians, and scientists who realize humans are rapidly approaching the planet's carrying capacity are advocating the use of the genuine progress indicator or similar measures. These indicators would factor biodiversity conservation into policies and decision-making frameworks for resource production and consumption while also increasing awareness of institutional and social changes needed to conserve biodiversity.
2. *Buy land and bar it from development.* This is happening increasingly in the United States and in underdeveloped countries strapped for currency, with the support of private foundations and philanthropists.
3. *Continued education and the instillation of a long-term environmental ethic in humans.* This may well be the most difficult approach. Short-term thinking might be part of humanity's evolutionary and cultural heritage. Before the great medical advances of the 20th century, high death rates encouraged high birth rates, increasing the chances of leaving children. Short-term thinking is also reinforced by the mass media and political and economic institutions. The effects of just one individual can be horrendous (**Box 17.4**) but positive effects can have just as much impact (**Figure 17.26**).

But if *Homo sapiens* ("prescient man") does indeed possess foresight and ingenuity, the world's biodiversity can get through what E. O. Wilson has called the narrow "bottleneck," or passageway, of the 21st century. This is when the increasing restriction and loss of habitats due to human population expansion will be at its greatest. After we pass through the bottleneck, human population expansion all over the world will have slowed considerably and extinction pressures should decrease. Even so, based on the fossil record, even after the biosphere has passed through the bottleneck, it will take many thousands of years or longer for new species to evolve and diversity to begin to increase.

CONCEPT AND REASONING CHECKS

1. How is habitat area related to other ecosystem functions such as energy flow, productivity, and biodiversity (see Chapter 2)?
2. How can one individual affect environmental change?

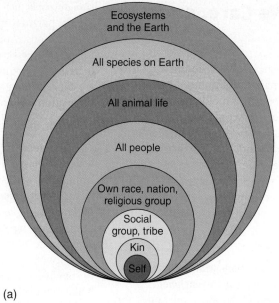

(a)

Data from: Noss, R. F. 1992. Issues of scale in conservation biology. In Fiedler, P. L., and Jains, S. K. (eds.) *Conservation Biology: The Theory and Practice of Nature Conservation, Preservation, and Management.* New York: Chapman and Hall.

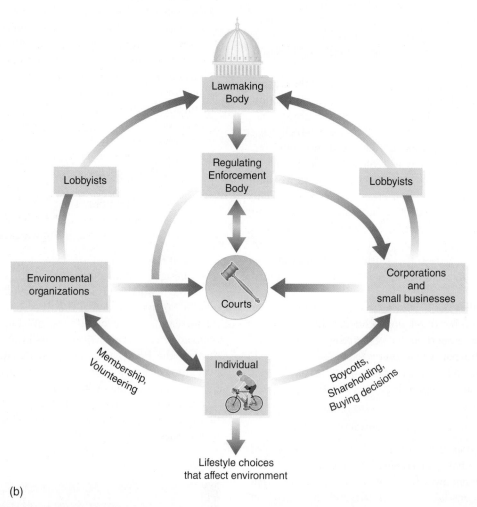

(b)

FIGURE 17.26 (a) An ethical sequence in which individual concerns reach outward from the self to larger and larger groups of persons. **(b)** The relationship of the individual to larger governing bodies, advocacy groups, and corporations.

BOX 17.4 The Individual and the Environment[1]

How much can one person affect the environment? A lot. Thomas Midgley (1889–1944) might have "had more impact on the atmosphere than any other single organism in Earth history."[2] Midgley was born into a family of inventors, grew up in Columbus and Dayton, Ohio, and received a degree in chemical engineering from Cornell University. By the time of his death Midgley held more than 100 patents and had won all the major prizes in chemistry in the United States. In 1921, while working for General Motors Research in Dayton, Midgley found that the addition of tetraethyl lead reduced engine knocking. A group of large corporations quickly formed the Ethyl Gasoline Corporation to market ethyl gasoline (or "ethyl," as it was widely known) beginning in 1923.

Lead was already known to be toxic in Midgley's time. Lead is a neurotoxin, affecting the brain and central nervous system, among other organs. With the introduction of ethyl, production workers suddenly began to exhibit symptoms of lead poisoning that included blindness, kidney failure, cancer, convulsions, and hallucinations that can lead to coma and death. Nevertheless, Midgley maintained that tetraethyl lead was perfectly safe. To allay the public's fears, he once conducted a public demonstration in which he washed his hands in tetraethyl lead and breathed the fumes for a full minute, knowing full well the consequences of long-term exposure because he had at one time exhibited symptoms of lead poisoning. The manufacturers of lead additives also funded studies devoted to showing tetraethyl lead was safe.

The downfall of leaded gasoline and lead-containing products began with another determined individual: Clair Patterson. This is the same Clair Patterson who was responsible for determining the age of Earth from radioactive isotopes (see Chapter 5). Patterson had continually noted unusually young ages in many of his rock samples due to lead. After many years he finally determined the source of the lead was the atmosphere and as much as 90% of the lead in the atmosphere might well come from automobile exhaust. Patterson then launched a campaign to stop the commercial use of lead, including in gasoline. For this, he was persecuted. Patterson had research contracts canceled and was excluded from committees that were ostensibly studying the health effects of lead, on which he was probably the foremost expert. The trustees of the California Institute of Technology, where Patterson worked, were also pressured by lead industry officials to fire Patterson.

Nevertheless, Patterson persevered. His efforts led to the passage of the *Clean Air Act* in 1970 and the banning of leaded gasoline in the United States in 1986. Today, we use unleaded gasoline instead. Almost immediately after the ban there was an 80% drop in the levels of lead in Americans' blood. Nevertheless, each American still carries about 625 times more lead in their blood than they would have a hundred years ago because of mining, smelting, and other industrial activities.

As for Midgley, he continued to wreak havoc on the environment. In the 1920s, he had discovered how to use *chlorofluorocarbons (CFCs)* such as Freon as coolants. By the 1930s, CFCs were being used in refrigerators and their use spread to air conditioners and deodorant sprays. Half a century later, CFCs were implicated in the destruction of Earth's ozone layer. A single kilogram of CFCs can destroy 70,000 kilograms of ozone. Ozone is also long-lasting, persisting in the atmosphere for up to 1 year, warming Earth. Thus, a single molecule of CFC is also about 10,000 times more efficient in greenhouse warming than a single molecule of carbon dioxide.

But Midgley never learned of this. Tragically, he contracted polio in 1940. He designed a system of ropes and pulleys to help him get in and out of bed but died of strangulation after entanglement in the rope network, "killed by the combination of bad luck and his own ingenuity."[3] Perhaps Midgley's fate is a metaphor for the affect of technologic achievement on the environment. Nevertheless, there is no turning back on technology, despite the potential unforeseen long-term environmental consequences.

The operative word here is "long term." No one had any idea that the lead in gasoline or CFCs would prove to be a problem after decades of use. More recently, in certain American markets where cuts in air pollution have been mandated, gasoline has been reformulated with a chemical called *methyl tert-butyl ether (MTBE)*. Although MTBE-formulated gasoline burns more cleanly, MTBE, which is a suspected carcinogen, is now being found in groundwater in these same markets. MTBE probably made its way into subsurface aquifers by spillage at gasoline stations and leaking tanks. As a result some states, including Ohio, New York, and Connecticut, have banned MTBE.

[1] Summarized from Bryson, B. 2003. *A short history of nearly everything*. London: Black Swan, pp. 193–205, and McNeill, J. R. 2000. *Something new under the sun: An environmental history of the twentieth-century world*. New York: W.W. Norton, pp. 111–113.

[2] McNeill, 2000, p. 111.

[3] McNeill, 2000, p. 113.

- Climate records strongly suggest that anthropogenic warming during the 20th century has accentuated natural global warming that began at the end of the Little Ice Age by roughly 0.5°C. The main cause of anthropogenic warming is thought to be the burning of fossil fuels and the production of CO_2, a greenhouse gas, although other gases such as methane (CH_4) have been implicated.
- The possible consequences of anthropogenic warming involve shifts in precipitation patterns, including drought and floods, and therefore shifts in agriculture; increased storm frequency and intensity; the spread of diseases that normally occur in the tropics; and political instability. Despite the development of alternative energy sources, humankind will continue to be heavily dependent on fossil fuels in coming decades. Moreover, the effects of fossil fuel combustion during the 20th century and coming years will persist for centuries.
- In the case of the ongoing Sixth Extinction, the main cause of habitat loss is explosive human population growth in underdeveloped countries, which tend to occur in the tropics, where the greatest biodiversity occurs. Based on the fossil record of extinction, the time it will take for the modern ecosystem to recover from its ongoing extinction due to humans might take seemingly countless human generations if the extinctions continue at the present pace.
- The value of biodiversity resides in two main factors: ecosystem services and the stability and productivity of the biosphere. The future of biodiversity resides in ethical issues that confront us regarding the distribution of wealth and poverty in the world, the lifestyles of developed countries, and humanity's view of its position in nature.

KEY TERMS

Arctic National Wildlife Refuge
biofuels
Chernobyl
CO_2 fertilization
CO_2 sequestration
demes
diapirs
directional drilling
Dust Bowl
ecosystem services
eutrophication
fracking

Fukushima
gas-hydrate deposits (clathrates)
genuine progress indicator
Gross Domestic Product
growth faults
high-head dams
hybrid cars
hydrogen fuel cells
iron hypothesis
Little Climate Optimum (Medieval Warm Period)
middens
nuclear accidents

Ogallala Formation
Organization of Petroleum Exporting Countries (OPEC)
peripheral isolates
permafrost
salt tectonics
Sixth Extinction
slash-and-burn
species–area relationship
Three Mile Island
vectors
water table
Yucca Mountain

REVIEW QUESTIONS

1. How are humans changing the Earth back to a more primordial state? How is energy use by humanity intertwined with natural climate change?
2. After the Middle East, where are the largest potential oil reserves found?
3. What are possible alternative energy sources to fossil fuels? What are the current limitations to these alternatives?
4. What are the potential consequences of anthropogenic warming? What are the limitations or uncertainties of the suggested remedies for anthropogenic warming?
5. How does nutrient-rich runoff affect coastal ecosystems?
6. What are some anthropogenic factors affecting biodiversity?
7. Describe the sequence of events that occurs as the habitat of a species is reduced.

FOOD FOR THOUGHT:
Further Activities In and Outside of Class

1. Why is there so much uncertainty in forecasting the effects of anthropogenic warming on future climate?
2. Comment on the role of causality in extinction based on your understanding of anthropogenic impacts on biodiversity.
3. What would you do to help the world pass through the biodiversity bottleneck of the 21st century?

SOURCES AND FURTHER READING

Autin, W. J., and Holbrook, J. M. 2012. Is the Anthropocene an issue of stratigraphy or pop culture? *GSA Today*, 22, 60–61.

Baker, A. C. 2001. Reef corals bleach to survive change. *Nature*, 411, 765–766.

Beaufort, L. et al. 2011. Sensivitiy of coccolithophores to carbonate chemistry and ocean acidification. *Nature* 476:80–83.

Bertness, M., Silliman, B. R., and Jefferies, R. 2004. Salt marshes under siege. *American Scientist*, 92(1): 54–61.

Blaustein, A. R., and Johnson, P. T. 2003. Explaining frog deformities. *Scientific American*, 288(2):60–65.

Cardinale, B. J., Palmer, M. A., and Collins, S. L. 2002. Species diversity enhances ecosystem functioning through interspecific facilitation. *Nature*, 415, 426–429.

Crabtree, G. W., Dresselhaus, M. S., and Buchanan, M. V. 2004. The hydrogen economy. *Physics Today*, 57(12): 39–44.

Davidson, E. A. et al. 2012. The Amazon basin in transition. *Nature*, 481, 321–328.

De Blib, H. 2005. *Why geography matters: Three challenges facing America: climate change, the rise of China, and global terrorism.* New York: Oxford University Press.

Eldredge, N. L. 1991. *The miner's canary: Unraveling the mysteries of extinction.* Englewood Cliffs, NJ: Prentice-Hall.

Evans, R. L. 2007. *Fueling our future: An introduction to sustainable energy.* Cambridge, UK: Cambridge University Press.

Finnegan, S. et al. 2015. Paleontological baselines for evaluating extinction risk in the modern oceans. *Science*, 348, 567–570.

Gibbs, W. W. 2001. On the termination of species. *Scientific American*, 285(5):40–49.

Hönisch, B. et al. 2012. The geologic record of ocean acidification. *Science*, 335, 1058–1063.

Houlahan, J. E. et al. 2001. Global amphibian population declines. *Nature*, 412, 499–500.

Jackson, J. B. C. et al. 2001. Historical overfishing and the recent collapse of coastal ecosystems. *Science*, 293, 629–638.

Kennett, D. J. et al. 2012. Development and disintegration of Maya political systems in response to climate change. *Science*, 338, 788–791.

Kiesecker, J. M., Blaustein, A. R., and Belden, L. K. 2001. Complex causes of amphibian population declines. *Nature*, 410, 681–684.

Lawler, A. 2012. Uncovering civilization's roots. *Science*, 335, 790–793.

Leakey, R., and Lewin, R. 1995. *The Sixth Extinction: Patterns of life and the future of humankind.* New York: Doubleday.

Liu, J., Daily, G. C., Ehrlich, P. R., and Luck, G. W. 2003. Effects of household dynamics on resource consumption and biodiversity. *Nature*, 421, 530–533.

McNeill, J. R. 2000. *Something new under the sun: An environmental history of the twentieth-century world.* New York: W.W. Norton.

Medina-Elizalde, M., and Rochling, E. J. 2012. Collapse of classic Maya civilization related to modest reduction in precipitation. *Science*, 335, 956–959.

Milne, G. A., Gehrels, R. W., Hughes, C. W., and Tamisiea, M. E. 2009. Identifying the causes of sea-level change. *Nature Geoscience*, 2, 471–478.

Parmesan, C., and Yohe, G. 2003. A globally coherent fingerprint of climate change impacts across natural systems. *Nature*, 421, 37–42.

Pauly, D., and Watson, R. 2003. Counting the last fish. *Scientific American*, 289(1):42–47.

Pimentel, D., and Ratzek, W. 2005. Ethanol production using corn, switchgrass, and wood: Biodiesel production using soybean and sunflower. *Natural Resources Research*, 289(1):65–76.

Raloff, J. 2004. Limiting dead zones: How to curb river pollution and save the Gulf of Mexico. *Science News*, 165, 378–380.

Rands, M. R. W. et al. 2010. Biodiversity conservation: challenges beyond 2010. *Science*, 329, 1298–1303.

Roberts, C. M. et al. 2002. Marine biodiversity hotspots and conservation priorities for tropical reefs. *Science*, 295, 1280–1284.

Rosenzweig, M. L. 1999. Heeding the warning in biodiversity's basic law. *Science*, 284, 276–277.

Ruddiman, W. R. 2005. *Plows, Plagues, and Petroleum*. Princeton, New Jersey: Princeton University Press, 202p.

Sachs, J. D., Mellinger, A. D., and Gallup, J. L. 2001. The geography of poverty and wealth. *Scientific American*, 284(3):70–75.

Sallenger, A. H., Doran, K. S., and Howd, P. A. 2012. Hotspot of accelerated sea-level rise on the Atlantic coast of North America. *Nature Climate Change*, 2, 884–888.

Scheffer, M. et al. 2001. Catastrophic shifts in ecosystems. *Nature*, 413, 591–596.

Scheffer, M. 2010. Forseeing tipping points. *Nature*, 467, 411–412.

Stenseth, N. C., Mysterud, A., Ottersen, G., Hurrell, J. W., Chan, K. S., and Lima, M. 2002. Ecological effects of climate fluctuations. *Science*, 297, 1292–1296.

Strain, D. 2011. 8.7 million: a new estimate for all the complex species on Earth. *Science*, 333, 1083.

Tsonis, A. A. 2004. Is global warming injecting randomness into the climate system? *Eos*, 85(38):361.

Vastag, B. 2010. Carbon-capture scheme could cause toxic blooms. *Nature*. Published online 15 March 2010. doi:10.1038/news.2010.124.

Vitousek, P. M., Mooney, H. A., Lubchenco, J., and Melillo, J. M. 1997. Human domination of Earth's ecosystems. *Science*, 277, 494–499.

Vörösmarty, C. et al. 2004. Humans transforming the global water system. *Eos*, 85(48):509.

Watchman, L. H., Groom, M., and Perrine, J. D. 2001. Science and uncertainty in habitat conservation planning. *American Scientist*, 89, 351–359.

Wellington, G. M. et al. 2001. Crisis on coral reefs linked to climate change. *Eos*, 82(1):1.

Wilson, E. O. 2002. The bottleneck. *Scientific American*, 286(2):82–91.

Wolanski, E., Richmond, R., McCook, L., and Sweatman, H. 2003. Mud, marine snow and coral reefs. *American Scientist*, 91(1):44–51.

Yergin, D. 2011. *The prize: the epic quest for oil, money, and power*. New York: Simon and Schuster.

GLOSSARY

Absaroka transgression: see Sequence, Absaroka.

absolute age: geologic age of a fossil, or a geologic event or structure expressed in units of time, usually years. See also radiometric date.

abyssal plain: vast, flat, sediment-covered area of the deep ocean floor.

acanthodians: a taxon of fish, believed to be predatory, that appeared during the Early Silurian. Characterized by much-reduced bony armor, scales, a streamlined body shape, and numerous paired fins supported by large spines, acanthodians also possessed jaws, a significant feature for the future of all ecosystems.

Acasta Gneiss: rock outcrop formed during the Hadean Eon, in the Northwest Territories, Canada, and estimated to be about 4.03 Ga.

accretion, continental: theory that continents have grown by the addition of new continental material around an original nucleus, mainly through the processes of sedimentation and orogeny.

accretionary wedge: prism consisting of thick "slices" of rock stacked on top of one another.

achondrite: group of meteorites, which has a complex origin involving asteroidal or planetary differentiation.

acritarch: microscopic organic fossil

active margin: margin of the continent along which plate boundaries move. Contrast with passive margin.

active site: the location on an enzyme molecule at which it binds and reacts with a substrate molecule. An active site is highly specific for the substrate molecule; thus, the substrate acts like a key into the lock represented by the enzyme.

actualism: philosophical doctrine that the study of modern processes tells us about the past. See also Uniformitarianism, Principle of.

adenosine triphosphate (ATP): energy storage and transfer molecule which is also a building block of the nucleotide adenine in DNA.

adaptive radiation: evolution of ecologic and phenotypic diversity within a rapidly multiplying lineage.

aerosol: gaseous suspension of fine solid or liquid particles.

Agnatha: a taxon of jawless fishes (the name means "without jaws"). Examples of which are placed in the ostracoderms, which possessed a bony skin or covering over their heads; however, the rest of the ostracoderm body lacked a hard internal skeleton and is thought to have been cartilaginous. Agnathans are still represented by the eel-like lamprey and hagfish.

albedo: Earth's surface reflectivity.

Allegheny Plateau: large dissected plateau area in New York, Pennsylvania, West Virginia, and eastern Ohio lying west of the Appalachians, formed from rocks on an ancient foreland basin.

allopatric speciation: the evolutionary process by which new species originate as a result of a local population (see also demes and peripheral isolates) of a given species being isolated from a main population by geographic obstructions such as rivers and mountain chains, or climatic conditions such as sunlight and moisture.

alpha (α) decay: type of radioactive decay in which an atomic nucleus emits an alpha particle consisting of two protons and two neutrons.

amber: fossilized tree resin.

amino acid: building block of protein.

ammonoid: cephalopod that evolved from straight-shelled nautiloids and differed in two respects: the shell was tightly coiled and the walls that separated the chambers started to become convoluted.

amniote egg: egg with compartmentalized sacs (a liquid-filled sac in which the embryo develops, a food sac, and a waste sac) that allowed vertebrates to reproduce on land.

amphibian: terrestrial vertebrate, including frogs, toads, and salamanders.

anagenesis: the gradual transition of one species into another, such that no more members of the original species remain (see also pseudoextinction), but members of a daughter species or subspecies carry on.

analog: A modern or ancient example.

analogous structure: feature that is similar in function but not in structure that evolved in response to a similar selective pressure. Contrast with homologous structure.

Ancestral Rockies: precursor to current-day mountain chain of the same name, believed to have been formed by high-angle faulting in response to compressive stresses generated during the collision of Gondwana with Laurasia during the Pennsylvanian.

andesite: dark, fine-grained, brown or grayish volcanic rock that is intermediate in composition between rhyolite and basalt.

angiosperm: flowering plant that produces seeds in specialized reproductive organs called flowers. Contrast with gymnosperm.

angular momentum: quantity of rotation of a body, which is the product of its mass and angular velocity.

angular unconformity: unconformity where horizontally parallel strata of sedimentary rock are deposited on tilted and eroded layers, producing an angular discordance with the overlying horizontal layers.

anhydrite: relatively common sedimentary mineral that forms massive rock layers.

ankylosaur: heavily built quadrupedal herbivorous dinosaur primarily of the Cretaceous period, armored with bony plates; thought to have walked with a sprawling gait resembling a lizard's.

anoxia: condition or environment where there is a total absence of oxygen.

Antarctic Bottom Water (AABW): Deep watermass produced off Antarctica and characterized by low temperature and high salinity and high nutrient levels.

Antarctic Circumpolar Current: ocean current that flows from west to east around Antarctica.

anthracosaur: extinct reptile-like amphibian that flourished during the Carboniferous and early Permian periods.

Anthropoidea: "higher primates" familiar to most people: the Old World monkeys and apes, and humans.

anthropogenic: relating to, or resulting from the influence of humans on nature.

anticline: formed when the strata dip away from the main axis of the fold, with the oldest strata found in the core of the anticline and the youngest strata on its crest and sides. Contrast with syncline.

aphanitic: refers to the texture of an igneous rock in which the crystalline components are not distinguishable by the unaided eye. Contrast with phaneritic.

Appalachian Mountains (Appalachians): system of mountains in eastern North America, first beginning to form roughly 480 million years ago.

apparent polar wandering curve: apparent path of magnetic poles actually produced by movement of the continents. Contrast with true polar wander.

aquifer: body of permeable rock that can contain or conduct groundwater.

aragonite sea: seas in which primarily aragonite or high-magnesium calcite are precipitated in inorganic cements and by organisms in shells.

Archaea: ancient domain of single-celled microorganisms. Contrast with Eubacteria and Eukarya.

archaeocyathids: sponge-like suspension feeder that lived in warm tropical and subtropical waters during the Early Cambrian; they diversified into hundreds of species of fascinating cup-like shapes, creating reefs.

Archean (Archeozoic): eon from about 3,800 million years to 2500 million years ago; Earth's crust formed; unicellular organisms are earliest forms of life.

Archaeopteryx: oldest known fossil bird, of the late Jurassic period. It had feathers, wings, and hollow bones like a bird, but teeth, a bony tail, and legs like a small bipedal dinosaur.

archosaur: reptile of the subclass Archosauria, which includes the dinosaurs, pterosaurs, and the modern crocodilians.

Arctic National Wildlife Refuge: national wildlife refuge in northeastern Alaska, consisting of over 77,000 km^2 (19,000,000 acres); the question of whether to drill for oil in this area has been an ongoing controversy since 1977.

arkose: sedimentary rock composed of sand-size fragments that contain a high proportion of feldspar in addition to quartz.

"arms race," biologic: competition between predator and prey, when each develops adaptations and counter-adaptations against each other for survival.

Artiodactyla: order of mammals that comprises the even-toed ungulates (artiodactyls), which today includes deer and many domesticated animals such as cows.

assimilation: process of magnetic differentiation that incorporates solid or fluid material that was originally in the rock wall into a magma.

asteroid: any of numerous small celestial bodies composed of rock and metal that move around the sun (mainly between the orbits of Mars and Jupiter).

asthenosphere: zone of the Earth's mantle that lies beneath the lithosphere and consists of several hundred kilometers of deformable rock. Compare with lithosphere.

astronomical unit (AU): unit of measurement equal to 149.6 million kilometers, the mean distance from the center of the Earth to the center of the sun.

Atlantic Coastal Plain: relatively flat landform extending 3,500 km (2,200 miles) from New York southward to the Eastern Continental Divide of Georgia/Florida; formed from sediment of the Mesozoic and Cenozoic age eroded from the Piedmont and Appalachian Mountains that lie west and north.

atmosphere: Earth system comprising the gaseous envelope surrounding Earth.

atmosphere, highly reducing: atmospheric condition in which free oxygen (O_2) is not present.

atmosphere, mildly reducing: condition in which increased oxygen combines with carbon and hydrogen to produce an atmosphere enriched in CO_2 and water vapor (H_2O), along with nitrogen (N_2).

atoll: ring-shaped reef, island, or chain of islands formed of coral.

ATP world hypothesis: theory that during the protometabolism phase leading to life, high-energy bond formation and energy transfer were taken over from thioester bonds by ATP. Contrast with clay world, pyrite world, and RNA world.

aulacogen: failed rift valley of the triple junctions of rift valleys.

australopithecine: extinct human-like bipedal primates with relatively small brains of the genus *Australopithecus;* from 1 to 4 million years ago.

autocatalysis: hypothesis in which reactants and products organize themselves into a more complex system.

autotroph (producer): organism capable of synthesizing its own food from inorganic substances, using light or chemical energy. Contrast with consumer.

autotrophic: requiring only carbon dioxide as a source of carbon for metabolic synthesis of organic molecules (as glucose).

Avalonia: Early Paleozoic terrane.

back-arc basin: submarine basin produced when the lithosphere behind a volcanic island arc is subjected to tensional forces. Compare with fore-arc basin.

back reef: area behind or to the landward of a reef.

balancing selection: state in which a relatively stable proportion of genotypes is maintained in a population by a selective force such as malaria or sickle cell anemia.

Baltica: late-Proterozoic, early-Palaeozoic continent that now includes mostly northern Europe west of the Ural Mountains of Russia.

banded iron formation (BIF): distinctive unit of sedimentary rock of Precambrian age. A typical BIF consists of repeated, thin layers of iron and a red form of chert called jasper.

barrier island: long narrow sandy island (wider than a reef) running parallel to the shore.

basalt: dark, fine-grained volcanic rock that sometimes displays a columnar structure.

Basin and Range Province: vast physiographic region in the American West and Southwest characterized by abrupt changes in elevation, alternating between narrow faulted mountain chains and flat arid valleys or basins.

bathyal zone: steep descent of the seabed from the continental shelf to the abyssal zone. See also continental slope.

bauxite: rock comprised of iron and aluminum hydroxides/oxides; the primary ore of aluminum.

belemnoid: extinct group of marine cephalopod, very similar in many ways to the modern squid and closely related to the modern cuttlefish.

benthic: lowest level of a body of water such as an ocean or a lake, including the sediment surface and some subsurface layers.

benthos: creatures that live on, in, or near the seabed, also known as the benthic zone.

beta (β) decay: type of radioactive decay in which a neutron breaks down into a proton and an electron and the proton is retained, whereas the electron is ejected.

Big Bang Theory: theory that the universe originated sometime between 10 billion and 20 billion years ago from the cataclysmic explosion of a small volume of matter at extremely high density and temperature. Contrast with Steady State Theory.

Big Five mass extinctions: five greatest mass extinctions over the past 500 million years, each of which is thought to have annihilated anywhere from 50 to 95 percent of all species on the planet; all took place in the Phanerozoic.

binomial nomenclature: scientific naming of species in which each species is given a unique name that consists of a generic and a specific term.

bioclast: skeletal fragment of marine or land organisms that are found in sedimentary rocks laid down in a marine environment.

bioeroder: animal, such as a sponge or clam, that bores into hard substances such as coral, weakening or killing them leading to decimation of the ecosystem (coral reef).

biofuel: type of fuel whose energy is derived from biologic carbon fixation; includes fuels derived from biomass conversion, as well as solid biomass, liquid fuels, and various biogases.

biogenetic law: theory that the stages in an organism's embryonic development and differentiation correspond to the stages of evolutionary development characteristic of the species.

biogenic ooze: pelagic sediments that have more than 30 percent skeletal material.

biogeochemical cycle: transport and transformation of chemical elements between reservoirs located in the biosphere and geologic reservoirs such as rocks, sediments, and soils.

biogeography: branch of biology that deals with the geographic distribution of plants and animals.

biologic classification: branch of biology dealing with the identification and naming of organisms. See also taxonomy.

Biological Continuity, Principle of: theory that, once established, all life comes from life and that biochemical pathways reflect their ancestry.

biome: large, naturally occurring community of flora and fauna occupying a major habitat.

biosphere: Earth system consisting of all living organisms and their dead remains.

biostratigraphy: science of dating rocks by using the fossils contained within them. See also chronostratigraphy and lithostratigraphy.

bioturbation: displacement and mixing of sediment particles by living things.

bipedal: ability to walk upright on two legs.

bipedalism: walking upright.

bird: warm-blooded egg-laying vertebrate distinguished by the possession of feathers, wings, and a beak and (typically) by being able to fly.

bivalve: mollusk of the class Bivalvia, such as an oyster or a clam, that has a shell consisting of two hinged valves.

Bivalvia: taxonomic class of marine and freshwater mollusks within the phylum Mollusca (common name bivalves), which includes clams, oysters, mussels, scallops, and many other families of mollusks that have two hinged shells.

blackbody radiation: randomized energy released when particles collide with each other very rapidly in a warm-to-hot object, such as the surface of a planet warmed by a star or the interior of the star itself.

black shale: fine-grained sedimentary rock containing 5% or more of organic carbon.

black smoker: type of hydrothermal vent found on the seabed, which appears as a black chimney-like structure that emits a cloud of black material (typically high levels of sulfur-bearing minerals, or sulfides).

blastoid: extinct stemmed echinoderm of the class Blastoidea, characterized by a persistent five-fold (pentagonal) symmetry.

blocking (or closure) temperature: refers to the temperature of a system, such as a mineral, at the time given by its radiometric date; typically hundreds of degrees cooler than the melting temperature of the mineral.

Blue Ridge Province: mountain range located in the eastern United States, starting in the south in Georgia and ending northward in Pennsylvania; produced as the result of thrusting, highly metamorphosed and deformed rocks that formed during the Precambrian, later brought to the surface and by thrusting over Early Paleozoic sediments.

blueschist: rock that forms by the metamorphism of basalt and rocks with similar composition at high pressures and low temperatures. The blue color of the rock comes from the presence of the mineral glaucophane.

body plan: blueprint for the way the body of an organism is laid out. See also groundplan.

bony fish: fish of the taxonomic class Osteichthyes that have bony, as opposed to cartilaginous, skeletons.

bottom-up approach, chemical evolution: strategy in which the fundamental building blocks of life are synthesized from simpler precursors and the building blocks then assembled to form living systems. Contrast with top-down approach.

brachiopod: suspension-feeding marine invertebrate of the phylum Brachiopoda, which has a bivalve dorsal and ventral shell enclosing a pair of tentacled, arm-like structures that are used to sweep minute food particles into the mouth.

braided streams and rivers: occur when a threshold level of sediment load or slope is reached.

breccia: sedimentary rock composed of sharp-angled fragments embedded in a fine-grained matrix.

brine: highly saline water.

brontothere: large ungulate mammal of the Eocene epoch with a horn-like bony growth on the nose.

browser: animal that feeds mainly on high-growing vegetation. Contrast with grazer.

bryozoans: geologically important phylum of small aquatic invertebrate animals; live for the most part in colonies of interconnected individuals.

Burgess Shale: bed of shale exposed in the Rocky Mountains in British Columbia, Canada, dating to about 530 million years ago, rich in well-preserved fossils of early marine invertebrates, many of which represent evolutionary lineages unknown in later times.

C3 grasses: grasses adapted to cool season establishment and growth in either wet or dry environments. Contrast with C4 grasses.

C4 grasses: grasses adapted to warm or hot seasonal conditions under moist or dry environments. Contrast with C3 grasses.

calcite compensation depth (CCD): depth in the ocean (~4,000–5,000 meters) below which solution of calcium carbonate occurs at a faster rate than its deposition.

calcite sea: seas in which primarily low-magnesium calcite is precipitated in inorganic cements and by organisms in shells. See aragonite sea.

calving: breaking off of a mass of ice from its parent glacier, iceberg, or ice shelf.

Cambrian: period from about 544 million to about 500 million years ago; earliest period in which fossils, notably trilobites, can be used in geologic dating.

Cambrian explosion: relatively rapid (over a period of many millions of years) appearance, around 530 million years ago, of diverse and abundant fossils.

cap limestone: limestone that lies on top of, or caps, a glacial deposit.

capture hypothesis: theory that the Moon was formed somewhere else in the solar system and was later captured by the gravitational field of the Earth.

carbonaceous chondrite: class of chondritic meteorite grouped according to distinctive compositions of the parent body from which it originated.

carbonic acid: weak acid that is created when carbon dioxide (CO_2) is dissolved in water (H_2O), resulting in the chemical formula H_2CO_3.

Carboniferous: period from about 360 million to 286 million years ago; first reptiles and seed-bearing plants appeared.

carbon isotope: stable isotope used to study the flux of carbon on geologic time scales between the different reservoirs of organic carbon.

Carnivora: order of mammals, including dogs, cats, bears, raccoons, and seals, which is distinguished by having powerful jaws and teeth adapted for stabbing, tearing, and eating flesh.

carnivore: member of the order Carnivora; animal that feeds on flesh.

carrying capacity: level at which populations or organisms are no longer sustainable.

cartilaginous fish: fish of the taxonomic class Chondrichthyes, distinguished by having a skeleton of cartilage rather than bone, including the sharks, rays, and chimeras.

cast: sedimentary structure representing the infilling of a mark or depression in a soft layer of sediment (or bed).

catastrophism: view of the Earth's geologic record as having resulted from a series of sudden global catastrophes.

cellulose: complex carbohydrate that forms the main constituent of the cell wall in most plants.

cementation: process by which sediment grains are bound together by precipitated minerals originally dissolved during the chemical weathering of preexisting rocks.

Cenozoic: most recent era of geologic time, beginning about 65 million years ago, during which modern plants and animals evolved.

Central Atlantic Magmatic Province: large connected magma flow formed during the breakup of Pangaea during the Mesozoic Era, providing a legacy of basaltic dikes, sills, and lavas over a vast area around the present central North Atlantic Ocean.

Central Dogma (of Cell Biology): theory that each gene in the DNA molecule carries the information needed to construct one protein, which, acting as an enzyme, controls one chemical reaction in the cell.

Cephalopoda: taxonomic class of marine mollusks within the phylum Mollusca (cephalopods), including octopuses, squids, cuttlefish, and pearly nautilus; characterized by bilateral body symmetry, a prominent head, and a set of arms or tentacles modified from the primitive molluscan foot.

ceratopsian: gregarious quadrupedal herbivorous dinosaur of a group found in the Cretaceous period, including triceratops. It had a large beaked and horned head and a bony frill protecting the neck.

chalicothere: horse-like fossil mammal of the late Tertiary period, with stout claws on the toes instead of hooves.

cheilostome bryozoan: exclusively marine, colonial invertebrate animal; cheilostome colonies are composed of calcium carbonate and grow on a variety of surfaces, including rocks, shells, seagrass, and kelps.

chemical evolution: formation of complex organic molecules from simpler inorganic molecules through chemical reactions in the oceans during the early history of the Earth; the first step in the development of life on this planet.

chemical weathering: breakdown of rock by chemical mechanisms, the most important ones being carbonation, hydration, hydrolysis, oxidation, and ion exchange in solution.

chemoautotrophic: being autotrophic and oxidizing an inorganic compound as a source of energy. Contrast with photoautotrophic.

chemosynthesis: synthesis of organic compounds by energy derived from chemical reactions, typically in the absence of sunlight.

chemosynthetic: product of the biologic conversion of one or more carbon molecules and nutrients into organic matter using the oxidation of inorganic molecules or methane as a source of energy, rather than sunlight. See also chemoautotrophic.

Chernobyl nuclear accident: result of the explosion of a Soviet nuclear reactor in 1986, which spewed radioactivity over a large part of Eastern Europe.

chert: sedimentary rock composed mostly of the mineral chalcedony—cryptocrystalline silica, or quartz in crystals of submicroscopic size.

Chicxulub crater: ancient impact crater buried underneath the Yucatán Peninsula in Mexico, dates from the end of the Cretaceous Period, roughly 65 million years ago.

chirality (handedness): refers to molecules. Two mirror images of a chiral molecule are called enantiomers or optical isomers. Pairs of enantiomers are often designated as "right-" and "left-handed."

Chiroptera: order of mammals that comprises the bats. There are over 900 living species of bats, and they are found on every continent except Antarctica.

chiton: primitive marine mollusk (Phylum Mollusca) that has an oval flattened body with a shell of overlapping plates.

chlorofluorocarbon (CFC): organic compound that contains carbon, chlorine, and fluorine, produced as a volatile derivative of methane and ethane. The chlorine in CFCs is harmful to the ozone layer in the Earth's atmosphere.

chloroplast: organelle found in plant cells and other eukaryotic organisms that conduct photosynthesis.

Chondrichthyes: taxonomic class of fish that includes those with a cartilaginous skeleton. See cartilaginous fish.

chondrite: stony meteorite that has not been modified due to melting or differentiation of the parent body.

chromosome: thread-like structure of nucleic acids and protein found in the nucleus of most living cells, carrying genetic information in the form of genes.

chronostratigraphy: science of dating rocks that uses age determination and time sequence of rock strata. See also biostratigraphy and lithostratigraphy.

clade: group of organisms, such as a species, whose members share homologous features derived from a common ancestor. Contrast with grade.

cladistics: method of classification of animals and plants according to the proportion of measurable characteristics that they have in common.

cladogram: branching diagram showing the cladistic relationship between a number of species.

clay world (or clay life) hypothesis: scenario in which clay particles served as scaffoldings or even catalysts for the production of more complex compounds serving as a source of information coding of prebiotic precursors. Contrast with ATP world, pyrite world, and RNA world.

Clean Air Act: federal law enacted by the United States Congress to control air pollution on a national level.

closure temperature: see blocking temperature.

CO$_2$ fertilization: enhanced growth of plants resulting from increased concentration of CO$_2$ in the atmosphere.

CO$_2$ sequestration: process of removing carbon from the atmosphere and depositing it in a reservoir.

coal: dark brown to black graphite-like material formed from fossilized plants and consisting of amorphous carbon with various organic and some inorganic compounds.

coccolithophorid: microscopic, planktonic marine alga, which secretes a calcite shell.

Cocos Plate: oceanic tectonic plate beneath the Pacific Ocean off the west coast of Central America; created by seafloor spreading along the East Pacific Rise and the Cocos Ridge.

coelacanth: branch of crossopterygian; large, bony marine fish with a three-lobed tail fin and fleshy pectoral fins. It is thought to be related to the ancestors of land vertebrates and was known only from fossils until one was found alive in 1938.

coevolution: change of a biologic object triggered by the change of a related object.

comet: celestial object consisting of a nucleus of ice and dust and, when near the sun, a "tail" of gas and dust particles pointing away from the sun.

community: interacting group of various species in a common location.

compaction: process by which the volume or thickness of sediment is reduced due to pressure from overlying layers of sediment.

compression: set of stresses directed toward the center of a rock mass. See also impression.

compressional tectonic regime: occur when pieces of crust are pushed together or over one another, such as during subduction or orogenesis. Contrast with extensional tectonic regime.

concurrent range zones: the overlap in stratigraphic ranges of different species, with each range representing variations within a species of a given fossil type.

concordant age: radiometric age data that are in agreement, confirming their accuracy. Contrast with discordant age data.

conglomerate: sedimentary rock consisting of individual gravel (>2 mm) clasts within a finer-grained matrix that have become cemented together.

conifer: gymnospermous tree or shrub bearing cones.

conodont: extinct marine animal of the Cambrian to Triassic periods, which was represented as a tooth-like structure composed of calcium phosphate and having a long worm-like body, numerous small teeth, and a pair of eyes.

consumer: living organism—omnivore, carnivore, or herbivore—that takes in energy by consuming another organism. Contrast with autotroph (producer).

contact metamorphism: alteration of rock that results in the formation of valuable minerals, such as garnet and emery, through the interaction of the hot magma with adjacent rock.

continent–continent plate boundary: Ocean crust of the leading edge of one plate bearing a continent is subducted beneath the margin of a second plate (also bearing a continent) as a volcanic island arc forms on the second plate.

continental drift: movement of the Earth's continents relative to each other.

continental shelf: area of seabed around a large landmass where the sea is relatively shallow compared with the open ocean. See also neritic zone.

continental shelf break: submerged border of a continent that slopes gradually and extends to a point of steeper descent to the ocean bottom.

continental slope: descent from the continental shelf to the ocean bottom. See also bathyal zone.

contingency: influence that historical circumstances have on the outcome of one or more processes or events but cannot be predicted with certainty.

continuous growth (of grasses): ability of grasses to grow back rapidly after being grazed.

control: sample in which a factor whose effect is being estimated is absent or is held constant, in order to provide a comparison.

convection cell: phenomenon of fluid dynamics that occurs in situations where there are density differences within a body of liquid or gas.

convergent evolution: the result of natural selection acting on evolutionarily unrelated taxa to produce or converge on a common evolutionary solution; for example, wings for flight or streamlined bodies for moving through water quickly.

convergent plate boundary: actively deforming region where two (or more) tectonic plates or fragments of lithosphere move toward one another and collide. Contrast with divergent plate boundary.

coral reef: erosion-resistant marine ridge or mound consisting chiefly of compacted coral that is a diverse biologic community.

cordaite: an extinct gymnosperm in the genus *Cordaite*, with very characteristic, long, strap-shaped leaves, which grew on wet ground similar to the Everglades in Florida. This group included large trees as well as shrub-like plants.

Cordillera: one of the longest mountain chains in the world, extending from Alaska to South America (about 600 miles).

core: central portion of the Earth below the mantle, beginning at a depth of about 2,900 kilometers (1,800 miles), and probably consisting of iron and nickel. It is made up of a liquid outer core and a solid inner core.

Coriolis effect: apparent force that as a result of the Earth's rotation deflects moving objects (as projectiles or air currents) to the right in the northern hemisphere and to the left in the southern hemisphere.

correlation: procedure that infers the age equivalence of rocks.

cosmic background radiation: radiation left over from an early stage in the development of the universe, and its discovery is considered a landmark test of the Big Bang model of the universe. See also microwave radiation.

cosmopolitan species: species with very large distribution in many, or all, parts of the world and ecosystems. Contrast with endemic species.

craton: large portion of a continent that has been relatively undisturbed since the Precambrian era and includes both shield and platform layers. Contrast with shield.

creodont: extinct carnivorous mammal of the early Tertiary period, ancestral to modern carnivores.

Cretaceous: period from about 146 million to 65 million years ago; first flowering plants emerged, and the domination of dinosaurs continued although they died out abruptly toward the end of it.

Cretaceous Interior Seaway (also called Western Interior Seaway): huge inland sea that split the North American continent into two halves during most of the Mid- and Late-Cretaceous Period. It was 760 meters (2,500 feet) deep, 970 kilometers (600 miles) wide, and over 3,200 kilometers (2,000 miles) long.

crinoid: echinoderm of the class Crinoidea, including sea lilies and feather stars, that is characterized by a cup-shaped body, feathery radiating arms, and either a stalk or claw-like structure with which they are able to attach to a surface.

cross-bedding: (near-) horizontal sedimentary structure that is internally composed of inclined layers deposited by wind or water.

cross cutting relationships, principle of: theory that an igneous intrusion is always younger than the rock it cuts across.

crossopterygian: type of lobe-finned fish with lungs that were ancestral to amphibians.

crust: outer layer of the Earth.

crust, primary: Earth's earliest crust that formed as the result of impact with meteors and other celestial bodies.

crust, secondary: crust that formed after destruction of the primary crust.

cryosphere: Earth system consisting of glaciers and related environments.

Cumberland Plateau: dissected plateau area in the southern part of the Appalachians extending from Kentucky, West Virginia, Tennessee, Alabama, and Georgia.

Curie point: temperature below which the polarity of the Earth's magnetic field is preserved in rocks.

cyanobacteria: early prokaryotes that are related to the bacteria but are capable of photosynthesis; the earliest known form of life on Earth existing approximately 3.5 billion years ago.

cycad: palm-like plant of tropical and subtropical regions, bearing large male or female cones. Cycads were abundant during the Triassic and Jurassic eras, but have since been in decline.

cycadeoid: gymnosperm seed plant not closely rated to, but superficially similar to, the cycad. Cycads and cycadeoids were dominant floristic elements of early and middle Mesozoic landscapes.

cyclothem: alternating stratigraphic sequence of marine and nonmarine sediments, sometimes interbedded with coal seams.

cynodont: therapsid which first appeared in the Late Permian (about 260 million years ago); characterized by its dog-like teeth. Compare with dicynodont.

daughter product (or isotope): compound remaining after the parent isotope (the original isotope) has undergone decay.

Deccan Traps: one of the largest volcanic provinces in the world, consisting of more than 2,000 meters (6,500 feet) of flat-lying basalt lava flows and covering an area of nearly 500,000 square kilometers (200,000 square miles) in west-central India.

decomposer: organism that breaks down dead organic matter to release nutrients trapped in the organic matter back to the ecosystem.

deduction: process of reaching a conclusion that is guaranteed to follow but which does not go beyond the information contained in the premises.

deep time: multimillion year time frame within which scientists believe the Earth has existed and which is supported by the observation of natural, mostly geologic phenomena.

dehydration–rehydration reaction: chemical reaction that involves the loss and then replenishment of water, possibly involved in the origin of life.

delta: triangular tract of sediment deposited at the mouth of a river.

Delta, Catskill: unit of mostly terrestrial sedimentary rock found in Pennsylvania and New York; formed during the Devonian period by an enormous complex of river environments that resulted from the Acadian Orogeny.

delta lobe-switching: occurs when delta shelf fills with sediment and the delta shifts position.

Delta, Queenston: 300-mile-wide clastic wedge of sediment deposited over what is now eastern North America during the late Ordovician period due to the erosion of mountains created during the Taconic orogeny.

delta, river-dominated: shape of a delta is controlled by flow of freshwater and sediment from river discharge.

delta, tide-dominated: shape of a delta is influenced by sediment input, wave energy, and tidal energy.

delta, wave-dominated: shape of a delta is controlled by wave erosion, although deposition still outweighs the amount of erosion and the delta is able to advance into the sea.

deme: term for a local population of organisms of one species that actively interbreed with one another and share a distinct gene pool. See also peripheral isolate.

dendrochronology: scientific discipline concerned with dating and interpreting past events, particularly paleoclimates and climatic trends.

dentary: one of a pair of membrane bones that in lower vertebrates form the distal part of the lower jaws and in mammals comprise the mandible.

depositional sequence: basic unit of sequence stratigraphy; conditions under which rocks or sediment is laid down.

descent with modification: refers to the passing on of traits from parent organisms to their offspring.

derived trait: trait that is present in an organism but was absent in the last common ancestor of the group being considered.

determinism: assumption that each effect (observation, outcome) has a particular cause and that we can predict an effect given a particular cause.

deuterium: hydrogen isotope that has an extra neutron.

Devonian: period from about 409 million to 363 million years ago; fish became abundant, the first amphibians evolved, and the first forests appeared.

diamictite: glacial deposit that consists of poorly sorted ice-rafted debris, including dropstones, deposited in seaways at the margins of melting sea ice.

diapir: domed rock formation in which a core of rock has moved upward to pierce the overlying strata.

diapsid: major amniote lineage that diverged about 300 million years ago, including the lizards, snakes, crocodiles, dinosaurs, and pterosaurs, characterized by two temporal openings in the skull; ultimately ancestral to mammals. Contrast with euryapsids and synapsids.

diastem: gap in the stratigraphic record so short it is virtually undetectable.

diatom: single-celled alga that has a cell wall of silica. Many kinds are planktonic, and extensive fossil deposits have been found.

diatomaceous ooze: pelagic, siliceous sediment composed of more than 30% diatom tests, up to 40% calcium carbonate, and up to 25% mineral grains.

dicynodont: herbivorous mammal-like reptile with two tusks. Compare with cynodont.

differential reproduction: idea that organisms best adapted to a given environment will be most likely to survive to reproductive age and have offspring of their own.

diluvial theory: interpretation of the geologic history of the Earth in terms of the global flood described in the story of Noah's Ark.

dinoflagellate: single-celled plankton with two flagella and, in some species, an external skeleton made of cellulose.

dinosaur: extinct terrestrial reptile of the Mesozoic era, often reaching an enormous size.

diorite: medium- to coarse-grained igneous rock that commonly is composed of about two-thirds plagioclase feldspar and one-third dark-colored minerals, such as hornblende or biotite.

directional drilling: technique that enables drilling at an angle, versus vertically, to reach a particular underground formation, for example an oil- or gas-bearing reservoir.

directional selection: a type of natural selection in which a particular genotype in a population is shifted toward dominance.

directionality: concept that Earth's systems change in certain directions through geologic time. See also secular change.

disarticulate: to separate at the joints.

disconformity: unconformity where the erosional surface is overlain by more horizontal layers of sediment.

discordant age: radiometric age data that are not in agreement. Contrast with concordant age data.

distinctive sequence: distinguishing arrangement (of rock units).

distributary: branch of a river that does not return to the main stream after leaving it (as in a delta).

divergent evolution: a process by which anatomic structures might evolve to be similar between taxa but the function of the structures is dissimilar. Compare this to convergent evolution.

divergent plate boundary: zone where two tectonic plates are pulling apart as they move away from each other. Contrast with convergent plate boundary.

dolostone: sedimentary rock composed primarily of dolomite, a mineral made up of calcium, magnesium, carbon, and oxygen.

domain: highest taxonomic rank of organisms.

Doppler effect: change in frequency of a sound wave for an observer moving relative to the source of the wave.

Drake equation: method of calculating the probability of intelligent life elsewhere in the universe.

drift: rock debris transported and deposited by or from ice, especially by or from a glacier.

dropstone: rock that was carried by a glacier or iceberg, and deposited as the ice melted.

dynamic equilibrium: the normal state in which Earth's natural systems are assumed to exist. These systems can oscillate, but in the long term, they maintain an average to which it is assumed they will return, even if disturbed away from its so-called equilibrium state.

earthquake: typically generated when one block of rock suddenly moves past another along a fault.

East Pacific Rise: mid-oceanic ridge located along the floor of the Pacific Ocean. It separates the Pacific Plate to the west from (north to south) the North American Plate, the Rivera Plate, the Cocos Plate, the Nazca Plate, and the Antarctic Plate.

eccentricity: extent to which the Earth's orbit around the Sun departs from a perfect circle.

echinoderm: suspension-feeding marine invertebrate of the phylum Echinodermata, which includes starfishes, sea urchins, and sea cucumbers, having an internal calcareous skeleton and often covered with spines.

echinoid: one of the more diverse and successful echinoderm groups today, including familiar echinoderms such as the sea urchins and sand dollars.

ecologic hypothesis: observation that primitive animals radiated into vacant ecologic space made suitable by changing physical environmental conditions. Contrast with genomic hypothesis.

ecologic replacement: gradual and orderly process of change in an ecosystem brought about by the progressive succession of one community by another until a stable climax is established.

ecosystem: complex of living organisms, their physical environment, and all their interrelationships in a particular unit of space.

ecosystem services: resources and processes supplied by ecosystems that benefit humankind; broken down into provisioning services (such as food and freshwater), regulating services (such as climate and disease regulation), supporting services (such as nutrient cycling), and cultural services (such as recreation).

ectothermic: "cold-blooded," refers to an animal that cannot regulate its own body temperature; rather it's regulated by the environment, so it often basks for heat, burrows, and hibernates.

electron capture: process in which a proton-rich nuclide absorbs an inner atomic electron (changing a nuclear proton to a neutron) and simultaneously emits a neutrino.

El Niño-Southern Oscillation (ENSO): quasiperiodic climate pattern that occurs across the tropical Pacific Ocean roughly every 5 years, when upwelling of cold, nutrient-rich water does not occur. It causes die-offs of plankton and fish and affects Pacific jet stream winds, altering storm tracks and creating unusual weather patterns in various parts of the world.

endemic species: species that is only found in a particular region. Contrast with cosmopolitan species.

endosymbiosis: symbiosis in which one of the symbiotic organisms lives inside the other, typically with the two organisms behaving as a single organism.

endosymbiotic theory of cell evolution: proposal that certain organelles, especially mitochondria and chloroplasts, originated as free-living bacteria that were taken inside another cell as endosymbionts (organism that lives within the body or cells of another organism).

endothermic: "warm-blooded," refers to an animal that generates body heat above ambient temperatures through various physiologic and anatomic specializations.

enzyme: protein that catalyzes (increases the rates of) biochemical reactions.

enzyme, active site: part of an enzyme at which catalysis of the substrate occurs.

Eocene: epoch from around 55 million to 38 million years ago, during which the ancestors of many modern animals appeared.

eon (of geologic time scale): major division of geologic time, subdivided into eras.

epicenter: surface location above the focus of an earthquake.

epifaunal: describes an animal lives upon the surface of sediments or soils, such as on the surface of a seabed. Contrast with infaunal.

epoch (of geologic time scale): unit of geologic time that is a subdivision of a period and is itself divided into ages.

equilibrium: condition where a system exhibits no net change.

era: major division of geologic time; an era is usually divided into 2 or more periods.

erratic: large boulder left behind as ice retreated.

estuary: tidal mouth of a river, where the tide meets the stream.

Eubacteria: large domain of prokaryotic microorganisms. Contrast with Archaea and Eukarya.

eugenics: science that deals with the improvement (as by control of human mating) of hereditary qualities of a race or breed.

Eukarya: domain of all of the organisms with eukaryotic cells—that is, those with membranous organelles (including mitochondria and chloroplasts). Contrast with Archaea and Eubacteria.

eukaryote: organism whose cells contain complex structures enclosed within membranes. Contrast with prokaryote.

euryapsids: marine reptiles of the Mesozoic (including the nothosaurs, plesiosaurs, and ichthyosaurs), which were characterized by a single upper temporal opening in the skull high above the cheek region. Compare with diapsids.

eurypterid: an extinct group of arthropods related to arachnids which include the largest known arthropods that ever lived.

eurytopic species: species able to tolerate a wide range of environmental changes. Contrast with stenotopic species.

eutrophication: excessive richness of nutrients in a lake or other body of water, frequently due to runoff from the land, which causes a dense growth of plant life and death of animal life from lack of oxygen.

evaporation: process by which any substance is converted from a liquid state into, and carried off in, vapor. Contrast with precipitation.

evaporites: a broad group of chemical sedimentary rocks which form by precipitation from water. Representative evaporites include gypsum, anhydrite, and halite (salt).

evolution: gradual process in which something changes into a different and usually more complex or better form.

evolution, convergent: describes the acquisition of the same biologic trait in unrelated lineages. Contrast with divergent pattern of evolution.

evolution, divergent: accumulation of differences between groups which can lead to the formation of new species. Contrast with convergent pattern of evolution.

evolutionary fauna, Cambrian: formed in the first 65 million years of the Paleozoic era, during which marine invertebrates, especially trilobites, early mollusks, and other arthropods, flourished.

evolutionary fauna, Paleozoic: began during the Ordovician Period, about 500 million years ago, dominated by taxa that fed on suspended organic matter and plankton, and marked by the adaptive radiation and diversification of marine invertebrates on the continental shelf.

evolutionary fauna, Modern: arose in the Triassic Period about 250 million years ago after the end-Permian extinction and dominated by the kinds of animals that still exist in the oceans and on land, particularly the vertebrates.

evolution, iterative: refers to repetitions of descendant types in taxonomic groups.

Exotic terrane: a terrane comprising lithologic, fossil, or paleomagnetic traits that differ dramatically from those of the continent with which it collided. See also terrane.

experimental science: science that uses what is known to try and prove ideas and concepts that are as yet untested.

extensional tectonic regime: associated with the stretching and thinning (pulling apart) of the crust or lithosphere. Contrast with compressional tectonic regime.

external mold: impression of the outside surface of shell, skeleton, or bone. Contrast with internal mode.

extinction: occurs when the last individual member of a species dies.

extinction, background: ongoing extinction of individual species due to environmental or ecologic factors such as climate change, disease, loss of habitat, or competitive disadvantage in relation to other species.

extinction, mass: extinction of one of more species in a relatively short period of geologic time, usually as a consequence of a catastrophic global event, natural disaster, or abrupt change in the environment.

extinction, minor: small-scale extinction that did not affect as many species as major mass extinctions but are critical to understanding the patterns of extinction.

extremophile: microorganism, especially an archaean, that lives in conditions of extreme temperature, acidity, alkalinity, or chemical concentration.

extrusive igneous rock: rock that formed from cooled lava of the Earth's surface. Compare with intrusive igneous rock.

facies: stratified body of rock that is distinguished from others by its appearance or composition. Contrast with formation.

facies change: lateral or vertical variation in the lithologic or paleontologic characteristics of contemporaneous sedimentary deposits.

Faint Young Sun Paradox: explains the apparent contradiction between observations of liquid water early in the Earth's history and high levels of greenhouse gases in the atmosphere as resulting from the existence of a faint young sun (70% to 75% as bright as today's sun).

Farallon Plate: ancient oceanic plate, which began subducting under the west coast of the North American Plate—then located in modern Utah— as Pangaea broke apart during the Jurassic Period.

fault: fracture along which rocks move past one another.

fauna: animals of a particular region, habitat, or geologic period.

Fauna, Cambrian: marine invertebrate community of relatively primitive taxa, including the trilobite, which were created in response to steep change in the diversity and composition of Earth's biosphere; dominated during the Cambrian, remained common in the Ordovician, and became progressively rarer in the Silurian and later.

Fauna, Ediacara: earliest known complex multicellular organisms that lived during the Ediacaran Period (about 635 to 542 million years ago), composed mostly of enigmatic tubular and frond-shaped, mostly sessile animals.

Fauna, Modern: marine invertebrates that radiated into the ecospace left vacant by the Late Permian extinctions. Forms included bivalves with longer siphons, carnivorous gastropods, sea urchins, crustaceans, and fish.

Fauna, Paleozoic: marine invertebrate community consisting primarily of bryozoans, mollusks, echinoderms, and brachiopods; present in the Cambrian, became more common in the Ordovician, and dominated the rest of the Paleozoic.

Faunal Succession, Principle of: observation that taxonomic groups of animals follow each other in time in a predictable manner.

feedback: process in which part of the output of a system acts on its input in order to regulate its further output.

felsic: igneous rock dominated by the light-colored, silicon- and aluminum-rich minerals feldspar and quartz, which give felsic rock its characteristic light gray color. Typical felsic rocks include granite and rhyolite. Contrast with mafic rock.

fenestrate bryozoan: microscopic sea animal that lived in colonies that are lace-like in construction. The individual bryozoan lived in microscopic tubes or pores on the lace branches.

First Appearance Datum (FAD): first evolutionary appearance of a key species. Contrast with Last Appearance Datum.

fish: limbless cold-blooded vertebrate animal with gills and fins that lives wholly in water.

fission: nuclear reaction in which the nucleus of an atom splits into two parts and releases a tremendous amount of energy. Contrast with fusion.

fission hypothesis: theory that the Moon was essentially thrown off from Earth because Earth was spinning so rapidly it ejected material that overcame gravitational attraction.

floodplain: area of low-lying ground adjacent to a river, formed mainly of river sediments and subject to flooding.

flora: plants of a particular region, habitat, or geologic period.

flux: flow of matter and energy through systems and their exchange of matter and energy with other systems and the surrounding environment.

flysch: sediments that are uplifted are shed into the deep waters of the foreland basin as shales and turbidites (term coined by geologists working in the Alps).

focal level: particular box or level of interest; its behavior is strongly constrained or controlled by the boxes or levels above and below it, which provide boundary conditions which constrain the behavior of the focal level. See hierarchy.

focus: site of energy release in an earthquake.

folding: when one or a stack of originally flat and planar surfaces, such as sedimentary strata, are bent or curved as a result of permanent deformation.

foliated: metamorphic rock that exhibits a layered structure as a result of differential stress that deforms the rock in one plane.

food chain: sequence of the transfer of food energy from one organism to another in an ecologic community.

food pyramid: graphic representation of human nutritional needs in the form of a pyramid; foods whose recommended daily intake is highest occupy the wider bottom part and foods whose recommended daily intake is lowest occupy the slender top part.

food web: system of interlocking and interdependent food chains or feeding relationships by which energy and nutrients are passed on from one species of living organisms to another.

foot (molluscan): muscle that acts to attach a mollusk to a hard surface (e.g., gastropods), burrow (bivalves), or is modified into a head-like region (cephalopods).

foraminifera: large group of single-celled protists which are among the most common plankton species. They have reticulating pseudopods, fine strands of cytoplasm that branch and merge to form a dynamic net.

fore-arc basin: depression in the sea floor located between a subduction zone and an associated volcanic island arc. Compare with back-arc basin.

foreland basin: depression that develops adjacent and parallel to a mountain belt.

formation: mappable unit of rock that is recognized based only on its lithology. Contrast with facies.

fossil: remnant or trace of an organism of a past geologic age, such as a skeleton or leaf imprint, embedded and preserved in the Earth's crust.

founder effect: loss of genetic variation that occurs when a new population is established by a very small number of individuals from a larger population.

fracking: use of liquid, sand, and chemicals shot through a bore hole to fracture oil-bearing rock formations, enhancing the recovery of oil and natural gas.

fractional crystallization: separation of a cooling magma into multiple minerals as the different minerals cool and congeal at progressively lower temperatures. Contrast with partial melting.

Franciscan mélange: product of the grinding and mixing of rocks (mélange is French for "mixture"), that consists of volcanic breccias, graywackes, mud-stones, cherts, and blueschist.

Franklin orogenic belt: an elongate trough that merges with the Caledonian belt In Northern Greenland. The Franklin orogenic belt was involved in the Ellesmere Orogeny.

freeze–thaw: type of physical weathering that occurs when the water inside of rocks freezes and expands, eventually breaking them apart.

Front Range: mountain range of the Southern Rocky Mountains formed around 300 million years ago as sediment-covered granite began to uplift, giving rise to the Ancestral Rocky Mountains.

Fukushima nuclear accident: series of equipment failures, nuclear meltdowns, and releases of radioactive materials at the Fukushima Nuclear Power Plant, following an earthquake and tsunami in March of 2011.

fungal spike: extraordinary abundance of fungal spores in sediments.

fusion: process by which two or more atomic nuclei join together to form a single heavier nucleus. This is usually accompanied by the release or absorption of large quantities of energy. Contrast with fission.

fusulinid: extinct order within the foraminifera in which the tests (shells) are composed of tightly packed, secreted microgranular calcite.

gabbro: medium- or coarse-grained rocks that consist primarily of plagioclase feldspar and pyroxene. Gabbros are found widely on the Earth and on the Moon.

gas hydrate (methane clathrate): crystalline solid; its building blocks consist of a gas molecule surrounded by a cage of water molecules or ice crystals.

gastropod: mollusk of the large class Gastropoda, such as a snail, slug, or whelk, typically having a one-piece coiled shell and flattened muscular foot with a head bearing stalked eyes.

Gastropoda: large taxonomic class within the phylum Mollusca (gastropods), which includes snails and slugs of all kinds and all sizes from microscopic to quite large.

gene: unit of heredity; a segment of DNA or RNA that is transmitted from one generation to the next, and that carries genetic information such as the sequence of amino acids for a protein.

gene, master control: first gene activated in a hierarchy that leads to differentiation along a particular pathway.

genetic drift: change in the frequency of a gene variant (allele) in a population due to random effects.

genetic engineering: deliberate modification of the characteristics of an organism by manipulating its genetic material.

genetic recombination: process of forming new allelic combination in offspring by exchanges between genetic materials (as exchange of DNA sequences between DNA molecules).

genomic hypothesis: explanation for the origin(s) of metazoans based on macroevolutionary innovation. Contrast with ecologic hypothesis.

genuine progress indicator: concept in green economics that has been suggested to replace, or supplement, gross domestic product as a metric of economic growth; based on whether a country's growth, increased production of goods, and expanding services have resulted in the improvement of the welfare of the people.

geologic time: time of the physical formation and development of the Earth, especially prior to human history.

geology: scientific study of the origin, history, and structure of the natural processes on Earth.

geosyncline: an earlier theory that proposed a downward movement of Earth's crust, forming long, linear troughs in which thick sequences of sediments could accumulate. The evidence in favor of continental drift and seafloor spreading has replaced this theory to the point that the term geosyncline has been abandoned.

ginkgo: deciduous Chinese tree related to the conifer, with fan-shaped leaves and yellow flowers. It has a number of primitive features and is similar to some Jurassic fossils.

glacial environment: where the land is covered with glaciers or masses of ice.

gneiss: foliated metamorphic rock in which the coarse mineral grains have been arranged into a banded structure.

gneiss terrane: zone of high grade Archaean rock that has been differentially exhumed relative to surrounding rock formations.

Golconda terrane: composite of several structurally bounded subterranes made of clastic, volcanic, and carbonate rocks, which was formed during the Late Paleozoic when the Golconda Arc collided with the North American continent.

Gondwana: southernmost of two supercontinents (the other being Laurasia) that later became parts of the Pangaea supercontinent. It existed from approximately 510 to 180 million years ago.

grade, evolutionary: Taxa that do not share a common ancestry but are mistakenly misclassified together. Contrast with clade.

graded bedding: stratification in which each stratum displays a gradation in the size of grains from coarse sediments at the bottom and become progressively finer upward.

Grand Unification Epoch: interval of the creation of the universe in which all known matter, radiation, fundamental forces of nature were one and the same.

granite: medium- or course-grained intrusive rock that is rich in quartz and alkali feldspar. One of the most common rocks of the Earth's crust, it is formed by the cooling of magma.

granitoid: component of the Earth's crust made up of a variety of coarse grained plutonic rock similar to granite which mineralogically are composed predominately of feldspar and quartz.

granodiorite: coarse-grained, plutonic rock containing quartz and plagioclase, between granite and diorite in composition.

graptolite: extinct marine invertebrate animal of the Paleozoic era, which consisted of colonies of tiny polyps with

tentacles and exoskeletons of a nitrogenous substance similar to fingernails.

gravel: Sediment grains >2 mm in size.

gravity slide: tectonic plate motion where plates slide down the side of the mid-ocean ridge by gravity.

grazer: animal which feeds on growing grass or other herbage on the ground. Contrast with browser.

Great American Interchange: important paleozoogeographic event in which land and freshwater fauna migrated from North America via Central America to South America and vice versa, as the volcanic Isthmus of Panama rose up from the sea floor and bridged the formerly separated continents.

Great Ordovician Biodiversification Event (GOBE): a diversification of the Paleozoic Fauna that occurred during the post-Cambrian. Marine fossil evidence Indicates that the GOBE biodiversification surpassed even that of the Cambrian.

greenhouse gas: any of the atmospheric gases that contribute to the greenhouse effect.

greenstone belt: zone of variably metamorphosed mafic to ultramafic volcanic sequences with associated sedimentary rocks that occur within Archean and Proterozoic cratons between granite and gneiss bodies.

green sulfur bacteria: family of obligately anaerobic photoautotrophic bacteria, which are nonmotile and come in spheres, rods, and spirals. See also purple sulfur bacteria.

graywacke: sedimentary rock consisting of angular fragments of quartz, feldspar, and other minerals set in a muddy base. Often includes volcanic fragments.

gross national product: total value of goods produced and services provided by a country during one year, equal to the gross domestic product plus the net income from foreign investments.

ground moraine: rock material carried and deposited in the base of a glacier.

groundplan: basic body plan.

growth fault: type of normal fault that develops and continues to move during sedimentation and typically has thicker strata on the downthrown, hanging wall side of the fault than in the footwall. Growth faults are common in the Gulf of Mexico and in other areas where the crust is subsiding rapidly or being pulled apart.

Gulf Stream: western boundary current of the North Atlantic subtropical gyre.

guyot: seamount of volcanic origin with a flat top. See also seamount.

gymnosperm: seed-bearing, vascular plant; seeds of these plants are not formed in an enclosed ovulary, but naked on the scales of a cone or cone-like structure. Contrast with angiosperm.

gypsum: soft white or gray mineral consisting of hydrated calcium sulfate.

gyre: circular or spiral motion or form; i.e., a giant circular oceanic surface current.

habitable zone (HZ): region within the solar system where life may exist.

hadal environment: deepest part of the ocean, below about 6,000 meters (20,000 feet).

Hadean: earliest eon in the history of the Earth from the 1st accretion of planetary material (around 4,600 million years ago) until the date of the oldest known rocks (about 3,800 million years ago).

hadrosaur ("duck-bill"): large bipedal ornithischian dinosaur having a horny duck-like bill and webbed feet; may have been partly aquatic.

half-life: period of time it takes for the amount of a radioactive substance undergoing decay to decrease by half.

halite: sodium chloride as a mineral, typically occurring as colorless cubic crystals; rock salt.

heat capacity: ratio of the heat energy absorbed by a substance to the substance's increase in temperature.

Heinrich event: periodic episode of rapid ice-rafted debris deposition as a result of massive discharge of icebergs from ice sheets.

herbivore: animal that feeds on plants.

hiatus: time represented by missing sediment.

hierarchy: arrangement or classification of things according to relative importance or inclusiveness, i.e., a set of boxes, each one of which is nested inside a larger box.

highstand systems tract: progradational deposits that form when sediment accumulation rates exceed the rate of increase in accommodation space.

historical science: study of the evolution of Earth and its life-forms by the interpretation of evidence from past events; based on the idea that history is just as important as physical laws.

Holocene: present epoch, which began 10,000 years ago; characterized by a warm climate and the development of modern human culture.

homeobox (*Hox*) gene: one of various similar homeotic genes that are involved in bodily segmentation during embryonic development.

homeostasis: the tendency of a system to maintain its internal stability in response to external disturbances.

hominid: primate of a family (Hominidae) that includes humans and their fossil ancestors.

***Homo*:** genus that includes modern humans and species closely related to them. The genus is estimated to be about 2.3 to 2.4 million years old.

homologous structure: feature that has common evolutionary ancestry but is dissimilar in function, such as the limbs of mammals. Contrast with analogous structure.

horst-and-graben: referring to regions that lie between normal faults and are either above or lower than the area beyond the faults. A horst represents a block pushed upward by the faulting, and a graben is a block that has dropped due to the faulting.

hotspot: area in the Earth's upper mantle, ranging from 100 to 200 kilometers in width, from which magma rises in a plume to form volcanoes.

Hubble Law: observation that there is a relationship between the distance to a galaxy and its red shift: the greater the distance to a galaxy, the greater the red shift.

humans: known taxonomically as *Homo sapiens* (Latin: "wise man" or "knowing man"). the only living species in the *Homo* genus of bipedal primates in Hominidae, the great ape family.

Hume's problem: philosophical question of whether inductive reasoning (established by observation, measurement, or calculation) leads to scientific knowledge.

humic acid: weak acid produced by biodegradation of dead organic (plant) matter.

humus: organic component of soil, formed by the decomposition of leaves and other plant material by soil microorganisms.

hybrid car: vehicle that uses both electricity and gasoline to run.

hybrid vigor: condition where the progeny of a cross (the hybrid) displays greater height, yield, resistance, etc., than the parents.

hydrogen: the lightest element and the most abundant element in the universe, making up about 90% of the universe by weight.

hydrogen fuel cell: cell that uses hydrogen gas and air to create an electrical current to power a vehicle, with only water as a byproduct.

hydrologic cycle: continual flow of water between land, sea, and atmosphere, through evaporation, condensation, and precipitation (rain).

hydrophilic: water-loving (tendency to interact with and be dissolved by water). Contrast with hydrophobic.

hydrophobic: water-hating (repelled by water). Contrast with hydrophilic.

hydrosphere: Earth system consisting of the oceans, rivers and streams, lakes, and ice contained in mountain glaciers and polar ice caps.

hyolith: worm-like animal from the Paleozoic Era that secreted a small conical shell with a lid composed of calcium carbonate.

hydrothermal metamorphism: alteration of rock by hot waters or gases associated with a magmatic source.

hydrothermal vent: fissure in the ocean floor especially at or near a mid-ocean ridge from which mineral-rich super-heated water issues.

hydrothermal weathering: occurs when seawater percolates through hot ocean crust at seafloor spreading centers, altering the ionic composition of seawater.

hyperthermophile: extreme heat-loving. Compare to thermophile.

hypothesis (pl: hypotheses): tentative explanation for an observation, phenomenon, or scientific problem that can be tested by further investigation.

hysteresis loop: rate-dependent feedback loop that results when the threshold to make a change in a system is different than the threshold to undo the change; useful tool to discuss the different possibilities for how the climate system can respond to changes in some controlling variable, for example the input of carbon into the Earth's atmosphere from fossil fuel combustion.

Iapetus Ocean: ancient ocean that was created as a rift began to widen in the Grenville super-continent about 575 million years ago and formed a wide oceanic basin between the continents of Laurentia to the west and Baltica to the east.

ice-rafted debris: objects lifted from a glacier bed and deposited out at sea as the iceberg melts.

ichnofossil: trace fossil, such as that of an animal's track or burrow.

ichthyosaur: extinct marine reptile of the Mesozoic era resembling a dolphin, with a long pointed head, four flippers, and a vertical tail.

impact: force transmitted by a collision, i.e., when an asteroid strikes Earth.

impact crater: crater on the Earth or Moon caused by the impact of a meteorite or other object, typically circular with a raised rim.

impact hypothesis: theory that a small planet the size of Mars struck the Earth just after the formation of the solar system, ejecting large volumes of heated material from the outer layers of both objects, and forming a disk of orbiting material, which eventually stuck together to form the Moon in orbit around the Earth.

impression: mark produced on a surface by pressure that exhibits the general outline of the original organism. See also compression.

index fossil: fossil species that is, ideally, easily identified, widespread, and abundant. Useful for general stratigraphic correlation.

induction: process of reaching a conclusion derived from particular facts or instances.

industrial melanism: adaptive melanism caused by anthropogenic alteration of the natural environment in terms of industrial pollution.

infaunal: describes an animal that lives within, rather than on, the substrate of a body of water, especially in a soft sea bottom. Contrast with epifaunal.

infiltration: slow passage of a liquid through a filtering medium; i.e., the percolation of rainwater through the soil.

inheritance, theory of, acquired characteristics: hypothesis that physiologic changes acquired over the life of an organism may be transmitted to offspring.

inheritance, theory of, blending: hypothesis that inherited traits are determined, randomly, from a range bounded by the homologous traits found in the parents.

inheritance, theory of, particulate: pattern of inheritance discovered by Gregor Mendel showing that characteristics can be passed from generation to generation through "discrete particles" (now known as genes).

insectivore: small insect-eating, mainly nocturnal terrestrial or fossorial (lives under ground) mammal.

insect metamorphosis: refers to the cycle that insects develop, grow, and change form.

interior drainage: system of streams that converge in a closed basin and evaporate without reaching the sea.

intermontane basin: basin between mountain ranges, often formed over a graben.

internal mold: formed when sediments or minerals fill the internal cavity of an organism, such as the inside of a bivalve or snail. Contrast with external mode.

intertidal zone: area between land and sea that is regularly exposed to the air by the tidal movement of the sea.

intrusive igneous rock: rock that formed from magma that cools and solidifies within the crust of a planet. Compare with extrusive igneous rock.

iridium: hard, brittle, silvery-white transition metal of the platinum family, the second-densest element, and the most corrosion-resistant metal, even at temperatures as high as 2,000°C. May occur in concentrations formed by meteor impact, volcanism, or low oxygen conditions.

iron hypothesis: theory that the biologic uptake of carbon dioxide could be accelerated by adding iron to the photic zone of the oceans.

isostasy: state of balance, or equilibrium, which sections of the Earth's lithosphere (whether continental or oceanic) are thought ultimately to achieve when the vertical forces upon them remain unchanged.

isostatic uplift (rebound): rise of land masses that were depressed by the huge weight of rock (in the case of mountains) or ice sheets by erosion or glacial retreat, respectively.

isotope, radioactive: natural or artificially created of a chemical element having an unstable nucleus that decays, emitting alpha, beta, or gamma rays until stability is reached. Contrast with isotope, stable.

isotope, stable: isotope that does not spontaneously undergo radioactive decay. Contrast with isotope, radioactive.

Isua Complex: Archean greenstone belt in southwestern Greenland aged between 3.7 and 3.8 Ga, making it among the oldest rock (crust) in the world.

jet stream: fast flowing, narrow air currents found in the atmosphere of Earth; the major jet streams on Earth are westerly winds (flowing west to east).

Juan de Fuca Plate: tectonic plate, generated from the Juan de Fuca Ridge, and still subducting under the northerly portion of the western side of the North American Plate at the Cascadia subduction zone.

Jurassic: period from about 208 million to 146 million years ago; large reptiles were dominant on both land and sea, ammonites were abundant, and the first birds appeared.

karst: area of irregular limestone in which erosion has produced fissures, sinkholes, underground streams, and caverns.

karst topography: landscape that is characterized by numerous caves, sinkholes, fissures, and underground streams.

Kaskaskia transgression: see Sequence, Kaskaskia.

Kazakhstania: small continental region in the interior of Asia, centered on Kazakhstan.

Kellwasser events: term given to the Late Devonian extinctions (one of the five mass extinctions), which occurred as two pulses of anoxia; associated with widespread shale deposition and negative carbon isotope excursions.

Klamath Island Arc: volcanic arc terrane accreted to the western margin of North America, which moved eastward to initiate the Antler Orogeny in the Late Devonian, depositing ophiolites, graywackes, and other submarine rocks and sediments in Nevada.

Labyrinthodont (temnospondyls): member of the extinct subclass of amphibians, which constituted some of the dominant animals of Late Paleozoic and Early Mesozoic times (about 390 to 210 million years ago).

lacustrine environment: lake or lake-like.

Lagerstätte (pl: Lagerstätten): sedimentary deposit that exhibits extraordinary fossil richness or completeness.

Lagomorpha: order of mammals, including hares, rabbits, and pikas, that are distinguished by the possession of double incisor teeth, and were formerly placed with the rodents.

Lake Bonneville: ice-age lake that formed in central Utah from melting mountain glaciers. The Great Salt Lake today is a much smaller remnant of Lake Bonneville.

La Niña: coupled ocean-atmosphere phenomenon characterized by unusually cold ocean temperatures in the eastern equatorial Pacific, as compared to El Niño, which is characterized by unusually warm ocean temperatures in the Equatorial Pacific.

large-number system: system composed of many components; studied primarily within experimental science. Contrast with small-number system.

Last Appearance Datum (LAD): last recorded appearance of a key species. Contrast with First Appearance Datum.

laterite: red residual soil produced by rock decay; contains insoluble deposits of ferric and aluminum oxides.

Laurentia: continental plate that was in existence from the Late Precambrian to the Silurian; today comprises North America and Greenland.

lava: hot molten or semifluid rock erupted from a volcano or fissure, or solid rock resulting from cooling of this.

law: concise relationship between phenomena that is considered a fundamental truth of science, for example Newton's law of motion.

limestone: hard sedimentary rock composed mainly of calcium carbonate or dolomite.

lipid bilayer: enclosing layer of phospholipid molecules found in a cell membrane. The hydrophobic parts of the molecules are in the middle of the bilayer, and the hydrophilic parts are on the inner and outer surfaces.

lithification: conversion of a newly deposited sediment into an indurated rock.

lithocorrelation: process of mapping rock units using lithology, or rock type, alone.

lithology: study of the general physical characteristics of rocks.

lithosphere: crust and upper mantle of the Earth. Compare with asthenosphere.

lithostratigraphy: science of dating rocks based on their physical and petrographic properties. See also biostratigraphy and chronostratigraphy.

Little Climate Optimum (Medieval Warm Period): time of warm climate in the North Atlantic region, which may have been related to other climate events around the world during that time, lasting from about 950 to 1250 A.D.

Little Ice Age: period of cooling that occurred after the Medieval Warm Period (Medieval Climate Optimum) about 1300 A.D. to about 1850 to 1880 A.D.

living fossil: informal term for an organism that lived during ancient times and still lives today, relatively unchanged, like the horseshoe crab, gingko tree, cycads, and other well-adapted organisms.

lobe-finned fish: fish of a largely extinct group having fleshy lobed fins, including the probable ancestors of the amphibians. See also labyrinthodont.

lowstand systems tract: deposits that accumulate after the onset of a relative sea-level rise.

lycopod: member of a group of plants that includes giant trees in the Carboniferous coal swamp forests and the spore-bearing clubmoss.

macroevolution: major evolutionary change. The term applies mainly to the evolution of whole taxonomic groups over long periods of time. Contrast with microevolution.

macroscopic behavior: collective behavior of a system used to understand the average, or bulk, behavior of a microscopic component. Contrast with microscopic behavior.

mafic: igneous rock that contains or is related to a group of dark-colored minerals, composed chiefly of magnesium and iron, that occur in igneous rocks. Common mafic rocks include basalt and gabbro. Contrast with felsic rock.

magma: body of molten rock that occurs below the surface of the Earth.

magma mixing: process by which two magmas meet, comingle, and form a magma of a composition somewhere between the two end-member magmas.

magma plume: upwelling of abnormally hot rock within the Earth's mantle.

magma ocean: global-scale ocean of magma, according to some calculations several hundred kilometers deep, thought to have existed during the final stages of accretion as the Earth was forming.

magmatic differentiation: process by which chemically different igneous rocks, such as basalt and granite, can form from the same initial magma.

magnetic reversal: change in the polarity of the Earth's magnetic field.

mammal: warm-blooded vertebrate animal of a class that is distinguished by the possession of hair or fur, the secretion of milk by females for the nourishment of the young, and (typically) the birth of live young.

mammal, archaic: mammal belonging to or characteristic of a much earlier ancient period.

mammal, modern: mammal that lives in present or recent times.

mammoth: large extinct elephant of the Pleistocene epoch, typically hairy with a sloping back and long curved tusks.

Manicouagan crater: one of the oldest known impact craters located in Québec, Canada; thought to have been caused by the impact of a 5 kilometer (3 mile) diameter asteroid about 215.5 million years ago.

mantle: layer of the Earth between the crust and the core.

mantle plume: upwelling of abnormally hot rock within the Earth's mantle.

marble: nonfoliated metamorphic rock composed of recrystallized carbonate minerals, most commonly calcite or dolomite.

marginal marine environment: where land and sea meet, the environments associated with coasts, bays, barrier islands, and estuaries that are often heavily influenced by the fresh waters and sediment brought by rivers and streams.

marine environment: relating to oceans, seas, bays, estuaries, and other major water bodies.

marine reptiles (euryapsids): reptiles which have become secondarily adapted for an aquatic or semiaquatic life in a marine environment.

marker bed: stratified unit with distinctive characteristics making it an easily recognized geologic horizon.

marsupial: mammal, such as a koala, kangaroo, or opossum, characterized by the presence of an abdominal pouch in which the young, which are born in a very undeveloped condition, are carried for some time after birth. Contrast with placental.

mass balance: observed measure of the change in mass of one reservoir at a certain point for a specific period of time to determine its effect on the mass of substances in other reservoirs.

mass extinction: sharp decrease in the diversity and abundance of life forms on Earth within a relatively short period of geologic time.

massif: massive topographic and structural feature, especially in an orogenic belt, commonly formed of rocks more rigid than those of its surroundings.

mastodon: large, extinct, elephant-like mammal of the Miocene to Pleistocene epochs, having teeth of a relatively primitive form and number.

maturity, compositional: type of maturity in sedimentary rocks in which a sediment approaches the compositional end product to which formative processes drive it.

maturity, textural: expression of the sorting, matrix content, and grain angularity in a sediment.

mechanistic view of causality: way of describing nature as if it is a machine.

megamonsoon: gigantic monsoon regime that occurred as the result of climatic extremes from the assembly of Pangaea in the Late Paleozoic.

mélange: jumbled mass of rock and sediment in deep water consisting of volcanic debris, muds, graywackes, turbidites, and pieces of ocean crust.

meridional ocean circulation: atmospheric circulation in a vertical plane oriented along a meridian; it consists, therefore, of the vertical and the meridional (north or south) components of motion only.

mesonychids: extinct order of medium to large-sized carnivorous mammals that were closely related to artiodactyls (eventoed ungulates) and to cetaceans (dolphins and whales).

Mesozoic: era from about 248 million to 65 million years ago, during which dinosaurs, birds, and flowering plants first appeared.

Mesozoic marine revolution: fundamental restructuring of marine ecosystems during the Mesozoic era, particularly in the Jurassic and Cretaceous, caused by increased predation pressure.

messenger RNA (mRNA): molecule of RNA encoding a chemical "blueprint" for a protein product. mRNA is transcribed from a DNA template, and carries coding information to the sites of protein synthesis: the ribosomes.

metabolism: chemical processes that occur within a living organism in order to maintain life.

metamorphic rock: rock that has been subjected to changing temperatures and pressures that physically and chemically transform it from one type to another.

metazoan: multicellular animal.

meteorite: natural object originating in outer space that survives impact with the Earth's surface.

meteorite, iron: meteorite composed of metallic material.

meteorite, stony: meteorite composed of rock material.

meterorite, stony-iron: meteorite composed of mixed material (rock and metals).

methanogenic (bacteria): bacteria that produce methane.

method of multiple working hypotheses: development of several hypotheses to explain the particular phenomenon being studied. See also multiple causation.

methyl tert-butyl ether (MTBE): fuel derived from methanol that has been discovered in groundwater supplies, thus leading to legislation banning its use in many states.

microcontinent: section of continental lithosphere that has broken off from a larger, distant continent, as by rifting.

microevolution: evolutionary change within a species or small group of organisms, especially over a short period of time. Contrast with macroevolution.

microfossil: very small fossil, best studied with the aid of a microscope.

microplate: fragment of crustal plate broken off from a larger, distant plate during rifting. Also called terrane.

microscopic behavior: individual behavior of each and every component at all times in a system composed of many parts. Contrast with macroscopic behavior.

microspherule: microscopic spherical crystalline or elliptical body resulting from meteor impact or volcanism.

microtektite: microscopic tektite found in ocean sediments and polar ice. See strewn field.

microwave radiation: electromagnetic radiation in the wavelength range 0.3 to 0.001 meters. See also cosmic background radiation.

Mid-Continent (Keweenawan) Rift: 2,000 km (1,200 mi) long geologic rift in the center of the North American continent and south-central part of the North American plate. It formed when the continent's crust began to split apart about 1.1 billion years ago.

mid-ocean ridge: elevated region with a central valley on an ocean floor at the boundary between two diverging tectonic plates where new crust forms from upwelling magma.

migmatite: rock composed of a metamorphic (altered) host material that is streaked or veined with granite rock; the name means "mixed rock."

Milankovitch cycles: Cycles of regular change over thousands of years in the shape of Earth's orbit, in the angle of tilt of its axis, and in its orientation toward other celestial bodies that change the intensity of solar radiation primarily at high latitudes.

Miller-Urey experiment: simulated hypothetical conditions thought at the time to be present on the early Earth and tested for the occurrence of chemical evolution.

Miocene: epoch from about 24 million to 5 million years ago, during which the modern ocean currents were established and Antarctica became frozen.

missing link: hypothetical fossil form intermediate between two living forms, i.e., between humans and apes.

Mississippian: period from about 363 to 323 million years ago.

Mississippi Embayment: physiographic feature that extends along the course of the Mississippi River from southern Illinois to the Gulf Mexico; thought to correspond to an ancient aulacogen buried deep beneath the surface.

mitochondrion (pl: mitochondria): organelle found in large numbers in most cells, in which the biochemical processes of respiration and energy production occur.

mobile (orogenic) belt: linear region that has undergone folding or other deformation during the orogenic cycle.

model: sophisticated hypothesis that tries to take into account all the components or processes of a system that are important to the system while excluding those that are not considered important.

Mohorovičić discontinuity (Moho): boundary between the Earth's crust and the mantle.

molasse: sediment consists of nonmarine sediments shed directly from the mountains onto river floodplains, swamps, and coastal and lagoon environments that may extend landward for hundreds of kilometers (term coined by geologists working in the Alps).

molecular clock: central concept of molecular evolution, which postulates that a gene evolves at a constant rate over time as long as its function does not change.

monoplacophoran: ancient animal possessing repeated sets of organs, possible ancestor of the mollusk.

monsoon: wind system that influences large climatic regions and reverses direction seasonally, characteristically accompanied by heavy rainfall.

mosasaur: large extinct marine reptile of the late Cretaceous period, with large-toothed jaws, paddle-like limbs, and a long flattened tail, related to the monitor lizards.

Mowry Sea: large inland sea (arm of the Arctic Ocean) that transgressed south over western North America during the mid-Cretaceous about 100 million years ago, leaving widespread deposits of black shale (clay and volcanic ash).

mud: Sediment grains <0.062 mm in size.

mud crack: irregular fracture formed by shrinkage of clay, silt, or mud under the drying effects of atmospheric conditions at the surface.

mudstone: fine-grained, dark gray sedimentary rock, formed from silt and clay and similar to shale but without laminations.

multiple causation: idea that a given response is likely to be the result of multiple variables and that any specific variable typically affects multiple responses. See also method of multiple working hypotheses.

mummification: mode of preservation in which dry environments inhibit organic decay.

nappe: formed by a sheet of rock that has moved sideways over neighboring strata as a result of an overthrust or folding.

natural selection, balancing: refers to a number of selective processes by which multiple alleles are actively maintained in the gene pool of a population at frequencies above that of gene mutation.

natural selection, directional: mode of natural selection in which a single phenotype is favored, causing the allele frequency to continuously shift in one direction.

nautiloid: early cephalopod that has a straight or coiled shell divided internally into a series of chambers of increasing size connected by a central tube; includes the modern nautiluses as well as numerous extinct species dating back as far as the Cambrian Period.

nebula: interstellar cloud of dust, hydrogen gas, helium gas, and other ionized gases.

nekton: aquatic animals that are able to swim and move independently of water currents.

neo-Darwinism: modern Darwinian theory that explains new species in terms of genetic mutations.

neogastropod: gastropod that evolved by the Cretaceous to become a predator with the ability to bore through calcareous shells and in some cases to inject poison into prey.

Neogene: period from about 23 million to 1.6 million years ago; mammals continued to evolve during this time, developing into the forms that are familiar today.

Neoproterozoic: ("new" Proterozoic) the last phase of the Precambrian, from 1.3 billion to about 450 million years ago. This period saw abrupt, radical climate changes involving runaway glaciations of most or all of Earth.

Neptunism: theory (erroneous) that rocks such as granite were formed by crystallization from the waters of a primeval ocean.

neritic zone: ocean waters from the low tide mark to a depth of about 200 meters. See also continental shelf.

niche: particular role or function of a species with the other members of an ecosystem.

nonconformity: unconformity where igneous and metamorphic rocks may also be eroded and sediments deposited on top.

nonfoliated: metamorphic rock that does not display planar patterns of strain.

nonlinear behavior: relatively small changes in a system produce an effect that is out of all proportion to the change.

normal fault: occurs when a block of rock—called the "hanging wall"—moves downward relative to the block on the other side of the fault, called the "footwall." Contrast with reverse fault.

North Atlantic Deep Water (NADW): Deep water mass that forms off Iceland and Greenland, flowing southward in the Atlantic Basin and eventually mixing with Antarctic Bottom Water.

nothosaur: extinct semiaquatic carnivorous reptile of the Triassic period, having a slender body and long neck, related to the plesiosaurs.

nuclear accident: an event that has led to significant consequences to people, the environment (i.e., damage from a large radioactivity release), or the facility (i.e., a reactor core melt).

nucleoid: region within the cell of prokaryotes, which has nuclear material without a nuclear membrane and where the genetic material is localized. Contrast with nucleus.

nucleosynthesis: fusion of atomic nuclei more complex than the hydrogen atom.

nucleotide: basic structural unit of nucleic acids (DNA and RNA).

nucleus: organelle present in most eukaryotic cells, typically a single rounded structure bounded by a double membrane, containing the genetic material. Contrast with nucleoid.

nummulitids: taxon of large, pancake-shaped benthic foraminifera that lived in shallow seaways; characterized by complex interiors, with smaller compartments joined by tunnel-like connections thought to have harbored algal symbionts.

nutrient: any substance required by an organism for normal growth and maintenance.

obliquity (tilt): angle between Earth's orbit plane and the plane of the Earth's equator.

oceanic anoxic event: occurs when the Earth's oceans become completely depleted of oxygen (O_2) below the surface levels.

oceanic burp hypothesis: theory that a great out-gassing of methane gas trapped under the ocean caused an explosion that deposited limestone and other organic-rich shale from the ocean floor onto the continental shelf.

oceanic-continent plate boundary: plate whose leading edge consists of ocean crust is subducted beneath another plate whose leading edge consists of continental crust.

oceanic conveyor: unifying concept that connects the ocean's surface and thermohaline (deep mass) circulation regimes, transporting heat and salt on a planetary scale. See also thermohaline circulation.

oceanic–oceanic plate boundary: forms by subduction of the leading edge of an oceanic crustal plate beneath the margin of another plate comprised of ocean crust.

Ockham's Razor (Principle of Parsimony): notion that the simplest explanation of a phenomenon is the one that is most likely to be correct.

Olbers' paradox: argument that the darkness of the night sky conflicts with the assumption of an infinite and eternal static universe.

Old Red Sandstone: minor supercontinent created in the Devonian as the result of a collision between the Laurentian, Baltica, and Avalonia cratons (Caledonian orogeny).

Oligocene: epoch from about 38 million to 24 million years ago, during which primates first appeared.

ontogeny: origin and development of an individual organism from embryo to adult. Compare with phylogeny.

oöid: A small calcium carbonate- or iron-coated grain generally found on the seafloor.

oölite: a particular type of limestone formed from oöids, which are spherical grains composed of concentric layers.

Oort cloud: immense spherical cloud surrounding the planetary system and extending approximately 3 light years, about 30 trillion kilometers from the sun.

ooze: soft deposit (as of mud, slime, or shells) on the bottom of a body of water.

ooze, calcareous: layers of muddy, soft rock sediment on the seafloor containing skeletons made of calcium carbonate.

ooze, siliceous (diatom): layers of soft pelagic sediment on the seafloor containing skeletons made of silica.

open system: physical system where reservoirs of the systems exchange matter (chemical substances) and energy (like sunlight) with their surrounding environment.

ophiolite: section of the Earth's oceanic crust and the underlying upper mantle that has been uplifted and exposed above sea level and often emplaced onto continental crustal rocks.

ordinary chondrite: class of stony chondritic meteorite that comprises almost 85% of all finds, hence the name.

Ordovician: period from about 510 million to 439 million years ago. It saw the diversification of many invertebrate groups and the appearance of the first vertebrates (jawless fish).

organelle: specialized subunit within a cell that has a specific function and is usually separately enclosed within its own lipid bilayer.

Organization of Petroleum Exporting Countries (OPEC): cartel of twelve countries (Algeria, Angola, Ecuador, Iran, Iraq, Kuwait, Libya, Nigeria, Qatar, Saudi Arabia, the United Arab Emirates, and Venezuela) formed in 1961 to agree on a common policy for the production and sale of petroleum.

Original Horizontality, Principle of: theory that layers of sediment are originally deposited horizontally under the action of gravity.

Ornithischia: one of the two orders of dinosaurs including armored dinosaurs; boneheaded and horned dinosaurs (marginocephalians); and duck-billed dinosaurs; distinguished by having a bird-like pelvis.

ornithopod: group of bird-hipped dinosaurs that started out as small, bipedal running grazers, and grew in size and numbers until they became one of the most successful groups of herbivores in the Cretaceous world, and dominated the North American landscape.

orogen: total mass of rock deformed during an orogeny.

Orogenic Belt, Cordilleran (Cordillera): western margin of the Americas, consisting of a mountain chain the extends 1,000 kilometers (600 miles) from Alaska down western North America as the Rocky Mountains into South America as the Andes.

Orogenic Belt, Franklin: located in Arctic Canada and northern Alaska and made up from Silurian and Devonian conglomerates and sandstone derived from Caledonian mountains, which were carried westward along a deep-water trough and deposited as a huge, elongate deep-sea fan.

orogenesis: mountain building. See also orogeny.

orogeny: process of mountain formation, especially by a folding and faulting of the Earth's crust. See also orogenesis.

Orogeny, Acadian: dating back 400 to 325 million years, which should not be regarded as a single event but rather as a chain of mountain building events; mostly responsible for the deformation in the northern Appalachians between New York and Newfoundland.

Orogeny, Alleghenian or Appalachian: occurred approximately 350 million to 300 million years ago, in the Carboniferous period. The orogeny was caused by Africa colliding with North America; it exerted massive stress on what is today the Eastern Seaboard of North America, forming a wide and high mountain chain.

Orogeny, Alpine: occurred in the Late Mesozoic when the continents of Africa and India and the small Cimmerian plate collided (from the south) with Eurasia in the north and formed the mountain ranges of the Alpine belt.

Orogeny, Ancestral Rockies: thought to have been formed by high-angle faulting in response to compressive stresses generated during the collision of Gondwana with Laurasia. The uplifts consisted of Precambrian crystalline (mainly metamorphic) basement rocks that were later eroded and rapidly deposited as conglomerates and arkoses along the margin of the modern Front Range.

Orogeny, Andean: series of orogeny occurring along the margins of the Pacific Ocean forming the modern Andes Mountains.

Orogeny, Antler: caused by the collision of the Antler volcanic island arc terrane with what was then the west coast of North America (between Utah and Nevada) during Late Devonian and Early Mississippian time, resulting in extensive deformation of Paleozoic rocks in Nevada and western Utah.

Orogeny, Caledonian: caused by the closure of the Iapetus Ocean when the continents and terranes of Laurentia, Baltica, and Avalonia collided.

Orogeny, Coast Ranges: period, mostly during the late Pliocene, of major deformation (mostly mountain building and faulting), metamorphism, and volcanic activity in the Coast Ranges in California.

Orogeny, Cordilleran: ranged from the Jurassic Period into the Eocene Epoch of the Cenozoic Era and uplifted the different regions that formed the Cordilleran Orogenic Belt.

Orogeny, Ellesmere: resulted from southward downward thrusting following a collision that occurred between northern Canada and possibly an arc or microcontinent as evidenced by middle Paleozoic granites, ultramafic and volcanic rocks.

Orogeny, Grenville: long-lived Mesoproterozoic mountain building event associated with the assembly of the supercontinent Rodinia, which spans a significant portion of the North American continent.

Orogeny, Hercynian or Variscan: caused by Late Paleozoic continental collision between Euramerica (Laurussia) and Gondwana to form the super-continent of Pangaea.

Orogeny, Himalayan: caused when the continent of India rifted away from Gondwanaland and collided into Asia in the Paleocene around 60 million years ago, causing the uplift of Tibet and the closure of the Tethys Sea, and leading to the formation of the Himalayan Mountains.

Orogeny, Laramide: third major mountain building episodes to transform western North America, affecting the topography of the central Rocky Mountains and adjoining Laramide regions (from central Montana to central New Mexico) during the early Cenozoic Era.

Orogeny, Mazatzal-Pecos: occurred around 1.68 to 1.65 Ga created by the collision of the Mazatzal island arc with the Yavapai continent accreted along the southern margin of North America into the western, southwestern, and central United States. See also Yavapai Orogeny.

Orogeny, Nevadan: first of three major mountain building episodes to transform western North America between the Late Mesozoic and Early Cenozoic Eras (about 180 to 140 million years ago); many of the plutons emplaced during this orogeny were later exposed by faulting to form the Sierra Nevada of eastern California.

Orogeny, Ouachita: occurred along the Gulf Coast as the result of a collision of South America with Laurentia; the collision that formed the Ouachita Mountains is thought to have been an extension of the Alleghenian Orogeny into the southwestern United States because the Ouachita Mountains lie on a line curving east and north into the Appalachians.

Orogeny, Pan African: series of major Neoproterozoic mountain-building events that related to the formation of the supercontinent Gondwana about 600 million years ago.

Orogeny, Sevier: second of three major mountain building episodes to transform western North America. This orogeny was the result of subduction of the oceanic Farallon Plate underneath the continental North American Plate that occurred between approximately 140 and 50 million years ago.

Orogeny, Sierra Nevada: formed from the uplift and intrusion of granitic plutons originally emplaced during the Jurassic Nevadan Orogeny.

Orogeny, Sonoma: period of mountain building in western North America, which occurred during the Permian/Triassic transition, around 250 million years ago; as a result the Sonomia terrane was added to what is now the western margin of the United States.

Orogeny, Taconic: mountain building period (responsible for the formation of the present Appalachian Mountains) that ended 440 million years ago and affected most of modern-day New England; involved the uplift of a volcanic island arc between the continents Laurentia and Baltica.

Orogeny, Trans-Hudson: occurred about 2.0 to 1.8 billion years ago and formed the Precambrian Canadian Shield, the North American craton (also called Laurentia), and the forging of the initial North American continent.

Orogeny, Uralian: involved the collision of Baltica and the smaller continent of Kazakhstan. Much of the rest of Asia during this time grew as a result of the collisions of microcontinents, such as Tibet, and terranes that had rifted northward from Gondwana.

Orogeny, Wopmay: occurred about 1.9 to 1.8 billion years ago in northwest Canada near the Arctic Circle.

Orogeny, Yavapai: major mountain building event around 1.68 to 1.65 Ga created by the collision of the Mazatzal island arc with the Yavapai continent accreted along the southern margin of North America into the western, southwestern, and central United States. See also Mazatzal-Pecos Orogeny.

orographic effect: atmospheric condition that results from or is enhanced by lifting of an air mass over mountains.

oscillation: climate pattern that can reach around the globe to affect the day-to-day weather, fluctuating on time scales ranging from days to decades.

Osteichthyes: taxonomic class of bony fish, characterized by a skeleton reinforced by calcium phosphate, the most diverse and abundant of all vertebrates.

ostracoderms: taxonomic class of early, small (10 to 20 cm) fish that possessed a bony skin or covering over their heads. The other parts of the ostracoderm body lacked a hard internal skeleton and is thought to have been cartilaginous. Ostracoderms lacked jaws and thus were likely not active predators.

outgassing (degassing): release of juvenile gases and water to the surface from a magma source.

outwash: material carried away from a glacier by meltwater and deposited beyond the moraine.

overkill hypothesis: argues that humans were responsible for the Late Pleistocene extinction of megafauna in northern Eurasia and North and South America.

overlapping (or concurrent) range zone: range of individual species within a biogeographic province. Contrast with total range zone.

overshoot: when a system exceeds equilibrium.

oviparous: describes an animal that produces its young by means of eggs that are hatched after they have been laid by the parent. Contrast with viviparous.

oxbow lake: crescent-shaped lake formed when a meander of a river or stream is cut off from the main channel.

oxygen isotope: stable isotope used to study the behavior of the hydrologic cycle, especially changes in ice volume, through time.

oxygen isotope, terminations: rapid shift to lower values than indicated rapid warming and melting of ice sheets; used to calibrate ages derived from reversals of Earth's magnetic field.

oxygen minimum zone (OMZ): zone in which oxygen saturation in seawater in the ocean is at its lowest, at depths of about 200 to 1,000 meters.

ozone: colorless, odorless reactive gas comprised of three oxygen atoms, found naturally in the Earth's stratosphere, where it absorbs the ultraviolet component of incoming solar radiation.

P-wave: type of elastic wave, also called seismic wave, which can travel through gases (as sound waves), solids, and liquids, including the Earth. Contrast with S-wave.

pachycephalosaur: bipedal herbivorous dinosaur of the late Cretaceous period with a thick bony covering over its skull.

Pacific Plate: oceanic tectonic plate that lies beneath the Pacific Ocean. At 103 million square kilometers, it is the largest tectonic plate. Northwest movement of the plate resulted in the origin of the San Andreas Fault.

Paleocene: epoch from about 63 million to 58 million years ago; appearance of birds and earliest mammals.

Paleocene–Eocene Thermal Maximum (PETM): climate change, which occurred over a few tens of thousands of years during the Paleocene and Eocene epochs, characterized by rapid global warming, profound changes in ecosystems, and major perturbations in the carbon cycle.

Paleogene: period from about 65 million to 23 million years ago; mammals diversified following the demise of the dinosaurs, and many new forms appeared.

paleosol: ancient soil preserved by burial underneath either sediments or volcanic deposits, which in the case of older deposits have lithified into rock.

Paleozoic: era from about 570 million to 248 million years ago, during which fish, insects, amphibians, reptiles, and land plants first appeared.

Pangaea: supercontinent including all the landmass of the Earth that existed during the Paleozoic and Mesozoic eras about 250 million years ago, before the component continents were separated into their current configuration.

Pannotia: supercontinent that existed from the Pan-African orogeny about 600 million years ago to the end of the Precambrian about 540 million years ago. Believed to have been produced when Laurentia (North America) collided with South America.

panspermia: theory that life on the Earth originated from microorganisms or chemical precursors of life present in outer space and able to initiate life on reaching a suitable environment.

Panthalassa: vast global ocean that surrounded the supercontinent Pangaea, during the late Paleozoic and the early Mesozoic years.

pantothere: animal of the extinct order Pantotheria that lived during the late Mesozoic Era, believed to be the ancestor of the marsupial and placental mammals.

ParaTethys Sea: large shallow sea formed about 34 million years ago as an extension of a rift that formed the Central Atlantic Ocean and was isolated during the Oligocene epoch; stretched north of the Alps over Central Europe to the Aral Sea in western Asia.

parent isotope: "starting point" of a radioactive decay series; decays by giving off radiation, changing into another element, or isotope, of the original element (the daughter isotope).

partial melting: incomplete melting of a rock composed of minerals with differing melting points. Contrast with fractional crystallization.

passerine (bird): perching bird, which belongs to the largest order birds; over half of the world's bird species are passerine.

passive margin: margin the of continent not presently associated with subduction or other types of large-scale tectonic activities such as orogenesis. Contrast with active margin.

peat: dark-brown or black soil material produced by the partial decomposition and disintegration of mosses, sedges, trees, and other plants that grow in marshes and other wet places.

pelagic: of, relating to, or living in open oceans or seas rather than waters adjacent to land or inland waters.

Pennsylvanian: period from about 323 to 290 million years ago.

period (of geologic time scale): unit of geologic time during which a system of rocks formed.

peripheral isolate: form of allopatric speciation in which a new species is formed from a small population isolated at the edge of the ancestral population's geographic range. See also deme.

Perissodactyla: order of mammals, including the horse, rhinoceros, and many extinct taxa, that comprises the odd-toed ungulates (perissodactyls).

permafrost: thick subsurface layer of soil that remains frozen throughout the year, occurring chiefly in polar regions.

Permian: period from about 290 million to 245 million years ago; climate was hot and dry during this period, which saw the extinction of many marine animals, including trilobites, and the proliferation of reptiles.

Permian Reef Complex: lays along the equator, in the lee of the Ouachita Mountains (New Mexico and Texas), along the margins of deep-water troughs that had developed in response to the collision of South America with the southern United States during the Ouachita Orogeny.

permineralization: process of fossilization in which mineral deposits fill voids of bone, wood, or shell and preserve structure in fine detail.

petrifaction: general term referring to the processes of fossilization in which dissolved minerals replace organic matter.

petroleum exploration and production: search for hydrocarbon deposits beneath the Earth's surface, such as oil and natural gas and methods for their extraction.

pH: measure of the acidity or alkalinity of a solution, numerically equal to 7 for neutral solutions, increasing with increasing alkalinity and decreasing with increasing acidity.

phaneritic: refers to the texture of an igneous rock in which the size of matrix grains in the rock are large enough to be distinguished with the unaided eye. Contrast with aphanitic.

Phanerozoic: current eon, beginning about 545 million years ago, and the one during which abundant animal life has existed.

phosphorite: phosphorus-rich sedimentary rock.

photic zone: region of the ocean through which light penetrates and where photosynthetic marine organisms live.

photoautotrophic: being autotropic and utilizing energy from light.

photolysis (photodissociation): chemical process by which molecules are broken down into smaller units through the absorption of light.

phyletic gradualism: model of evolution that theorizes that most speciation is slow, uniform, and gradual. Contrast with punctuated equilibrium.

phyllite: intermediate-grade, foliated metamorphic rock type that resembles its sedimentary parent rock, shale, and its lower-grade metamorphic counterpart, slate.

phylogeny: sequence of events involved in the evolutionary development of a species or taxonomic group of organisms. Compare with ontogeny.

phylum: taxonomic rank below kingdom and above class.

Phylum Arthropoda: jointed-foot invertebrates including arachnids, crustaceans, insects, millipedes, centipedes, and the extinct trilobites.

Phylum Annelida: segmented worms including earthworms, lugworms, and leeches.

Phylum Brachiopoda: marine invertebrates that resemble mollusks.

Phylum Chordata: comprises true vertebrates and animals having a notochord, a dorsal (or upper) nerve cord, and paired gill slits like those found in fish.

Phylum Cnidaria: includes hydras, polyps, jellyfish, sea anemones, and corals.

Phylum Echinodermata: radially symmetric marine invertebrates including starfish, sea urchins, and sea cucumbers.

Phylum Mollusca: includes gastropods, bivalves, cephalopods, and chitons.

Phylum Porifera: multicellular organisms having less-specialized cells than in the Metazoa; includes sponges.

physical weathering: process by which rocks break down through natural, physical means.

phytolith: microscopic granules of silica found in plants.

phytoplankton: plankton consisting of free-floating algae, protists, and cyanobacteria. Phytoplankton form the beginning of the food chain for aquatic animals and fix large amounts of carbon, which would otherwise be released as carbon dioxide.

Piedmont: plateau region located in the eastern United States between the Atlantic Coastal Plain and the main Appalachian Mountains, stretching from New Jersey in the north to central Alabama in the south; also known as the crystalline Appalachians for the predominance for igneous and metamorphic rocks.

pillow lava: lava that has solidified as rounded masses, characteristic of eruption under water.

placental: mammal whose young develops inside the mother (within a placenta). Contrast with marsupial.

placoderm: fish-like vertebrate with bony plates on head and upper body; dominant in seas and rivers during the Devonian; considered the earliest vertebrate with jaws.

placodont: extinct marine shellfish-eating reptile of the Triassic period, having short flat grinding palatal teeth and sometimes a turtle-like shell.

plane of the ecliptic: plane of the Earth's orbit around the sun.

planet, Jovian: large planet that is not primarily composed of rock or other solid matter. Contrast with terrestrial planet.

planet, terrestrial (inner): planet that is composed primarily of silicate rocks and/or metals. Contrast with Jovian planet.

planetesimal: remains of a small planet thought to have orbited the sun during the formation of the planets.

plankton: small or microscopic organisms, including algae and protozoans, that float or drift in great numbers in fresh- or saltwater.

plant: Any of various photosynthetic, eukaryotic, multicellular organisms of the kingdom Plantae characteristically producing embryos, containing chloroplasts, having cellulose cell walls, and lacking the power of locomotion.

plant, terrestrial: plant that grows on land.

plasma: charged particles ejected from the upper atmosphere of the sun. See solar wind.

plate: section of the Earth's lithosphere, constantly moving in relation to the other sections.

plate boundary: place where two or more plates in the Earth's crust meet.

plate tectonics: theory that the Earth's outer rigid shell is composed of several dozen "plates," or pieces, that float on a ductile mantle, like slabs of ice on a pond.

Pleistocene: epoch from about 2 million to 11 thousand years ago; extensive glaciation of the northern hemisphere; the time of human evolution.

plesiosaur: large extinct marine reptile of the Mesozoic era, with a broad flat body, large paddle-like limbs, and typically a long flexible neck and small head.

Pliocene: epoch from about 3 million to 2 million years ago; growth of mountains; cooling of climate; more and larger mammals.

pluton: body of intrusive igneous rock that crystallized from magma slowly cooling below the surface of the Earth. Plutons include batholiths, dikes, sills, laccoliths, and other igneous bodies.

polar easterlies: winds that originate at the polar highs and blow to the subpolar lows in a east to west direction.

polypeptide: protein or part of a protein made of a chain of amino acids joined by a peptide bond.

preadaptation: characteristic evolved by an ancestral species or population that serves an adaptive though different function in a descendant.

prebiotic: existing or occurring before the emergence of life.

Precambrian: time from 4,650 million years ago to the base of the Phanerozoic Eon, during which the Earth's crust consolidated and primitive life first appeared.

precession of the equinoxes: gradual shift in the orientation of Earth's axis of rotation, which, like a wobbling top, traces out a pair of cones joined at their apices in a cycle of approximately 26,000 years.

precipitation: any form of water, such as rain, snow, sleet, or hail, which falls to the Earth's surface. Contrast with evaporation.

primary productivity, gross: initial production of organic carbon by photosynthesis.

primary productivity, net: amount of carbon dioxide vegetation takes in during photosynthesis minus the amount of carbon dioxide the plants release during respiration (metabolizing sugars and starches for energy).

primates: order of mammals, including lemurs, monkeys, apes, and humans, which is distinguished by having hands, hand-like feet, and forward-facing eyes; with the exception of humans, primates are typically agile tree dwellers.

primitive trait: trait inherited from distant ancestors.

principle: rule or law concerning a natural phenomenon or the function of a complex system.

Proboscidea: order of mammals, including elephants and their extinct relative, the mastodon, distinguished by the possession of a trunk and tusks.

productid brachiopod: spiny marine animal that characteristically has a straight hinge line, a brachial valve that is flat or concave, as well as an outward-flaring shell in the area of the hinge that gave it a scallop-like appearance.

progradation: growth of a river delta farther out into the sea over time.

prokaryote: organism that lacks a cell nucleus (=karyon) or any other membrane-bound organelles. Contrast with eukaryote.

protein: basic chemical that makes up the structure of cells and directs their activities.

proteinoid: protein-like molecules formed inorganically from amino acids. Contrast with protein.

protein sequencing: type of molecular clock used to determine the amino acid sequence of a protein, as well as which conformation the protein adopts and the extent to which it is complexed with any nonpeptide molecules.

Proterozoic: eon from about 2.5 billion to 570 million years ago, during which sea plants and animals first appeared.

protometabolism: series of linked chemical reactions, in a prebiotic environment, that has characteristics of true metabolism. Compare with metabolism.

provenance: where something originated or was nurtured in its early existence.

province: area of land, less extensive than a region, having a characteristic plant and animal population.

pseudoextinction: occurs where there are no more living members of a species, but members of a daughter species or subspecies remain alive.

pterosaur: extinct flying reptile of the Jurassic and Cretaceous periods, with membranous wings supported by a greatly lengthened fourth finger.

punctuated equilibrium: hypothesis that evolutionary development is marked by isolated episodes of rapid speciation between long periods of little or no change. Contrast with phyletic gradualism.

purple sulfur bacteria: group of Proteobacteria capable of photosynthesis, which are anaerobic or microaerophilic, and are often found in anoxic zones like hot springs or stagnant water. See also green sulfur bacteria.

pyrite world hypothesis: proposes that early life may have formed on the surface of iron sulfide (pyrite) minerals, hence the name. Contrast with ATP world, clay world, and RNA world.

pyritization: development of pyrite in a solid rock.

quark: elementary particle and a fundamental constituent of matter.

quartzite: hard nonfoliated metamorphic rock that formed from the metamorphism of sandstone.

Quaternary: current period, beginning about 1.8 million years ago; considered the "age of man."

Queenston Delta: a clastic wedge formed from the molasse shed from the Taconic Mountains.

race: population within a species that is distinct in some way. Contrast with species; see also subspecies.

radioactive decay: spontaneous disintegration of a radioactive substance along with the emission of ionizing radiation.

radiolarian: single-celled aquatic animal that has a spherical, amoeba-like body with a spiny skeleton of silica. Their skeletons can accumulate as a silica-rich chert on the seabed.

radiometric date: determination of the age of materials (typically rocks) by analyzing the decay of radioactive isotopes in the material.

radula: layer of serially arranged teeth within the mouth that is used by mollusks for feeding.

Rare Earth Elements (REEs): set of 14 chemical elements that are relatively insoluble in water and not easily incorporated into the crystals of other minerals.

recovery phase: period when new taxa arise through adaptive radiation.

recrystallization: occurs where crystals form within the original structure, eventually changing the original into a crystal replica.

recycling: conversion of waste into reusable material.

redbed: stratum of reddish-colored sedimentary rocks such as sandstone, siltstone, or shale that were deposited in hot climates under oxidizing conditions.

red clay: brown to red, widely distributed deep-sea deposit consisting chiefly of microscopic particles and tinted red by iron oxides and manganese.

red shift: happens when light seen coming from an object is proportionally increased in wavelength, or shifted to the red end of the spectrum. Attributable to the Doppler effect.

red tide (algal bloom): an event in which estuarine, marine, or freshwater algae accumulate rapidly in the water column and results in discoloration of the surface water.

reductionism: doctrine that nature can be understood by taking it apart, like a machine, into smaller parts that are more easily analyzed.

reef: ridge of jagged rock, coral, or sand just above or below the surface of the sea.

reef, fringing: coral reef formed close to a shoreline.

reef, patch: relatively small, isolated coral formation that is nearer to shore than the reef itself.

region: part of a country or the world having defining characteristics but not always fixed boundaries.

regional continuity (multiregional) hypothesis: model that holds that humans evolved from the beginning of the Pleistocene two million years ago within a single, continuous human species.

regional metamorphism: alteration of rock that occurs over a very large area in response to increased temperature and pressure.

regression: relative fall in sea level. Contrast with transgression.

relative age: whether something is older or younger than something else. Contrast with absolute age.

relict sediment: sediment that was in equilibrium with its environment when first deposited but which is unrelated.

remineralization: restoration of minerals to demineralized structures or substances.

replacement: growth of a mineral within another of different chemical composition by gradual simultaneous deposition and removal.

reproductive isolation: inability of a species to breed successfully with related species due to geographic, behavioral, physiologic, or genetic barriers or differences.

reptile: cold-blooded, usually egg-laying vertebrate of the class Reptilia, such as a snake, lizard, crocodile, turtle, or dinosaur.

reservoir: compartment within a system which may communicate with its external environment by fluxes of matter and energy into and out of the reservoir.

reverse fault: occurs when hanging wall moves up relative to the footwall. Contrast with normal fault.

reverse transcription: process in which RNA is first transcribed into DNA before protein synthesis.

rhipidistian: branch of crossopterygian; lobe-finned fishes that are the ancestors of the tetrapods who were the dominant predators during the late Paleozoic.

Rhynie Chert: a fossil site in Aberdeenshire in the north of Scotland. The Rhynie Chert represents one of the earliest terrestrial ecosystems preserved in the fossil record, dating back to the Early Devonian.

rhyolite: pale fine-grained volcanic rock of granitic composition.

ribonucleic acid (RNA): nucleic acid that is used in key metabolic processes for all steps of protein synthesis in all living cells and carries the genetic information of many viruses.

ribosomal RNA (rRNA): RNA component of the ribosome, the enzyme that is the site of protein synthesis in all living cells. Ribosomal RNA provides a mechanism for decoding mRNA into amino acids.

ribosome: minute particle consisting of RNA and associated proteins, found in large numbers in the cytoplasm of living cells.

ribozyme: specialized RNA molecule that can catalyze a chemical reaction and replicate itself.

ridge push: tectonic plate motion where plates move by pushing from either magma upwelling at spreading centers.

rift: place where the Earth's crust and lithosphere are being pulled apart.

rifting: long narrow zone of faulting resulting from tensional stress in the Earth's crust.

Rio Grande Rift: north-trending continental rift zone that separates the Colorado Plateau in the west from the interior of the North American craton on the east; began forming between 35 and 29 million years ago when Earth's crust began to spread apart.

ripple mark: small ridge produced in sand by water currents or by wind.

RNA virus: virus that has RNA (ribonucleic acid) as its genetic material.

RNA world hypothesis: theory that life based on ribonucleic acid (RNA) predates the current world life of deoxyribonucleic acid (DNA). Contrast with ATP world, clay world, and pyrite world.

Roberts Mountains terrane: sliver of exotic terrane accreted to the western margin of North America (now Nevada) as a result of the Antler Orogeny, which entailed the collision of the Klamath Arc with the western margin of North America during Devonian and Early Carboniferous time.

rock cycle: interrelated sequence of events by which rocks are initially formed, altered, destroyed, and reformed as a result of magmatism, erosion, sedimentation, and metamorphism.

Rodentia: order of mammals, including rodents such as rats, mice, squirrels, and porcupines, characterized by two continuously growing incisors in the upper and lower jaws which must be kept short by gnawing.

Rodinia: oldest known supercontinent formed approximately 1 billion years ago due to the subduction of ocean basins followed by a series of continental collisions.

rudist: cone-shaped extinct bivalve mollusk that formed reefs in the Cretaceous period.

rugose (or tetra) coral: extinct coral abundant in Paleozoic seas that was characterized by anemone-like polyps that secreted a small cup subdivided by partitions or septa arranged in fours; some species were colonial and could form large structures.

ruminant: cud-chewing hoofed mammal having a stomach divided into four (occasionally three) compartments; includes cattle, sheep, antelopes, deer, giraffes, and their relatives.

runaway glaciation: cooling of a planet's surface sufficient for massive glaciation. Thought to have occurred on Earth during Snowball Earth episodes and on Mars.

runaway greenhouse: where positive feedback increases the strength of its greenhouse effect until Earth's water evaporates into space.

runoff: draining away of water (or substances carried in it) from the surface of an area of land.

S-wave: type of elastic wave, also called seismic wave, which can travel through solid bodies but not liquid ones. Contrast with P-wave.

salinity: refers to the water's "saltiness."

San Andreas Fault: continental strike-slip fault that runs a length of roughly 1,300 kilometers (810 miles) through California. The fault's motion is right-lateral strike-slip (horizontal motion). It forms the tectonic boundary between the Pacific Plate and the North American Plate.

sand: sediment grains ranging in size from 0.062 to 2 mm in size.

sandstone: sedimentary rock formed by the cementing together of grains of sand.

Sauk Sea: a shallow sea that covered most of North America, except for the Transcontinental Arch and the Canadian Shield, during the Sauk Sequence of the Cambrian. The Sauk transgression left a succession of Tapeats Sandstone, Bright Angel Shale, and Muav Limestone, among other formations.

Saurischia: one of the two orders of dinosaurs including sauropods and theropods; distinguished by having a three-pronged pelvis like that of a crocodile ("lizard-hipped").

sauropod: herbivorous dinosaur of the Jurassic and Cretaceous having a small head, a long neck and tail, and five-toed limbs; largest known land animal.

schist: metamorphic rock form characterized by strong foliation so that it is readily split into thin flakes or slabs.

Scleractinia (stony corals): exclusively marine animals similar to sea anemones but which generate a hard skeleton. They first appeared in the Middle Triassic and replaced tabulate and rugose corals that went extinct at the end of the Permian.

sclerosponge: sponge with a soft body that covers a hard, often massive skeleton made of calcium carbonate, either aragonite or calcite. Modern representatives of the ancient Paleozoic reef-builders, stromatoporoids.

seafloor spreading: process that occurs at mid-ocean ridges, where new oceanic crust is formed through volcanic activity and then gradually moves away.

seafloor spreading center: see mid-ocean ridge.

seafloor trench: long, narrow depression of the deep-sea floor having steep sides and containing the greatest ocean depths; formed by depression, to several kilometers depth, of the high-velocity crustal layer and the mantle.

seamount: mountain rising from the ocean seafloor that does not reach to the water's surface (sea level) and thus is not an island. See also guyot.

seaway: way or route by sea.

secondary consumers: in the food chain—these animals are carnivores that feed on herbivores.

secular change: change that occurs slowly over a long period of time. See also directionality.

sediment grain: small particle of naturally occurring material that is broken down by processes of weathering and erosion.

sedimentary rock: rock formed by consolidated sediment deposited in layers.

sedimentary rock, biogenic: created when organisms use materials dissolved in air or water to build their tissue.

sedimentary rock, chemical evaporites: forms when mineral constituents in a solution become supersaturated and inorganically precipitate.

sedimentary rock, terrigenous: composed of silicate minerals and rock fragments transported by moving fluids and were deposited when these fluids came to rest.

seed fern: primitive gymnosperm (not fern) that lived in swampy areas from the Mississipian Epoch through the Mesozoic Era; the plant was topped with a fern-like frond which bore seeds.

seismic tomography: methodology for estimating the Earth's properties.

self-organizing system: system in which simpler components organize themselves into more complex systems.

sequence boundary: significant erosional unconformity and its correlative conformities.

Sequence, Absaroka: major sea-level transgression that extended from the end of the Mississippian through the Permian periods; it is the unconformity between this sequence and the preceding Kaskaskia that divides the Carboniferous into the Mississippian and Pennsylvanian periods in North America.

Sequence, Kaskaskia: major sea-level transgression that began in the mid-Ordovician and peaked in the Mid-Mississippian. Like earlier transgressions, basal sediments of the Kaskaskia consist of clean, well-sorted sandstones that give way upward to limestones, including reefs.

Sequence, Sauk: first major sea-level transgression onto Laurentia, starting in the Late Proterozoic and ending with a regression in the Early Ordovician; in western North America, the Sauk transgression was embodied by the succession of the Tapeats Sandstone, Bright Angel Shale, and Muav Limestone of the Cambrian.

sequence stratigraphy: subdivides the sedimentary record along continental margins and in interior basins into a succession of depositional sequences as regional and interregional correlative units.

Sequence, Tippecanoe: major sea-level transgression that followed the Sauk sequence, extending from the Middle Ordovician to the Early Devonian; the transgression resulted in the deposition of clean sandstones across the craton, followed by abundant carbonate deposition.

Sequence, Zuni: transgression during the Cretaceous that resulted in the formation of the Cretaceous Interior Seaway.

sessile: permanently attached or fixed; not free-moving.

sexual selection: natural selection arising through preference by one sex for certain characteristics in individuals of the other sex.

shales: mudstones that exhibit a layered appearance. Mudstone itself is composed of mud particles that are bonded together.

shield: ancient, stable, interior layer of continents composed of primarily Precambrian igneous or metamorphic rocks.

shock metamorphism: alteration of rock that occurs as a result of shock-wave related deformation and heating during impact events.

Siberia: extremely ancient craton that formed an independent continent before the Permian period craton; today it comprises the Central Siberian Plateau.

Siberian Traps: large region of volcanic rock, known as a large igneous province, in the Russian region of Siberia; formed as the result of a massive volcanic eruption that spanned the Permian-Triassic boundary about 250 million years ago.

siderophilic: iron-loving.

siderophilic element: chemical element such as iridium or gold that tends to bond with metallic iron.

Silurian: period from about 439 million to 409 million years ago; first jawed fish and land plants appeared.

simultaneous accretion (double planet) hypothesis: theory that the Moon formed through accretion just as Earth did.

single origin ("Out of Africa") hypothesis: model that holds that modern humans came from or evolved from a single origin (i.e., in Africa).

single species hypothesis: proposes that there was only sufficient carrying capacity on Earth to accommodate one species of human and that ancient hominids evolved slowly through time by phyletic gradualism.

singularity: hypothesis that all the matter in the universe was initially compressed into a single "primeval atom."

siphon: tubular organ used in feeding and respiration by an infaunal animal, such as a bivalve, that lives beneath the sediment surface.

Sirenia: order of large aquatic plant-eating mammals, including manatees, or sea cows, and dugong, that are distinguished by paddle-like forelimbs and a tail flipper replacing hind limbs.

Sixth Extinction (Holocene): widespread, ongoing human-caused extinction of species during the present Holocene epoch; the term implies that the loss in biodiversity resembles in magnitude the other "Big Five" mass extinctions of the Phanerozoic.

slab gap hypothesis: proposes that, as the crustal block moved northwest, it caused the Basin and Range province to expand like a spreading fan, causing the surrounding mountain ranges to swing to the west around the "hinge" of the fan; hypothesis is named for the spreading and corresponding extension of the range.

slab pull: tectonic plate motion where the descending portion of the plate remains attached to the rest of the plate pulling the rest of the plate behind it.

slash-and-burn: method of agriculture in which existing vegetation is cut down and burned off before new seeds are sown, typically used as a method for clearing forest land for farming.

slate: fine-grained, foliated homogeneous metamorphic rock derived from an original shale-type sedimentary rock composed of clay or volcanic ash through low-grade regional metamorphism.

slate belt: fine-grained metamorphic rock that splits into thin, smooth-surfaced layers.

Slushball Earth hypothesis: contends that Earth was not completely frozen over during Precambrian times but instead included large areas of thin ice or open ocean that allowed the exchange of water vapor between the oceans and atmosphere.

small-number system: system composed of a few components; studied primarily within historical science. Contrast with large-number system.

small shelly fossils: fossils found all over the world which are represented by tubes and scale, shield-like, or spiny structures composed of calcium phosphate or calcium carbonate.

Snowball Earth hypothesis: posits that the Earth's surface became entirely or nearly entirely frozen at least once, sometime earlier than 650 million years ago.

social Darwinism: theory, now largely discredited, that society should improve by the action of natural selection on human efforts: some persons naturally succeed, whereas others fail.

soft-bottom community: comprised of unconsolidated, soft sediment occurring in freshwater, estuarine, and marine systems, which supports a large number of organisms.

soil: top layer of the Earth's surface, consisting of rock and mineral particles mixed with organic matter.

solar nebula hypothesis: hypothesis of the formation of the solar system according to which a rotating nebula cooled and contracted, throwing off rings of matter that contracted into the planets and their moons, while the great mass of the condensing nebula became the sun.

solar wind: stream of charged particles ejected from the upper atmosphere of the sun.

sole mark: irregularity or penetration on the undersurface of a sedimentary stratum.

solid Earth system: nonliving, solid Earth, from its center to its surface, including the continents and the seafloor.

Sonomia: microcontinent, represented today by Nevada, southeast Oregon, and northern California, which added a large portion of landmass to the western margin of the United States during the Sonoma Orogeny.

sorting: separation of sediment grains according to size.

speciation, allopatric: formation of reproductively isolated species due to the divergence of populations that are geographically isolated from each other.

species: taxonomic group of living organisms consisting of similar individuals capable of interbreeding and producing fertile offspring. Compare with race, subspecies.

species–area relationship: relationship between the area of a habitat, or of part of a habitat, and the number of species found within that area.

species sorting: theory which states that evolution can proceed by the natural selection of traits found at the species level rather than the organismal or genetic level.

spectroscopy: study of the interaction between matter and radiated energy.

spectrum, electromagnetic: the entire spectrum of all kinds of electric, magnetic, and visible radiation ranging from zero to infinity.

spectrum, visible: portion of the electromagnetic spectrum that is visible to (can be detected by) the human eye.

speleothem: structure formed in a cave by the deposition of minerals from water, e.g., a stalactite or stalagmite.

sphenopsid: member of a group of plants that includes trees in the Carboniferous coal swamp forests as well as the living horsetail (*Equisetum*).

spontaneous generation: theory that living entities are generated by the power of nature, and new living forms are constantly being generated from nonliving.

spore-bearing fern: fern that reproduces by means of spores.

stalactite: circle-shaped mineral deposit, usually calcite or aragonite, hanging from the roof of a cavern, formed from dripping water.

stalagmite: conical mineral deposit, usually calcite or aragonite, built up on the floor of a cavern, formed from the dripping of calcareous water.

Steady State Theory: theory that the universe maintains a constant average density with matter created to fill the void left by galaxies that are receding from each other. Contrast with Big Bang Theory.

stegosaur: small-headed quadrupedal herbivorous dinosaur of the Jurassic and early Cretaceous periods, with a double row of large bony plates or spines along the back.

stenotopic species: species able to tolerate only a narrow range of environmental factors. Contrast with eurytopic species.

Strangelove Ocean: phrase used to describe the condition of the Earth's ocean immediately following the mass extinction of the dinosaurs (around 65 million years ago), where the carbon cycle was shut down for many thousands of years killing every living thing in the ocean's surface waters.

stratified: horizontally layered.

stratigraphy: geologic science associated with the study of strata or rock layers.

stratum (pl: strata): sheet-like mass of sedimentary rock of one kind lying between beds of other kinds.

strewn field: area where tektites formed when a large meteoroid enters the atmosphere and fragments into many pieces before touching the ground due to thermal shock.

strike-slip fault: see transcurrent fault.

stromatolite: large mineral structures formed in shallow water by microorganisms, especially cyanobacteria. Some stromatolites are among the most ancient fossils known, dating to about 3.5 billion years ago.

stromatoporoid: extinct, sessile, coral-like marine organism of uncertain relationship that built up calcareous masses composed of laminae and pillars, occurring from the Cambrian to the Cretaceous. See also sclerosponge.

subduction: sideways and downward movement of the edge of a plate of the Earth's crust into the mantle beneath another plate.

subduction zone: convergent plate boundary where one plate subducts beneath the other, usually because it is denser.

submarine canyon: steep-sided valley on the sea floor of the continental slope.

submarine fan: accumulation of land-derived sediment on the deep seafloor.

sub-Milankovitch time scales: measurement of rapid climate change that occurred over time spans less than a few thousand years all the way down to durations as short as decades or less.

subspecies: taxonomic group that is a division of a species; usually arises as a consequence of geographic isolation within a species. See also race.

Sundance Sea: epeiric sea that existed in North America during the Mid to Late Jurassic Period. It was an arm of what is now the Arctic Ocean and extended through what is now western Canada into the central western United States.

sunspot minima: cycle of reduced sunspot frequency.

sunspots: relatively cool regions of the sun (that are still intensely hot) that appear visibly as dark spots compared to surrounding regions. They are caused by intense magnetic activity forming areas of reduced surface temperature.

supernova (pl: supernovae): explosive death of a massive star whose energy output causes its expanding gases to glow brightly for weeks or months.

supraorbital ridge: refers to a bony ridge located above the eye sockets of primates.

Superposition, Principle of: theory that in a sequence of rocks, younger rocks lie on top of older rocks.

suturing: major fault zone through an orogen or mountain range.

SWEAT hypothesis: theorizes that the southwestern United States was at one time connected to East Antarctica (SWEAT stands for South West U.S. and East Antarctica).

symbiotic algae: relationship where algae supply photo-synthates (organic substances) to the host organism providing protection to the algal cells and the host derives energy requirements from the algae.

synapsid: major amniote lineage that diverged about 300 million years ago, which showed increasingly mammalian characteristics and have a single temporal opening in the each side of the skull; ultimately ancestral to mammals. Contrast with diapsid.

syncline: formed when the strata dip toward the center of the fold and the strata found along the axis of the syncline are younger than strata farther away from the axis. Contrast with anticline.

system: series of parts or components that interact together to produce a larger, more complex whole.

systems tract: discrete package of distinctive sediment types (facies) that are laid down during different phases of a cycle of sea-level change.

tabulate coral: extinct coral that formed colonies of individual hexagonal cells known as corallites defined by a skeleton of calcite, similar in appearance to a honeycomb, and characterized by horizontal partitions.

taeniodont: early group of mammals who lived from the Paleocene to the Eocene, which evolved quickly into highly specialized digging animals.

taxonomy: science of biologic classification.

tectonics: the processes that cause the movement of the plates that comprise Earth's lithosphere.

tectonic (Wilson) cycle: period of geologic history that is characterized by a certain sequence of tectonic and other geologic events.

Tejas Seaway: waterway formed from the regression that led to the retreat of the Cretaceous Interior Seaway.

tektite: small dark glassy object found in several areas around the world, thought to be a product of meteorite impact.

teleconnection: Causal link between patterns of weather in two locations, or between two atmospheric occurrences, which are very far apart.

Teleologic: life forms fulfill a preordained plan and design based on final goals and purposes.

teleost fish: largest group of bony fish; includes ray-finned fish that have a caudal fin, scales, and a swim bladder.

temnospondyl: extinct amphibian theoretically resembling the crocodile, with a flattened head, long snout, limbs positioned at right angles to its body, eyes on top of its head, and separate vertebrae. Previously referred to as a labyrinthodont because of its highly folder (labyrinthine) teeth.

terminal moraine: moraine deposited at the point of furthest advance of a glacier or ice sheet.

terrane: fault-bounded area or region with a distinctive stratigraphy, structure, and geologic history.

terrestrial environment: Earth's land area, including its human-made and natural surface and subsurface features, and its interfaces and interactions with the atmosphere and the oceans.

test: shell of certain microorganisms, such as testate foraminifera, testate amoebae, and sea urchins.

Tertiary: period from about 65 million to 1.8 million years ago; considered the "age of mammals."

texture, igneous: texture occurring in igneous rocks. Igneous textures are used by geologists in determining the mode of origin igneous rocks and are used in rock classification.

theory: set of statements or principles devised to explain a group of facts or phenomena, especially one that has been repeatedly tested or is widely accepted and can be used to make predictions about natural phenomena.

therapsids: "mammal-like reptiles" considered the direct ancestor of mammals.

thermohaline circulation: refers to a part of the large-scale ocean circulation that is driven by global density gradients created by surface heat and freshwater fluxes. See also oceanic conveyor.

thermophile: heat-loving. Compare to hyperthermophile.

theropod: carnivorous dinosaur of the Triassic to Cretaceous with short forelimbs that walked or ran on strong hind legs.

thioester: compound with a functional group characterized by a sulfur atom flanked by one carbonyl group and one carbon of any hybridization.

Three Mile Island nuclear accident: Pennsylvanian nuclear power plant that suffered a reactor meltdown in 1979 due to a cooling malfunction. While some radioactive gas was released a couple of days after the accident, no injuries or adverse health effects resulted.

thrust fault: particular type of reverse fault in which the angle of the fault is quite low or nearly horizontal.

tiering: suspension feeding at different levels within the sediment above and below the seafloor.

till: boulder clay or other unstratified sediment deposited by melting glaciers or ice sheets.

tillite: sedimentary rock composed of compacted glacial till.

tillodont: extinct order of mammals that may be related to the pantodont. They were widespread across North America and Eurasia during the late Paleocene and most of the Eocene.

tilt: see obliquity.

Tippecanoe Sea: the second epeiric sea of the Paleozoic (following the Sauk Sea), the Tippecanoe Sea formed during the Tippicanoe Sequence, transgressing the craton from the Middle Ordovician to the Early Devonian; the transgression resulted in the deposition of clean sandstones across the craton, followed by widespread carbonate deposition.

tonalite-trondhjemite-granodiorite (TTG): aggregation of rocks that are formed by melting of hydrous mafic crust at high pressure.

top carnivore: animal that is highest on the food chain and consumes all lower levels and is not consumed by any other animal.

top-down approach, chemical evolution: strategy in which existing biologic systems and biochemical pathways are extrapolated backward to earlier, simpler components and systems. Contrast with bottom-up approach.

total range zone: range of a species in all the provinces in which it occurs. See also index fossil. Contrast with overlapping range zone.

trade winds: winds that blow steadily from east to west and toward the equator The trade winds are caused by hot air rising at the equator, with cool air moving in to take its place from the north and from the south.

Transantarctic Mountains: located in Antarctica and one of the world's longest continental rift flank uplifts, with a total length of about 3,500 kilometers.

transcription: organic process whereby the DNA sequence in a gene is copied into mRNA; the process whereby a base sequence of messenger RNA is synthesized on a template of complementary DNA.

transcurrent fault: formed by horizontal movement of rocks relative to each other along the fault. Also known as strike-slip fault.

transform plate boundary: marks the location where plates or segments of plates move past one another horizontally.

transgression: spreading of the sea over land as evidenced by the deposition of marine strata. Contrast with regression.

transgressive systems tract: deposits that accumulated from the onset of coastal transgression until the time of maximum transgression of the coast.

translation: process whereby genetic information coded in messenger RNA directs the formation of a specific protein at a ribosome in the cytoplasm.

transpiration: emission of water vapor from the leaves (stomata) of plants.

travertine: white or light-colored calcareous rock deposited from mineral springs.

trepostome: extinct order of bryozoans characterized by delicate to massive colonies that consisted of countless numbers of microscopic polyps that together secreted calcareous skeletons that held the colonies above bottom or encrusted hard surfaces.

Triassic: period from about 245 million to 208 million years ago; many new organisms appeared following the mass extinctions of the end of the Paleozoic era, including the earliest dinosaurs and ammonites and the first primitive mammals.

trilobite: extinct marine arthropod that occurred abundantly during the Paleozoic era, with a carapace over the forepart and a segmented hind part divided longitudinally into three lobes; the earliest animal to exhibit evidence of eyesight in the fossil record.

triple junction: point where the boundaries of three tectonic plates meet.

trophic relationship: feeding relationship of organisms in communities and ecosystems.

true polar wander: involves the rotation of the entire crust and mantle as a single unit around Earth's core. Contrast with apparent polar wandering curve.

turbidite: sediment or rock deposited by a turbidity current.

turbidity current: swift downhill current in water, air, or other fluid, triggered by the weight of suspended material such as silt in a current.

turnover pulse hypothesis: proposes that a major pulse of mammal evolution worldwide beginning about 2.7 million years ago led to early hominid forms in Africa as the result of adaptation to climate change.

tyrannosaurid: large-headed carnivorous dinosaur with short, two-fingered hands, that most likely functioned as top predator in the Late Cretaceous ecosystem.

ultramafic: igneous rock consisting dominantly of mafic minerals, containing less than 10 percent feldspar. Includes dunite, peridotite, amphibolite, and pyroxenite.

unconformity: buried erosion surface separating two rock masses or strata of different ages, indicating that sediment deposition was not continuous.

undershoot: when a system falls short of equilibrium.

uintathere: hoofed North American mammal of the extinct genus *Dinoceras,* of the Eocene Epoch, having a massive body and three pairs of horns.

ungulate: animal which uses the tips of its toes, usually hoofed, to sustain its whole body weight while moving.

Uniformitarianism, Principle of: doctrine that the processes and the rates of those processes we observe today are the same as those that have always operated.

uniformity of nature: see Uniformitarianism, Principle of.

upwelling: process by which warm, less-dense surface water is drawn away from along a shore by offshore currents and replaced by cold, denser water brought up from the subsurface.

uraninite, detrital: mineral readily oxidized in the presence of oxygen and thus does not survive weathering processes during erosion, transport, and deposition in an oxygenated atmosphere.

Valley and Ridge province: westernmost province of the Appalachian Mountains formed when sedimentary rocks were deposited on the shallow continental shelves of epeiric seas; characterized by long, even ridges (folds), with long, continuous valleys in between.

varve: annual layer of sediment or sedimentary rock in a body of still water (lake).

vascular tissue: complex conducting tissue, formed of more than one cell type, found in vascular plants.

vector (disease): Organisms (e.g., rodents, flies, mosquitoes) that carry pathogens between humans, or from infected animals to humans.

Vendozoa: distinct phylum, or even kingdom, of metazoans from the Ediacara fauna that eventually became extinct and were replaced by more modern metazoans.

vestigial structure: structure in an organism that has lost all or most of its original function in the course of evolution, such as human appendixes.

viviparous: describes animal that gives birth to living off-spring that develop within the mother's body. Most mammals and some other animals are viviparous. Contrast with oviparous.

volcanic ash: fine particles of mineral matter from a volcanic eruption, which can be dispersed long distances by winds aloft.

walking the outcrop: method of mapping rock units in the field visually over the area in which they are exposed.

Walther's Law: states that the vertical succession of facies reflects lateral changes in environment.

water table: level below which the ground is permanently saturated with water.

weathering: mechanical and chemical breakdown of rocks by the action of rain, snow, cold, etc.

well-log: method of gathering and recording information about subsurface geologic formations (strata and their pore liquids) using remote sensing apparatus lowered down a well.

westerlies: belt of prevailing westerly winds in medium latitudes in the southern hemisphere.

Wilson cycle: see tectonic cycle.

wobble: see precession of the equinoxes.

Wrangellia terrane: large exotic terrane, extending from Washington into Alaska, which accreted to the north during the Mesozoic Era.

Yellowstone hotspot: volcanic hotspot responsible for large scale volcanism in Oregon, Nevada, Idaho, and Wyoming.

Younger Dryas: geologically brief period of cold climatic conditions and drought (referred to as the "Big Freeze") that occurred between approximately 12,800 and 11,500 years ago as the result of the collapse of North American ice sheets.

Yucca Mountain: proposed federal repository for nuclear waste in Nevada.

zircon, detrital: silicate mineral found in sedimentary rocks that was eroded and transported from previously existing crust.

zooplankton: plankton consisting of small animals and the immature stages of larger animals.

Figures are indicated by *f*. Tables are indicated by *t*. Boxes are indicated by *b*.

Grenville Orogeny, 228
"grid-independent" hybrid cars, 538
gross domestic product (GDP), 530*f*, 550
gross primary production, 77
groundplans, 137
growth faults, 531*b*
gulf and Atlantic coastal plains
    Paleogene Period, 429
Gulf Coast, 430*f*
Gulf of Mexico
    dead zone in, 553*f*
    oil in, 531
    oil rig in, 534*f*
    simultaneous meltwater pulses in, 510*f*
Gulf Stream, 471
guyots, 31, 32*f*
gymnosperms, 330, 365
gypsum, 93
gyres, circular, 73

H

habitable zone (HZ), 190, 218
    atmospheric evolution of Venus, Earth, and Mars, 191*f*
    evidence for water on Mars, 191*f*
    spectroscopic data, 192
hadal environment, 99
Hadean Eon, 179
    Earth's earliest evolution, 179–182
    origin of Moon, 182–185
hadrosaurs, 399
Haldane, J. B. S., 204
half-life concept, 143, 144*f*
halite, 93
Halkieria, 258
Hallucigenia, 260, 262*f*, 266
halophiles, 220
haplorhines, 483
Harpid trilobites, 302
Hebertella, 306*f*
Heinrich events, 511, 512*f*
Hemichordata, 256
Hemicyclaspis, 314, 315*f*
herbivores, 76
Hercynian Orogeny, 332
Hess's hypotheses, 32
hexacorals, 389*f*
hiatus, 159
high-head plants, 530
highstand systems tract, 161
Himalayan Orogeny, 423, 424*f*
historical sciences, 16
    geology as, 16–17
H.M.S. Beagle, 99, 117
Holmes, Sir Arthur, 29
Holocene, 150, 474, 499, 502
    climatic events, 502*f*
    Noah's flood, 504–505*b*

rapid climate change
    on centennial time scales, 513–514
    on interdecadal to multidecadal time scales, 514–518
    on millennial time scales, 507–513
sea level and Antarctic ice sheets and ice shelves, 506*f*
sea-level rise, 502–503
Younger Dryas, 503–506
Homeobox, 131, 269
homeostasis, 8*b*
hominids, 483
hominins, 483, 491–492
Homo sapiens (*H. sapiens*), 115, 481, 557
homologous structures, 126
horizontally layered rocks, 10–11
horn corals, 305, 307
horst-and-graben structures, 39, 39*f*, 467
hotspot, 32, 33*f*, 542
    stationary, 33–34*b*
Hox genes, 131, 132, 133*f*, 269, 405
Hubble, Edwin P., 171
Hubble Law, 172
human–climate interactions, 526–527
human disturbance, 552–556*b*
Human-generated activities. *See* Anthropogenic activities
human impacts on global hydrologic cycle, 546*f*
Hume's problem, 18
humic acids, 86
humus, 87
Hutton, James, 11, 11*f*, 12*f*, 13, 18, 143
hybrid cars, 537–538
hybrid vigor, 125
Hydnoceras, 309*f*
hydroelectric power, 530
    exploration for petroleum, 531–534*b*
    projected oil production for scenarios, 531*f*
hydrogen (H$_2$), 174–175, 179, 181, 204
    fuel cells, 538
hydrogen cyanide (HCN), 204
hydrogen sulfide (H$_2$S), 208
hydrologic cycle, 69–73, 72*f*
hydrophilic, 207
hydrophobic, 207
hydrosphere, 7, 69
    See also biosphere
    hydrologic cycle, 69–73, 72*f*
    ocean circulation, 73–75
hydrothermal metamorphism, 65
hydrothermal vents, 208–209
    unusual communities, 210*b*
hydrothermal weathering, 65, 66*f*
Hylonomus, 349, 349*f*
hyoliths, 260
hyperthermals, 432–433
hyperthermophilic bacteria, 205
hypothesis, 170, 208
hysteresis loop, 518, 519*f*

vascular tissue, 319
vectors, 546
Vendobionta, 252
Vendozoa, 252
Vernadsky, Vladimir Ivanovich, 8*b*
vertebrates, 308, 348–353, 393, 435–437
    dinosaurs warm-blooded, 402*b*
    evolution of dinosaurs, 400*f*
    evolutionary relationships of diapsid reptiles, 398*f*
    ornithischian dinosaurs, 403*f*
    Proterosuchus, 399*f*
    saurischian dinosaurs, 404*f*
    structure of pelvis, 402*f*
    tremendous diversity of ornithischians and
        saurischians, 401*f*
vestigial structures, 134, 136*f*
Vine, Fred, 35
viscosity, 61
volcanic
    ash, 60, 377
    ash layer, 61*f*
    eruption styles, 61
volcanism, 357, 367, 428, 429
    and uplift, 467–468
volcanoes, 60, 186, 448–451
    types, 63

W

Walcott, Charles, 264
walking outcrop, 150
Wallace, Alfred Russel, 118*f*
Walther's Law, 152
warm-blooded reptiles. *See* endothermic reptiles
water, 69, 86
    cures, 119
    table, 535–536
Watson, James, 122
wave-dominated deltas, 97
wave ripples, 105

weathering, 9, 144
    of continental rocks, 78
    on inner planets, 193
        fate of Earth, 193–194*b*
        relative distribution of surface elevations, 195*f*
    processes, 86–88
Wegener, Alfred, 29, 29*f*
well-logs, 150, 152*f*
West Coast of North America, Neogene Period, 462–465
West Gondwana, collision of, 278
westerlies, 69
Western North America, Neogene Period, 458, 462
"whaleback" anticline, 331*f*
White Cliffs of Dover, 13*f*, 381
William of Ockham, 17
Wilson cycle, 48
wind energy, 530
wind systems of Earth, 73
winter solstice, 474
wobble, 474
Wöhler, Friedrich, 203
wood-burning stoves, 537
Wopmay Orogen, 227
Wrangellia, 373

X

xylem, 319

Y

Yavapai, 228
Yellowstone National Park, 219
Younger Dryas, 503–506, 507*f*
Yucca Mountain nuclear accidents, 535

Z

zircons, 185
zooplankton, 102
Zuni Sequence, 367